T0388035

Brewing Microbiology

Woodhead Publishing Series in Food Science, Technology and Nutrition

Brewing Microbiology

Managing Microbes, Ensuring Quality and Valorising Waste

Second Edition

Edited by

Annie E. Hill

Woodhead Publishing is an imprint of Elsevier
50 Hampshire Street, 5th Floor, Cambridge, MA 02139, United States
125 London Wall, London EC2Y 5AS, United Kingdom

ISBN: 978-0-323-99606-8

For information on all Woodhead Publishing publications visit our website at https://www.elsevier.com/books-and-journals

Publisher: Zoe Kruze
Acquisitions Editor: Kelsey Connors
Editorial Project Manager: Deepak Vohra
Production Project Manager: Sruthi Satheesh
Cover Designer: Vicky Pearson Esser

Typeset by TNQ Technologies

Contents

Part II
Spoilage bacteria and other contaminants

Part III
Reducing microbial spoilage: Design and technology

Part IV
Impact of microbiology on sensory quality

18. Biotransformation of wort components for appearance, flavour and health

Mei Z.A. Chan and Shao Quan Liu

19. Sensory analysis as a tool for microbial quality control in the brewery

Gary Spedding and Tony Aiken

Part V
The recycling and valorisation of brewing residues

20. Anaerobic treatment of brewery wastes

Joseph C. Akunna

21. Water treatment and reuse in breweries

Geoff S. Simate

Contributors

Tony Aiken, Data Collection Solutions, Lexington, KY, United States

Joseph C. Akunna, Abertay University, Dundee, United Kingdom

Anne Anstruther, The International Centre for Brewing & Distilling, Heriot-Watt University, Edinburgh, Scotland, United Kingdom

Jan Biering, VLB Berlin e.V., Berlin, Germany

Chris A. Boulton, Boulton Brewing Consultancy Ltd., Staffordshire, United Kingdom

Alex Bowler, Faculty of Environment, School of Food Science and Nutrition, University of Leeds, Leeds, United Kingdom

Scott J. Britton, Brouwerij Duvel Moortgat, Puurs-Sint-Amands, Belgium; The International Centre for Brewing & Distilling, Heriot-Watt University, Edinburgh, United Kingdom

Mei Z.A. Chan, National University of Singapore, Singapore

Ben Connolly, Briggs of Burton Plc, Staffordshire, United Kingdom

Scott Davies, Briggs of Burton Plc, Staffordshire, United Kingdom

Jan Fischer, VLB Berlin e.V., Berlin, Germany

Gary J. Freeman, baftec, Surrey, United Kingdom

Brian Gibson, Department of Brewing and Beverage Technology, Institute of Food Technology and Food Chemistry, Technische Universität Berlin, Berlin, Germany

John Hancock, Briggs of Burton Plc, Staffordshire, United Kingdom

Annie E. Hill, The International Centre for Brewing & Distilling, Heriot-Watt University, Edinburgh, United Kingdom

Ruslan Hofmann, PureMalt Products Ltd., Berlin, Germany

Mathias Hutzler, Technical University of Munich (TUM), Research Center Weihenstephan for Brewing and Food Quality, Freising, Germany

Riikka Juvonen, VTT Technical Research Centre of Finland, Espoo, Finland

Ulf-Martin Kohlstock, Scanbec, Bitterfeld-Wolfen, Germany

Yohannes N. Kurniawan, Asahi Group Holdings, Ltd., Tokyo, Japan

Ajja Laitila, VTT Technical Research Centre of Finland Ltd., Espoo, Finland

Shao Quan Liu, National University of Singapore, Singapore; National University of Singapore (Suzhou) Research Institute, Suzhou, Jiangsu, China

Andrew J. MacIntosh, Food Science and Human Nutrition Department, University of Florida, Gainesville, FL, United States

Kathleen Merx, Scanbec, Bitterfeld-Wolfen, Germany

Ashtavinayak D. Paradh, Technical Innovation (R&D) Center, Nashik, Pernod Ricard India, Nashik, Maharashtra, India

Mark Phillips, Briggs of Burton Plc, Staffordshire, United Kingdom

David E. Quain, International Centre for Brewing Science, School of Biosciences, University of Nottingham, Sutton Bonington Campus, Leicestershire, United Kingdom

Tuija Sarlin, VTT Technical Research Centre of Finland Ltd., Espoo, Finland

Jvo Siegrist, Merck Group, St Gallen, Switzerland

Geoff S. Simate, School of Chemical and Metallurgical Engineering, University of the Witwatersrand, Johannesburg, South Africa

Gary Spedding, Brewing and Distilling Analytical Services, LLC, Lexington, KY, United States

Graham G. Stewart, GGStewart Associates, Cardiff, Wales, United Kingdom

Koji Suzuki, Asahi Quality & Innovations, Ltd., Ibaraki, Japan

Trevor Sykes, Briggs of Burton Plc, Staffordshire, United Kingdom

Kathleen Vetter, Scanbec, Bitterfeld-Wolfen, Germany

Frank Vriesekoop, Harper Food Innovation, Harper Adams University, Newport, United Kingdom

Nicholas Watson, Faculty of Environment, School of Food Science and Nutrition, University of Leeds, Leeds, United Kingdom

Edward Wray, Hepworth and Co. Ltd, West Sussex, United Kingdom

Preface

The purpose of brewing is to produce beer through the hydrolysis of starch from barley malt, together with wheat, maize, rice, sorghum, unmalted barley and other cereals, and with the incorporation of hops that contribute characteristic flavours to the product. These raw materials are mashed into a sugary nitrogenous fermentable liquid called wort. This medium is converted into an alcoholic, carbonated beverage by yeast. The brewing process is essentially a microbiological/biochemical series of reactions. This involves a number of complementary disciplines including plant breeding and cultivation, chemistry, chemical/civil/mechanical/electrical engineering, which laterally have come under the auspices of computer control. This volume not only focuses on the positive microbiological aspects of brewing but also considers in detail the microbiological contamination of the process starting with its raw materials. This volume concludes with a discussion of the quality of the drinkable beer (fresh and not so fresh) in both bottles and cans and on draft.

Although there are many excellent text books on brewing, their primary focus continues to be a consideration of the whole brewing process with microbiological aspects being integrated into its syntax. As a consequence, the discussion of fermentation tends to be more biochemical in its emphasis than microbiological. This volume's focus is decidedly microbiological! This applies to both brewer's yeast strains and contaminating microorganisms – bacteria, wild yeasts and mycelial fungi.

Distinct areas are devoted to brewer's yeast and its metabolism and consider, in appropriate detail, their taxonomy and related areas such as identification and its characterisation. Wort fermentation and metabolism are discussed, and in particular, the metabolic engineering of these organisms is considered. The fact that brewer's yeast cultures are routinely recycled through a number of wort fermentations is emphasised, and details of yeast management between fermentations are discussed.

Contaminating fungi, both yeast and mycelial fungi, are discussed in the context of their influence on beer characteristics and quality. It is emphasised, in a number of chapters, that brewing is usually a sterile process (unlike distilling). This is due to the fact that the wort is boiled and in many situations (not all) benefits from the antiseptic properties of hop acids. It is appreciated that often wild yeasts can contaminate pitching yeast cultures and that acid washing of the yeast does not always cleanse the brewing yeast culture of such microorganisms. Also, the stimulation of beer gushing by mycotoxins is discussed.

Most of the remainder of the text focuses on a detailed discussion of contaminating bacteria (both Gram positive and Gram negative) that can occur in brewing. Sometimes, these bacteria are welcome (e.g., in Lambic beer), but usually this is not always the case. This unwelcome contamination can occur on raw materials (particularly malt and water), during fermentation and maturation and in the final beer. Contamination in all these production stages will influence beer flavour (and its overall characteristics) and stability (physical and flavour). The implications of these bacterial infections are considered in detail.

Graham G. Stewart

Heriot Watt University, Edinburgh, Scotland, August 2024.

Introduction to brewing microbiology 2E

Brewers are optimists. Anyone relying on the behavior of a living organism needs to be! Fortunately, our understanding of industrial strains of brewing yeast, in terms of the typical flavour and aroma profiles they impart, handling and parameters required for optimal fermentation, is well developed, and through the appliance of science, brewers can create the conditions most conducive to success. There is also an appreciation of the vulnerability of the process and product to contamination, particularly for no and low alcohol beer, with published methods to mitigate against spoilage and to ensure quality and consistency.

As stated in the first edition of this text, more knowledge brings the realisation of how much we still must learn, but also the wherewithal to alleviate risks, solve problems, and manipulate microbes to improve and develop new products and processes. Significant strides have been made in the development of tools to characterise microbes and determine their value in terms of production, including valorisation of waste streams. These aspects are covered in detail within this volume, but to provide some background:

Brewing yeast

Yeast species are found in a range of fungal lineages and are predominantly single-celled microorganisms able to grow in both the presence and absence of oxygen. Over 2000 yeast species are currently accepted (Boekhout et al., 2022). Of these, there are basically two major strains used in brewing: *Saccharomyces cerevisiae* (ale) and *Saccharomyces pastorianus* (lager), a hybrid of *S. cerevisiae* and *Saccharomyces eubayanus* (Libkind et al., 2011). Ale yeast operate at around room temperature (18–22°C), ferment quickly and produce the 'fruitiness' characteristic of most ales. Lager yeast work at colder temperatures (8–15°C), ferment slowly and utilise more wort sugars, leaving a cleaner, crisp taste. Ale and lager yeast are the most commonly used worldwide, but there is increased use of other yeast strains such as *Brettanomyces* spp., which are traditionally used in Lambic beer production.

In the early 2000s, the discovery and whole-genome sequencing of *S. eubayanus* caught the imagination of both brewers and research microbiologists alike (Libkind et al., 2011). It was known for some time that *S. pastorianus* was a hybrid organism involving *S. cerevisiae*, but the other parent(s) were unknown until the isolation of *S. eubayanus*. Genome sequencing revealed that it was an almost exact genetic match of the non-*S. cerevisiae* subgenome of lager yeast. First isolated in Patagonia, it was thought that the parent *S. eubayanus* strain had its origin in South America and had been introduced into an environment containing *S. cerevisiae*. Recent surveys have, however, recovered *S. eubayanus* from China, Argentina, Chile, Tibet, New Zealand, the United States and Ireland, and it has been proposed that the critical event that led to hybridisation was in fact the introduction of *S. cerevisiae* into an environment where *S. eubayanus* was present (Hutzler et al., 2023).

The increased intensity in genomics research led to the realisation that, as with *S. eubayanus*, many commercial yeast strains are natural yeast hybrids (Gibson & Liti, 2015). High genetic diversity within yeasts used in the wine industry suggests that hybridisation events are common. The ability of yeasts to adapt to changing conditions through hybridisation not only confers evolutionary advantage but also presents us with an opportunity to manipulate mating to create novel strains without resorting to genetic modification (GM).

Advances in yeast research, using both GM and non-GM approaches, have continued apace and led to the development of a wide range of new production strains, most recently using CRISPR technology. Several commercial yeast companies offer strains with attributes such as increased production of esters, thiols or terpenes, and strains that mitigate against hop creep, or produce beer that is diacetyl-free. An entirely synthetic functional *S. cerevisiae* genome has also been developed, paving the way for even further novel yeast development.

Process and product integrity

There are literally millions of food spoilage organisms. However, those responsible for beer spoilage are limited to only a few species of bacteria and 'wild' yeast. Beer has a range of properties that hinder microbial growth including low pH, high alcohol concentration, low nutrient level, antiseptic action of hop acids, low oxygen concentration and carbonation. Its production is a microbiological process though, meaning that the medium into which the brewing yeast is pitched is an ideal environment for the growth of a range of microorganisms. The increasing number of no and low alcohol beers that are being produced, and the wider use of more unusual raw materials and adjuncts also increase the likelihood of microbial beer spoilage.

Most brewers take a proactive approach to beer-spoilage organisms beginning with brewhouse design: use of closed vessels, avoiding dead legs in pipework and use of cleaning-in-place (CIP). These are all methods that are designed-in to new plants but can also be retrofitted or integrated into existing breweries. A second improvement in tackling spoilage is to carry out ATP testing on brewing liquor, CIP rinse water and vessel surfaces; this rapid method of microbiological testing does not identify bacteria or yeasts but gives a very quick indication of plant cleanliness and the success of CIP cycles.

Raw materials and final product testing are still predominantly carried out using traditional methods, but rapid methods, such as PCR, are becoming standard practice. Improvements in methodology mean that tests previously consigned to research laboratories or dedicated microbiology services are now possible without extensive training or specialised facilities. As equipment and consumables costs fall, we will see further take up in tools to tackle microbial spoilage within breweries and further improvements in product quality and consistency.

Waste valorisation

Breweries no longer produce waste; the term 'coproduct' has been adopted to cover all nonbeer outputs such as spent grain and yeast. As our understanding of microbial metabolism has increased and tools to manipulate specific biochemical pathways have been developed, a range of new applications have been identified, including methods of converting 'waste' to either new products or energy.

Brewery waste streams often contain high-value chemicals that can be extracted and reused in other industries. Spent yeast and grain are most commonly used in animal feed and human nutrition, but both can also be used as a flavouring agent, as a source of enzymes and single cell protein, or as a filter element for beverage clarification. Yeast (and bacteria) may also be used in wastewater treatment and biogas production, and as such represent an important tool within the drive to more sustainable brewing practices.

Since the first edition of this text, we have certainly learned of the challenges that are posed by microorganisms. Global pandemic aside, the increased interest in exploration of novel yeasts and the potential for microbes to enhance products and processes is infectious, making this an exciting time to be brewing!

Annie E. Hill

References

Boekhout, T., Amend, A. S., El Baidouri, F., Gabaldón, T., Geml, J., Mittelbach, M., Robert, V., Tan, C. S., Turchetti, B., Vu, D., & Wang, Q. M. (2022). Trends in yeast diversity discovery. *Fungal Diversity, 114*(1), 491–537.

Gibson, B., & Liti, G. (2015). *Saccharomyces pastorianus*: genomic insights inspiring innovation for industry. *Yeast, 32*(1), 17–27.

Hutzler, M., Morrissey, J. P., Laus, A., Meussdoerffer, F., & Zarnkow, M. (2023). A new hypothesis for the origin of the lager yeast *Saccharomyces pastorianus*. *FEMS Yeast Research, 23*, Article foac023.

Libkind, D., Hittinger, C. T., Valério, E., Gonçalves, C., Dover, J., Johnston, M., Gonçalves, P., & Sampaio, J. P. (2011). Microbe domestication and the identification of the wild genetic stock of lager-brewing yeast. *Proceedings of the National Academy of Sciences, 108*(35), 14539–14544.

Acknowledgements

Many thanks to all of the contributors to this volume, both new and returning, for sharing their great breadth of knowledge and experience. I am indebted to past, present and future students of the International Centre for Brewing and Distilling – a constant source of enthusiasm and inspiration! Thanks also to past and present colleagues, in particular Emeritus Professor Graham Stewart for his unstinting support that has continued throughout his retirement.

In memory of Professor Fergus Priest, a great role model when I started at Heriot-Watt University and an all-round good human.

Part I

Yeast: Properties and management

Chapter 1

Brewing yeasts: An overview

Brian Gibson
Department of Brewing and Beverage Technology, Institute of Food Technology and Food Chemistry, Technische Universität Berlin, Berlin, Germany

Beer has been brewed for thousands of years (Hirschfelder & Trummer, 2022), and in humankind's journey towards civilisation, the brewing yeasts have been faithful travelling companions. However, for much of the time we have been brewing beer, the role and nature of yeast were unknown or poorly understood. It has only been since the early to mid-19th century, when microscope technology had advanced and biologists, rather than chemists, focussed their attention on fermentation, that a clearer understanding of brewing yeast was possible (Barnett & Barnett, 2011). More recently, advances in genome analysis techniques have greatly increased our appreciation of brewing yeasts as organisms genetically and physiologically distinct from other yeast groups. This improved understanding of the nature of brewing yeasts as well as their potential applications has fortuitously coincided with a period in the brewing industry when innovation and experimentation are required to satisfy consumer demand for novel and diverse beer styles. A less conservative approach to brewing yeast, and brewing generally, has also been necessitated by challenges related to climate change, resource-use efficiency and raw material insecurity. Advances in our understanding have facilitated the development and selection of new brewing yeasts and new approaches to yeast application in breweries. Some of the most significant recent findings and applications are summarised in the following text.

1.1 Yeasts

Yeasts are a diverse group of fungi characterised primarily by their occurrence as single-celled organisms. This morphological simplicity belies an incredible genetic diversity. Outwardly similar yeasts may possess genomes that are as disparate as any seen in the animal or plant kingdoms (Shen et al., 2018). Of the many thousands of yeast species known, only a very restricted number are utilised in brewing. Of these, *Saccharomyces cerevisiae* is the most significant. This is one of eight recognised species in the *Saccharomyces* genus and is the only species routinely found associated with food or beverage fermentations. In addition to beer fermentation, the species is associated with bread, chocolate, coffee, olive, sausage and wine fermentation, as well as many others. For a number of reasons, some which remain obscure, the species appears to be uniquely adaptable to man-made environments. Additionally, *Saccharomyces* species, including *S. cerevisiae*, are found in nature, typically associated with woodland environments and often with trees of the family Fagaceae (Mozzachiodi et al., 2022). In its wild state, *S. cerevisiae* typically occurs as a diploid organism, with duplicate copies of 16 chromosomes consisting of 6000 genes and an approximately 12 Mb genome size (Goffeau et al., 1996). The yeasts are capable of sporulation and sexual reproduction in addition to asexual reproduction involving budding.

1.2 Ale yeasts

Brewing strains of *S. cerevisiae*, which are used for ale production, possess a number of features that differentiate them from their wild relatives and also from production yeast used for nonbeer beverages. Two seminal papers published in 2016 highlight some of the main differences (Gallone et al., 2016; Gonçalves et al., 2016). Perhaps most importantly, brewing yeasts can efficiently ferment the most abundant wort sugar maltose and in most cases also maltotriose. This allows for efficient fermentation with little residual sugar remaining in the beer. The ability of brewing yeasts to utilise these sugars is determined by the presence of transmembrane sugar transporters such as *MTT1* and *AGT1* (Vidgren &

Brewing Microbiology. https://doi.org/10.1016/B978-0-323-99606-8.00004-3

Londesborough, 2012) that are not found in other yeast groups. In addition to the ethanol production and carbonation that result as a consequence of fermentable sugar use, the yeasts also influence beer flavour by producing volatile flavour compounds. In particular, higher alcohols and esters are important for beer aroma and essential for certain styles of beer, e.g., wheat beers and Belgian ales. It should be noted that production of these volatile compounds is not restricted to brewing yeasts. What is however particular to many brewing yeast strains is the absence of 4-vinylguaiacol production. This phenolic acid derivative is created by decarboxylation of ferulic acid (Richard et al., 2015) and imparts a characteristic clove/spice aroma to beer when present. The majority of beers are produced however with yeast strains that do not produce 4-vinylguaiacol due to loss-of-function mutations in the key genes involved in this biochemical step. The mutation does not offer any obvious competitive advantage to yeasts in the brewing environment, and it is likely that its persistence is purely related to artificial selection, i.e., due to brewers retaining batches of yeast with little or no production of this compound. The trait, like efficient maltotriose use, can be considered a clear sign of domestication. Another characteristic of brewing yeasts is their tendency to aggregate towards the end of fermentation, in a process termed flocculation. This property allows for their efficient removal and hence clarification of beer. While flocculation is not unique to brewing yeasts, the timeliness and strength of this activity are more pronounced in brewing yeasts than in other groups. Again, this can be considered a domestication trait unique to the group. The brewing yeasts are not only phenotypically distinct but also genetically distinct from other *S. cerevisiae strains*. Two major clades have so far been identified. The two groups, Beer 1 and Beer 2, appear to represent two separate historical lineages, with the Beer 1 group representing most of the ale yeasts used in modern brewing (Gallone et al., 2016; Kerruish et al., 2024). Like *S. cerevisiae* strains encountered in the wild, brewing strains possess approximately 6000 genes. However, unlike their wild counterparts, genome sizes are typically larger, and rather than being diploid, the strains' genomes are characteristically polyploid and aneuploid. Aneuploidy, basically an uneven number of chromosomes, has rendered these strains sterile or almost sterile (Gallone et al., 2016).

Typically, ale brewing yeasts are accessible via culture collections. However, recent studies have highlighted the fact that many brewing strains may never have been deposited in an official collection. For example, the 'kveik' strains used in Norwegian farmhouse ale brewing had been used for generations, perhaps centuries, before being deposited in a culture collection or being made available commercially. Rather, these yeasts had been maintained using traditional techniques, mainly drying. Their survival is due solely to the persistence of traditional brewing practices in Norway, in particular in rural areas (Garshol, 2020). In other parts of the Europe, it is likely that unique strains have been lost as a result of modernisation and urbanisation replacing traditional brewing traditions. The few studies on kveik strains have shown that they are both genetically and phenotypically distinct from strains used in industrial breweries. A notable feature is the ability to ferment at very high temperatures, while in many cases still producing a clean tasting beer without excessive aroma production (Preiss et al., 2018). The kveik strains have shown themselves to be suitable for production of many beer styles, even lager beers. The recent 'discovery' of kveik strains and their successful application in industrial brewing has emphasised the importance of strain diversity and also the vulnerability of strains that have not yet been deposited in culture collections for safety. Indeed, traditional brewing is carried out still in many parts of the world—chicha in Latin American and sorghum-based beers in Africa for example. Only very rarely have yeasts been isolated from these environments and characterised with respect to their brewing performance. Preliminary studies suggest that these strains too possess signatures of domestication and could be valuable additions to the brewing yeast canon (Grijalva-Vallejos et al., 2020; Paraíso et al., 2023). Existing collections of brewing yeasts are influenced by a clear Eurocentric bias and do not at all reflect the diversity of this group globally.

Brewing yeasts are generally considered to be highly domesticated, so much so that they are considered incapable of surviving in the wild. This, however, does not mean necessarily that wild yeasts could not be used in brewing. In recent years, there has been a greater focus on non-conventional yeasts used for brewing beer. This has been inspired in part by an increased interest in beer diversity, development of new beer styles and a greater spirit of experimentation in the craft brewing industry, which has, to some extent, crossed over to mainstream brewing. A number of studies have, for example, evaluated the wort fermentation performance of alternative *Saccharomyces* species. As stated above, *S. cerevisiae* is accompanied by seven sister species in the genus. Many of these are maltose positive and have potential for brewing. *Saccharomyces eubayanus*, first discovered in Argentinian Patagonia (Libkind et al., 2011) has, for example, been shown to be capable of wort fermentation (Gibson et al., 2013) and has been used in commercial beer production. *Saccharomyces paradoxus*, a common species in the Northern Hemisphere (Mozzachiodi et al., 2022), also possesses wort fermentation ability and has likewise been used in commercial brewing (Nikulin et al., 2020). A number of other *Saccharomyces* species have shown potential for application in brewing. These include *S. uvarum* and *S. jurei*, both of which can utilise maltose, with the latter also showing ability to consume maltotriose, despite this being seen as a typical trait of domesticated yeasts (Hutzler et al., 2021; Nikulin et al., 2018). Though apparently unable to utilise maltose, the remaining species

(*S. arboricola* and *S. mikatae*) could have potential for low-alcohol brewing. Though typically less efficient at fermentation compared to their domesticated cousins, the wild yeast offer other advantages, including unique flavour profiles. *S. eubayanus*, for example, produces a distinct floral aroma due to phenylethanol production (Gibson et al., 2013; Mardones et al., 2020), while *S. jurei* creates more tropical fruit aromas due to strong ethyl ester production (Hutzler et al., 2021). In all cases, the wild yeast produce the spice-like 4-vinylguaiacol aroma, which may limit their application to certain beer styles. It should however be noted that this is a malleable trait and elimination is possible (Diderich et al., 2018). Also relevant is that all of the wild species have a greater cold-tolerance than *S. cerevisiae* (Magalhães et al., 2021) and could potentially be used in low-temperature lager fermentations.

While it may seem unusual to utilise wild yeast in brewing, the traditional view that brewing strains are highly adapted to the brewing environment, and unable to exist in the wild, has been recently challenged. A lager strain has for example been isolated from the wild (Tafer et al., 2018) and two *S. cerevisiae* strains belonging to the Beer 2 group have been isolated from the fruits in Southern Africa (Paraíso et al., 2023). This suggests that, either brewing yeasts have the ability to escape the brewing environment, essentially becoming feral or, alternatively, that wild strains of brewing yeast survive in natural niches and may historically have been an inoculum source for brewing. This semi-wild, semi-domesticated existence has been seen in wine yeasts, which must survive for long periods between the seasonal wine production periods. The view that brewing yeasts are dependent on humans for their survival has also been challenged by the discovery that contaminant *S. cerevisiae* strains often found in breweries are, in fact, genetically identical to brewing yeasts. Specifically, diastatic strains of *S. cerevisiae*, long believed to be wild strains, were recently found to belong to the Beer 2 clade (Krogerus & Gibson, 2020) and many show good potential for brewing (Meier-Dörnberg et al., 2018). The ubiquity of these strains in breweries (including breweries that do not utilise Beer 2 strains for production) raises the question of their origin. Again, one may speculate about the existence of wild or feral populations of brewing yeasts living independently in the natural environment.

1.3 Lager yeasts

The vast majority of beer produced globally is of the lager type. This style is characterised by a clean flavour profile, that is, limited aroma volatiles produced during fermentation. This is achieved by fermenting at lower temperatures using cold-tolerant lager yeasts. This unique ability is due to the lager yeasts' genetic makeup. In contrast to the *Saccharomyces* yeasts mentioned above, the lager yeast (*S. pastorianus*) is not a true species, but rather an interspecies hybrid created from a cross between an *S. cerevisiae* ale strain and a wild strain of *S. eubayanus* (Gibson & Liti, 2015). Efficient wort sugar fermentation was inherited from the former species and cold tolerance from the latter species (Krogerus et al., 2015). After the hybrid was created, a number of other traits such as inability to convert ferulic acid to 4-vinylguaiacol, and strong flocculation, would have been introduced via artificial selection in breweries. How and when the lager yeast was first created is uncertain, but it seems reasonable to suggest that this occurred in Central Europe, possibly in the early 17th century (Hutzler et al., 2023). During this time, brewing during the warmer months of the year was prohibited, meaning that both brewers and brewer's yeasts were under pressure to ferment at lower temperatures. Under such conditions, it is understandable that a yeast with enhanced cold-tolerance achieved via hybridisation or otherwise would have a competitive advantage over other yeasts. Relative to ale yeasts, the lager yeasts possess a low level of diversity. All extant lager yeasts are derived from two hybridisation events (Okuno et al., 2016), with one leading to the Saaz yeast and the other leading to the Frohberg yeasts. The former group is rarely used in modern brewing due to limited maltotriose use (Gibson et al., 2013; Magalhães et al., 2016). The availability of lager yeasts today is directly influenced by advances in yeast handling in the 19th century, with the existence of the two lager groups being directly related to isolations carried out by the German scientist Paul Linder (Barnett & Barnett, 2011). The availability of strains is also directly related to the work of Emil Christian Hansen at the Carlsberg brewery in Denmark, who pioneered the practice of purifying yeast cultures in brewing. This innovation led to more reliable and consistent lager fermentations and is still standard practice today. Despite the industrial benefits of isolation and purification of lager yeast strains, these developments likely led to a severe loss of lager yeast diversity – what one might call a microbial mass extinction event—as brewers discarded their own mixed cultures in favour of the limited number of purified strains that became available. It is quite certain that more hybrids were in circulation prior to these innovations becoming standard. Hybrid yeasts are still quite common in ale brewing, particularly in Belgian breweries (Langdon et al., 2019).

The lack of diversity in lager yeasts is something that can be corrected. The recent discovery of *S. eubayanus* has facilitated the creation of new lager yeasts. A number of studies have demonstrated how this could be achieved through interspecies mating, that is, crossing cells or spores of an ale yeast with those of *S. eubayanus* (Krogerus et al., 2015; Mertens et al., 2015). Resultant hybrids possessed the key features of lager yeasts, that is, efficient fermentation of wort sugars coupled with psychrotolerance. Predictably, these new hybrids possess the wild type phenotype of 4-vinylguaiacol production, but this can be eliminated through the screening and selection of spore clones from fertile hybrids (Krogerus et al., 2021). Studies also demonstrated that the *S. eubayanus* contribution to the hybrid was limited to enhancing cold tolerance (Krogerus et al., 2015). Further studies demonstrated that a number of different interspecies hybrids could replicate the lager yeast phenotype. Relative to *S. cerevisiae*, all other *Saccharomyces* species are cold tolerant and can be used instead of *S. eubayanus* to create new lager yeasts (Nikulin et al., 2018). Indeed, a recent study has suggested that even *S. cerevisiae* × *S. kudriavzevii* hybrids may have been used in the past for lager brewing (Guan et al., 2024). Hybridisation allows for an increased diversity of lager yeasts and possibly even the introduction of novel traits not previously seen in lager yeasts. This was demonstrated, for example, by Krogerus et al. (2023), who created a novel lager strain with high β-lyase activity by crossing an ale strain with an *S. uvarum* strain possessing an active *IRC7* gene. β-Lyase activity markedly increased the new lager strain's biotransformation capability, with resultant beers having higher levels of the thiols that impart tropical aromas. This is a trait that is not typical for traditional lager yeasts, and its absence may be considered a domestication trait (Ruiz et al., 2021). Brewers now have the possibility to create new lager yeast strains tailored to their own requirements rather than simply selecting from the historically available strains.

1.4 Non-*saccharomyces* brewing yeasts

Many of the studies mentioned above have shown how a less restrictive definition of 'brewing yeast' may be helpful in increasing the diversity of *Saccharomyces* yeasts employed in breweries and consequently the diversity of beers available to consumers. Similarly, this approach may usefully be applied to non-*Saccharomyces* yeasts. Many such yeasts have been utilised in traditional brewing, with *Brettanomyces* species being good examples. So-called 'Brett' yeasts (strains of *B. anomalus* or *B. bruxellensis*) contribute to traditional Lambic and Berliner Weiβe fermentations, in particular by adding flavours that are not encountered when *Saccharomyces* strains alone are used in fermentation (Colomer et al., 2019). Typically, Brett fermentations lead to complex flavour profiles, with acetic, goaty, phenolic and tropical flavours being present at varying levels. The *Brettanomyces* strains are a good example of the utility of non-*Saccharomyces* yeasts in creating beers that could not be produced with standard brewing yeasts. Likewise, brewing yeasts are not the best option if a low-alcohol beer is to be produced. Here, *Saccharomycodes ludwigii*, a yeast unable to utilise the main wort sugars, is used to restrict alcohol production, typically below 0.5% ABV. The species also benefits from the generation of a clean flavour profile, with no clear off-flavours like 4-vinylguaiacol produced (Johansson et al., 2021). The advantage of *S. ludwigii* was recognised early, and strains of the species have been used for almost a century (Haehn & Glaubitz, 1933). Inability to use maltose is however a relatively common trait and a number of other yeasts, some with interesting flavour production have likewise been proposed for low-alcohol brewing. These include amongst others *Cyberlindnera* spp., *Kazachstania servazzii*, *Mrakia gelida*, *Pichia fermentans*, *Pichia kluyveri*, *Scheffersomyces shehatae*, *Toruslaspora delbrueckii*, and *Wickerhamomyces anomalus* (Bellut et al., 2019; De Francesco et al., 2018; Johansson et al., 2021; Li et al., 2011; Linnakoski et al., 2023; Methner et al., 2022; Walker, 2011). The different species offer different advantages with respect to fermentation characteristics and beer quality. In some cases, the strains produce relatively high levels of volatile aromas, compensating for the limited fermentation. Other phenotypes, such as tolerance to low temperatures, could help in introducing strains to low-temperature lager fermentation systems, where also the temperature limits the possibility of contamination by other microorganisms (Linnakoski et al., 2023). In other cases, poor tolerance to low temperature could be advantageous. Nikulin et al. (2022), for example, showed that a cold-sensitive strain of *T. delbrueckii* could be used in a cold-contact fermentation process. In this case, alcohol production is negligible, but reduction of the aldehydes that cause 'worty' off-flavours in beer is still effective.

The value of novel non-*Saccharomyces* yeasts is not limited to production of low-alcohol beers. Strains are also available for production of sour beers. A number of yeast species are capable of producing lactic acid during fermentation. In particular, *Lachancea* species are effective in this regard. Co-production of ethanol and lactic acid is termed 'primary souring' and allows for production of sour beers without the use of lactic acid bacteria, albeit with some reduction in fermentation efficiency expected (Domizio et al. 2016). Such strains are now commercially available.

As with alternative *Saccharomyces* strains, non-*Saccharomyces* strains for brewing can be obtained from different sources. Interestingly, many of the species suitable for brewing application have been obtained from natural locations similar to those of *Saccharomyces* species (Mozzachiodi et al. 2022). In other cases, fermentation environments have been found to be potentially important reservoirs for yeast with brewing potential. These include bioethanol, sourdough, and kombucha fermentations (Cubillos et al., 2019). Contaminant yeasts found in breweries have also been found to have potential brewing capabilities (Krogerus et al., 2022).

1.5 Perspectives

The brewing industry has in recent years faced unprecedented challenges, ranging from energy crises to radically altered customer expectations for beer. The market demands more product diversity and simultaneously requires these products to be created as efficiently and sustainably as possible. Such challenges have inspired new approaches to brewing. These novel approaches also extend to the yeasts employed in fermentation; in many cases, it is no longer sufficient to simply select a traditional brewing strain from a culture collection. Here, a good example is low-alcohol brewing. Clearly, the strains historically deposited in culture collections were done so because of their superior fermentation ability. This advantage is however clearly a disadvantage when the objective is to limit alcohol production. There is also a growing interest in novel flavour profiles, with alcohol content being of secondary importance. More than ever, it is necessary for the industry to be open-minded about the use of yeast in brewing, and more yeasts should be screened for their potential application in brewing. This applies also to non-European brewing strains, for example, those of the Global South that have been utilised for centuries, but overlooked by the brewing industry, and never deposited in culture collections. The kveik story is a good example of how valuable such strains can be and also how vulnerable such strains are to becoming extinct through neglect and ignorance. Clearly, consumers have been embracing more diverse beers. We have the opportunity now to maintain this diversity by redefining how we think of 'brewing yeast', and by embracing a more diverse collection of strains for application in breweries.

References

Barnett, J. A., & Barnett, L. B. (2011). *Yeast research: A historical overview*. USA: American Society for Microbiology Press.

Bellut, K., Michel, M., Zarnkow, M., Hutzler, M., Jacob, F., Atzler, J. J., et al. (2019). Screening and application of *Cyberlindnera* yeasts to produce a fruity, non-alcoholic beer. *Fermentation, 5*, 103. https://doi.org/10.3390/fermentation5040103

Colomer, M. S., Funch, B., & Forster, J. (2019). The raise of *Brettanomyces* yeast species for beer production. *Current Opinion in Biotechnology, 56*, 30–35. https://doi.org/10.1016/j.copbio.2018.07.009

Cubillos, F. A., Gibson, B., Grijalva-Vallejos, N., Krogerus, K., & Nikulin, J. (2019). Bioprospecting for brewers: Exploiting natural diversity for naturally diverse beers. *Yeast, 36*, 383–398. https://doi.org/10.1002/yea.3380

De Francesco, G., et al. (2018). *Mrakia gelida* in brewing process: An innovative production of low alcohol beer using a psychrophilic yeast strain. *Food Microbiology, 76*, 354–362.

Diderich, J. A., Weening, S. M., van den Broek, M., Pronk, J. T., & Daran, J.-M. G. (2018). Selection of POF- *Saccharomyces eubayanus* variants for the construction of *S. cerevisiae* × *S. eubayanus* hybrids with reduced 4-vinyl guaiacol formation. *Frontiers in Microbiology, 9*, 1640. https://doi.org/10.3389/fmicb.2018.01640

Domizio, P., House, J. F., Joseph, C. M. L., et al. (2016). *Lachancea thermotolerans* as an alternative yeast for the production of beer. *Journal of the Institute of Brewing, 122*, 599–604.

Gallone, B., Steensels, J., Prahl, T., Soriaga, L. B., Saels, V., Herrera-Malaver, B., et al. (2016). Domestication and divergence of *Saccharomyces cerevisiae* beer yeasts. *Cell, 166*, 1397–1410.

Garshol, L. M. (2020). *Historical brewing techniques: The lost art of farmhouse brewing*. Colorado, USA: Brewers Pubn.

Gibson, B., & Liti, G. (2015). *Saccharomyces pastorianus*: Genomic insights inspiring innovation for industry. *Yeast, 32*, 17–27.

Gibson, B. R., Storgårds, E., Krogerus, K., & Vidgren, V. (2013). Comparative physiology and fermentation performance of Saaz and Frohberg lager yeast strains and the parental species *Saccharomyces eubayanus*. *Yeast, 30*, 255–266. https://doi.org/10.1002/yea.2960

Goffeau, A., Barrell, B. G., Bussey, H., et al. (1996). Life with 6000 genes. *Science, 274*.

Gonçalves, M., Pontes, A., Almeida, P., et al. (2016). Distinct domestication trajectories in top-fermenting beer yeasts and wine yeasts. *Current Biology, 26*, 2750–2761.

Grijalva-Vallejos, N., Krogerus, K., Nikulin, J., Magalhães, Aranda, A., Matallana, E., et al. (2020). Potential application of yeasts from Ecuadorian chichas for controlled beer and chicha production. *Food Microbiology*. https://doi.org/10.1016/j.fm.2020.103644

Guan, Y., Li, Q., Liu, C., & Wang, J. (2024). Assess different fermentation characteristics of 54 lager yeasts based on group classification. *Food Microbiology, 120*, 104479.

Haehn, H., & Glaubitz, M. (1933). *Beer manufacture*. Patent US1898047 A.

Hirschfelder, G., & Trummer, M. (2022). *Bier: Die ersten 13.000 jahre*. Darmstadt, Germany: WBG.

Hutzler, M., Michel, M., Kunz, O., Kuusisto, T., Magalhães, F., Krogerus, K., et al. (2021). Unique brewing-relevant properties of the non-domesticated yeast *Saccharomyces jurei* isolated from ash (*Fraxinus excelsior*). *Frontiers in Microbiology, 12*, Article 645271. https://doi.org/10.3389/fmicb.2021.645271

Hutzler, M., Morrissey, J. P., Laus, A., Meussdoerffer, F., & Zarnkow, M. (2023). A new hypothesis for the origin of the lager yeast Saccharomyces pastorianus. *FEMS Yeast Research, 23*. https://doi.org/10.1093/femsyr/foad023

Johansson, L., Nikulin, J., Juvonen, R., Krogerus, K., Magalhães, F., Mikkelson, A., et al. (2021). Sourdough cultures as reservoirs of maltose-negative yeasts for low-alcohol beer brewing. *Food Microbiology, 94*, Article 103629. https://doi.org/10.1016/j.fm.2020.103629

Kerruish, D. W. M., Cormican, P., Kenny, E. M., Kearns, J., Colgan, E., Boulton, C. A., et al. (2024). The origins of the Guinness stout yeast. *Communications Biology, 7*, 68. https://doi.org/10.1038/s42003-023-05587-3

Krogerus, K., Eerikäinen, R., Aisala, H., & Gibson, B. (2022). Repurposing brewery contaminant yeast as production strains for low-alcohol beer fermentation. *Yeast, 39*, 156–169. https://doi.org/10.1002/yea.3674

Krogerus, K., & Gibson, B. (2020). A re-evaluation of diastatic *Saccharomyces cerevisiae* strains and their role in brewing. *Applied Microbiology and Biotechnology, 104*, 3745–3756.

Krogerus, K., Magalhães, F., Castillo, S., et al. (2021). Lager yeast design through meiotic segregation of a *Saccharomyces cerevisiae* × *Saccharomyces eubayanus* hybrid. *Frontiers in Fungal Biology, 2*. https://doi.org/10.3389/ffunb.2021.733655

Krogerus, K., Magalhães, F., Vidgren, V., & Gibson, B. (2015). New lager yeast strains generated by interspecific hybridization. *Journal of Industrial Microbiology and Biotechnology, 42*, 769–778. https://doi.org/10.1007/s10295-015-1597-6

Krogerus, K., Rettberg, N., & Gibson, B. (2023). Increased volatile thiol release during beer fermentation using constructed interspecies yeast hybrids. *European Food Research and Technology, 249*, 55–69. https://doi.org/10.1007/s00217-022-04132-6

Langdon, Q. K., Peris, D., Baker, E. P., et al. (2019). Fermentation innovation through complex hybridization of wild and domesticated yeasts. *Nature Ecology and Evolution, 3*, 1576–1586. https://doi.org/10.1038/s41559-019-0998-8

Li, H., Liu, Y., & Zhang, W. (2011). *Method for preparing non-alcoholic beer by* Candida shehatae. Patent CN102220198 B.

Libkind, D., Hittinger, C. T., Valério, E., et al. (2011). Microbe domestication and the identification of the wild genetic stock of lager-brewing yeast. *Proceedings of the National Academy of Sciences of the United States of America, 108*, 14539–14544. https://doi.org/10.1073/pnas.1105430108

Linnakoski, R., Veteli, P., Cortina-Escribano, M., Eerikäinen, R., Magalhães, F., Järvenpää, E., et al. (2023). Brewing potential of strains of the boreal yeast *Mrakia gelida*. *Frontiers in Microbiology*. https://www.frontiersin.org/articles/10.3389/fmicb.2023.1108961/abstract.

Magalhães, F., Calton, A., Heiniö, R. L., & Gibson, B. (2021). Frozen-dough baking potential of psychrotolerant *Saccharomyces* species and derived hybrids. *Food Microbiology, 94*, Article 103640. https://doi.org/10.1016/j.fm.2020.103640

Magalhães, F., Vidgren, V., Ruohonen, L., & Gibson, B. (2016). Sugar utilization by group 1 strains of the hybrid lager yeast *Saccharomyces pastorianus*. *FEMS Yeast Research, 16*, Article fow053.

Mardones, W., Villarroel, C. A., Krogerus, K., et al. (2020). Molecular profiling of beer wort fermentation diversity across natural *Saccharomyces eubayanus* isolates. *Microbial Biotechnology, 13*, 1012–1025. https://doi.org/10.1111/1751-7915.13545

Meier-Dörnberg, T., Kory, O. I., Jacob, F., Michel, M., & Hutzler, M. (2018). Saccharomyces cerevisiae variety diastaticus friend or foe?—spoilage potential and brewing ability of different Saccharomyces cerevisiae variety diastaticus yeast isolates by genetic, phenotypic and physiological characterization. *FEMS Yeast Research, 18*(4), foy023.

Mertens, S., Steensels, J., Saels, V., et al. (2015). A large set of newly created interspecific *Saccharomyces* hybrids increases aromatic diversity in lager beers. *Applied and Environmental Microbiology, 81*, 8202–8214. https://doi.org/10.1128/AEM.02464-15

Methner, Y., Magalhães, F., Raihofer, L., Zarnkow, M., Jacob, F., & Hutzler, M. (2022). Metabolic properties of the yeast strain *Saccharomycopsis fibuligera* with emphasis on sugar uptake and its performance during fermentation of brewer's wort. *Frontiers in Microbiology, 13*, Article 1011155. https://doi.org/10.3389/fmicb.2022.1011155

Mozzachiodi, S., et al. (2022). Yeasts from temperate forests. *Yeast, 39*, 4–24. https://doi.org/10.1002/yea.3699

Nikulin, J., Aisala, H., & Gibson, B. (2022). Non-alcoholic beer production via cold-contact fermentation with *Torulaspora delbrueckii*. *Journal of the Institute of Brewing, 128*, 28–35. https://doi.org/10.1002/jib.681

Nikulin, J., Krogerus, K., & Gibson, B. (2018). Alternative *Saccharomyces* interspecies hybrid combinations and their potential for low-temperature wort fermentation. *Yeast, 35*, 113–127. https://doi.org/10.1002/yea.3246

Nikulin, J., Vidgren, V., Krogerus, K., Magalhães, F. Valkeemäki, S., Kangas-Heiska, T., et al. (2020). Brewing potential of the wild yeast species *Saccharomyces paradoxus*. *European Food Research and Technology, 246*, 2283–2297. https://doi.org/10.1007/s00217-020-03572-2

Okuno, M., Kajitani, R., Ryusui, R., Morimoto, H., Kodama, Y., & Itoh, T. (2016). Next-generation sequencing analysis of lager brewing strains reveals the evolutionary history of interspecies hybridization. *DNA Research, 23*, 67–80.

Paraíso, F., Pontes, A., Neves, J., Lebani, K., Hutzler, M., Zhou, N., et al. (2023). Do microbes evade domestication? - evaluating potential ferality among diastatic *Saccharomyces cerevisiae*. *Food Microbiology, 115*, Article 104320.

Preiss, R., Tyrawa, C., Krogerus, K., Garshol, L. M., & van der Merwe, G. (2018). Traditional Norwegian kveik are a genetically distinct group of domesticated *Saccharomyces cerevisiae* brewing yeasts. *Frontiers in Microbiology, 9*. https://doi.org/10.3389/fmicb.2018.02137

Richard, P., Viljanen, K., & Pentilla, M. (2015). Overexpression of *PAD1* and *FDC1* results in significant cinnamic acid decarboxylase activity in *Saccharomyces cerevisiae*. *AMB Express, 5*, 12. https://doi.org/10.1186/s13568-015-0103-x

Ruiz, J., Celis, M., Martín-Santamaría, M., Benito-Vázquez, I., Pontes, A., Lanza, V. F., et al. (2021). Global distribution of *IRC7* alleles in *Saccharomyces cerevisiae* populations: A genomic and phenotypic survey within the wine clade. *Environmental Microbiology, 23*, 3182–3195. https://doi.org/10.1111/1462-2920.15540

Shen, X. X., Opulente, D. A., Kominek, J., et al. (2018). Tempo and mode of genome evolution in the budding yeast subphylum. *Cell, 175*, 1533–1545.

Tafer, H., Sterflinger, K., & Lopandic, K. (2018). Draft genome sequence of the interspecies hybrid *Saccharomyces pastorianus* strain HA2560, isolated from a municipal wastewater treatment plant. *Genome Announcements, 6*. https://doi.org/10.1128/genomea.00341-18

Vidgren, V., & Londesborough, J. (2012). Characterization of the *Saccharomyces bayanus*-type *AGT1* transporter of lager yeast. *Journal of the Institute of Brewing, 118*, 148–151.

Walker, G. M. (2011). *Pichia anomala*: Cell physiology and biotechnology relative to other yeasts. *Antonie van Leeuwenhoek, 99*, 25–34.

Chapter 2

Yeast quality assessment, management and culture maintenance

Graham G. Stewart[1] and Anne Anstruther[2]
[1]GGStewart Associates, Cardiff, Wales, United Kingdom; [2]The International Centre for Brewing & Distilling, Heriot-Watt University, Edinburgh, Scotland, United Kingdom

2.1 Introduction

For many decades, brewer's yeast strains have been selected empirically by brewers. Strains were required in order to suit the brewing process and the final product (the beer) and also sometimes the whim of the brewer! However, during the past 40 years, or so, research and development has focussed on an understanding of the objectives of yeast performance during wort fermentation in order to produce beer that possesses the necessary flavour, stability and drinkability. Detailed studies have been encouraged by an increasing knowledge of both the brewing microbiology and biochemistry of brewing wort fermentation by yeast (Stewart, Hill, & Russell, 2013). Allied to this development is the advent of sophisticated analytical methodology together with appropriate instrumentation. Principal examples of this development are gas chromatography together with mass spectrometry (GC-MS) (Boulton & Quain, 2001), which is employed for the identification and quantification of beer volatile flavour compounds (Boulton & Quain, 2001). High-performance liquid chromatography (HPLC) has also been employed in order to determine the concentration of wort sugars and dextrins (D'Amore, Russell, & Stewart, 1989), amino acids and small peptides (Lekkas, Stewart, Hill, & Taidi, 2007), together with hop constituents (Roberts & Wilson, 2006). Also, thermal energy analysis (TEA) for N-nitrosodimethylamine determinations (NDMA) – (Goff & Fine, 1979) and bioluminescence for the detection of ATP in biological materials (Boulton & Quain, 2001) has been used to detect possible contaminants that may still be present after process cleaning (but they are not directly relevant in the context of this particular chapter). The initial analytical two methods (GC and HPLC) are an integral part of yeast strain selection and overall brewing fermentation research (Hutzler, Jacob, & Geiger, 2010).

2.2 Objectives of wort fermentation

The objectives of brewer's wort fermentation are: To 'consistently metabolise wort constituents into: ethanol, carbon dioxide and other fermentation products in order to produce beer with satisfactory quality and stability'. It is also important to produce and harvest yeast crops by their flocculation properties that can be confidently re-pitched into subsequent brews (Stewart, 2018).

This is unlike distiller's yeast strains where a yeast culture is only used once during the production process – it is not repitched (Russell & Stewart, 2014). During the brewing process, overall yeast performance is controlled by a plethora of factors which include:

- The yeast strain(s) employed—ale or lager, flocculent or non-flocculent, top or bottom cropping, etc.;
- The condition of a yeast culture at pitching and throughout wort fermentation, particularly its viability and vitality features (Younis & Stewart, 1998);
- The wort's fermentable sugar spectrum (such as: glucose, fructose, sucrose, maltose and maltotriose) (Panchal & Stewart, 1980);
- The concentration of wort's free amino nitrogen (FAN) and the category of assimilable nitrogen (Jones & Pierce, 1964); Table 2.1.

Brewing Microbiology. https://doi.org/10.1016/B978-0-323-99606-8.00022-5

TABLE 2.1 Order of wort amino acids and ammonia uptake during fermentation (Jones & Pierce, 1964).

Group A	Group B	Group C	Group D
Fast absorption	**Intermediate absorption**	**Slow absorption**	**Little or No. absorption**
Glutamic acid	Valine	Glycine	Proline
Aspartic acid	Leucine	Phenylalanine	
Asparagine	Isoleucine	Tyrosine	
Glutamine	Histidine	Tryptophan	
Serine		Alanine	
Methionine		Ammonia	
Threonine			
Lysine			
Arginine			

- The concentration of a plethora of ions in wort, that includes: calcium, zinc, sodium, potassium, carbon, chlorine, etc., to list just a few ions (Goldhammer, 2002);
- The tolerance of yeast cultures and individual cells to stress factors such as: osmotic pressure, ethanol, specific enzymes, temperature, desiccation and mechanical stress (Stewart, 2018); Fig. 2.1.
- The gravity (concentration) of the wort at yeast pitching (Stewart, 1999);
- The wort's dissolved oxygen (DO) level at yeast pitching (Stewart, 2007);
- A yeast culture's flocculation characteristics throughout the fermentation cycle (Fig. 2.2);
- The geometry of fermenters being employed (horizontal, vertical, conical or flat bottom, overall capacity, etc.) and the type of fermentation being conducted, for example batch or continuous (Stewart, Hill, & Russell, 2013).

All of these factors, individually or more often in combination with one another, permit the definition of the requirements of an acceptable brewer's yeast strain (Stewart & Russell, 2009) which is 'in order to achieve beer of consistent quality, it is appropriate that not only must the correct yeast strain be effective in metabolising the required nutrients from the growth/fermentation medium (such as the wort), able to tolerate the prevailing environmental conditions (for example, the wort osmotic pressure, pH, temperature and ethanol tolerance), and impart the desired flavour to the beer in question, but the microorganisms (mostly yeast) themselves must be effectively removed from the fermented wort by: flocculation, centrifugation and/or filtration after they have fulfiled their metabolic roles' (Stewart, 2018).

It has already been discussed here (and this is worthy of further emphasis) that the major difference between brewer's yeast strains and other alcohol producing yeasts is that brewers (unlike distillers) recycle their yeast cultures from one fermentation into a subsequent one. It is, therefore, important to jealously protect the quality of the cropped yeast culture because it will be used later in order to be repitched into a subsequent fermentation and this will, consequently, have a profound effect on the quality of the resulting beer produced by it. Distillers (for example, Scotch whisky producers) only use a yeast culture once (Russell & Stewart, 2014). This introduces a separate series of selection criteria for a particular yeast culture which is beyond the scope of this chapter to discuss this matter in greater detail.

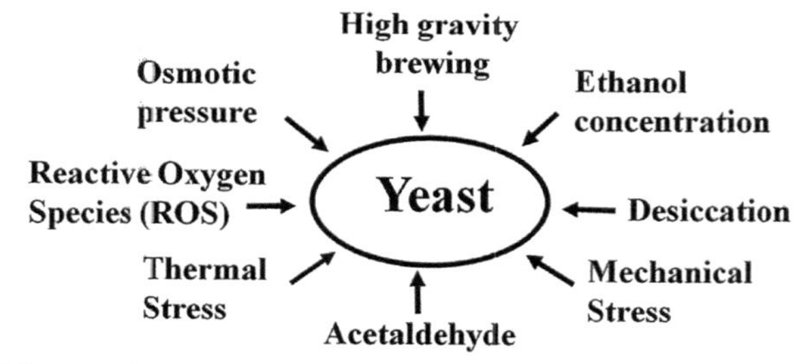

FIGURE 2.1 Stress factors which promote yeast proteinase A release.

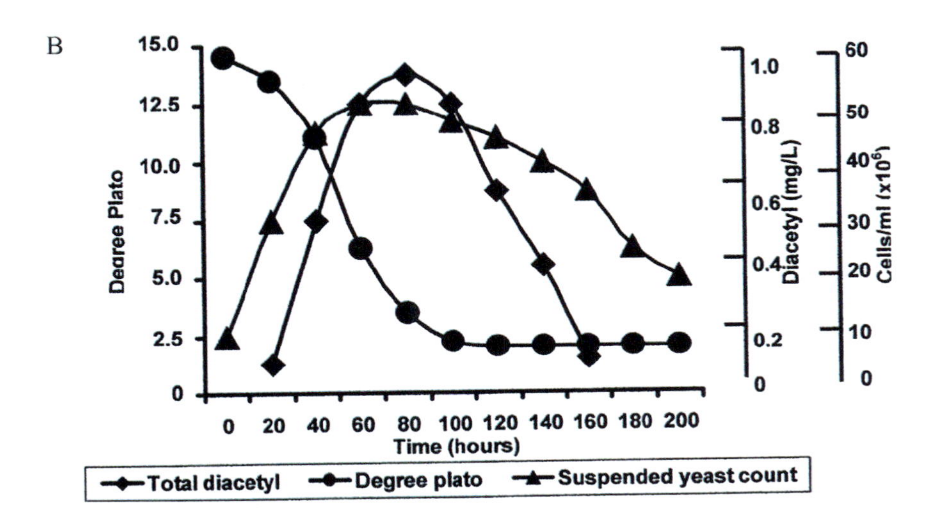

FIGURE 2.2 (A) Giant colony morphologies of co-flocculent ale yeast strains and the influence on wort diacetyl levels. (B) Flocculation characteristics of a brewer's yeast strain during static fermentation and its influence on wort diacetyl levels.

2.3 Brewer's yeast species

There are basically three different types of beer: lager, ale and stout (dark beer). In reality, stout is a form of ale. The commercial worldwide production volume of ale has always been considerably less than that of lager and over the years, this yeast difference has grown worldwide (Stewart, 2013). However, this difference between ale and lager beer production volumes has recently narrowed slightly, particularly in the USA. during the past decade, and it is currently approximately 4.6% ale, largely due to the increasing popularity of the craft brewing sector (Lorenzoni, 2023).

Although there are several differences between the production methods of these two types of beer, one of the main differences is the characteristics of the ale and lager yeast strains employed primarily during the production process. Consequently, research by many breweries and research institutions on this topic has been extensively conducted (Stewart, Hill, & Lekkas, 2013) and typical differences between ale (*Saccharomyces cerevisiae*) and lager (*Saccharomyces pastorianus*) yeast strains has been established, details later (Table 2.2).

With the advent of novel molecular biology-based technologies, the gene sequencing of ale and lager brewing strains has shown that they are interspecies hybrids with homologous relationships to one another and also to the appropriate yeast genus *Saccharomyces bayanus*, a yeast species that is employed in wine fermentations and has been identified as a wild yeast for brewing fermentations (Sofie, Saerens, Duong, & Nevoigt, 2010). The gene homology between *S. pastorianus* and *S. bayanus* strains is relatively high at 72% (Fig. 2.3) whereas the homology between *S. pastorianus* and *S. cerevisiae* is much lower at 50% (Pederson, 1995) (Fig. 2.3).

Libkind et al. (2011), working in Argentina, Portugal and the U.S.A. published a paper entitled Microbe Domestication and the Identification of the Wild Genetic Stock of Lager-Brewing Yeast' (Libkind et al., 2011). This publication confirmed that *S. pastorianus* is a domesticated yeast species that has been developed by the fusion of *S. cerevisiae* with a

TABLE 2.2 Differences between ale and lager yeast strains.

Ale yeast	Lager yeast
Saccharomyces cerevisiae (ale type)	*Saccharomyces carlsbergensis*
Saccharomyces cerevisiae (ale and distillers yeast)	*Saccharomyces uvarum* (*carlsbergensis*)
	Saccharomyces cerevisiae (Lager type)
	Saccharomyces pastorianus (current taxonomic name)
Fermentation temperature (18–25°C)	Fermentation temperature (8–15°C)
Cells can grow at 37°C and higher	Cells cannot grow above 34°C
Cells cannot ferment the disaccharide melibiose (galactose – glucose)	Ferments melibiose
Strains with distinctive colonial Morphology on wort-gelatin medium	Strains do not have a distinctive morphology on wort-gelatin medium (Stewart, Garrison Russell)
'Top' fermentation	'Bottom' fermentation

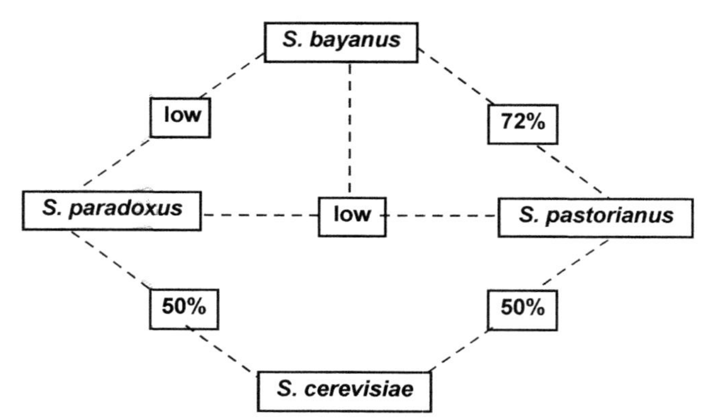

FIGURE 2.3 The *Saccharomyces sensu stricto* group for ale and lager strains.

previously unknown species that has now been designated *Saccharomyces eubayanus* because of its close relationship to *S. bayanus* (Libkind et al., 2011). They also reported that *S. eubayanus* exists in the forests of Patagonia and was not found in Europe until the advent of trans-Atlantic trade between Argentina and Europe. Libkind et al. (2011) paper contains a draft genome sequence of *S. eubayanus*. It is 99.5% identical to the non-*S. cerevisiae* portion of the *S. pastorianus* genome sequence and suggests specific changes in fermentable wort sugars, such as glucose, fructose, maltose and maltotriose and sulphite metabolism (Stewart & Ryder, 2019) compared to the ale strains that are critical for determining lager beer characteristics. More recent studies suggest that the *S. cerevisiae* parent of *S. pastorianus* originated in Europe, and the development of pure starter cultures led to the global spread of the *S. pastorianus* lineages (Hutzler, Morrissey, Laus, Meussdoerffer, & Zarnkow, 2023).

2.4 Yeast management

The overall process between fermentations is collectively described as yeast management. This process includes: strain storage (in a culture collection), culture propagation, cropping, culture storage, acid washing (if required) and, subsequently, wort fermentation itself. These latter brewing procedures are not usually regarded as yeast management as such and will not be discussed further here. The use of dried brewer's yeast is becoming increasingly popular in some brewing operations, and, as a consequence, it will be discussed later in this chapter (Finn & Stewart, 2002).

It is a normal procedure, in most breweries, to propagate fresh yeast (particularly with lager yeast strains) every 8–10 generations (fermentation cycles) or less. Prolonged yeast cycles can result in sluggish fermentations, usually due to slower reduced rates of both wort maltose and maltotriose uptake, higher levels of sulphur dioxide and hydrogen sulphide production and prolonged diacetyl formation (Fig. 2.4 – formation of diacetyl) and reduction times (Fig. 2.5), also, increased yeast flocculation and culture sedimentation rates (Fig. 2.6).

The long-term preservation of a brewing yeast culture requires that not only is optimal self-survival important, but it is also imperative that no changes in the character of the yeast culture occurs. Many yeast strains are difficult to maintain in a stable state and long-term preservation by lyophilisation (freeze drying), which has proven useful with mycelial fungi and bacteria.

Schlee, Miedl, Leiper, and Stewart (2006) has also been found to give poor fermentation results with most brewing yeast strains (Stewart & Russell, 2009). Storage studies have been conducted with a plethora of both ale and lager brewing strains—details later (Russell & Stewart, 1981).

It is important to emphasise that considerable information is available regarding brewer's yeast fermentation per se (for example, Boulton & Quain, 2001; Sofie et al., 2010; Stewart & Russell, 2009). However, by comparison, basic detailed information on yeast management processes between wort fermentations has been lacking. Indeed, although overall fermentation procedures and its control have become very sophisticated, yeast management was, until recently, the 'poor relation' of the overall process!

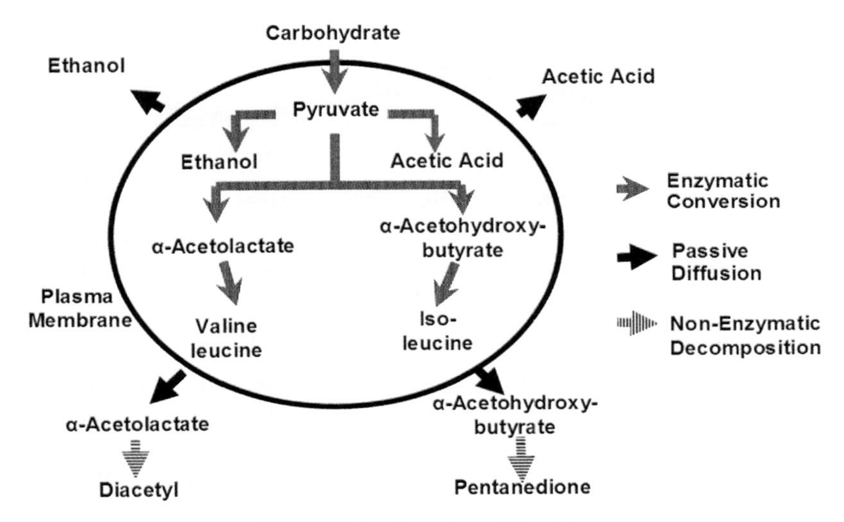

FIGURE 2.4 Formation of diacetyl.

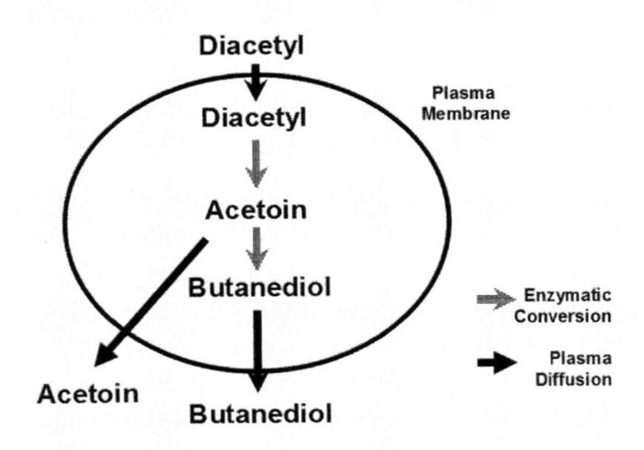

FIGURE 2.5 Reduction of diacetyl.

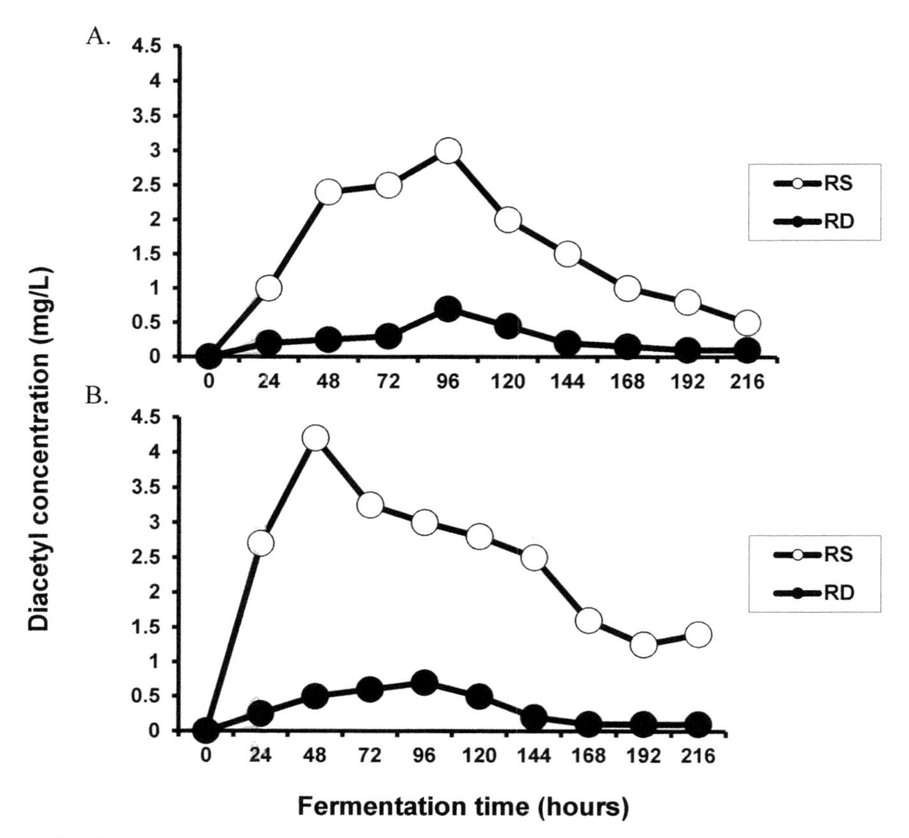

FIGURE 2.6 Production of free diacetyl (a) and total diacetyl (diacetyl plus α-acetolactate) (b) during fermentation of 16°Plato wort by a lager strain RS (●) and its spontaneously generated RD mutant (○). Fermentation was conducted in 30 L static fermentation at 15°C.

Yeast management can be divided into a number of overlapping procedures:

- Prior to propagation (the production of yeast biomass) after fermentation and cropping most (not all) yeast strains are stored under standard conditions in a brewery or in an accredited culture collection—sometimes, both for security (Russell & Stewart, 1981);
- Yeast propagation (biomass formation) in wort under aerobic conditions (Stewart & Russell, 1981);
- Following propagation, the yeast is pitched into wort. This is the first generation (cycle) of a multi-generation procedure;
- At the end of wort fermentation (attenuation), yeast cropping occurs followed by storage prior to re-pitching. Cropping occurs using the flocculation characteristics of the appropriate yeast strain or with a centrifuge (Fig. 2.7)— details later;
- In order to eliminate contaminating bacteria, the yeast slurry can be acid washed (Chlup & Stewart, 2011). Also, sometimes, but less frequently these days, the yeast slurry is also sieved to remove contaminating trub (coagulated protein-phenol solid materials) (Bamforth, 2020).

2.5 Storage of yeast stock cultures between propagations

It has already been discussed here (Bamforth & Lentini, 2009) that the advent of pure yeast strain fermentation dates from the studies of Emil Christian Hansen (1883), who worked in the Copenhagen Carlsberg Laboratory during the latter decades of the nineteenth century. Hanson isolated four separate yeast strains from the Carlsberg lager yeast culture (Holter & Moller, 1976). He studied these four lager strains from the standpoint of overall brewery performance and only one of them proved to be suitable for lager beer fermentations. This strain, designated as ‘Carlsberg Yeast No. One’, was introduced into the Carlsberg Brewery in Copenhagen for its use on a production scale on 13 May 1883 and pure strain brewing of lager beer can be considered to have commenced from this date (Holter & Moller, 1976). Carlsberg Yeast No.1 was named *Saccharomyces carlsbergensis* (Hansen, 1883) and is now known as *S. pastorianus* (Pederson, 1995).

FIGURE 2.7 The disk stack centrifuge (5 hL/h).

With the advent of the use of pure yeast strain fermentations in brewing, Hansen soon found that it was necessary to furnish the Carlsberg brewery with production quantities of pure yeast cultures of the single lager strain and that it would be more convenient to develop a specific apparatus for the purpose of large scale yeast propagation. Consequently, in association with a coppersmith (W.E. Jansen), Hansen developed an apparatus specific for this purpose (Fig. 2.8).

At the beginning of 1886, this apparatus was effectively working in the Carlsberg brewery and was also operating in a number of other breweries, including Heineken in the Netherlands and in a number of breweries in the United States.

As a result of Hansen and Jansen's efforts, the practice of employing a strain for lager production was soon adopted by many breweries worldwide, particularly in the USA. However, ale-producing regions met this 'radical innovation' with severe opposition and scepticism! This method was merely regarded (by some) as a means of reducing wild yeasts and bacterial infections. It was not until the middle of the 19th century that the pure strain fermentation methods were adopted. Indeed, a few ale producing brewers have yet to adopt this procedure! Anderson (2012) has published a paper entitled: 'One yeast or two? Pure yeast and top fermentations' which focusses on the reluctance of British ale brewers to introduce pure yeast ale strains into the production of top fermentation beers until comparatively recently. Currently, yeast propagation equipment is available for both large and small breweries worldwide 'off the shelf' (Nielsen, 2010).

2.6 Preservation of yeast strains

The long-term preservation of a brewing yeast strain requires that not only is optimal survival important, it is imperative that no change in the characteristics of the yeast strain occurs. Hansen's studies resulted in storage of his yeast strains in liquid nutrient media prior to culture propagation. This propagation technique has evolved in many breweries with independent culture collections who maintained their yeast strains on nutrient media solidified initially with gelatine and subsequently with agar (Stewart, 2017). Some yeast strains are difficult to maintain in a stable state and long-term preservation by lyophilisation (freeze drying) and has proven useful for mycelial fungi and bacteria (Kirsop & Doyle, 1990), but has been found to give poor survival results with brewing yeast strains (Kirsop, 1955). However, it has already been described here that the use of dried yeast cultures for pitching into wort is rapidly increasing in popularity (Finn & Stewart, 2002).

Storage studies have been conducted with a number of both ale and lager brewing yeast strains (Russell & Stewart, 1981). The following yeast storage conditions have been investigated:

- Low temperature as a result of storage in liquid nitrogen (−196°C). With the advent of −70°C refrigerators in the 1980s, liquid nitrogen has been largely replaced for this purpose with similar results (Russell & Stewart, 1981);

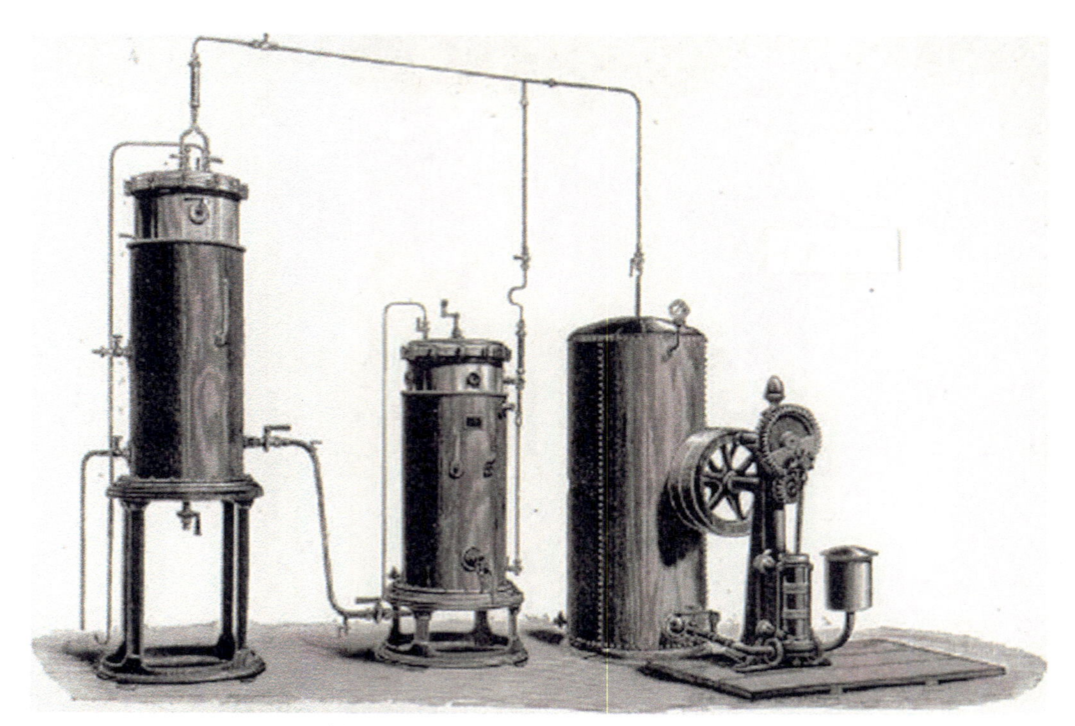

FIGURE 2.8 Hansen and Jansen's yeast propagation apparatus.

- Lyophilisation (freeze drying);
- Storage in distilled water;
- Storage under oil;
- Repeated direct transfer on solid culture media—subcultured once a week for 2 years;
- Long-term storage at 21°C on a solid nutrient medium—subcultured every 6 months for 2 years;
- Long-term storage at 4°C on a solid nutrient medium—subcultured every 6 months.

After a 2-year storage period, wort fermentation tests that included the fermentation rate and the wort sugar uptake efficiency (Fig. 2.9), flocculation characteristics (Fig. 2.10), sporulation ability, formation of respiratory deficient colonies (Ernandes et al., 1993) (Figs 2.11 and 2.12) and the rate of culture survival were conducted, and the results were compared to the characteristics of a stored control culture. Low-temperature storage appears to be the storage method of choice. However, there are capital and ongoing cost considerations connected with this method. Storage at 4°C on nutrient agar slopes, subcultured every 6 months, or so, was the next method of preference to low temperature storage. This method is simple to perform and relatively inexpensive. Lyophilisation and other storage methods revealed yeast's instability which

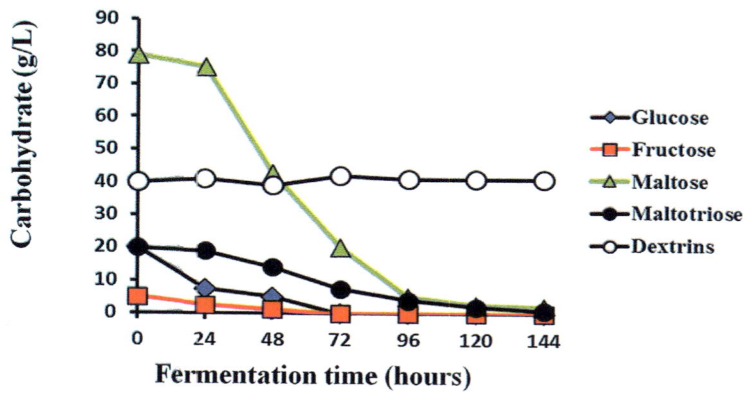

FIGURE 2.9 Order of the uptake of wort sugars by brewer's yeast cultures.

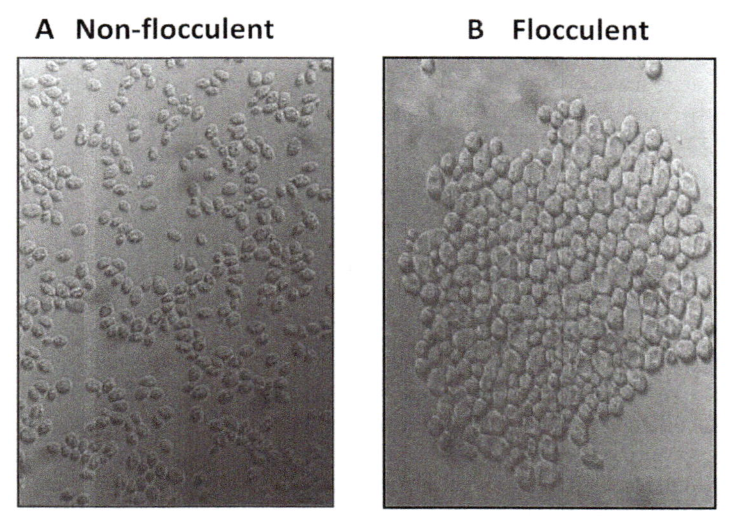

FIGURE 2.10 Flocculation characteristics of a brewer's yeast strain.

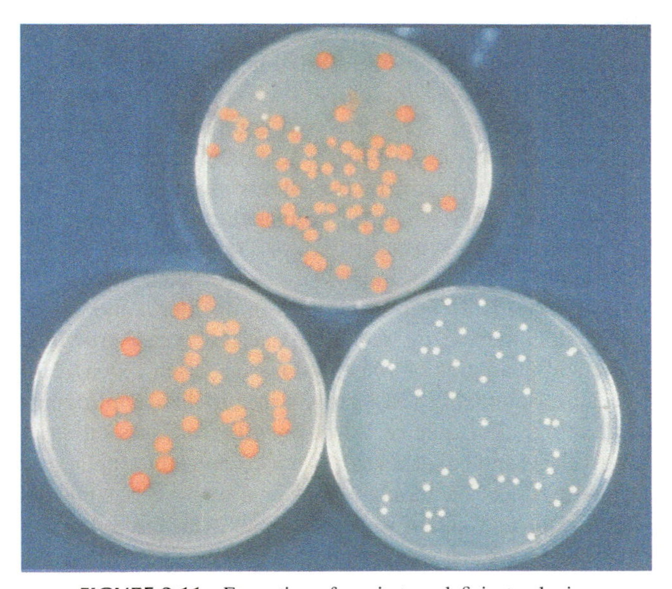

FIGURE 2.11 Formation of respiratory deficient colonies.

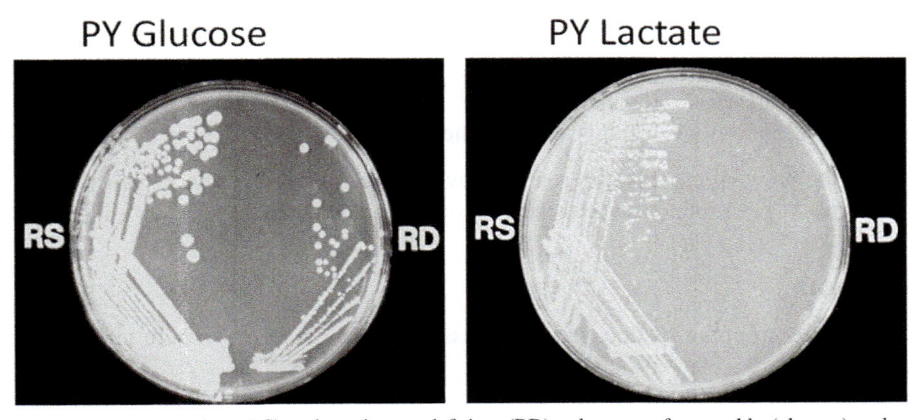

FIGURE 2.12 Growth of respiratory sufficient (RS) and respiratory deficient (RD) cultures on fermentable (glucose) and non-fermentable (lactate) carbon sources. *PY*: Peptone-yeast extract.

varied from strain to strain. Many breweries currently store their yeast strains (or contract store them) at −70°C. Routine subculturing of yeast cultures on solid media every 6 months or so, although being a less desirable storage procedure, is still an acceptable method. Freeze drying should be avoided as a yeast storage method (Finn & Stewart, 2002) although, its use for a pitching culture is becoming increasingly popular!

2.7 Yeast propagation

Yeast propagation is a traditional and well-established process in most large breweries (Nielsen, 2010). Also, some multibrewery operations propagate their yeast centrally and distribute the yeast cultures to individual breweries internationally. Nevertheless, propagation development equipment is constantly ongoing, and questions remain to be answered (Stewart, Hill, & Lekkas, 2013, Stewart, Hill, & Russell, 2013). The requirements for a freshly propagated yeast culture are that it is not stressed, is highly vital and viable, and is free from contaminating organisms. The way to this objective involves a carefully designed sanitary propagation plant with an aeration (oxygenation) system that is able to supply sufficient oxygen to all yeast cells during the propagation phase, without causing mechanical stress to the cells, which are in a wort consisting of the appropriate nutrient composition (further details of oxygenation during brewing later).

No matter how many of these conditions are optimised, it is still only possible to obtain relatively low cell numbers (approximately 100−200 million cells/mL which is equivalent to 2.5−5.0 g dry matter per litre). In order to avoid losing time during the delay for the yeast to consume all of the wort sugars, a complimentary process should be employed. This process has been adapted from the baker's yeast propagation process and is conducted in a fed batch reactor, whereby the sugar concentration is maintained at a consistently low level, but not too low, in order to avoid the yeast growing aerobically and, thereby, potentially losing some of its fermentation characteristics during the propagation procedure. Consequently, a hybrid process employed between traditional brewery propagations and the aerobic yeast propagation process, employed for baker's yeast propagation, is the preferred compromise (Boulton & Quain, 1999).

In a typical brewery, propagation is carried out in a batch reactor, with wort as the growth culture medium. This is basically the same medium that will be used later in order to ferment wort into beer. Although wort gravities have increased for fermentation, weaker worts are still more appropriate for propagation. The propagation medium used to produce yeast for both the distilling and baker's yeast industries is usually molasses (sometimes hydrolysed whey) where the major sugar is sucrose plus a nitrogen source (usually ammonium ions). Also, a fed batch reactor with a continuous supply of dilute substrate and intense aeration (oxygenation) is used to produce both distiller's and baker's yeast. When propagation in a brewery is carried out in a batch reactor, the use of wort limits aerobic yeast growth in a concentrated sugar solution making it difficult to produce theoretical quantities of yeast biomass (Nielsen, 2005). However, the brewing industry has chosen to tolerate this problem, because optimising yeast growth in a molasses/nitrogen medium could jeopardise wort's fermentation properties and lead to poorer beer quality. Also, brewing focusses on strict sanitary conditions in order to avoid infection (the production of distiller's and baker's yeast is not completely aseptic) and to minimise yeast stress during propagation in order to avoid negative effects on fermentation. It is worth repeating that brewer's yeast propagation is based on aerobic conditions and the extensive use of sterile air or oxygen throughout this process. This process differs extensively from brewing fermentations where oxygen is only required at the beginning of the procedures in order for the lag phase cells to begin to synthesise both unsaturated fatty acids and sterols which are both important membrane constituents.

This culture synthesis occurs largely from glycogen as the substrate (Fig. 2.13). Parenthetically, it is interesting to note that oxygen is only required during the following stages for the malting and brewing process:

- During barley germination for malting;
- For biomass formation during yeast propagation;
- At the beginning of fermentation when the yeast is pitched into wort.

At any other point in the brewing process, oxygen can have a negative effect on beer quality particularly when there is dissolved oxygen in the packaged product that can lead to stale characteristics in the beer (Stewart, 2004).

2.8 Yeast collection

Yeast collection (also termed cropping or flocculation) (Russell & Stewart, 1981) where techniques vary depending on whether one is dealing with:

- a traditional ale top-cropping fermentation system;

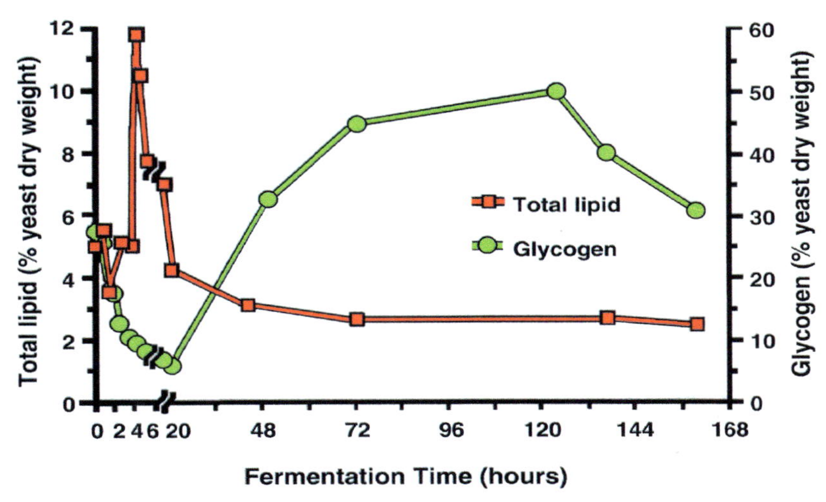

FIGURE 2.13 Intracellular concentration of glycogen and lipids in a lager yeast strain during fermentation of a 15°Plato wort.

- a traditional lager bottom fermentation system;
- a cylindroconical fermentation system (Stewart & Russell, 2009);
  or
- a non-flocculent yeast culture where the cells are cropped with a centrifuge.

With traditional top ale cropping fermentation systems, there are many variations to this system. For example, a simple, dual or multi-strain yeast system, can be employed (Stewart & Garrison, 2018). The timing of the skimming process can be critical in order to maintain the flocculation characteristics of the ale strains. Traditionally, the first skim or 'dirt skim', with the trub present, is discarded, as is the final skim usually with the middle skim being kept for repitching.

With the traditional lager bottom fermentation system, the yeast is deposited on the bottom of the vessel at the conclusion of fermentation. This type of yeast collection is essentially non-selective and the yeast will normally contain entrained trub.

The cylindroconical fermentation system has now been widely adopted for both ale and lager fermentations, with the angle at the bottom of the fermenter allowing for the effective removal of the yeast plug.

The use of centrifuges for the removal of yeast from the fermented wort is complex, and the collection of the flocculated pitching yeasts is now commonplace. There are a number of advantages to the use of centrifuges, which includes shorter process times, cost reductions (after significant initial capital costs), increased productivity and reduced wort shrinkage (Chlup & Stewart, 2011).

Care must be taken to ensure that elevated temperatures (above 20°C) are not generated during centrifugation and that the design ensures low dissolved oxygen pickup and a high culture throughput (Chlup, Bernard, & Stewart, 2007). In addition, centrifugation can (under certain circumstances) cause physical damage to yeast cells and, consequently, can negatively affect beer physical stability (haze). This is dependent on the centrifuge's operating parameters.

Hydrodynamic forces and yeast cell interactions within the gap of the centrifuge disc stack can create collisions amongst yeast cells producing kinetic energy that causes cellular damage. Also, release of cell wall mannan during mechanical agitation of yeast slurries in conjunction with an increase in beer haze has been well documented (Chlup, Conery, & Stewart, 2007, 2008).

2.9 Yeast storage

At the end of fermentation, the brewing yeast is cropped for further use employing the flocculating characteristics of the yeast strain or with a centrifuge. However, in this discussion, yeast cropping is considered to be part of fermentation but not yeast management between fermentations. It has already been described here that one method of yeast cropping that is increasing in popularity, is the use of centrifuges, although their use has not been without its problems (Table 2.3) (Chlup & Stewart, 2011).

If a cropped yeast culture is not stored properly, cell consistency will suffer which can adversely affect batch fermentation and beer quality. Following cropping, the yeast is stored in a separate facility that is conveniently sanitised,

TABLE 2.3 Characteristics of yeast cells recycled nine times in 20°P wort and beer before and after centrifugation at high g-force.

Characteristic	Before Centrifugation	After Centrifugation
Viability (%)	85	42
Extracellular pH	4.2	6.0
Intracellular pH	5.8	5.3
Damaged cells (%)	4	15
Glycogen (ppm)	18	8
Trehalose (ppm)	22	6
Mannan released (counts)	400	1000
Proteinase A (U/mL)	3.1	6.2
Hydrophobic polypeptides (mg/L)	48	25
Beer foam stability (NIBEM)	110	82

contains a plentiful supply of sterile water and a separate filtered air supply with positive pressure in order to prevent the entry of contaminants at a temperature of 0°C. Alternatively, insulated tanks in a dehumidified room can be employed. Also, 'off the shelf' yeast storage facilities are currently available in a plethora of working capacities.

Yeast is predominantly stored under six inches of beer (sterile water has also been employed in the past but its use is currently unpopular!). When high gravity brewing procedures are practised (Murray & Stewart, 1991), it is important to ensure that the ethanol level of the stored beer is decreased to 4%–6% (v/v) ethanol in order to maintain the viability and vitality of the stored yeast. As more sophisticated systems become available, storage tanks with external cooling (0–4°C), equipped with low shear stirring devices, have become popular. The need for low shear stirring systems has been shown to be important. With high velocity agitation in a yeast storage tank, the yeast cell surface can become disrupted and intracellular proteases (particularly proteinase A [PrA]) is excreted from the yeast cells and this can result in unfilterable mannan hazes in the beer (Stoupis, Stewart, & Stafford, 2002) together with poor head retention due to protease hydrolytic activity on foam stability enhancing peptides (Cooper, Stewart, & Bryce, 2000). There are procedures where the yeast is not stored between fermentations. In this case, the yeast is pitched directly from one fermenter to another. This yeast handling procedure occurs with cylindroconical (vertical) fermenters and is termed 'cone to cone yeast pitching'. This procedure was employed by some breweries in the 1980s and 1990s but currently has limited application because of lack of opportunity and time to conduct quality and contamination studies on the yeast between fermentations.

One of the factors that will affect the fermentation rate is the condition under which the yeast culture is stored between fermentations. Of particular importance in this regard is the influence of temperature during these storage conditions on the cell's intracellular glycogen level. Glycogen is the major reserve carbohydrate stored within the yeast cell and is similar in structure to plant amylopectin (Fig. 2.14).

It has already been discussed here that glycogen serves as a store of biochemical energy during the lag phase of fermentation when the yeast energy demand is intense for the synthesis of such compounds as sterols and unsaturated fatty acids. Consequently, it is important that appropriate levels of glycogen and trehalose are maintained during culture storage so that during the initial stages of fermentation, the yeast cell is able to synthesise appropriate levels of sterols and unsaturated fatty acids. Trehalose is a non-reducing disaccharide that plays a protective role during osmoregulation, for the protection of cells during conditions of nutrient depletion and starvation and improving a cell's resistance to high and low temperatures and elevated ethanol concentrations.

Storage temperatures (Fig. 2.15) do have a direct influence on the rate and extent of glycogen dissimulation, as might be expected, considering the effect that temperature has upon metabolic rates in general. Although yeast strain dependent, of particular interest, is the fact that within 48 h, the yeast stored semi-aerobically at 15°C and has only 15% of the original glycogen concentration remaining. Glycogen reductions to this extent will have a profound overall effect on wort's fermentation.

The number of times that a yeast crop (generations or cycles) is used for wort fermentations is usually standard practice in a particular brewery. Typically, a yeast culture is used 6–10 times prior to reverting to a fresh culture of the same yeast

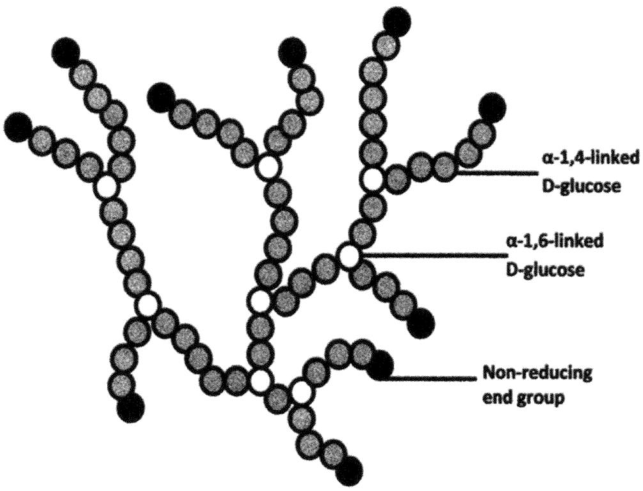

FIGURE 2.14 Structure of glycogen.

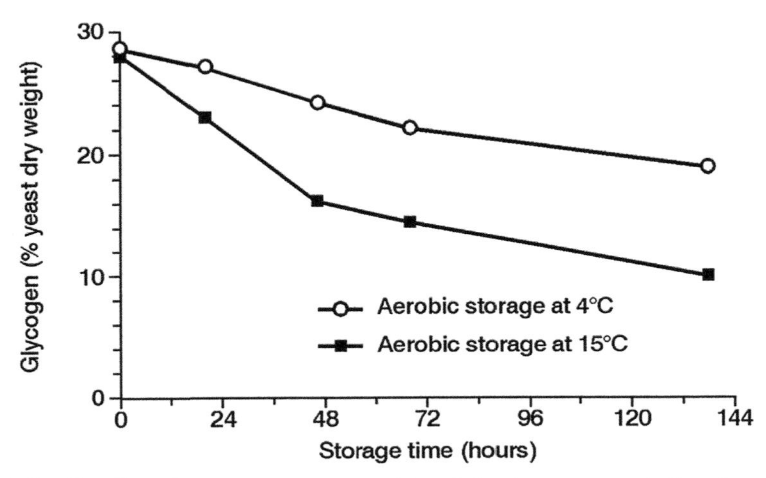

FIGURE 2.15 The effect of storage temperatures on intracellular glycogen concentration of yeast cultures.

strain obtained from the pure yeast culture plant. If a particular yeast culture is used beyond its crop specification, fermentation difficulties (fermentation rate and extent are typical examples) are often encountered! An example of this effect is when a brewery increases its wort gravity in order to adopt high gravity brewing procedures (Stewart, Russell, & Garrison, 1975). A particular brewing operation, over a 15 year period, has increased its wort gravity incrementally. In order to avoid fermentation difficulties, it reduced the number of yeast cycles from a single propagation:

12 degrees Plato wort — > 20 yeast cycles
14 degrees Plato wort — 16 yeast cycles
16 degrees Plato wort — 12 yeast cycles
18 degrees Plato wort — eight yeast cycles

Currently, some breweries have adopted a yeast re-use specification as few as four to six cycles!

The reason why multiple yeast generations can exhibit a negative effect on a culture's fermentation performance is unclear. However, multiple generations will result in reduced levels of intracellular glycogen and an increase in trehalose concentrations indicating additional stress conditions on the culture as the cycles progress (Table 2.4) (Boulton & Quain, 2001).

Yeast storage conditions between brewing fermentations can affect fermentation efficiency and result in agreeable beer quality. Good yeast handling practices should encompass collection and storage procedures, should avoid the inclusion of oxygen in the slurry. Cooling the slurry to 0–4°C, soon after its collection, and perhaps most importantly, ensuring that intracellular glycogen levels are maintained because of its critical property at the start of wort fermentation!

TABLE 2.4 Concentration of trehalose and glycogen in a lager yeast culture following one, four and eight cycles after fermentation in 15°Plato wort.

	Fermentations (cycles)		
	One	Four	Eight
Trehalose[a]	8.8	9.2	11.6
Glycogen[b]	14.6	12.6	9.2

[a]*mg/g dry weight of yeast.*
[b]*mg/g dry weight of yeast.*

2.10 Yeast washing

Acid washing pitching yeast at pH 2–2.2 (with either phosphoric, tartaric, hydrochloric, sulphuric and nitric acid solutions) usually during the later stages of storage just prior to being pitched into wort for fermentation, has been employed by many breweries for the past 100 years (and longer) as an effective method to eliminate contaminating bacteria (not wild yeasts) without adversely affecting the physiological quality of the yeast culture. The acid washing regime differs between breweries with some brewers routinely acid washing their yeast after each fermentation cycle, whereas other brewers only acid wash their pitching yeast when there is significant bacterial contamination and some breweries (not all) refrain from acid washing their yeast culture completely! Brewer's yeast strains are normally resistant to acidic conditions when the washing procedure is conducted properly. However, if other environmental and operating conditions are modified, then the acid resistance of the yeast culture will vary. Simpson and Hammond (1989) demonstrated that if the temperature of acid washing was greater than 5°C and/or the ethanol concentration was greater than 8%(v/v), acid washing had a detrimental effect on the culture, causing decreases in the culture's viability and their overall fermentation performance (Pratt-Marshall, Brey, de Costa, Bryce, & Stewart, 2002). The physiological condition of the yeast prior to acid washing is an important factor for acid tolerance with yeast, in poor physiological condition prior to washing, being more adversely affected by acid washing than a healthy yeast!

Acid washing primarily affects the yeast cell envelope with the physiological systems associated with both the cell wall and the plasma membranes, subsequently, decreasing a yeast culture's vitality is measured by the acidification power test (Kara, Simpson, & Hammond, 1988). Studies in this laboratory (Cunningham & Stewart, 2000) have reported that acid washing pitching yeast cropped from high gravity (20 degree Plato) wort fermentations did not affect the fermentation performance of cropped yeast if it was maintained in good physiological condition. Oxygenation of the yeast at the start of fermentation stimulated yeast growth leading to a more efficient wort fermentation and equally important in the context of yeast management between fermentations produced yeast that was in good physiological condition permitting the culture to tolerate exposure to acid washing conditions (phosphoric acid solution at pH 2.2). These data support the findings of Simpson and Hammond (1989), who concluded that yeast in poor physiological condition should not be acid washed.

In summary, the do's and do not's for yeast acid washing, listed by Simpson and Hammond (1989), are still appropriate.

The Do's of acid washing are:

- Use food grade acid;
- Chill the acid and the yeast slurry before its use to less than 5°C;
- Wash the yeast as a beer slurry or as a slurry in water;
- Ensure constant stirring whilst the acid is added to the yeast slurry and preferably throughout the washing procedure;
- Ensure that the temperature of the yeast slurry does not exceed 5°C during the washing procedure;
- Routinely verify the pH of the yeast slurry throughout the washing procedure; and
- Pitch the yeast immediately after washing.

The Do Nots of acid washing are:

- Do not wash a yeast culture for more than 2 h—very important!;
- Do not store a washed yeast culture for more than 2 h—equally important;
- Do not wash unhealthy yeast; and
- Avoid washing yeast from high-gravity wort fermentations prior to dilution.

There are also a number of options to acid washing brewer's yeast:

- Never acid wash yeast;
- Low yeast generation (cycle) specification;
- Discard yeast when there is evidence of contamination (bacteria and/or wild yeast);
- Acid wash yeast every fermentation cycle, this procedure can have adverse effects on yeast or;
- Acid wash the yeast slurry when bacterial infection levels warrant this procedure.

2.11 Yeast stress

During wort fermentations, a yeast culture is exposed to a number of stress conditions and the primary stress factor (not the only one) is the use of high gravity worts (Pratt-Marshall et al., 2002). During these circumstances, yeast cells are exposed to numerous stresses, including osmotic stress at the beginning of fermentation due to high concentrations of wort sugars and ethanol stress at the end of fermentation (Stewart, 1999). Other forms of yeast stress are: desiccation (details later), mechanical stress and thermal stress (hot and cold). The yeast is expected to maintain its metabolic activity during stressful conditions by not only surviving these stresses but by rapidly responding in order to ensure continued cell viability and vitality (Casey, Chen, & Ingledew, 1985).

Stress can have profound and varied effects on yeast cells, including:

- A negative effect on overall yeast fermentation performance resulting in decreased attenuation rates, sluggish fermentation and a marked reduction in cell volumes with a concomitant loss of cell viability and vitality.
- Cell autolysis can occur with a loss in viability and vitality with their cell contents being excreted into the fermenting wort. This effect has a number of consequences on beer flavour and stability—especially both beer foam and its overall stability.
- Stress can also result in the excretion of intracellular enzymes, particularly PrA which will also negatively affect beer foam stability.

The balance between PrA occurring in the fermenting wort as a result of cell autolysis and/or enzyme excretion of whole cells is still an unanswered question. Nevertheless, the occurrence of active PrA is important because of its negative effects on beer foam stability (Bamforth, 2012).

2.12 Dried yeast

Dried yeast has been employed in the baking and distilling industries for over 70 years (Pyke, 1958). However, its use in brewing is relatively recent (Fels, Reckelbus, & Gosselin, 1998). One of the reasons for this delay and the differences in drying characteristics is because ale yeast strains dry relatively well, whereas lager yeast cultures, when dried, have comparatively lower cell viability. The reasons for the drying differences between these two yeast species is still not fully understood but levels of the storage carbohydrates glycogen and trehalose have been implicated (Gadd, Chalmers, & Reed, 1987). Another reason for the delay in adopting dried yeast in brewing is that this yeast is often contaminated with various bacteria and wild yeasts. However, this problem is not as prevalent today as it was 25 years ago (Pyke, 1958).

The use of dried yeast in brewing has several advantages and similarities compared to the use of fresh yeast (Fels et al., 1998) which are:

- It is easier to handle and convenient to store;
- It can replace yeast propagation in breweries;
- In some cases, it can replace the need for wort aeration during pitching;
- Recent studies have shown that dried yeast often has characteristics similar to those of its fresh counterpart with analytical and flavour profiles, rates of fermentation and final attenuation all matching favourably to those of fresh yeast (Debourg & Van Nedervelde, 1999);
- Although the average viability determined by methylene violet or methylene blue staining of dried yeast is 20%−30% lower than that of freshly propagated yeast (Pratt-Marshall, 2002). This problem can be accommodated by pitching according to viable cell number.

It has also been reported (Finn & Stewart, 2002) that dried yeast samples exhibited different flocculation and haze formation characteristics when compared to fresh yeast samples. The flocculation rate with fresh and dried ale cultures was rapid with most of the yeast sedimenting out of suspension within the first minute of a Helm's Sedimentation Test with 80% of the culture eventually flocculating out of suspension (Fig. 2.16).

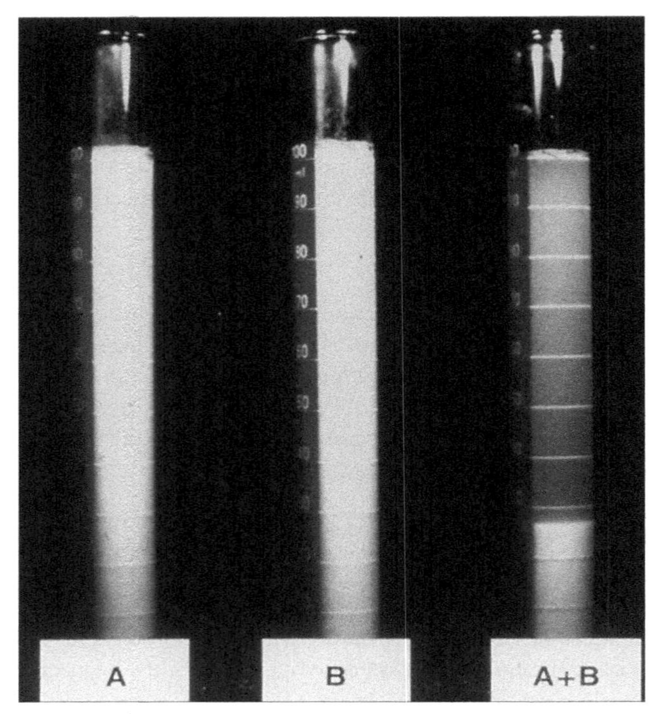

FIGURE 2.16 Co-flocculation: Helm's sedimentation in vitro test.

The flocculating differences between fresh lager and dried yeast samples were more pronounced than with ale strains. Virtually no flocculation took place within a 10-minute test period with the dried yeast samples. This test indicated that the lager dried yeast samples were modified in some way, and, as a consequence, the culture exhibited non-flocculent characteristics (Chlup & Stewart, 2011).

During the studies on flocculation, it was observed that the dried yeast fermentation often left a haze in suspension (Finn & Stewart, 2002). Even during the ale yeast fermentation, although the yeast flocculated, a haze remained in suspension. This may have been due, in part, to the number of dead cells pitched into the wort. In addition, the fresh yeast fermentation exhibited a foam head, but a foam head was absent on the dried yeast samples (both ale and lager cultures). PrA (and other proteinases) is released by dried yeast into the wort in much greater quantities than fresh yeast under similar fermentation conditions (Finn & Stewart, 2002). PrA release into the wort will have an impact on beer foaming characteristics (Cooper, Stewart, & Bryce, 2000; Osmond, Lebor, & Sharpe, 1991). The decreased foam stability is due to the hydrolysis of hydrophobic polypeptides by PrA. Hydrophobic polypeptides are known to be mainly responsible for beer foam stability (Bamforth, 2012). Leakage of intracellular proteinases from living brewer's yeast cells has been demonstrated by Dreyer, Biedermann, and Otteson (1983) particularly when they are under stress (Stewart, 1999). Indeed, the addition of dead cells (as could be the case with a dried culture) would greatly increase the levels of PrA.

2.13 Conclusions

During most (not all) lager fermentation ethanol production procedures, a yeast culture is used only once for a single fermentation cycle. However, in brewing, the yeast culture is harvested at the end of a fermentation for reuse in a subsequent wort fermentation. In between fermentations, the cropped yeast is normally appropriately stored prior to being repitched. As a consequence of this procedure, the yeast culture must be carefully managed between each fermentation in order to maintain its quality. The steps in this yeast management procedure have been discussed in other sections of this chapter.

Acknowledgements

I want to express my sincere thanks to Anne Anstruther for her hard work and support during the development of this manuscript. In addition, Olga Stewart is also thanked for her patience during the development of this chapter. Part of this paper was first presented at the 14th Africa Convention of the Institute of Brewing & Distilling held in Accra, Ghana, 2013.

References

Anderson, R. (2012). One yeast or two? Pure yeast and top fermentation. *Brewery History, 149*, 30–38.

Bamforth, C. W. (2012). The chemistry of foaming. In *Foam* (pp. 15–28). St. Paul, Minn: American Society of Brewing Chemists.

Bamforth, C. W. (2020). *The everyday guide to beer*. The Great Courses, DVD and CD.

Bamforth, C. W., & Lentini, A. (2009). The flavour instability of beer. In C. W. Bamforth (Ed.), *Beer: A quality perspective* (pp. 85–109). Burlington MA: Academic Press.

Boulton, C., & Quain, D. (1999). A novel system for propagation of brewing yeast. In *Proceedings of the European brewing convention congress* (pp. 647–654). Cannes.

Boulton, C., & Quain, D. (2001). *Brewing yeast and fermentation*. Oxford, UK: Pub. by Blackwell Science Ltd.

Casey, P. H., Chen, E. C. H., & Ingledew, W. M. (1985). High-gravity brewing: Production of high levels of ethanol without excessive concentrations of esters and fusel alcohols. *Journal of the American Society of Brewing Chemists, 43*, 178–182.

Chlup, P. L., Bernard, D., & Stewart, G. G. (2007a). The disc stack centrifuge and its impact on yeast and beer quality. *Journal of the American Society of Brewing Chemists, 65*, 29–37.

Chlup, P. L., Bernard, D., & Stewart, G. G. (2008). Disc stack centrifuge operating parameters and their impact on yeast physiology. *Journal of the Institute of Brewing, 114*, 45–61.

Chlup, P. L., Conery, J., & Stewart, G. G. (2007b). Detection of mannan from *Saccharomyces cerevisiae* by flow cytometry. *Journal of the American Society of Brewing Chemists, 65*, 151–155.

Chlup, P. H., & Stewart, G. G. (2011). Centrifuges in brewing. *Technical Quarterly - Master Brewers Association of the Americas, 48*, 46–50.

Cooper, D. J., Stewart, G. G., & Bryce, J. H. (2000). Yeast proteolytic activity during high and low gravity wort fermentations and its effect on head retention. *Journal of the Institute of Brewing, 106*, 197–201.

Cunningham, S., & Stewart, G. G. (2000). Acid washing and serial repitching a brewing ale strain of *Saccharomyces cerevisiae* in high gravity wort and the role of wort oxygenation conditions. *Journal of the Institute of Brewing, 106*, 389–402.

D'Amore, T., Russell, I., & Stewart, G. G. (1989). Sugar utilization by yeast during fermentation. *Journal of Industrial Fermentation, 4*, 315–344.

Debourg, A., & Van Nedervelde, L. (1999). The use of dried yeast in the brewery industry. In *Proceedings of the European brewery convention congress* (pp. 751–760). Cannes.

Dreyer, T., Biedermann, K., & Otteson, M. (1983). Yeast proteinase in beer. *Carlsberg Research Communications, 48*, 249–253.

Ernandes, J. R., Williams, J. W., Russell, I., & Stewart, G. G. (1993). Respiratory deficiency in brewing yeast strain – effects on fermentation, flocculation and beer flavour components. *Journal of the American Society of Brewing Chemists, 51*, 16–20.

Fels, S., Reckelbus, B., & Gosselin, Y. (1998). Why use dried yeast for brewing your beers? *Brewing and Distilling International, 29*(5), 17–19.

Finn, D., & Stewart, G. G. (2002). Fermentation characteristics of dried brewers' yeast, the effect of drying on flocculation and fermentation. *Journal of the American Society of Brewing Chemists, 60*, 135–139.

Gadd, G. M., Chalmers, K., & Reed, R. H. (1987). The role of trehalose in dehydration resistance of *Saccharomyces cerevisiae*. *Federation of European Microbiological Societies Microbiology Letters, 48*, 249–254.

Goff, E. U., & Fine, D. H. (1979). Analysis of volatile N-nitrosamine in alcoholic beverages. *Food and Cosmetics Toxicology, 17*(6), 569–573.

Goldhammer, T. (September 2002). Chapter 5, brewing water, minerals in brewing water, page 68. In *The brewer's handbook* (3rd ed.). Apex Publishers.

Hansen, E. C. (1883). *Undersøgelser over alkoholgjaersvampenes fysiologi og morfologi*. II Om askosposedann elsen hos slaegten *Saccharomyces*. *Meddelelser fra Carlsberg Laboratoriet, 2*, 29–104.

Holter, H., & Moller, K. M. (1976). *The Carlsberg laboratory 1876-1976*. Copenhagen: Pub. by International Service and Art Publishers.

Hutzler, M., Jacob, F., & Geiger, E. (2010). Use of PCR-DHPLC (polymerase chain reaction—denaturing high performance liquid chromatography) for the rapid differentiation of industrial *Saccharomyces pastorianus* and *Saccharomyces cerevisiae* strains. *Journal of the Institute of Brewing, 116*, 464–474.

Hutzler, M., Morrissey, J. P., Laus, A., Meussdoerffer, F., & Zarnkow, M. (2023). A new hypothesis for the origin of the lager yeast Saccharomyces pastorianus. *FEMS Yeast Research, 23*, Article foad023.

Jones, M., & Pierce, J. S. (1964). Absorption of amino acids from wort by yeasts. *Journal of the Institute of Brewing, 70*(4), 307–315.

Kara, B. V., Simpson, W. J., & Hammond, J. R. M. (1988). Prediction of the fermentation performance of brewing yeast with the acidification power test. *Journal of the Institute of Brewing, 94*, 153–158.

Kirsop, B. (1955). Maintenance of yeasts by freeze drying. *Journal of the Institute of Brewing, 61*, 466–471.

Kirsop, B., & Doyle, A. (1990). Maintenance of microorganisms and cultured cells. In *A manual of food practice* (2nd ed.). London: Academic Press.

Lekkas, C., Stewart, G. G., Hill, A. E., Taidi, B., & Hodgson, J. (2007). Elucidation of the role of nitrogenous wort components in yeast fermentation. *Journal of the Institute of Brewing, 113*, 3–8.

Libkind, D., Hittinger, C. T., Valério, E., Gonçalves, C., Dover, J., Johnston, M., et al. (2011). Microbe domestication and the identification of the wild genetic stock of lager-brewing yeast. *Proceedings of the National Academy of Science, United States of America, 108*, Article 1105430108.

Lorenzoni, L. (2023). The future of craft beer – part one. *Brewers Journal*. https://www.brewersjournal.info/the-future-of-craft-beer-part-one/.

Murray, C. R., & Stewart, G. G. (1991). Experience with high gravity lager brewing. *Birra et Malto, 44*, 52–64.

Nielsen, O. (2005). Control of the yeast propagation process – how to optimize oxygen supply and minimize stress. *Technical Quarterly - Master Brewers Association of the Americas, 42*, 128–132.

Nielsen, O. (2010). Status of the propagation process and some aspects of propagation for refermentation. *Cerevisiae: Belgian Journal of Brewing Biotechnology, 35*, 71–74.

Osmond, I. H. L., Lebor, E. F., & Sharpe, F. R. (1991). Yeast proteolytic enzyme activity during fermentation. In *Proceedings of the European brewing convention congress* (pp. 457–464). Lisbon: IRL Press: Oxford. Copenhagen.

Panchal, C. J., & Stewart, G. G. (1980). The effect of osmotic pressure on the production and excretion of ethanol and glycerol by a brewing yeast strain. *Journal of the Institute of Brewing, 86*, 207–210.

Pederson, M. B. (1995). Recent views and methods for the classification of yeasts. *Cerevisia – Belgian Journal of Brewing Biotechnology, 20*, 28–33.

Pratt-Marshall, P. L. (2002). *High gravity brewing – an inducer of yeast stress: Its effect on cellular morphology and physiology*. PhD Thesis. Heriot-Watt University.

Pratt-Marshall, P. L., Brey, S. E., de Costa, S. D., Bryce, J. H., & Stewart, G. G. (2002). High gravity brewing – an inducer of stress. *Brewers' Guardian, 131*, 22–26.

Pyke, M. (1958). The technology of yeast. In *The chemistry and biology of yeasts* (pp. 535–586). New York: Academic Press Inc.

Roberts, T. R., & Wilson, R. H. H. (2006). Hops. In F. G. Priest, & G. G. Stewart (Eds.), *Handbook of brewing* (pp. 177–279). New York: Taylor & Francis.

Russell, I., & Stewart, G. G. (1981). Liquid nitrogen storage of yeast cultures compared to more traditional methods. *Journal of the American Society of Brewing Chemists, 39*, 19–24.

Russell, I., & Stewart, G. G. (Eds.). (2014). *Whisky: Technology, production and marketing* (2nd ed.). Boston, USA: Pub. by Academic Press (Elsevier), 978-0-12-401735-1.

Schlee, C., Miedl, M., Leiper, K. A., & Stewart, G. G. (2006). The potential of confocal imaging for measuring physiological changes in brewer's yeast. *Journal of the Institute of Brewing, 112*, 134–147.

Simpson, W. J., & Hammond, J. R. M. (1989). The response of brewing yeasts to acid washing. *Journal of the Institute of Brewing, 95*, 347–354.

Sofie, M. G., Saerens, C., Duong, C. T., & Nevoigt, E. (2010). Genetic improvement of brewer's yeast: Current state, perspectives and limits. *Applied Microbiology and Biotechnology, 86*, 1195–1212.

Stewart, G. G. (1999). High gravity brewing. *Brewers' Guardian, 128*, 31–37.

Stewart, G. G. (2004). Chemistry of beer instability. *Journal of Chemical Education, 81*, 963–968.

Stewart, G. G. (2007). High gravity brewing – the pros and cons. *New Food, 1*, 42–46.

Stewart, G. G. (2013). Biochemistry of brewing. In N. A. M. Eskin, & F. Shahidi (Eds.), *Biochemistry of foods* (pp. 291–318). London, U.K: Academic Press.

Stewart, G. G. (2017). *Brewer's yeast propagation – the basic principles* (Vol 54, pp. 125–131). Tech. Q. Master Brew. Assoc. Am..

Stewart, G. G. (2018). Yeast flocculation—sedimentation and flotation (2018). *Fermentation, 4*(2), 28. https://doi.org/10.3390/fermentation4020028

Stewart, G. G., & Garrison, I. F. (2018). *Some observations on Co-flocculation in* Saccharomyces cerevisiae (pp. 118–131). Published online: 31 Jul 2018 Proceedings. Annual meeting - American Society of Brewing Chemists.

Stewart, G. G., Hill, A., & Lekkas, C. (2013). Wort FAN – its characteristics and importance during fermentation. *Journal of the American Society of Brewing Chemists, 71*, 179–185.

Stewart, G. G., Hill, A. E., & Russell, I. (2013). 125th Anniversary review: Developments in brewing and distilling strains. *Journal of the Institute of Brewing, 119*, 202–220.

Stewart, G. G., & Russell, I. (2009). *An introduction to brewing science and technology. Series lll, Brewer's yeast* (2nd ed.). London: Pub The Institute of Brewing and Distilling, 0900498-13-8.

Stewart, G. G., & Russell. (1981). Chapter 4 Yeast culture collections, strain maintenance and propagation, 4.3 propagation of yeast cultures. In *Handbook of yeast.*

Stewart, G. G., Russell, I., & Garrison, I. F. (1975). Some considerations of the flocculation characteristics of ale and lager yeast strains. *Journal of the Institute of Brewing, 81*, 248–257.

Stewart, G. G., & Ryder, D. S. (2019). Sulfur metabolism during brewing. *Technical Quarterly - Master Brewers Association of the Americas, 56*, 39–46.

Stoupis, T., Stewart, G. G., & Stafford, R. A. (2002). Mechanical agitation and rheological considerations of ale yeast slurry. *Journal of the American Society of Brewing Chemists, 60*, 58–62.

Younis, O. S., & Stewart, G. G. (1998). Sugar uptake and subsequent ester and alcohol production in *Saccharomyces cerevisiae. Journal of the Institute of Brewing, 104*, 255–264.

Chapter 3

Theory and practice of fermentation modeling

Andrew J. MacIntosh
Food Science and Human Nutrition Department, University of Florida, Gainesville, FL, United States

3.1 Introduction

For thousands of years, the fermentation of cereal wort has challenged our ability to explain, predict and control the behaviour of yeast. This chapter discusses some common tools used to describe and predict fermentation behaviour and how our understanding of this phenomenon has improved over time. This will be accomplished through the use of statistical techniques to fit models to fermentation data. Within the brewing industry, the term 'model' is often applied to any equation (empirical or theoretical) that is fit to fermentation data; it is this definition that will be used throughout this chapter. The advantages and disadvantages of modelling techniques (simple and complex) will be discussed, as will models commonly used by industry. A distinction is also made between theoretical models that attempt to explain behaviour and empirical models that mathematically attempt to follow expected trends as closely as possible (where the parameters describe only the physical shape of an equation over time). As fermentation is influenced by many parameters, an overview of how some of these affect this process will be discussed. Finally, applications will be discussed, as will several advanced brewing techniques (such as high-gravity brewing) in which accurate prediction of behaviour is critical. The chapter will end with a discussion of future trends.

3.2 Parameters influencing yeast growth and fermentation of barley malt

During brewing operations, there are numerous factors affecting the growth of yeast cells and subsequent fermentation of wort. Therefore, when attempting to model and explain this phenomenon, it is imperative to know which parameters affect fermentation and in what manner. A typical fermentation will be affected in many ways, such as the rate of fermentation, the degree to which the media fermented and the ratio of products (and by-products) that are formed. These are affected not only by extrinsic parameters such as temperature and headspace composition but also by numerous intrinsic properties that are covered in other chapters. Most modern prediction methods use models that describe the expected behaviour of the fermentation while occasionally take into consideration the expected behaviour of a major parameter (such as the original density of the wort, number of yeast, or occasionally temperature).

Brewing fermentations typically use either *Saccharomyces cerevisiae* (ale yeast) or *Saccharomyces pastorianus* (lager yeast), with the latter species producing $>90\%$ of the global product (Canadean, 2011; Goncalves et al., 2016). Over the course of 4–20 days, the fermentable sugars within the wort are consumed, and fermentation products (predominantly ethanol and CO_2) are produced. The density of the media (commonly expressed as specific gravity or apparent extract) is often used as an easily measured analogue for the concentration of sugar within the media (although this must be corrected for alcohol concentration). The vast majority of industrial brewing operations use batch fermentations, in which yeast is added (pitched) at concentrations of approximately $12-15 \times 10^6$ cells/mL (Briggs et al., 2004). Although continuous industrial fermentations do exist (i.e., Morton Coutts' method used in New Zealand ([Virkajärvi & Kronlöf, 1998]), these introduce additional complexity and are difficult to run economically. Continuous fermentations have very different behaviour and characteristics from those described within this chapter.

Brewing Microbiology. https://doi.org/10.1016/B978-0-323-99606-8.00012-2

TABLE 3.1 Typical sugar components of brewing wort.

Saccharide:	Chemical formula	Typical percent composition[a] (%)
Glucose	$C_6H_{12}O_6$	10–15
Fructose	$C_6H_{12}O_6$	1–2
Sucrose	$C_{12}H_{22}O_{11}$	1–2
Maltose	$C_{12}H_{22}O_{11}$	50–60
Maltotriose	$C_{18}H_{32}O_{16}$	15–20
Higher Saccharides	$H_2O+(C_6H_{10}O_5)_n$	20–30

[a]*Typical composition as a percentage of total sugars (Stewart, 2006).*

The initial concentration of each fermentable sugar will define many characteristics of the fermentation. This parameter is highly dependent upon the malt and mashing style, as every wort will comprise a different configuration of fermentable and nonfermentable sugars. The sugars present in wort (and typical concentrations) are described by Stewart (2006) and are listed in Table 3.1. Although various brewing yeast strains are able to metabolise different sugars, Table 3.1 highlights those sugars most commonly found and metabolised from brewers wort (Stewart, 2006). During brewing operations, the uptake of fermentable sugars by yeast is a highly ordered process; glucose and fructose are consumed first, with any sucrose present being hydrolysed extracellularly via the enzyme β-fructosidase (invertase) excreted by yeast (Briggs et al., 2004). The presence of glucose in sufficient quantities has been shown to inhibit respiration and the uptake of maltose in brewing yeast. Once the concentration of glucose is sufficiently low, maltose is sequentially utilised by the yeast, followed by maltotriose (Stewart, 2006). Both maltose and maltotriose are hydrolysed intracellularly into glucose via the enzyme α-glucosidase (maltase) (Briggs et al., 2004; de Godoy, Muller, & Stambuk, 2014). Most brewing strains cannot metabolise chains of sugar in excess of three glucose units (Stewart & Russell, 1998).

As wort is an aqueous mixture high in fermentable sugars, wort density is often used as an easily measured indicator of fermentation progress. This is directly related to consumption of sugar and subsequent production of alcohol, which results in density attenuation. This decline in density (commonly measured in either units of degree Plato [°P], or specific gravity), characteristically follows a sigmoidal (s-shaped) curve (Corrieu et al., 2000; Speers et al., 2003; Trelea, Latrille, et al., 2001). Similarly, each individual fermentable sugar follows a sigmoidal decline. However, these consumption curves are influenced by a variety of factors, such as yeast state, species and sugar type. The consumption of total sugar (as shown in Fig. 3.1), as well as individual sugar attenuation, is often lagged before consumption and may be symmetrical or asymmetrical.

Although the concentration of sugars will ultimately define the maximum and minimum parameters of any equation that attempts to model fermentation behaviour, the shape will be influenced by the intrinsic and extrinsic parameters. Novel research on how both extrinsic and intrinsic parameters affect fermentation is completed at many notable institutions and is discussed by other researchers mentioned in this text.

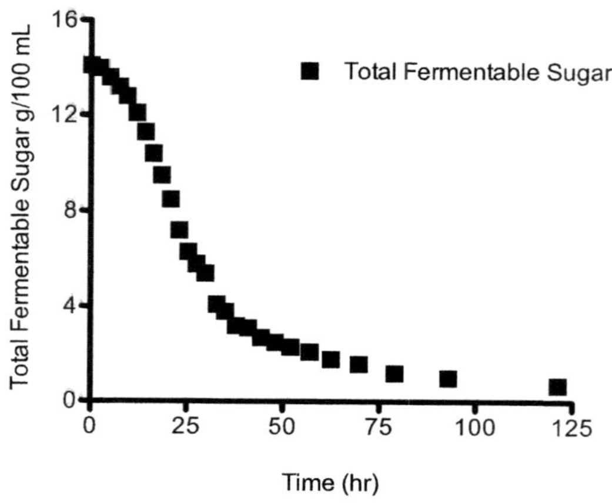

FIGURE 3.1 Typical attenuation of sugars during brewing fermentations.

3.3 Modelling: techniques and applications

Possibly the most well-known early attempt to model brewing fermentation was completed in 1865 by the chemist Carl Balling. Using beer with original wort extract of 10−14 °P (°P—a measure of density), Balling reported that from 2.0665 g of fermented extract, the following products were generated: 1.000 g alcohol, 0.9565 g CO_2 and 0.11 g dry yeast matter (Balling, 1845−1865). Upon analysis, it appears as though this formula is a combination of the theoretical conversion of glucose to ethanol and CO_2, combined with empirical assessments of yeast mass generation measured at the end of fermentation. This formula, and associated calculation of original extract (OE), is used worldwide and endorsed by both the European Brewing Convention (EBC Method 9.4) and the American Society of Brewing Chemists (ASBC Beer-6B). However, that is not to say that the formula has remained unchallenged. In the ensuing years since its derivation, the formula has been disputed on multiple grounds. Subsequent researchers have noted that although it is not perfect, the formula is a good approximation that is well known and widely used (Nielsen et al., 2007). Additionally, several issues with Balling's formula can be corrected for, as summarised by Nielsen et al. (2007). The aforementioned studies have assessed the accuracy of Balling's formula when used to model fermentations, specifically the relationship between final density and OE. However, the ratio of fermentation products is known to vary throughout the fermentation. For example, the majority of yeast propagation is completed during the first half of fermentation, whereas the initial CO_2 produced is dissolved within the wort and does not evolve. The result is that if Balling's formula is used to relate the products of fermentation to the OE of a fermentation halfway through, it will not be accurate. Modern methods of analysis now allow researchers to predict the parameters of Balling's formula over the entire fermentation and to examine how the product ratios likely change with time. Unfortunately, not every fermentation has each variable monitored in real time; in industrial settings, measurements are usually taken intermittently and when convenient for scheduling purposes. It is impractical to precisely monitor every aspect of fermentation performance and to understand how this will affect the overall fermentation and real-world concerns of economics and scheduling often supersede that of rigour. Therefore, assumptions must be made that allow the brewer or researcher to roughly predict fermentation behaviour. Often, only one or two parameters are actually monitored to observe whether, or how, a fermentation deviates from previously completed fermentations that had identical (or similar) initial parameters.

Throughout the brewing process, sugars are metabolised into alcohol and carbon dioxide, resulting in wort density attenuation. When plotted with respect to time, this decline follows a sigmoidal curve, from an initial sugar concentration of anywhere from 10% to 20% (or higher when using high-gravity brewing techniques) to 2%−4% over the course of a typical fermentation. Mathematical models can be fit to this data allowing brewers to predict, assess and more accurately compare fermentations. Within the brewing industry, there are several models that can be applied, each with advantages and disadvantages. Some models are theoretically derived, whereas others are fully or semi-empirical. In modelling sugar attenuation, brewing researchers use simpler models; however, these may not accurately characterise real-world fermentations (particularly at the onset and latter half of fermentation). With a limited number of data points, important trends can be missed, and small errors in measurement can greatly affect alcohol and extract calculations. With the development of computer-aided modelling, scientists have applied non-linear fitting techniques to model and to more precisely determine interpolated values of variables (Speers et al., 2003). Since then, advances in other scientific fields have introduced novel models that may be more adept at modelling the patterns observed during fermentation (Reid et al., 2021; Rudolph et al., 2020).

3.3.1 Modelling fermentations

Modelling total sugar consumption has many advantages, such as predicting the final density/sugar content (Defernez et al., 2007; Reid et al., 2021), approximating the time until completion (Speers et al., 2003), characterising yeast behaviour (Guadalupe-Daqui et al. 2023), detecting aberrant fermentations (Armstrong et al., 2018; Rudolph et al., 2020) and assessing the effects of mixed cultures (Nemenyi et al. 2024). Non-linear models are already promoted for use in various analytical methods within the brewing industry, such as the 'nearest neighbour' and 'predictive modelling' techniques (Trelea, Titica, et al., 2001), in which easily measured parameters are related to others such as using evolved CO_2 to estimate sugar consumption. The most common functions used to predict density decline in brewing fermentations are the logistic model (Speers et al., 2003; ASBC Yeast-14, 2011; Armstrong et al., 2018), the regularised incomplete β-function (IBF) (Trelea, Latrille, et al., 2001) and the modified Gompertz function (Gibson et al., 1988). Differences in reported and predicted density can significantly influence the decision-making process in large breweries and can make comparing metrics (such as fermentability of grain) very problematic. Because of this, using the correct function for a given application is often of the upmost importance. The following section discusses several commonly used sigmoidal models; the first—Richard's model—is based upon theoretical principles (Richards, 1959), but is not often used in the brewing industry. The second—the incomplete β distribution—is an applied empirical distribution that is often reported in the brewing literature (Trelea, Latrille, et al., 2001). The final—Gompertz's model—is an empirical model that is widely used in microbiology (Gibson et al., 1988) to describe growth curves (analogous to consumption curves). Each model is described in detail below.

3.3.2 Logistic models

The logistic model is a family of nested equations that describe a sigmoidal (s-shaped) curve. The two-parameter logistic model has parameters, which describe the midpoint and slope of the curve. As additional parameters are introduced, the curve gains more flexibility to fit data such as a lower and upper asymptote which describe the final attenuation and OE of the fermentation respectively. The ideal number of parameters for a particular dataset must be statistically determined via an F-test. The four-parameter logistic function (4P logistic model) is often used to describe changes in a population, as it effectively models autocatalytic behaviour (Eq. 3.1). This curve is commonly used in the brewing industry to model the decline in apparent extract and is the basis of ASBC Yeast-14 (ASBC, 2011). Here, it is used to assess malt for premature yeast flocculation behaviour and to compare the fermentability of yeast wort under controlled and repeatable conditions. The generalised logistic model is a five-parameter variant of the logistic model (Eq. 3.2) that expands the theoretical basis to an asymmetrical curve (Richards, 1959), required for modelling sugar attenuation. The five-parameter (5P) logistic model is commonly used in other fields for applications such as population growth modelling and dosage calculations (Gottschalk & Dunn, 2005). The generalised logistic curve is equal to the symmetrical 4P logistic when the parameter $s = 1$: Eqs. (3.1 and 3.2) where s is a variable that modifies the point of inflection (M).

$$P_{(t)} = P_e + \frac{P_i - P_e}{\left(1 + e^{-B(t-M)}\right)} \tag{3.1}$$

Where P_i is the initial asymptotic density value (in °P) for the density attenuation regression, B is a function of the slope at the inflection point, t is time, P_t is the density at time t, M is the time at point B, and P_e is the equilibrium asymptotic density value.

$$P_{(t)} = P_e + \frac{P_i - P_e}{\left(1 + se^{-B(t-M)}\right)^{1/s}} \tag{3.2}$$

Where s is a variable that modifies the point of inflection (M).

As shown in Fig. 3.2, the consumption of sugar follows an asymmetrical sigmoidal curve. As the consumption of sugar follows the same shape as microbial growth, the use of a common biological growth curve has often been used for this

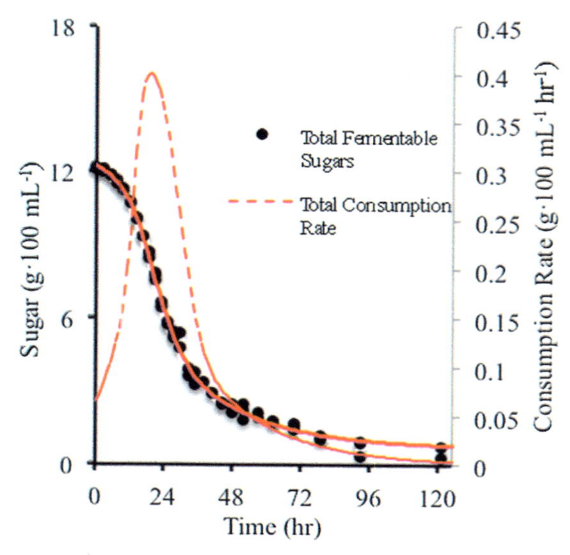

FIGURE 3.2 Total fermentable sugar data and total sugar consumption rate as calculated by the summation of each individual sugar consumption rate.

application. The Gompertz model (GM) is an empirical model named after Benjamin Gompertz (1825) that is widely used in the field of microbiology to predict the growth curves of bacteria (Buchanan et al., 1997). This model is a special case of the generalised logistic formula and describes a sigmoidal curve in which the latter half of the curve approaches the asymptote more slowly than the initial half. This model is often used when one expects an asymmetrical curve, such as when working with microorganisms. A modified version of the Gompertz curve mentioned in the brewing literature to describe density attenuation is described in Eq. (3.3) (Speers et al., 2003)

$$P_t = P_e + (P_i - P_e) \cdot e^{\left(e^{-B(t-M)}\right)} \tag{3.3}$$

where P_i and P_e are the upper and lower asymptotes, respectively, M is the time of the inflection point of the curve, B is the consumption rate factor and t is the time at $P_{(t)}$.

This version of the Gompertz model is an empirical model, not one derived from theory (Speers et al., 2003). An advantage of this model is the low number of parameters (four) required to fit the model while still allowing for an asymmetrical shape; this is particularly advantageous with a limited number of data points. However, in testing of data from more than 50 industrial brewing fermentations, Speers et al. (2003) showed that the 4P logistic model fit the data more accurately than the modified Gompertz model.

3.3.3 The incomplete β-function

The incomplete β-function (IBF) can be used to describe an asymmetric curve as described by Eq. (3.4). The full name for this equation is the regularised incomplete β-function; however, the name is often shortened in literature to the incomplete β-function. The IBF has been modified for describing the attenuation of extract by Trelea, Latrille, et al. (200) and has been used by several researchers (i.e., Defernez et al., 2007) to model and predict the end parameters of fermentation. Eq. (3.5) is the aforementioned modified version of the IBF with two additional terms (P_i and P_e) added to fit experimental data (describing the upper and lower boundaries of the sugar consumption curve). With the additional variables, the IBF can be used to describe brewing fermentations that follow a symmetrical sigmoidal curve. However, as with the modified Gompertz model, the fit is purely empirical and the shape parameters do not describe biological functions (Eq. 3.4).

$$IBF = \frac{\beta(x: \propto, \beta)}{\beta(\propto, \beta)} = \frac{\int_0^x u^{\propto-1}(1-u)^{\beta-1}du}{\int_0^1 u^{\propto-1}(1-u)^{\beta-1}du} \tag{3.4}$$

where β and α are shape parameters and:

$$P_t = P_i - (P_i - P_e) \cdot IBF(x \cdot t: \propto, \beta) \tag{3.5}$$

3.3.4 Additional models

Although the 4P logistic, IBF and Gompertz models are all discussed within brewing literature, there are many additional models used to describe sigmoidal curves outside of this industry. The fields of predictive microbiology, medical science and biology all offer additional models that may be useful in describing sugar attenuation; however, those described are most prevalent in the brewing literature. It is noteworthy that a review of the literature will reveal many forms of the logistic model that all effectively describe sigmoidal curves.

3.3.5 Application of models

For this chapter, each model was applied to sugar attenuation data taken from multiple brewing fermentations (assessed using high-pressure liquid chromatography). Three techniques: Akaike's Information Criterion (corrected) (AICc), comparison of the coefficients of determination (r^2) and absolute residual sum of squares (RSS), were used to compare the fit of each model. Ideally, the data would adhere to a simplistic, theoretically derived formula such as a low parameter symmetric model. Unfortunately, the variability in both shape and lag time for each individual sugar attenuation necessitated a more flexible model. Table 3.2 details the fit of each model through examination of the residuals, coefficients of determination and absolute residual sums of squares. As shown in Fig. 3.3, each sugar was found (not surprisingly) to follow a sigmoidal attenuation. Noteworthy is that although the attenuation of glucose was immediate, the attenuation of other sugars (maltose, glucose, maltotriose, and fructose (Figs. 3.4–3.7), respectively) were delayed (lagged) to varying degrees.

TABLE 3.2 Residual analysis for each sugar attenuation modelled using the modified Gompertz, IBF and 5P logistic models.

Sugar:	Modified Gompertz Residual pattern r^2 - RSS		IBF Residual pattern r^2 - RSS		5P logistic Residual pattern r^2 - RSS	
Glucose	Random[a]	0.998–0.308	Pattern	0.996–0.678	Random[a]	0.998–0.299
Fructose	Random[a]	0.987–0.015	Random[a]	0.989–0.013	Random[a]	0.988–0.014
Maltose	Pattern	0.979–3.473	Pattern	0.991–1.436	Random[a]	0.996–0.660
Maltotriose	Pattern	0.986–0.210	Pattern	0.994–0.084	Random[a]	0.996–0.061

[a]*The heteroscedasticity caused by known variance inherent to the assay was common to all models.*

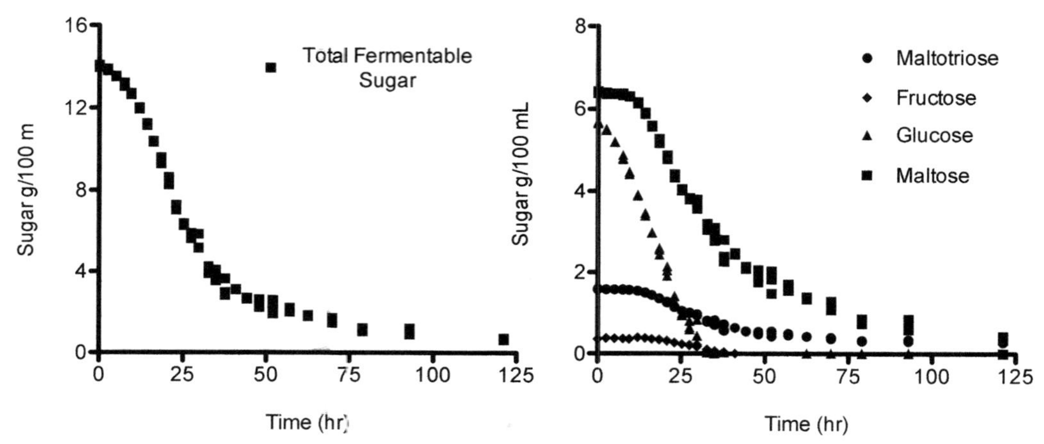

FIGURE 3.3 Raw total (left) and individual (right) sugar attenuation values taken throughout an experimental fermentation.

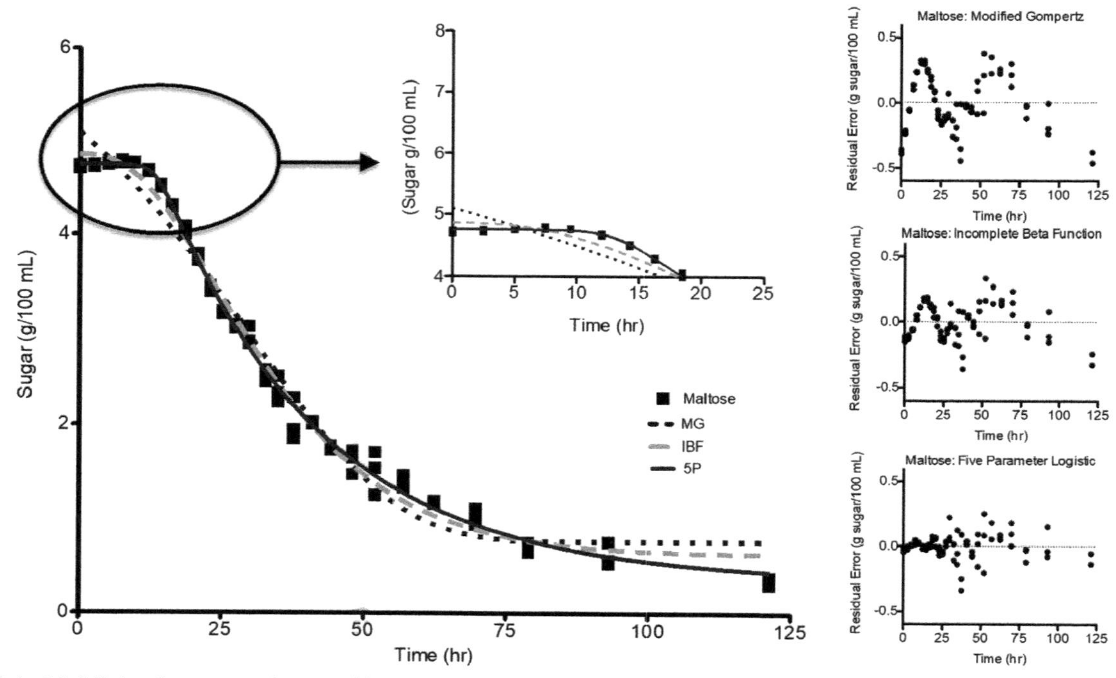

FIGURE 3.4 Modelled maltose attenuation data (MG, modified Gompertz; IBF, incomplete β-function; 5P, five-parameter logistic). The residuals for each model are depicted on the right.

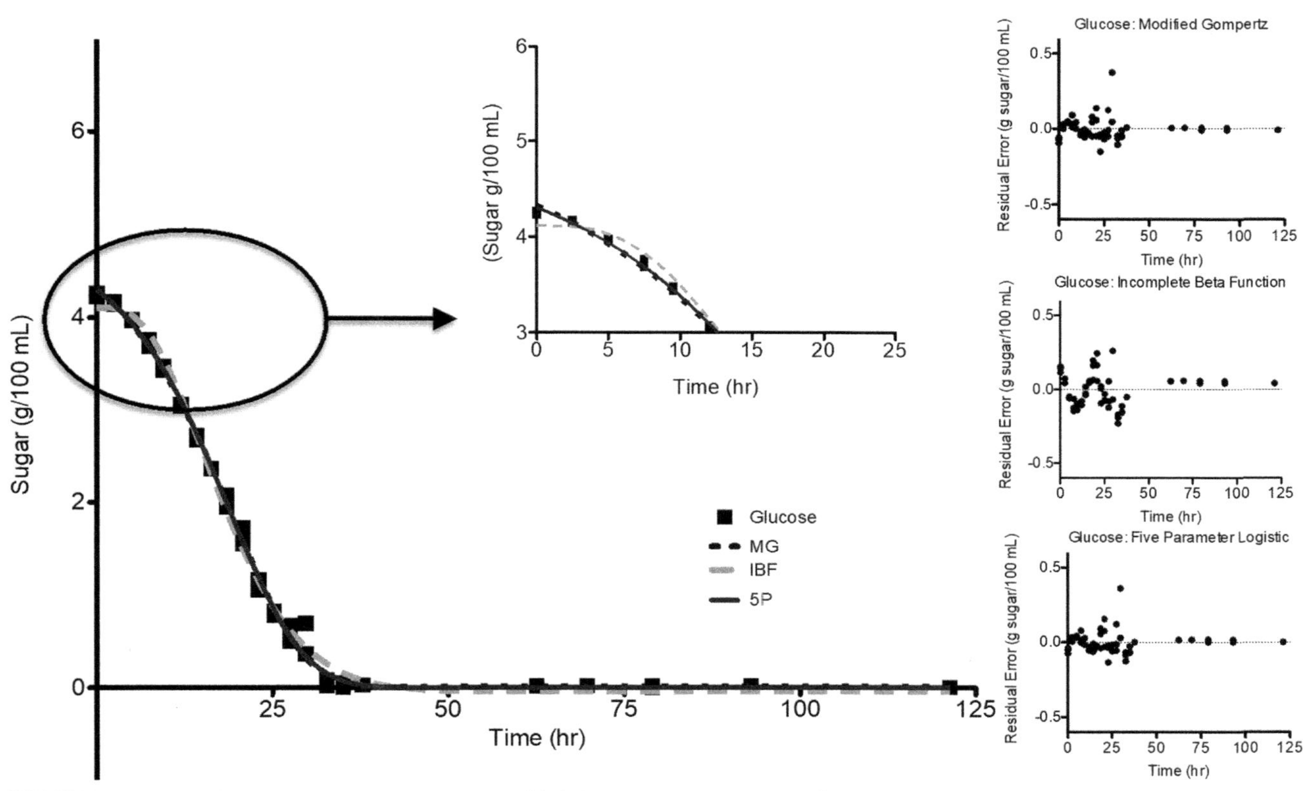

FIGURE 3.5 Modelled glucose attenuation data (MG, modified Gompertz; IBF, incomplete β-function; 5P, five-parameter logistic). The residuals for each model are depicted on the right.

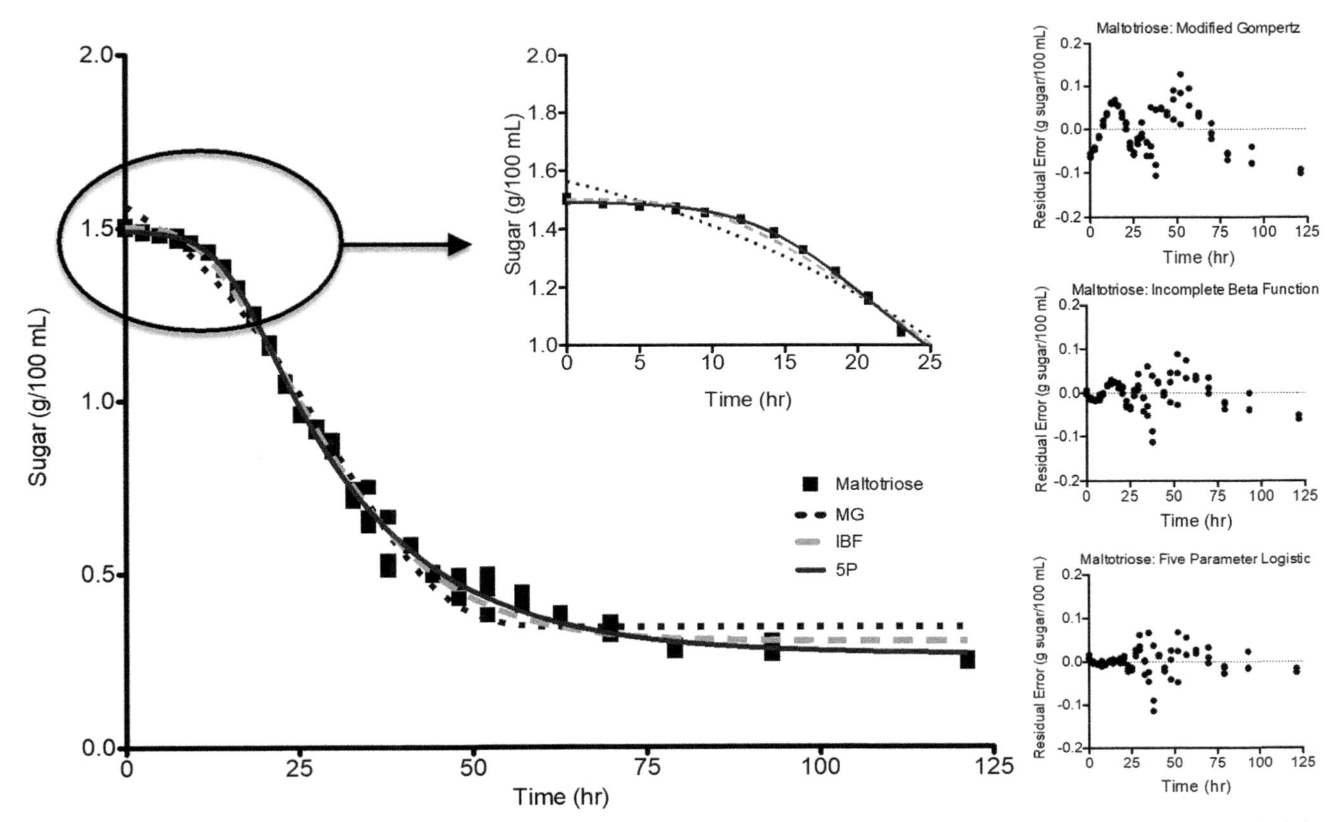

FIGURE 3.6 Modelled maltotriose attenuation data (MG, modified Gompertz; IBF, incomplete β-function; 5P, five-parameter logistic). The residuals for each model are depicted on the right.

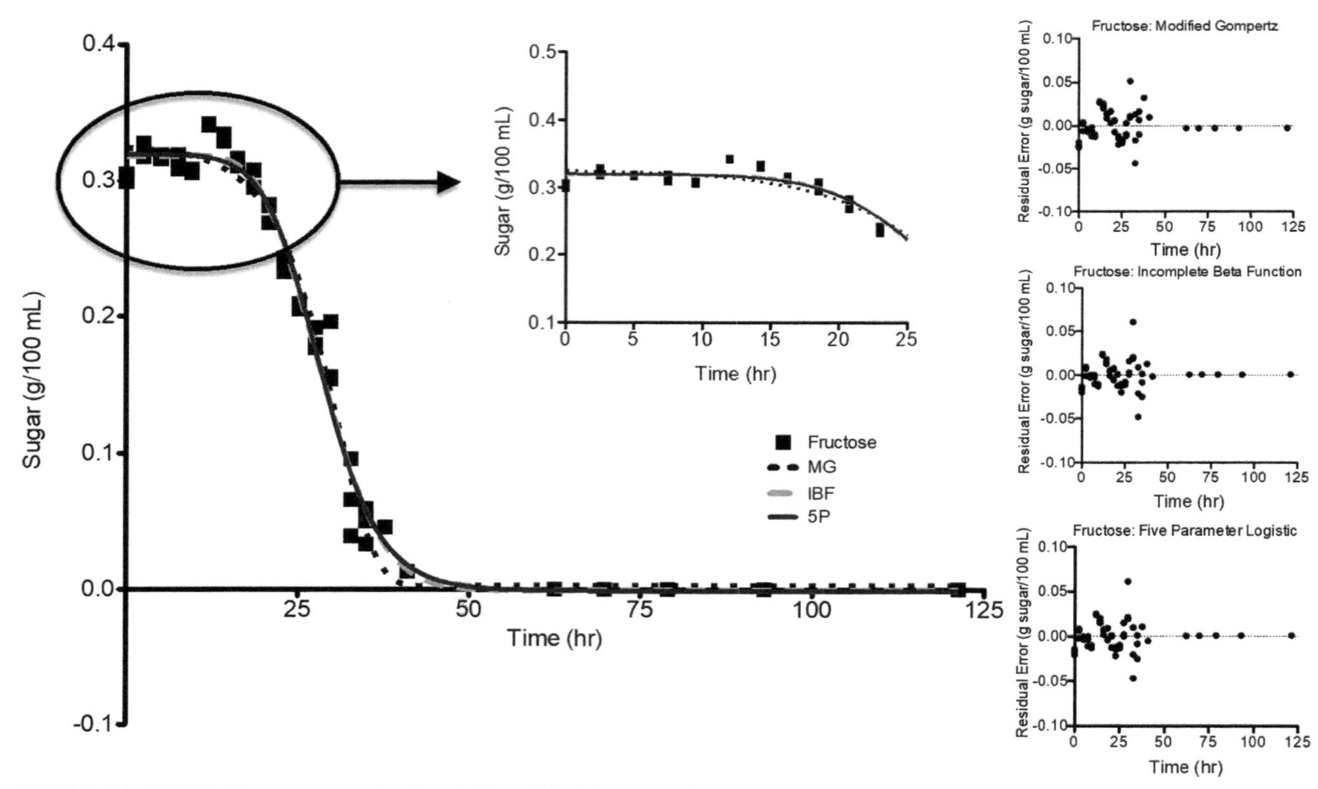

FIGURE 3.7 Modelled fructose attenuation data (MG, modified Gompertz; IBF, incomplete β-function; 5P, five-parameter logistic). The residuals for each model are depicted on the right.

The Gompertz model, although widely used for modelling the growth of many organisms (Buchanan & Cygnarowicz, 1990), has limited potential in modelling brewing fermentations. This model fits sugar attenuation well, provided that the sugar does not undergo consumption 'lag'. However, as we can see with maltose and maltotriose (Figs 3.4 and 3.6, respectively), this model deviates from the data near the beginning of fermentation, creating a trend in the residual error. Therefore, care should be taken when using this model, as it may fail to adequately describe brewing data. Next, although the IBF describes a versatile sigmoidal curve, it is an expanded mathematical distribution that is not designed to model biological behaviour. The limitations of this approach are apparent when modelling consumption data for sugars without a lag period. This is especially evident with glucose attenuation data (Fig. 3.5), in which the derivative of the curve (the rate of consumption) at time zero, will be zero by definition.

The theoretical basis behind the logistic model is that the primary variable (sugar concentration or density) will have an autocatalytic effect upon the rate of change. Although this is shown to be likely true (as attested to by the sigmoidal shape of the sugar curves), the non-symmetrical nature of the data alludes to additional factors beyond substrate consumption that slows attenuation during the second half of fermentation (such as flocculation or alcohol concentration). Therefore, the semi-empirical logistic model, which allows for asymmetry within the curve (i.e., 5P logistic), produced the most accurate fit, conforming to the actual shape of the attenuation curves. Additionally, the biological significance of the parameters described by Richard's curve provides a means of comparison between trials. As evidenced by the lack of pattern in every residual chart, this model can be used to accurately describe sugar attenuation in brewing operations. That being said, the 5P logistic model may suffer from 'overparameterisation' should the number of data points fall sufficiently low (the exact number depends upon when the samples are taken), which may often be the case in an industrial setting. This can be assessed by comparing the 4P logistic to the nested 5P using an F-test.

In summary, each model has advantages and disadvantages with respect to a particular situation. For example, the logistic models and IBF require numerous parameters to accurately model asymmetrical fermentations, whereas the IBF and Gompertz models each shows pattern residual deviation under specific circumstances (with and without consumption lag). Furthermore, in specific circumstances, the simpler Gompertz model was chosen by statistical rigour to be superior. This example illustrates the importance of understanding the capability and limitations of a chosen model for brewing applications, as the differences reported by each model are significant at scale.

3.4 Advanced fermentation techniques

Highlighting the importance of proper modelling and fermentation prediction is the phenomenon of high-gravity brewing. This increasingly common brewing technique can be simply described as the brewing of wort with a higher than normal specific gravity (the result of additional sugars). The resulting product is diluted or blended, allowing greater capacity within the brewery (more beer is produced without the need to expand fermentor capacity/plant size). Although the use of this technique poses unique challenges to the brewer, this procedure often has potential to deliver large benefits, while also granting a great deal of control over the final product. The necessity of diluting post fermentation allows the brewer to blend the high ethanol liquor into a variety of brands and to achieve greater consistency. The main economical benefits of this technique come from the increased capacity without the need for additional equipment. Additionally, a slightly higher ethanol yield per unit sugar can be expected as the number of yeast produced during fermentation is lower than would be expected from multiple fermentations.

There are, however, numerous considerations that the brewer must understand before committing to high-gravity brewing. Some of the drawbacks include greater stress upon the yeast (greater osmotic stress, ethanol content, etc.) and difficulty in matching the flavour of low-gravity (and undiluted) beer, often due in part to extensive use of adjuncts in high-gravity worts). Therefore, when contemplating high-gravity brewing, yeast selection and care is critical, as is quality control. Other noteworthy considerations include additional demands on the kettle (with higher carbohydrate loads), longer fermentations (partially yeast dependent) and potentially reduced foam stability (Stewart & Russell, 1998).

Due to the large volumes typically used for high gravity brewing, small changes in fermentation kinetics can have substantial economic impacts. In these circumstances, being able to fit models to ongoing fermentations allows brewers to detect deviations and take corrective actions. Furthermore, if incomplete fermentation data are fit using the models mentioned in this chapter, brewers will be able to get a rough prediction of final attenuation values (Reid et al., 2021). The accuracy of this prediction will depend upon many factors outside of the model.

3.5 Future trends and sources for further information

Like most industries, the brewing industry will ultimately follow the will of the consumer. For many years, this led to the development of lighter beer (one of the most popular products in North America). However, a perceived lack of choice has (among other factors) contributed to the success of craft markets in North America. Looking forward, the modern brewer must be keenly aware of consumer perception, especially in an increasing global market, in which nearly all growth is in emerging markets.

3.6 Closing remarks

The tradition of brewing is thousands of years old and has played an important social and economic role in many cultures. With this historical significance, it is not surprising that brewing has been highly scrutinised and that many of the processes and mechanisms that take place during fermentations are well understood and documented. However, as scientific methods and tools evolve, there are opportunities to re-evaluate and improve our understanding even of topics that are well understood. Often, apparent discrepancies observed between theoretical and observed results can be explained with a greater understanding of the process. As concluded by many researchers, modelling fermentations is a powerful tool that is easily used by modern brewers. The examples in this chapter shows that, for brewery applications, the common models used in the fermentation industry each have advantages and disadvantages. The additional accuracy resulting from the use of more complex models will be more easily utilised as modern instrumentation becomes more commonplace, allowing brewers to accurately model and predict their fermentations.

References

American Society of Brewing Chemists ASBC. (2011). Methods of analysis. In *Yeast-14 Miniature fermentation assay* (12th ed.). St. Paul, MN: The Society.

Armstrong, M., MacIntosh, A. J., Josey, M., & Speers, R. A. (2018). Examination of premature yeast flocculation in U.K. malts. *MBAA TQ, 55*(3), 54–60.

Balling, C. I. N. (1845–1865). *Die Gärungschemie (I-II, I-III, I-IV)*. Prague: Czech Polytechnical Institute.

Briggs, D. E., Boulton, C. A., Brookes, P. A., & Stevens, R. (2004). *Brewing science and practice*. Boca Raton, FL: CRC Press LLC.

Buchanan, R. L., & Cygnarowicz, M. L. (1990). A mathematical approach toward defining and calculating the duration of the lag phase. *Food Microbiology, 7*, 237–240.

Buchanan, R. L., Whiting, R. C., & Damert, W. C. (1997). When is simple good enough: A comparison of the Gompertz, barany, and three-phase linear models for fitting bacterial growth curves. *Food Microbiology, 14*, 313–326.

Canadean. (2011). *Beer, cider and flavored alcoholic beverages market.* Retrieved 2012 from Canadean Wisdom database . Retrieved 2012 from Canadean Wisdom database.

Corrieu, G., Trelea, I. C., & Perret, B. (2000). Online estimation and prediction of density and ethanol evolution in the brewery. *Master Brewers Association of Americas Technical Quarterly, 37*, 173–181.

de Godoy, V. R., Muller, G., & Stambuk, B. (2014). Efficient maltotriose fermentation through hydrolysis mediated by the intracellular invertase of *Saccharomyces cerevisiae. BMC Proceedings, 8*(Suppl. 4), Article P181.

Defernez, M., Foxall, R. J., O'Malley, C. J., Montague, G., Ring, S. M., & Kemsley, E. K. (2007). Modeling beer fermentation variability. *Journal of Food Engineering, 83*, 167–172.

Gibson, A. M., Bratchell, N., & Roberts, T. A. (1988). Predicting microbial growth: Growth responses of salmonellae in a laboratory medium as affected by pH, sodium chloride and storage temperature. *International Journal of Food Microbiology, 6*, 155–178.

Gompertz, B. (1825). On the nature of the function expressiveness of the law of human mortality, and a new mode of determining the value of life contingencies. *Philosophical Transactions of the Royal Society, 115*, 513–585.

Gonçalves, M., Pontes, A., Almeida, P., Barbosa, R., Serra, M., Libkind, D., et al. (2016). Distinct domestication trajectories in top-fermenting beer yeasts and wine yeasts. *Current Biology, 26*, 2750–2761.

Gottschalk, P. G., & Dunn, J. R. (2005). The five-parameter logistic: A characterization and comparison with the four-parameter logistic. *Analytical Biochemistry, 343*, 54–65.

Guadalupe-Daqui, M., Goodrich-Schneider, R. M., Sarnoski, P. J., Carriglio, J. C., Sims, C. A., Pearson, B. J., et al. (2023). The effect of CO2 concentration on yeast fermentation: Rates, metabolic products, and yeast stress indicators. *Journal of Industrial Microbiology and Biotechnology, 50*(1).

Nemenyi, J., Pitts, E. R., Martin Ryals, A., Boz, Z., Zhang, B., Jia, Z., et al. (2024). The effect of mixed culture fermentation of *Saccharomyces cerevisiae* and *Saccharomyces cerevisiae* var. *diastaticus* on fermentation parameters and flavor profile. *Journal of Food Science, 89*, 513–522.

Nielsen, H., Kristiansen, A. G., Lassen, K. M. K., & Ericstrøm, C. (2007). Balling's Formula – scrutiny of a brewing dogma. *Brauwelt International, 11*, 90–93.

Reid, S. J., Josey, M., MacIntosh, A. J., Maskell, D. L., & Alex Speers, R. (2021). Predicting fermentation rates in ale, lager and whisky. *Fermentation, 7*, 13.

Richards, F. J. (1959). A flexible growth function for empirical use. *Journal of Experimental Botany, 10*, 290–301.

Rudolph, A., MacIntosh, A. J., Speers, R. A., & St Mary, C. (2020). Modeling yeast in suspension during laboratory and commercial fermentations to detect aberrant fermentation processes. *Journal of the American Society of Brewing Chemists, 78*(1), 63–73.

Speers, R. A., Rogers, P., & Smith, B. (2003). Non-linear modeling of industrial brewing fermentations. *Journal of the Institute of Brewing, 109*, 229–235.

Stewart, G. G. (2006). Studies on the uptake and metabolism of wort sugars during brewing fermentations. *Master Brewers Association of Americas Technical Quarterly, 43*, 265–269.

Stewart, G. G., & Russell, I. (1998). *An introduction to brewing science & technology series III: Brewers yeast.* London, England: The Institute of Brewing.

Trelea, I. C., Latrille, E., Landaud, S., & Corrieu, C. (2001). Reliable estimation of the key variables and of their rates in the alcoholic fermentation. *Bioprocess and Biosystems Engineering, 24*, 227–237.

Trelea, I. C., Titica, M., Landaud, S., Latrille, E., Corrieu, G., & Cheruyb, A. (2001). Predictive modeling of brewing fermentation: From knowledge-based to black-box models. *Mathematics and Computers in Simulation, 56*, 405–424.

Virkajärvi, I., & Kronlöf, J. (1998). Long-term stability of immobilized yeast columns in primary fermentation. *Journal of the American Society of Brewing Chemists, 56*, 70–75.

Further reading

MacIntosh, A. J., Josey, M., & Speers, R. A. (2016). An examination of substrate and product kinetics during brewing fermentations. *Journal of the American Society of Brewing Chemists, 74*(4), 250–257.

Chapter 4

Advances in metabolic engineering of yeasts

Chris A. Boulton
Boulton Brewing Consultancy Ltd., Staffordshire, United Kingdom

4.1 Introduction

The realisation is that, in nature, a reservoir existed of strains of brewing yeast with differing properties coincided with the isolation of the first pure cultures by Hansen in 1883. Following the widespread adoption of these techniques brewing companies rapidly adopted and jealously guarded their own proprietary strains. A logical extension of this was that it should be possible to apply the principles of breeding programmes used elsewhere and not only to choose strains from those already existing but actively to create new ones possessing even more desirable properties and eliminating disadvantageous traits.

Early strain improvement programmes were hampered in several ways. There was insufficient knowledge of yeast genetics, especially of brewing strains and a lack of tools for manipulating the genome with the necessary degree of precision for the creation of new strains with desired new traits and no undesirable changes. Linked to this was a paucity of detailed knowledge of the relationships between the yeast genome and the results of its expression. In consequence, there was a lack of predictability as to how the phenotype of engineered yeast strains would be expressed in the conditions of commercial brewing.

The huge growth in knowledge of the yeast genome acquired over the last 20 years, or so, coupled with the development of methods allowing precise genetic manipulation have superseded the much more difficult approaches of classical breeding or random mutagenesis. These have combined to provide excellent methods for the construction of novel strains. For example, the *Saccharomyces* Genome Database (www.yeastgenome.org), is an excellent on-line resource providing up to date information on this topic. The genetics and evolution of brewing yeast strains are described in Nakao et al., 2009; Rainieri, 2009; Olauno et al., 2016, de Vries et al., 2019. Sequencing data for brewing and some other industrial strains can be found in Gonçalves et al. (2016) and Gallone et al. (2016). The latter two references provide much data regarding the patterns of domestication and its consequences.

Good as these techniques are they can only be used profitably if they are underpinned with a precise knowledge of how the make-up and regulation of the yeast genome impacts on the phenotype. This aspect of yeast behaviour has lagged behind the development of techniques for genetic manipulation. The regulation of metabolism at the level of biochemical pathways is complex but essential to understand if the benefits of genetic engineering are to be gainfully exploited.

This need is addressed by the concept of metabolic engineering, which can be defined as the application of genetic techniques for the manipulation of metabolic pathways to bring about desirable changes in the activities of an organism. In this chapter may be found a discussion of metabolic engineering, and the tools that are used to apply it. Developments are described in modern brewing practice and how these are providing the impetus for developing new brewing yeast strains.

4.2 Metabolic engineering

The concept of metabolic engineering was introduced by Bailey in the early 1990s (Bailey, 1991; Bailey *et al.*, 1990). Several other terms have been used to describe the similar and allied concepts; these include metabolic pathway

Brewing Microbiology. https://doi.org/10.1016/B978-0-323-99606-8.00019-5

engineering (Tong et al., 1991), cellular engineering (Nerem, 1991) and in vitro evolution (Timmis et al., 1988). In each case, as the names suggest, the emphasis is on the adoption of a systematic engineering approach to the characterisation of cellular activities in order that these can be manipulated to achieve an altered and desired outcome. In other words, rather than adopt the random approach of selecting desirable strains from a pool of natural or induced variants, the likelihood of success is increased, provided there is sufficient prior knowledge of the physiological basis of the trait that is under consideration. Implicit in this is the realisation that although the phenotype is driven by the genome, there are several additional layers of control operating at the pathway level and involving regulation of enzyme activity by effector molecules and the concentrations of substrates and end-products. If these are not quantified and understood, strain improvement programmes are unlikely to succeed.

The original definition of Bailey (1991) defined metabolic engineering as ‘the improvement of cellular activities by manipulation of enzymatic, transport and regulatory functions of the cell with the use of recombinant DNA technology’. In two later excellent reviews, Stephanopoulis (1999) and Ostergaard et al. (2000) refined the definition to ‘the directed improvement of product formation or cellular properties through the modification of specific biochemical reaction(s) or the introduction of new one(s) with the use of recombinant DNA technology’. Key to the approach is the use of the word ‘directed’. Unlike earlier attempts at strain improvement, the genetic techniques that are used allow more precise manipulation without the introduction of unwanted and potentially undesirable additional changes.

Metabolic engineering of a desired trait is a two-stage process. The cellular basis of the trait which is to be modified is first analysed to identify specific targets and secondly, these target sites are then subjected to genetic modification in a highly focussed manner to make the necessary changes. Several operations are possible. These include amplification, inhibition, deletion, or deregulation of native genes and in addition, heterologous genes may be introduced. The genetics of brewing yeasts are discussed in Section 4.5. The target traits that have been identified with respect to brewing yeast are described in Sections 4.6 and 4.7.

Metabolic engineering seeks to close the gap in understanding between a basic knowledge of biochemical pathways and the over-arching mechanisms that regulate cellular functions. Information such as the sequences of chemical steps that make up biochemical pathways, the enzymes which catalyse the individual steps and the genes responsible for their synthesis is essential but gives relatively little information as to how the individual pathways are regulated and how the control mechanisms are integrated with other pathways at any given time. Without this higher-level information, it is unlikely that attempts to steer changes to provide an altered outcome will be entirely successful.

With regard to brewing, this can be illustrated by considering the flow of metabolites through pyruvate and acetyl-CoA, as shown in Fig. 4.1. Clearly, this is not a complete picture, and many steps in the pathways shown have been omitted; nevertheless, it can be seen that there are major branch-points linking the uptake of major wort nutrients such as sugars, free amino nitrogen and oxygen and the formation of key products including ethanol, organic acids, higher alcohols, vicinal diketones, esters, fatty acids and sterols. Many of these compounds will be important participants in potential strain-improvement programmes, and therefore, the pathways they form part of will be targets for genetic manipulation. Not shown in the figure are the proportions of carbon metabolised via the glycolytic and pentose phosphate pathways, the roles of anaplerotic and gluconeogenic pathways, neither are the issues surrounding cell compartmentalisation. It is obvious that unless the bigger picture is considered, changes made in one pathway could affect carbon flow through others with unexpected and perhaps undesirable results.

It is helpful to consider cellular activities in terms of a hierarchy which collectively provides a framework within which the various levels of function can be viewed as an ordered whole. These are the genome, the transcriptome, the metabolome and the fluxome (Fig. 4.2).

The genome delineates the potential capabilities of the cell, and strain improvement programmes may seek to delete or amplify parts of it or augment it with heterologous genes. As discussed later in this chapter (Section 4.5), the genomes of many brewing strains are by no means simple, and these complexities make them not always easily manipulated. By definition, the transcriptome is that part of the genome that at any given instant has been transcribed and is therefore represented by molecules of RNA. An analysis of the transcriptome gives a snapshot of gene activity at any given instant. The products of translation collectively constitute the proteome, and this provides a picture of the immediate results of transcription. Many of these will be enzymes, and the proteome describes the types and quantities present but provides no indication of activity. Some components of the proteome interact directly with the genome by acting as transcriptional factors. The longevity of enzymes in the proteome is of interest since mechanisms for regulating protein turn-over will have an obvious impact on the results of genetic manipulations, particularly where heterologous genes are involved.

The metabolome describes the concentrations of the entire complement of small molecules present in the cell at any given time (Zamboni & Sauer, 2009). Advances in the power of analytical techniques, as discussed in the next Section, 4.3, now allow detailed scrutiny to be made such that it is possible to monitor the whole of the metabolome. The compounds

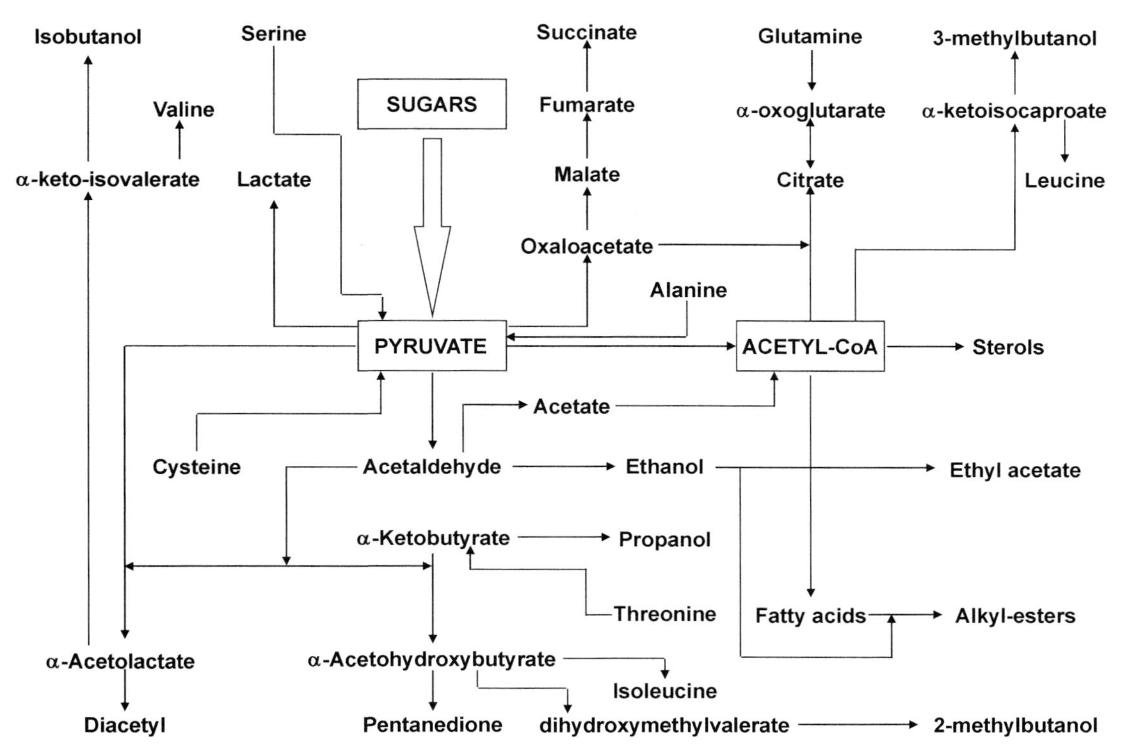

FIGURE 4.1 An overview of some of the metabolic pathways flowing to and from pyruvate and acetyl-CoA in yeast.

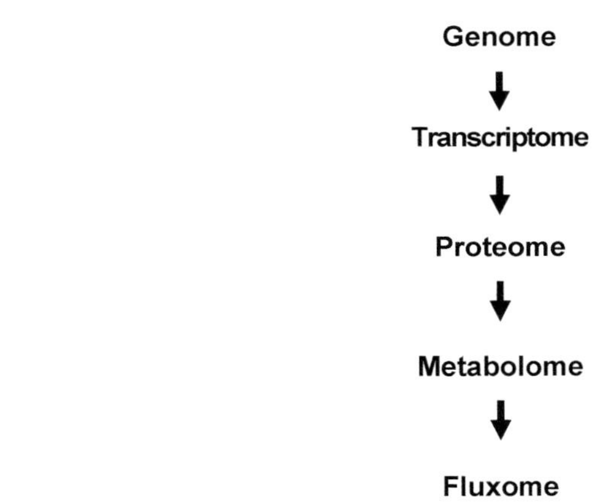

FIGURE 4.2 Hierarchy of cellular functions (see text for discussion).

involved are direct participants in metabolic pathways, and they can also function as modulators of enzyme activity. They can be used to infer the presence of heretofore unknown pathways, and they provide information as to how nutrients become distributed throughout the cell and how and where excreted products arise.

Metabolomic information is obviously very helpful, but it does not provide any information regarding flow rates of metabolites through individual pathways. For this, it is necessary to measure changes in the concentrations of components of the metabolome as a function of time. Collectively these changes are termed the fluxome (Stephanopoulis, 1999; Wiechart, 2001; Zamboni, 2011). Studies which seek to investigate changes in the metabolome are referred to as metabolic flux analysis (Christiansen & Nielsen, 2000; Zamboni, 2011). Thus, Stephanopoulis (1999) defines a biochemical pathway as a 'sequence of feasible and observable biochemical steps connecting a specified set of input and output metabolites', whereas the metabolic flux is 'the rate at which material is processed through a metabolic pathway'. The flux is considered with measurement of the concentrations of metabolites, and collectively, these are considered to be the minimum information required to define the physiology of a cell under a given set of environmental conditions. Study of metabolic fluxes provides the necessary level of detail for a proper appreciation of the effect of changes in physiological condition on cellular function. With reference to the genome and transcriptome, the effect of genetic engineering can be assessed.

The totality of cellular functions needs to be considered. This includes the basic pathways and also global regulatory functions. In the context of brewing, the growth of yeast on wort, usually with an initial aerobic phase, followed by a transition to anaerobiosis, requires the cells to have mechanisms for dealing with the complex mixture of nutrients available and the constantly changing conditions. Apart from short-term regulation of metabolic pathways via interactions between the metabolome and enzyme activity global effects on the genome must also be considered. These signal transduction pathways regulate the cell cycle (Nishida & Gotsh, 1993; Posas et al. 1998), responses to applied stresses, for example, the osmo-sensing (HOG) pathway (Brewster et al., 1993) and responses to mixtures of nutrients as embodied by carbon (Gancedo, 1998; Schüller, 2003) and nitrogen catabolite repression (Wiame et al., 1985; Boczo et al., 2005). Of particular note in brewing, where serial fermentation is the norm, is the transition of cells between the G_0 phase of the cell cycle, which occurs at the end of fermentation and persists through storage and back into G_1 at the end of the lag phase when pitched into the next fermentation (Wei *et al.,* 1998; Gibson *et al.,* 2005).

Initiation of these pathways requires communication between the cell and the external environment, and therefore, the role of receptor sites and uptake mechanisms must be considered. The onset of flocculation would be another response, which requires responses with cells and the environment and with each other. The appearance of metabolic end-products in the medium requires mechanisms for intracellular transport and excretion. Within the cell, trafficking of metabolites and enzymes is of importance, particularly in the case of eukaryotic cells, where the impact of compartmentalisation has to be considered in strategies for strain improvement.

4.3 Tools for metabolic engineering

Successful metabolic engineering of microbial cells requires a stringent set of criteria to be satisfied. A high degree of knowledge of the organisation and function of the target cell is required. It is preferable that the genome has been sufficiently characterised to allow annotation of gene to cellular function. Knowledge of the underlying metabolic pathways is needed together with information as to how these are linked together and regulated.

It is unlikely that it will be possible to make some measurements in batch fermentations, especially commercial large-scale fermentations featuring the growth of yeast on a relatively uncharacterised medium such as wort. Batch cultures are inherently difficult to work with since by definition, conditions are in a state of continuous change. The corollary is that physiological condition is also in a state of continuous change in response to the altered environment. The conditions that yeast is exposed to in commercial brewing fermentations are very different to those where the same strains are used in laboratory studies. This poses a dilemma since it is clearly difficult to impose the appropriate degree of experimental stringency in production scale trials where ultimately, the results of metabolic engineering must be assessed. However, when carrying out the initial studies into cellular function and subsequent genetic manipulation, it is essential to have a properly controlled experimental system. Where batch fermentations are used at laboratory scale, the apparatus should be capable of both controlling and monitoring all basic parameters such as temperature, gas analysis, pH and agitation. For more controlled experimentation, it is desirable to use a chemostat or related cultivation system (Pirt, 1975). Chemostats form the basis of continuous cultivation and therefore allow establishment of steady states in which growth rate is proportional to the rate of supply of nutrients and physiological state is constant. Many of the techniques used in flux analysis require that steady-state conditions are maintained. The ability to perturb growth conditions in chemostats by allowing transitions between different steady states is a particularly powerful tool for elucidating metabolic fluxes (Lapujade et al., 2003; Kuyper et al., 2005).

Many metabolic engineering initiatives where the cell may be viewed as a factory have some choice regarding the host organism. It is best that this is stable and capable of growing on a simple and preferably reasonably well-characterised medium. In the case of brewing yeast, these niceties are of course irrelevant, and it is certainly true that the genome of lager yeasts in particular poses some problems (see Section 5.5). Nevertheless, a range of tools are available to probe and modify the different levels of cellular activities summarised in Fig. 4.2.

The yeast genome can now be sequenced and manipulated with relative ease, and there is huge literature describing the techniques that can be used and the information obtained; for example, Oliver et al., 1998; Fromont-Racin et al., 1991. The complete sequencing of several genomes, including that of yeast (Goffeau et al. 1996), has allowed comparisons of coding regions of genes with known function such that a large proportion of the component parts of genomes can be associated with specific cellular functions (Liti & Louis, 2012; Schilling et al., 1999).

Individual genes can be amplified or deleted, and heterologous genes can be inserted using recombination or with suitable vectors. Vectors must be stable, form the desired product and be present in all transformed cells with a consistent copy number (Nevoight, 2008). Single or multi-copy vectors can be used to regulate copy number. In the case of brewing yeast, the most appropriate tool is to increase copy number by integrating several gene copies into chromosomes. Successful attempts have made use of the delta regions which occur in several locations in yeast genomes and therefore provide a means of inserting multiple gene copies (Kudla & Nicolas, 1992; Lee & da Silva, 1997).

The transcriptome can be monitored using a variety of techniques which allow the quantification of the degree of expression of entire genomes. Techniques such as serial analysis of gene expression (SAGE) and array analysis have been applied to yeast (Gibson et al., 2008; Oleson et al., 2002; Velculescu et al., 1997), in the latter two papers, lager yeasts during the course of commercial large capacity fermentations.

Control of the proteome is obviously of primary importance since control at the cellular level is exerted via the activities of enzymes either via catalysis of individual steps in pathways, performing transport functions or by acting as regulatory proteins. The proteome is amenable to analysis via extractive techniques followed by chromatographic or electrophoretic separation and analysis usually by mass spectrometry (Gavin et al., 2002; Picotti et al., 2013; Washburn et al., 2001). These methods have been particularly useful for gene annotation.

Transcription of individual genes can be modified by the addition of a suitable promoter (Mumberg et al., 1995). It is possible to regulate transcription efficiency by choosing alternative yeast promoters with the appropriate strength. Verstrepen and Thievelein, (2004) successfully applied this technique to control the transcription of homologous genes in *S. cerevisiae*. However, this may have to be used in conjunction with methods for manipulating gene copy number. Another approach is to delete or modify the function of transcription factors by targeting the encoding genes. This has been applied to the alleviation of glucose repression allowing constitutive maltose uptake (Nielsen, 2001) and modulation of ethanol formation by manipulation of the activity of pyruvate decarboxylase (Nevoight & Stahl, 1996). Omura et al., (2005) introduced efficient proline assimilation into a lager yeast strain using an alternative strategy by inhibiting ubiquitination of an amino acid permease and thereby rendering the protein more long lasting.

Metabolomic and flux analysis are the most challenging. The components of the metabolome can be characterised by separation via liquid or gas chromatography followed by analysis using mass spectrometry or nuclear mass spectrometry. Typically, several methods of separation and detection are combined to allow quantification of several classes of compounds (Büscher et al., 2009; (Dettmer et al., 2008); Ohashi et al., 2008; van der Werf et al., 2007). These approaches have managed to characterise nearly 400 individual compounds although only in bacteria. Sampling remains a problem since after removal from the growth vessel, all cellular activities must be rapidly quenched in order to avoid artefacts. In the case of bacteria, the process disrupts the membrane and corrections must be made for leakage into the medium. In the case of eukaryotes, such as yeast cells, quenching will disrupt intracellular membranes making probing of compartmentalisation difficult.

The dynamic nature of flux analysis introduces much more complexity. Studies are dependent on measurement of the relevant components of the metabolome followed by the application of mathematical modelling and statistical analysis to unravel the huge quantities of data generated. The usual approach is to use a combination of tracer studies using isotopes of carbon, mainly ^{13}C together with gas chromatography mass spectrometry (GC-MS) or nuclear magnetic resonance (NMR) (Whitmann, 2007). The ^{13}C-labelled substrate is fed into the system, and the label becomes distributed into the intermediates and products of subsequent metabolism. The identity of the compounds formed and the position of the label in the molecule are of significance. These combinations of isotopes and isomers are referred to as isotopomers (Wiechart, 2001). The use of isotopomers dramatically increases the value of the information supplied. A compound containing n carbon atoms, each of which may or may not be labelled, can form 2^n different isotopomers. From the position of the label in a product, it may be possible to identify the pathway by which it was formed, or the presence of unknown pathways can be identified. Where several isotopomers of the same compound are detected, the presence of multiple pathways can be inferred. At steady states, the concentration of each labelled intermediate provides flux information.

Zamboni (2011) describes flux analysis via two approaches, termed stationary or non-stationary. In both cases, the basic requirements are a model metabolic network, a closed carbon balance measured in terms of rates of substrate uptake and product excretion and information regarding the patterns of ^{13}C labelling in product and intermediates. The stationary strategy relies on measurement of the proportions of isotopomers of a common compound formed at steady state from which converging alternate pathways can be resolved. The non-stationary method monitors the spread of the ^{13}C label with time. Together with analyses of the concentrations of relevant components of the metabolome, the latter approach is capable of quantifying fluxes. It is much more demanding in terms of data handling.

Undoubtedly, flux analysis offers great promise as a tool to be used in metabolic engineering; however, there remain significant challenges, and these are very pertinent to applications in brewing. The problems of compartmentalisation in eukaryotic cells are problematic (Niklas et al. 2010). With regard to brewing the enzymes associated with branched chain amino acid metabolism, such as valine, which are presumed to be implicated in diacetyl formation and reduction in beer are located in mitochondria and by inference two transport steps must be involved between this compartment and the cell exterior. In order to understand properly and manipulate metabolic fluxes and how these relate to events such as free amino nitrogen (FAN) uptake and vicinal diketone formation, it seems probable that compartmentalisation will be pertinent. Nevertheless, as a result of metabolic engineering, the sub-cellular localisation of enzymes has been successfully moved from mitochondria to the cytosol (Moreira dos Santos et al., 2004).

It is interesting to note that petite mutants which can arise during the course of brewing are known to generate higher concentrations of diacetyl and are slow to remove it compared with the wild type. It has been shown (Desari & Kolling, 2011) that at least a partial explanation may be that the acetohydroxy acid synthase, the enzyme which forms part of the ILV pathway and which catalyses the step that produces the immediate precursor of diacetyl, fails to be transported from the cytosol, where it is synthesised, to the mitochondria, the location of the ILV pathway. This enzyme is coded by a nuclear gene, and as a necessary precursor for its relocation to the mitochondria, it is labelled with a targeting sequence. This relocation process is defective in petites which presumably is implicated in disruption of the usual system for diacetyl formation and ultimate removal. Apart from indicating the complexity of the pathways leading to diacetyl arising in beer, the observations show that engineering must take factors such as cellular compartmentalisation into account.

Much work remains to be done to begin to understand the control of cellular functions in non-steady states during growth on complex media. The great promise of metabolic engineering and the analytical methods on which strategies for genetic manipulation are based is that it allows a much more integrated view of cellular function from metabolic pathways up to the genome. An output from these studies has been the appreciation of the role of regulation networks, in particular, the importance of non-transcriptional events such as phosphorylation of proteins (Heinemann & Sauer, 2010).

4.4 Strategies for metabolic engineering

The complete appreciation of cellular functions continues to lag behind the ability to analyse and manipulate the genome. As a result of this, strategies for metabolic engineering still remain largely reliant on successive rounds of genetic manipulation and analysis in order to hone the acquisition of the desired phenotype. Metabolic engineering is based on what has been defined as a rational metabolic engineering in which genetic manipulation is based on prior knowledge. The results are often disappointing because of the appearance of unpredicted side-effects and the fact that the advances made in laboratory studies fail to translate to commercial fermentations (Nevoight, 2008).

For these reasons, other strategies have been devised. Evolutionary engineering (Sauer, 2001) relies on a random approach in which the genome is altered via mutagenesis or DNA shuffling followed by screening for variants with a desired phenotype. The obvious problem is the random nature of the process and the need for precise screening methods. The latter can be a particular problem with some traits. Bayer and Smolke (2005) describe a work-around in which the activation of riboregulators by a target ligand is connected to the expression of a gene that generates a fluorescent dye. In this way, any phenotype should be amenable to analysis.

The major drawback of evolutionary metabolic engineering is that is does not produce an understanding of the metabolic networks involved nor does it relate this to the relevant genomic information. Global transcription machinery engineering (gTME) targets random mutations involving protein transcription factors. Mutants showing phenotypic improvements are selected, and the gene mutations are identified. The phenotype modifications can be transferred to other cells, and via analysis of the transcriptome, the associated changes in gene expression can be identified. An advantage of this strategy is that it can be applied to polygenic traits such as stress responses (Alper et al., 2006). In another strand, the concept of reverse metabolic engineering has been introduced (Bailey et al. 2002). This requires selection of systems in which the desired phenotype is expressed differently. These might be different but related strains or the same strain in which the trait is expressed differently under varying environmental conditions. The genetic basis of the observed differences is then resolved, and hopefully, this provides the necessary information required to reproduce the desired phenotype in another host. Confirmation that the genetic basis of the phenotype has been correctly identified is made via deletion or amplification of the target portion of the genome.

The strengths and weaknesses of this approach are embodied by the foregoing discussion regarding the different levels of cellular function (Fig. 4.2). Assessing the phenotype based on genome analysis is unlikely to reveal the precise source of phenotypic variations. The metabolome and fluxome provided the greatest level of detail but are the most difficult to relate to the genome. Nevoight (2008) claims several advantages for inverse metabolic engineering: it is not necessary to have prior knowledge of the relevant pathways or their regulation; industrial strains can be used under production conditions; heterologous genes are not involved and the transformants can be considered self-cloned; it may lead to the chance discovery of novel genetic targets.

The introduction of highly focussed gene editing techniques such as clustered relatively interspaced short palindromic repeats (CRISPR) makes the process of genetic manipulation relatively easy. In an interesting paper, Krogerus et al. (2021) used the technique as a demonstration of concept to allow efficient breeding of brewing stains, a trait in which most commercial strains are deficient. The authors demonstrated that the method can be used to make intraspecific hybrids and thereby introduce new and industrially useful traits. Additional information on the utility of hybrids may also be found in an earlier paper by the same research group in Krogerus et al. (2017). Approaches such as this have the advantage that they permit genetic engineering in a way which is perhaps more 'natural' and therefore acceptable to the public. The perception of public hostility to genetic engineering remains a bar which many brewers are reluctant to scale.

4.5 Brewing yeast genetics

Haploid cells of *S. cerevisiae* contain 16 chromosomes, which contain more than 12 million base pairs comprising more than 6000 genes. Additionally, there is a separate mitochondrial genome, which contains genes mainly coding for components of the respiratory electron transport chain. Sequencing data of mitochondrial DNA for several strains together with information regarding their origins and evolution can be found in Wolters et al. (2015). Mitochondrial and nuclear genomes are not independent, and indeed many mitochondrial functions are coded for by nuclear genes. The cooperation provides an example of intra-organelle trafficking and communication (Rodley et al., 2012). In addition, it exemplifies the extra levels of complexity, which must be considered when embarking on programmes of genetic engineering.

Brewing yeast strains have much more complex genetics compared to the haploid types used in most laboratory studies.

Brewing yeasts are differentiated into ale and lager types. The former are considered older in evolutionary terms and are classified as *S. cerevisiae*. They express considerable genetic variability (Pedersen, 1985, 1986a, 1986b, 1994). Lager strains are currently classified as *S. pastorianus* (Vaughan-Martini & Martini, 1987) and form a less diverse group of strains compared with ale types. The genome of lager strains differs significantly from ale types. Ale strains are polyploid whereas lager types are allotetraploids (Smart, 2007). Studies have shown that the lager genome is hybrid in nature in which chromosomes may show homology with those of *S. cerevisiae*, with no homology with *S. cerevisiae* or mosaics of the two. Individual genes may be *S. cerevisiae* (Sc-) -type or non-*cerevisiae*, termed *S. pastorianus* (Sp-), lager (lg-) or *S. carlsbergensis* (CA-) type. Lager strains apparently arose as a result of one or more hybridisation events between two closely related *Saccharomyces* strains. One parental type was *S. cerevisiae* the identity of the other remains subject to debate but is currently considered most likely to be *S. bayanus* (Smart, 2007). The mitochondrial genome is circular and in lager yeast strains shows most homology with that of *S. bayanus* (Smart, 2007).

The hybrid nature of lager strains introduces much complexity. Kodama et al. (2001) investigated uptake of branched chain amino acids in a lager strain. Two permeases were present coded for by *BAP2* genes. Of the two genes one (*cer-BAP1*) was identical to that of *S. cerevisiae*, whereas the second (*lg-BAP2*) wAs identical with that from *S. bayanus*. The genes show 88% homology with each other but were regulated differently. Generally two copies of each gene were present.

Nakao et al. (2009) sequenced the entire genome of a lager strain and found that a 25 Mb genome could be divided into two distinct nuclear sub-genomes with homologies with *S. cerevisiae* and *S. bayanus*. The size was roughly double that of ale strains. Some 36 different chromosomes could be distinguished of which 8 were mosaic types where the breakpoints were within open reading frames.

The tetraploid nature of the lager yeast genome appears to confer inherent genetic instability. The application of applied stresses of the types now commonplace in commercial brewing such as the use of fermentations at elevated temperatures with very high gravity worts are causes of genetic changes (James et al. 2008).

4.6 Metabolic engineering and the complexities of metabolic regulation

Metabolism is undeniably complex in terms of the multiplicity of pathways of which it is comprised. Most of the individual reactions are well-characterised as are the enzymes which catalyse them, and in many cases, the genes that code for them are known. This should give a sound basis for devising successful strategies for manipulating metabolism to deliver brewing yeast strains with new and desirable properties. However, this is only a small part of the complexities of cellular functions.

The impact of intra-cellular compartmentalisation has been alluded to already. Cells use compartmentalisation as a method for maintaining different conditions in separate parts of the cell, for example, regulation of pH at values appropriate for the reactions being conducted. In addition, it provides a method for ensuring that enzymes come into contact with desired substrates and not those where modification would be undesirable. An example would be targeting selected proteins for breakdown and turnover carried out in vacuoles. Individual compartments are associated with specific functions. These must be supported by systems for communication and translocation. Gene transcription largely in the nucleus, cytosolic translation into protein synthesis and assembly and subsequent transport of proteins to their intended site of operation must be underpinned by an exquisitely balanced system of regulation. It is obvious that attempts to engineer these processes to deliver a change from the normal outcome will be fraught with difficulty if the intricacies of these highly nuanced systems are ignored. For example, amplifying or deleting a single gene without considering the context within which it operates is unlikely to deliver the desired outcome.

A universal theme in metabolic regulation is the use of signalling hierarchies in which the cell is able to respond to the external environment by adjusting its genome to take best advantage of particular sets of circumstances. Molecular cues from the environment trigger metabolic cascades of gene up and down regulation that elicit an appropriate global cellular response. The TOR and Ras/PKA systems act as the primary signalling receivers. In yeast, TORC 1 is responsible for nitrogen nutrient sensing and linking this, via the RAS/PKA system, to detection of sugar availability and an appropriate growth response. The TORC 2 system has numerous roles including membrane lipid sensing, cell cycle control and regulation of pathways such as the pentose phosphate pathway (Gonzales & Hall, 2017; Tamanoi, 2011).

These sensing systems are responsible for the control of global responses including amongst others, glucose repression and de-repression, ordered assimilation of nutrients, cellular growth and proliferation, stress responses and the spectrum and yields of numerous metabolic by-products. Obviously, many of these have direct relevance to the performance of brewing yeast and beer quality. A facet of this, not always appreciated, is that many of these global responses are initiated by relatively simple signalling molecules. These include nutrients, as would be expected, but also many other compounds that are metabolic intermediates. Knowledge of these triggering molecules their sites of action and the responses that are elicited via signalling cascades will be likely starting points for programmes of metabolic engineering designed to modify and improve brewing yeast performance.

A common theme and complicating factor with potential signalling molecules is that many have already well-established other metabolic functions. Some examples are given in Table 4.1. No doubt there are many others. This multiplicity of roles introduces complexities which must be considered when pathways in which they are involved are chosen as candidates for metabolic engineering. On the other hand, proper understanding of this facet of metabolism is likely to identify many new routes for improving brewing yeast performance via metabolic engineering.

TABLE 4.1 Multiple functions of potential signalling molecules.

Metabolite	Metabolic function	Signalling role	References
Glucose	Energy generation and carbon source for anabolism (amongst others).	Transition from quiescence to growth. Trigger of regulated cell death in target cells	Liko et al. (2010) Ruckenstuhl et al. (2020)
Higher alcohols	Redox role	Transitions from planktonic to biofilms Flocculation	Chauhan and Karuppayil (2020)
Organic acids Acetate specifically	Acidification of medium. pH homoeostasis	Trigger of regulated cell death in target cells.	Chaves et al. (2021)
Esters	Regulation of Coenzyme A availability and lipid synthesis.	Dispersal agents via insect vectors.	Malcorps et al. (1991) Christaens et al. (2014)

4.7 Impact of brewing yeast population dynamics

The literature devoted to *S. cerevisiae* is enormous largely because it is used as a model eukaryotic cell that is easily cultivated and amenable to study. In particular, it has been widely used to study mammalian ageing, regulated cell death and cancer biogenesis (Gershon & Gershon, 2000; Vanderwaeren et al., 2022). Implicit in much of this work is the fact that the same, or very similar metabolic machinery, is shared by budding yeast and higher eukaryotes. Furthermore, it has become clear that individual yeast cells in many ways are non-identical and rather than existing in splendid isolation, populations function in concert in ways which are analogous to cells that comprise individual organs or whole organisms of higher eukaryotes.

This concept is hardly new but in terms of brewing practice little note seems to have been taken of this fundamental aspect of brewing yeast populations. Yeast is treated as a bulk resource, for good practical reasons, where individual cells are analysed tests are quite rudimentary, for example, distinguishing live from dead cells via the use of vital stains. So-called yeast vitality tests (Kwolek-Mirek & Tecza, 2014) have been developed which probe the physiology of the live fraction of yeast populations but again many of these test bulk yeast samples to give an 'averaged' result. Typically, yeast considered to have high vitality is that which is predicted to give vigorous fermentations, a quality not necessarily of great importance since it does not correlate with fermentation efficiency or give a useful measure of beer quality.

Section 4.6 of this chapter described the idea that relatively simple metabolites can function as signalling molecules via initiating cascades in which multiple pathways are up or down regulated. An extrapolation of this is that within yeast populations distinct sub-groups communicate with each other by acting either as signal generators or receivers (or both).

Brewing yeast cells age in two ways, termed chronological and replicative ageing. The latter describes the number of times that an individual cell can undergo budding. This is strain-specific and is typically 10 to 30 times (Powell et al., 2000). Budding is unequal such that daughter cells are always smaller than their parents and sharing of assets such as cellular organelles and error-free DNA always ensures that daughters receive the best of the bargain. In this sense, new virgin cells are reset as having a youthful status. Chronological lifespan is defined as the time that cells can survive under non-growth conditions.

Cell death can occur as a result of exposure to un-survivable trauma, but more usually, it is a programmed and deliberate process referred to as regulated cell death. This is the same system that underpins cellular homoeostasis of the organs of higher eukaryotes in which older cells with reduced function must die to make way for newly formed replacements.

The process requires a differential response to signalling systems in which specific sub-populations are targeted for death, whereas others are able to modify modulators of the same pathways in a manner that is pro-life (Zimmerman et al., 2020). The usual agents that trigger death are reactive oxygen species (ROS). Susceptible cells cannot recover from the damage caused but non-targeted cells have ameliorating protective systems that allow them to survive. The underlying principle is that chosen cells which have outlived their useful contribution are sacrificed for the benefit of the whole population. In so doing, they lyse and release their contents for recycling by survivors.

Implicit in this is that communication between individual cells results in the differentiated response in a manner that requires cooperation. This is not restricted to decisions regarding cell death or survival. Other differentiated global changes in cellular physiology in selected sub-populations have been shown to occur, for example, in yeast colonies growing on solid media (Vachova et al., 2012). Predictably, cell proliferation is greatest at the peripheries of colonies where nutrients are more readily available and less amongst cells in the middle of colonies. These studies revealed marked differences in gene expression of cells based on their spatial location, both radially and vertically, within the colony. Primary signals initiating these changes were underpinned by regulated shifts in pH, both acidification and alkalinisation, the latter via the formation of ammonia.

These results suggest complex systems of communication between sub-populations. Differences in gene expression in cells within flocs analogous to those described for colonies have also been recorded. This is discussed in more detail with specific reference to brewing yeast and fermentation in Boulton (2021).

An ineluctable conclusion of these studies is that cooperative behaviour between individual yeast cells is the norm. Responses are mediated by relatively simple signalling molecules. Provided we make efforts to identify the individual words and learn the meaning of the language and its effects, this must give new avenues for genetic manipulation.

4.8 Targets for engineering of brewing yeast

Traits chosen for manipulation in brewing strains fall into the following categories.

i. Process improvement
ii. Altered spectrum of substrate utilisation
iii. Improved control of beer flavour and stability
iv. Production of novel beers by fermentation
v. Altered resistance to spoilers

Many examples of the development of engineered brewing strains with traits falling into the categories given above have been published. Space does not permit a complete listing of these, but some examples are given in Table 4.2. From the references, it may be seen that most of these pieces of work are now comparatively old, and this possibly reflects the fact that the lack of commercial take-up has made many researchers switch to more receptive alternative industries such as bioenergy.

TABLE 4.2 Targets for metabolic engineering of brewing yeast.

Trait	Driver	Trait	Target	References
Process improvement	Increased yield	Increased ethanol yield	Overexpression of *GLT1*	Cao et al. (2007)
		Improved ethanol tolerance	Altered transcription via mutation of Spt15p transcription factor	Alper et al. (2006)
		Improved osmotolerance	NAD^+-dependent glycerol 3-phosphate dehydrogenases	Ansell et al. (1997) and Siderius et al. (2000)
		Improved performance at high gravity	Selection of variants from UV-treated brewing yeast	Blieck et al. (2007)
		Dextrin utilisation	Insertion of glucoamylase and α-amylase from *Lipomyces kononenkoae*	Eksteen et al. (2003)
			Addition of *DEX* gene from *Saccharomyces diastaticus*	Perry and Meaden (1988)
		Improved sugar utilisation	Constitutive expression of *MAL* genes	Kodama et al. (1995)
	Reduced fermentation cycle time	Increased thermo-tolerance	Selection and isolation of variants in distillery yeasts	Abdel-Fattah et al. (2000)
		Increased or altered diacetyl metabolism	Disruption of *ILV2* via self-cloning	Zhang et al. (2008) and Wang, He, Liu, and Zhang (2008)
			Disruption of *ILV* genes	Gjermansen et al. (1988)
			Introduction of α-acetolactate decarboxylase	Fujii et al. (1990)
	Enhanced substrate utilisation	Dextrin utilisation	As above	Eksteen et al. (2003) and Perry and Meaden (1988)
		Pentose utilisation	Incorporation of xylose-utilising pathway, manipulation of redox control	Jeffries and Jin (2004)
	Better beer filterability	β-glucan utilisation	Incorporation of β-glucanase from *Trichoderma reesei*	Penttilä et al. (1988)
	Resistance to contamination	Acquisition of killer phenotype	Transfer of killer factor via rare mating	Young (1981)
	Cropping behaviour	Altered flocculation	Manipulation of expression of *FLO1*	Verstrepen et al. (2001)
Beer quality	Altered flavour	Altered volatile spectrum	Alcohol acetyltransferase	Fujii et al. (1994)
			Manipulation of *BAP2* to modulate higher alcohols	Kodama et al. (2001)
		Reduced hydrogen sulphide	Increased copy number of MET25 gene	Omura et al. (1995)
		Reduced dimethyl sulphide	Removal of dimethyl sulphide oxidase by deletion of *MXR1*	Hansen et al. (2002)
	Enhanced flavour stability	Elevated sulphur dioxide	Over-expression of *MET14* and *SSU1*	Donalies and Stahl (2002)
Novel beers	Low or zero alcohol beers	Increased glycerol production	Manipulation of *GDP1*	Nevoight et al. (2002)
	Low carbohydrate beers	Dextrin utilisation	As above	Eksteen et al. (2003) and Perry and Meaden (1988)

4.9 Future perspective

The global brewing industry continues to be dominated by a relatively small number of large companies. They are internationals operating in some markets that are flat or declining and others where volume growth is very high. Competition is very fierce and growth via acquisition is likely to continue. Pale pilsner-style lager beers vastly out-sell all other beer styles. Against this backdrop, there is a burgeoning craft brewing sector that has a relatively small but growing volume. Although small, in many countries, the craft segment has a loud voice, and they have done much to polarise arguments over what is and what is not 'real beer'. Undoubtedly, this will colour the views of many consumers regarding the probity of using genetically engineered yeast strains. Hammond (1995) summarised the then current situation regarding the use of genetically modified yeast for brewing. Several strains had been developed successfully, and as a test case, one had been granted approval for commercial use in the UK. To date, no brewers have chosen to use modified strains, and this situation does not show signs of changing. Interestingly, there is less resistance to using commercial exogenous enzymes some of which have been derived from transgenic organisms.

The major brewers face several challenges. They must minimise costs to maintain a competitive edge. Sustainability is a big driver and in many markets minimising water usage is important. The move from draught to small-pack continues and coupled with the need to export, there is a much focus on extending beer shelf life and maximising beer flavour stability. Large international brands and multi-site brewing must be backed up by excellent control of beer flavour and quality. In the case of developments in selection of brewing yeast, many of these needs are reflected in the strategies for metabolic engineering summarised in Table 4.2.

A common wish is to brew at very high gravity (>20 degrees Plato) using high fermentation temperatures for short vessel residence times and with very large batch sizes. If this is to be combined with serial re-pitching, as would be the norm, many currently used brewing yeast strains are barely fit-for-purpose, and there is a very real danger that the stresses imposed on yeast will lead to greater than normal losses in viability and consequent problems with beer quality. The usual response namely to reduce the number of acceptable serial fermentations before introduction of a new culture is not a particularly tenable strategy since many propagation facilities are unable to cope with the additional demands. It may be argued therefore that there is a real need to develop brewing strains that are able to tolerate these conditions. These may not occur in nature, and therefore, they will need to be constructed. Implicit in this is that for the first time, it may not be possible to isolate suitable new strains from nature.

Another current trend is renewed interest in low and zero alcohol beers, typically made via de-alcoholisation of full-strength beers using membrane technology. Organoleptically, these beers are superior to most of their forebears. A simpler and probably less costly approach would be to revisit the development of new strains via genetic engineering.

Public acceptance of genetically engineered strains is not forthcoming in many countries. The brewing industry has a hard task in this regard since many of the goals of yeast genetic engineering are good for brewers' profits but perhaps less obviously of direct benefit to the consumer. The trend towards an increased need for ingredient labelling may become problematic in many markets. The development of engineered strains via self-cloning techniques where the introduced material is not viewed as 'foreign' is likely to have importance.

Although the brewing industry will exercise its usual caution with regard to the adoption of engineered strains, it is essential that development work continues in which the principles of metabolic engineering are exploited. A major failing of many previous attempts to engineer yeasts have failed, or at least achieved only partial success, because there was insufficient knowledge of the physiological basis of the traits under investigation. The use of metabolomics and metabolic flux control studies will greatly assist in filling this gap. This basic knowledge is bound to provide benefits whether or not the information is ultimately used for yeast strain modification.

4.10 Further sources of further information

For more information on brewing and the role of yeast in fermentation and beer quality see Boulton and Quain (2006). Metabolic engineering and the tools available for its application are described comprehensively by Wittmann and Lee (2012). The Journal of Metabolic Engineering (www.journals.elsevier.com/metablic-engineering) is an excellent resource for up to date information regarding all aspects of metabolic engineering.

References

Abdel-Fattah, W. R., Fadil, M., Nigam, P., & Banat, I. M. (2000). Isolation of thermotolerant ethanologenic yeasts and use of selected strains in industrial scale fermentation in an Egyptian distillery. *Biotechnology and Bioengineering, 68*, 531–535.

Alper, H., Moxley, J., Nevoight, E., Fink, G. R., & Stephanopoulos, G. (2006). Engineering yeast transcription machinery for improved ethanol tolerance and production. *Science, 314*, 1565–1568.

Büscher, J. M., Czernik, D., Ewald, J. C., Sauer, U., & Zamboni, N. (2009). Cross-platform comparison of methods for quantitative metabolomics of primary metabolism. *Analytical Chemistry, 81*, 2135–2143.

Ansell, R., Granath, K., Hohmann, S., Thevelein, J. M., & Adler, L. (1997). The two isoenzymes for yeast NAD+-dependent glycerol 3-phosphate dehydrogenase encoded by GPD1 and GPD2 have distinct roles in osmo-adaptation and redox regulation. *EMBO Journal, 16*, 2179–2187.
Bailey, J. E. (1991). Towards a science of metabolic engineering. *Science, 2521*, 1668–1675.
Bailey, J. E., Birnbaum, S., Galazzo, J. L., Khosla, C., & Shanks, J. V. (1990). Strategies and challenges in metabolic engineering. *Annals of the New York Academy of Sciences, 5891*, 1–15.
Bailey, J. E., Sburlati, A., Hatzimanikatis, V., Lee, K., Renner, W. A., & Tsai, P. S. (2002). Inverse metabolic engineering: A strategy for directed genetic engineering of useful phenotypes. *Biotechnology and Bioengineering, 79*, 568–579.
Bayer, T. S., & Smolke, C. D. (2005). Programmable-ligand controlled riboregulators of eukaryotic gene expression. *Nature Biotechnology, 233*, 337–343.
Blieck, L., Toye, G., Dumortier, F., Verstrepen, K. J., Delvaux, F. R., Thevelein, J. M., et al. (2007). Isolation and characterization of brewer's yeast variants with improved fermentation performance under high-gravity conditions. *Environmental Microbiology, 73*, 815–824.
Boczo, E.,M., Cooper, T. G., Gedeon, T., Mischaikow, K., Murdock, D. G., Pratrap, S., et al. (2005). Structure theorems and the dynamics of nitrogen catabolite repression in yeast. *Proceedings of National Academic Sciences of United States of America, 102*, 5647–5652.
Boulton, C. A. (2021). All yeasts are born equal, some grow to be more equal than others. *Journal of the Institute of Brewing, 127*, 83–106.
Boulton, C. A., & Quain, D. E. (2006). *Brewing yeast and fermentation*. Oxford, UK: Wiley-Blackwell.
Brewster, J. L., de Valoir, T., Dwyer, N. D., Winter, E., & Gustin, M. C. (1993). An osmo-sensing signal transduction pathway in yeast. *Science, 259*, 1760–1763.
Cao, L., Zhang, A., Kong, Q., Xu, X., Josine, T. L., & Chen, X. (2007). Over-expression of GLT1 in fps1DeltagpdDelta mutant for optimum ethanol formation by *Saccharomyces cerevisiae*. *Biomolecular Engineering, 24*, 638–646.
Chauhan, N., & Karuppayil, S. M. (2020). Dual identities for various alcohols in two different yeasts. *Mycology, 12*, 1–14.
Chaves, C. R., Rego, A., Martins, V. M., Santos-Pereira, C., Sousa, M. J., & Corte-Real, M. (2021). Regulation of cell death induced by acetic acid in yeasts. *Frontiers in Cell and Developmental Biology, 9*, 1–20.
Christaens, J. F., Franco, L. M., Cools, T. L., De Meester, L., Michiels, J., Wenseleers, T., et al. (2014). The aroma gene, *ATF1*, promotes dispersal of yeast cells through insect vectors. *Cell Reports, 9*, 425–432.
Christiansen, B., & Nielsen, J. (2000). Metabolic network analysis. A powerful tool in metabolic engineering. *Advances In Biochemical Engineering, 66*, 209–223.
De Vries, A. R. G., Pronk, J. T., & Daran, J.-M.,G. (2019). Lager brewing yeasts in the era of modern brewing. *FEMS Yeast Research, 19*, 1093.
Desari, S., & Kolling, R. (2011). Cytosolic localisation of acetohydroxyacid synthase (ilv2) and its impact on diacetyl formation during beer fermentation. *Applied and Environmental Microbiology, 77*, 727–731.
Dettmer, K., Aronov, P. A., & Hammock, B. D. (2007). Mass spectrometry based metabolomics. *Mass Spectrometry Reviews, 26*, 51–78.
Donalies, U. E., & Stahl, U. (2002). Increasing sulphite formation in *Saccharomyces cerevisiae* by over-expression of MET14 and SSU1. *Yeast, 19*, 475–484.
Eksteen, J. M., Van Rensburg, P., Cordero Otero, R. R., & Pretorius, I. S. (2003). Starch fermentation by recombinant *Saccharomyces cerevisiae* strains expressing the alpha-amylase and glucoamylase genes from *Lipomyces kononenkoae* and *Saccharomycopsis fibuligera*. *Biotechnology and Bioengineering, 84*, 639–646.
Fromont-Racin, M., Rain, J.-C., & Legrain, P. (1991). Toward a functional anlaysis of the yeast genome through exhaustive two-hybrid systems. *Nature Genetics, 16*, 277–282.
Fujii, T., Kondo, K., Shimizu, F., Sone, H., Tanaka, J., & Inoue, T. (1990). Application of a ribosomal DNA integration vector in the construction of a brewer's yeast having alpha-acetolactate decarboxylase activity. *Applied and Environmental Microbiology, 56*, 997–1003.
Fujii, T., Nagasawa, N., Iwamatsu, A., Bogaki, T., Tamai, Y., & Hamachi, M. (1994). Molecular cloning, sequence analysis, and expression of the yeast alcohol acetyltransferase gene. *Applied and Environmental Microbiology, 60*, 2786–2792.
Gallone, B., Steensels, J., Prahl, T., Soriaga, L., Saels, V., Herrera-Malaver, B., et al. (2016). Domestication and divergence of Saccharomyces cerevisiae beer yeasts. *Cell, 166*(6), 1397–1410.
Gancedo, J. M. (1998). Yeast carbon catabolite repression. *Microbiology and Molecular Biology Reviews, 62*, 334–361.
Gavin, A.-C., Bösche, M., Krause, R., Grandi, P., Marzioch, M., Bauer, A., et al. (2002). Functional organisation of the yeast proteome by systematic analysis of protein complexes. *Nature, 415*, 141–147.
Gershon, H., & Gershon, D. (2000). The budding yeast, *Saccharomyces cerevisiae*, as a model for ageing research: A critical review. *Mechanism of Ageing and Development, 120*, 1–22.
Gibson, B. R., Boulton, C. A., Box, W. G., Graham, N. S., Lawrence, S. J., Linforth, R. S., et al. (2008). Carbohydrate utilisation and the lager yeast transcriptome during brewery fermentation. *Yeast, 8*, 549–562.
Gibson, B. R., Lawrence, S. J., Leclaire, J. P. R., Powell, C. D., & Smart, K. S. (2005). Yeast responses to stresses associated with industrial brewery handling. *FEMS Microbiology Reviews, 31*, 535–569.
Gjermansen, C., Nilsson-Tillgren, T., Litske Petersen, J. G., Kielland-Brandt, M. C., Sigsgaard, P., & Holmberg, S. (1988). Towards diacetyl-less brewers' yeast. Influence of *ilv2* and *ilv5* mutations. *Journal of Basic Microbiology, 28*, 175–183.
Goffeau, W., Barrell, B. G., Bussey, H., Davis, R. W., Dujon, B., Feldmann, H., et al. (1996). Life with 6000 genes. *Science, 274*, 546–567.
Gonçalves, M., Pontes, A., Almeida, P., Barbosa, R., Serra, M., Libkind, D., et al. (2016). Distinct domestication trajectories in top-fermenting beer yeasts and wine yeasts. *Current Biology, 26*(20), 2750–2761.
Gonzales, A., & Hall, M. N. (2017). Nutrient and TOR signalling in yeast and mammals. *EMBO Journal, 36*, 397–408.

Hammond, J. R. M. (1995). Genetically-modified yeast for the 21st century. Progress to date. *Yeast, 11*, 1613–1627.

Hansen, J., Bruun, S. V., Bech, L. M., & Gjermansen, C. (2002). The level of *MXR1* gene expression in brewing yeast during beer fermentation is a major determinant for the concentration of dimethyl sulfide in beer. *FEMS Yeast Research, 2*, 137–149.

Heinemann, M., & Sauer, U. (2010). Systems biology of microbial metabolism. *Current Opinion in Microbiology, 13*, 337–343.

James, T. C., Usher, J., Campbell, S., & Bond, U. (2008). Lager yeasts possess dynamic genomes that undergo re-arrangements and gene amplification in response to stress. *Current Genetics, 53*, 139–152.

Jeffries, T. W., & Jin, Y.-S. (2004). Metabolic engineering for improved fermentation of pentoses by yeasts. *Applied Microbiology and Biotechnology, 63*, 495–509.

Kodama, Y., Fukui, N., Ashikari, T., Shibano, Y., Morioka-Fujimoto, K., Hiraki, Y., et al. (1995). Improvement of maltose fermentation efficiency: Constitutive expression of *MAL* genes in brewing yeasts. *Journal of the American Society of Brewing Chemists, 53*, 24–29.

Kodama, Y., Omura, F., Miyajima, K., & Ashikari, T. (2001). Control of higher alcohol production by manipulation of the *BAP2* gene in brewing yeast. *Journal of the American Society of Brewing Chemists, 59*, 157–162.

Krogerus, K., Fletcher, E., Rettberg, N., Gibson, B., & Priess, R. (2021). Efficient breeding of industrial brewing yeast strains using CRISPR/Cas9-aided mating-type switching. *Applied Microbiology and Biotechnology, 105*, 8359–8376.

Krogerus, K., Magalhaes, F., Vidgren, V., & Gibson, B. (2017). Novel brewing yeast hybrids: Cfreation and application. *Applied Microbiology and Biotechnology, 101*, 65–78.

Kudla, B., & Nicolas, A. (1992). A multi-site integrative cassette for the yeast *Saccharomyces cerevisiae*. *Gene, 119*, 49–56.

Kuyper, M., Toirkens, M. J., Diderich, J. A., Winkler, A. A., van Dijken, J. P., & Pronk, J. (2005). Evolutionary engineering of mixed-sugar utilisation by a xylose-fermenting *Saccharomyces cerevisiae* strain. *FEMS Yeast Research, 5*, 925–934.

Kwolek-Mirek, M., & Tecza, R. Z. (2014). Comparison of methods used for assessing the viability and vitality of yeast cells. *FEMS Yeast Research, 14*, 1068–1079.

Lapujade, D., Jansen, M. L. A., Daran, J.-M., van Gluik, W., de Winde, J. H., & Pronk, J. (2003). Role of transcriptional regulation in controlling fluxes in central carbon metabolism of *Saccharomyces cerevisiae*, a chemostat study. *Journal of Biological Chemistry, 91*, 9125–9138.

Lee, F. W., & da Silva, N. A. (1997). Improved efficiency and stability of multiple cloned gene insertions at the delta sequences of *Saccharomyces cerevisiae*. *Applied Microbiology and Biotechnology, 48*, 339–345.

Liko, D., Conway, M. K., Grunwald, D. S., & Heideman, W. (2010). Sbt3 plays a role in the glucose-induced transition from quiescence to growth in *Saccharomyces cerevisiae*. *Genetics, 185*, 797–810.

Liti, G., & Louis, E. J. (2012). Advances in quantitative trait analysis in yeast. *PLoS Genetics, 8*, 1–7.

Malcorps, P., Cheval, J. M., Jamil, S., & Dufour, J. P. (1991). A new model for the regulation of synthesis by alcohol acetyl transferase in Saccharomyces cerevisiae in fermentation. *Journal of the American Society of Brewing Chemists, 49*, 47–53.

Moreira dos Santos, M., Raghevendran, V., Kotter, P., Olsson, L., & Nielsen, J. (2004). Manipulation of malic enzyme in Saccharomyces cerevisiae for increasing NADPH production capacity aerobically in different cellular compartments. *Metabolic Engineering, 6912*, 363.

Mumberg, D. R., Muller, D. R., & Funk, M. (1995). Yeast vectors for the controlled expression of heterologous proteins in different genetic backgrounds. *Gene, 156*, 119–122.

Nakao, Y., Kanamori, T., Itoh, T., Kodama, Y., Rainieri, S., Nakamura, N., et al. (2009). Genome sequence of the lager brewing yeast, an interspecies hybrid. *DNA Research, 16*, 115–129.

Nerem, R. M. (1991). Cellular engineering. *Annals of Biomedical Engineering, 19*, 529–545.

Nevoight, E. (2008). Progress in metabolic engineering of *Saccharomyces cerevisiae*. *Microbiology and Molecular Biology Reviews, 72*, 379–412.

Nevoight, E., Pilger, R., Mast-Gerlach, E., Schmidt, U., Freihammer, S., Eschenbrenner, M., et al. (2002). Genetic engineering of brewing yeast to reduce the content of ethanol in beer. *FEMS Yeast Research, 2*, 225–232.

Nevoight, E., & Stahl, U. (1996). Reduced pyruvate decarboxylase and increased glycerol 3-phosphate dehydrogenase (NAD+) levels enhance glycerol production in *Saccharomyces cerevisiae*. *Yeast, 121*, 331–337.

Nielsen, J. (2001). Metabolic engineering. *Applied Microbiology and Biotechnology, 552*, 263–283.

Niklas, J., Schneider, K., & Heinzle, E. (2010). Metabolic flux analysis in eukaryotes. *Current Opinion in Biotechnology, 21*, 63–69.

Nishida, E., & Gotsh, Y. (1993). The MAP kinase cascade is essential for driving signal transduction pathways. *Trends in Biochemical Sciences, 18*, 128–131.

Ohashi, Y., Hirayama, A., Ishikawa, T., Nakamura, S., Shimizu, K., Ueno, Y., Tomita, M., & Soga, T. (2008). Depiction of metabolome changes in histidine-starved *Escherichia coli* by CE-TOFMS. *Molecular BioSystems, 4*, 135–147.

Olauno, M., Kaijitani, R., Ryusui, R., Morimoto, H., Kodamo, Y., & Itoh, T. (2016). Next generation sequencing analysis of lager brewing yeast strains reveals the evolutionary history of interspecies hybridization. *DNA Research, 23*, 67–80.

Oliver, S. G., Winson, M. T., Kell, D. B., & Bayang, F. (1998). Systematic functional analysis of the yeast genome. *Trends in Biotechnology, 16*, 373–378.

Omura, F., Fujita, A., Miyajima, K., & Fukui, N. (2005). Engineering of yeast Put4 permease and its application to lager yeast for efficient proline assimilation. *Bioscience, Biotechnology, and Biochemistry, 69*, 1162–1171.

Omura, Y. S., Fukui, N., & Nakatani, K. (1995). Reduction of Hydrogen sulfide production in brewing yeast by constitutive expression of MET25 gene. *Journal of the American Soceity of Brewing chemists, 53*, 58–62.

Ostergaard, S., Olsson, L., & Nielsen, J. (2000). Metabolic engineering of *Saccharomyces cerevisiae*. *Microbiology and Molecular Biology Reviews, 64*(1), 34–50.

Pedersen, M. B. (1986b). DNA sequence polymorphism in the genus *Saccharomyces* IV. Homologous chromosomes III in *Saccharomyces bayanus*, S. carlbergensis and S. uvarum. *Carlsberg Research Communications, 51*, 185–202.

Pederson, M. B. (1985). DNA sequence polymorphisms in the genus *Saccharomyces* II. Analysis of the gene *RDN1, HIS4, LEU2* and Ty transposable elements in Carlsberg, Tuborg and 22 Bavarian brewing strains. *Carlsberg Research Communications, 50*, 263–272.

Pederson, M. B. (1986a). DNA sequence polymorphisms in the genus *Saccharomyces* III. Restriction endonuclease fragment patterns of chromosomal regions in brewing and other yeast strains. *Carlsberg Research Communications, 51*, 163–183.

Pederson, M. B. (1994). Molecular analyses of yeast DNA, tools for pure yeast maintenance in the brewery. *Journal of the American Society of Brewing Chemists, 52*, 23–27.

Penttilä, M. E., Suihko, M. L., Lehtinen, U., Nikkola, M., & Knowles, J. K. C. (1988). Construction of brewer's yeasts secreting fungal endo-ß-glucanase. *Current Genetics, 12*, 413–430.

Perry, C., & Meaden, P. (1988). Properties of a genetically-engineered dextrin-fermenting strain of brewers' yeast. *Journal of the Institute of Brewing, 94*, 64–67.

Picotti, P., Clément-Ziza, M., Lam, H., Campbell, D. S., Schmidt, A., Deutsch, E. W., et al. (2013). A complete mass-spectrometric map of the yeast proteome applied to quantitative trait analysis. *Nature, 494*, 266–270.

Pirt, S. J. (1975). *Principles of microbe and cell cultivation*. Oxford, UK: Blackwell Scientific Publications.

Posas, F., Takekawa, M., & Saito, H. (1998). Signal transduction by MAP kinase cascades in budding yeast. *Current Opinion in Microbiology, 1*, 175–182.

Powell, C. D., Van Zandycke, S. M., Quain, D. E., & Smart, K. A. (2000). Replicative ageing and senescence in *Saccharomyces cerevisiae* and the impact on brewing fermentations. *Microbiology, 146*, 1023–1034.

Rainieri, S. (2009). The brewing yeast genome: From its origin to our current knowledge. In *Beer and health and disease prevention* (pp. 89–101). Academic Press.

Rodley, C. D. M., Grand, R. S., Gehlen, L. R., Greyling, G., Jones, M. B., & O'Sullivan, J. M. (2012). Mitochondrial-nuclear DNA interactions contribute to the regulation of nuclear transcript levels as part of the inter-organelle communication system. *PLoS, 7*(1), Article e30943. https://doi.org/10.1371/journal.pone.0030943

Ruckenstuhl, C., Carmona-Guiterez, D., & Madeo, F. (2020). The sweet taste of death: Glucose triggers apoptosis during yeast chronological ageing. *Aging, 10*, 643–649.

Sauer, U. (2001). Evolutionary engineering of industrially important microbial phenotypes. *Advances in Biochemical Engineering, 73*, 129–169.

Schüller, H. J. (2003). Transcriptional control of non-fermentative metabolites in yeast. *Current Genomics, 43*, 139–160.

Schilling, C. H., Schuster, S., Palsson, B. O., & Heinrich, R. (1999). Metabolic pathway analysis: Basic concepts and scientific applications in the post-genomic era. *Biotechnology Progress, 15*, 296–303.

Siderius, M., Van Wuytswinkel, O., Reijenga, K. A., Kelders, M., & Mager, W. H. (2000). The control of intracellular glycerol in *Saccharomyces cerevisiae* influences osmotic stress response and resistance to increased temperature. *Molecular Microbiology, 36*, 1381–1390.

Smart, K. S. (2007). Brewing yeast genomes and genome-wide expression and proteome profiling during fermentation. *Yeast, 24*, 993–1013.

Stephanopoulis, G. (1999). Metabolic fluxes and metabolic engineering. *Metabolic Engineering, 1*, 1–11.

Tamanoi, T. (2011). Ras signalling in yeast. *Genes and Cancer, 2*, 210–215.

Timmis, K. N., Rojo, F., & Ramos, J. L. (1988). 'Prospects for laboratory engineering of bacteria to degrade pollutants. *Basic Life Sciences, 45*, 61–79.

Tong, L.-T., Liao, H. H., & Cameron, D. C. (1991). 1,3-propanediol production by *Escherichia coli* expression genes from the *Klebsiella pneumonia dha* regulon. *Applied and Environmental Microbiology, 57*, 3541–3546.

Váchová, L., Cáp, M., & Paiková, Z. (2012). Yeast colonies: A model for studies of ageing, environmental adaption and longrevity. *Oxid. Med. Cell, 10*.

Van der Werf, M. J., Overkamp, K. M., Muilwijk, B., Coulier, L., & Hankemeier, T. (2007). Microbial metabolomics: Toward a platform with full metabolome coverage. *Analytical Biochemistry, 370*, 17–25.

Vanderwaeren, L., Dok, R., Voordeckers, K., Nuyts, S., & Verstrepen, K. J. (2022). *Saccharomyces cerevisiae* as a model system for eukaryotic cell biology, from cell cycle control to DNA damage response. *International Journal of Molecular Sciences, 23*, 11665–11678.

Vaughan-Martini, A., & Martini, A. (1987). Three newly delineated species of *Saccharomyces sensu stricto*. *Antonie Van Leeuwenhoek, 53*, 77–84.

Velculescu, V. E., Zhang, L., Zhou, W., Vogelstein, J., Basrai, M. A., Basset, D. E., et al. (1997). Characterisation of the yeast transcriptome. *Cell, 88*, 243–251.

Verstrepen, K. J., Derdelinckx, G., Delvaux, F. R., Winderickx, J., Thevelein, J. M., Bauer, F. F., et al. (2001). Late fermentation expression of *FLO1* in. Saccharomyces cerevisiae. *Journal of the American Soceity of Brewing chemists, 59*, 69–71.

Verstrepen, K. J., & Thievelein, J. M. (2004). Controlled expression of homologous genes by genomic promoter replacement in the yeast *Saccharomyces cerevisiae*. *Methods in Molecular Biology, 267*, 259–266.

Wang, Z.-Y., He, X.-P., Liu, N., & Zhang, B. R. (2008). Construction of self-cloning industrial brewing yeast with high glutathione and low diacetyl production. *International Journal of Food Science and Technology, 43*, 989–994.

Washburn, M. P., Worters, D., & Yates, J. R. (2001). Large-scale analysis of the yeast proteome by multi-dimensional protein identification technology. *Nature Biotechnology, 19*, 242–247.

Wei, W., Nurse, P., & Broek, D. (1998). Yeast cells can enter a quiescent state through G_1, S, G_2 or M phase of the cell cycle. *Cancer Research, 53*, 1867–1870.

Whitmann, C. (2007). Fluxome analysis using GC-MS. *Microbial Cell Factories, 8*, 6–22.

Wiame, J.-M., Grenson, M., & Arst, H. N. (1985). Nitrogen catabolite repression in yeasts and filamentous fungi. *Advances in Microbial Physiology, 26*, 1−38.

Wiechart, W. (2001). ^{13}C Metabolic flux analysis. *Metabolic Engineering, 3*, 195−206.

Wittmann, C., & Lee, S. Y. (2012). *Systems metabolic engineering*. New York and London: Springer.

Wolters, J. F., Chui, K., & Fiumera, H. L. (2015). Population structure of mitochondrial genomes in *Saccharomyces cerevisiae. BMC Genomics, 16*, 451−467.

Young, T. W. (1981). The genetic manipulation of killer factor into brewing yeast. *Journal of the Institute of Brewing, 87*, 292−295.

Zamboni, N. (2011). ^{13}C metabolic flux analysis in complex systems. *Current Opinion in Biotechnology, 22*, 103−108.

Zamboni, N., & Sauer, U. (2009). Novel insights through metabolomics and ^{13}C-flux analysis. *Current Opinion in Microbiology, 12*, 553−558.

Zhang, Y.-H., Wang, Z.-Y., He, X.-P., Liu, N., & Zhang, B. R. (2008). New industrial brewing yeast strains with ILV2 disruption and LSD1 expression. *International Journal of Food Microbiology, 123*, 18−24.

Zimmerman, A., Tadic, J., Kainz, K., Hofer, S. J., Bauer, M. A., Carmona-Guiterez, et al. (2020). Transcriptional and epigenetic control of regulated cell death in yeast. *International Review of Cell Molecular Biology, 352*, 55−82.

Further reading

Timmischi, B., Dettmer, K., Kaspar, H., Thieme, M., & Oefner, P. J. (2008). Development of a quantitative validated capillary electrophoretic-time of flight-mass spectrometry method with high confidence analyte identification for metabolomics. *Electrophoresis, 29*, 2203−2214.

Wainwright, T. (1973). Diacetyl − a review. *Journal of the Institute of Brewing, 79*, 451−470.

Chapter 5

Yeast identification and characterisation

Mathias Hutzler

Technical University of Munich (TUM), Research Center Weihenstephan for Brewing and Food Quality, Freising, Germany

5.1 Biodiversity and characterisation of yeast species and strains from a brewing environment

More than 2000 yeast species are currently known (Boekhout et al., 2022). Estimations indicate that an additional 669,000 extant yeast species have not yet been described (Verstrepen et al., 2006). The most important yeast species for fermentation technology belong to the genus *Saccharomyces* and were formerly taxonomically grouped in the *Saccharomyces sensu stricto* complex (Rainieri et al., 2003; Vaughan-Martini & Martini, 2011). The genus *Saccharomyces* consists of the following: *Saccharomyces cerevisiae,* the yeast used for the production of top-fermented beers (often referred to as 'ale'), wine, distillers' mash, sake and many other alcoholic beverages; *Saccharomyces bayanus* (an *S. uvarum* x *S. eubayanus* hybrid), applied in wine, cider and apple wine production and a brewing contamination species; *Saccharomyces pastorianus,* the starter culture for bottom-fermented beer (lager) and apple wine production; as well as seven additional *Saccharomyces* species (*S. kudriavzevii, S. mikatae, S. paradoxus, S. arboricola, S. uvarum, S. jurei* and *S. eubayanus*) that are not used as common industrial brewing starter cultures, but some of those species (*S. eubayanus, S. jurei, S. paradoxus*) were already used in studies to investigate their brewing potential or were used in brewing trials or in smaller breweries or limited production scales and periods (Alsammar & Delneri, 2020; Bamforth, 2005; Hutzler et al. 2021; Libkind et al., 2011; Rainieri et al., 2003, 2006; Wang & Bai, 2008). Strains of the former species *S. cariocanus* are now regarded as a subgroup of *S. paradoxus*. In total it is 8 *Saccharomyces* species and 2 *Saccharomyces* hybrid species. Libkind et al. reported that the bottom-fermenting (BF) strains of the species *Saccharomyces pastorianus* used in lager beer production are genetic hybrids of *Saccharomyces cerevisiae* and the Patagonian wild yeast *S. eubayanus* (Libkind et al., 2011). Dunn and Sherlock postulated that all *Saccharomyces pastorianus* lager strains consist of at least two types (Dunn & Sherlock, 2008). The *S. pastorianus* strains that they studied were divided into the groups: Saaz and Frohberg. Some industrial strains exhibiting strong fermentation performance belong to the Frohberg group.

Rapid species identification within the *Saccharomyces sensu stricto* group is of great importance for verifying the purity of a species in a beer starter culture and for detecting cross-contaminations. In addition, there are some non-*Saccharomyces* yeast species that are used as starter cultures for special beer styles. *Schizosaccharomyces pombe* is found in some traditional African beers (which are often produced using cereals such as Sorghum) and *Dekkera bruxellensis* in Belgian beers and in German Berliner Weiβe. *Saccharomycodes ludwigii* is used for the production of low- and non-alcoholic beer styles, whereas *Torulaspora delbrueckii* can be used in the production of top-fermented wheat beer as a supplemental yeast strain to generate a distinct fruity aroma. In spontaneous beer fermentations, other non-*Saccharomyces* species can be involved, such as *Debaryomyces* spp., *Meyerozyma guilliermondii, Pichia membranifaciens, Candida friedrichii, Naumovia castellii, Dekkera anomala, Priceomyces* spp. in Lambic beer (Spitaels et al., 2014) and *Cryptococcus keutzingii, Rhodotorula mucilaginosa, Candida krusei, Pichia fermentans* and *Pichia opuntiae* in American coolship ale (Bokulich et al., 2012). Table 5.1 provides an overview of the yeast species common in beer production.

Breweries either maintain individual brewing strains or they order yeast strains from yeast strain providers or culture collections. Yeasts are available as pure liquid or solid cultures or in dried form. For dry yeast, a rehydration process in a tank or a reactor is necessary before the yeast can be pitched. Pure culture yeast must be cultivated in a laboratory until the required volume is reached. Afterwards, the yeast can be transferred from the Carlsberg flask to the propagator. In the

Brewing Microbiology. https://doi.org/10.1016/B978-0-323-99606-8.00018-3

TABLE 5.1 Brewing strains of yeast species used for different beer types.

	Fermentation/flocculation characteristic		Beer type	Genus/species
Brewing yeast strains	Bottom-fermenting	Strong flocculation	Lager, pilsener, export, bottom-fermented special beers, bottom-fermented low alcohol beer, etc.	*S. pastorianus* (spp. *carlsbergensis*)
		Low flocculation		
	Top-fermenting	Low flocculation	German wheat beer, ale, stout, kolsch, alt, Belgian special beer styles (Witbier, Trapist beer), African indigenous beer styles, etc.	*S. cerevisiae*
	–	–	Non- and low-alcohol beer styles	*Saccharomycodes ludwigii*
	–	–	Berliner Weiβe, Belgian special beer styles (e.g., Lambic)	*Dekkera bruxellensis*
	–	–	African indigenous beer styles	*Schizosaccharomyces pombe*
	–	–	Special beer styles with fruity character	*Torulaspora delbrueckii*

propagation system, once the target cell concentration for yeast growth under ideal conditions has been achieved, the yeast is grown to the appropriate volume and cell concentration for pitching in wort for beer production. It must be said that handling pure culture yeast is more demanding than working with dry yeast, but it is generally safer from a microbiological point of view. The greater the number of yeast strains maintained in one brewery, the greater the danger for cross-contamination. A brewing yeast strain should be taxonomically classified by means of molecular biological methods at species and strain levels. In addition, its propagation and fermentation performance, as well as its aroma profile, should be characterised. Classification with molecular biological methods is described in sections 5.1 and 5.2. Fig. 5.1 shows a brief

Description of **yeast strain TUM 34/70**
***Saccharomyces pastorianus ssp. carlsbergensis*,**
bottom-fermenting, flocculent yeast

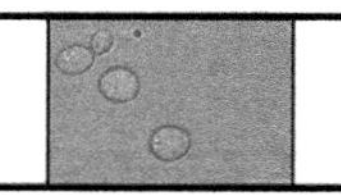

This yeast strain is excellently suited for the production of bottom-fermented beers of all types. The resulting beer possesses an extremely pure flavor, a fine, subtle aroma, and a mild overall impression. Fermentation is rapid with optimal yeast flocculation. The length of time that the yeast remains in suspension is dependent on conditions present in the brewery. The degree to which the color lightens during fermentation is sufficient.

Beer analysis after 6 days of primary fermentation (original gravity: 11.6 %)

Beer analysis	
Apparent degree of attenuation (%)	73
Cells in suspension (millions/ml)	12.5
pH value	4.6
Difference between final and apparent degree of attenuation (%)	0
Diacetyl (mg/L)in green beer	0.7
Diacetyl (mg/L)in matured beer	0.1
Acetaldehyde (mg/L)	6
Higher (aliphatic) alcohols (mg/L)	58
Esters (mg/L)	18.5
Foam, Ross & Clark method (s)	132

TUM 34/70 overview of attributes

Fermentation rate	High
pH reduction	Normal
Flocculation	Optimal
Diacetyl reduction	Very good
Foam	Very good
Difference between final and apparent degree of attenuation	Very low
Acetaldehyde	Normal
Higher alcohols	Very low
Esters	Pronounced

FIGURE 5.1 Description of *Saccharomyces pastorianus* ssp. *carlsbergensis* TUM 34/70, a bottom-fermenting lager yeast strain, in terms of fermentation parameters, beer-quality parameters and aroma (Hutzler et al., 2014). *Author's own image.*

description of the BF lager yeast strain TUM 34/70 according to parameters relevant for fermentation, beer quality and aroma.

Strain TUM 34/70 is one of the most abundant lager yeast strains in the brewing industry and is a strain to which all others are compared regarding fermentation performance and the pure flavour of lager beers produced with it. The genome of TUM 34/70 was also the first of the BF strains to be sequenced and published (Nakao et al., 2009). A recent study conducted by Mueller-Auffermann used the characteristics of TUM 34/70 as a reference for developing a method to rapidly compare the performance of lager yeast strains (Mueller-Auffermann, 2014b). Data comparing six BF lager yeast strains are provided in Fig. 5.2, which shows 9 of 17 properties of yeast analysed and characterised using this method. Strain TUM 66/70 does not flocculate well and possesses a 'powdery' character, whereas the other five strains in Fig. 5.2 are flocculent.

The pilot plant used to conduct these trials consisted of 27 2-L small-scale fermentation tanks. In each trial, two strains ($2 \times 9 = 18$ tanks) were compared to the reference strain TUM 34/70 (in nine tanks). The values measured during fermentation (Fig. 5.2) for the two strains characterised in the trials were compared to the values for strain TUM 34/70. An arrow in Fig. 5.2 represents a shift of 5% in the value for a specific parameter relative to the same one for strain TUM 34/70, whereas two arrows represent a shift of 10% (Mueller-Auffermann, 2014b). TUM 193 produces more SO_2 than TUM 34/70 as well as more acetaldehyde, esters and fusel alcohols. Fermentation performance is similar to that of TUM 34/70M; however, the pH drops more slowly at the beginning of fermentation (data not shown). TUM 193 is advantageous for improving the flavour stability of beers (SO_2) and also produces a slight estery, fruity note (esters, acetaldehyde, fusel alcohols) relative to TUM 34/70. These kinds of trials are very useful for breweries wishing to replace their yeast strain or introduce a second one to develop speciality beers with particular properties or to modify or improve existing lager beer styles.

The booming North American craft beer scene is now conquering Europe, and a lot of speciality beers with distinctive flavours are appearing on the market, especially beers fermented with top-fermenting (TF) *Saccharomyces cerevisiae* strains. These strains produce intense flavours, and they are in the focus of many craft- and microbreweries. These include Bavarian wheat beer, ales and Belgian speciality beers. The wide biodiversity and availability of different strains of *Saccharomyces cerevisiae* offer brewers enormous possibilities to create beers with unique attributes and flavour profiles.

	Saccharomyces pastorianus ssp. carlsbergensis strains					
Fermentation Characteristics	**TUM 34/78**	**TUM 193**	**TUM 194**	**TUM 66/70**	**TUM 44**	**TUM 69**
Physical characteristics of yeast cells						
Low flocculation strength (powdery character)	↑	-	↓	↑↑	-	↑
Sedimentation	-	-	-	-	↑	-
Turbidity	↑↑	-	↑↑↑	-	-	↑
Fermentation by-products						
Diacetyl	↓	-	-	↓↓	↓	↓
SO_2	↓↓	↑↑↑	↓↓	↑	-	-
Acetaldehyde	↓	↑	↓↓	-	-	↑
Esters	↓	↑	↓	↓	-	↓
Fusel alcohols	↓	↑	↓	↓	↑	-
Other quality parameters						
Foam	↑↑	-	↓↓	-	↓↓	↓

FIGURE 5.2 Characterisation of six lager stains in comparison to the reference TUM 34/70 by a new pilot-scale approach (Mueller-Auffermann, 2014b). *Author's own image.*

Therefore, descriptions of these TF speciality yeast strains are of great importance in selecting suitable strains for developing special products. Fig. 5.3 provides a description of the most widely used Bavarian wheat beer strain TUM 68.

TUM 68 is a phenolic off-flavour (POF)-positive Bavarian wheat beer strain. Depending on the production process, Bavarian wheat beers can exhibit very strong fruity, clove-like, estery flavours or a more neutral, yeasty, top-fermented character with a decent fruity note or they can fall somewhere in between the two. In addition to the process parameters, the strain and how it is handled play a prominent role in determining the aroma of the finished beer. Schneiderbanger described the impact of the different wheat beer yeast strains on fermentation performance and their respective aroma profiles (Schneiderbanger, 2014). These authors found that the Bavarian wheat beer strain TUM 127 used to ferment the first batch does not ferment maltotriose, which results in a differing mouthfeel and aroma compared to wheat beer strains without this maltotriose gap (Schneiderbanger et al., 2013). Describing both existing and new brewing yeast strains will aid in our understanding of their characteristics and will open doors to experimentation for innovative brewers around the world to create novel products for the beer market. The potential for increasing the biodiversity among brewing yeast strains is more or less infinite.

5.2 Microbiological, physiological, identification and typing methods

5.2.1 Differences in top- and bottom-fermenting brewing yeast strains

Saccharomyces pastorianus ssp. *carlsbergensis* strains are BF lager yeast strains, whereas *Saccharomyces cerevisiae* brewing yeast strains are TF. Both differ significantly from one another with regard to numerous characteristics. The former strains are able to ferment at lower temperatures, able to flocculate well during primary fermentation and are harvested from the bottom of a fermentation tank, for example, a cylindro-conical tank (CCT). The latter strains cease to function or they are only able to ferment very slowly at low temperatures (6–10°C). They ferment most readily at temperatures between 15°C and 25°C, depending on the strain and type of wort. Cell growth is more vigorous among TF brewing yeast strains, and the cells do not flocculate out as rapidly compared to BF yeast strains. A large portion of the yeast population remains suspended in the liquid phase for a long time and is even buoyed to the top of the fermentation

Description of **yeast strain TUM 68**
Saccharomyces cerevisiae, top-fermenting,
Bavarian wheat beer yeast, POF positive

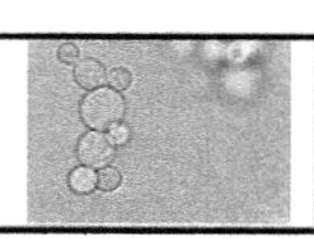

Color:	light brown, naturally cloudy
Foam (visual):	good
Aroma:	pure, pleasant top-fermented aroma, pleasant clove aroma, trace of banana
Flavor:	pure, pleasant top-fermented flavor, pleasant clove aroma, trace of banana, full-bodied, mild, well-balanced aftertaste

Analysis (fermentation and beer parameters)	
Degree of attenuation	0.83
ΔpH (pH reduction)	1.0
Cells in suspension — value at maximum	39.67 million/mL
ΔFAN	131.1 mg FAN/L
Sugar spectrum, total	79.54 g/L → 0.77 g/L
Isoamyl acetate after 96 h	3.16 mg/L
Ethyl acetate after 96 h	30.16 mg/L
4-Vinylguaiacol after 96 h	2.71 mg/L
Isoamyl alcohol after 96 h	63.91 mg/L
Diacetyl after 96 h	0.51 mg/L
Total score according to the DLG rating scheme	4.45

Abbreviations:
POF = phenolic off flavor
FAN = free amino nitrogen
DLG = Deutsche-Landwirtschafts-Gesellschaft (German agricultural society)

FIGURE 5.3 Description of *Saccharomyces cerevisiae* TUM 68, a top-fermenting Bavarian wheat beer strain, in terms of fermentation parameters, beer-quality parameters and aroma (Hutzler et al., 2014). *Author's own image.*

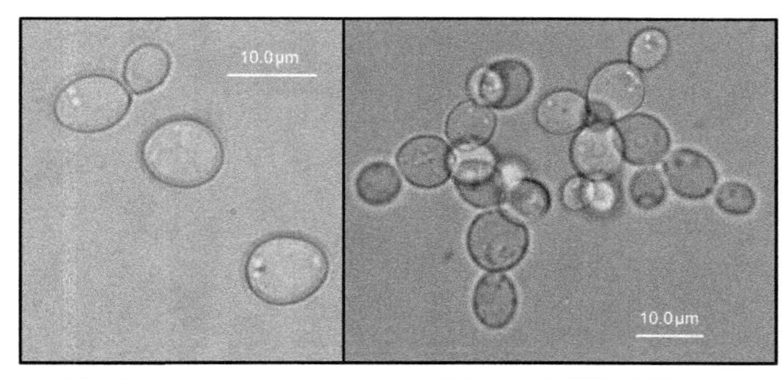

FIGURE 5.4 Microscopic picture of *Saccharomyces pastorianus* ssp. *carlsbergensis* TUM 34/70 and *Saccharomyces cerevisiae* TUM 127.

vessel by bubbles of CO_2. If it is possible to crop the yeast from the top, a positive selection of the most vital yeast can be carried out. In this way, many Bavarian wheat beer brewers select their yeast over numerous generations and can continue to repitch it, virtually a countless number of times. In ale and kölsch beer production, most yeast strains have been selected by means of bottom cropping from CCTs and have therefore lost their vigorous TF character to some extent. A further distinguishing feature of BF and TF yeast becomes apparent through microscopic analysis. Bottom-fermenting yeast strains occur as single cells and perform unilateral budding. Most TF yeast strains occur in groups consisting of more than 2 cells and perform multilateral budding. The BF lager strain *Saccharomyces pastorianus* ssp. *carlsbergensis* TUM 34/70 is depicted on the left in Fig. 5.4, both as a single cell and as a cell having undergone unilateral budding. The Bavarian wheat beer strain *Saccharomyces cerevisiae* TUM 127 (right) forms two three-dimensional stellar patterns through multilateral budding.

Large cell formations or large star-like clusters are not formed by all *Saccharomyces cerevisiae* strains, however. Most *Saccharomyces cerevisiae* strains used to produce Altbier and Kölsch as well as certain ale strains form only very small cell clusters consisting of one to four cells. These are often very difficult to differentiate from *Saccharomyces pastorianus* lager strains, especially when they appear close together. Additional characteristics differentiating BF and TF yeast are as follows:

- Respiration rate (higher in TF)
- Metabolism rate (higher in TF)
- Optimum growth temperature (higher in TF)
- Bands of cytochrome spectrum (TF four bands, BF two bands)
- Optimum catalase pH (TF 6.2–6.4, BF 6.5–6.8)
- Complete use of melibiose and raffinose (BF positive)
- Yeast crop cell mass (higher in TF if cropped from above)
- Sulphite production (BF positive)
- Flocculation (stronger in BF)
- Fructose transport via active transporter (BF positive)
- Growth at 37°C (TF positive)
- Ascospore formation on acetate agar according to a defined procedure (TF positive)
- Glucose effect, carbon catabolite repression (BF positive)
- Repitching cycles (higher in TF)
- Growth on pantothenate agar, a culture medium containing no pantothenic acid (BF positive)

Detailed descriptions of these differences and the related methods are described in several publications and overviews (Back, 1994; Dufour et al., 2003; Hutzler, 2009; Priest & Campell, 2003; Röcken & Marg, 1983).

5.2.2 Overview of identification methods

Over recent decades, the classical standard methods for microbiology and physiology have been modified and, in part, replaced by sophisticated molecular microbiological, physical–chemical methods, such as polymerase chain reaction (PCR)–based DNA techniques or chemotaxonomic or spectrometric methods. Methods commonly used in the brewing industry for identifying and differentiating yeast species can be found in Table 5.2, including the degree of differentiation possible with each method.

TABLE 5.2 Overview of methods for yeast identification and differentiation that can be applied for yeast species from a brewing environment.

Methods	Degree of differentiation	References source
Physiological, morphological methods		
Standard methods	Genus, species	
Miniature commercial systems (e.g., API 20C AUX, rapid IDyeast plus)	Genus, species	
Chemotaxonomic methods		
Total fatty acids analysis (FAME = determination of fatty acid methyl ester compounds)	Species	Timke et al. (2008)
Protein fingerprinting (e.g., 2D protein map)	Species, strain	Abdel-Aty (1991), Kobi et al. (2004)
Mass spectrometry methods (e.g., MALDI-TOF MS, Py-MS, DIMS, GC-TOF MS)	Species, strain	Blattel et al. (2013), Pope et al. (2007), Schuhegger et al. (2008), Timmins et al. (1998), Usbeck et al. (2013), Usbeck et al. (2014)
Fourier transform infrared spectroscopy (FT-IR)	Species, strain	Buchl et al. (2008), Buchl et al. (2010), Timmins et al. (1998), Wenning et al. (2002)
Immunological methods		
Technique based on monoclonal antibodies (e.g., enzyme-linked immunosorbent assay)	Species, strain	Abdel-Aty (1991), Kuniyuki et al. (1984)
Molecular genetics methods		
Sequencing	Species	Arias et al. (2002), Hutzler (2009), Laitila et al. (2006), Timke et al. (2008), Van Der Aa Kuhle et al. (2001)
Karyotyping	Species	Antunovics et al. (2005), Demuyter et al. (2004), Esteve-Zarzoso et al. (2001), Guerra et al. (2001), Naumov et al. (2000), Naumov et al. (2002), Naumov et al. (2001), Martinez et al. (2004), Rodriguez et al. (2004)
Restriction fragment length polymorphism (RFLP) mt DNA	Species	Beltran et al. (2002), Cappello et al. (2004), Comi et al. (2000), Esteve-Zarzoso et al. (2004), Esteve-Zarzoso et al. (2001), Fernandez-Gonzalez et al. (2001), Granchi et al. (2003), Lopes et al. (2002), Martinez et al. (2004), Pramateftaki et al. (2000), Rodriguez et al. (2004), Torija et al. (2001), Torija et al. (2003)
Fluorescence/chemoluminescence in-situ hybridisation (FisH/CisH)	Genus, species	Roder et al. (2007), Stender et al. (2001), Xufre et al. (2006)
Polymerase chain reaction (PCR)-based methods		
PCR (specific primers)	Species	Muir et al. (2011)
PCR-RFLP of the 5.8s ITS rDNA region	Species	Arias et al. (2002), Beltran et al. (2002), Esteve-Zarzoso et al. (2001), Ganga and Martinez (2004), Las Heras-Vazquez et al. (2003), Morrissey et al. (2004), Pramateftaki et al. (2000), Rodriguez et al. (2004), Torija et al. (2001), Van Der Aa Kuhle et al. (2001)

PCR-DGGE, PCR TGGE	Species, strain	Cocolin et al. (2000), Prakitchaiwattana et al. (2004)
PCR-DHPLC	Species, strain	Buchl et al. (2010), Hutzler (2009), Hutzler et al. (2010)
Real-time PCR	Species, subspecies	Bleve et al. (2003), Brandl (2006), Brandl et al. (2005), Casey and Dobson (2004), Delaherche et al. (2004), Dörries (2006), Hutzler (2009), Phister and Mills (2003)
RAPD-PCR	Strain	Gomes et al. (2002), Guerra et al. (2001), Scherer (2002)
SAPD-PCR	Species, strain	Blattel et al. (2013)
Microsatellite PCR	Strain	Howell et al. (2004), Scherer (2002)
AFLP-PCR	Strain	Schöneborn (2001)
δ-Sequence PCR	Strain	Ciani et al. (2004), Cappello et al. (2004), Demuyter et al. (2004), Legras and Karst (2003), Lopes et al. (2002), Pramateftaki et al. (2000), Scherer (2002), Tristezza et al. (2009)

Some promising new methods are not provided in this table; however, they are discussed in Section 5.2.4 below. The variety of identification and differentiation methods is extensive; only a few are implemented in the routine analysis conducted in commercial laboratories and even fewer in the microbiology laboratories of breweries. A list of some of the established and reliable routine methods for rapid yeast identification is found in Section 5.2.3. Which method is applied in brewery laboratories or combined with the established routine methods depends on the practicability and the degree of acceptance for new techniques, which require special training and knowledge of yeast handling.

5.2.3 Selection of successful standard and recently introduced methods

For the detection of contamination by wild yeast, culture media are still the state-of-the-art in brewery laboratories. Many, however, do not even test for wild yeast, only for beer spoilage bacteria. More than one medium is necessary to detect a broad range of these yeasts. Differentiation of strains in the *Saccharomyces sensu stricto* complex and top- and BF brewing yeasts using culture media is particularly challenging. In the differentiation of *Saccharomyces cerevisiae* (TF) and *Saccharomyces pastorianus* ssp. *carlsbergensis* (BF) strains, media such as WLN agar, X-α-GAL, YM agar or wort agar at 37°C, melibiose/bromocresol purple agar, or pantothenate agar may be used.

WLN agar is based on a colour reaction of bromocresol green. *S. cerevisiae* (TF) strains are not able to reduce bromocresol green and therefore form dark green colonies. Lager strains (BF) as well as *Saccharomyces* and non-*Saccharomyces* wild yeasts reduce bromocresol green, forming pale green, bluish or white colonies (Jespersen & Jakobsen, 1996). For this reason, WLN agar is very useful in analysing *S. cerevisiae* (TF) starter cultures for contamination with lager yeast and varieties of wild yeast. The method is described in Analytica-Microbiologica EBC 3.3.2.2 (Analytica-EBC, 2014b).

The X-α-GAL medium exploits the fact that BF yeast strains secrete α-galactosidase (melibiase) and TF strains do not. Whereas TF brewing yeast colonies remain white, BF brewing yeast colonies turn blue-green. A description of this method is given in the ASBC methods of analysis, microbiology, Yeast-10 A (ASBC, 2014). BF colonies grow on melibiose/bromocresol purple agar. This agar, described by back, turns yellow when BF colonies are cultured on it, whereas TF colonies exhibit no growth or develop only micro-colonies (Back, 1994). However, a study has shown that the discriminative power of the method, that is, the capability of differentiating real colonies and micro-colonies, is insufficient (Anonymous, 1994).

As stated in Section 5.2.1, pantothenate agar does not contain pantothenic acid. Top-fermenting *Saccharomyces cerevisiae* strains do not grow on pantothenate agar, unlike BF strains, which flourish on it. Röcken and Marg found that some *S. cerevisiae* strains could also be cultivated on the pantothenate agar. This agar nevertheless provides a valuable tool for breweries to test for the presence of TF strain *S. cerevisiae*, which in contrast to a broad variety of wild yeasts, does not grow on this medium (Röcken & Marg, 1983). Before performing this test, the yeast sample must be thoroughly washed to avoid transferring of pantothenic acid to the pantothenate agar (Back, 1994; Röcken & Marg, 1983).

The 37°C method has been approved by the ASBC and is found in its Methods of Analysis, Microbiology, Yeast-10 B (ASBC, 2014) and Analytica-Microbiologica EBC 4.2.5.2 (Analytica-EBC, 2014b). Hutzler tested the YM medium at an incubation temperature of 37°C for differentiation of BF and TF brewing yeasts and for detection of wild yeast. In addition, he compared YM medium with three other media (YM + $CUSO_4$, CLEN and XMACS) developed for the detection of a wide range of wild yeasts (Hutzler, 2009). The results are shown in Table 5.3.

The results in Table 5.3 provide proof of the differentiation potential of this agar for TF and BF brewing yeast strains. YM agar is also an adequate medium for detecting wild yeast cells in BF lager yeast at 37°C, with the exception of some *Saccharomyces bayanus/pastorianus* strains that do not grow on YM agar at 37°C. In Table 5.3, wort agar was used as a positive control.

YM + 195 ppm $CuSO_4$ can serve as a good culture medium for the detection of non-*Saccharomyces* and *Saccharomyces cerevisiae* wild yeast cells among BF and TF brewing yeast strains. It has also been successfully used in the form of YM + $CuSO_4$ broth for real-time PCR pre-enrichment of *S. cerevisiae* var. *diastaticus*, to directly detect these super-attenuating strains in beer samples containing brewing yeast (Brandl et al., 2005).

Use of CLEN is an improved method over simple lysine agar for the detection of non-*Saccharomyces* wild yeasts (Anonymous, 1997, 1998). Most non-*Saccharomyces* yeast species grow on this medium. The disadvantages inherent to this medium are that it necessitates washing the sample and that brewing yeasts are able to grow in micro-colonies on the medium, making it difficult to distinguish them from other yeasts.

The XMACS medium exhibits similar disadvantages. De Angelo and Siebert proposed the XMACS with five carbon sources for the detection of *Saccharomyces* and non-*Saccharomyces* wild yeast strains (De Angelo & Siebert, 1987). They also demonstrated that more *Saccharomyces* wild yeast strains were more readily detectable with this medium than with YM + $CuSO_4$ and YM agar at 37°C (De Angelo & Siebert, 1987). This was also tested by Hutzler, whose results are

TABLE 5.3 Growth spectra of brewing and wild yeast strains on five different cultivation media.

Yeast strain/media	Wort agar	YM+ 195 ppm $CuSO_4$	YM + 37°C	CLEN	XMACS
***S. pastorianus* bottom-fermenting brewing strains**					
S. pastorianus TUM 34/70	+/1/<6	-/7/-	-/7/-	+/3/<1	+/3/<1
S. pastorianus TUM 34/78	+/1/<5	-/7/-	-/7/-	+/3/<1	+/2/<2
S. pastorianus TUM 44	+/1/<6	-/7/-	-/7/-	+/3/<1	+/3/<1
S. pastorianus TUM 66	+/1/<6	-/7/-	-/7/-	+/3/<1	+/2/<1
***S. cerevisiae* top-fermenting brewing strains**					
S. cerevisiae TUM 68	+/1/<6	-/7/-	+/2/<5	-/7/-	+/3/<1
S. cerevisiae TUM148	+/1/<6	-/7/-	+/2/<5	+/3/<1	+/2/<1
S. cerevisiae TUM 175	+/1/<11	-/7/-	+/2/<8	-/7/-	+/3/<2
S. cerevisiae TUM 184	+/1/<6	-/7/-	+/2/<5	+/3/<1	+/3/<1
***S. cerevisiae* wild yeasts**					
S. cerevisiae DSM 70451	+/1/<3	+/5/<2	+/4/<2	+/6/<1	-/7/-
S. c. var. *diastaticus* TUM K 3-D-2	+/1/<4	+/5/<2	+/1/<5	+/2/<1	+/2/<1
S. c. var. *diastaticus* TUM K 1-H-7	+/1/<4	+/1/<2	+/1/<2	+/2/<1	+/2/<3
S. c. var. *diastaticus* TUM K 1-B-8	+/1/<5	-/7/-	+/1/<3	+/2/<1	+/2/<3
***S. bayanus/pastorianus* wild yeasts**					
S. bayanus DSM 70411	+/1/<4	-/7/-	-/7/-	+/2/<1	+/2/<2
S. bayanus DSM 70412T	+/2/<9	-/7/-	+/1/<1	+/2/<1	+/2/<4
S. bayanus DSM 70508	+/1/<6	-/7/-	-/7/-	+/5/<1	-/7/-
S. bayanus DSM 70547	+/1/<6	-/7/-	-/7/-	+/2/<1	+/3/<4
S. bayanus TUM K 1-C-3	+/1/<5	-/7/-	-/7/-	+/2/<1	+/3/<3
S. pastorianus DSM 6580NT	+/3/<4	-/7/-	-/7/-	-/7/-	-/7/-
Non-*Saccharomyces* wild yeasts					
C. sake TUM K 1-B-3	+/1/<3	+/1/<3	+/3/<1	+/1/<2	+/2/<4
C. tropicalis TUM K 1-A-3	+/1/<11	+/1/<6	+/3/<1	+/1/<5	+/1/<7
D. bruxellensis CBS 2797	+/3/<5	+/4/<2	+/3/<1	+/5/<2	–/7/–
L. kluyveri CBS 3082T	+/1/<8	+/1/<6	+/1/<6	+/2/<1	+/2/<4
N. castellii TUM K 3-I-1	+/1/<2	+/1/<2	+/4/<2	+/2/<1	+/3/<1
P. membranifaciens CBS 107	+/1/<3	+/1/<3	+/3/<1	+/2/<2	+/2/<2
Sch. pombe CBS 356	+/1/<5	+/1/<2	+/3/<1	+/1/<4	+/1/<3
Z. bailii CBS 1097	+/1/<7	+/3/<2	+/3/<1	+/1/<3	+/2/<4

**Growth [+/−]/incubation time until positive result [days]/
Colony diameter after 7 days [mm]**

shown in Table 5.3 (Hutzler, 2009). *Saccharomyces bayanus* wild yeasts were able to grow on XMACS medium, and a number of brewing strains grew to form larger micro-colonies with diameters of approximately 2 mm (strains TUM 34/78 and strain TUM 175 in Table 5.3). Therefore, the detection of wild yeasts among these brewing strains is difficult.

In selecting the most suitable medium for wild yeast, it quickly becomes apparent that one must first know which brewing strains a brewery uses and the number of different media that a brewery laboratory is inclined to keep on hand. YM agar at 37°C is often used for analysis procedures involving BF yeast strains, whereas YM + 195 ppm $CuSO_4$ is typically used for TF yeast strains. WLN agar, X-α-GAL medium, pantothenate agar and XMACS agar can be effective in conjunction with additional media, depending on which microbes are to be targeted. YM agar + bromophenol blue - + coumaric acid at a pH of 6.0 offer an effective means for detecting some *Saccharomyces* and non-*Saccharomyces* wild yeasts present in brewing yeast and can be used to detect phenolic off-flavour-positive yeast strains (Hutzler, 2009). They can be distinguished based on the different colours produced by the colonies.

The real-time PCR is well established internationally as a method in brewing microbiology, primarily in relation to the detection of beer spoilage bacteria (Hutzler, Schuster, & Stettner, 2008). Large brewing companies, the central laboratories of brewing groups and commercial service laboratories use real-time PCR for the detection and identification of beer spoilage bacteria and, to some extent, also for wild yeast as well as brewing yeast. It is very common among breweries and laboratories of this size to use commercially available PCR kits; however, in small-to-mid-sized breweries, real-time PCR is rarely used. It provides a rapid and reliable means for identifying and differentiating *Saccharomyces* and non-*Saccharomyces* brewing species. Real-time PCR can be used to identify single colonies of an unknown yeast strain at the species level and can also serve as a tool for finding trace contaminations in mixed populations at concentrations of one contaminating cell in 1000 culture yeast cells, for example, 1 cell of *Saccharomyces cerevisiae* in 1000 cells of *Saccharomyces pastorianus* ssp. *carlsbergensis* (Hutzler, 2009; Hutzler, Schoenenberg, et al., 2008). Identifying the correct species to which a brewing yeast strain belongs can rapidly be carried out. Schönling et al., Brandl, Hutzler and Dörries have published information pertaining to PCR systems for the detection of wild and beer spoilage yeast (Brandl, 2006; Brandl et al., 2005; Dörries, 2006; Hutzler, 2009; Schönling et al., 2009). Table 5.4 lists the primer sets and target specificities of real-time PCR systems for the detection and identification of the *Saccharomyces sensu stricto* and non-*Saccharomyces* species that are used in brewing.

The real-time PCR probes belonging to the systems and primers listed in Table 5.4 can be found in Table 5.5.

All real-time PCR systems in Tables 5.4 and 5.5 are compatible, operate using the same temperature protocol (95°C 10 min; 40 × 95°C 10 s, 60°C 55 s; 20°C ∞) and detect the products in the FAM channel, which is available in most real-time PCR cyclers. The composition of the PCR mix as well as the DNA sequences and the application of the internal positive/amplification control (IPC or IAC) in the HEX channel are described by Brandl and Hutzler (Brandl, 2006; Hutzler, 2009). Qualitative results for certain strains of the *Saccharomyces sensu stricto* complex analysed with the real-time PCR systems in Tables 5.4 and 5.5 are shown in Table 5.6.

Saccharomyces cariocanus, paradoxus, kudriavzevii and *mikatae* can be identified using the specific real-time PCR systems presented in Tables 5.4 and 5.5. No specific real-time PCR system has been developed for *S. arboricolus*. Up to now, *S. arboricolus* has not been observed in a brewing environment. The only strain that produced a negative result with this specific system was *S. cariocanus* CBS 5313, but it produced a positive result with all *Saccharomyces cerevisiae*-specific systems and was subsequently sequenced on both IGS rDNA regions as *S. cerevisiae* (data not shown). *S. cerevisiae* can be distinguished from other species with TF-COXII and SC real-time PCR systems. *Saccharomyces pastorianus* ssp. *carlsbergensis* strains (BF, lager type) can be discerned using two different real-time PCR systems. The hybrid character of lager strains can be differentiated by using, for example, a combination of Sc-GRC3 and BF-LRE1 or Sce and BF-300. *S. bayanus/eubayanus/pastorianus* strains produce negative results with Sc-GRC3 or Sce systems and therefore can be differentiated (in single colonies). *S. bayanus* consists of two types (varying BF-300 and BF-LRE1 results). *S. cerevisiae* brewing strains cannot be distinguished from *S. cerevisiae* wild yeast strains using these methods. Contaminations by all relevant *Saccharomyces sensu stricto* species can be detected directly in beer or starter cultures containing *S. cerevisiae* brewing yeast through use of the systems listed in Tables 5.4–5.6. An analysis scheme for contaminations in mixed populations is shown in Fig. 5.5.

Contamination by all relevant *Saccharomyces sensu stricto* species can be detected directly in *Saccharomyces pastorianus* ssp. *carlsbergensis* brewing yeast containing beer or starter cultures with the exception of *Saccharomyces bayanus/eubayanus/pastorianus* contaminations. For this specific problem, an additional quantitative approach, which is shown in Fig. 5.6, can be implemented.

The PCR signal of the Sbp system for *Saccharomyces pastorianus* ssp. *carlsbergensis* brewing yeast is weak, and therefore, the C_t value (cycle number with positive fluorescence signal) is high (~30), as can be seen in Fig. 5.6. This means that only a small number of target DNA copies of this gene were in the sample. A DNA isolate consisting of approximately 50 million cells/mL from a sample of fermented beer was analysed. Pure cultures of *Saccharomyces bayanus, eubayanus* or *pastorianus* produce C_t values of approximately 20 using the Sbp system. This indicates that a large number of target DNA copies of this gene were present in the sample. Hence, the C_t value of an *S. pastorianus* ssp.

TABLE 5.4 Primer sequences of real-time polymerase chain reaction (PCR) systems to differentiate *Saccharomyces sensu stricto* and other brewing culture species.

Target-specificity	Primer	Probe	System name	Primer sequence (5′→3′)	References
D. bruxellensis	Db-f	Y58	Dbr	TGCAGACACGTGGATAAGCAAG	Brandl (2006)
	Db-r			CACATTAAGTATCGCAATTCGCTG	
S. bayanus, S. pastorianus	Sbp-f	Y58	Sbp	CTTGCTATTCCAAACAGTGAGACT	Josepa et al. (2000), Brandl (2006)
	Sbp-r1Sbp-r2			TTGTTACCTCTGGGCGTCGAGTTTGTTACCTCTGGGCTCG	
S. cariocanus	Sca-f	Scar	Sca	TTAGACTTACGTTTGCTCCTCTCATG	Hutzler (2009)
	Sca-r			TGCAAATGACAAATGGATGGTTAT	
S. cerevisiae, *S. pastorianus* ssp. *carlsbergensis*,*S. paradoxus*, *S. cariocanus*	Sc-f	Scer	Sce	CAAACGGTGAGAGATTTCTGTGC	Josepa et al. (2000), Brandl (2006)
	Sc-r			GATAAAATTGTTTGTGTTTGTTACCTCTG	
S. cerevisiae *S. pastorianus* ssp. *carlsbergensis*	Sc-GRC-f	Sc-GRC	Sc-GRC3	CACATCACTACGAGATGCATATGCA	Hutzler (2009)
	Sc-GRC-r			GCCAGTATTTTGAATGTTCTCAGTTG	
S. cerevisiae	TF-f	TF-MGB	TF-COXII	TTCGTTGTAACAGCTGCTGATGT	Hutzler (2009)
	TF-r			ACCAGGAGTAGCATCAACTTTAATACC	
S. cerevisiae	SCF1	SCTM	SC	GGACTCTGGACATGCAAGAT	Salinas et al. (2009)
	SCR1			ATACCCTTCTTAACACCTGGC	
S. cerevisiae var. *diastaticus*	Sd-f	Sdia	Sdi	TTCCAACTGCACTAGTTCCTAGAGG	Scherer (2002), Brandl (2006)
	Sd-r			GAGCTGAATGGAGTTGAAGATGG	
S. kudriavzevii	Sk-f	Skud	Sku	TCCTTACCTTATTCATCATATTCTCCAC	Hutzler (2009)
	Sk-r			CGATATTTGGTAAGGGGAGGTAGA	
S. mikatae	Sm-f	Smik	Smi	ACAACCGCCTCCCCAATT	Hutzler (2009)
	Sm-r			AAATGACAAGTAGTGGGTTGGAAGT	
S. paradoxus	Sp-f	Spar	Spa	CATACTATCAATACTGCCGCCAAA	Hutzler (2009)
	Sp-r			GGCGGATGTGGGTGGTAA	
S. pastorianus, *S. bayanus* (partially) Main target: Bottom-fermenting culture yeast	BF300E	BF	BF300	CTCCTTGGCTTGTCGAA	Brandl (2006)
	BF300M			GGTTGTTGCTGAAGTTGAGA	

Continued

TABLE 5.4 Primer sequences of real-time polymerase chain reaction (PCR) systems to differentiate *Saccharomyces sensu stricto* and other brewing culture species.—cont'd

Target-specificity	Primer	Probe	System name	Primer sequence (5′→3′)	References
S. pastorianus, *S. bayanus* (partially) Main target: Bottom-fermenting culture yeast	UG-LRE-f	UG-LRE	UG-LRE1	ACTCGACATTCAACTACAAGAGTAAAATTT	Hutzler (2009)
	UG-LRE-r			TCTCCGGCATATCCTTCATCA	
Saccharomycodes ludwigii	Sl-f	Y58	Slu	GACGAGCAATTGTTCAAGGGTC	Brandl (2006)
	Sl-r			ACTTATCGCAATTCGCTACGTTC	
T. delbrueckii	Td-f	Y58	Tde	AGATACGTCTTGTGCGTGCTTC	Hutzler (2009)
	Td-r			GCATTTCGCTGCGTTCTT	

TABLE 5.5 Probe sequences of real-time PCR systems to differentiate *Saccharomyces sensu stricto* and other brewing culture species.

Probe name	Reporter	Quencher	Sequence 5′→3	References
TF-MGB	FAM	BHQ1	ATGATTTTGCTATCCCAAGTT	Hutzler (2009)
SC	FAM	BHQ1	CCCTTCAGAGCGTTTTCTCTAAATTGATAC	Salinas et al. (2009)
Sbp	FAM	BHQ1	ACTTTTGCAACTTTTTCTTTGGGTTTCGAGCA	Brandl (2006)
Scar	FAM	BHQ1	TCACCAAAACTGCACCATACGTACAAAATACC	Hutzler (2009)
Scer	FAM	BHQ1	ACACTGTGGAATTTTCATATCTTTGCAACTT	Brandl (2006)
Sc-GRC	FAM	BHQ1	TCCAGCCCATAGTCTGAACCACACCTTATCT	Hutzler (2009)
Sdia	FAM	BHQ1	CCTCCTCTAGCAACATCACTTCCTCCG	Brandl (2006)
Skud	FAM	BHQ1	TGCTATTACTTTTGCTTTTTCACTCACCACACCCT	Hutzler (2009)
Smik	FAM	BHQ1	AACATCCATCATCTATGTGCTCTAAATCCTCACTTTATCA	Hutzler (2009)
Spar	FAM	BHQ1	CTGCACCATACGTACAAAATCTCCCTCCTTC	Hutzler (2009)
BF	FAM	BHQ1	TGCTCCACATTTGATCAGCGCCA	Brandl (2006)
BF-LRE	FAM	BHQ1	ATCTCTACCGTTTTCGGTCACCGGC	Hutzler (2009)
Y58	FAM	BHQ1	AACGGATCTCTTGGTTCTCGCATCGAT	Brandl (2006)

carlsbergensis brewing yeast sample contaminated with *S. bayanus/eubayanus/pastorianus* should be between 20 and 30. The Sbp C_t value is strain dependent and is ≥ 30 for the strain *S. pastorianus* ssp. *carlsbergensis* TUM 34/70, given that DNA is extracted from cell concentrations of $\leq$100 million cells/mL. To obtain a reliable result for contamination, the C_t shift should be at least 3 to 4 C_t values. For example, if yeast TUM 34/70 is to become contaminated, the Sbp C_t value should be $\leq$ 26 to 27. At the moment, this is the only rapid method that can directly detect contamination by *S. bayanus/ eubayanus/pastorianus* in *S. pastorianus* ssp. *carlsbergensis* brewing yeast.

Identifying the species of brewing yeast as well as the strain must be possible in order to maintain reproducible beer quality. Therefore, frequently monitoring of strain identity and purity is of great consequence. Table 5.2 in Section 5.2.2 lists a number of methods applicable for differentiating yeasts on the strain level. Karyotyping and other PCR typing or PCR fingerprint techniques are standard for brewing strains, and karyotyping can be regarded as the reference method or gold standard. For the most part, *Saccharomyces cerevisiae* strains are more heterogeneous than lager strains, and one suitable typing method is sufficient to differentiate between two strains of this species. If they are very closely related, two typing methods have to be combined, for example, when one strain is the ancestor of the other strain and was used for a long period of time under different conditions. *S. pastorianus* ssp. *carlsbergensis* lager strains are very homogenous genetically, and a set of lager strains can generally be distinguished using merely two or more typing techniques (Van Zandycke et al., 2007). Fig. 5.7 shows the two PCR typing methods IGS2_314 and Interdelta PCR system delta12-delta21 (δ12-δ21), both of which are combined with capillary electrophoresis (lab on a chip, Bioanalyser, Agilent, Santa Clara, CA, USA) and subsequent data analysis with Bionumerics (Applied Maths, Ghent, Belgium). The IGS2_314 system was originally developed for adjacent gel electrophoresis and was subsequently combined with denaturing high performance liquid chromatography (DHPLC; Buchl et al., 2010; Hutzler, 2009; Hutzler & Goldenberg, 2007; Hutzler et al., 2010). The system can differentiate most lager strains. The PCR system δ12-δ21 was developed by Legras and Karst and then combined with capillary electrophoresis and subsequent data analysis with Bionumerics by Hutzler et al. (Hutzler et al., 2014; Legras & Karst, 2003).

Fig. 5.7 shows the homogeneity of *Saccharomyces pastorianus* brewing strains and the heterogeneity of *Saccharomyces cerevisiae* brewing strains, as well as the discriminatory power of combining the two methods. Further combinations of typing methods are also possible and can be evaluated by using Bionumerics, for example.

Aside from rapid identification and differentiation of brewing yeast strains and common wild yeast species, a reliable and easy-to-handle identification technique for 'unknown' yeast species from a brewing environment would also be advantageous. The reference method entails sequencing the 26S rDNA D1/D2 domain (Kurtzman & Robnett, 1998). Other useful regions for species identification are, for example, the ITS1-ITS4 rDNA and the IGS2 rDNA (Fernandez-Espinar et al., 2006; Hutzler, 2009).

TABLE 5.6 Comparison of qualitative results of different strains analysed with real-time PCR systems for differentiation of *Saccharomyces sensu stricto* species with focus on the differentiation of brewing species (Hutzler, 2009).

Species	Strain examples	PCR system							
		Sc-GRC3	Sce	SC	TF-COX II	Sbp	BF-LRE1	BF-300	Sdia
S. bayanus	DSM 70412T, 70547, TUM K 1-C-3 (type B)	–	–	–	–	+	–	–	–
	DSM 70411, 70508 (type A)	–	–	–	–	+	+	+	–
S. bayanus/pastorianus	CBS 2440, 6017	–	–	–	–	+	+	+	–
S. eubayanus	CBS 12537	–	–	–	–	+	+	+	–
S. pastorianus	CBS 1503, 1513 DSM 6580NT, 6581	–	–	–	–	+	+	+	–
S. pastorianus (bottom-fermenting brewing yeast)	TUM 34/70, 34/78, 69, 128, 168,194 (Flocculating yeasts) TUM 71, 144 (non-flocculating yeasts) CBS 1484, 5832, CBS 6903, NBRC 2003, TUM B-I-4, B-J-4, B-J-5	+	+	–	–	+	+	+	–
	TUM 120 (flocculating yeast) TUM 66/70, 204 (non-flocculating yeasts), CBS 5832, CBS 6903	+	+	–	–	+/–	+	+	–
S. cerevisiae	DSM 70424, 70449T, 70451, CBS 1464, 8803, TUM K 3-A-1, 3-C-3	+	+	+	+	–	–	–	–
S. cerevisiae (Top-fermenting brewing yeast)	TUM 68, 127, 149, 175, 205, TUM K 5-A-8 (wheat beer yeasts)	+	+	+	+	–	–	–	–
	TUM 148, 184, 208 (Alt beer yeasts)	+	+	+	+	–	–	–	–
	TUM 165, 177 (Kolsch beer yeasts)	+	+	+	+	–	–	–	–
	TUM 210, 211, 213 (ale yeasts)	+	+	+	+	–	–	–	–
S. cerevisiae (Top-fermenting yeast from other fermentation industries)	TUM V Bingen, V Bordeaux, V Epernay, V Wädensvil, V Loureiro (wine yeast)	+	+	+	+	–	–	–	–
	TUM D4 (distillery yeast)	+	+	+	+	–	–	–	–
	TUM S2 (sparkling wine yeast)	+	+	+	+	–	–	–	–
S. cerevisiae var. *diastaticus*	CBS 1782, DSM 70487, TUM 1-B-8, 1-H-7, 2-A-7, K 2-F-1, 3-D-2, 3-H-2	+	+	+	+	–	–	–	+
S. cariocanus	CBS 7995, 8841	–	+	–	–	–	–	–	–
	CBS 5313	+	+	+	+	–	–	–	–
S. kudriavzevii	CBS 8840	–	–	–	–	–	–	–	–
S. mikatae	CBS 8839	–	–	–	–	–	–	–	–
S. paradoxus	CBS 406, 432, 2908, 5829, 7400, 8436	–	+	–	–	–	–	–	–

Culture collection abbreviations. TUM, Technische Universität München, Forschungszentrum Weihenstephan BLQ, Freising, Germany; CBS, Centraalbureau voor Schimmelcultures, Utrecht, Netherlands; DSMZ, Deutsche Stammsammlung für Mikroorganismen und Zellkulturen, Braunschweig, Germany; NBRC, Biological Resource Centre, Tokyo, Japan.
Author's own table.

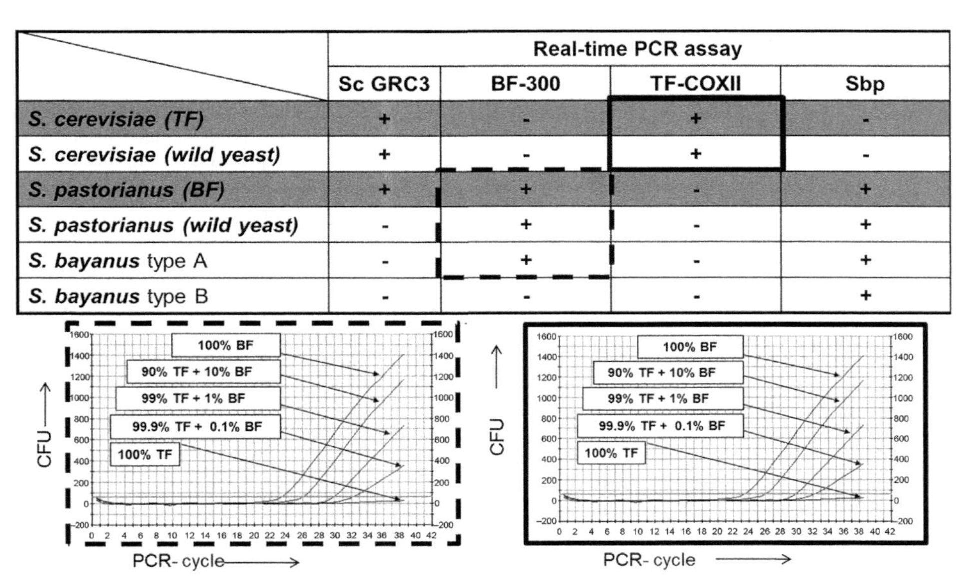

	Real-time PCR assay			
	Sc GRC3	BF-300	TF-COXII	Sbp
S. cerevisiae (TF)	+	-	+	-
S. cerevisiae (wild yeast)	+	-	+	-
S. pastorianus (BF)	+	+	-	+
S. pastorianus (wild yeast)	-	+	-	+
S. bayanus type A	-	+	-	+
S. bayanus type B	-	-	-	+

FIGURE 5.5 Real-time polymerase chain reaction detection scheme and results for contaminations of *S. cerevisiae* brewing yeast in *S. pastorianus* ssp. *carlsbergensis* brewing yeast and vice versa (Hutzler, 2009). *Author's own image.*

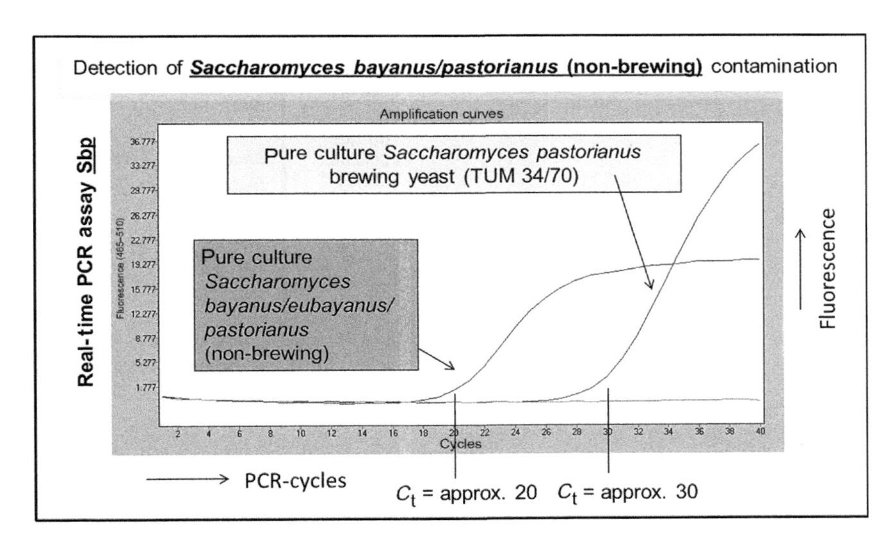

FIGURE 5.6 Real-time polymerase chain reaction basis for the detection of *S. bayanus/eubayanus/pastorianus* contaminations in *S. pastorianus* ssp. *carlsbergensis* brewing yeast using the C_t shift of the Sbp system (Hutzler et al., 2012). *Author's own image.*

5.2.4 Promising new methods

Yeast databases for both brewing yeast and wild yeast were established for the methods matrix-assisted laser desorption/ ionisation time-of-flight mass spectrometry (MALDI-TOF-MS) and Fourier transform infrared spectroscopy (Usbeck et al., 2012, 2013; Hutzler & Wenning, 2009; Timmins et al., 1998; Wenning et al., 2002). Strains stored in databases can easily be rechecked with the relevant methods, and databases can be rapidly expanded to allow mapping of culture collections or microhabitats. Identification on the basis of large database analyses is inexpensive compared to the use of molecular biological methods. Multi-locus sequencing and whole-genome/next-generation sequencing will be of immense importance in describing lineages of brewing yeasts and for exploring the potential of as-yet unproven brewing yeast strains from other industries or from the environment for applications in brewing. Metagenomics, microbiome analysis and droplet PCR can play important roles in the clarification of mixed species populations, such as those found in spontaneously fermented beer.

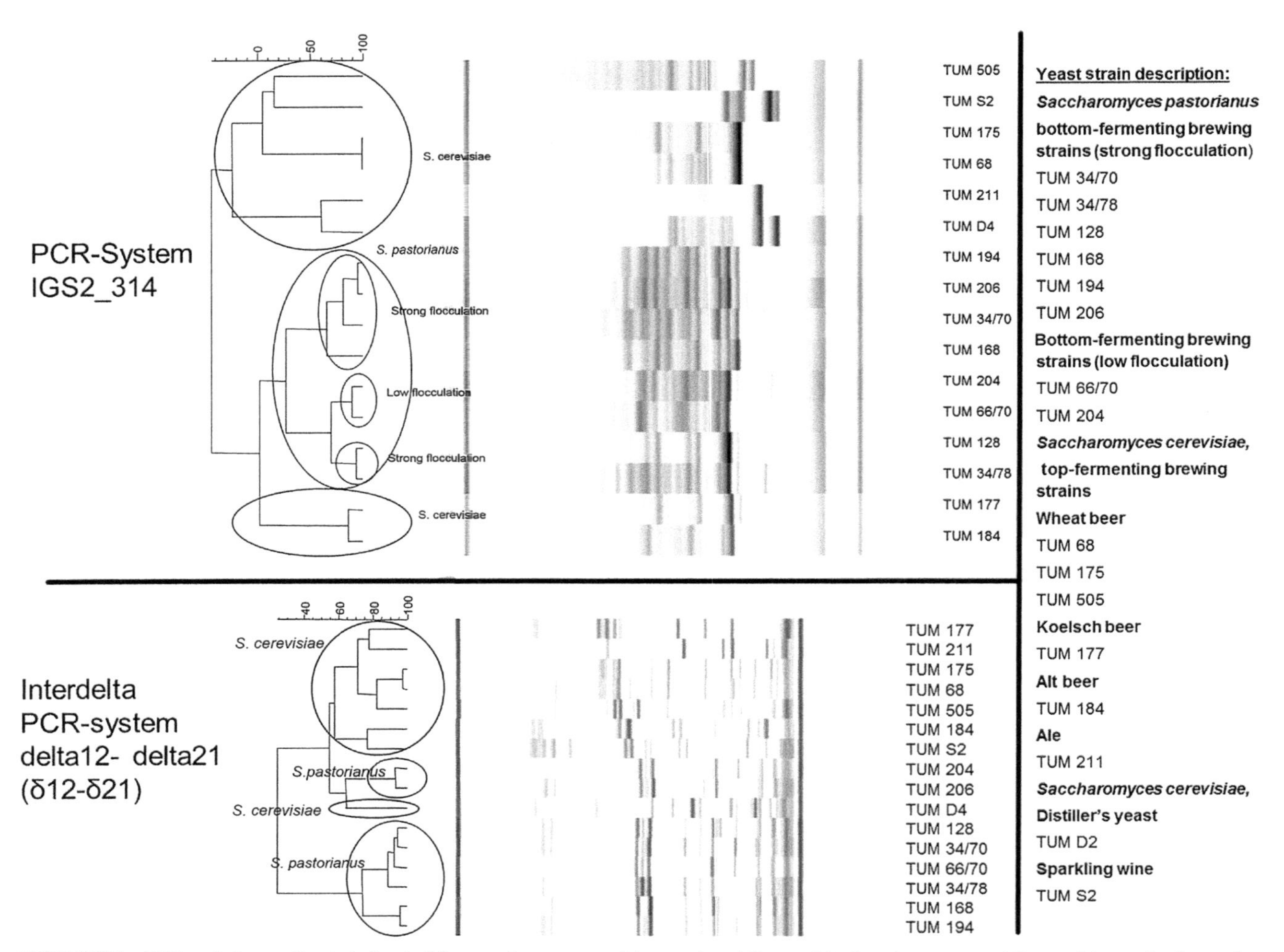

FIGURE 5.7 Differentiation on the strain level of *S. pastorianus* ssp. *carlsbergensis* and *S. cerevisiae* brewing yeasts and *S. cerevisiae* strains from other beverage fermentations using polymerase chain reaction systems IGS2_314 and Interdelta δ12-δ21 (Hutzler et al., 2014). *Author's own image.*

5.3 Brewing yeast cell count/viability/vitality methods

5.3.1 Cell-counting methods

Yeast cell concentration is an important parameter in beer fermentation technology. It is usually expressed in millions of cells per millilitre and should be able to be determined rapidly to ensure that, for example, pitching yeast volumes or harvest yeast dilutions can be calculated and adjusted expeditiously. The most common method used in brewery laboratories for determining yeast concentration is to count the cells under a microscope in a counting chamber, for example, with a Thoma cell-counting chamber. Fig. 5.8 shows *Saccharomyces pastorianus* ssp. *carlsbergensis* TUM 34/70 cells on the grid of a Thoma cell-counting chamber square, which consists of 16 small squares.

There are 16 large grid squares in a Thoma cell-counting chamber, and each one of these is divided into 16 smaller squares, for a total of 256 small squares. The Thoma cell-counting chamber is designed to hold a defined volume so that the cell count can be converted to a cell concentration using a formula provided with each specific type of chamber. A Thoma cell-counting chamber suitable for counting yeast cells is 0.1 mm deep with an area of 0.00,025 mm^2. Concentrations are calculated with the following formula: Yeast cell number/mL = (average yeast cell number per single grid square) $\times$ 4×10^6

Manual cell-counting methods are described in MEBAK III 10.4.3.1/10.11.4.4 (Pfenniger, 1996), Analytica-Microbiologica EBC 3.1.1.1 (Analytica-EBC, 2014b) and in ASBC Methods of Analysis, Microbiology, Yeast-4 (ASBC, 2014). Currently, there are automated cell chambers, for example, Cellometer (Nexcelom, Lawrence, KS,

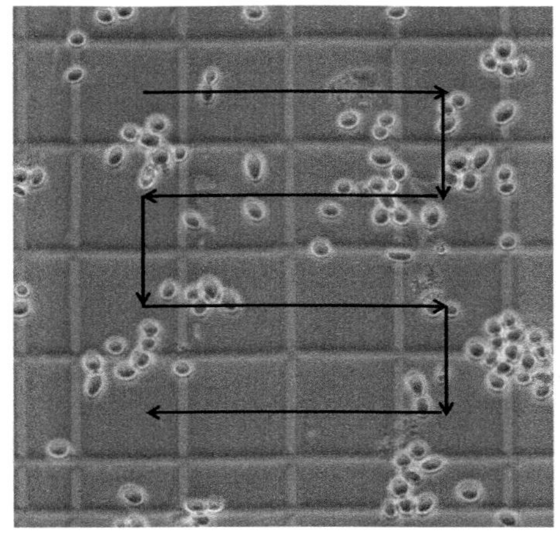

FIGURE 5.8 *Saccharomyces pastorianus* ssp. *carlsbergensis* TUM 34/70 in a Thoma cell-counting chamber, which is counted following the arrows over 16 small grid squares.

USA), available that are very practical for rapidly counting a large number of yeast samples. Manual counting in such cases can be very time consuming. Automated cell chambers can also be combined with fluorescence staining to simultaneously measure viability (see 6.3.2). Other non-microscope-based automated yeast counting systems such as Coulter counter (Beckman Coulter, Brea, CA, USA) or the flow cell-based system Nucleocounter (IUL Instruments, Königswinter, Germany) are also used in some breweries. The system Nucleocounter can also be used to measure viability. An exemplary electronic automated yeast counting technique is described in MEBAK III 10.11.4.5 (Pfenniger, 1996) and in Analytica-Microbiologica EBC 3.1.1.2 (Analytica-EBC, 2014b). Photometric determination of the yeast cell concentration is of less practical relevance and is outlined in MEBAK III 10.4.3.2 (Pfenniger, 1996) and in Analytica-Microbiologica EBC 3.1.1.3 (Analytica-EBC, 2014b). The correct yeast cell concentration is of great importance for reproducibility in fermentation processes. Automated methods should be optimally calibrated before introducing them into a brewery laboratory. They should also be thoroughly compared with existing manual methods. Top-fermenting *Saccharomyces cerevisiae* brewing strains, for example, Bavarian wheat beer strains TUM 68 and 127, which form larger cell clusters, can cause cell concentrations to vary in automated systems. In such cases, the introduction of a correction factor can be necessary. Other microbiological methods, for example, pour plate technique, surface spread-plate technique, can also be used to determine the number of colony-forming units (CFU) per millilitre. These methods are described in MEBAK III 10.11.4.1/10.11.4.2 (Pfenniger, 1996), Analytica-Microbiologica EBC 2.3.3.1 (Analytica-EBC, 2014b) and ASBC Methods of Analysis, Microbiology, Microbiological Control-2 (ASBC, 2014). The microbiological method for determining the CFU per defined volume is far too slow for the breweries to measure yeast concentration because results are needed quickly so that adjustments to brewing process parameters can be made. However, for many microbiological analyses, the CFU per defined volume is highly practical. Inline systems for the analysis of the yeast cell concentrations are described in Section 5.4.2.

5.3.2 Viability methods

Yeast viability describes the percentage of living cells in a yeast population. For a population of brewing yeast, this value should be at least >95% living cells. One of the objectives of yeast management and yeast handling in a brewery should be to achieve a yeast viability of 99%–100%. Viability is the chief analysis criterion, which must be determined in order to rapidly evaluate the condition and quality of the yeast. The standard method for measuring yeast viability is methylene blue or methylene violet staining, described in MEBAK III 10.4.4.1/10.11.3.3 (Pfenniger, 1996), Analytica-Microbiologica EBC 3.2.1.1 (Analytica-EBC, 2014b) and in ASBC Methods of Analysis, Microbiology, Yeast-3 A (ASBC, 2014). Additionally, various fluorophores are used for staining living or dead cells to facilitate counting them under a fluorescence microscope. A standard stain for living cells is fluorescein diacetate (FDA). Dead cells can be stained with propidium iodide (PI) and 1,8-ANS (1-anilinonaphthalene-8-sulphonic acid). The viability method in Analytica-Microbiologica EBC 3.2.1.1 describes dead yeast cell staining with 1,8-ANS. The substance 4′,6-diamidino-2-phenylindole (DAPI) is able to cross the cell membrane and thus can be used for staining both living and dead cells. This technique can be applied as

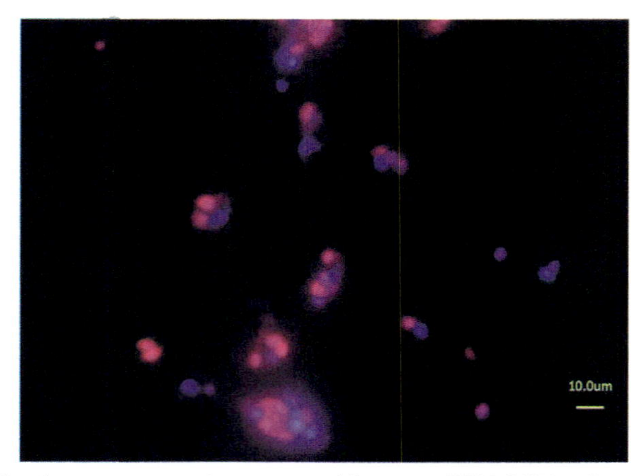

FIGURE 5.9 Microscopic picture of *Saccharomyces pastorianus* ssp. *carlsbergensis* TUM 34/70 cells stained with DAPI (*blue*, living cells) and propidium iodide (*red* to *purple*, dead cells).

counterstaining against PI. Fig. 5.9 shows a microscopic image of *Saccharomyces pastorianus* ssp. *carlsbergensis* TUM 34/70 cells stained with DAPI (blue, living cells) and PI (red to purple, dead cells).

Another combination that is routinely used is FDA (green, living cells) and PI (red, dead cells). Other fluorescent dyes that can be used for this purpose include DiBAC4(3), berberine, acridine orange, primuline, Sytox Orange, cFDA, trypan blue or Hoechst stains. A study by Van Zandycke et al. found that fluorophore staining was perceived to be less subjective by individuals conducting the analysis than bright field dye staining because of the lack of intermediate colour variations (Van Zandycke et al., 2003). If a fluorescence-based viability method has been established in a brewing microbiology laboratory, then a comparison with the standard method methylene blue staining is recommended. Back described the difference in viability using FDA/PI fluorescence staining and methylene blue staining (Back, 1994). Automated yeast counting systems (outlined in Section 5.3.1), which offer a fluorescence measurement option, also find applications in automated yeast viability measurements. Flow cytometry can be used to measure viability as well. A standard fluorescent stain for this application is PI.

5.3.3 Vitality methods

Yeast vitality expresses the activity of the yeast metabolism or in simple words the 'fitness' of a yeast population. Vital brewing yeast should exhibit an excellent growth rate and a strong fermentation performance. Different characteristics or parameters depend on or influence the vitality of the yeast. Table 5.7 presents various methods to quantify yeast vitality, which are based on the measurement of metabolic activity, cellular components or fermentation capacity.

Only a few of the methods in Table 5.7 are actually used in brewing laboratories. Most breweries do not measure yeast vitality at all. A few use time-consuming small fermentation trials yielding delayed results or have in-house methods to evaluate the fermentation capacity of brewing yeast. It is exceedingly rare for service laboratories and for breweries around the world to use the intracellular pH (ICP) method. Weigert et al. modified the ICP method to be based on a flow cytometry platform and developed a rapid technique to determine yeast vitality (Weigert et al., 2009). Flow cytometry can also be used to determine the life cycle of yeast cells and to measure specific cell components, which provide an indication of the yeast vitality. Various flow cytometry protocols and applications for these methods with brewing yeast have been described by Boyd et al., Hutter, Kobayashi et al., Novak et al. and Schönenberg (Boyd et al., 2003; Hutter, 2001; Kobayashi et al., 2007; Novak et al., 2007; Schönenberg, 2011). Methods that have little practical relevance are not discussed further in this chapter but can be reviewed in the publication by Heggart et al. (2000). Recently, Mueller-Auffermann developed two methods based on measurement of the CO_2 volume produced by a defined yeast cell concentration within a specified time period (Mueller-Auffermann, 2014a; Mueller-Auffermann et al., 2011). One method is based on the volumetric measurement of CO_2 in an Einhorn saccharometer, whereas the other system is a CO_2 gas pressure control system. Both methods are shown in Fig. 5.10.

Both methods can easily be transferred to brewery laboratories because the equipment is small and compact and can be placed in incubators. For the Einhorn fermentation saccharometer measurement, the protocol was optimised and is subject to the following conditions: maltose substrate consisting of a 10% ([m/v]) solution; centrifugation of the yeast sample at 750 g for 5 min; adjustment of the yeast cell concentration to 200 million cells/mL with water; mixing according to a ratio

TABLE 5.7 Classification and evaluation of yeast vitality methods according to Heggart et al., Thiele and Mueller-Auffermann (Mueller-Auffermann et al., 2011; Heggart et al., 2000; Thiele, 2006).

Method based on	Examples	Direct	Practical relevance
Metabolic activity	Vitality staining	–	–
	Microcalorimetry	–	–
	Reduction of vicinal diketones (VDK)	–	–
	Protease activity of yeast	–	–
	Magnesium ion release test (MRT)	–	–
	Specific oxygen uptake	–	–
	Acidification power test	–	–
	Intracellular pH value (ICP)	–	X
Measurement of cellular components	Adenosin triphosphate (ATP)	–	–
	Adenylate energy charge (AEC)	–	–
	NADH (fluorometry)	–	–
	Glycogen and trehalose	–	–
	Sterols und unsaturated fatty acids	–	–
Fermentation capacity or glycolytic flow rate	Glycolytic flow rate	X	–
	CO_2 measurement	X	X
	Rapid fermentation trials	X	X

Author's own table.

of 6 mL yeast suspension (200 million cells/mL) + 14 mL maltose solution; mixing in Einhorn saccharometer and equilibration for 60 min in an incubator at 28°C; after 120 min of fermentation, readings are taken for the produced CO_2 volume every 10 min. Conditions of the CO_2 gas pressure control system have been described by Mueller-Auffermann (2014a).

Empirically obtained, sound values for yeast vitality using the two methods are given in Fig. 5.10. ICP, which serves as a kind of reference method, provides the following range of values in assessments of yeast quality: good vitality $\geq$5.8; average vitality 5.8–5.4 and poor vitality $\leq$5.4.

5.4 Monitoring yeast and fermentation

5.4.1 Wort and yeast specifications

A rapid start by the yeast and a 'nicely progressing' fermentation necessitate that the wort composition and the condition of the yeast meet certain requirements. Wort and yeast specifications important for favourable propagation and good fermentation performance are presented in Table 5.8.

High-gravity worts with extract concentrations of $\geq$15 degrees Plato are not generally recommended for yeast propagation processes, due to the osmotic stress on the yeast. High-gravity worts blended with de-aerated water to extract levels of approximately 11–12 degrees Plato must be carefully checked according to the wort specifications in Table 5.8. Here, the parameters free amino nitrogen (FAN) and zinc are the most essential. Under normal conditions, the wort specifications in Table 5.8 guarantee that enough nitrogen and carbon sources for growth and fermentation are available for the yeast. Table 5.8 also contains microbiological specifications for wort and yeast. Brewing yeast should be free of other microorganisms that can negatively influence propagation and fermentation processes negatively. In propagation systems, aerobic or facultative anaerobic microorganisms such as Gram-negative acetic acid bacteria or aerobic and facultative anaerobic wild yeasts may cause problems. During the initial phase of fermentation, until the brewing yeast has reduced the

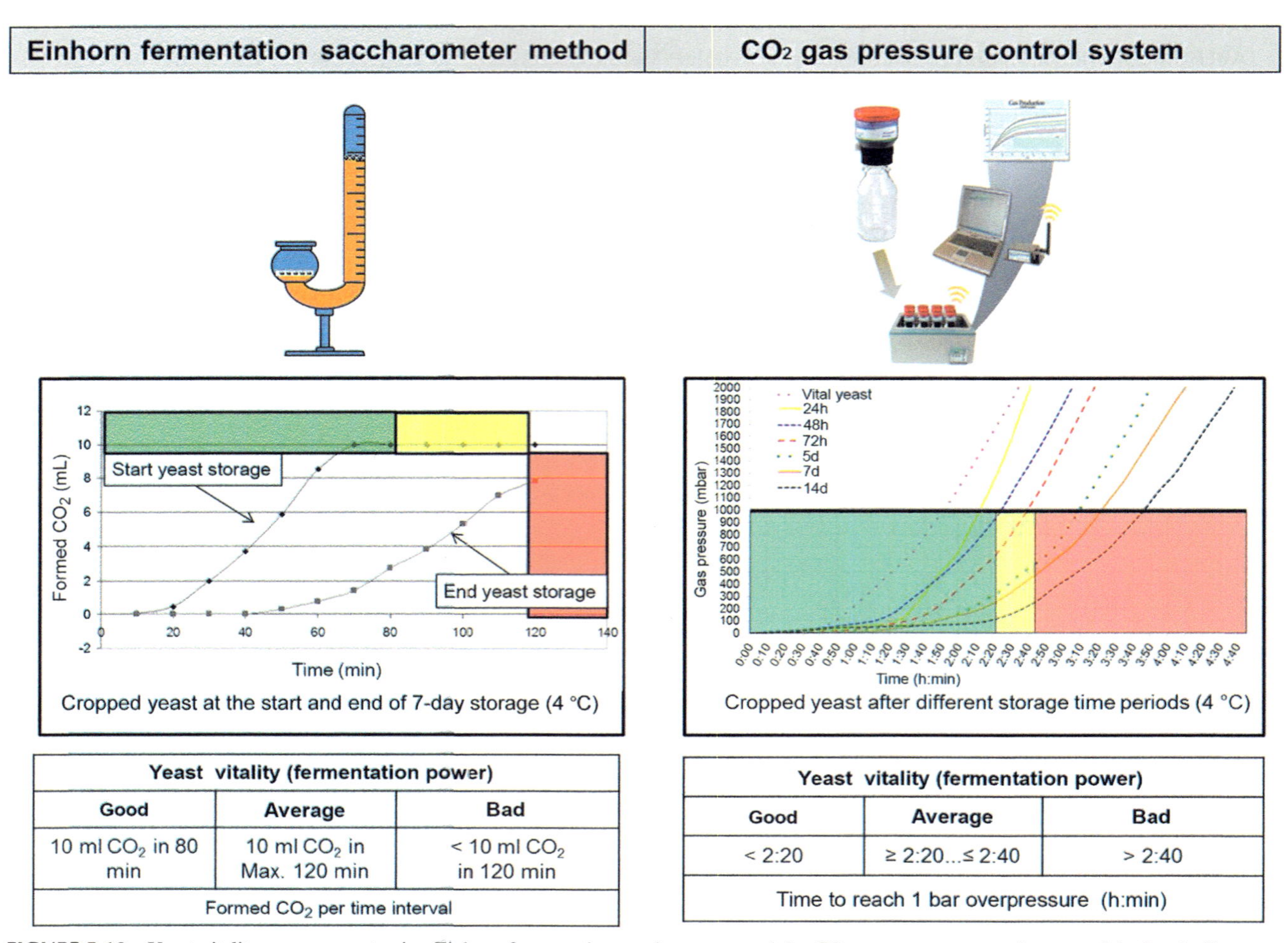

FIGURE 5.10 Yeast vitality measurement using Einhorn fermentation saccharometer and the CO_2 gas pressure control system (Mueller-Auffermann, 2014a; Mueller-Auffermann et al., 2011). *Author's own image.*

pH to below 4.8, and the ethanol concentration rises above 1.5%–2%, the pitched wort is exceedingly susceptible to microbial spoilage. Background flora associated with wort, mostly Gram-negative bacteria such as *Enterobacter* sp., *Rahnella* sp., *Citrobacter* sp., *Klebsiella* sp., *Pantoea* sp. and *Serratia* sp., are particularly apt to cause spoilage and generate off-flavours (Hutzler et al., 2013). Therefore, yeast should be monitored using wort agar and Wallerstein differential agar (WLD agar) and wort on wort agar. A test for Enterobacteriaceae may also be performed on Violet Red Bile Dextrose (VRBD) agar. Wild yeasts may also compromise the quality of the fermenting wort by competing with brewing yeasts in the initial stage of fermentation. For this reason, wort should be tested for wild yeast growth using wort agar or any other universal nutrient medium for yeast. Brewing yeast should be monitored for the presence of wild yeast, for example, with the 37°C test (for BF) or YM + 200 ppm $CuSO_4$ (for BF and TF). There are also other media for the detection of wild yeasts. Furthermore, wort and yeast should also be checked for beer spoilage microorganisms. The total colony count is a good comparative value that enables CFU/mL comparisons with other foods, beverages, water, or intermediate products. Methods for determining and assigning values to yeast viability and vitality are described in Sections 5.3.2 and 5.3.3. If the pH of the supernatant of a centrifuged yeast suspension is above 5.8, the yeast population most likely contains residues of autolysed yeast cells, which increase the alkalinity of the suspension. Preliminary examination of the yeast with a microscope provides quick information on the amount of nonyeast particulates present. Especially bottom-cropped brewing yeast that has been repitched many times may contain higher percentages of nonyeast particulates, for example, oxalate.

TABLE 5.8 Yeast and wort specifications for sufficient yeast growth and fermentation performance.

Wort specifications			
Total nitrogen	900–1100		mg/L
FAN free amino nitrogen cast wort	200–250		mg/L
pH value (depending on additional acidification)	5.0–5.6		
Zinc	0.1–0.3		mg/L
Attenuation limit of wort (apparent)	≥78		%
Photometric iodine method (MEBAK)	<0.45		
Wort-associated microbial background flora (wort agar)	≤1		per mL
Beer spoilage bacteria (NBB)	Negative		In 1 mL
Yeast specifications			
Target cell concentration in propagation end sample	80–100		Million cells/mL
Target cell concentration in fermenter full sample	≥10 (BF)	≥3–10 (TF)	Million cells/mL
Viability (methylene blue)	≥95		%
Vitality (Einhorn saccharometer, 80 min runtime)	≥10		mL
Vitality (CO_2 pressure system, 1 bar over pressure)	≤2:20		h:min
pH of yeast supernatant	≤5.8		
Bacterial flora in fermentation systems (WLD)	≤1		per mL
Beer spoilage bacteria (NBB)	Negative		In 1 mL
Wort-associated bacterial background flora (wort agar + actidione)	≤1		per mL
Acetic acid bacteria	Negative		In 0.1 mL
Wild yeast ($CuSO_4$ broth)	Negative		In 0.01 mL
Wild yeast (37°C method)	Negative		In 0.01 mL
Microscopic pre-check	No spoilage organisms,		
	No or few non-yeast particles		

BF, bottom fermenting; *TF*, top fermenting.

5.4.2 Monitoring of fermentation and maturation parameters and their application with yeast

Measurement of the fermentation and maturation parameters allows the fermentation performance, as well as any problems or any other developments that may arise, to be monitored and rectified if necessary. Table 5.9 shows a list of analysis methods for monitoring the principal fermentation parameters and also for selected fermentation by-products approved by the ASBC, MEBAK and EBC.

The principal fermentation parameters are useful for creating fermentation diagrams and therefore for comparing different fermentation processes with each other. The same is true for propagation and maturation processes. The process can be tailored to fit specific parameters, according to an individual brewery's needs. Hence, a fruity beer character can be augmented by increasing ester production, or the flavour stability can be improved by enhancing SO_2 production. In Fig. 5.11, the values for various fermentation by-products from two different beers are discussed with regard to the errors in the brewing process that may have produced them.

The following errors in the brewing process listed in Fig. 5.11 can result in extreme values for the parameters measured: high fermentation/maturation temperature, insufficient aeration, long maturation time (yeast excretion), very strong yeast growth/propagation, FAN deficiency, low yeast cell concentration and short maturation.

In-process or in-line measurements throughout the entire brewing process, and for yeast and fermentation parameters in particular, are growing in significance and have already become well established in some areas. Boulton published, in a review article, a list of in-line analyses for all stages of the brewing process. In this article, he described the types of sensors

TABLE 5.9 Approved analysis methods for fermentation main parameters and selected fermentation by-products of the analysis commissions of ASBC, MEBAK and EBC (ASBC, 2014; Analytica-EBC, 2014a; Jacob, 2012).

	Methods		
Fermentation main parameters	**ASBC**	**MEBAK**	**EBC**
Extract	Beer 2, 3, 5	2.9	8.3, 9.43
Ethanol	Beer 4	2.9	9.2–9.4
CO_2	Beer 13	2.26	9.28
Final attenuation of beer	Beer 16	2.8	9.7
pH	Beer 9	2.13	9.35
Selected fermentation by-products	**ASBC**	**MEBAK**	**EBC**
Volatile fermentation by-products (headspace)	Beer 48	2.21.1	–
Fermentation by-products (other methods)	–	2.21.2	–
Fatty acids	–	2.21.4	–
Aromatic alcohols and phenolic acids	–	2.21.3	–
Vicinal diketones	Beer 25	2.21.5	9.24
Acetoin	–	2.21.5.4	–
Higher alcohols and esters	–	2.21.6	–
Organic acids	–	2.21.7	9.32, 9.34
SO_2	Beer 21	2.21.8	9.25
Glycerol	–	–	9.33

Author's own table.

	Beer 1	Beer 2	
Diacetyl	0.05	0.13 E	mg/L
Pentandione-2,3 7	0.02	0.08 E	mg/L
Acetoin	2.8 A	7.9 A	mg/L
Ethyl acetate	12.4 D	25.1 F	mg/L
Isobutyl acetate	0.01 D	0.05 F	mg/L
Isoamyl acetate	0.6 D	1.8	mg/L
2-Phenylethanol	31.5 A	15.5	mg/L
Decanoic acid	1.9 D	1.0	mg/L
Ethyl decanoate	0.11 D	0.05	mg/L
Ethyl hexanoate (ethyl caproate)	0.16 B	0.23 B	mg/L

Possible fermentation/process errors causing out-of-range values:
A High fermentation/ maturation temperature
B Insufficient aeration
C Long maturation time (yeast excretion)
D Strong yeast growth/propagation
E Deficient FAN, low yeast concentration, short maturation
F Low yeast growth/propagation

FIGURE 5.11 Interpretation of selected fermentation by-products of two beers in terms of process errors (Jacob, 2014). *Author's own image.*

necessary for monitoring the corresponding processes, including all those involving yeast (Boulton, 2012). Boulton also discussed the equipment required for determining the concentration or cell mass of the yeast, turbidity measurements (Optek-Danulat, Germany) and capacitance measurement/dielectric spectroscopy (Aber Instruments, UK/Hamilton Messtechnik, Germany/Fogale Nanotech, France). Capacitance measurement/dielectric spectroscopy can also be used to measure the amount of viable yeast cells. Many breweries, especially small- and mid-sized breweries, adjust their yeast cell concentrations, pitching rates and quantities of yeast cropped by means of volumetric measurements, sight glass inspection and manual process control without knowing the exact yeast concentration and the percentage of viable cells in the pitching yeast population. Essentially there is a room for improvement in the in-line yeast process control at most breweries. Even in breweries where in-line measurement and control systems for yeast are already in place, optimal adjustment of the finer points of these processes according to the parameters of each individual process is necessary and may require some time for sufficient optimisation. Selected publications by Tibayrenc et al., Sandhar et al., Mas et al., Kiviharju et al., and Krause et al., discuss the principles, applications and scientific approaches pertaining to the topic of in-line yeast measurement (Kiviharju et al., 2008; Krause et al., 2011; Mas et al., 2001; Sandhar, 2014; Tibayrenc et al., 2011).

Aside from yeast concentration, ethanol, CO_2 and specific gravity/extract are considered to be principal fermentation parameters and thus are of great importance. The relevant methods for in-line measurement of these parameters were also included by Boulton review (Boulton, 2012). Boulton and Nordkvist and Biering and Bockisch cited different methods for in-tank measurement of fermentation parameters that are used to observe certain conditions present in fermentation tanks (e.g., layer effects, homogeneous distribution, etc.) (Biering & Bockisch, 2014; Boulten & Nordkvist, 2014). Grassi et al. demonstrated that FT-NIR spectroscopy, when combined with locally weighted regression, is a perfectly suitable quantitative method for measuring pH, biomass and Brix (extract) and can be readily implemented in the beer production process (Grassi, Amigo, Lyndgaard, Foschino & Casiraghi, 2014a, 2014b). This method offers a potential in-line measurement tool for recording the values of principal fermentation parameters.

References

Abdel-Aty, L. M. (1991). *Immunchemische und molekularbiologische Untersuchungen an verschiedenen Saccharomyces- und Schizosaccharomyces Hefen* (Dissertation, TU München).

Alsammar, H., & Delneri, D. (2020). An update on the diversity, ecology and biogeography of the *Saccharomyces* genus. *FEMS Yeast Research, 20*(3), Article foaa013. https://doi.org/10.1093/femsyr/foaa013

Analytica-EBC. (2014a). *Chemical/physical*. Nürnberg: Hans Carl Fachverlag. http://www.analytica-ebc.com. (Accessed 24 April 2014).

Analytica-EBC. (2014b). *Microbiology*. Hans Carl Fachverlag. http://www.analytica-ebc.com. (Accessed 29 April 2014).

Anonymous. (1994). Differentiation of ale and lager yeast. *Journal of the American Society of Brewing Chemists, 52*, 184–188.

Anonymous. (1997). CLEN medium for the detection of wild yeast. *Journal of the American Society of Brewing Chemists, 55*, 185–189.

Anonymous. (1998). CLEN medium for the detection of wild yeast. *Journal of the American Society of Brewing Chemists, 56*, 202–208.

Antunovics, Z., Irinyi, L., & Sipiczki, M. (2005). Combined application of methods to taxonomic identification of *Saccharomyces* strains in fermenting botrytized grape must. *Journal of Applied Microbiology, 98*, 971–979.

Arias, C. R., Burns, J. K., Friedrich, L. M., Goodrich, R. M., & Parish, M. E. (2002). Yeast species associated with orange juice: Evaluation of different identification methods. *Applied and Environmental Microbiology, 68*, 1955–1961.

ASBC. (2014). *Methods of analysis*. St. Paul: ASBC. Available http://www.asbcnet.org/moa/.

Back, W. (1994). *Farbatlas und Handbuch der Getränkemikrobiologie*. Nürnberg: Hans Carl Fachverlag.

Bamforth, C. W. (2005). *Food, fermentation and microorganisms*. Oxford: Blackwell Publishing.

Beltran, G., Torija, M. J., Novo, M., Ferrer, N., Poblet, M., Guillamon, J. M., et al. (2002). Analysis of yeast populations during alcoholic fermentation: A six year follow-up study. *Systematic & Applied Microbiology, 25*, 287–293.

Biering J.; Bockisch A. (April 13–17, 2014) Oscillating conditions in yeast fermenation—multiposition sensors for process control and optimization. In: 11th international trends in brewing. KU Leuven: Ghent, Belgium.

Blattel, V., Petri, A., Rabenstein, A., Kuever, J., & Konig, H. (2013). Differentiation of species of the genus *Saccharomyces* using biomolecular fingerprinting methods. *Applied Microbiology and Biotechnology, 97*, 4597–4606.

Bleve, G., Rizzotti, L., Dellaglio, F., & Torriani, S. (2003). Development of reverse transcription (RT)-PCR and real-time RT-PCR assays for rapid detection and quantification of viable yeasts and molds contaminating yogurts and pasteurized food products. *Applied and Environmental Microbiology, 69*, 4116–4122.

Boekhout, T., Amend, A. S., El Baidouri, F., Gabaldón, T., Geml, J., Mittelbach, M., et al. (2022). Trends in yeast diversity discovery. *Fungal Diversity, 114*, 491–537. https://doi.org/10.1007/s13225-021-00494-6

Bokulich, N. A., Bamforth, C. W., & Mills, D. A. (2012). Brewhouse-resident microbiota are responsible for multi-stage fermentation of American coolship ale. *PLoS One, 7*, Article e35507.

Boulten, C., & Nordkvist, M. (2014). A stable platform for control of fermentation. In *11th international trends in brewing, 13th–17th April 2014*. Ghent, Belgium: KU Leuven.

Boulton, C. (2012). 125th anniversary review: Advances in analytical methodology in brewing. *Journal of the Institute of Brewing, 118*, 255–263.

Boyd, A. R., Gunasekera, T. S., Attfield, P. V., Simic, K., Vincent, S. F., & Veal, D. A. (2003). A flow-cytometric method for determination of yeast viability and cell number in a brewery. *FEMS Yeast Research, 3*, 11–16.

Brandl, A. (2006). *Entwicklung und Optimierung von PCR-Methoden zur Detektion und Identifizierung von brauereirelevanten Mikroorganismen zur Routine-Anwendung in Brauereien* (Dissertation, TU München)).

Brandl, A., Hutzler, M., & Geiger, E. (2005). Optimisation of brewing yeast differentiation and wild yeast identification by real-time PCR ([CD-ROM], Nürnberg). In *Proc. 30th ebc congr. Prague. Fachverlag hans carl* ([CD-ROM], Nürnberg).

Buchl, N. R., Hutzler, M., Mietke-Hofmann, H., Wenning, M., & Scherer, S. (2010). Differentiation of probiotic and environmental *Saccharomyces cerevisiae* strains in animal feed. *Journal of Applied Microbiology, 109*, 783–791.

Buchl, N. R., Wenning, M., Seiler, H., Mietke-Hofmann, H., & Scherer, S. (2008). Reliable identification of closely related *issatchenkia* and *pichia* species using artificial neural network analysis of fourier-transform infrared spectra. *Yeast, 25*, 787–798.

Cappello, M. S., Bleve, G., Grieco, F., Dellaglio, F., & Zacheo, G. (2004). Characterization of *Saccharomyces cerevisiae* strains isolated from must of grape grown in experimental vineyard. *Journal of Applied Microbiology, 97*, 1274–1280.

Casey, G. D., & Dobson, A. D. (2004). Potential of using real-time PCR-based detection of spoilage yeast in fruit juice—a preliminary study. *International Journal of Food Microbiology, 91*, 327–335.

Ciani, M., Mannazzu, I., Marinangeli, P., Clementi, F., & Martini, A. (2004). Contribution of winery-resident *Saccharomyces cerevisiae* strains to spontaneous grape must fermentation. *Antonie van Leeuwenhoek, 85*, 159–164.

Cocolin, L., Bisson, L. F., & Mills, D. A. (2000). Direct profiling of the yeast dynamics in wine fermentations. *FEMS Microbiology Letters, 189*, 81–87.

Comi, G., Maifreni, M., Manzano, M., Lagazio, C., & Cocolin, L. (2000). Mitochondrial DNA restriction enzyme analysis and evaluation of the enological characteristics of *Saccharomyces cerevisiae* strains isolated from grapes of the wine-producing area of Collio (Italy). *International Journal of Food Microbiology, 58*, 117–121.

Dörries. (2006). *Entwicklung von real-time PCR Nachweissystemen für getränkerelevante Hefen* (Dissertation, TU Berlin).

De Angelo, J., & Siebert, K. J. (1987). A new medium for the detection of wild yeast in brewing culture yeast. *Journal of the American Society of Brewing Chemists, 45*, 135–140.

Delaherche, A., Claisse, O., & Lonvaud-Funel, A. (2004). Detection and quantification of *Brettanomyces bruxellensis* and 'ropy' *Pediococcus damnosus* strains in wine by real-time polymerase chain reaction. *Journal of Applied Microbiology, 97*, 910–915.

Demuyter, C., Lollier, M., Legras, J. L., & Le Jeune, C. (2004). Predominance of *Saccharomyces uvarum* during spontaneous alcoholic fermentation, for three consecutive years, in an Alsatian winery. *Journal of Applied Microbiology, 97*, 1140–1148.

Dufour, J.-P., Verstrepen, K., & Derdelinckx, G. (2003). Brewing yeasts. In T. Boekhout, & V. Robert (Eds.), *Yeasts in food*. Hamburg: Behr's Verlag.

Dunn, B., & Sherlock, G. (2008). Reconstruction of the genome origins and evolution of the hybrid lager yeast *Saccharomyces pastorianus*. *Genome Research, 18*, 1610–1623.

Esteve-Zarzoso, B., Fernandez-Espinar, M. T., & Querol, A. (2004). Authentication and identification of *Saccharomyces cerevisiae* 'flor' yeast races involved in sherry ageing. *Antonie van Leeuwenhoek, 85*, 151–158.

Esteve-Zarzoso, B., Peris-Toran, M. J., Garcia-Maiquez, E., Uruburu, F., & Querol, A. (2001). Yeast population dynamics during the fermentation and biological aging of sherry wines. *Applied and Environmental Microbiology, 67*, 2056–2061.

Fernandez-Espinar, M. T., Martorell, P., De Llanos, R., & Querol, A. (2006). Molecular methods to identify and charcterize yeasts in foods and beverages. In A. Querol, & G. H. Fleet (Eds.), *Yeasts in food and beverages*. New York: Springer Inc.

Fernandez-Gonzalez, M., Espinosa, J. C., ubeda, J. F., & briones, A. I. (2001). Yeasts present during wine fermentation: Comparative analysis of conventional plating and PCR-TTGE. *Systematic & Applied Microbiology, 24*, 634–638.

Ganga, M. A., & Martinez, C. (2004). Effect of wine yeast monoculture practice on the biodiversity of non-*Saccharomyces* yeasts. *Journal of Applied Microbiology, 96*, 76–83.

Gomes, F. C., Pataro, C., Guerra, J. B., Neves, M. J., Correa, S. R., Moreira, E. S., et al. (2002). Physiological diversity and trehalose accumulation in *Schizosaccharomyces pombe* strains isolated from spontaneous fermentations during the production of the artisanal Brazilian cachaca. *Canadian Journal of Microbiology, 48*, 399–406.

Granchi, L., Ganucci, D., Viti, C., Giovannetti, L., & Vincenzini, M. (2003). *Saccharomyces cerevisiae* biodiversity in spontaneous commercial fermentations of grape musts with 'adequate' and 'inadequate' assimilable-nitrogen content. *Letters in Applied Microbiology, 36*, 54–58.

Grassi, S., Amigo, J. M., Lyndgaard, C. B., Foschino, R., & Casiraghi, E. (2014b). Beer fermentation: Monitoring of process parameters by FT-NIR and multivariate data analysis. *Food Chemistry, 155*, 279–286.

Grassi, S., Amigo, J. M., Lyndgaard, C., Foschino, R., & Casiraghi, E. (2014). A comprehensive study of beer fermentation by using NIR and MIR spectroscopy and advanced data analysis. In *11th international trends in brewing*. Ghent, Belgium: KU Leuven.

Guerra, J. B., Araujo, R. A., Pataro, C., Franco, G. R., Moreira, E. S., Mendonca-Hagler, L. C., et al. (2001). Genetic diversity of *Saccharomyces cerevisiae* strains during the 24 h fermentative cycle for the production of the artisanal Brazilian cachaca. *Letters in Applied Microbiology, 33*, 106–111.

Heggart, H. M., Margaritis, A., Stewart, R., Pikington, J. H., Sobczak, J., & Russell, I. (2000). Measurement of brewing yeast viability and vitality: A review of methods. *Technical Quarterly - Master Brewers Association of the Americas, 37*, 409–430.

Howell, K. S., Bartowsky, E. J., Fleet, G. H., & Henschke, P. A. (2004). Microsatellite PCR profiling of *Saccharomyces cerevisiae* strains during wine fermentation. *Letters in Applied Microbiology, 38*, 315–320.

Hutter, K. J. (2001). Flusszytometrische prozesskontrolle untergäriger bierhefen. *Brewing Science, 1*(2), 13–27.

Hutzler, M. (2009). *Entwicklung und Optimierung von Methoden zur Identifizierung und Differenzierung von getränkerelevanten Hefen* (Dissertation, TU München).

Hutzler, M., Geiger, E., & Jacob, F. (2010). Use of PCR-DHPLC (Polymerase chain reaction—denaturing high performance liquid chromatography) for the rapid differentiation of industrial *Saccharomyces pastorianus* and *Saccharomyces cerevisiae* strains. *Journal of the Institute of Brewing, 116*, 464–474.

Hutzler, M., & Goldenberg, O. (2007). PCR-DHPLC: A potential novel method for rapid screening of mixed yeast populations. In *31th Congress European brewery convention. Venice: Fachverlag hans carl, nürnberg.*

Hutzler, M., Michel, M., Kunz, O., Kuusisto, T., Magalhães, F., Krogerus, K., et al. (2021 Mar 31). Unique brewing-relevant properties of a strain of Saccharomyces jurei isolated from ash (fraxinus excelsior). *Frontiers in Microbiology, 12*, Article 645271. https://doi.org/10.3389/fmicb.2021.645271

Hutzler, M., Mueller-Auffermann, K., & Jacob, F. (2012). Yeast quality control—standards and novel approaches. In *EBC Symposium 2012-from filler to chiller*. Copenhagen: EBC: Carlsberg.

Hutzler, M., Mueller-Auffermann, K., Koob, J., Riedl, R., & Jacob, F. (2013). Beer spoiling microorganisms—a current overview. *Brauwelt International, 31*, 23–25.

Hutzler, M., Schoenenberg, S., Koetke, H., Geiger, E., & Rainieri, S. (2008). Rapid brewing yeast differentiation by real-time PCR. In *First international symposium for young scientists and technologists in malting, brewing and distilling, 6th November 2008*. Cork: UCC.

Hutzler, M., Schuster, E., & Stettner, G. (2008). Ein Werkzeug in der Brauereimikrobiologie - real-time PCR in der Praxis. *Brauindustrie, 4*, 52–55.

Hutzler, M., Stretz, D., Schneiderbanger, H., Mueller-Auffermann, K., & Riedl, R. (2014). Yeast strain information sheets. Available from: Research Center Weihenstephan for Brewing and Food Quality, TU München http://www.blq-weihenstephan.de/en/yeast-center/microorganisms.html.

Hutzler, M., & Wenning, M. (2009). Yeast identification using infrared spectroscopy, friend oder foe? *Brauwelt, 27*, 74–77.

Jacob, F. (2012). *MEBAK brautechnische analysenmethoden, würze, bier biermischgetränke*. Freising: MEBAK.

Jacob, F. (April 3, 2014). *Hefe und Bierqualität. Hefe und Mikrobiologie—Forschung und Praxis*. Freising-Weihenstephan.

Jespersen, L., & Jakobsen, M. (1996). Specific spoilage organisms in breweries and laboratory media for their detection. *International Journal of Food Microbiology, 33*, 139–155.

Josepa, S., Guillamon, J. M., & Cano, J. (2000). PCR differentiation of *Saccharomyces cerevisiae* from *Saccharomyces bayanus/Saccharomyces pastorianus* using specific primers. *FEMS Microbiology Letters, 193*, 255–259.

Kiviharju, K., Salonen, K., Moilanen, U., & Eerikainen, T. (2008). Biomass measurement online: The performance of in situ measurements and software sensors. *Journal of Industrial Microbiology and Biotechnology, 35*, 657–665.

Kobayashi, M., Shimizu, H., & Shioya, S. (2007). Physiological analysis of yeast cells by flow cytometry during serial-repitching of low-malt beer fermentation. *Journal of Bioscience and Bioengineering, 103*, 451–456.

Kobi, D., Zugmeyer, S., Potier, S., & Jaquet-Gutfreund, L. (2004). Two-dimensional protein map of an "ale"-brewing yeast strain: Proteome dynamics during fermentation. *FEMS Yeast Research, 5*, 213–230.

Krause, D., Birle, S., Hussein, M. A., & Becker, T. (2011). Bioprocess monitoring and control via adaptive sensor calibration. *Engineering in Life Sciences, 11*, 402–416.

Kuniyuki, A. H., Rous, C., & Sanderson, J. L. (1984). Enzymatic-linked immunosorbent assay (ELISA) detection of *Brettanomyces* contaminants in wine production. *American Journal of Enology and Viticulture, 35*, 143–145.

Kurtzman, C. P., & Robnett, C. J. (1998). Identification and phylogeny of ascomycetous yeasts from analysis of nuclear large subunit 26s partial sequences. *Antonie Van Leeuwenhoek, 73*, 331–371.

Laitila, A., Wilhelmson, A., Kotaviita, E., Olkku, J., Home, S., & Juvonen, R. (2006). Yeasts in an industrial malting ecosystem. *Journal of Industrial Microbiology and Biotechnology, 33*, 953–966.

Las Heras-Vazquez, F. J., Mingorance-Cazorla, L., Clemente-Jimenez, J. M., & Rodriguez-Vico, F. (2003). Identification of yeast species from orange fruit and juice by RFLP and sequence analysis of the 5.8S rRNA gene and the two internal transcribed spacers. *FEMS Yeast Research, 3*, 3–9.

Legras, J. L., & Karst, F. (2003). Optimisation of interdelta analysis for *Saccharomyces cerevisiae* strain characterisation. *FEMS Microbiology Letters, 221*, 249–255.

Libkind, D., Hittinger, C. T., Valerio, E., Goncalves, C., Dover, J., Johnston, M., et al. (2011). Microbe domestication and the identification of the wild genetic stock of lager-brewing yeast. *Proceedings of the National Academy of Sciences USA, 108*, 14539–14544.

Lopes, C. A., Van Broock, M., Querol, A., & Caballero, A. C. (2002). *Saccharomyces cerevisiae* wine yeast populations in a cold region in Argentinean Patagonia. A study at different fermentation scales. *Journal of Applied Microbiology, 93*, 608–615.

Martinez, C., Gac, S., Lavin, A., & Ganga, M. (2004). Genomic characterization of *Saccharomyces cerevisiae* strains isolated from wine-producing areas in South America. *Journal of Applied Microbiology, 96*, 1161–1168.

Mas, S., Ossard, F., & Ghommidh, C. (2001). On-line determination of flocculating *Saccharomyces cerevisiae* concentration and growth rate using a capacitance probe. *Biotechnology Letters, 23*, 1125–1129.

Morrissey, W. F., Davenport, B., Querol, A., & Dobson, A. D. (2004). The role of indigenous yeasts in traditional Irish cider fermentations. *Journal of Applied Microbiology, 97*, 647–655.

Mueller-Auffermann, K. (2014). Bestimmung der Vitalität und Viabilität von Hefen. In *Hefe und Mikrobiologie—Forschung und Praxis*. Freising-Weihenstephan.

Mueller-Auffermann, K. (2014). Neuartige charakterisierung untergäriger hefestämme. In *Hefe und Mikrobiologie—Forschung und Praxis*. Freising-Weihenstephan. Research Center Weihenstephan for Food and Beverage Quality, TU München.

Mueller-Auffermann, K., Schneiderbanger, H., Hutzler, M., & Jacob, F. (2011). Scientific evaluation of different methods for the determination of yeast vitality. *Brewing Sciences, 64*, 107–118.

Muir, A., Harrison, E., & Wheals, A. (2011). A multiplex set of species-specific primers for rapid identification of members of the genus *Saccharomyces*. *FEMS Yeast Research, 11*, 552–563.

Nakao, Y., Kanamori, T., Itoh, T., Kodama, Y., Rainieri, S., Nakamura, N., et al. (2009). Genome sequence of the lager brewing yeast, an interspecies hybrid. *DNA Research: International Journal for Rapid Publication of Reports on Genes and Genomes, 16*, 115–129.

Naumov, G. I., Masneuf, I., Naumova, E. S., Aigle, M., & Dubourdieu, D. (2000). Association of Saccharomyces bayanus var. uvarum with some French wines: Genetic analysis of yeast populations. *Research in Microbiology, 151*, 683–691.

Naumov, G. I., Naumova, E. S., Antunovics, Z., & Sipiczki, M. (2002). *Saccharomyces bayanus* var. uvarum in Tokaj wine-making of Slovakia and Hungary. *Applied Microbiology and Biotechnology, 59*, 727–730.

Naumov, G. I., Nguyen, H. V., Naumova, E. S., Michel, A., Aigle, M., & Gaillardin, C. (2001). Genetic identification of *Saccharomyces bayanus* var. uvarum, a cider-fermenting yeast. *International Journal of Food Microbiology, 65*, 163–171.

Novak, J., Basarova, G., Teixeira, J. A., & Vicente, A. A. (2007). Monitoring of brewing yeast propagation under aerobic and anaerobic conditions employing flow cytometry. *Journal of the Institute of Brewing, 113*, 249–255.

Pfenniger, H. (1996). *MEBAK brautechnische analysenmethoden band III*. Freising: MEBAK.

Phister, T. G., & Mills, D. A. (2003). Real-time PCR assay for detection and enumeration of Dekkera bruxellensis in wine. *Applied and Environmental Microbiology, 69*, 7430–7434.

Pope, G. A., Mackenzie, D. A., Defernez, M., Aroso, M. A., Fuller, L. J., Mellon, F. A., et al. (2007). Metabolic footprinting as a tool for discriminating between brewing yeasts. *Yeast, 24*, 667–679.

Prakitchaiwattana, C. J., Fleet, G. H., & Heard, G. M. (2004). Application and evaluation of denaturing gradient gel electrophoresis to analyse the yeast ecology of wine grapes. *FEMS Yeast Research, 4*, 865–877.

Pramateftaki, P. V., Lanaridis, P., & Typas, M. A. (2000). Molecular identification of wine yeasts at species or strain level: A case study with strains from two vine-growing areas of Greece. *Journal of Applied Microbiology, 89*, 236–248.

Priest, F. G., & Campell, I. (2003). *Brewing microbiology*. New York: Kluver Academic/Plenum Publishers.

Röcken, W., & Marg, C. (1983). Nachweis von Fremdhefen in der obergärigen Brauerei - Vergleich verschiedener Nährböden. *Monatsschrift für Brauwissenschaft, 7*, 276–279.

Rainieri, S., Kodama, Y., Kaneko, Y., Mikata, K., Nakao, Y., & Ashikari, T. (2006). Pure and mixed genetic lines of *Saccharomyces bayanus* and *Saccharomyces pastorianus* and their contribution to the lager brewing strain genome. *Applied and Environmental Microbiology, 72*, 3968–3974.

Rainieri, S., Zambonelli, C., & Kaneko, Y. (2003). *Saccharomyces sensu stricto*: Systematics, genetic diversity and evolution. *Journal of Bioscience and Bioengineering, 96*, 1–9.

Roder, C., Konig, H., & Frohlich, J. (2007). Species-specific identification of Dekkera/Brettanomyces yeasts by fluorescently labeled DNA probes targeting the 26S rRNA. *FEMS Yeast Research, 7*, 1013–1026.

Rodriguez, M. E., Lopes, C. A., Van Broock, M., Valles, S., Ramon, D., & Caballero, A. C. (2004). Screening and typing of Patagonian wine yeasts for glycosidase activities. *Journal of Applied Microbiology, 96*, 84–95.

Salinas, F., Garrido, D., Ganga, A., Veliz, G., & Martinez, C. (2009). Taqman real-time PCR for the detection and enumeration of *Saccharomyces cerevisiae* in wine. *Food Microbiology, 26*, 328–332.

Sandhar, P. (2014). The evalution of the yeast monitor as a critical control processin brewing. In *11th international trends in brewing, 13th–17th April 2014*. Ghent, Belgium: KU Leuven.

Schöneborn, H. (2001). *Differenzierung und Charakterisierung von Betriebshefekulturen mit genetischen und physiologischen Methoden*. Dissertation, TU München).

Schönenberg, S. (2011). *Der physiologische Zustand und der Sauerstoffbedarf von Bierhefen unter brautechnologischen Bedingungen*. Dissertation, TU München).

Schönling, J., Koetke, H., Wenning, M., & Hutzler, M. (May 10–14, 2009). Spoilage yeasts in breweries and their detection by realtime multiplex PCR. In *EBC congress*. Hamburg: EBC.

Scherer, A. (2002). *Entwicklung von PCR-Methoden zur Klassifizierung industriell genutzter Hefen*. München: TU.

Schneiderbanger, H. (April 3, 2014). *Weizenbierhefen—Charakterisierung, Aroma, Technologie. Hefe und Mikrobiologie - Forschung und Praxis*. Freising-Weihenstephan: Research Center Weihenstephan for Food and Beverage Quality, TU München.

Schneiderbanger, H., Strauß, M., Hutzler, M., & Jacob, F. (2013). Aroma profiles of selected wheat beer yeast strains. *Brauwelt International, 31*, 219–222.

Schuhegger, R., Skala, H., Maier, T., & Busch, U. (April 9–11, 2008). Identifizierung von Mikroorganismen mit MALDI TOF Massenspektrometrie in der Routineanalytik. (Stuttgart). In *Dghm fachgruppe lebensmittelmikrobiologie und -hygiene, F.L.D.V*, 10. Fachsymposium lebensmittelmikrobiologie. . (Stuttgart).

Spitaels, F., Wieme, A. D., Janssens, M., Aerts, M., Daniel, H. M. A., et al. (2014). The microbial diversity of traditional spontaneously fermented lambic beer. *PLoS One, 9*, Article e95384.

Stender, H., Kurtzman, C., Hyldig-Nielsen, J. J., Sorensen, D., Broomer, A., Oliveira, K., et al. (2001). Identification of Dekkera bruxellensis (Brettanomyces) from wine by fluorescence in situ hybridization using peptide nucleic acid probes. *Applied and Environmental Microbiology, 67*, 938–941.

Thiele, F. (2006). *Einfluss der Hefevitalität und der Gärparameter auf die Stoffwechselprodukte der Hefe und auf die Geschmacksstabilität* (Dissertation, TU München).

Tibayrenc, P., Preziosi-Belloy, L., & Ghommidh, C. (2011). On-line monitoring of dielectrical properties of yeast cells during astress-model alcoholic fermentation. *Process Biochemistry, 46*, 193–201.

Timke, M., Wang-Lieu, N. Q., Altendorf, K., & Lipski, A. (2008). Identity, beer spoiling and biofilm forming potential of yeasts from beer bottling plant associated biofilms. *Antonie van Leeuwenhoek, 93*, 151–161.

Timmins, E. M., Quain, D. E., & Goodacre, R. (1998). Differentiation of brewing yeast strains by pyrolysis mass spectrometry and Fourier transform infrared spectroscopy. *Yeast, 14*, 885–893.

Torija, M. J., Beltran, G., Novo, M., Poblet, M., Guillamon, J. M., Mas, A., et al. (2003). Effects of fermentation temperature and *Saccharomyces* species on the cell fatty acid composition and presence of volatile compounds in wine. *International Journal of Food Microbiology, 85*, 127–136.

Torija, M. J., Rozes, N., Poblet, M., Guillamon, J. M., & Mas, A. (2001). Yeast population dynamics in spontaneous fermentations: Comparison between two different wine-producing areas over a period of three years. *Antonie van Leeuwenhoek, 79*, 345–352.

Tristezza, M., Gerardi, C., Logrieco, A., & Grieco, F. (2009). An optimized protocol for the production of interdelta markers in *Saccharomyces cerevisiae* by using capillary electrophoresis. *Journal of Microbiological Methods, 78*, 286–291.

Usbeck, J. C., Kern, C. C., Vogel, R. F., & Behr, J. (2013). Optimization of experimental and modelling parameters for the differentiation of beverage spoiling yeasts by matrix-assisted-laser-desorption/ionization-time-of-flight mass spectrometry (MALDI-TOF MS) in response to varying growth conditions. *Food Microbiology, 36*, 379–387.

Usbeck, J. C., Kern, C., Vogel, R., & Behr, J. (July 28–August 1, 2012). *Fast and reliable identification and differentiation of beverage spoiling yeasts by MALDI-TOF MS*. Portland: WBC. ASBC.

Usbeck, J. C., Wilde, C., Bertrand, D., Behr, J., & Vogel, R. F. (2014). Wine yeast typing by MALDI-TOF MS. *Applied Microbiology and Biotechnology, 98*, 3737–3752.

Van Der Aa Kuhle, A., Jesperen, L., Glover, R. L., Diawara, B., & Jakobsen, M. (2001). Identification and characterization of *Saccharomyces cerevisiae* strains isolated from West African sorghum beer. *Yeast, 18*, 1069–1079.

Van Zandycke, S. M., Bertrand, D., & Powell, C. (May 6–10, 2007). The use of genetic tools to differentiate related lager yeast strains. In *EBC congress. EBC: Venice.*

Van Zandycke, S. M., Simal, O., Gualdoni, S., & Smart, K. A. (2003). Determination of yeast viability using fluorophores. *Journal of the American Society of Brewing Chemists, 61*, 15–22.

Vaughan-Martini, A., & Martini, A. (2011). Saccharomyces meyen ex rees. In C. P. Kurtzman, & J. W. Fell (Eds.), *The yeasts, a taxonomic study* (5th ed.). Amsterdam: Elsevier.

Verstrepen, K., Chambers, P. J., & Pretorius, I. S. (2006). The development of superior yeast strains for the food and beverage industries: Challenges, opportunities and potential benefits. In A. Querol, & G. H. Fleet (Eds.), *Yeasts in food and beverages*. New York: Springer Inc.

Wang, S. A., & Bai, F. Y. (2008). Saccharomyces arboricolus sp. nov., a yeast species from tree bark. *International Journal of Systematic and Evolutionary Microbiology, 58*, 510–514.

Weigert, C., Steffler, F., Kurz, T., Shellhammer, T. H., & Methner, F. J. (2009). Application of a short intracellular pH method to flow cytometry for determining *Saccharomyces cerevisiae* vitality. *Applied and Environmental Microbiology, 75*(17), 5615–5620.

Wenning, M., Seiler, H., & Scherer, S. (2002). Fourier-transform infrared microspectroscopy, a novel rapid tool for identification of yeasts. *Applied and Environmental Microbiology, 68*, 4717–4721.

Xufre, A., Albergaria, H., Inacio, J., Spencer-Martins, I., & Girio, F. (2006). Application of fluorescence in situ hybridisation (FISH) to the analysis of yeast population dynamics in winery and laboratory grape must fermentations. *International Journal of Food Microbiology, 108*, 376–384.

Part II

Spoilage bacteria and other contaminants

Chapter 6

Toxigenic fungi and mycotoxins in the barley-to-beer chain

Ajja Laitila and Tuija Sarlin
VTT Technical Research Centre of Finland Ltd., Espoo, Finland

6.1 Introduction

Contamination by toxigenic fungi of cereals used as raw materials in brewing is a great concern. Harmful fungal metabolites can cause failures during both malting and the brewing process. Furthermore, toxic metabolites may have severe adverse effects on human and animal health. Mycotoxins are toxic, low-molecular-weight natural compounds produced as secondary metabolites by various different filamentous fungi (El-Sayed, Jebur, Kang, & El-Demerdash, 2022; Moretti, Logrieco, & Susca, 2017; Rubio-Armendáriz et al., 2022). Mycotoxins are considered as climate-dependent plant-, storage- and process-associated problems (Paterson & Lima, 2010). The growth of toxigenic fungi and subsequent toxin production vary considerably from year to year and place to place, depending especially on climatic conditions.

Fig. 6.1 shows the transmission of toxigenic fungi and their metabolites in the barley-to-beer chain. Mycotoxins can enter human and animal food chains through direct or indirect contamination (Edite, da Chagas Oliveira, Erlan Feitosa, Florindo Guedes, & Rondina, 2014; Tolosa, Rodríguez-Carrasco, Ruiz, & Vila-Donat, 2021). Products can be directly contaminated with toxigenic fungi with the concomitant toxin production. More often mycotoxins enter the food and feed chain indirectly. Raw materials used in the barley-to-beer chain can be contaminated with toxigenic fungi, and even though the fungi have been eliminated in the process, the mycotoxins survive and remain in the final product. Mycotoxin present in grain dust can also be transmitted via air. Mycotoxins can be transmitted to by-products used as animal feed. Barley rootlets and brewers' spent grains are important by-products used as animal feed. When contaminated grains or by-products are fed to livestock, mycotoxins can be transferred into milk, eggs and meat and then back to human consumption.

Health hazards associated with mycotoxins in humans are rarely seen in western countries, but mycotoxins are recognised as a serious food safety issue, especially in developing countries, due to a combination of subsistence farming, poor post-harvest handling and storage and unregulated local markets (Chakraborty & Newton, 2011; Kang'ethe et al., 2017). The public health risks related to mycotoxins in beer are generally regarded as low for moderate consumers. This is mainly because malting and brewing raw materials are carefully selected and inspected for quality prior their use. However, clear risk has been identified for by-products from malting and brewing processes as well as for rejected barley batches used as animal feed. In addition to mycotoxin production, the activity of barley-associated toxigenic fungi may lead to serious quality problems (discussed in Section 6.4). Occurrence of mycotoxins in the food and feed chain is expected to increase due to climate change (Moretti, Pascale, & Logrieco, 2019; Singh et al., 2023; Zingales, Taroncher, Martino, Ruiz, & Caloni, 2022). Thus, toxigenic fungi in cereal processing will be a great concern in the future.

This chapter begins by discussing the evolution and impacts of fungi in the barley-to-beer chain. It then gives an overview of the current knowledge on toxigenic fungi and mycotoxins. Regulation and emerging mycotoxin issues, such as modified mycotoxins, will also be discussed. Finally, this chapter gives a review on preventive actions.

Brewing Microbiology. https://doi.org/10.1016/B978-0-323-99606-8.00010-9

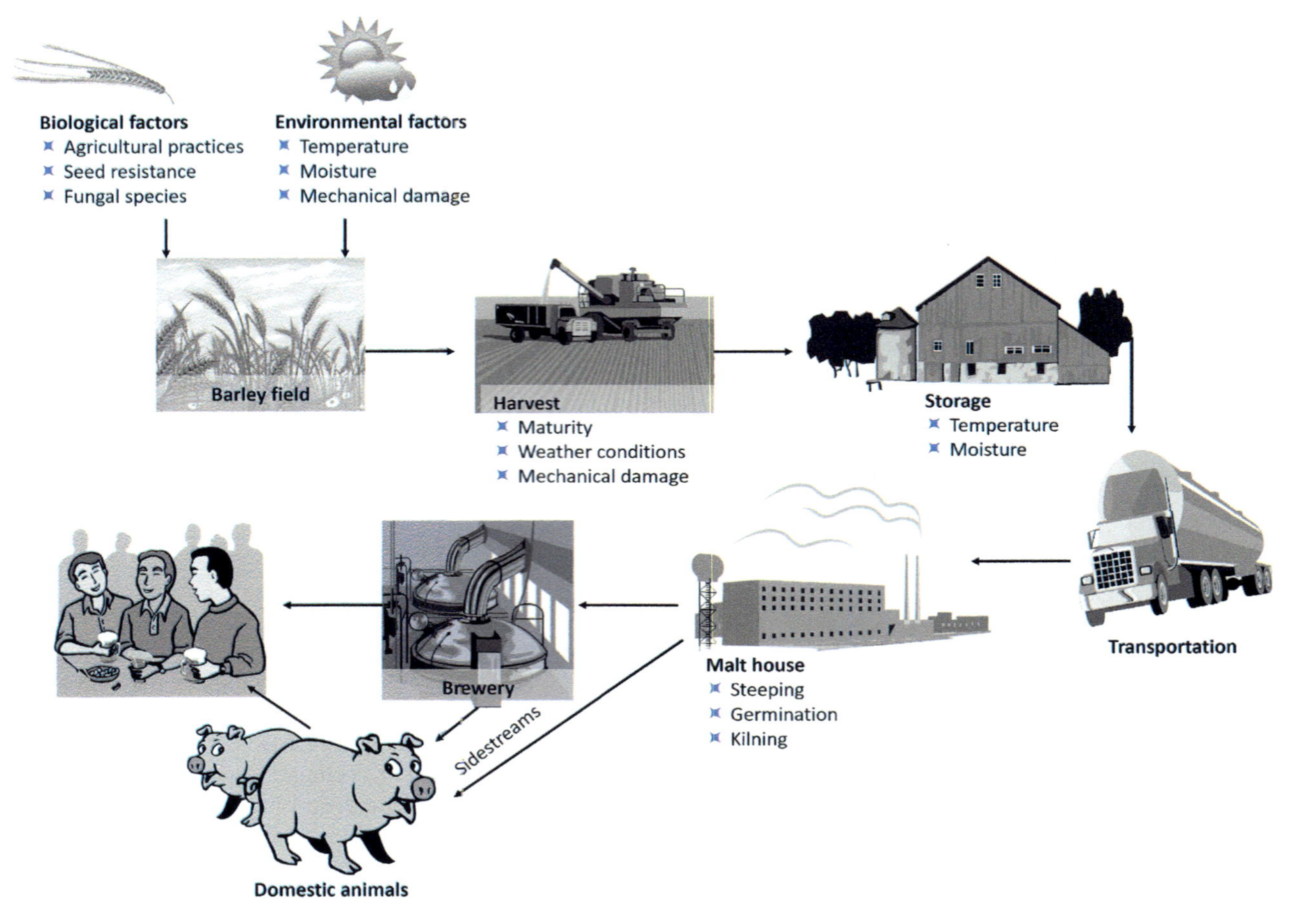

FIGURE 6.1 Transport of toxigenic fungi and their metabolites in the barley-to-beer chain.

6.2 Barley malt: A key raw material in brewing

Barley (*Hordeum vulgare* L.) is one of the most important cereals worldwide. Barley can be cultivated in highly diverse geographical regions from subarctic to subtropical. In 2020, global barley production reached 157 million tonnes (FAOSTAT, 2022). The largest value-added use for barley is the production of malt (Schwarz & Li, 2011). Malted barley provides the basis of most beers in the world. Approximately 20% of the barley produced worldwide is used by malting, brewing and distilling industries (International Grains Council (igc.int), 2022; Tricase, Amicarelli, Lamonaca, & Rana, 2018). The rest is consumed mostly as animal feed (approx. 70%) and to a lesser extent as food (approx. 5%).

Barley malt is produced by germinating grains under controlled moisture and temperature conditions. In 2020, the global malting capacity was around 30 million tons, with approximately 1/3 located in EU countries (www.euromalt.be, 2022). Malting is a natural, biological process involving a wide range of biochemical and physiological reactions. Malting traditionally involves three stages: steeping, germination and kilning. The main goal is to produce various enzymes capable of degrading the grain macromolecules into soluble compounds. The outward appearance of the final malt resembles that of the unmalted barley, but the physical, biochemical and microbiological composition is changed.

In addition to germinating grain, the malting process includes another metabolically active component: a diverse microbial community that naturally colonises the barley grains (Chen et al., 2022; Laitila, 2007; Laitila et al., 2018; Østlie, Porcellato, Kvam, & Wicklund, 2021; Raulio, Wilhelmson, Salkinoja-Salonen, & Laitila, 2009). The indigenous microbial community of barley harbours a wide range of microbes, including numerous species of bacteria, yeasts and filamentous fungi (moulds). Malting can be considered as a complex ecosystem involving two metabolically active groups: the barley grains and the diverse microbial community. The grain ecosystem is greatly influenced by the whole history experienced by the grain during the growth period, harvesting and storage. Furthermore, the behaviour of both barley and microbes during the malting process is influenced by multiple interactive factors, such as moisture, temperature, gaseous atmosphere and time.

In the brewery, malt is milled and mashed with water. In the mashing stage, malt enzymes break down the grain components into fermentable sugars and other yeast nutrients. The liquid fraction (wort) is then separated from the grain insoluble parts (spent grains). The spent grains are most often used as animal feed. After boiling with hops, cooling and aeration wort is ready for beer fermentation. In addition to barley malt, many other cereals, such as maize, millet, oats, rice, rye, sorghum and wheat, can be applied to bring additional sources of carbohydrates and proteins into wort (Cadenas, Caballero, Nimubona, & Blanco, 2021; Goode & Arent, 2006).

6.3 Evolution of fungi in the barley–malt ecosystem

The microbial community characteristics of barley products develop in the field, under storage and during processing. Many intrinsic and extrinsic factors, including plant variety, climate, soil type, agricultural practices, storage and transport, influence the diversity and structure of the microbial community present in the barley grains (Flannigan, 2003; Hoheneder et al., 2022; Noots, Delcour, & Michiels, 1999; Petters, Flannigan, & Austin, 1988; Singh et al., 2023). Of these, climate plays a particularly important role. Therefore, barleys cultivated in different geographic locations have different microbial communities. The composition of the microbial community on barley grains changes dramatically as a result of post-harvest operations. Some of the grain-associated microbes and their metabolites are removed during the processing of grains, whereas every process step in the barley-malt-beer chain can be a source of additional populations.

Fungi that contaminate barley grains may come from air, dust, soil, water, insects, rodents, birds, animals, humans, storage and transport containers and handling and processing equipment. Hundreds of different fungal species have been found on grains as surface contaminants or as internal invaders. Novel culture independent methods based on the genetic diversity enable the characterisation of the entire microbial communities in detail. Chen et al. (2022) investigated the microbiota of western Canadian barley and corresponding malt samples by high-throughput sequencing. They recovered 758 fungal operational taxonomic units (OTUs) representing 48 fungal genera from the samples.

Grain-associated filamentous fungi are mainly divided into two main rather distinct ecological groups, namely, field and storage fungi. In addition, third fungal group called intermediate fungi can be included in classification. This group comprises fungi that continue to develop in storage if water activity (a_w) remains high. This group includes species of the genera *Cladosporium*, *Fusarium* and *Trichoderma* (Mannaa & Kim, 2017).

Field fungi invade the kernels while the plant is growing or during the harvest. Among the most common and widespread field fungi are species of *Alternaria*, *Cladosporium*, *Cochliobolus*, *Epicoccum*, *Fusarium* and *Pyrenophora* (Ackermann, 1998; Chen et al., 2016; Flannigan, 2003; Kazartsev, Gagkaeva, Gavrilova, & Gannibal, 2020; Noots et al., 1999). These fungi require relatively high water availability for growth ($a_w > 0.85$, moisture content $>18\%$). Thus, their growth is restricted during storage by appropriate drying of barley. However, dormant spores can survive in normal storage conditions for years.

After harvest, barley grains are stored for about 2 months to 1 year to allow the breakup of the normal dormancy before malting. Microbes are not usually active, and their number generally decreases during storage under appropriate conditions. Microbial growth and spoilage of stored barley are determined especially by water activity and temperature (Cao, Lou, Jiang, Zhang, & Liu, 2022; Frisvad, Andersen, & Samson, 2007; Mannaa & Kim, 2017). Xerophilic *Aspergillus*, *Eurotium* and *Penicillium* are the most characteristic fungi found in the storage environment (Mannaa & Kim, 2017; Samson, Hoekstra, Frisvad, & Filtenborg, 2000). Storage fungi are able to grow on kernels of moisture content as low as 13%–15% ($a_w \sim 0.70$). Storage fungi are habitually present in the dust and air of the storage environment and can also be found in different farm and malting equipment such as harvesters and elevators (Halstensen, Nordby, Elen, & Eduard, 2004; Sauer, Meronuck, & Christensen, 1992; Yashiro, Savova-Bianchi, & Niculita-Hirzel, 2019). Thus, it is almost impossible to avoid contamination of cereals by these fungi. The best way of controlling fungal activity in stored barley is to ensure that conditions do not allow their growth. It is well known that temperature and moisture content together determines the length of safe storage.

It must be noted that this differentiation into field and storage fungi is applicable only in temperate climates, since in warmer regions, some species normally considered as storage fungi may be found already in the developing barley (Mannaa & Kim, 2017; Medina et al., 2006; Noots et al., 1999). *Aspergillus* species capable of producing mycotoxins have been detected in malting barley in southern Europe where the temperatures during ripening can be high (Medina et al., 2006; Nikolić et al. 2020).

The microbial ecology of barley changes again during malting. Before entering the malting process, barley is cleaned and graded in order to remove foreign material, dust and small and broken kernels. Cleaning procedures also diminish the microbial load. However, malting conditions are favourable for microbial growth in terms of available nutrients, temperature, moisture content and gaseous atmosphere. Steeping of barley leads to leakage of nutrients into steeping water and rapidly activates the dormant microbes present in barley grains (Laitila, 2007). Although some of the microbes and soluble nutrients are washed away along with steep water draining, the viable microbial numbers increase markedly during the steeping period (Flannigan, 2003; O'Sullivan, Walsh, O'Mahony, Fitzgerald, and van Sinderen et al., 1999). Steeping is generally regarded as the most critical step in malting with respect to microbiological safety (Laitila, 2007; Noots et al., 1999).

Microbial activity remains high throughout the germination period. Germination significantly influences both the bacterial and fungal composition in barley grain (Chen et al., 2022; Østlie et al., 2021).

Furthermore, microbial growth is accelerated during the first hours of kilning (Wilhelmson et al., 2003). The kilning regime has been identified as a significant factor in controlling microbial communities. Although high temperatures effectively restrict the growth and activity of microbes, kilning appears to have little effect on the viable counts of bacteria and fungi. The viable counts of microbes are generally higher in the finished malt than in the native barley (Noots et al., 1999). The barley bed dries progressively from the bottom to the top of the grain bed. Temperature and moisture gradients are formed in the grain bed. The conditions that prevail during the first hours of kilning before the temperature break-through, especially in the top layers of the grain bed, are rather favourable for microbial growth and activity (Laitila et al., 2006; Wilhelmson et al., 2003).

The microbial community is also significantly influenced by the malthouse operations, and it has been shown that a specific microbial community develops in each malting plant (O'Sullivan et al., 1999). Great variation in fungal communities has been observed due to the differences in malting practices in different geographic locations (Ackermann, 1998; Flannigan, 2003). The microbial community of final malt reaching the brewery is naturally influenced by the handling and storage operations after the malting process as well as during the transport of malt. In addition to barley malt, adjuncts used in breweries are potential sources for fungi and their metabolites; thus, they may contribute to the mycotoxin levels in the final product.

6.4 Impacts of barley-associated fungi on malt quality

It is evident that fungi as well as other microbes associated with barley actively interact with the grain and thus greatly influence barley, malt and beer quality and safety (Table 6.1). Deterioration or damage caused by intensive fungal proliferation during storage or processing include decrease of germination, postharvest dormancy, discolouration, off-flavours and off-odours, dry matter losses, changes in chemical and nutritional composition of grains, heating and caking of cereal lots during storage, production of gushing factors and formation of toxic metabolites (Laitila, 2007; Noots et al., 1999; Sarlin, Laitila, Pekkarinen, & Haikara, 2005; Sarlin, Nakari-Setälä, Linder, Penttilä, & Haikara, 2005; Sauer et al., 1992; Schwarz, 2017; Wolf-Hall, 2007).

Fusarium fungi are considered as perhaps the most important group of filamentous fungi with respect to malt and beer quality. Many barley-associated fusaria are plant pathogens causing devastating infections and thus lead to quality and yield reduction. Small cereal grains, such as barley, are greatly influenced by the plant disease *Fusarium* head blight (FHB), also known as scab (Karlsson, Persson, & Friberg, 2021; McMullen, Jones, & Gallenberg, 1997; Parry, Jenkinson, & McLeod, 1995; Paulitz & Steffenson, 2011). FHB is a disease complex in which several *Fusarium* and *Microdochium* species are involved. *Fusarium graminearum* is the most common causal agent of FHB, especially in the temperate and warmer regions of the USA, China and the southern hemisphere. *Fusarium culmorum* is traditionally more frequently

TABLE 6.1 Overview of reported negative impacts of filamentous fungi associated with barley and malt during storage and processing.

Quality reduction	Process failures	Health hazards
Plant diseases	Spontaneous heating of grain batch in silos	Allergens
Qualitative and quantitative changes in grain carbohydrates, proteins, lipids	Reduced grain germination	Mycotoxins
Off-odours and off-flavours	Factors inducing premature yeast flocculation (PYF)	
Discolouration of kernels	Production of gushing inducers	

found in cooler regions such as the UK, Northern Europe and Canada. A number of other species are also reported with this disease complex, especially in Europe, including *Fusarium avenaceum*, *Fusarium poae*, *Fusarium sporotrichioides*, *Fusarium tricinctum* as well as *Microdochium nivale* and *Microdochium majus* (Beccari et al., 2017; Bottalico & Perrone, 2002; Karlsson, Friberg, Kolseth, Steinberg, & Persson, 2017; Karlsson et al., 2021; Osborne & Stein, 2007). Climate change and changes in agricultural practices are altering *Fusarium* mycobiota in cereals (Parikka, Hakala, & Tiilikkala, 2012; Valverde-Bogantes et al., 2020). Toxigenic *F. graminearum* and *F. langesthiae* species have become more common in Northern Europe that poses an increased mycotoxin risk in the cereal production chain (Hietaniemi et al., 2016; Sohlberg et al., 2022). A shift in the predominant species from *F. graminearum* to *F. poae* has been observed in Italy and Canada recently (Beccari et al., 2017; Valverde-Bogantes et al., 2020).

Fusarium damaged barley cannot be processed in the malting plant. They produce a wide range of enzymes and a diverse array of secondary metabolites with a range of biological activities, including pigmentation, plant growth regulation and toxicity to other microbes, humans and animals (Brown & Proctor, 2013). The main problem associated with using *Fusarium*-infected barley malt in brewing is in the alteration of wort-soluble nitrogen compounds. This will have an impact on colour, flavour, texture and foaming properties of the beer (Sarlin, Laitila, et al., 2005; Schwarz, Horsley, Steffenson, Salas, & Barr, 2006; Wolf-Hall, 2007). Yeast fermentation failures may also be due to fungal activity in barley malt. Premature yeast flocculation (PYF) has been associated with fungal activity in barley (Van Nierop, Rautenbach, Axcell, & Cantrell, 2006). PYF is a phenomenon in which the brewing yeast prematurely settles at the bottom of the fermentation tank leading to an incomplete fermentation and undesirable beer flavour (Blechova, Havlova, & Havel, 2005; Van Nierop et al., 2006). Heavy contamination of the barley crop by fusaria or other filamentous fungi may increase the gushing (beer overfoaming) propensity of beer (Sarlin, 2012; Sarlin, Nakari-Setälä, et al., 2005). Fungi may produce gushing factors during the cultivation period in the field or during the malting process. It has been shown that small secreted fungal proteins called hydrophobins act as gushing factors (Sarlin, 2012; Sarlin, Nakari-Setälä, et al., 2005).

Fungi present in barley and malt or in grain dust are also potent sources of allergens to the workers in malthouses and breweries. Diseases such as farmer's or maltworker's lung and brewer's asthma are results of allergic responses to high concentrations of inhaled spores (Flannigan, 1986; Heaney, McCrea, Buick, & MacMahon, 1997; Rylander, 1986). Mycotoxins can also be concentrated on grain dust (Nordby et al., 2004).

Most plant-pathogenic or spoilage fungi can produce a wide range of toxic metabolites, mycotoxins, that are toxic to other microbes, plants, animals and humans. The most important fungal genera producing mycotoxins include *Aspergillus*, *Fusarium* and *Penicillium*.

6.5 *Aspergillus, Penicillium* and *Fusarium* mycotoxins

The problems associated with moulds and concomitant mycotoxin production are worldwide. The Food and Agricultural Organization (FAO, 2008) of the United Nations estimated that 25% of the world's crops are contaminated by mycotoxins each year. Recently this estimation was evaluated and confirmed to be true for the mycotoxin occurrence above the European Union legislation and Codex Alimentarius limits for food and feed (Eskola et al., 2020). However, the prevalence of detectable levels of mycotoxins seems to be much higher than this; up to 60%–80% of the global crops are estimated to contain mycotoxins (Eskola et al., 2020). Mycotoxins are fungal metabolites that cause sickness or death in people and other animals when ingested, inhaled and/or absorbed (Furian Flávia, Fighera, Royes, & Oliveira, 2022; Paterson & Lima, 2010; Rubio-Armendáriz et al., 2022). Mycotoxins include a very large, heterogeneous group of substances, and toxigenic species can be found in all major taxonomic groups. Thousands of mycotoxins exist, but only a few present significant food safety challenges (Murphy, Hendrich, Landgren, & Bryant, 2006). The relevant mycotoxins related to foods and beverages are mainly produced by species in the genera *Aspergillus*, *Penicillium* and *Fusarium* and include aflatoxins (AFs), ochratoxin A (OTA) and *Fusarium* toxins such as trichothecenes, zearalenone and fumonisins. When present in high levels, mycotoxins can have toxic effects ranging from acute (for example, kidney or liver damage) to chronic symptoms (increased cancer risk and suppressed immune system).

Production of a particular mycotoxin is a species- or strain-specific property, and usually a toxigenic fungus can produce several toxins. Furthermore, several different toxins are often present in the contaminated raw materials, and they have poorly understood synergistic effects.

In addition to health hazards, several mycotoxins have phytotoxic impacts on host plants and may cause loss of viability and reduced quality of plant seed (Nishiuchi, 2013). Several mycotoxins have antimicrobial activity and thus may also influence the behaviour of other microbes present in the same surrounding. Mycotoxins may have adverse effects on animal health if they are transmitted to side streams used as animal feed. Consumption of contaminated batches at farms

TABLE 6.2 Some mycotoxins most commonly associated with particular fungi.

Mycotoxin	Major producer fungi	Common food and beverage source
Aflatoxins B_1 (AFB_1), AFB_2 AFB_1, AFB_2 AFG_1, AFG_2	*Aspergillus flavus, Aspergillus nomius, Aspergillus paracitus*	Cereals, nuts, seeds, dried fruits, spices
Ochratoxin A (OTA)	*Aspergillus carbonarius, Aspergillus ochraceus, Penicillium nordicum, Penicillium verrucosum*	Dried fruits, cereals, grape juice, wine, coffee
A-type trichothecenes T-2 and HT-2, diacetoxyscirpenol (DAS)	*Fusarium acuminatum, Fusarium langsethiae, Fusarium poae, Fusarium sambucinum, Fusarium sporotrichoides*	Cereals and cereal products
B-type trichothecenes[a] deoxynivalenol (DON), nivalenol (NIV)	*Fusarium cerealis, Fusarium culmorum, Fusarium graminearum, Fusarium poae*	Cereals and cereal products
Zearalenone (ZEA)	*Fusarium crookwellense, Fusarium culmorum, Fusarium equiseti, Fusarium graminearum, Fusarium semitectum*	Cereals and cereal products, other food commodities
Fumonisin B_1 (FB_1)	*Fusarium verticillioides, Fusarium proliferatum, Fusarium nygamai*	Corn, corn meal, grits

[a]*Over 200 compounds are included in the trichothecenes. In addition to* Fusarium *fungi, species belonging to the genera* Myrothecium, Phomopsis, Stachybotrys, Trichoderma *and* Trichothecium *can also produce trichothecenes.*
Bennett and Klich (2003), El-Sayed et al. (2022), Frisvad et al. (2007), Murphy et al. (2006), Paterson and Lima (2010).

can lead to reduced livestock productivity and serious illness or even death (Murphy et al., 2006). Some of the most common mycotoxins associated with foods and beverages are presented in Table 6.2.

6.5.1 Aflatoxins

Afs are the most important mycotoxins worldwide (Agriopoulou, Stamatelopoulou, & Varzakas, 2020; Bennett & Klich, 2003; El-Sayed et al., 2022; Murphy et al., 2006). They are typical toxins in tropical and subtropical regions. They are mainly produced by *Aspergillus* species, particularly *Aspergillus flavus* and *Aspergillus parasiticus*. Afs occur in several chemical forms, and four compounds are commonly produced in foods: aflatoxins B_1, B_2, G_1 and G_2. Letters B and G refer to the blue or green fluorescence observed under ultraviolet light. Furthermore, aflatoxin M1, which is a metabolite of aflatoxin B1, can be found in milk and milk products. When cows consume aflatoxin-contaminated feed, they metabolise aflatoxin B_1 into a hydroxylated form called M_1. Aflatoxins are considered to be the most toxic natural compounds and are classified as proven human carcinogens (Edite Bezerra da Rocha et al., 2014). The International Agency for Research on Cancer (IARC) has classified aflatoxins as human carcinogens Class 1. They have been detected as natural contaminants in brewing materials and may pass from contaminated raw materials or adjuncts even into final beer (Mably et al., 2005; Scott, 1996). However, malting barley and malt are not the major source for these toxins if barley is properly dried and secondary contamination is restricted during storage (Benesova, Belakova, Mikulikova, & Svoboda, 2012). It must be noted that aflatoxin contamination poses an increasing risk especially in maize in the future due to the climate change (Moretti et al., 2019).

6.5.2 Ochratoxin A

OTA can be found in a large variety of products since it is produced by several fungal strains of *Aspergillus* and *Penicillium*. OTA is a derivative of isocumarin linked via peptide bond with L-phenylalanine. It is nephrotoxic, teratogenic, immunotoxic and carcinogenic (El-Sayed et al., 2022). Based on the IARC classification, OTA is considered as a possible human carcinogen (Group 2B). Cereals are the major source of human OTA exposure. Cereals contribute 50%–80% of the OTA intake among European consumers. OTA has been found in barley, oats, rye, wheat, coffee beans and other plant products, with barley having a particularly high likelihood of contamination (Anli & Mert Alkis, 2010; Bennett & Klich, 2003; El-Sayed et al., 2022; Kuruc et al. 2015). OTA is considered as a post-harvest problem and not produced in the field in Europe. Toxin production is often related to improper storage conditions. It is mainly produced by *Penicillium verrucosum* in cool and temperate zones and by *Aspergillus ochraceus* and *Aspergillus carbonarius* in warmer regions. OTA occurrence in malting barley has been associated with *P. verrucosum* (Mateo, Medina, Mateo, Mateo, & Jimenez, 2007). OTA producers require rather high water activity for growth. Rapid growth occurs at a_w 0.98–0.99 (≥27%–30% moisture content) over the temperature range 10–25°C. Growth and toxin production is almost completely inhibited at about 0.80–0.83 (=17.5%–18% m.c.) (Anli & Mert Alkis, 2010; Magan & Aldred, 2005). OTA can also be present in other adjuncts used in breweries. OTA has been detected in beers usually at low levels (Bertuzzi, Rastelli, Mulazzi, Donadini, & Pietri, 2018; Schabo, Alvarenga, Schaffner, & Magnani, 2021; Silva, Teixeiram, Pereira, Pena, & Lino, 2020).

6.5.3 *Fusarium* toxins

Production of mycotoxins is probably the most negative consequence associated with heavy contamination of barley and malt by *Fusarium* fungi.

Trichothecenes are tricyclic sesquiterpenes, and they can be classified into four major types (A-D) based on their chemical structure. More than 200 trichothecenes have been identified (Nishiuchi, 2013; Polak-Śliwińska & Paszczyk, 2021). Although a high number of molecules have been characterised, only a few of them have been characterised from barley. Types A and B are frequent contaminants in cereal grains and cereal-based products. Type A includes T-2 toxin, neosolaniol (NEO) and diacetoxyscirpenol (DAS). Type B includes fusarenon-x, nivalenol (NIV) and deoxynivalenol (DON).

Trichothecenes bind to eukaryotic ribosomes and inhibit protein synthesis (Foroud et al., 2019; Pestka, Zhou, Moon, & Chung, 2004). Different trichothecenes interfere with initiation, elongation and termination stages of protein synthesis. They are also immunosuppressive. Acute trichothecene mycotoxicosis is rare, but when ingested in high doses by farm animals, they cause nausea, vomiting and diarrhoea. DON is also called a vomitoxin or food refusal factor (Bennett & Klich, 2003). Trichothecenes are not classifiable as to their carcinogenicity to humans (Class 3) (IARC).

Deoxynivalenol (DON) is the most important trichothecene worldwide and is often detected in small cereal grains such as barley, oats and wheat. Due to relatively good thermal stability, DON can be transmitted from contaminated barley into the final product (Habler, Geissinger, et al., 2017; Schwarz, 2017). DON is frequently detected in barley and in commercial beer (discussed in Section 6.6.2). The occurrence of DON is largely dependent on weather conditions in the particular location and year. DON is predominantly produced by *F. culmorum* and *F. graminearum* species.

The mycotoxin T-2 and its deacetylated form HT-2 toxins are type a trichothecenes and perhaps the second most important *Fusarium* toxins. They are often treated as a pair when considering incidence and regulatory aspects as these closely related mycotoxins have equivalent toxicity. The major producers of T-2 and HT-2 are *F. sporotrichioides* and *F. langsethiae*. T-2 is a potent inhibitor of protein synthesis, and it is considered to be significantly more toxic than DON. The LD50 value (mg/kg for mice) for DON is 70, and for T-2, it is only 5.2. The data available show that T-2 and HT-2 levels have been an increasing problem, especially in oats (Edwards, 2009a; van der Fels-Klerx & Stratakou, 2010; Hietaniemi et al., 2016; Kiš et al., 2021). Since 2004, the occurrence of T-2 and HT-2 producers and toxin incidences appear to also be increasing in barley (Kiš et al., 2021; Strub, Pocaznoi, Lebrihi, Fournier, & Mathieu, 2010; van der Fels-Klerx & Stratakou, 2010). Studies have indicated that T-2 and HT-2 production in malting barley is rather unpredictable (Euromalt, 2013). Modelling based on the weather data can be used for prediction of DON. Currently, forecasting systems for T-2 and HT-2 production are not available. More knowledge is required for the understanding of the biological role and induction of the T-2 and HT-2 toxin synthesis.

Zearalenone (ZEA) is a nonsteroidal oestrogenic mycotoxin produced by several *Fusarium* species, including *F. graminearum*, *F. equiseti*, *F. culmorum*, *F. tricinctum* and *F. crookwellense* (Zinedine, Soriano, Molto, & Manes, 2007). Chemically, it is a phenolic resorcyclic acid lactone. The detrimental effects caused by consumption of zearalenone-contaminated grains include impaired reproduction and altered sexual development in farm animals (Murphy et al., 2006). ZEA often coexist with DON, as *F. graminearum* and *F. culmorum* may produce both compounds (Richard, 2007). Most often, this toxin is found in maize but also in other important cereal crops, including barley. ZEA was evaluated by the IARC (1993). Based on inadequate evidence in humans and limited evidence in experimental animals, it was categorised in Group 3 (not classifiable as to its carcinogenicity to humans) together with trichothecenes.

Fumonisins are produced by a number of *Fusarium* species, notably *Fusarium verticillioides* (formerly known as *Fusarium moniliforme*), *Fusarium proliferatum*, *Fusarium nygamai* and also *Alternaria alternata*. Fumonisins in brewing processes are seldom related to barley but rather maize adjuncts (Pietri, Bertuzzi, Agosti, & Donadini, 2010). *F. verticillioides* has economic importance, since it is present in almost all maize samples (Bennett & Klich, 2003). However, not all strains are toxigenic, so the presence of this fungus does not necessarily mean that toxin is formed. Fumonisins have also been detected in wheat (Cendoya, Chiotta, Zachetti, Chulze, & Ramirez, 2018). The presence of fumonisins in maize has been associated with oesophageal cancer in regions of Africa, China and Italy (Edite Bezerra da Rocha et al., 2014). IARC has categorised fumonisins as Class 2B (possible human carcinogen).

6.6 Fate of mycotoxins in the barley-to-beer chain

Toxigenic fungi and their metabolites are a natural part of the barley—malt ecosystem. Table 6.3 compiles the fate of mycotoxins in the malting and brewing process. The majority of fungal metabolites may be produced during the following steps:

1. during crop cultivation in the field,
2. while cereal awaits drying after harvest,
3. if cereal is inadequately dried or becomes damp during storage or transport and
4. during the malting process.

The various processing steps along the malting and brewing chain such as sorting, cleaning and grading, malting, roasting, milling, mashing and fermentation will influence the final mycotoxin levels (Bullerman & Bianchini, 2007). Generally mycotoxin concentrations significantly decrease in the brewing process but are not completely eliminated.

6.6.1 Mycotoxins in barley and malting

Barley naturally contains *Fusarium* mycotoxins at harvest, whereas *Aspergillus* and *Penicillium* toxins are seldom detected in good-quality malting barley (Baxter, Slaiding, & Kelly, 2001; Benesova et al., 2012; Bolechová et al., 2015; Parry et al., 1995). Mycotoxin production in the field is a complex biological process and is influenced by several different interrelated factors such as fungus type, crop resistance, cultivation practices and climatic conditions. Surveys on the presence of *Fusarium* toxins in barley have been carried out in several countries. During the past decade, the amounts of *Fusarium* toxins in malting barley have been below the EU acceptance limits (Belakova, Benesova, Caslavsky, Svoboda, & Mikulikova, 2014; Benesova et al., 2012; Dohnal et al., 2010; Drakopoulos, Sulyok, Krska, Logrieco, & Vogelgsang, 2021; Edwards, 2009b; Euromalt, 2013; 2023; Ibanez-Vea, Lizarraga, Gonzalez-Penas, & Lopez de Cerain, 2012; Ksieniewicz-Woźniak et al., 2019; van der Fels-Klerx & Stratakou, 2010). However, some single samples occasionally may exceed the legal limits. DON is the prevalent toxin in malting barley. Type-A trichothecenes such as T-2 and HT-2 are detected less frequently and at lower concentrations than DON.

TABLE 6.3 Mycotoxins in the malting and brewing process.

Process phase	Potential change in mycotoxins	Cause
Barley cleaning and grading	Reduction	Removal of infected grains, fungal spores and dust
Steeping	Reduction	Removal of water-soluble toxins
Germination	No effect/increase/decrease	Growth of toxigenic fungi suppressed Fungal growth and concomitant mycotoxin production Liberation of conjugated forms/deconjugation
Kilning	Increase/no effect	Fungi capable of mycotoxin production during early hours
Mashing	Increase	Enzymatic release of toxins from protein conjugates/mycotoxins transmitted via contaminated adjuncts
Wort boiling	No effect/decrease	Mycotoxins stable/removal with trub
Fermentation	Increase/decrease	Increase due to deconjugation/decrease due to absorption to yeast cell or bioconversion
Final beer	No effect/dilution	
Side-streams from malting and brewing	Increase	Accumulation

There has been increasing consumer interest in organic products. Bernhoft, Clasen, Kristoffersen, and Torp (2010) and Ibanez-Vea et al. (2012) found less *Fusarium* infestation and mycotoxin production in organically grown barley compared to conventional farming, obviously due to better crop rotation and soil management. Based on the studies by Edwards (2009a, 2009c), organic cultured wheat and oats appear to have lower mycotoxin contamination compared to conventional cultivation, but there were no significant differences in mycotoxin levels of organic and conventional barley samples in the UK (Edwards, 2009b). Pleadin et al. (2017) reported that mycotoxin contamination detected in Croatian organic cereals and cereal products including barley does not significantly differ from that in their conventional counterparts. Due to contradictory results, more systematic studies are needed on the farming practices and mycotoxin production.

Harvesting causes a major change in the balance in fungal ecology. The availability of water decreases and new fungal populations can spread by the machinery, field dust and crop residues (Noots et al., 1999). If the grain has been dried before storage, the risk of enhancing fungal activity is low. Baxter et al. (2001) reported that in normal conditions, OTA was undetectable in malting barley. However, slight changes in temperature and moisture parameters can lead to rapid deterioration of the barley. Fungi are unevenly distributed in silos. If the conditions at one spot allow the proliferation of xerophilic fungi their activity changes the microenvironment so that it becomes more favourable for toxigenic fungi (Noot et al., 1999; Magan & Aldred, 2005). If *P. verrucosum* spores are present at hot spots, they will proliferate and produce OTA (Anli & Mert Alkis, 2010).

The malting process has typically a significant cleaning function. The majority of the water-soluble toxins are washed out during the steeping step. Euromalt, the trade association representing the malting industry in the European Union, is carefully following the occurrence of mycotoxins in malting barley. They have been performing annual surveys of barley and malt since 2002. Samples ($\sim$100) are collected from all EU Member States with significant malt production (Figs 6.2 and 6.3; Laitila, 2015).

Several studies have revealed that the levels of type a trichothecenes decreased during malting (Euromalt, 2013, 2023; Habler et al., 2016; Malachova et al., 2010; Prusova et al., 2022). Approximately 60%−80% reduction in the sum of T-2 and HT-2 toxins was detected during the malting process (Euromalt, 2013, 2023; Laitila, 2015). In the years 2013−2021, a mean value of the sum of T-2 and HT-2 toxin was below 50 μg/kg in barley and below 10 μg/kg in malt (Fig. 6.2). Surveys also revealed that the levels of T-2 and HT-2 in winter grown barleys were lower than in spring grown varieties (Euromalt, 2013; Karlsson, Mellqvist, & Persson, 2023; Linkmeyer et al., 2016; van der Fels-Klerx & Stratakou, 2010). However, the initial differences in T-2 and HT-2 contamination of winter and spring varieties disappeared during malting, resulting in comparable levels in the final malt (Euromalt, 2013).

The impact of malting on type B trichothecenes is different. Processing barley into malt has generally little effect on the overall DON level. Both a decrease and production of DON can occur during malting. Dohnal et al. (2010) studied DON production in laboratory scale malting with 20 different barley varieties. They observed a decrease of DON levels during malting in 10% of samples and an increase in 20% of samples. A statistically significant impact of malting to DON content was not found. A similar trend was reported by Malachova et al. (2010). In the Euromalt survey, the mean levels of DON in malt were generally lower than in barley (Fig. 6.3; Laitila, 2015). However, little correlation between individual barley and malt samples was observed. Schwarz et al. (2006) reported that barley with a DON level <1.0 ppm produced acceptable malt. Substantial increase in DON content during malting of artificially inoculated barleys was detected by

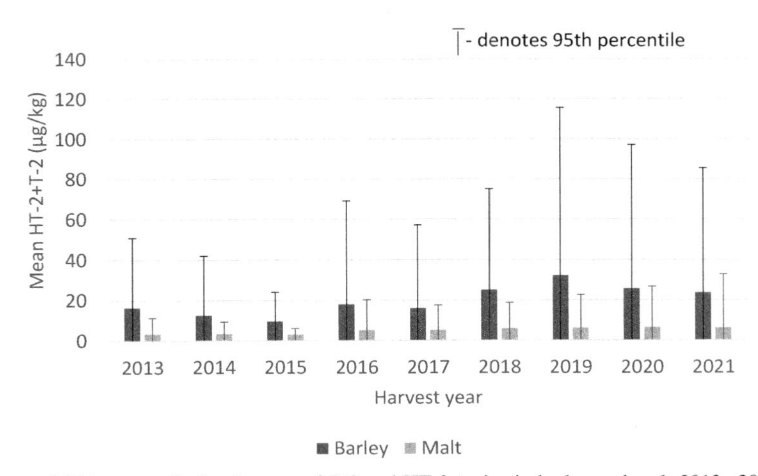

FIGURE 6.2 Mean and 95th percentile for the sum of T-2 and HT-2 toxins in barley and malt 2013−2021 (Euromalt, 2023).

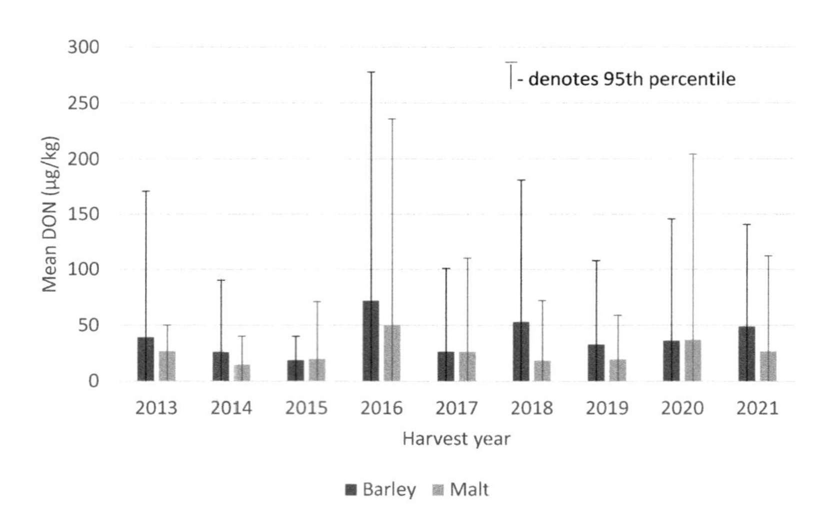

FIGURE 6.3 Mean and 95th percentile for DON in barley and malt 2013–2021 (Euromalt, 2023).

Habler et al. (2016). Recently, Jin et al. (2021) observed that extensive internal fungal colonisation in *Fusarium* head blight–infected grains contributes to the elevated DON production during the malting process. Image analysis showed that the internal *Fusarium* infection of barley typically occurred in the regions that are likely protected from removal during grain cleaning and steeping and, thus, can start to proliferate during malting. Hence, reported differences in the fate of DON during malting may be associated with the physical location of *Fusarium* spp. within kernels. Besides trichothecenes, small amounts of zearalenone have been detected in European malting barley and malt occasionally (Euromalt, 2023).

It is well known that malting conditions are favourable for fungal growth. Dormant spores present in barley are activated during the steeping. Reformation of toxins may occur with the growth of toxigenic fungi during steeping, germination and early hours of kilning (Jin et al., 2021; Prusova et al., 2022; Sarlin, Laitila, et al., 2005; Schwarz, 2017; Wolf-Hall, 2007). Stress conditions for fungi during the kilning phase can trigger the mycotoxin synthesis. It has been shown that the temperature rise can activate mycotoxin production (Prusova et al., 2022; Wolf-Hall, 2007). Thus, kilning can be regarded as an important step with regard to mycotoxin production and safety. The elevated temperatures at malt kilning are not enough to destroy mycotoxins.

6.6.2 Mycotoxins in brewing and beer

Mycotoxins are stable compounds and can therefore survive throughout the brewing process and enter the final product. Other raw materials used in beer production can be additional sources for mycotoxins. Many factors influence the concentration in the final product (raw materials, production steps in the breweries such as the mashing step, fermentation, yeast type). Thus, prediction of transmission is difficult. Several mycotoxins, like trichothecenes, are water soluble and are extracted into wort during mashing. Cantrell (2008) concluded that the majority (65%–100%) of T-2 and HT-2 toxins present in malt persist into beer. There was little or no significant losses of these toxins into by-products such as spent grains or brewers' yeast. Prusova et al. (2022) reported that the transfer of mycotoxins from mash to sweet wort and subsequently to beer corresponded to their polarity; polar mycotoxins such as NIV, HT-2 and T-2 passed almost completely into beer. Furthermore, mashing conditions and enzymatic activity may contribute to the release of mycotoxins from the grain matrix during mashing (Wolf-Hall, 2007). Mashing may increase toxin levels such as DON due to the release of additional DON from conjugates (see Section 6.8.1 for modified mycotoxins). Mycotoxins are relatively stable at temperatures used in mashing and wort boiling. For example, DON is stable at 120°C (Hazel & Patel, 2004).

Mycotoxins may cause process failures in beer production. They have been shown to disturb yeast metabolism during fermentation (Boeira, Bryce, Stewart, & Flannigan, 1999a, 1999b, 2000; Koshinsky, Crosby, & Khachhatourians, 1992; Whitehead & Flannigan, 1989). Mycotoxins may have negative impacts on the viability, biomass growth and metabolic activity of yeasts. AFB_1 and trichothecenes are known to inhibit the alcohol hydrogenase activity resulting in decreased fermentation activity and lower CO_2 liberation (Klosowski, Mikulski, Grajewski, & Blajet-Kosicka, 2010). Mycotoxin contamination of raw materials has also led to significant reduction of ethanol yield (Klosowski et al., 2010). The degree of growth inhibition has been shown to depend on the toxin concentration, yeast type, fermentation conditions and length (Boeira et al., 1999a, 1999b; Boeira, Bryce, Stewart, & Flannigan, 2002). Table 6.4 illustrates some mycotoxins and their

TABLE 6.4 Mycotoxin concentration (ppm = μg/mL) causing significant inhibitory effects on the growth and activity of brewing yeast.

Mycotoxin	Lager yeast	Ale yeast
Deoxynivalenol	100	50
Nivalenol	50	50
Fumonisin B1	10	50
T2	10	nd
Zearalenone	50	50
Diacetoxyscirpenol (DAS)	5–10	nd

nd, not determined.
Boeira et al. (1999a, 1999b, 2000), Whitehead and Flannigan (1989).

levels that have shown to cause significant adverse impacts. A combination of various toxins at low concentrations may be one of the reasons for unexplained unfinished fermentations. Synergistic effects of various toxins are still poorly understood.

Although mycotoxins have been shown to influence yeast behaviour, they do not form immediate concern for brewing. Even if highly contaminated grains were used in the brewing process, it is unlikely that any of the toxins would be found at high concentrations during fermentation (Wolf-Hall, 2007). Studies have indicated that DON has little importance in conventional brewing or distilling fermentations (Boeira et al., 2000; Kostelanska et al., 2009; Nathanail et al., 2016; Wolf-Hall, 2007). The inhibitory DON level is 10,000 times higher than that measured in the survey of commercial beers (Kostelanska et al., 2009).

The fermentation process can also decrease the amount of mycotoxins. Part of the toxins can be bound to yeast cells and thus removed from the wort. In addition, brewers' yeast also has the ability to detoxify mycotoxins by bioconversion of mycotoxins to less toxic derivatives (Halady Shetty & Jespersen, 2006; Inoue, Nagatomi, Uyama, & Mochizuki, 2013; Nathanail et al., 2016). Mizutani, Nagatomi, and Mochizuki (2011) showed that the major part of ZEA (89.5%) was converted to β-zearalenol, which has lower oestrogenic activity than ZEA.

Although part of the mycotoxins are removed or destroyed during brewing, some toxins can be transmitted into beer. Surveys on mycotoxins in beer are continuously being carried out in different countries. Detectable amounts of mycotoxins have been found in commercial beer samples regardless of the market where they have been collected. In all surveys the mycotoxin levels have been rather low. DON, fumonisin, nivalenol, T-2, HT-2, diacetoxyscirpenol, enniatins, zearalenone, aflatoxins and OTA have been detected in beers at trace (ppb) levels (Anli & Mert Alkis, 2010; Bertuzzi et al., 2018; Cantrell, 2008; Habler, Gotthardt, et al., 2017; Kostelanska et al., 2009; Ksieniewicz-Woźniak, Bryła, Waskiewicz, Yoshinari, & Szymczyk, 2019; Mably et al., 2005; Mateo et al., 2007; Schabo et al., 2021; Scott, 1996; Silva et al., 2020; Tangni, Ponchaut, Maudoux, Rozenberg, & Larondelle, 2002).

Aflatoxins are rarely found in beers. However, some warmer climates, such as South America and equatorial Asia, can have a higher incidence for AFs in raw materials and thus contamination levels in final products (Mably et al., 2005; Scott, 1996). Mably et al. (2005) carried out a survey of aflatoxins in beer sold in Canada. Both domestic and imported beers from 36 countries were included and 12 samples were positive for AFs. The highest incidence (4/5) and levels (max 230 ng/L) occurred in beers from India. Pietri et al. (2010) studied the transfer of AFB_1 and fumonisin B_1 from naturally contaminated raw materials to beer during a full-scale industrial brewing process. The content of AFB_1 in maize grit ranged from 0.31 to 14.85 μg/kg. AFB_1 was also found in malted barley at levels of 0.20–4.07 μg/kg. Approximately 0.6%–2.2% of AFB_1 was recovered from the final beer. Benesova et al. (2012) analysed AFs from different brewing materials (61 malting barley, 77 malt, 54 hop, 12 brewers' yeast and 12 spent grain) and 117 beers obtained from EU malting plants and breweries. AFs in trace concentration were found in approximately 3% of the material samples and in 5% of beer samples coming from European countries. More recently, a meta-analysis of global prevalence of mycotoxins in beers revealed a low prevalence of AFs in beer (12%), which is probably due to their slight solubility in water (Schabo et al., 2021).

OTA present in contaminated grains can be transmitted to beer, although the brewing process decreases the concentration (Baxter et al., 2001; Mateo et al., 2007; Tangni et al., 2002). Baxter et al. (2001) reported that substantial loss of OTA (up to 40%) was observed during mashing. This was obviously due to the proteolytic degradation and conversion of

OTA to non-toxic ochratoxin α. OTA structure includes a peptide bond, thus the cleavage is possible. Approximately 16% of OTA was removed along with the spent grain. The final beer contained 13%−32% of the toxin present in the contaminated malt sample. OTA has been found in beers all over the world in surveys carried out since 1998. A worldwide systematic review and meta-analysis revealed that OTA was the second most prevalent mycotoxin detected in beers (52% positive samples), and the global estimated concentration of OTA was 0.59 μg/L (95% confidence interval = 0.51−0.68 μg/L) (Schabo et al. 2021). Mateo et al. (2007) concluded that beer is not a relevant contributor to OTA exposure in human consumption. However, highly contaminated batches may occur in the beer production chain. Thus, it is important to follow and control OTA in brewery products to maintain OTA intake at the lowest achievable level (Silva et al., 2020; Tangni et al., 2002).

Fusarium toxins can be frequently found in beers. For DON, the transfer from malt in finished beer was between 80% and 93% (Schwarz, Casper, & Beattie, 1995). Accordingly, DON and its derivatives were found to be the most prevalent mycotoxins in beers globally (Schabo et al., 2021). European beers were surveyed for the presence of T-2 and HT-2 toxins during the years 2006 (including 195 samples) and 2007, (including 196 samples) from 27 countries (Cantrell, 2008). The majority (65%−100%) of T-2 and HT-2 toxins present in malt persists into beer. The mean level was 0.098 μg/L in 2006 (maximum 0.73) and 0.053 μg/L in 2007 (maximum 2.67). Schwarz et al. (1995) studied the fate of ZEA in brewing. They reported that the majority of ZEA (60%) was detected from the spent grains. ZEA was not detected in the final beer. This might be due to conversion of ZEA to zearalenol by the brewers' yeast. Moreover, Pascari, Rodriguez-Carrasco, Juan, Mañes, Marín, Ramos, et al. (2019) reported a reduction effect of wort boiling on ZEA as well as fumonisin B1and B2 levels.

Overall, it can be concluded that mycotoxins can be transmitted to beer. However, mycotoxin in commercial beers does not form a significant health risk for moderate consumers (Ibanez-Vea et al., 2012; Ksieniewicz-Woźniak et al., 2019; Varga, Malachova, Schwartz, Krska, & Berthiller, 2013). However, rather high incidences of aflatoxins (AFs), ochratoxin A (OTA) and zearalenone (ZEA) have been found in locally brewed commercial and home-brewed beers (maize−or sorghum-based) in warm climates, particularly in Africa (Mably et al., 2005; Odhav, 2002). Increased toxin levels were often due to improper storage of raw materials.

6.6.3 Mycotoxins in by-products

The potential health risks related to beer consumption are low, but mycotoxins present in grain dust or in by-products used as food ingredients or as animal feed are a concern (Mastanjević et al., 2019; Nordby et al., 2004; Wolf-Hall, 2007). High levels of DON and its derivatives have been detected in the rootlets after malting of *Fusarium* infected barleys (Habler et al., 2016). Mycotoxins present in contaminated rootlets and spent grains have caused serious mycotoxicoses in production animals (Flannigan, 2003). Cavaglieri et al. (2009) reported that potential toxin producers such as *A. flavus* (potential aflatoxin producer) and *F. verticilloides* (potential fumonisin producer) were frequently found in barley rootlets. Poor management of by-products during storage and transportation can lead to fungal growth and mycotoxin production. Several mycotoxins have been shown to accumulate in spent grains and thus form a health risk to animals (Gonzalez, Rosa, Dalcero, & Cavaglieri, 2011).

Aspergillus clavatus is considered to be one of the major causes of allergic alveolitis amongst malt workers (Flannigan, 2003). It can also produce various mycotoxins, including patulin and cytochalasin E. Lopez-Diaz and Flannigan (1997) showed that *A. clavatus* could produce these toxins during laboratory-scale malting. *A. clavatus* and toxins have been associated with mycotoxicoses in animals fed with by-products from malting houses and breweries (Lorretti et al., 2003).

6.7 Regulation of mycotoxins in Europe

Management of mycotoxin contamination in cereals is a global objective for farmers, breeders, manufacturers, regulatory agencies and the research community (Cheli, Battaglia, Gallo, & Dell'Orto, 2014). Mycotoxin regulations have been established in about 100 countries. The European Union Commission Regulation (EU) 2023/915 sets maximum levels for certain contaminants including several mycotoxins in food in order to protect consumer health (European Commission, 2023a). Beer must conform to legal limits for raw materials. Mycotoxins originating from raw materials are diluted in the brewing process. It has been estimated that the toxin levels in the final beer are decreased by one order of magnitude compared to the levels of raw material (Kostelanska et al., 2009).

Legal limits for aflatoxin B1 and for the sum of aflatoxins B1, B2, G1 and G2 in cereals and products derived from cereals are 2.0 and 4.0 μg/kg, respectively.

The legal limits for ochratoxin A in cereals in the EU are 5.0 μg/kg for raw grain and 3.0 μg/kg for processed cereal products (which includes malt). In addition, the Commission Regulation 2023/915 (European Commission, 2023a) set a separate maximum OTA level of 3 μg/kg for non-alcoholic malt beverages.

Maximum limits for *Fusarium* toxins DON and ZEA in cereals have been operated in the EU. The Commission Regulation (EU) 2024/1022 (European Commission, 2024a) amending the Regulation (EU) 2023/915 lowered the maximum DON levels in food. The current DON limits are 1000 μg/kg for unprocessed cereals except oat, durum wheat and maize, 750 μg/kg for cereals placed on the market for the final consumer and 600 mg/kg for milling products. The maximum ZEA limits for cereals other than maize are as follows: 100 mg/kg for unprocessed cereal grains and 75 μg/kg for cereals placed on the market for the final consumer. Legislative limits have not been set for nivalenol (NIV) as it is considered to be a co-contaminant of DON and as such can be controlled by controlling DON.

The maximum levels for fumonisins (sum of B_1 and B_2) are set for maize and maize products and vary from 200 to 4000 μg/kg.

The maximum levels for the sum of T-2 and HT-2 in food were set recently in EU (Commission Regulation 2024/1038, European Commission, 2024b). The sum limits of T-2 and HT-2 for unprocessed grains vary form 50 to 1250 μg/kg depending on the cereal type being 200 mg/kg for unprocessed malting barley. Corresponding sum limits for cereals placed on the market for the final consumer and for cereal products are set between 10 to 100 μg/kg.

The Panel on Contaminants in the Food Chain of the European Food Safety Authority (EFSA) (http://www.efsa.europa.eu/en/panels/contam.htm) deals with contaminants in the food chain. Scientific opinions on health risks are prepared by this panel. It is anticipated that the number of mycotoxins with regulatory status will increase in the future. Other *Fusarium* mycotoxins with possible regulatory interest in the future are fusarenone-x (an acetylated form of nivalenol), fusarin C, enniatins, beauvericin, diacetoxyscirpenol and moniliformin. More scientific opinions and risk assessments on mycotoxins can be found in the EFSA webpages (http://www.efsa.europa.eu/en/topics/topic/mycotoxins.htm).

The EU Commission has also set a regulation sampling. Regulation (EU) 2023/2782 on the sampling and analysis of mycotoxins in food entered into force the 1st of April, 2024 (European Commission, 2023b). However, a transitional period until the 1st of January, 2029 was imposed for the adaptation of analytical methods. Representative sampling of biological samples is always a great challenge due to the sporadic distribution of the target molecules. Differences in the mycotoxin contamination pressure can be seen between different regions and also inside one field or in stored barley batches. Thus, sampling is a key factor in mycotoxin control. Regardless of the mycotoxin detection method chosen, the final results will be only as good as the sample taken.

6.8 Emerging mycotoxin issues

6.8.1 Modified mycotoxins

Food or feed are not necessarily safe due to the absence or low concentrations of well-known mycotoxins, as these toxic compounds can be present in disguise as a result of plant, mammal or fungal metabolism or even food processing (Angioni, Russo, La Rocca, Pinto, & Mantovani, 2022; Berthiller, Schuhmache, Adam, & Krska, 2009; Berthiller et al., 2013; Nathanail, 2019, p. 123). Modified mycotoxins (also called masked mycotoxins) are mainly conjugation products due to detoxification mechanisms of living organisms (Angioni et al., 2022; Rychlik et al., 2014). The great challenge is that modified mycotoxins are often undetectable by conventional analytical techniques since their structure has been changed due to conjugation with sugars, amino acids or proteins. They can be either in soluble form (then called masked or preferably modified mycotoxins) or in non-extractable form attached to macromolecules (bound mycotoxins). Bound mycotoxins can be considered to be detoxified compounds as long as they are not released from the cereal matrix during processing or digestion (Berthiller et al., 2013). However, possible hydrolysis of modified mycotoxins back to their toxic forms within the mammalian gastrointestinal tract is a great concern. A study by Dall'Erta et al. (2013) demonstrated that modified forms of DON and ZEA could be deconjugated by the human colonic microbiota.

Several *Fusarium* toxins such as DON, fumonisins, fusarenon-x, fusaric acid, nivalenol, T-2, HT-2 and ZEA are prone to masking biotransformations or binding by plants (Berthiller et al., 2013; Nathanail, 2019, p. 123). In addition, other mycotoxins such as OTA and patulin have been found in conjugated forms. The major form of modified DON is deoxynivalenol-3-β-D-glucopyranoside (D3G), which is perhaps the most widely studied cereal-associated modified mycotoxin so far.

Food and beverage processing can alter mycotoxins chemically. For example, microbes used in fermentations may transform mycotoxins to less toxic compounds. So far, only a few studies have been carried out concerning the modified mycotoxins in the barley-to-beer chain. The studies have indicated that bound DON appears to be fairly common in barley

(Lancova et al., 2008; Nathanail et al., 2015; Zhou, He, & Schwarz, 2008). Zhou et al. (2008) reported that bound DON in naturally infected barley was detected in almost 40% of samples and represented an additional 6%–21% of DON determined by the standard gas chromatography method. Even higher incidence of modified DON was detected in a nationwide survey of Finnish barley, oats and wheat grains, which revealed that 81% of the samples contained deoxynivalenol-3-glucoside (Nathanail et al., 2015).

D3G has been found at levels comparable or even higher than DON in malt and beer (Lancova et al., 2008; Schwarz, 2017). Both an increase and a decrease of modified or bound mycotoxins can occur during malting and brewing (Habler et al., 2016; Kostelanska et al., 2011; Lancova et al., 2008; Pascari, Gil-Samarra, Marín, Ramos, Sanchis, 2019). The formation and detoxification mechanisms still need further studies. However, it has been suggested that barley could metabolise the *Fusarium* toxin produced during the malting process. In addition, bound mycotoxins originally present in the plant cell walls could be liberated due to enzymatic actions during processing. Maul et al. (2012) reported that the germination process has a significant impact on biotransformation of DON. Approximately 50% of DON was transformed to D3G during 5 days of germination. Substantial increases in D3G content during malting have also been reported by Habler et al. (2016), Vaclavikova et al. (2013). D3G has been detected in all types of malts, including light, caramel, Munich and wheat malt (Kostelanska et al., 2011), except roasted malt used in dark beer production. This was obviously due to thermal degradation under roasting temperatures. Kostelanska et al. (2011) reported that D3G was not detected in brewing intermediates, including spent grains, indicating that extractable mycotoxins were effectively transferred into wort. In summary, three critical steps can be identified in the barley-to-beer chain with respect to transformation of modified mycotoxins: (1) barley germination, (2) liberation of bound toxins during mashing and (3) yeast fermentation.

Traces of modified mycotoxins can be found in beers worldwide. Kostelanska et al. (2009) analysed 176 commercial beers for DON and D3G. Samples were collected from European and North American markets in 2007. Almost 74% of the samples contained the modified form of DON exceeding the detection limit (1 μg/L). The highest level of D3G was 37 μg/mL. Varga et al. (2013) analysed 374 beer samples from 38 countries for the presence of DON, D3G and 3-acetyldeoxynivalenol (3ADON). They reported that trace amounts of DON were found in 77% of beers and D3G in 93% of beers. 3ADON was not detected in beer. The majority of the samples contained DON and D3G less than 10 μg/L. The highest concentrations of both toxins (>80 μg/L) were found in the same beer sample. It was shown that stronger beers with higher ethanol concentrations contained higher DON and DON-conjugate levels (Kostelanska et al., 2009; Varga et al., 2013). This was obviously due to the use of larger wort extract volumes in strong beer production. DON and its conjugates have also been detected in non-alcoholic beers (Kostelanska et al., 2011; Varga et al., 2013). However, the non-alcoholic products have shown the lowest contamination levels (Varga et al., 2013).

The extractable conjugated or bound mycotoxins are not currently regulated by legislation (Berthiller et al., 2013; European Commission regulation [EC] 1881/2006). Further studies are required on determination of modified mycotoxins and on their stability, transformation, toxic properties and bioavailability. With respect to regulation, a possible approach in the future could be the definition of the sum of all relevant forms of each mycotoxin, including its relevant derivatives.

6.8.2 Enniatins and beauvericin

Barley-associated fungi are also responsible for the production of bioactive compounds called emerging or 'minor' mycotoxins. This group includes toxins such as enniatins (ENNs) and beauvericin (BEA). Recently, the occurrence and fate of these toxins in the barley-to-beer chain have gained attention (Hu, Gastl, Linkmeyer, Hess, & Rychlik, 2014; Prusova et al., 2022; Vaclavikova et al., 2013).

ENNs and BEA are cyclic hexadepsipeptides consisting of three D-2-hydroxycarboxylic acid and *N*-methylamino acid moieties (Jestoi, 2008). They are often found in *Fusarium*-infected cereals, including barley. ENNs appear in nature as mixtures of four main variants: enniatin A, A_1, B_1 and B_2 and the minor variants C, D, E and F. BEA is produced by a relatively wide range of *Fusarium* spp., including F. *avenaceum*, *Fusarium oxysporum*, *F. poae*, *F. proliferatum*, *Fusarium subglutinans* and *Fusarium semitectum* and the relatively closely related fungus *Beauveria bassiana*, which is a natural soil fungus (Jestoi, 2008; Logrieco, Rizzo, Ferracane, & Ritieni, 2002). Jestoi (2008) summarised that 26 *Fusarium* species are known to produce BEA. ENNs are also produced by several different *Fusarium* species (17 reported enniatins producers), including *Fusarium acuminatum*, *F. avenaceum*, *Fusarium sambucinum* and *F. tricinctum*. A review published by Jestoi (2008) gives an overview of the properties of these emerging toxins. They are of interest due to their wide range of biological activity.

The ENNs act as ionophores, and thus, they can serve as antimicrobial compounds (Uhlig, Ivanova, Petersen, & Kristensen, 2009). They may disrupt the membrane permeability of other microbes. Hiraga, Yamamotoa, Fukuda, Hamanakaba, and Oda (2005) showed that ENNs can influence functions of *Saccharomyces cerevisiae*. They were identified as inhibitors of major drug efflux pumps. Thus, they could potentially cause failures with metabolism of brewers' yeast. However, the impacts of these emerging mycotoxins on the behaviour and activity of brewers' yeast still need further investigation.

Hu et al. (2014) and Prusova et al. (2022) studied the fate of ENNs and BEA during the malting and brewing process. They showed that a part of the preformed toxins were leached out during steeping, but the major part of ENNs and BEA remained in barley, obviously due to their low water solubility. Furthermore, toxin production occurred during barley germination. A further increase in ENNs and BEA during kilning was observed by Prusova et al. (2022). Vaclavikova et al. (2013) reported that ENN levels were decreased during malting by approximately 30% of the initial content in barley. Some losses of ENNs and BEA could also be seen due to thermal degradation or biodegradation during kilning. Significant proportions of these toxins were removed along with the rootlets (reduction range 28%−59% including the losses during kilning and rootlet removal) (Hu et al., 2014). In the brewing process, these toxins were mostly retained spent grains (53% −98%) due to low water solubility (Hu et al., 2014; Prusova et al., 2022; Vaclavikova et al., 2013). Additionally, some part of the toxins remained in the wort and were discarded along with trub. Hu et al. (2014) reported that less than 0.2% of the ENNs and BEA present in barley were detected in the final beer. Prusova et al. (2022) did not detect ENNs and BEA in green or final beers. Thus, they do not form an immediate health risk for consumers. However, high concentrations of the toxins could be concentrated in rootlets and spent grains used as animal feed and should be taken into consideration. Further studies are needed on their toxicity in mammals.

6.9 Preventive actions

Toxigenic fungi and mycotoxins are present in brewing raw materials and recognised as a risk in the beer production chain. Therefore, it is highly important to set up specific procedures to assure the safety of the products.

Management of fungi and their metabolites in the entire barley-malt-beer chain relies on good agricultural practices (GAP) as well as on good malting and brewing practices. The first line of defence is always at farm level. GAP are general procedures to reduce hazards already at the farm level. The selection of barley cultivars and agricultural practices such as crop rotation, tillage practices and fungicide use will influence fungal dynamics and their mycotoxin production.

EU Regulation (No. 852/2004) requires food business operators to establish, implement and maintain permanent procedures based on Hazard Analysis and Critical Control Points (HACCP) (European Commission, 2004). It has also been implemented in the malting and brewing industry (Davies, 2006; Erzetti et al., 2009; Rush, 2006). HACCP involves identifying all points in the manufacturing process where biological, chemical and physical hazards could occur and then controlling and monitoring those risks. It also covers the cereal coproducts such as malt rootlets and spent grains of the malting and brewing process used as animal feed. Erzetti et al. (2009) reviewed how to develop HACCP programmes for mycotoxins, nitrosoamines and biogenic amines in the brewing production chain. They highlighted that small enterprises (microbreweries, micropubs, etc.) should also pay attention to safety issues, since they often use nonstandardised barley and malt lots.

Preventive actions are highly important in maintaining the quality of malting barley and in assuring safety throughout the malting and brewing process. These procedures must be implemented for both pre-harvest and post-harvest actions. Generally, the preventive actions can be divided into three categories (Wolf-Hall, 2007):

1. removal and/or separation of infected kernels (for example, cleaning, grading, peeling);
2. treatments intended to prevent mould growth and
3. decontamination or elimination of mycotoxins present.

Controlling the harmful fungi and their metabolites is extremely challenging because the procedures carried out in the barley-to-beer chain should not have negative impacts on the grain germination performance or on malt and beer properties.

6.9.1 Pre-harvest management

Management of toxigenic fungi in field conditions requires an integrated approach including proper agricultural practices to minimise the risk for fungal growth, development and use of resistant cultivars, use of fungicides and/or biocontrol to protect the susceptible host, minimising insect infestation and utilisation of weather-based risk assessment for disease forecasting.

The use of high-quality seeds and seed dressing, weed management, careful timing of harvest as well as crop rotation are well-known GAPs to reduce the incidence of toxigenic fungi and mycotoxins in cereals (Hietaniemi et al., 2016). The use of resistant varieties is the most cost-effective management strategy for toxigenic fungi. The development of resistant varieties has benefitted from the advances in next-generation sequencing technologies including whole genome sequencing, genome-wide association studies, genotyping by sequencing and RNA sequencing that have facilitated the selection of resistant breeding lines through marker-assisted selection (Fernando et al., 2021). Many quantitative trait loci (QTL) associated with moderate disease resistance against *Fusarium* head blight have been identified in wheat and barley.

Previous crops and the amount of crop residue on the soil surface are considered major factors in spreading the pathogen (Leplat, Friberg, Abid, & Steinberg, 2012; Osborne & Stein, 2007; Vogelgsang et al., 2019). For example, the use of maize as a previous crop for barley, wheat and oats may increase the risk for FHB and increased mycotoxin production. Maize is an important host for a number of *Fusarium* fungi, including *Fusarium graminearum*. The survival of fungi is enhanced with reduced tillage systems. Fungi are present in crop residues and may survive in plant debris over winter, while burial of plant residues speeds decomposition and reduces pathogen survival (Hofgaard et al. 2016; Osborne & Stein, 2007). On the other hand, reduced and no-tillage systems have been shown to increase the accumulation of organic carbon and microbial biomass in the soil surface layer, which tends to improve soil suppressiveness against toxigenic fungi (Palojärvi, Kellock, Parikka, Jauhiainen, Alakukku, 2021). Harnessing of the microbial related soil ecosystem functions is a potential natural mean to manage mycotoxin risk. More research needs to be conducted on this topic.

A lot of attention is paid to finding effective fungicide treatments against FHB. However, contradictory results have been obtained with fungicides. The effects are highly dependent on cultivar resistance, fungicide efficacy, fungicide coverage, application method, timing and the aggressiveness of the pathogen. Previous studies have reported no or only a small impact of fungicides on the contamination of cereals with *Fusarium* toxins (Henriksen & Elen, 2005; van der Fels-Klerx & Stratakou, 2010). Henriksen and Elen (2005) showed that fungicide treatment even increased the *Fusarium* infection level in spring barley when treatment had been applied to control other fungal diseases. The suppression of competing moulds in the barley ecosystem may lead to increased *Fusarium* growth. Even increased mycotoxin production has been observed when fungicides have been applied. Malachova et al. (2010) reported that a combination of two fungicide preparations led to increased NIV production and in some crop years also enhanced DON production in field trials of 12 malting barley cultivars. Fungicides primarily in the demethylation inhibitor class and one in the succinate dehydrogenase inhibitor are nowadays regarded as the most effective against FHB and DON (Crop Protection Network, 2021). A novel fungicide, pydiflumetofen co-formulated with prothioconazole, has been shown to have a good activity against FHB resulting in reduced DON contamination in wheat (Edwards, 2022). It is recommended to use fungicides as part of an integrated FHB management strategy, not as the only control method. Especially, the combination of moderate cultivar resistance and fungicide can reduce FHB substantially (Cowger, Arellano, Marshall, & Fitzgerald, 2019). Several forecasting systems have been developed to monitor the risk of FHB outbreak and to optimise the use of fungicide only when warranted (Wegulo et al., 2015; Van der Lee et al., 2018). Recently Wang, Liu, and van der Fels-Klerx (2022) developed a machine learning algorithm to predict the risk levels of one or more mycotoxins in wheat on a regional basis in Europe using crop phaenological data, weather data and satellite images as input.

Biological control of plant pathogens using antagonistic bacteria or fungi is a sustainable option suitable for both organic and conventional farming. Several biocontrol agents (BCAs) showing antagonistic activity against *Fusarium* species have been identified and tested including bacteria *Bacillus* spp., lactic acid bacteria, *Pseudomonas* spp., *Streptomyces* spp. and fungi *Aureobasidium* spp., *Cryptococcus* spp., *Clonostachys rosea*, *Phoma glomerata*, *Pythium oligandrum* and *Trichoderma* spp. (Byrne, Thapa, Doohan, & Burke, 2022; Comby et al., 2017; Palazzini et al., 2018; Pellan et al., 2020; Wegulo et al., 2015). The application of atoxigenic strain of *Aspergillus flavus* has been shown to displace aflatoxin-producers in the fields resulting in reduced aflatoxin contamination (Agbetiameh et al., 2019; Mauro, Garcia-Cela, Pietri, Cotty, & Battilani, 2018). Some microbes have been shown to have DON detoxification ability; they can metabolise DON into less toxic or nontoxic compounds (Tian et al., 2022). Application of BCAs is a potential mycotoxin mitigation mean as a part of integrated pest management strategy. Currently only few commercial BCA products applicable in cereal production exist (Ouadhene, Ortega-Beltran, Sanna, Cotty, & Battilani, 2023; Pellan et al., 2020).

6.9.2 Post-harvest management

Storage of cereal lots on the farm level or on manufacturing silos can be regarded as the most important postharvest phase with respect to mycotoxin production. Water is the most important single factor limiting microbial growth. Barley should be dried immediately after harvest at least to a moisture content $<14\%$ if it is stored for any period of time and to $<12.5\%$ to exclude the fungal growth during storage (Flannigan, 2003; Hietaniemi et al., 2016). During storage, the barley moisture content is in equilibrium with the moisture content of the air. Therefore, grains may be further dried or they can absorb water from the surrounding air during storage. The storage life of stored grains is increased by cooling. Barley and malt should always be stored in a dry and cool environment to avoid the potential risks associated with fungal growth and possible mycotoxin accumulation. Furthermore, empty silos should be cleaned to remove the grain residues and occasionally fumigated in order to eliminate the contaminants. Pest control is highly important since they are vectors for toxigenic fungi. Metabolic activity of insects and mites increases the moisture content and temperature of contaminated grains and thus creates conditions favourable for fungal growth. Sanitation of empty malting vessels and air-conditioning systems is also carried out in malting houses in order to avoid harmful process contaminants. Fumigation of stored grains with ozone, peracetic acid and allyl isothiocyanate, a bioactive compound characteristic of the plants of the Brassicaceae family, have been shown to reduce the amounts of insects and toxigenic fungi as well as mycotoxin contamination (Christ, Savi & Scussel, 2016; Luz et al., 2021; McDonough et al., 2011; Quiles et al. 2019). Application of the fumigation treatment on an industrial scale needs further studies.

The best preventive method is to avoid highly contaminated material in malt and beer production. Heavily infected barley lots are then discarded before entering the malting process. Cleaning and grading of grains during the harvest and prior processing are crucial steps and significantly reduce fungal contamination and also mycotoxins. Grain deterioration due to plant-pathogenic fungi often leads to poor kernel size and highly contaminated grains can be removed by sorting. By rejection of the smallest sized kernels (<2.5 mm), a significantly reduced level of *Fusarium*-contaminated grains and mycotoxins can be obtained (Brodal, Aamot, Almvik, & Hofgaard, 2020; Perkowski, 1998). Lancova et al. (2008) reported that cleaning of kernels reduced DON content by 30%–50%. However, there are differences between the removal of mycotoxins during cleaning and grading procedures. Although cleaning significantly influences *Fusarium* toxins, only 2%–3% reduction of OTA in barley/wheat was obtained by cleaning (Scudamore, 2005). Typical grain cleaning steps include separation based on size and density difference, optical sorting of discoloured or defective grain kernels and if applicable debranning/dehulling as well as washing. Fluorescence and near-infrared technologies have been applied for more advanced grain sorting devices (Kautzman, Wickstrom, & Scott, 2015; Slettengren et al., 2018).

6.9.3 Mitigation measures during processing

The treatments carried out pre- or post-harvest on barley should not significantly influence the seed vigour. Various common practices are routinely applied to reduce adverse effects of fungi and other microbes during malting, especially during the steeping phase, such as changing the steeping water in order to remove microbes and leached nutrients, balancing the temperature or modifying aeration. Furthermore, steep water must be warm enough to allow rapid water uptake and germination of the grains but cool enough to avoid extensive microbial growth. Therefore, steeping is normally carried out at 10–20°C. It is also important to provide sufficient aeration and to pulse the circulation throughout the immersion period in order to keep the grains moving and to avoid anaerobic, hot pockets in the grain bed which would lead to increased microbial activity and poor grain germination (Davies, 2006).

It has been estimated that approximately 70%–100% of the toxins present in grains can be removed during steeping (Sarlin, Laitila, et al., 2005; Schwarz et al., 1995; van der Fels-Klerx & Stratakou, 2010). The following factors influence the removal of toxins: (1) amount of water, (2) temperature, (3) number of steeps, (4) duration of steeping periods and (5) extent of mixing during steeping (van der Fels-Klerx & Stratakou, 2010).

It has been shown that a variety of chemical agents such as acids, bases and oxidising reagents could be applied to intensify the washing effect during steeping. Papadopoulou, Wheaton, and Muller (2000) suggested that fungal proliferation could be restricted by adding hop beta-acids in the malting process. Moreover, they demonstrated that the growth of fungi was inhibited by washing barley first with sodium hypochlorite (alkaline wash) followed by an acid wash with hypochloric acid. Lake, Browers, and Yin (2007) reported that soaking of barley in sodium bisulphite (10 g/L) enhanced the removal of DON during steeping without influencing grain germination. Most studies have been carried out in a laboratory or pilot scale, and the feasibility and safety of treatments in large scale remain to be confirmed. Although different additives may effectively improve processing, their use in industrial processes is often limited by legislation. Furthermore, the industry has a strong emphasis towards natural processing without chemicals. Ozonation is one of the potential technologies to reduce fungal growth and mycotoxins during malting since it would not leave residual chemicals (Kottapalli, Wolf-Hall, & Schwarz; Tiwari et al., 2010; Zuluaga-Calderón,González, Alzamora, & Coronel, 2022). Ozone has been reported to be effective in inactivation of the toxigenic fungi as well as in detoxification and degradation of mycotoxins such as aflatoxin, OTA, DON and ZEA (Li, Guan, & Bian, 2019; Tiwari et al., 2010). Kottapalli et al. (2005) concluded that gaseous ozone and hydrogen peroxide are potential means for reducing *Fusarium* survival during malting. Residues and undesirable reaction products in germinating barley and in the subsequent malt are of concern especially with chemical treatments, since they may have a negative impact on malt properties and yeast fermentation performance

(Laitila, 2007). Furthermore, precautions must be taken as some of the antimicrobial treatments in sublethal doses may stimulate the production of harmful metabolites such as gushing factors and mycotoxins. Barley- and malt-derived microbes, especially lactic acid bacteria and certain fungi, offer a potential alternative as natural, food-grade biocontrol agents (Dalie, Deschamps & Richard-Forget, 2010; Laitila, 2007; Lowe & Arendt, 2004; Rouse & van Sinderen, 2008). Biocontrol candidates isolated from the brewing raw materials will most likely persist in the habitat from which they have been isolated. LAB and fungi with antagonistic properties have been shown to restrict fungal growth and prevent mycotoxin formation during malting (Boivin & Malanda, 1997; Laitila, 2007; Laitila, Alakomi, Raaska, Mattila-Sandholm, & Haikara, 2002; Laitila, Tapani, & Haikara, 1997; Postulkova et al., 2018). Natural biocontrol agents are attractive as they have a better public image, and they could potentially be used as starter cultures in bioprocesses in which the use of chemicals is considered undesirable. Starter technology, in which well-characterised microbes are added to the barley during processing, has been introduced into the malting and brewing industry.

Biodegradation of mycotoxins has become an area of great interest. Biological detoxification involves the enzymatic degradation or transformation of toxins to less toxic compounds and is often a detoxification or resistance mechanism used by microbes or plants for protection from adverse impacts of toxins. It has been shown that many bacterial and fungal species including *Bacillus* spp., *Pseudomonas* spp., *S. cerevisiae* and lactic acid bacteria are potential candidates for mycotoxin decontamination (Halady Shetty & Jespersen, 2006; Liu et al., 2022).

Biocontrol combined with other physical and chemical preventive actions along the barley-to-beer chain could result in a successful strategy for controlling toxigenic fungi. Several new potential techniques have been studied in the laboratory or pilot scale and transfer of technologies into the industrial scale requires further studies.

6.9.4 Sampling and on-site mycotoxin detection methods

Representative sampling is the most critical step in mycotoxin analysis. Mycotoxins can be heterogeneously distributed throughout grain lots, which makes them reliable determination challenging. European Commission Regulation for sampling (EU) 2023/2782 (European Commission, 2023b) lays down the methods of sampling and analysis for the official control of the levels of mycotoxins in foodstuffs. Codex Alimentarius Commission (2019, pp. 193−1995) has published recommendations of sampling plans for various mycotoxins and commodities. The online Mycotoxin Sampling Tool of Food and Agriculture Organization of United Nations (FAO) provides support in analysing performance of sampling plans and determining the most appropriate plan to meet user's defined objectives (http://tools.fstools.org/mycotoxins/). Representative sampling procedures need workforce and time. Innovative solutions like dust sampling have been proposed to overcome laborious sampling procedures (Ciasca et al., 2022; Reichel, Mänz, & Biselli, 2018; Reichel, Staiger, & Biselli, 2014). Good correlation of mycotoxin levels between corresponding dust and bulk-ware samples has been shown for several *Fusarium* toxins, aflatoxins and OTA in cereals. Dust samples can directly be extracted and analysed without additional milling or homogenisation step. However, further studies are needed to standardise the dust sampling approach.

Developed simplified and rapid test methods enable on-site screening of mycotoxins. Antibody-based assays are the most widely used methods (Lattanzio et al., 2019). Besides classical enzyme linked immune sorbent assays (ELISAs), user-friendly lateral flow devices (LFD) have appeared on the market. Some LFD kits are equipped with a specific application (app) to scan the test strip with a smartphone. Molecularly imprinted polymers (MIPs) and oligonucleotide aptamers are widely investigated alternatives to antibodies with improved performance characteristics (Tittlemier et al., 2020). Liquid chromatography-mass spectrometry (LC-MS) techniques are regarded as successful solution for simultaneous screening of a large number of mycotoxins including their transformation products. Development of rapid methods and biosensors for simultaneous determination of multiple mycotoxins is a fast-growing research field. Infrared spectroscopy methods have been studied as a non-destructive fast approach for mycotoxin screening, but these tools need further improvement in respect of detection accuracy and sensitivity (Levasseur-Garcia, 2018).

6.10 Future trends

It is clear that contamination of brewing raw materials with toxigenic fungi cannot be completely avoided, especially in crop years when bad weather conditions favour the growth of gushing active and toxigenic species in large barley production areas. Recognising, understanding and management of toxigenic fungi and mycotoxin production require close cooperation and communication between different sectors along the food, beverage and feed production chain. Control of toxigenic fungi in changing climatic conditions together with changing agricultural practices is a global future challenge. Occurrence of mycotoxins is expected to increase due to climate change. Climate represents the key agro-ecosystem driving force of fungal colonisation and mycotoxin production. Climate change will have direct effects on the fungal host interactions. Mycotoxin production is also influenced by non-infectious factors in the field, e.g., plant stress, bioavailability of (micro) nutrients, insect damage and other pest attacks, which are in turn driven by climatic conditions. In addition, indirect effects will be due to the changes in agricultural crop production systems. Changes in farming practices and new crop varieties in different cereal production areas are expected due to climate change. The need for sharing information and practices related to food and feed safety issues has been globally recognised.

Currently, reduction of water and energy consumption in all industrial processes is an economical and environmental challenge. Simultaneously, the environment is becoming more favourable for toxigenic fungi. Food safety issues should be taken into account when changes are made in the malting and brewing industry.

According to Codex Alimentarium (2003), complete elimination of mycotoxins in cereals and cereal-based products may not be achievable, but reduction of toxins in every step along the ecosystem functions is needed to control the growth of fungi and removal of mycotoxins at the pre-harvest level as well as during processing. Novel molecular biological tools will speed up the breeding process of resistant cultivars. In addition, these tools enable research to investigate the interactions between growing plants and the whole microbiome around them in the field that enhances the exploitation potential of natural ecosystem functions to suppress the growth of toxigenic fungi and mycotoxin production. Utilisation of the novel ICT systems for mycotoxin regional prediction will facilitate the timely use of appropriate mitigation means in the fields. Further development of fast and reliable screening approaches is required for multitoxin detection, since several mycotoxins and their derivatives can co-occur in the cereal production chain. The multitarget control strategies in combination with novel monitoring tools will open up new possibilities for ensuring safety along the barley-to-beer chain.

6.11 Sources of further information and advice

Food and Agriculture Organization of the United Nations provides a lot of information and relevant links related to food safety and quality, including mycotoxins. Available from: http://www.fao.org/food/food-safety-quality/home-page/en/.

EFSA collects and evaluates occurrence data on mycotoxins in food and feed. It provides scientific advice and risk assessments on mycotoxins for EU risk managers to help them assess the need for regulatory measures as regards to the safety of mycotoxin-contaminated food and feed. Further information can be found on the EFSA homepages: http://www.efsa.europa.eu/en/topics/topic/mycotoxins.htm.

References

Ackermann, A. (1998). Mycoflora of South African barley and malt. *Journal of the American Society of Brewing Chemists, 56*, 169–176.

Agbetiameh, D., Ortega-Beltran, A., Awuah, R. T., Atehnkeng, J., Islam, M.-S., Callicott, K. A., et al. (2019). Potential of atoxigenic Aspergillus flavus vegetative compatibility groups associated with maize and groundnut in Ghana as biocontrol agents for aflatoxin management. *Frontiers in Microbiology, 10*, 2069.

Agriopoulou, S., Stamatelopoulou, E., & Varzakas, T. (2020). Advances in occurrence, importance, and mycotoxin control strategies: Prevention and detoxification in foods. *Foods, 9*, 137.

Alimentarium, C. (2003). *Code of practice for the prevention and reduction of mycotoxin contamination in cereals, including annexes on ochratoxin a, zearalenone, fumonisins and trichothecenes*. CAC/RCP 51-2003 13.

Angioni, A., Russo, M., La Rocca, C., Pinto, O., & Mantovani, A. (2022). Modified mycotoxins, a still unresolved issue. *Chemistry, 4*, 1498–1514.

Anli, E., & Mert Alkis, I. (2010). Ochratoxin and brewing technology: A review. *Journal of the Institute of Brewing, 116*, 23–32.

Baxter, E., Slaiding, I., & Kelly, B. (2001). Behaviour of ochratoxin a in brewing. *Journal of the American Society of Brewing Chemists, 59*, 98–100.

Beccari, G., Prodi, A., Tini, F., Bonciarelli, U., Onofri, A., Oueslati, S., Limayma, M., & Covarelli, L. (2017). Changes in the Fusarium head blight complex of malting barley in a three-year field experiment in Italy. *Toxins, 9*, 120.

Belakova, S., Benesova, K., Caslavsky, J., Svoboda, Z., & Mikulikova, R. (2014). The occurrence of selected *Fusarium* toxins in Czech malting barley. *Food Control, 37*, 93–98.

Benesova, K., Belakova, S., Mikulikova, R., & Svoboda, Z. (2012). Monitoring of selected aflatoxins in brewing materials and beer by liquid chromatography/mass spectrometry. *Food Control, 25*, 626–630.

Bennett, J. W., & Klich, M. (2003). Mycotoxins. *Clinical Microbiology Reviews, 16*, 497–516.

Bernhoft, A., Clasen, P.-E., Kristoffersen, A., & Torp, M. (2010). Less *Fusarium* infestation and mycotoxin contamination in organic than in conventional cereals. *Food Additives & Contaminants, 27*, 842–852.

Berthiller, F., Crews, C., Dall'Asta, C., De Saeger, S., Haesaert, G., Karlovsky, P., et al. (2013). Masked mycotoxins: A review. *Molecular Nutrition & Food Research, 57*, 165–186.

Berthiller, F., Schuhmacher, R., Adam, G., & Krska, R. (2009). Formation, determination and significance of masked and other conjugated mycotoxins. *Analytical and Bioanalytical Chemistry, 395*, 1243–1252.

Bertuzzi, T., Rastelli, S., Mulazzi, A., Donadini, G., & Pietri, A. (2018). Known and emerging mycotoxins in small- and large-scale brewed beer. *Beverages, 4*, 46. https://doi.org/10.3390/beverages4020046

Blechova, P., Havlova, P., & Havel, J. (2005). The study of premature yeast flocculation and its relationship with gushing of beer. *Monatsschrift für Brauwissenschaft*, 56–70.

Boeira, L., Bryce, J., Stewart, G., & Flannigan, B. (1999a). Inhibitory effect of *Fusarium* mycotoxins on growth of brewing yeasts. 1 Zearalenone and Fumonisin B1. *Journal of the Institute of Brewing, 105*, 366–374.

Boeira, L., Bryce, J., Stewart, G., & Flannigan, B. (1999b). Inhibitory effect of *Fusarium* mycotoxins on growth of brewing yeasts. 2 Deoxynivalenol and nivalenol. *Journal of the Institute of Brewing, 105*, 376–381.

Boeira, L., Bryce, J., Stewart, G., & Flannigan, B. (2000). The effect of combinations of *Fusarium* mycotoxins (deoxynivalenol, zearalenone and fumonisin B1) on growth of brewing yeast. *Journal of Applied Microbiology, 88*, 388–403.

Boeira, L., Bryce, J., Stewart, G., & Flannigan, B. (2002). Influence of cultural conditions on sensitivity of brewing yeasts growth to *Fusarium* mycotoxins zearalenone, deoxynivalenol and fumonisin B1. *International Biodeterioration & Biodegradation, 50*, 69–81.

Boivin, P., & Malanda, M. (1997). Improvement of malt quality and safety by adding starter culture during the malting process. *Technical Quarterly - Master Brewers Association of the Americas, 34*, 96–101.

Bolechová, M., Benesová, K., Beláková, S., Cáslavský, J., Pospíchalová, M., & Mikulíková, R. (2015). Determination of seventeen mycotoxins in barley and malt in the Czech Republic. *Food Control, 47*, 108–113.

Bottalico, A., & Perrone, G. (2002). Toxigenic *Fusarium* species and mycotoxins associated with head blight in small grain. *European Journal of Plant Pathology, 108*, 611–624.

Brodal, G., Aamot, H. U., Almvik, M., & Hofgaard, I. S. (2020). Removal of small kernels reduces the content of Fusarium mycotoxins in oat grain. *Toxins, 12*(5), 346.

Brown, D., & Proctor, R. (2013). Diversity of polyketide synthases in *Fusarium*. In D. Brown, & R. Proctor (Eds.), *Fusarium, genomics, molecular and cellular biology* (pp. 143–163). Norfolk: Caister Academic Press.

Bullerman, L., & Bianchini, A. (2007). Stability of mycotoxins during food processing. *International Journal of Food Microbiology, 119*, 140–146.

Byrne, M. B., Thapa, G., Doohan, F. M., & Burke, J. I. (2022). Lactic acid bacteria as potential biocontrol agents for Fusarium head blight disease of spring barley. *Frontiers in Microbiology, 13*, Article 912632.

Cadenas, R., Caballero, I., Nimubona, D., & Blanco, C. A. (2021). Brewing with starchy adjuncts: Its influence on the sensory and nutritional properties of beer. *Foods, 10*, 1726. https://doi.org/10.3390/foods10081726

Cantrell, I. (2008). EBC/the brewers of Europe survey of *Fusarium* toxins in European beers. In *The 4th* Fusarium-*toxin forum*. held 10–11 January in Brussels.

Cao, D., Lou, Y., Jiang, X., Zhang, D., & Liu, J. (2022). Fungal diversity in barley under different storage conditions. *Frontiers in Microbiology, 13*, Article 895975.

Cavaglieri, L., Keller, K., Pereyra, C., Gonzalez, P. M., Alonso, V., Rojo, F., et al. (2009). Fungi and natural incidence of selected mycotoxins in barley rootlets. *Journal of Stored Products Research, 45*, 147–150.

Cendoya, E., Chiotta, M. L., Zachetti, V., Chulze, S. N., & Ramirez, M. L. (2018). Fumonisins and fumonisin-producing Fusarium occurrence in wheat and wheat by products: A review. *Journal of Cereal Science, 80*, 158–166.

Chakraborty, S., & Newton, A. (2011). Climate change, plant diseases and food security: An overview. *Plant Pathology, 60*, 2–14.

Cheli, F., Battaglia, D., Gallo, R., & Dell'Orto, V. (2014). EU legislation on cereal safety: An update with a focus on mycotoxins. *Food Control, 37*, 315–325.

Chen, W., Cheung, H. Y. K., McMillan, M., Turkington, T. K., Izydorczyk, M. S., & Gräfenhan, T. (2022). The dynamics of indigenous epiphytic bacterial and fungal communities of barley grains through the commercial malting process in Western Canada. *Current Research in Food Science, 5*, 1352–1364.

Chen, W., Turkington, T. K., Lévesque, C. A., Bamforth, J. M., Patrick, S. K., Lewis, C. T., et al. (2016). Geography and agronomical practices drive diversification of the epiphytic mycoflora associated with barley and its malt end product in western Canada. *Agriculture, Ecosystems & Environment, 226*, 43–55.

Christ, D., Savi, G. D., & Scussel, V. M. (2016). Effectiveness of ozone gas in raw and processed food for fungi and mycotoxin decontamination - a review. *Journal of Chemical, Biological and Physical Sciences, 6*(2), 326–348.

Ciasca, B., De Saeger, S., De Boevre, M., Reichel, M., Pascale, M., Logrieco, A. F., et al. (2022). Mycotoxin analysis of grain via dust sampling: Review, recent advances and the way forward: The contribution of the MycoKey project. *Toxins, 14*(6), 381.

Codex Alimentarius Commission. (2019). *Codex general standard for contaminants and toxins in food and feed*. General standard.

Comby, M., Gacoin, M., Robineau, M., Rabenoelina, F., Ptas, S., Dupont, J., et al. (2017). Screening of wheat endophytes as biological control agents against Fusarium head blight using two different in vitro tests. *Microbiological Research, 202*, 11–20.

Cowger, C., Arellano, C., Marshall, D., & Fitzgerald, F. (2019). Managing Fusarium head blight in winter barley with cultivar resistance and fungicide. *Plant Disease, 103*(8), 1858–1864.

Crop Protection Network. (2021). *An overview of Fusarium head blight*. https://doi.org/10.31274/cpn-20211109-0. Published: 11/03/2021 https://cropprotectionnetwork.org/publications/an-overview-of-fusarium-head-blight.

Dalie, D., Deschamps, A., & Richard-Forget, F. (2010). Lactic acid bacteria — potential control of mould growth and mycotoxins: A review. *Food Control, 21*, 370–380.

Dall'Erta, A., Cirlini, M., Dall'Asta, M., Del Rio, D., Galaverna, G., & Dall'Asta, C. (2013). Masked mycotoxins are effectively hydrolysed by human colonic micorbiota releasing their aglycones. *Chemical Research in Toxicology, 26*, 305–312.

Davies, N. (2006). Malt and malt products. In C. Bamforth (Ed.), *Brewing, new technologies* (pp. 68–101). Cambridge, UK: Woodhead Publishing Limited and CRC Press LLC.

Dohnal, V., Jezkova, A., Pavlikova, L., Musilek, K., Jun, D., & Kuca, K. (2010). Fluctuation in the ergosterol and deoxynivalenol content in barley and malt during malting process. *Analytical and Bioanalytical Chemistry, 397*, 109–114.

Drakopoulos, D., Sulyok, M., Krska, R., Logrieco, A. F., & Vogelgsang, S. (2021). Raised concerns about the safety of barley grains and straw: A Swiss survey reveals a high diversity of mycotoxins and other fungal metabolites. *Food Control, 125*, Article 107919.

Edite, B.da R. M., da Chagas Oliveira, F., Erlan Feitosa, M. F., Florindo Guedes, M., & Rondina. (2014). Mycotoxins and their effects on human and animal health. *Food Control, 36*, 159–165.

Edwards, S. (2009a). *Fusarium* mycotoxin content of UK organic and conventional oat. *Food Additives & Contaminants, 26*, 1063–1069.

Edwards, S. (2009b). *Fusarium* mycotoxin content of UK organic and conventional barley. *Food Additives & Contaminants, 26*, 1185–1190.

Edwards, S. (2009c). *Fusarium* mycotoxin content of UK organic and conventional wheat. *Food Additives & Contaminants, 26*, 496–506.

Edwards, S. G. (2022). Pydiflumetofen co-formulated with prothioconazole: A novel fungicide for Fusarium head blight and deoxynivalenol control. *Toxins, 14*(1), 34.

El-Sayed, R. A., Jebur, A. B., Kang, W., & El-Demerdash, F. M. (2022). An overview on the major mycotoxins in food products: Characteristics, toxicity, and analysis. *Journal of Future Foods, 2–2*, 91–102.

Erzetti, M., Marconi, O., Bravi, E., Perretti, G., Montanari, L., & Fantozzi, P. (2009). HACCP in the malting and brewing production chain: Mycotoxin, nitrosamine and biogenic amine risks. *Italian Journal of Food Science, 21*, 211–230.

Eskola, M., Kos, G., Elliott, C. T., Hajšlová, J., Mayar, S., & Krska, R. (2020). Worldwide contamination of food-crops with mycotoxins: Validity of the widely cited 'FAO estimate' of 25%. *Critical Reviews in Food Science and Nutrition, 60*.

Euromalt. (2013). T-2 and HT-2 toxins in barley and malt. In *EU Commission's mycotoxins forum*. held 5th September in Brussels.

Euromalt. (2023). *Data provided by Euromalt 14.4.2023*. Unpublished data.

European Commission. (2004). Commission regulation (EC) 852/2004 of April 29, 2004 on the hygiene of foodstuffs. *Official Journal of the European Union*.

European Commission. (2023a). Commission regulation (EU) 2023/915 of April 25, 2023 on maximum levels for certain contaminants in food and repealing regulation (EC) No. 1881/2006. *Official Journal of the European Union*.

European Commission. (2023b). Commission implementing regulation (EU) 2023/2782 of December 14, 2023 laying down the methods of sampling and analysis for the control of the levels of mycotoxins in food and repealing regulation (EC) No 401/2006. *Official Journal of the European Union*.

European Commission. (2024a). Commission regulation (EU) 2024/1022 of April 8, 2024 amending regulation (EU) 2023/915 as regards maximum levels of deoxynivalenol in food. *Official Journal of the European Union*.

European Commission. (2024b). Commission regulation (EU) 2024/1038 of April 9, 2024 amending regulation (EU) 2023/915 as regards maximum levels of T-2 and HT-2 toxins in food. *Official Journal of the European Union*.

FAO. (2008). Food and agriculture organisation of the United Nations, climate change: Implications for food safety, 49 Food and Agriculture Organization of the United Nations: Rome 49 http://www.fao.org/docrep/010/i0195e/i0195e00.htm

FAOSTAT. (2022). *World barley production in 2020*. UN Food and Agriculture Organization Corporate Statistical Database. https://www.fao.org/faostat/en/#data/QCL. (Accessed 15 November 2022).

Fernando, W. G. D., Oghenekaro, A. O., Tucker, J. R., & Badea, A. (2021). Building on a foundation: advances in epidemiology, resistance breeding, and forecasting research for reducing the impact of Fusarium head blight in wheat and barley. *Canadian Journal of Plant Pathology, 43*(4), 495–526.

Flannigan, B. (1986). *Aspergillus* clavatus – an allergenic, toxigenic deteriogen of cereals and cereal products. *International Biodeterioration & Biodegradation, 22*, 79–89.

Flannigan, B. (2003). The microbiota of barley and malt. In F. Priest, & I. Campell (Eds.), *Brewing microbiology* (3rd ed., pp. 113–180). New York: Kluwer Academic/Plenum Publishers.

Foroud, N. A., Baines, D., Gagkaeva, T. Y., Thakor, N., Badea, A., Steiner, B., et al. (2019). Trichothecenes in cereal grains – an update. *Toxins, 11*, 634.

Furian Flávia, A., Fighera, M. R., Royes, L. F. F., & Oliveira, M. S. (2022). Recent advances in assessing the effects of mycotoxins using animal models. *Current Opinion in Food Science, 47*, Article 100874.

Frisvad, J., Andersen, B., & Samson, R. (2007). Association of moulds to food. In J. Dijksterhuis, & R. Samson (Eds.), *Food mycology – a multifaceted approach to fungi and food* (pp. 199–239). New York: CRC Press, Taylor & Francis Group.

Gonzalez, P. M., Rosa, C., dalcero, A., & Cavaglieri, L. (2011). Mycobiota and mycotoxins in malted barley and brewer's spent grain from Argentinean breweries. *Letters in Applied Microbiology, 53*, 649–655.

Goode, D., & Arent, E. (2006). Developments in the supply of adjunct materials for brewing. In C. Bmforth (Ed.), *Brewing new technologies* (pp. 30–67). Cambridge: Woodhead Publishing.

Habler, K., Geissinger, C., Hofer, K., Schüler, J., Moghari, S., Hess, M., et al. (2017). Fate of Fusarium toxins during brewing. *Journal of Agricultural and Food Chemistry, 65*(1), 190–198.

Habler, K., Gotthardt, M., Schüler, J., & Rychlik, M. (2017). Multi-mycotoxin stable isotope dilution LC–MS/MS method for fusarium toxins in beer. *Food Chemistry, 218*, 447–454.

Habler, K., Hofer, K., Geißinger, C., Schüler, J., Hückelhoven, R., Hess, M., et al. (2016). Fate of Fusarium toxins during the malting process. *Journal of Agricultural and Food Chemistry, 64*(6), 1377–1384.

Halady Shetty, P., & Jespersen, L. (2006). *Saccharomyces cerevisiae* and lactic acid bacteria as potential mycotoxin decontaminating agents. *Trends in Food Science and Technology, 17*, 48–55.

Halstensen, A. S., Nordby, K. C., Elen, O., & Eduard, W. (2004). Ochratoxin A in grain dust — estimated exposure and relations to agricultural practices in grain production. *Annals of Agricultural and Environmental Medicine, 11*, 245—254.

Hazel, C., & Patel, S. (2004). Influence of processing on trichothecene levels. *Toxicology Letters, 153*, 51—59.

Heaney, L., McCrea, P., Buick, B., & MacMahon, J. (1997). Brewer's asthma due to malt contamination. *Occupational Medicine, 47*, 397—400.

Henriksen, E., & Elen, O. (2005). Natural *Fusarium* grain infection level in wheat, barley and oat after early application of fungicides and herbicides. *Journal of Phytopathology, 153*, 214—220.

Hietaniemi, V., Rämö, S., Yli-Mattila, T., Jestoi, M., Peltonen, S., Kartio, M., et al. (2016). Updated survey of *Fusarium* species and toxins in finnish cereal grains. *Food Additives & Contaminants: Part A, 33*, 831—848.

Hiraga, K., Yamamotoa, S., Fukuda, H., Hamanakaba, N., & Oda, K. (2005). Enniatin has a new function as an inhibitor of Pdr5p, one of the ABC transporters in Saccharomyces cerevisiae. *Biochemical and Biophysical Research Communications, 328*, 1119—1125.

Hofgaard, I. S., Seehusen, T., Aamor, H. U., Riley, H., Razzghian, J., Le, V. H., et al. (2016). Inoculum potential of Fusarium spp. relates to tillage and straw management in Norwegian fields of spring oats. *Frontiers in Microbiology, 7*, 556.

Hoheneder, F., Biehl, E. M., Hofer, K., Petermeier, J., Groth, J., Herz, M., et al. (2022). Host genotype and weather effects on Fusarium head blight severity and mycotoxin load in spring barley. *Toxins, 14*, 125. https://doi.org/10.3390/toxins14020125

Hu, L., Gastl, M., Linkmeyer, A., Hess, M., & Rychlik, M. (2014). Fate of enniatins and beauvericin during malting and brewing process determined by stable isotope dilution assays. *LWT - Food Science and Technology, 56*, 469—477.

IARC. (1993). Some naturally occurring substances: Food items and constituents, heterocyclic aromatic amines and mycotoxins. *Monographs on the evaluation of carcinogenic risks to humans, 56*, 489—521.

Ibanez-Vea, M., Lizarraga, E., Gonzalez-Penas, E., & Lopez de Cerain, A. (2012). Co-occurrence if type A and type B trichothecenes in barley from a northern region of Spain. *Food Control, 25*, 81—88.

Inoue, T., Nagatomi, Y., Uyama, A., & Mochizuki, N. (2013). Fate of mycotoxins during beer brewing and fermentation. *Bioscience, Biotechnology, and Biochemistry, 77*, 1410—1415.

International Grains Council. (2022). *Market information, supply & demand, barley and world total* Accessed via International Grains Council (igc.int) on November 16th, 2022.

Jestoi, M. (2008). Emerging Fusarium-mycotoxins fusaproliferin, beauvericin, enniatins, and moniliformin — a review. *Critical Reviews in Food Science and Nutrition, 48*, 21—49.

Jin, Z., Solanki, S., Ameen, G., Gross, T., Sharma Poudel, R., Borowicz, P., et al. (2021). Expansion of internal hyphal growth in *Fusarium* head blight infected grains contribute to the elevated mycotoxin production during the malting process. *Molecular Plant-Microbe Interactions, 34*, 1—38.

Kang'ethe, E. K., Gatwiri, M., Sirma, A. J., Ouko, E. O., Mburugu-Musoti, C. K., Kitala, P. M., et al. (2017). Exposure of Kenyan population to aflatoxins in foods with special reference to Nandi and Makueni counties. *Food Quality and Safety, 1*, 131—137.

Karlsson, I., Friberg, H., Kolseth, A.-K., Steinberg, C., & Persson, P. (2017). Agricultural factors affecting Fusarium communities in wheat kernels. *International Journal of Food Microbiology, 252*, 53—60.

Karlsson, I., Mellqvist, E., & Persson, P. (2023). Temporal and spatial dynamics of Fusarium spp. and mycotoxins in Swedish cereals during 16 years. *Mycotoxin Research, 39*, 3—18. https://doi.org/10.1007/s12550-022-00469-9

Karlsson, I., Persson, P., & Friberg, H. (2021). *Fusarium* head blight from a microbiome perspective. *Frontiers in Microbiology, 12*, Article 628373. https://doi.org/10.3389/fmicb.2021.628373

Kautzman, M. E., Wickstrom, M. L., & Scott, T. A. (2015). The use of near infrared transmittance kernel sorting technology to salvage high quality grain from grain downgraded due to Fusarium damage. *Animal Nutrition, 1*(1), 41—46.

Kazartsev, I. A., Gagkaeva, T. Y., Gavrilova, O. P., & Gannibal, P. B. (2020). Fungal microbiome of barley grain revealed by NGS and mycological analysis. *Foods and Raw Materials, 8*(2), 286—297.

Kiš, M., Vulić, A., Kudumija, N., Šarkanj, B., Jaki, T. V., Aladić, K., et al. (2021). Two-year occurrence of Fusarium T-2 and HT-2 toxin in Croatian cereals relative of the regional weather. *Toxins, 13*, 39.

Klosowksi, G., Mikulski, D., Grajewski, J., & Blajet-Kosicka, A. (2010). The influence of raw material contamination with mycotoxins on alcoholic fermentation indicators. *Bioresource Technology, 101*, 3147—3152.

Koshinsky, H., Crosby, R., & Khachhatourians, G. (1992). Effect of T-2 toxin on ethanol production by *Saccharomyces cerevisiae. Biotechnology and Applied Biochemistry, 16*, 275—286.

Kostelanska, M., Hajslova, J., Zachariasova, M., Malachova, A., Kalachova, K., Poustka, J., et al. (2009). Occurrence of deoxynivalenol and its major conjugate, deoxynivalenol-3 glucoside, in beer and some beer intermediates. *Journal of Agricultural and Food Chemistry, 57*, 3187—3194.

Kostelanska, M., Zachariasova, M., Lacina, O., Fenclova, M., Kollos, A.-L., & Hajslova, J. (2011). The study of deoxynivalenol and its masked metabolites fate during brewing process realised by UPLC-TOFMS method. *Food Chemistry, 126*, 1870—1876.

Kottapalli, B., Wolf-Hall, C., & Schwarz, P. (2005). Evaluation of gaseous ozone and hydrogen peroxide treatments for reducing Fusarium survival in malting. *Journal of Food Protection, 68*, 1236—1240.

Ksieniewicz-Woźniak, E., Bryła, M., Waskiewicz, A., Yoshinari, T., & Szymczyk, K. (2019). Selected trichothecenes in barley malt and beer from Poland and an assessment of dietary risks associated with their consumption. *Toxins, 11*, 715.

Kuruc, J., Hegstad, J., Lee, H. J., Simons, K., Ryu, D., & Wolf-Hall, C. (2015). Infestation and quantification od ochratoxigenic fungi in barley and wheat naturally contaminated with ochratoxin A. *Journal of Food Protection, 78*(7), 1350—1356.

Laitila, A. (2007). *Microbes in the tailoring of barley malt properties (PhD thesis)* (Vol. 645). VTT Publications, 107 pp.

Laitila, A. (2015). Toxigenic fungi and mycotoxins in the barley-to-beer chain. In A. E. Hill (Ed.), *Brewing microbiology* (1st ed., pp. 107–139). Woodhead Publishing/Elsevier.

Laitila, A., Alakomi, H.-L., Raaska, L., Mattila-Sandholm, T., & Haikara, A. (2002). Antifungal activities of two *Lactobacillus plantarum* strains against *Fusarium* moulds in vitro and in malting of barley. *Journal of Applied Microbiology, 93*, 566–576.

Laitila, A., Sweins, H., Vilpola, A., Kotaviita, E., Olkku, J., Home, S., et al. (2006). *Lactobacillus plantarum* and *Pediococcus pentosaceus* starter cultures as a tool for microflora management in malting and for enhancement of malt processability. *Journal of Agricultural and Food Chemistry, 54*, 3840–3851.

Laitila, A., Manninen, J., Priha, O., Smart, K., Tsitko, I., & James, S. (2018). Characterisation of barley-associated bacteria and their impact on wort separation performance. *Journal of the Institute of Brewing, 124*, 314–324.

Laitila, A., Tapani, K.-M., & Haikara, A. (1997). Lactic acid starter cultures for prevention of the formation of *Fusarium* mycotoxins during malting. In *Proc 26th congr Eur brew conv* (pp. 137–144). Maastricht: Oxford University Press.

Lake, J., Browers, M., & Yin, X. (2007). Use of sodium bisulfite as a method to reduce DON levels in barley during malting. *Journal of the American Society of Brewing Chemists, 65*, 172–176.

Lancova, K., Hajslova, J., Poustka, J., Krplova, A., Zachariasova, M., Sachambula, L., et al. (2008). Transfer of *Fusarium* mycotoxins and 'masked' deoxynivalenol (deoxynivalenol-3-glucoside) from field barley through malt to beer. *Food Additives & Contaminants, 25*, 732–744.

Lattanzio, V. M. T., von Holst, C., Lippolis, V., De Girolamo, A., Logrieco, A. F., Mol, H. G. J., et al. (2019). Evaluation of mycotoxin screening tests in a verification study involving first time users. *Toxins, 11*, 129.

Leplat, J., Friberg, H., Abid, M., & Steinberg, C. (2012). Survival of Fusarium graminearum, the causal agent of Fusarium head blight. A review. *Agronomy for Sustainable Development, 33*(1), 97–111.

Levasseur-Garcia, C. (2018). Updated overview of infrared spectroscopy methods for detecting mycotoxins on cereals (corn, wheat, and barley). *Toxins, 10*, 38.

Li, M., Guan, E., & Bian, K. (2019). Structure elucidation and toxicity analysis of the degradation products of deoxynivalenol by gaseous ozone. *Toxins, 11*(8), 474.

Linkmeyer, A., Hofer, K., Rychlik, M., Herz, M., Hausladen, H., Hückelhoven, R., et al. (2016). Influence of inoculum and climatic factors on the severity of Fusarium head blight in German spring and winter barley. *Food Additives & Contaminants: Part A, 33*(3), 489–499.

Liu, M., Zhao, L., Gong, G., et al. (2022). Invited review: Remediation strategies for mycotoxin control in feed. *Journal of Animal Science and Biotechnology, 13*, 19.

Logrieco, A., Rizzo, A., Ferracane, R., & Ritieni, A. (2002). Occurrence of beauvericin and enniatins in wheat affected by *Fusarium avenaceum* head blight. *Applieed and Environmental Microbiology, 68*, 82–85.

Lopez-Diaz, T., & Flannigan, B. (1997). Production of patulin and cytochalasin E by *Aspergillus clavatus* during malting of barley and wheat. *International Journal of Food Microbiology, 35*, 129–136.

Loretti, A., Colodel, E., Driemeier, D., Correa, A., Bange, J., & Ferreiro, L. (2003). Neurological disorder in dairy cattle associated with consumption of beer residues contaminated with *Aspergillus clavatus*. *Journal of Veterinary Diagnostic Investigation, 15*, 123–132.

Lowe, D., & Arendt, E. (2004). The use and effects of lactic acid bacteria on malting and brewing with their relationship to antifungal activity, mycotoxins and gushing: A review. *Journal of the Institute of Brewing, 110*, 163–180.

Luz, C., Carbonell, R., Quiles, J. M., Torrijos, R., de Melo, N. T., Manes, J., et al. (2021). Antifungal activity of peracetic acid against toxigenic fungal contaminants of maize and barley at the postharvest stage. *LWT - Food Science and Technology, 148*, Article 111754.

Mably, M., Mankotia, M., Cavlovic, P., Tam, J., Wong, L., Pantazopoulos, P., Scott, P., et al. (2005). Survey of aflatoxins in beer sold in Canada. *Food Additives & Contaminants, 22*, 1252–1257.

Magan, N., & Aldred, D. (2005). Conditions of formation of ochratoxin A in drying, transport and in different commodities. *Food Additives & Contaminants, 22*(Suppl. 1), 10–16.

Malachova, A., Cerkal, R., Ehrenbergerova, J., Dzuman, Z., Vaculova, K., & Hajslova, J. (2010). *Fusarium* mycotoxins in various barley cultivars and their transfer into malt. *Journal of the Science of Food and Agriculture, 90*, 2495–2505.

Mannaa, M., & Kim, K. D. (2017). Influence of temperature and water activity on deleterious fungi and mycotoxin production during grain storage. *Mycobiology, 45*(4), 240–254.

Mastanjević, K., Lukinac, J., Jukić, M., Šarkanj, B., Krstanović, V., & Mastanjević, K. (2019). Multi-(myco)toxins in malting and brewing by-products. *Toxins, 11*, 30. https://doi.org/10.3390/toxins11010030

Mateo, R., Medina, A., Mateo, E., Mateo, F., & Jimenez, M. (2007). An overview of ochratoxin in beer and wine. *International Journal of Food Microbiology, 119*, 79–83.

Maul, R., Müller, C., Riess, S., Koch, M., Methner, F.-J., & Nehls, I. (2012). Germination induces the glycosylation of the *Fusarium* mycotoxin deoxynivalenol in various grains. *Food Chemistry, 131*, 274–279.

Mauro, A., Garcia-Cela, E., Pietri, A., Cotty, P. J., & Battilani, P. (2018). Biological control products for aflatoxin prevention in Italy: Commercial field evaluation of atoxigenic Aspergillus flavus active ingredients. *Toxins, 10*, 30.

McDonough, M. X., Campabadal, C. A., Mason, L. J., Maier, D. E., Denvir, A., & Woloshuk, C. (2011). Ozone application in a modified screw conveyor to treat grain for insect pests, fungal contaminants, and mycotoxins. *Journal of Stored Products Research, 47*, 249–254.

McMullen, M., Jones, R., & Gallenberg, D. (1997). Scab of wheat and barley: A re-emerging disease of devastating impact. *Plant Disease, 81*, 1340–1348.

Medina, A., Valle-Algarra, F., Mateo, R., Gimeno-Adelantado, J., Mateo, F., & Jimenez, M. (2006). Survey of the mycobiota of Spanish malting barley and evaluation of the mycotoxin producing potential of species of *Alternaria, Aspergillus* and *Fusarium. International Journal of Food Microbiology, 108*, 196–203.

Mizutani, K., Nagatomi, Y., & Mochizuki, N. (2011). Metabolism of zearalenone in the course of beer fermentation. *Toxins, 3*, 134–141.

Moretti, A., Logrieco, A. F., & Susca, A. (2017). Mycotoxins: An underhand food problem. In A. Moretti, & A. Susca (Eds.), *Mycotoxigenic fungi. Methods in molecular biology* (Vol. 1542). New York, NY: Humana Press.

Moretti, A., Pascale, M., & Logrieco, A. F. (2019). Mycotoxin risks under a climate change scenario in Europe. *Trends in Food Science & Technology, 84*, 38–40.

Murphy, P., Hendrich, S., Landgren, C., & Bryant, C. (2006). Food mycotoxins: An update. *Journal of Food Science, 71*, 51–65.

Nathanail, A. V. (2019). *Modified Fusarium mycotoxins: A threat in disguise* (PhD thesis) (p. 123). Helsinki: Unigrafia pdf http://ethesis.helsinki.fi.

Nathanail, A. V., Gibson, B., Han, L., Peltonen, K., Ollilainen, V., Jestoi, M., et al. (2016). The lager yeast *Saccharomyces pastorianus* removes and transforms *Fusarium* trichothecene mycotoxins during fermentation of brewer's wort. *Food Chemistry, 203*, 448–455.

Nathanail, A. V., Syvähuoko, J., Malachová, A., Jestoi, M., Varga, E., Michlmayr, H., et al. (2015). Simultaneous determination of major type A and B trichothecenes, zearalenone and certain modified metabolites in Finnish cereal grains with a novel liquid chromatography-tandem mass spectrometric method. *Analytical and Bioanalytical Chemistry, 407*, 4745–4755.

Nikolić, M., Savić, I., Obradović, A., Srdić, J., Stanković, G., Stevanović, M., et al. (2020). First report of *Aspergillus parasiticus* on barley grain in Serbia. *Plant Disease, 104*(3), 987.

Nishiuchi, T. (2013). Plant responses to *Fusarium* metabolites. In D. Brown, & R. Proctor (Eds.), Fusarium, *genomics, molecular and cellular biology* (pp. 165–178). Norfolk: Caister Academic Press.

Noots, I., Delcour, J., & Michiels, C. (1999). From field barley to malt: Detection and specification of microbial activity for quality aspects. *Critical Reviews in Microbiology, 25*, 121–153.

Nordby, K. C., Halstensen, A. S., Elen, O., Clasen, P. E., Langseth, W., Kristensen, P., & Eduard, W. (2004). Trichothecene mycotoxins and their determinats in settled dust related to grain production. *Annals of Agricultural and Environmental Medicine, 11*(1).

Nordby, K. C., Straumfors, H. A., Elen, O., Clasen, P. E., Langseth, W., Kristensen, W., et al. (2004). Trichothecene mycotoxins and their determinants in settled dust related to grain production. *Annals of Agricultural and Environmental Medicine, 11*, 75–83.

Odhav, N. V. (2002). Mycotoxins in South African traditionally brewed beers. *Food Additives & Contaminants, 19*, 55–61.

Østlie, H. M., Porcellato, D., Kvam, G., & Wicklund, T. (2021). Investigation of the microbiota associated with ungerminated and germinated Norwegian barley cultivars with focus on lactic acid bacteria. *International Journal of Food Microbiology, 341*, Article 109059.

O'Sullivan, T., Walsh, Y., O'Mahony, A., Fitzgerald, G., & van Sinderen, D. (1999). A comparative study of malthouse and brewhouse microflora. *Journal of the Institute of Brewing, 105*, 55–61.

Osborne, L., & Stein, J. (2007). Epidemiology of *Fusarium* head bligh on small grains. *International Journal of Food Microbiology, 119*, 103–108.

Ouadhene, M. A., Ortega-Beltran, A., Sanna, M., Cotty, P. J., & Battilani, P. (2023). Multiple year influences of the aflatoxin biocontrol product AF-X1 on the A. flavus communities associated with maize production in Italy. *Toxins, 15*(3), 184.

Palazzini, J., Roncallo, P., Cantoro, R., Chiotta, M., Yerkovich, N., Palacios, S., et al. (2018). Biocontrol of Fusarium graminearum sensu stricto, reduction of deoxynivalenol accumulation and phytohormone induction by two selected antagonists. *Toxins, 10*(2), 88.

Palojärvi, A., Kellock, M., Parikka, P., Jauhiainen, L., & Alakukku, L. (2021). Tillage system and crop sequence affect soil disease suppressiveness and carbon status in boreal climate. *Frontiers in Microbiology, 11*, Article 534786.

Papadopoulou, A., Wheaton, L., & Muller, R. (2000). The control of selected microorganisms during the malting process. *Journal of the Institute of Brewing, 106*, 179–188.

Parry, D., Jenkinson, P., & McLeod, L. (1995). *Fusarium* ear blight (scab) in small grain cereals — a review. *Plant Pathology, 44*, 207–238.

Parikka, P., Hakala, K., & Tiilikkala, K. (2012). Expected shifts in Fusarium species' composition on cereal grain in Northern Europe due to climatic change. *Food Additives & Contaminants: Part A, 29*(10), 1543–1555.

Pascari, X., Gil-Samarra, S., Marín, S., Ramos, A. J., & Sanchis, V. (2019). Fate of zearalenone, deoxynivalenol and deoxynivalenol-3-glucoside during malting process. *LWT - Food Science and Technology, 99*, 540–546.

Pascari, X., Rodriguez-Carrasco, Y., Juan, C., Mañes, J., Marín, S., Ramos, A. J., et al. (2019b). Transfer of Fusarium mycotoxins from malt to boiled wort. *Food Chemistry, 278*, 700–710.

Paterson, R. R. M., & Lima, N. (2010). Toxicology of mycotoxins. In A. Luch (Ed.), *Molecular, clinical and environmental toxicology. Clinical toxicology* (Vol. 2, pp. 31–63). Basel: Springer.

Paulitz, T., & Steffenson, B. (2011). Biotic stress in barley: Disease problems and solutions. In S. Ullrich (Ed.), *Barley: Production, improvement and uses* (pp. 307–353). Chichester: Wiley-Blackwell.

Pellan, L., Durand, N., Martinez, V., Fontana, A., Schorr-Galindo, S., & Strub, C. (2020). Commercial biocontrol agents reveal contrasting comportments against two mycotoxigenic fungi in cereals: Fusarium graminearum and Fusarium verticillioides. *Toxins, 12*(3), 152.

Perkowski, J. (1998). Distribution of deoxynivalenol in barley kernels infected by *Fusarium. Nahrung, 42*, 81–83.

Pestka, J., Zhou, H., Moon, Y., & Chung, Y. (2004). Cellular and molecular mechanisms for immune modulation by deoxynivalenol and other trichothecenes: Unraveling a paradox. *Toxicology Letters, 153*, 61–73.

Petters, H., Flannigan, B., & Austin, B. (1988). Quantitative and qualitative studies of the microflora of barley malt production. *Journal of Applied Bacteriology, 65*, 279–297.

Pietri, A., Bertuzzi, T., Agosti, B., & Donadini, G. (2010). Transfer of aflatoxin B1 and fumonisin B1 from naturally contaminated raw materials to beer during an industrial brewing process. *Food Additives & Contaminants, 27*, 1431–1439.

Pleadin, J., Staver, M. M., Markov, K., Frece, J., Zadravec, M., Jaki, V., et al. (2017). Mycotoxins in organic and conventional cereals and cereal products grown and marketed in Croatia. *Mycotoxin Research, 33*, 219–227.

Polak-Śliwińska, M., & Paszczyk, B. (2021). Trichothecenes in food and feed, relevance to human and animal health and methods of detection: A systematic review. *Molecules, 26*, 454.

Postulkova, M., Rezanina, J., Fiala, J., Ruzicka, M. C., Dostalek, P., & Branyik, T. (2018). Suppression of fungal contamination by Pythium oligandrum during malting of barley. *Journal of the Institute of Brewing, 124*, 336–340.

Prusova, N., Dzuman, Z., Jelinek, L., Karabin, M., Jana Hajslova, J., Rychlik, M., et al. (2022). Free and conjugated *Alternaria* and *Fusarium* mycotoxins during Pilsner malt production and double-mash brewing. *Food Chemistry, 369*, Article 130926.

Quiles, J. M., de Melo Nazareth, T., Luz, C., Luciano, F. B., Mañes, J., & Meca, G. (2019). Development of an antifungal and antimycotoxigenic device containing allyl isothiocyanate for silo fumigation. *Toxins, 11*, 137.

Raulio, M., Wilhelmson, A., Salkinoja-Salonen, M., & Laitila, A. (2009). Ultrastructure of biofilms formed on barley kernels during malting with and without starter culture. *Food Microbiology, 26*, 437–443.

Reichel, M., Mänz, J. S., & Biselli, S. (2018). Advanced and representative sampling for mycotoxin - the most critical step. In A. Laitila, P. Vahala, & T. Sarlin (Eds.), *VTT technical research centre of Finland, Helsinki, FinlandThe 1st MycoKey technological workshop: Integrated preventive actions to avoid mycotoxins in malting and brewing*.

Reichel, M., Staiger, S., & Biselli, S. (2014). Analysis of Fusarium toxins in grain via dust: A promising field of application for rapid test systems. *World Mycotoxin Journal, 7*(4), 465–477.

Richard, J. (2007). Some major mycotoxins and their mycotoxicises — an overview. *International Journal of Food Microbiology, 119*, 3–10.

Rouse, S., & van Sinderen, D. (2008). Bioprotective potential of lactic acid bacteria in malting and brewing. *Journal of Food Protection, 71*, 1724–1733.

Rubio-Armendáriz, C., Revert, C., Paz-Montelongo, S., Gutiérrez-Fernández, Á. J., Luis-González, G., & Hardisson, A. (2022). Mycotoxins. In *Reference module in biomedical sciences*. Elsevier.

Rush, B. (2006). Developing HACCP programs in grain-based brewing and food ingredient production facilities. *Technical Quarterly - Master Brewers Association of the Americas, 1*, 26–30.

Rychlik, M., Humpf, H.-U., Marko, D., Dänicke, S., Mally, A., Berthiller, F., et al. (2014). Proposal of a comprehensive definition of modified and other forms of mycotoxins including "masked` mycotoxins. *Mycotoxin Research, 30*, 197–205.

Rylander, R. (1986). Lung diseases caused by organic dusts in the farm environment. *American Journal of Industrial Medicine, 10*, 221–227.

Samson, R., Hoekstra, E., Frisvad, J., & Filtenborg, O. (2000). *Introduction to food- and borne fungi*, 389 pp. (6th ed.). Wageningen: The Netherlands: Ponsen & Looyen

Sarlin, T. (2012). Detection and characterization of *Fusarium hydrophobins* inducing gushing in beer (Doctoral dissertation). *VTT Science, 13*, 82.

Sarlin, T., Laitila, A., Pekkarinen, A., & Haikara, A. (2005). Effects of three *Fusarium* species on the quality of barley and malt. *Journal of the American Society of Brewing Chemists, 43*, 43–49.

Sarlin, T., Nakari-Setälä, T., Linder, M., Penttilä, M., & Haikara, A. (2005). Fungal hydrophobins as predictors of the gushing activity of malt. *Journal of the Institute of Brewing, 111*, 105–111.

Sauer, D., Meronuck, R., & Christensen, C. (1992). Microflora. In D. B. Sauer (Ed.), *Storage of cereal grains and their products* (pp. 313–340). St Paul, MN, USA: American Association of Cereal Chemists, Inc.

Schabo, D. C., Alvarenga, V. O., Schaffner, D. W., & Magnani, M. (2021). A worldwide systematic review, meta-analysis, and health risk assessment study of mycotoxins in beers. *Comprehensive Reviews in Food Science and Food Safety, 20*, 5742–5764.

Schwarz, P. B. (2017). Fusarium head blight and deoxynivalenol in malting and brewing: Successes and future challenges. *Tropical Plant Pathology, 42*, 153–164.

Schwarz, P., Casper, H., & Beattie, S. (1995). Fate and development of naturally occurring *Fusarium* toxins during malting and brewing. *Journal of the American Society of Brewing Chemists, 53*, 121–127.

Schwarz, P., Horsley, R., Steffenson, B., Salas, B., & Barr, M. (2006). Quality risks associated with the utilisation of *Fusarium* head blight infected malting barley. *Journal of the American Society of Brewing Chemists, 64*, 1–7.

Schwarz, P., & Li, Y. (2011). Malting and brewing uses of barley. In S. Ullrich (Ed.), *Barley: Production, improvement and uses* (pp. 478–521). Chichester: Wiley-Blackwell.

Scott, P. (1996). Mycotoxins transmitted into beer from contaminated grains during brewing. *Journal of AOAC International, 79*, 875–882.

Scudamore, K. (2005). Prevention of ochratoxin in commodities and likely effects of processing fractionation animal feeds. *Food Additives & Contaminants, 22*(Suppl. 1), 17–25.

Silva, L. J. G., Teixeiram, A. C., Pereira, A. M. P. T., Pena, A., & Lino, C. M. (2020). Ochratoxin a in beers marketed in Portugal: Occurrence and human risk assessment. *Toxins, 12*(4), 249.

Singh, B. K., Delgado-Baquerizo, M., Egidi, E., Guirado, E., Leach, J. E., Liu, H., et al. (2023). Climate change impacts on plant pathogens, food security and paths forward. *Nature Reviews Microbiology, 21*, 640–656.

Slettengren, K., Pascale, M., Reichel, M., Vega, A., Rittenauer, M., Deefholts, B., et al. (2018). Advanced grain cleaning solutions for mycotoxin reduction. In A. Laitila, P. Vahala, & T. Sarlin (Eds.), *VTT technical research centre of Finland, Helsinki, FinlandThe 1st MycoKey technological workshop: Integrated preventive actions to avoid mycotoxins in malting and brewing*.

Sohlberg, E., Virkajärvi, V., Parikka, P., Rämö, S., Laitila, A., & Sarlin, T. (2022). Taqman qPCR quantification and *Fusarium* community analysis to evaluate toxigenic fungi in cereals. *Toxins, 14*, 45. https://www.mdpi.com/2072-6651/14/1/45/pdf.

Strub, C., Pocaznoi, D., Lebrihi, A., Fournier, R., & Mathieu, F. (2010). Influence of barley malting operating parameters on T-2 and HT-2 toxinogenesis of *Fusarium langsethieae*, a worrying contaminant of malting barley in Europe. *Food Additives & Contaminants, 27*, 1247–1252.

Tangni, E., Ponchaut, S., Maudoux, M., Rozenberg, R., & Larondelle, Y. (2002). Ochratoxin in domestic and imported beers in Belgium: Occurrence and exposure assessment. *Food Additives & Contaminants, 19*, 1169–1179.

Tian, Y., Zhang, D., Cai, P., Lin, H., Ying, H., Hu, Q.-N., et al. (2022). Elimination of Fusarium mycotoxin deoxynivalenol (DON) via microbial and enzymatic strategies: Current status and future perspectives. *Trends in Food Science & Technology, 124*, 96–107.

Tittlemier, S. A., Cramer, B., Dall'Asta, C., Iha, M. H., Lattanzio, V. M. T., Maragos, C., et al. (2020). Developments in mycotoxin analysis: An update for 2018–19. *World Mycotoxin Journal, 13*(1), 3–24.

Tiwari, B., Brennan, C., Curran, T., Gallagher, E., Cullen, P., & O'Donnell, C. (2010). Application of ozone in grain processing. *Journal of Cereal Science, 51*, 248–255.

Tolosa, J., Rodríguez-Carrasco, Y., Ruiz, M. J., & Vila-Donat, P. (2021). Multi-mycotoxin occurrence in feed, metabolism and carry-over to animal-derived food products: A review. *Food and Chemical Toxicology, 158*, Article 112661.

Tricase, C., Amicarelli, V., Lamonaca, E., & Rana, R. L. (2018). Economic analysis of the barley market and related uses. In Z. Tadele (Ed.), *Grasses as food and feed*. London: IntechOpen. https://doi.org/10.5772/intechopen.78967. Available from: https://www.intechopen.com/chapters/62190.

Uhlig, S., Ivanova, L., Petersen, D., & Kristensen, R. (2009). Structural studies on minor enniatins from *Fusarium* sp. VI 03441: Novel *N*-methyl-threonine containing enniatins. *Toxicon, 53*, 734–742.

Vaclavikova, M., Malachova, M., Veprikova, Z., Dzuman, Z., Zachariasova, M., & Hajslova, J. (2013). Emerging mycotoxins in cereals processing chain: Changes of enniatins during beer and bread making. *Food Chemistry, 136*, 750–757.

Valverde-Bogantes, E., Bianchini, A., Herr, J. R., Rose, D. J., Wegulo, S. N., & Hallen-Adams, H. E. (2020). Recent population changes of *Fusarium* head blight pathogens: Drivers and implications. *Canadian Journal of Plant Pathology, 42*, 3 315–329.

van der Fels-Klerx, H., & Stratakou, I. (2010). T-2 and HT-2 toxin in grain and grain-based commodities in Europe: Occurrence, factors affecting occurrence, co-occurrence and toxicological effects. *World Mycotoxin Journal, 3*, 349–376.

Van der Lee, T. A. J., Molendijk, L. P. G., Audenaert, K., Landschoot, S., Verwaeren, J., Leggieri, M. C., et al. (2018). MycoKey App: An ICT solution to facilitate mitigation of mycotoxin risks. In A. Laitila, P. Vahala, & T. Sarlin (Eds.), *VTT technical research centre of Finland, Helsinki, FinlandThe 1st MycoKey technological workshop: Integrated preventive actions to avoid mycotoxins in malting and brewing*.

Van Nierop, S., Rautenbach, M., Axcell, B., & Cantrell, I. (2006). The impact of microorganisms on barley and malt – a review. *Journal of the American Society of Brewing Chemists, 64*, 69–78.

Varga, E., Malachova, A., Schwartz, H., Krska, R., & Berthiller, F. (2013). Survey of deoxynivalnol and its conjugates deoxynivalenol-3-glucoside and 3 acetyl-deoxynivalenol in 374 beer samples. *Food Additives & Contaminants, 30*, 137–146.

Vogelgsang, S., Beyer, M., Pasquali, M., Jenny, E., Musa, T., Bucheli, T. D., et al. (2019). An eight-year survey of wheat shows distinctive effects of cropping factors on different Fusarium species and associated mycotoxins. *European Journal of Agronomy, 105*, 62–77.

Wang, X., Liu, C., & van der Fels-Klerx, H. J. (2022). Regional prediction of multi-mycotoxin contamination of wheat in Europe using machine learning. *Food Research International, 159*, Article 111588.

Wegulo, S. N., Baenziger, P. S., Nopsa, J. H., Bockus, W. W., & Hallen-Adams, H. (2015). Management of Fusarium head blight of wheat and barley. *Crop Protection, 73*, 100–107.

Whitehead, M., & Flannigan, B. (1989). The *Fusarium* mycotoxin deoxynivalenol and yeast growth and fermentation. *Journal of the Institute of Brewing, 95*, 411–413.

Wilhelmson, A., Laitila, A., Heikkilä, J., Räsänen, J., Kotaviita, E., Olkku, J., et al. (2003). Changes in gas atmosphere during industrial scale malting. *Proceedings of the Congress European Brewery Convention Dublin, 29*, 226–233.

Wolf-Hall, C. (2007). Mold and mycotoxin problems encountered during malting and brewing. *International Journal of Food Microbiology, 119*, 89–94.

Yashiro, E., Savova-Bianchi, D., & Niculita-Hirzel, H. (2019). Major differences in the diversity of mycobiomes associated with wheat processing and domestic environments: Significant findings from high-throughput sequencing of fungal barcode ITS1. *International Journal of Environmental Research Public Health, 16*, 2335. https://doi.org/10.3390/ijerph16132335

Zhou, B., He, G.-Q., & Schwarz, P. (2008). Occurrence of bound deoxynivalenol in *Fusarium* head blight-infected barley (*Hordeum vulgare* L.) and malt as determined by solvolysis with trifluoroacetic acid. *Journal of Food Protection, 71*, 1266–1269.

Zinedine, A., Soriano, J., Molto, J., & Manes, J. (2007). Review on the toxicity, occurrence, metabolism, detoxification, regulations and intake of zearalenone: An oestrogenic mycotoxin. *Food and Chemical Toxicology, 45*, 1–18.

Zingales, V., Taroncher, M., Martino, P. A., Ruiz, M.-J., & Caloni, F. (2022). Climate change and effects on molds and mycotoxins. *Toxins, 14*, 445. https://doi.org/10.3390/toxins14070445

Zuluaga-Calderón, B., González, H. H. L., Alzamora, S. M., & Coronel, M. B. (2022). Multi-step ozone treatments of malting barley: Effect on the incidence of *Fusarium graminearum* and grain germination parameters. *Innovative Food Science & Emerging Technologies, 83*(7), Article 103219.

Further reading

Euromalt. (2022). http://www.euromalt.be.

Peters, J., Van Dam, R., Van Doorn, R., Katerere, D., Berthiller, F., Haasnoot, W., et al. (2017). Mycotoxin profiling of 1000 beer samples with a special focus on craft beer. *PLoS One, 12*(10), Article e0185887. https://doi.org/10.1371/journal.pone.0185887

Röder, O., Jahn, M., Schröder, T., Stahl, M., Kotte, M., & Beuermann, S. (2009). E-ventus technology – an innovative treatment method for sustainable reduction in the use of pesticides with recommendation for organic seed. *Journal für Verbraucherschutz und Lebensmittelsicherheit, 4*, 107–117.

Chapter 7

Gram-positive spoilage bacteria in brewing

Koji Suzuki[1] and Yohannes N. Kurniawan[2]
[1]*Asahi Quality & Innovations, Ltd., Ibaraki, Japan;* [2]*Asahi Group Holdings, Ltd., Tokyo, Japan*

7.1 Introduction

Beer has been recognized as a microbiologically stable beverage. This is due to the presence of ethanol (0.5%−10% w/w), hop bitter compounds (*ca* 17−55 ppm of iso-α-acids) and high carbon dioxide content (approximately 0.5% w/v), as well as low pH (3.8−4.7) and reduced concentration of oxygen (generally less than 0.3 ppm) (Suzuki, Iijima, Sakamoto, et al., 2006). Beer is also a poor medium because nutrients are almost depleted by the fermentative activities of brewing yeast. As a result, most Gram-positive bacteria, such as *Bacillus* and *Staphylococcus*, do not grow or survive in beer (Bunker, 1955). Despite these hostile features, a limited number of Gram-positive species are able to grow in beer. The two genera *Lactobacillus* and *Pediococcus* are predominant beer spoilers in Gram-positive bacteria (note: The genus *Lactobacillus* has been reclassified into 25 genera, including 23 novel genera since 2020 (Zheng et al., 2020)). Nevertheless, brewing microbiologists are more familiar with the conventional nomenclature, and, to avoid the confusion, the former genus name, *Lactobacillus*, will be used for the rod-shaped beer-spoilage lactic acid bacteria (LAB). For the current nomenclature of major beer spoilage lactobacilli, see Table 7.1. These beer-spoilage LAB exhibit strong tolerance to hop bitter acids, a distinguishing feature that is not observed with non-spoilage Gram-positive counterparts. Hop tolerance of beer-spoilage LAB is considered as the interplay of several distinct mechanisms that collectively counteract the toxic effects of hop bitter acids.

In this chapter, the first Section 7.2 describes the taxonomy and history of beer-spoilage LAB, in addition to other features concerning these spoilage microorganisms. The second Section 7.3 is specifically devoted to the hop tolerance mechanisms in LAB, as well as the identification of hop tolerance genes and their significance in the brewing industry. The preservation and subculture methods of beer-spoilage LAB strains are described in the third Section 7.4, to which attention should be paid to maintaining the original states of beer-spoilage LAB when detected as primary isolates from brewing environments. In addition, beer spoilage ability of LAB largely depends on the species status, as well as the presence or absence of particular genetic markers related to hop tolerance (diagnostic marker genes), the identification methods for beer-spoilage LAB strains will be presented in the fourth Section 7.5. Other Gram-positive bacteria relevant for the brewing will be also briefly summarised in the final Section 7.6.

7.2 Beer-spoilage LAB

7.2.1 Historical backgrounds and taxonomy

Beer-spoilage LAB were found by Pasteur in 1871 through microscopic examinations of spoiled beer (Pasteur, 1876). Initially, beer-spoilage LAB were grouped in rods and cocci. Rod-shaped LAB strains were originally designated *Saccharobacillus pastorianus* by van Laer (1892). This species was named in honour of Pasteur and later redesignated *Lactobacillus pastorianus*. van Laer also reported that *L. pastorianus* did not show culturability on ordinary nutrient media and therefore used unhopped beer solidified with gelatin for isolation. Due to its extremely low culturability on ordinary culture media, *L. pastorianus* had been poorly characterised, despite the fact that this species exhibits very strong beer-

Brewing Microbiology. https://doi.org/10.1016/B978-0-323-99606-8.00020-1

TABLE 7.1 Current nomenclature for major beer-spoilage lactobacilli.

Former classification	Current nomenclature
Lactobacillus brevis	*Levilactobacillus brevis*
Lactobacillus lindneri	*Fructilactobacillus lindneri*
Laactobacillus backii	*Loigolactobacillus backii*
Lactobacillus (para)collinoides	*Secundilactobacillus (para)collinoides*
Lactobacillus (para)buchneri	*Lentilactobacillus (para)buchneri*
Lactobacillus rossiae	*Furfurilactobacillus rossiae*
Lactobacillus perolens	*Schleiferilactobacillus perolens*
Lactobacillus harbinensis	*Schleiferilactobacillus harbinensis*
Lactobacillus (para)casei	*Lacticaseibacillus (para)casei*
Lactobacillus plantarum	*Lactiplantibacillus plantarum*
Lactobacillus coryniformis	*Loigolactobacillus coryniformis*
Lactobacillus acetotolerans	*Lactobacillus acetotolerans*
Lactobacillus curtus	*Furfurilactobacillus curtus*
Lactobacillus nagelii	*Liquorilactobacillus nagelii*

spoilage ability (Suzuki, Iijima, Sakamoto, et al., 2006). However, the recent development of new culture techniques has enabled *L. pastorianus* to be isolated from brewing environments (Suzuki, 2012). Since then, the insights into this species have been accumulated, and it was reported that *L. pastorianus* is a much more common beer spoiler than previously assumed (Iijima, Suzuki, Asano, Kuriyama, & Kitagawa, 2007; Shimokawa & Suzuki, 2021). *L. pastorianus* is now considered as a synonym of *Lactobacillus paracollinoides*, and *L. paracollinoides* has been accepted as a formal species name (Ehrmann & Vogel, 2005; Suzuki, 2020; Suzuki, Asano, Iijima, & Kitamoto, 2008). Through the subsequent development of phylogenetic studies, the taxonomy of rod-shaped lactobacilli has been changed a lot since the end of the nineteenth century, and beer-spoilage lactobacilli are now divided into *Lactobacillus brevis*, *Lactobacillus lindneri*, *L. paracollinoides*, *Lactobacillus backii* and several other *Lactobacillus* species (Hutzler et al., 2013). On the other hand, coccal strains were originally named *Pediococcus cerevisiae* by Blacke in 1884 (Kitahara, 1974). *Ped. cerevisiae* is now redesignated as *Pediococcus damnosus*, a species name proposed by Claussen (1903). *P. claussenii* has been described as a novel beer-spoilage LAB species since 2002 (Dobson et al., 2002). *P. inopinatus* is also recognized as a potential beer spoiler (Back, 2005a; Iijima et al., 2007).

7.2.2 General features of beer-spoilage LAB

LAB contain a large group of genera and species of Gram-positive bacteria, including *Lactobacillus* and *Pediococcus*. In the period 1980–2002, approximately 60%–90% of the microbiological spoilage incidents in Germany were caused by *Lactobacillus* and *Pediococcus* (Table 7.2; Back, 1994a, 1994b, 2003). A similar trend was observed in the more recent studies conducted during the 2010–16 period, using polymerase chain reaction (PCR) analysis (Table 7.3; Hutzler et al., 2012; Koob et al., 2014; Schneiderbanger, Grammer, Jacob, & Hutzler, et al., 2018). Among these LAB, *L. brevis*, *L. lindneri* and *Ped. damnosus* are considered major beer spoilers. *L. brevis* has been reported as the most frequently detected LAB species in spoiled beer products, as well as in fermentation and maturation processes (Back, 2005a), and hence most extensively studied in brewing microbiology. *L. brevis* is widespread in the food industry and natural environments and is generally known to be physiologically versatile in that this species grows relatively well on many laboratory culture media and in temperature ranges wider than most other beer-spoilage LAB species. However, the beer-spoilage ability of *L. brevis* varies considerably depending on the strain and the source of isolation (Back, 2005a; Suzuki, Iijima, Sakamoto, et al., 2006). Some strains spoil almost all kinds of beer, causing turbidity, sediment and acidification but produce no diacetyl off-flavour. In contrast, *L. brevis* strains isolated from sources other than brewing environments generally exhibit no or very weak beer-spoilage ability (Kern, Vogel, & Behr, 2014; Nakagawa, 1978; Suzuki, 2009). For

TABLE 7.2 Percentages of beer-spoilage microorganisms in incident reports in Germany during 1980–2002 periods.[a]

Genus/species[b]	1980–90	1992[c]	1993[c]	1997	1998	1999	2000	2001	2002
L. brevis	40	39	49	38	43	41	51	42	51
L. lindneri	25	12	15	5	4	10	6	13	11
L. plantarum	1	3	2	1	4	2	1	1	2
L. casei/paracasei	2			6	9	5	8	4	4
L.coryniformis	3			4	11	4	1	3	6
Ped. damnosus	17	4	3	31	14	12	14	21	12
Pectinatus	4	28	21	6	3	6	5	10	7
Megasphaera	2	7	3	2	2	4	4	4	2
Saccharomyces wild yeasts	N.A.	5	5	7	6	11	5	2	3
Non-*Saccharomyces* wild yeasts	N.A.	0	0	0	3	4	5	0	2
Others	N.A.	2	2	0	1	1	0	0	0

[a]*This table is adapted from the studies conducted by Back (1994b, 2003, 2005a) during 1980–2002. N.A.: not available.*
[b]Lactobacillus brevis *includes* Lactobacillus brevisimilis, *which exhibits morphological similarities to* L. brevis. *According to Back,* L. brevis *in this table consists of several types on the basis of carbohydrate fermentation profiles, arginine utilization pattern, and morphological features, suggesting that this group of lactic acid bacteria may be further divided into separate species.*
[c]*In 1992 and 1993 studies,* Lactobacillus plantarum, Lactobacillus casei, Lactobacillus paracasei, *and* Lactobacillus coryniformis *were consolidated into one group.*

TABLE 7.3 Percentages of beer-spoilage microorganisms in incident reports in Europe during 2010–16 periods.[a]

Genus/species[b]	2010	2011	2012	2013	2014	2015	2016
L. brevis	49.2	35.4	40.0	44.7	36.8	31.1	25.6
L. lindneri	10.2	10.8	4.3	0.0	3.5	4.9	6.8
L. backii	5.1	13.8	8.6	10.6	8.8	11.5	9.4
L. collinoides/paracollinoides	0.0	1.5	7.1	2.1	5.3	1.6	6.0
L. buchneri/parabuchneri	3.4	7.7	1.4	10.6	5.3	1.6	6.8
L. rossiae	1.7	0.0	0.0	2.1	1.8	3.3	6.8
L. perolens/harbinensis	3.4	3.1	1.4	4.3	7.0	6.6	9.4
L. casei/paracasei	8.5	10.8	14.3	8.5	7.0	13.1	8.5
L. (para)plantarum/coryniformis	6.8	4.6	1.4	8.5	8.8	8.2	8.5
Ped. damnosus	1.7	4.6	14.3	6.4	8.8	8.2	6.8
Ped. inopinatus	0.0	0.0	1.4	2.1	0.0	0.0	0.0
Ped. claussenii	0.0	1.5	0.0	0.0	0.0	0.0	0.0
Other *Pediococcus* spp.	0.0	0.0	0.0	0.0	0.0	1.6	0.9
Pectinatus spp.	3.4	4.6	4.3	0.0	5.3	8.2	1.7
Megasphaera cerevisiae	6.8	1.5	1.4	0.0	1.8	0.0	2.6

[a]*This table is adapted from the studies conducted by Hutzler et al. (2012), Koob et al. (2014) and Schneiderbanger et al. (2018).*
[b]*The identification of the genus/species was performed with polymerase chain reaction (PCR) analysis.*

these reasons, intraspecies differentiation of beer-spoilage ability in *L. brevis* is important in the brewing industry. Some strains of beer-spoilage *L. brevis*, formerly known as *L. frigidus*, produce extracellular capsules and show resistance to disinfectants used in breweries and may tolerate up to 25 pasteurisation units (Back, 2005a). This particular subgroup of *L. brevis* strains causes severe hazes, sediments and ropiness in beer. In a few cases, another subgroup of *L. brevis* strains, previously known as *L. diastaticus*, utilise dextrin, causing the superattenuation of worts during fermentation (Briggs et al., 2004). On the other hand, *L. lindneri* is highly tolerant to hop compounds and grows optimally at 19−23°C (Back, 2005a). It is also reported that *L. lindneri* is unable to grow at temperatures higher than 28°C. Nonetheless, this species is known to tolerate relatively high thermal treatment (up to 15 pasteurisation units) and sometimes survive suboptimal pasteurisation process (Back, Leibhard, & Bohak, 1992). Furthermore, *L. lindneri* grows poorly on many laboratory detection media described in the brewing industry and often causes spoilage incidents without being detected by microbiological quality control (QC) tests (Suzuki, Asano, Iijima, Kuriyama, & Kitagawa, 2008). *L. lindneri* causes relatively faint haze and sediment with very little off-odour formation in beer (Back, 2005a). The occurrence outside brewing environments has rarely been reported for this species, although it is suggested that an LAB species closely related to *L. lindneri* was isolated from wine grapes and wine-making processes (Back, 2005b; Suzuki, 2012). One striking observation is that *L. brevis* and *L. lindneri* strains grown in beer exhibit reduced cell size and more easily penetrate membrane filters used for the removal of microorganisms (sterile filtration) in the brewing industry (Asano et al., 2007).

Beer spoilage caused by *Ped. damnosus* is characterised by acid formation and buttery off-flavour of diacetyl (Back, 2005a). The amount of diacetyl produced by *Ped. damnosus* is high and often noticeable even with the low level of contamination (as low as 20,000 cells/mL (Suzuki, 2020)). Some strains of *Ped. damnosus* produce exopolysaccharides, making beer ropy and gelatinous. *Ped. damnosus* is found commonly as a contaminant in pitching yeast and beer but not found in brewing raw materials, suggesting that this species is particularly well adapted to the brewing environment (Priest, 2003). *Ped. damnosus* has a long association with brewing microbiology and was originally known as sarcina because their cell morphology was confused with the cubical packets of 8 cells of *Sarcinia* spp. (Briggs et al., 2004). *Ped. damnosus* is also known as one of the most frequent contaminants in fermentation and maturation processes, due partly to its ability to grow at low temperatures (Back, 2005a). The unexpected rise in diacetyl level during the fermentation and maturation process is often caused by the presence of *Ped. damnosus*. In addition, *Ped. damnosus* is reported to adhere to brewing yeast and sometimes induce premature sedimentation of yeast cells, resulting in retardation of the fermentation process (Priest, 2003; Suzuki, Shinohara, & Kurniawan, 2021). The adherence to the brewing yeast has been observed for *L. lindneri*, as well (Storgårds, Pihlajamäki, & Haikara, 1997), suggesting that these two species tend to be latent in fermentation and maturation processes. Furthermore, *Ped. damnosus* is known as a slow grower on laboratory detection media and often requires some beer-specific components for growth (Back, 2005a). *Ped. damnosus* grows at rather low temperature, and its optimum temperature lies around 22−25 °C. Therefore, the incubation temperature of laboratory detection media used in quality control (QC) tests should be kept relatively low (typically 25−28°C) to comprehensively detect beer-spoilage LAB species, including *Ped. damnosus* and *L. lindneri*. In addition, the *Ped. damnosus* species is known to preferentially grow under relatively CO_2-rich environments and almost exclusively isolated from beer-brewing and wine-making environments (Back, 2005b). On the other hand, *Ped. inopinatus* is detected in pitching yeast but rarely in other stages of brewing processes. This species is reported to grow in beer at pH values above 4.2 and with low contents of hop bitter acids and ethanol (Sakamoto & Konings, 2003). The production of diacetyl by *Ped. inopinatus* is generally weak and less noticeable than that of *Ped. damnosus* (Priest, 2003).

L. paracollinoides, *L. backii* and *Lactobacillus paucivorans* have been proposed as novel species since the 2000s (Hutzler et al., 2013; Suzuki, Funahashi, et al., 2004). The genetic characterisation indicates that *L. paracollinoides* and *L. backii* are closely related to *L. collinoides* and *L. coryniformis*, respectively. Accordingly, some of the strains belonging to *L. paracollinoides* and *L. backii* might have been misidentified as *L. collinoides* and *L. coryniformis* in the past. In addition to the above-mentioned species, *L. acetotolerans* has been recently recognized as a beer-spoilage species (Deng et al., 2014). Similar to the case with *L. lindneri*, *L. paracollinoides* and *L. acetotolerans* show very poor culturability on many conventional culture media, which is especially true upon primary isolation from brewing environments (Shimokawa & Suzuki, 2021; Suzuki, Asano, Iijima, Kuriyama, & Kitagawa, 2008). This is probably the main reason that these *Lactobacillus* species had remained uncharacterised and underreported until recently. *L. acetotolerans* and/or its closely related species are also isolated from vinegar and sake brewing environments and known as hard-to-cultivate LAB in those industries (Suzuki, 2012). In addition, *Ped. claussenii* has been reported as a beer-spoilage species (Dobson et al., 2002). Some strains of *Ped. claussenii* produce exopolysaccharides. All the strains of *L. paracollinoides*, *L. backii*, *L. paucivorans* and *Ped. claussenii* characterised to date have been isolated from brewing environments and therefore considered as unique LAB species to the brewing industry. Other LAB species, such as *L. curtus* and *L. cerevisiae*, have been proposed as novel beer-spoilage species (Suzuki, 2020). In addition, the conventionally established LAB species, such as *Lactobacillus rossiae* and *Lactobacillus nagelii*, have been recognized to include beer-spoilage strains (Suzuki, 2020).

In contrast, *Lactobacillus casei/paracasei*, *L. coryniformis* and *Lactobacillus plantarum* are ubiquitously found in nature and exhibit relatively weak hop tolerance. Therefore, these *Lactobacillus* species spoil only weakly hopped beers or those with elevated pH values (Back, 2005a). Although the frequencies of spoilage incidents by these relatively hop-sensitive lactobacilli are generally low, they are known to cause diacetyl off-flavour in beer. One trend to be noted, however, is that the spoilage incidents by *L. (para) casei* appear to have increased since 2010 (Table 7.3). This trend should be watched more closely in future surveys. *Lactobacillus curvatus*, *Lactobacillus malefermentans* and *Pediococcus dextrinicus* were also recognized as beer-spoilage LAB species (Farrow, Phillips, & Collins, 1988; Sakamoto & Konings, 2003), but the spoilage incidents by these LAB species appear to be rare and are now considered as potential beer-spoilage LAB species of less importance (Hutzler et al., 2013). Currently, approximately 20 LAB species have been recognized as obligate or potential beer spoilers (Table 7.4), but the strain-dependent differences in beer-spoilage ability within the identical species are often observed (Suzuki, 2012). Additionally, the beer-spoilage ability of LAB strains is substantially affected by their physiological conditions (the degree of adaptation to hop bitter acids) and the beer types (bitterness units, pH values, ethanol contents and other antibacterial factors). These aspects of beer-spoilage LAB will be discussed later in this chapter.

Apart from spoilage incidents of beer products, certain thermophilic lactobacilli, including *Lactobacillus delbrueckii*, have been noted as contaminants of sweet wort. They are killed by the boiling process, but if the wort is kept sweet for an extended period, even stored hot (*ca* 60°C or below), thermophilic lactobacilli spoil sweet wort by producing lactic acid (Priest, 2006).

7.2.3 Association of beer-spoilage LAB with their habitat

L. brevis, the most frequent beer-spoilage species in the brewing industry, is isolated from a diverse source of environments, including milk, cheese, silage, faeces and the intestinal tracts of mammals (Kandler & Weiss, 1986). Nonetheless, it seems that beer-spoilage ability is not an innate character for *L. brevis*. In fact, *L. brevis* strains, isolated from sources other than beer-brewing environments, were reported to generally lack beer-spoilage ability (Nakagawa, 1978). In contrast, beer-spoilage *L. brevis* strains are typically isolated from brewing environments and are suggested to possess numerous layers of hop tolerance mechanisms that appear to have been acquired in a stepwise manner during their long history of beer adaptation processes (Behr, Gänzle, & Vogel, 2006; Behr, Israel, Gänzle, & Vogel, 2007; Suzuki, 2009). Interestingly, the sequencing analysis of *gyrB* indicated that beer-spoilage *L. brevis* strains form a distinct subgroup within this species (Nakakita, Maeba, & Takashio, 2003). In addition, the comparative study on electrophoretic mobilities of D-lactate dehydrogenase (LDH) supported that beer-spoilage *L. brevis* is a phylogenetically distinct subgroup that can be discriminated from non-spoilage *L. brevis* (Takahashi, Nakakita, Sugiyama, Shigyo, & Shinotsuka, 1999). The recent study using the matrix-assisted laser desorption ionization time-of-flight mass spectrometry (MALDI-TOF MS) showed that the MS spectrum profiles of strongly beer-spoilage strains cluster closely together and share remarkable similarities among this specific *L. brevis* group (Kern et al., 2014), further supporting that beer-spoilage *L. brevis* forms a phylogenetically distinct subgroup within the species of *L. brevis*. From a phenotypic viewpoint, it has been shown that beer-spoilage *L. brevis* strains tend to show preference for maltose over glucose as a fermentable sugar (Rainbow, 1981), suggesting beer-spoilage *L. brevis* is well adapted to brewing environments where maltose is a more abundant sugar source. From these findings, it is conceivable that a particular subgroup of *L. brevis* strains chose brewing environments for their habitats and evolved along the history of brewing. In addition, one proteomic study of hop-tolerant *L. brevis* TMW 1.465 showed that up to 84% of the investigated proteins were identified based on the genome sequence data of *L. brevis* ATCC 367, a non-spoilage strain isolated from silage (Behr et al., 2007), suggesting that approximately 20% of the genome was not shared between these two strains. From an evolutional standpoint, these observations suggest that considerable portions of the genome have been acquired (or lost) through the long association with brewing environments, which led to the diversification of beer-spoilage *L. brevis* from non-spoilage counterparts of this species.

L. lindneri, *L. paracollinoides* and *Ped. damnosus* are also frequently encountered species that exhibit strong beer-spoilage ability (Back, 2005a; Iijima et al., 2007). With the exception for *Ped. damnosus*, which is occasionally found in wineries, these species have been almost exclusively isolated from beer and related environments (Back, Bohak, Ehrmann, Ludwig, & Schleifer, 1996; Storgårds & Suihko, 1998; Suzuki, 2012), indicating that *L. lindneri*, *L. paracollinoides* and *Ped. damnosus* are brewery-associated microorganisms. Furthermore, it has also been reported that certain components in beer promote the growth of *L. lindneri*, *L. paracollinoides* and *Ped. damnosus* strains (Back, 2005a; Suzuki, 2012), suggesting the strong adaptation of these species to brewing environments. In addition to these

TABLE 7.4 Beer-spoilage gram-positive bacteria and their microbiological characteristics.[a]

Species	Beer-spoilage ability[b]	Primary/secondary contamination[c]	Exopolysaccharide formation[d]	Diacetyl production[e]	Culturability on MRS agar[f]
L. acetotolerans	+	s > p	−	N.A.	Poor
L. backii	++	p > s	−	−	Presumably good
L. brevis	++	s > p	+	−	Relatively good
L. buchneri/ parabuchneri	+	p > s	+	−	Presumably good
L. casei/paracasei	+	s > p	−	+	Good
L. coryniformis	+	s > p	−	+	Good
L. collinoides/ paracollinoides	++	s > p	−	+	Poor
L. lindneri	++	p > s	−	−	Poor
L. perolens/ harbinensis	+	s > p	−	+	Presumably good
L. paucivorans	++	p	−	N.A.	Presumably good if fructose is supplemented
L. plantarum	+	s > p	−	+	Good
L. rossiae	+	s > p	+	−	Presumably good
Lactococcus lactis	−/+	s > p	−	+	Good
Leuc. mesenteroides/ paramesenteroides	−/+	s > p	+	+	Good
Kocuria kristinae	−/+	s	−	+	N.A.
Ped. claussenii	+	p > s	+	+	Presumably good
Ped. damnosus	++	p > s	+	+	Poor
Ped. inopinatus	+	p > s	−	−/+	N.A.

[a]*This table is adapted from the review authored by Hutzler et al. (2013) with some modifications (Back, 1981, 2005a; Garg et al., 2010). See the review (Hutzler et al., 2013) and relevant literature for more details.*
[b]*++: strong beer-spoilage ability, +: intermediate beer-spoilage ability, −/+: weak or negative beer-spoilage ability.*
[c]*p: primary contamination, s: secondary contamination, p > s: more cases with primary contaminations observed, s > p: more cases with secondary contaminations observed.*
[d]*+: strain-dependent production of exopolysaccharides may be observed, which makes beer ropy.*
[e]*N.A.: information not available, −/+: most strains produce less noticeable amount of diacetyl.*
[f]*N.A.: sufficient information is not available. 'Presumably good' indicates that at least some strains have been reported to grow on MRS agar. It is possible that some others belonging to the same species show hard-to-cultivate characteristics on MRS agar.*

observations, the sugar utilization profiles of *L. lindneri*, *L. paracollinoides* and *Ped. damnosus* are relatively narrow (Suzuki, Asano, Iijima, & Kitamoto, 2008). The narrow sugar utilization profile was also noted for *L. paucivorans,* and *L. backii* has been reported to ferment fewer sugars than its closest species, *L. coryniformis*. This is a feature often noted with those highly adapted to a particular environmental niche. For instance, *L. delbrueckii* sub. *bulgaricus*, a yogurt-producing LAB species, is able to utilise only a few sugars, including lactose (Suzuki, Asano, Iijima, & Kitamoto, 2008). This is in contrast to other subgroups of *L. delbrueckii*, such as *L. delbrueckii* subsp. *lactis,* which can ferment a much wider spectrum of sugars. These differences are regarded as an indication that *L. delbrueckii* sub. *bulgaricus* is deeply adapted to milk environments, in which lactose is the predominant sugar source. Therefore, beer-spoilage LAB, such as *L. lindneri*, *L. paracollinoides* and *Ped. damnosus*, might have lost the ability to utilise a wide variety of sugars due to the deep associations with a particular environmental niche. Taken together, these findings strongly suggest the close associations of *L. lindneri*, *L. paracollinoides* and *Ped. damnosus* with brewing environments.

Furthermore, strains belonging to these three species show hard-to-cultivate characteristics on primary isolation, and often fail to grow on de Man, Rogosa and Sharpe (MRS) agar and Raka-Ray agar, the laboratory detection media widely recommended for beer-spoilage LAB by major brewery associations, such as European Brewery Convention, American Society of Brewing Chemists and Brewery Convention of Japan (Suzuki, Kuriyama et al., 2008). Interestingly, recent studies have shown that repeated passages in beer gradually reduce the culturability of originally easy-to-cultivate strains of *L. lindneri* and *L. paracollinoides*, eventually leading to the acquisitions of hard-to-cultivate strains that mimic the state of primary isolates of these species (Shimokawa & Suzuki, 2021; Suzuki, 2012). Strikingly, it was observed that highly beer-adapted strains of *L. lindneri* and *L. paracollinoides* die swiftly on MRS agar, a behaviour that is in sharp contrast to that of the easy-to-cultivate counterparts of the same species that grow well on MRS agar. It has been increasingly recognized that microorganisms too deeply associated with a particular environment tend to exhibit hard-to-cultivate characteristics (Suzuki, 2012). These lines of evidence therefore suggest once again the profound association of these species with brewing environments and further indicate that the physiological characteristics of beer-spoilage LAB living in brewing environments are drastically different from those of laboratory strains maintained in nutrient-rich media (Shimokawa & Suzuki, 2021; Suzuki, Kuriyama et al., 2008). It should be also noted that because of the hard-to-cultivate nature of *L. lindneri*, *L. paracollinoides* and *Ped. damnosus*, these LAB species may have been underreported as causative agents of microbiological incidents in the brewing industry.

Taken collectively, these observations suggest that the beer-spoilage LAB species have long been associated with brewing environments, and a deeply beer-adapted status presumably represents their intrinsic state in nature. Therefore, these features should be taken into account when brewing microbiologists develop any QC methods for beer-spoilage LAB. This aspect of beer-spoilage LAB will be further discussed in the third section of this chapter.

7.2.4 Factors affecting the growth of LAB in beer

Growth capability of beer-spoilage LAB depends on the strain and the type of beer. In one study, the ability of 14 strains of hop-tolerant LAB (*Lactobacillus* spp. and *Pediococcus* spp.) to grow in 17 different beers was assessed using a biological challenge test (Fernandez & Simpson, 1995). A statistical analysis of the relationship between spoilage potential and 56 parameters of beer composition revealed a correlation with eight parameters: pH, beer colour, the content of free amino nitrogen, total soluble nitrogen, a range of individual amino acids, maltotriose and the undissociated forms of SO_2 and hop bitter acids. Among them, the correlation coefficient of pH value and undissociated hop bitter acids were found to be −0.72 and 0.70, respectively, suggesting that these two factors have strong influence on the sensitivity of beers to spoilage by LAB. Hop bitter acids are hypothesised to act as mobile ionophores, and their activity is pH dependent. Low pH favours antibacterial activity, but high pH reduces it. Small changes in beer pH are known to cause large changes in the antibacterial activity of hop bitter acids in beer. For instance, it has been shown that an increase in pH value of as little as 0.2 can reduce the protective effect of hop compounds by 50% (Simpson, 1993). However, it seems evident that factors other than pH and hop bitter acids are also influential in determining the susceptibility of beer to spoilage by LAB.

As growth inhibitors, the phenolic compounds, such as phytic acid and ferulic acid, have been shown to be antimicrobial in beer, and the antimicrobial effects of ferulic acid are significantly enhanced when it is converted enzymatically to 4-vinyl guaiacol (Hammond, Brennan, & Price, 1999). In addition, undissociated SO_2 seems to have a negative effect on the growth of LAB in beer (Fernandez & Simpson, 1995). Carbon dioxide, which is considered as a growth promoter for *Lactobacillus* and *Pediococcus* at low concentrations (<0.3 g/L), was also found to be inhibitory at the concentrations found typically in beer, indicating that beers with low carbon dioxide can be more prone to spoilage by LAB. Therefore, there is a need for particular attention to be given to hygiene when dealing with cask-conditioned beers of low carbon dioxide content and beers dispensed with nitrogen gas. It has also been suggested that the use of LAB in bioacidification of wort is beneficial in enhancing the microbiological stability of finished beer. This is because LAB produce lactic acid that lowers pH value of finished beer products (often below 4.3), and possibly form antibacterial compounds, including bacteriocins (Gänzle, 2004; Vaughan, O'Sullivan, & van Sinderen, 2005).

As growth promoters, citrate, pyruvate, malate and arginine in beer were shown to be utilised by beer-spoilage LAB (Geissler, Behr, von Kamp, & Vogel, 2016; Suzuki et al., 2005b). These four components were also found to yield ATP in beer-spoilage LAB, indicating that these energy sources help them grow in beer where nutrients are almost depleted by the fermentative activities of brewing yeast. As other nutrient sources in beer, varying amounts of maltose, maltotriose, maltotetraose, lysine and tyrosine were found to be consumed by beer-spoilage LAB strains. In some cases, dextrins up to 14 or 15 glucose units were hydrolysed (Lawrence, 1988). Taken together, the sensitivities of beer to spoilage by LAB are determined by various growth inhibitors and promoters present in beer, although pH value and undissociated hop bitter acids are predominant factors.

7.2.5 Probiotic potential of beer-spoilage LAB

Increasingly, *L. brevis* is recognized to possess beneficial effects on human health as probiotics. For instance, *L. brevis* KB290 was reported to be useful for early intervention in irritable bowel syndrome and to improve gut health (Waki et al., 2013), and particularly relevant for brewers are the studies conducted on *L. brevis* SBC8803 (Segawa, Nakakita, et al., 2008; Segawa, Wakita, Hirata, & Watari, 2008). This strain was isolated from barley malt for brewing and, according to the authors, exhibits growth capability in beer. The oral intake of *L. brevis* SBC8803 was suggested to alleviate not only allergic symptoms related to type I allergy but also alcoholic liver diseases, especially the development of alcohol-induced fatty liver. Although the use of LAB is currently limited to special beers such as Lambic beers, these studies indicate the promising potential of beer-spoilage LAB as probiotics. It is generally known that LAB surviving harsh environments are able to tolerate acidic conditions and bile acids encountered in human digestive systems and are more likely to show probiotic effects there. Beer can be considered as one of those harsh environments, and beer-spoilage LAB are able to survive in the brewing environments. In the future, these aspects of beer-spoilage LAB should be more vigorously studied in the brewing industry.

7.3 Hop tolerance mechanisms in beer-spoilage LAB

Hop tolerance in beer-spoilage LAB is a progressively evolving area of research, and many studies have been conducted to elucidate the inhibitory effects of hops and the tolerance to these inhibitory effects. In this section, recent progress in this area of research is briefly summarised and, in relevant cases, the examples for spoilage LAB in sake (Japanese rice wine) and wine are described for references. This is because these alcoholic beverages represent harsh environments characterised by low pH value and high ethanol content, and, in this sense, spoilage LAB in sake and wine exhibit responses similar to those of beer-spoilage LAB. More information concerning this aspect of spoilage LAB in alcoholic beverages is available from the previous literature (Suzuki, 2012).

7.3.1 Antibacterial effects of hop bitter acids

The antibacterial activities of α-acid (humulone) and β-acid (lupulone) have been studied since before 1950. Their antibacterial activities are higher than those of iso-α-acids, but these nonisomerised hop acids dissolve to a lesser extent in beer and water, so iso-α-acids are considered as a principal antibacterial agent in beer (Sakamoto & Konings, 2003). Antibacterial effects of hop bitter acids were extensively investigated by Simpson (1993) and Simpson and Fernandez (1994). According to a series of meticulous studies, hop bitter acids were found to act as protonophores and inhibit the growth of hop-sensitive LAB strains by dissipating the transmembrane pH gradient. In LAB, transmembrane pH gradient is an important component of proton motive force (PMF), providing mechanisms by which generation of energy (ATP) and its utilization for nutrient transport can be coupled. In addition, the intracellular pH influences nutrient transport and metabolic processes. It was thus suggested that the hop-induced reduction in intracellular pH leads to the inhibition of nutrient transport and thereby the starvation of hop-sensitive LAB strains. Accompanied by the dissipation of transmembrane pH gradient, the loss of intracellular Mn^{2+} was observed, and it was suggested that hop bitter acids exchange protons for cellular divalent cations, such as Mn^{2+}. In this hypothetical model, hop anions bind with intracellular divalent cations including Mn^{2+} and diffuse them out of the cell. Thus, the ionophoric action of hop bitter acids, together with the diffusion of the hop—metal complex, results in an electroneutral exchange of cations across the cytoplasmic membrane, leading to the growth inhibition of LAB (Sakamoto & Konings, 2003). Although Mn^{2+} is required for growth and survival of most bacteria, many LAB are known to have higher requirements of Mn^{2+} and accumulate high intracellular levels of Mn^{2+} (Groot et al., 2005; Vogel, Preissler, & Behr, 2010). Therefore, the loss of divalent cations, Mn^{2+} in particular, is presumably detrimental for the survival of LAB in hop-containing environments. On the other hand, it has been recently reported that another mode of antibacterial effects by hop bitter acids is a transmembrane redox reaction that occurs in the presence of intracellular Mn^{2+} and causes cellular oxidative damage (Behr & Vogel, 2010). Although more studies are needed to elucidate a comprehensive picture of antibacterial mechanisms by hop bitter acids, intracellular levels of Mn^{2+} and its interactions with hop bitter acids appear to play a major role in their antibacterial effects.

7.3.2 Hop tolerance mechanisms associated with cytoplasmic membrane

Because hop bitter acids are assumed to intrude into the cells as proton ionophores, it is important for beer-spoilage LAB to alleviate the intrusion of hop compounds into the cell. The *horA* and *horC* genes, originally identified in *L. brevis*, have been shown to confer hop tolerance on LAB. HorA, a product of the *horA* gene, was demonstrated to act as an ATP-binding cassette (ABC) transporter and to efflux hop bitter acids out of the cells (Fig. 7.1). It was also shown that HorA confers resistance to multiple drugs that are structurally unrelated to hop bitter acids, making this protein the second member of the multidrug ABC transporters ever discovered in bacteria (Sakamoto, Margolles, van Veen, & Konings, 2001). On the other hand, the presumed secondary structure of HorC is similar to those of PMF-dependent multidrug transporters belonging to the resistance—nodulation—cell division (RND) superfamily (Suzuki et al., 2005a). The functional expression of HorC in *L. brevis* demonstrated that this protein confers tolerance to hop bitter acids, as well as other structurally unrelated drugs. Therefore, HorC was postulated to function as a PMF-dependent multidrug efflux pump, and a defence mechanism similar to that of HorA was hypothesised (Fig. 7.1; Iijima et al., 2006). Accordingly, the activities of HorA and HorC presumably result in a reduced net influx of the undissociated and membrane-permeable hop bitter acids into the cytoplasm and thereby limit the antibacterial protonophoric effect of hop-derived compounds. However, since beer-spoilage LAB strains develop tolerance against rather high concentrations of hop bitter acids, the question arises as to whether functional expression of HorA and HorC is sufficient to confer hop tolerance or whether other activities could contribute to defence mechanisms against hop bitter acids.

Hop compounds are weak acids, which can cross the cytoplasmic membrane in the undissociated form (Simpson, 1993). Due to the higher intracellular pH, hop bitter acids dissociate internally, thereby dissipating the transmembrane pH gradient. As a result of this protonophoric action of hop bitter acids, the viability of the exposed bacteria decreases. On the other hand, microorganisms have been found to increase PMF-generating activities in their cytoplasmic membranes when they are confronted with a high influx of protons (Suzuki, 2012). Therefore, it is conceivable that, to defend against the antibacterial effects of hop bitter acids, beer-spoilage LAB strains respond by increasing the rate at which protons are expelled out of cells. In fact, the hop-tolerant LAB strains were found to maintain a larger transmembrane pH gradient than hop-sensitive strains (Simpson, 1993), and *L. brevis* was demonstrated to increase the activity and expression level of proton-translocating ATPase upon acclimatization to hop bitter acids (Sakamoto et al., 2002). These findings indicate that the extrusion of protons by proton-translocating ATPase counteracts the ionophoric effects of hop compounds and helps beer-spoilage LAB strains maintain the transmembrane pH gradient.

Given that the above defence mechanisms are energy consuming in nature, beer-spoilage LAB strains require substantial energy sources to grow in beer. Nevertheless, beer is generally considered as a poor medium to support the growth of bacteria because most of the nutrients have been depleted by brewing yeast. Furthermore, it has been reported that the protonophoric action of hop compounds inhibits the uptake of nutrients by bacteria (Simpson, 1993). Despite these disadvantages, beer-spoilage LAB strains are still capable of growing in beer. Indeed, three beer-spoilage LAB species, *L. brevis*, *L. lindneri* and *L. paracollinoides*, were found to exhibit strong ATP-yielding ability in beer (Suzuki et al., 2005b). The inoculation tests into beer indicated that citrate, pyruvate, malate and arginine were consumed to support the growth of spoilage LAB strains in beer. The four components induced considerable ATP production even in the presence of hop compounds, accounting for the ATP-yielding ability of the spoilage LAB strains observed in beer. In general, the metabolism of organic acids and amino acids in LAB is known to directly or indirectly enhance the energy production and PMF generation in conditions where nutrients are otherwise scarce. The putative metabolic pathways of these substrates have been discussed in the previous literature (Geissler et al., 2016; Suzuki et al., 2005b; Suzuki, Iijima, Sakamoto, et al., 2006).

In contrast to these active hop tolerance mechanisms described so far, passive defence mechanisms are also important, in which energy sources are not required once they are established. In *L. brevis*, the membrane composition was reported to change towards the incorporation of more saturated fatty acids, such as C16:0, rendering the membrane less fluid and

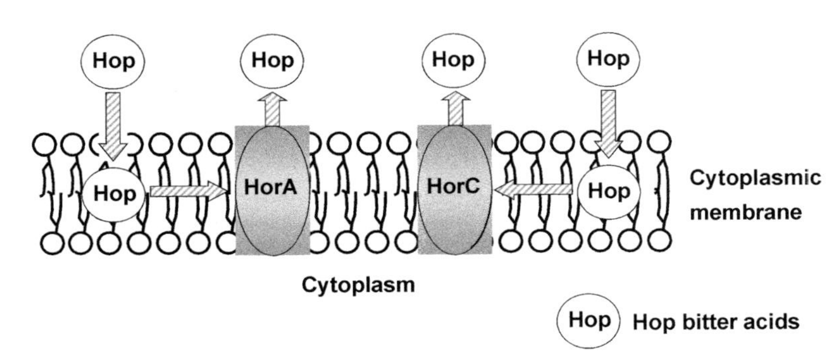

FIGURE 7.1 Efflux activities of hop bitter acids by HorA and HorC. HorA was shown to act as an ABC multidrug transporter and alleviate the intrusion of hop bitter acids into the cytoplasm. On the other hand, HorC was suggested to function as a proton motive force (PMF)—dependent multidrug transporter and to extrude hop bitter acids in a manner similar to that of HorA. In addition, HorC was postulated to act as a homodimer (Iijima et al., 2009). The secondary structures of HorA and HorC were described previously (Suzuki, 2012).

protecting the cell against the intruding hop bitter acids (Behr et al., 2006). This phenomenon is reminiscent of sake-spoilage *Lactobacillus fructivorans* that possesses long-chain fatty acids exceeding 24 carbons in length, which are not observed in ordinary LAB (Ingram, 1986). The proportion of these long-chain fatty acids reaches 30%–40% in the entire fatty acid compositions of the membrane, when sake-spoilage *L. fructivorans* is grown in an environment containing a high concentration of ethanol. It is presumed that these unusually long-chain fatty acids prevent the intrusion of ethanol into the cytoplasmic membrane. In a wine-associated lactic acid bacterium *Oenococcus oeni*, the changes in membrane fluidity coupled with the upregulation of heat-shock proteins lead to the reduction in the permeability of membrane and reinforcement of membrane structures, thereby protecting cells from the bactericidal effects of ethanol (Graça da Silveira, Vitória San Romão, Loureiro-Dias, Rombouts, & Abee, 2002; Grandvalet et al., 2008). From these observations, defence mechanisms associated with cytoplasmic membrane are generally important for LAB living in harsh environments.

7.3.3 Hop tolerance mechanisms associated with cell wall

In beer-spoilage *L. brevis*, it has been shown that higher-molecular-weight lipoteichoic acids (LTAs) in cell wall increase in response to the presence of hop bitter acids (Behr et al., 2006). These changes in the compositions of LTAs are suggested to reduce the intrusion of hop bitter acids into cells by increasing the barrier functions of the cell wall against hop bitter acids. LTAs are also hypothesised to act as reservoirs of divalent cations, such as Mn^{2+}, which are otherwise scarce as a result of complexation with hop compounds (Behr et al., 2006; Vogel et al., 2010). The altered LTAs have an increased potential to bind with divalent cations and compete for them with hop bitter acids, thus reducing the detrimental effects of hops towards the cells. This type of tolerance can also be considered as a passive defence mechanism that requires little energy once established. In relation to the reservoir function of LTAs for Mn^{2+}, Hayashi, Ito, Horiike, and Taguchi (2001) proposed HitA as one of the mediators of hop tolerance in *L. brevis* and suggested that HitA plays a role in the uptake of divalent cations, such as Mn^{2+}, whereas hop bitter acids reduce the intracellular divalent cations. Thus, HitA may modulate and maintain the levels of intracellular divalent cations in beer-spoilage LAB. In fact, many of the proteins involved in energy generation and redox homoeostasis are dependent on Mn^{2+}; therefore, intracellular Mn^{2+} plays an important role in LAB (Behr et al., 2007). Accordingly, these mechanisms relating to the cell envelope may function in concert for beer-spoilage LAB to counteract the loss of intracellular Mn^{2+}.

Defence mechanisms against toxic compounds involving the cell wall are also known for LAB in other alcoholic beverage industries. In sake-spoilage *L. fructivorans* and *Lactobacillus homohiochi*, for instance, the presence of ethanol has been reported to induce the increase in cell wall thickness (Suzuki, 2012). It was thus suggested that the increase in cell wall thickness is involved in ethanol tolerance observed in sake-spoilage LAB. In another instance, the *gtf* gene that encodes glucosyltransferase is known to exist in some strains of wine-associated *O. oeni* (Dols-Lafargue et al., 2008). The presence of this gene induces the formation of the cell envelope, consisting mainly of β-glucans, and elevates the ethanol tolerance of *O. oeni* strains that possess the *gtf* gene. Accordingly, the defence mechanisms associated with the cell wall appear to play a vital role for various LAB in alcoholic beverages.

7.3.4 Other hop tolerance mechanisms

It has been reported that Mn^{2+}-dependent enzymes are induced by hop bitter acids in *L. brevis* (Behr et al., 2007). These hop-inducible enzymes are suggested to be involved in energy generation and redox homoeostasis. One explanation for this phenomenon is that the cells respond to Mn^{2+} limitations by upregulating these enzymes, thus compensating for the loss of Mn^{2+}-dependent enzyme activities caused by the reduced intracellular manganese availability. It has been therefore suggested that beer-spoilage LAB can cope with low intracellular manganese levels, where hop-sensitive LAB cannot maintain metabolic activities. Relatively recently, the antibacterial mechanisms of hop compounds have been suggested to involve proton ionophoric actions and redox-reactive uncoupler activities occurring in parallel (Behr & Vogel, 2010). Accordingly, it is plausible that beer-spoilage LAB have to cope with oxidative stress induced by hop compounds, in addition to PMF and intracellular Mn^{2+} depletion. Thus, the observed upregulation of Mn^{2+}-dependent enzymes responsible for redox homoeostasis is most likely part of defensive responses to the oxidative stress caused by hop bitter acids (Vogel et al., 2010). In this hypothetical model, intracellular Mn^{2+} can be a target for hop-induced oxidative stress; thus hop-tolerant LAB may attempt to alleviate the oxidative stress by actively lowering the concentrations of its intracellular target. Therefore, it is possible that beer-spoilage LAB modulate and maintain the appropriate levels of intracellular Mn^{2+} for survival in hop-containing environments. However, this aspect of hop tolerance mechanisms has not been fully explored and the potential control of intracellular Mn^{2+} levels by beer-spoilage LAB will have to be further examined in future studies.

On the other hand, the morphological shifts into smaller rods were observed in beer-adapted *L. brevis* and *L. lindneri* cells (Fig. 7.2; Asano et al., 2007). The diminished cell size is presumably due to the efforts by beer-spoilage LAB to reduce surface area that is in contact with beer. This is conceivable because beer contains many bactericidal factors, including hop compounds. In addition to reducing the defense perimeters, the minimized cell surface area presumably helps beer-spoilage LAB deploy membrane-bound tolerance mechanisms more efficiently (Suzuki, 2012). In similar cases, it has been observed that sake-spoilage LAB remain morphologically compact in the presence of high ethanol content, whereas ethanol-sensitive LAB tend to exhibit elongated cell forms (Suzuki, Asano, Iijima, & Kitamoto, 2008). From these observations, the reduced surface area that is in contact with external environments is probably advantageous for spoilage LAB that must survive in hostile milieu.

The hop tolerance mechanisms described above are summarized in Fig. 7.3. However, it should be noted that hop tolerance mechanisms are more complex than previously assumed. Presumably these multiple layers of defense systems in beer-spoilage LAB have been acquired progressively through centuries of brewing history. Undoubtedly these are only part of the whole tolerance mechanisms of beer-spoilage LAB, and novel defense mechanisms will be found in future. In addition, the inhibitory actions of hop compounds have recently been shown to involve oxidative stress. This newly found inhibitory mechanism goes beyond the proton ionophoric actions and Mn^{2+} depletion activities of hop compounds that have been traditionally accepted. It is hoped that more comprehensive pictures will emerge concerning the hop tolerance of beer-spoilage LAB, which eventually leads to the development of more accurate QC tests in breweries.

7.3.5 Hop tolerance genes and their distribution in beer-spoilage LAB

The distribution of hop tolerance genes, *horA* and *horC*, has been investigated, using 183 strains that consist of various species of LAB and frequent brewery isolates. As a result of PCR and Southern blot analysis, *horA* and *horC* homologues have been detected widely and almost exclusively in beer-spoilage LAB strains (Iijima et al., 2007; Sami et al., 1997; Suzuki et al., 2005a; Suzuki, Shinohara, & Kurniawan, 2021). Among the 104 beer-spoilage LAB strains investigated, 92 strains possessed *horA* homologues, whereas *horC* homologues were detected in 97 strains (Fig. 7.4). When LAB strains with weak beer-spoilage ability were included, the presence of *horA* and *horC* homologues was found to be almost completely exclusive in beer-spoilage LAB strains, indicating that *horA* and *horC* are uniquely associated with beer-spoilage LAB. Equally interestingly, PCR and Southern blot analysis indicated that the flanking DNA regions of the hop tolerance genes are detected simultaneously with *horA* and *horC* homologues in these beer-spoilage LAB strains (Suzuki, Iijima, Sakamoto, et al., 2006; Suzuki, Shinohara, & Kurniawan, 2021). From these observations, it is quite conceivable that the flanking open reading frames (ORFs) in *horA*- and *horC*-carrying gene clusters collectively confer hop tolerance on LAB, and indeed some of the ORFs in these gene clusters potentially encode proteins that may be involved in cell wall synthesis (Fig. 7.5).

From a practical point of view, it is interesting to note that all of the beer-spoilage LAB strains examined in the studies were found to possess at least one of the homologues of hop tolerance genes (Fig. 7.4). This insight indicates that *horA* and *horC* are excellent genetic markers for the species-independent determination of beer-spoilage ability of LAB strains. In addition, the combined use of *horA* and *horC* was proposed for the detection of as yet uncharacterised beer-spoilage LAB species, as well as the established spoilage species (Suzuki et al., 2005a). Therefore, *horA*- and *horC*-specific detection methods may find potential applications in microbiological QC in breweries. This aspect of *horA* and *horC* was more thoroughly reviewed in previous literature (Suzuki, 2012; Suzuki, Iijima, Sakamoto, et al., 2006). Other genetic markers for differentiating beer-spoilage ability of LAB are listed in Table 7.5. These species-independent methods have been

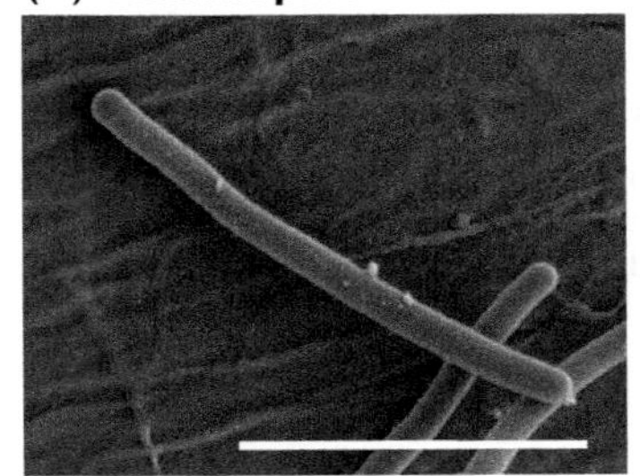

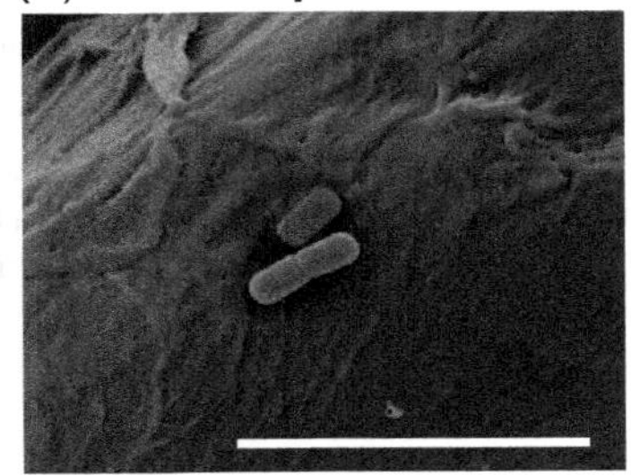

FIGURE 7.2 Effects of beer adaptation on morphological features of beer-spoilage lactic acid bacteria (LAB). Beer-spoilage *Lactobacillus brevis* was grown in MRS broth (a) and degassed beer (b). Cells were trapped on a membrane filter and the morphological features of beer-adapted and non-adapted strains were compared using scanning electron microscopy. Bar, 5 μm. Similar tendencies were also observed for beer-adapted *Lactobacillus lindneri* (data not shown).

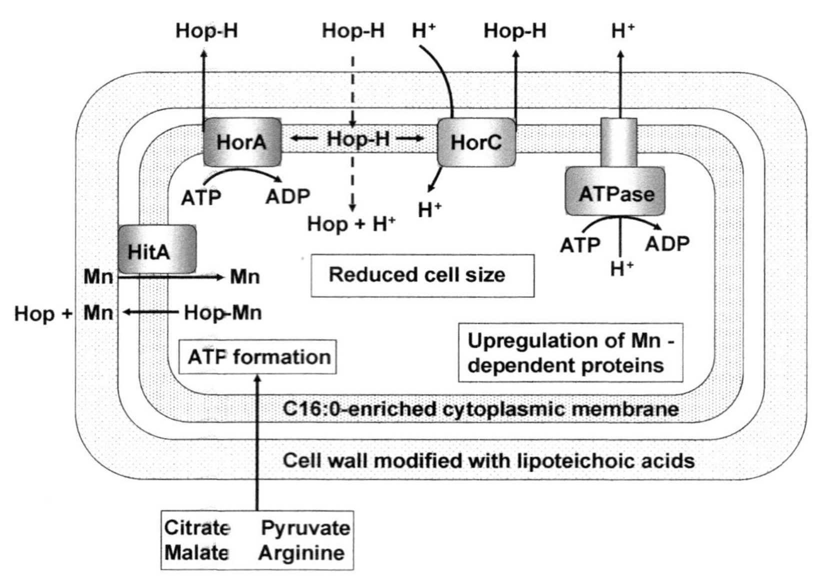

FIGURE 7.3 Complex hop tolerance mechanisms in beer-spoilage *Lactobacillus brevis*. Hop tolerance mechanisms so far reported are comprised of the following defence systems. (1) alleviation of ionophoric actions of hop compounds: Mechanisms for prevention of hop incursions involve HorA and HorC as efflux transporters, cytoplasmic membrane modifications, and cell wall modifications. These systems presumably function together to reduce the incursion of undissociated and membrane-permeable hop compounds (Hop-H). Proton-translocating ATPase also counteracts the proton ionophoric actions of hop compounds by pumping out intruding protons. (2) Mn^{2+} homoeostasis and countermeasures against the diffusions of intracellular Mn^{2+}: Intracellular Mn^{2+} levels are maintained by the actions of the putative Mn^{2+} transporter HitA. In addition, the modified cell wall functions as Mn^{2+} reservoirs, and presumably counteracts the loss of intracellular Mn^{2+}. Furthermore, Mn^{2+}-dependent proteins are upregulated in response to hop compounds. The upregulation of these proteins presumably compensates for the loss of their activities caused by reduced intracellular Mn^{2+} levels. It is also hypothesised that the upregulated Mn^{2+}-dependent proteins that are involved in redox homoeostasis counteract the oxidative stress conferred by hop compounds. (3) energy supply: Metabolisms with citrate, pyruvate, malate, and arginine supply ATP and proton motive force (PMF) for active defence mechanisms involving HorA, HorC, and proton-translocating ATPase. (4) other defence systems: Morphological shifts into smaller rods reduce the contact areas against hostile external milieu and help beer-spoilage lactic acid bacteria (LAB) to more efficiently deploy membrane-bound defence mechanisms, such as those driven by HorA, HorC, and proton-translocating ATPase.

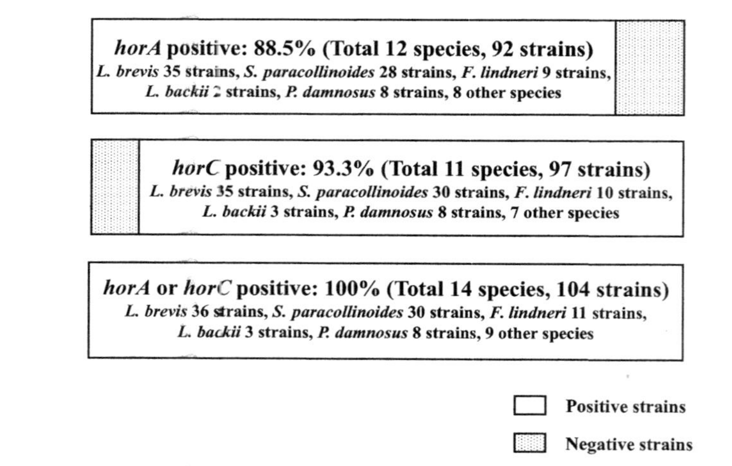

FIGURE 7.4 Compensatory relationship between *horA*- and *horC*-specific determination methods for beer-spoilage ability of lactic acid bacteria (LAB) strains. A total of 104 beer-spoilage strains belonging to various LAB species were examined by polymerase chain reaction and Southern blot analysis. It was shown that beer-spoilage LAB strains surveyed in this study possess at least one of the genetic markers, indicating that *horA* and *horC* are excellent genetic markers for comprehensibly determining beer-spoilage ability of LAB.

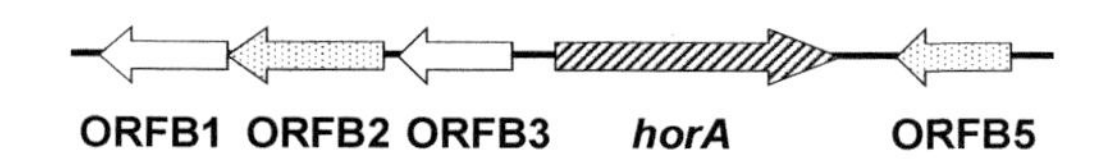

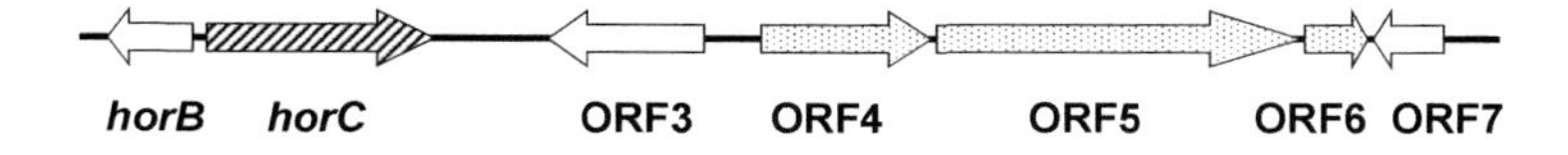

FIGURE 7.5 ORF structures of the gene clusters carrying *horA* and *horC*. (a) *horA* and its flanking DNA regions (*ca* 5.6 kb); (b) *horC* and its flanking DNA regions (*ca* 8.2 kb). The hop tolerance genes, *horA* and *horC*, are indicated by the striped arrows, and the open reading frames potentially involved in cell wall synthesis are shown by the dotted arrows.

TABLE 7.5 Hop tolerance(-related) genes and diagnostic marker genes for the determination of beer-spoilage ability of LAB.

Genes/genetic markers	Functions	Correlations with beer-spoilage ability	References
horA	ABC multidrug transporter	*Lactobacillus* spp. and *Pediococcus* spp.	Sami et al. (1997); Suzuki et al. (2005a); Haakensen et al. (2007); Ehrmann et al. (2010); Pittet et al., 2012; Deng et al. (2014)
horB	Putative transcriptional regulator of *horC*	*Lactobacillus* spp. and *Pediococcus* spp.	Suzuki et al. (2005a); Fujii et al. (2005); Iijima et al. (2006, 2007, 2008)
horC	Putative PMF-dependent multidrug transporter	*Lactobacillus* spp. and *Pediococcus* spp.	Suzuki et al. (2005a); Fujii et al. (2005); Iijima et al. (2006, 2007, 2008)
hitA	Putative Mg2+ transporter	*Lactobacillus brevis*	Hayashi et al. (2001); Behr et al. (2006)
bsrA	Putative ABC multidrug transporter	*Pediococcus* spp.	Haakensen et al. (2009)
gtfD15/ORFB5	Cell wall modification and fortification	*Lactobacillus* spp.	Suzuki, Sami, Ozaki, & Yamashit (2004); Feyereisen et al. (2020)
fabZ	Fatty acid biosynthesis	*Ped. damnosus* and possibly *L. backii*	Behr et al. (2016)
M37	Putative gene for DNA modification	*L. brevis*	Geissleer (2016); Suzuki et al. (2021)
ORF5 and adjacent DNA regions	Putative genes associated with cell wall synthesis	*Lactobacillus* spp. and *Pediococcus* spp.	Suzuki et al. (2004b, 2005a), Suzuki, Asano, Iijima, & Kitamoto, 2008; Fujii et al. (2005)

known to be useful for detecting previously unencountered beer-spoilage species and differentiating intraspecies differences in beer-spoilage ability of LAB, but it is possible that they have some false-positive and false-negative results (Suzuki, Shinohara, & Kurniawan, 2021). Therefore, these species-independent genetic markers should be used in conjunction with conventional species identification methods to conduct comprehensive QC tests.

7.3.6 Hypothetical origin of beer-spoilage LAB

From a phylogenetic standpoint, however, beer-spoilage LAB species are not closely related to each other (Fig. 7.6). As a matter of fact, beer-spoilage LAB species do not closely cluster together relative to other non-spoilage LAB species, on the basis of the phylogenetic comparison of their housekeeping genes. In contrast, based on a fragmented all-vs-all alignment of their total plasmid DNA sequences derived from a given strain, beer-spoilage LAB strains isolated from breweries cluster together, independent of their species status, indicating that beer-spoilage potential is more related to the plasmid compositions and isolation sources than to the species status (Geissler, Behr, Schmid, Zehe, & Vogel, 2017; Suzuki, Shinohara, & Kurniawan, 2021). These observations indicate that the relevant plasmid-localised genetic elements have

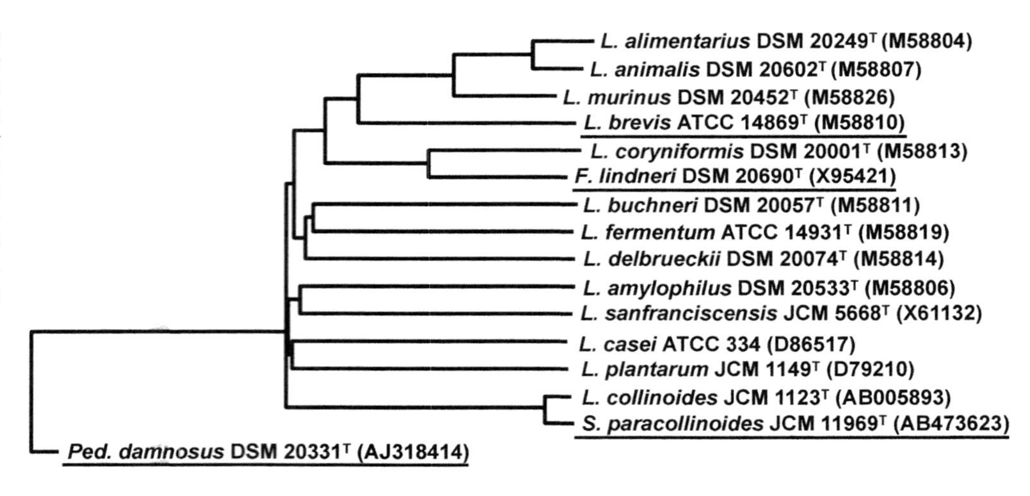

FIGURE 7.6 Phylogenetic tree of *Lactobacillus* and *Pediococcus* species derived from 16S rRNA gene sequence data, using neighbour-joining method for calculation. The bar indicates the number of inferred substitutions per 100 nucleotides. The accession numbers of 16S rRNA gene sequence are shown in parentheses, and the strong beer-spoilage species are underlined.

been horizontally acquired beyond the species boundaries by LAB strains inhabiting the brewing environments. The recent developments in this area of research suggest that hop tolerance genes, such as *horA* and *horC*, transformed originally nonspoilage LAB strains into beer-spoilage strains through plasmid- and transposon-mediated horizontal gene transfer (Fig. 7.7; Haakensen et al., 2007; Suzuki, Iijima, Sakamoto, et al., 2006). In addition, the comparative analysis of the gene clusters harbouring *horA* and *horC* indicates that their nucleotide sequences are approximately 99% identical among distinct beer-spoilage LAB species. These findings suggest that the acquisitions of hop tolerance genes by LAB strains were relatively recent events in the long history of LAB evolution (Fig. 7.8).

Beer-spoilage LAB strains have been almost invariably isolated from beers and related environments and therefore have been regarded as the microorganisms closely associated with brewing environments. Hop tolerance genes *horA* and *horC* have also been found almost exclusively in beer-spoilage LAB (Suzuki, Iijima, Sakamoto, et al., 2006). Therefore, beer-spoilage LAB and hop tolerance genes are unique to beers and related environments, although exceptions have been occasionally found (Suzuki, Shinohara, & Kurniawan, 2021). The insight that the presence of hop bitter acids is required in culture media for the maintenance of hop tolerance genes further indicates that beer-spoilage LAB and hop tolerance genes

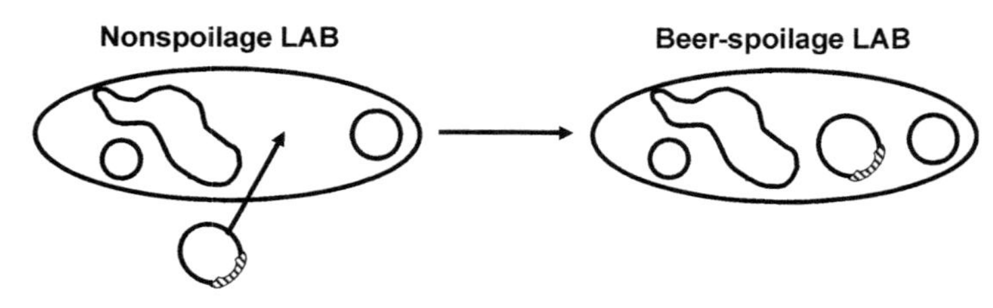

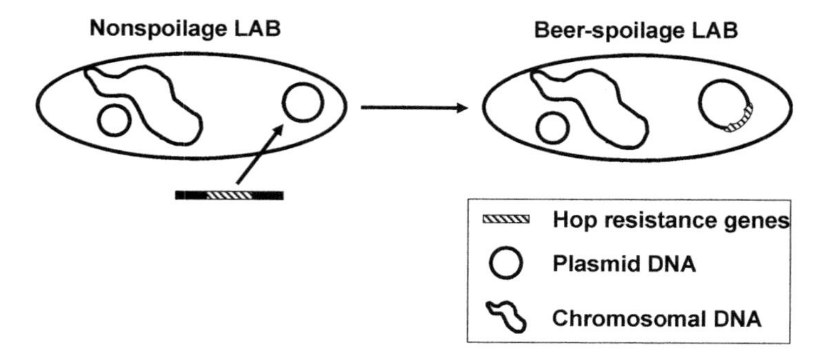

FIGURE 7.7 Hypothetical horizontal transfer of *horA* and *horC*. Two modes of horizontal transfer of hop tolerance genes, plasmid-mediated (a) and transposon-mediated (b) types, have been postulated on the basis of the nucleotide sequence identities and open reading frame analysis of *horA*- and *horC*-containing DNA regions identified in *Lactobacillus brevis*, *Lactobacillus lindneri*, *Lactobacillus paracollinoides*, *Lactobacillus backii*, *Pediococcus damnosus* and *Pediococcus inopinatus*. The exact mechanisms underlying the horizontal gene transfer of *horA* and *horC* are currently unknown, but several mechanisms, including conjugative transmission of hop tolerance genes, are postulated.

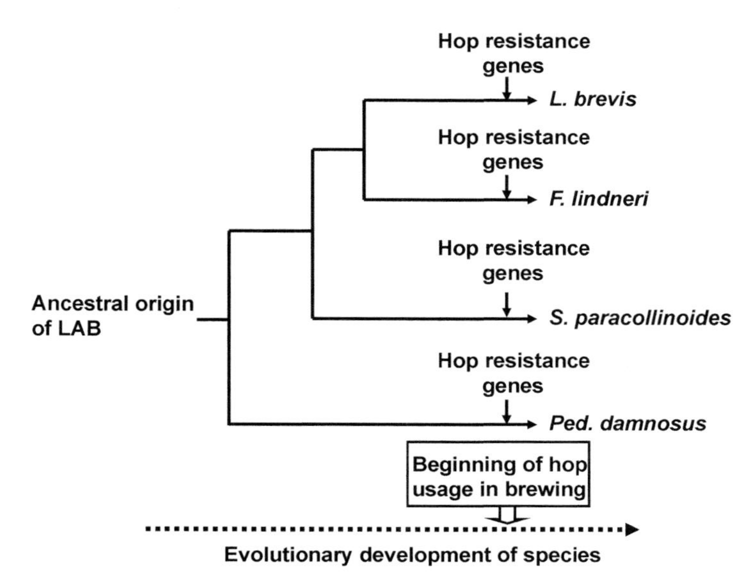

FIGURE 7.8 Hypothetical origin of beer-spoilage lactic acid bacteria (LAB) species. Based on the extraordinarily high sequence identities observed with various species of beer-spoilage LAB, the acquisitions of the hop tolerance genes are considered to be relatively recent events. Although the exact point of these events cannot be determined in the history of species evolution, it may have occurred as brewers widely adopted hop for a raw material in brewing.

have been inextricably linked with the historical development of hop use in brewing. Conceivably, hop tolerance genes chose LAB as companions for their own survival, and conversely LAB allowed the states of symbiosis with hop tolerance genes to continue in order to gain decisive advantages in brewing environments in which almost no other microorganisms can survive. In other words, beer-spoilage LAB and hop tolerance genes have developed mutually beneficial relationships along the long history of brewing.

At this time, it is difficult to determine exactly when beer-spoilage LAB emerged in the history of brewing. However, it seems increasingly likely that the progressively widespread use of hops in brewing has been responsible for the emergence and spread of beer-spoilage LAB and hop tolerance genes. The origin of the use of hops in brewing is still surrounded by controversy, and many aspects of the early cultivation of hops are unclear. It is, however, reasonable to assume that the cultivation of hops began in Central Europe sometime between the fifth and seventh centuries (Barth, Klinke, & Schmidt, 1994). On the other hand, the records of hop use in brewing has existed since around 1079 (Moir, 2000), so it is plausible that the first use of hops in brewing occurred between these periods. It is therefore tempting to imagine that beer-spoilage LAB and hop tolerance genes have emerged and spread with the increasingly widespread use of hops in brewing worldwide for the past 10–15 centuries.

7.4 Culture and preservation methods for beer-spoilage LAB

As discussed earlier in this chapter, beer-spoilage LAB strains have long been associated with brewing environments, and the deeply beer-adapted status presumably represents their intrinsic states in nature. On the basis of this hypothesis, beer-spoilage LAB strains to be used for the development of QC tests in breweries should be maintained and preserved as a culture stock so that they remain in a state as if they were living in the brewing environments. Otherwise, important physiological and genetic traits of beer-spoilage LAB might be changed during the subculture and preservation processes. In this section, it will be illustrated that the subculture and preservation methods affect various aspects of beer-spoilage LAB, and a new approach will be proposed to minimise the artefacts caused by the conventional subculture and preservation methods.

7.4.1 Stability of hop tolerance ability

Induction of hop tolerance ability in beer-spoilage LAB strains is important, especially when biological challenge tests are conducted to evaluate the microbiological stability of beer products. The levels of hop tolerance ability in beer-spoilage LAB depend largely on the preculture conditions. For example, hop tolerance ability of *L. brevis* strains can be elevated by adding the subinhibitory concentrations of hop bitter acids to culture media used for preculture. It has been reported that hop-adapted strains of beer-spoilage *L. brevis* exhibit 4- to 12-fold increased levels of hop tolerance ability compared with those of preadapted *L. brevis* strains (Simpson, 1993; Suzuki, Iijima, Sakamoto, et al., 2006). Conversely, strongly hop-tolerant *L. brevis* strains were shown to exhibit gradually diminished hop tolerance ability when subcultured with hop-free culture media. Therefore, subculture conditions substantially affect the hop tolerance ability; in particular, the presence or absence of hop bitter acids in preculture media plays an important role in determining the hop tolerance ability and beer-spoilage ability of the LAB strains tested. However, when laboratory culture media are used to induce hop adaptation, caution should be exercised because some components in media, such as Tween 80, inhibit or retard the hop adaptation process of beer-spoilage LAB strains (Simpson & Smith, 1992). Accordingly, MRS broth without Tween 80 was proposed as a base medium for hop adaptation. It is also the experience of the authors that the use of beer with somewhat weaker microbiological stability as a preculture medium is one good way to induce hop tolerance ability/beer-spoilage ability of a wide variety of spoilage LAB strains.

7.4.2 Stability of hop tolerance genes

It has been reported that the repeated subculture of beer-spoilage LAB strains often leads to the loss of the hop tolerance genes *horA* and *horC* (Suzuki, Iijima, Sakamoto, et al., 2006). This phenomenon typically occurs when laboratory culture media are used without hop bitter acids for subculture. It has also been observed that the loss of these hop tolerance genes is accompanied by the reduced levels of hop tolerance ability and/or beer-spoilage ability. This phenomenon was originally reported to occur in beer-spoilage *L. brevis* strains, but similar observations were subsequently made in beer-spoilage *L. paracollinoides*, *L. lindneri* and *Ped. damnosus* (Table 7.6; Suzuki, Iijima, Sakamoto, et al., 2006). The addition of hop bitter acids to laboratory culture media prevents or retards the loss of hop tolerance genes. In a few cases, however, the loss of hop tolerance genes was still observed after the repeated subcultures, even when laboratory culture media were supplemented with subinhibitory levels of hop bitter acids. In contrast, it was observed that the hop tolerance genes are stably maintained in degassed beer even after more than 100 subcultures, suggesting that *horA* and *horC* are required for growth in beer and may be responsible not only for hop tolerance but also for the tolerance to other hostile factors in beer. This notion is supported by the case with OmrA, a protein found in wine-associated *O. oeni* that shows 54% identity with HorA. OmrA has been demonstrated to confer tolerance not only to ethanol but also to multiple stress factors found in wine-making environments (Bourdineaud et al., 2004), suggesting that HorA and possibly HorC have much wider functions upon the survival in beer other than the roles as efflux pumps of hop bitter acids. At any rate, the use of beer itself as culture media seems important for maintaining the hop tolerance genes and concomitantly the beer-spoilage ability of LAB strains. However, it should be noted that the loss of hop tolerance genes occasionally occurs when the strains are repeatedly subcultured in beers to which laboratory culture media are added for supplemental nutrient sources. One example of this failure is MRS broth prepared with beer instead of water. Presumably, nutrients contained in laboratory culture media counteract the hostile factors in beer that are important for the maintenance of hop-tolerance genes.

Another important observation to be noted is that the loss of *horA* and *horC* regions sometimes occurs with freeze drying (lyophilisation), a method that is typically used for preserving bacterial cultures (Suzuki, Iijima, Sakamoto, et al., 2006). This phenomenon was observed when the type strain of *L. paracollinoides* was deposited in the Japan Collection of Microorganisms (JCM) and in the Deutsche Sammlung von Mikroorganismen und Zellkulturen GmbH (DSMZ), and their stock cultures showed declined beer-spoilage ability. Currently, JCM and DSMZ have restocked the strain at −80°C in a frozen state with suitable protectants to avoid the loss of the hop tolerance genes. Therefore, brewing microbiologists may have to be careful with the culture stock storage conditions of important beer-spoilage strains. In our experience, examining for the presence of *horA* and *horC* homologues is a useful indicator to determine whether the stock conditions of beer-spoilage LAB strains are suitable.

7.4.3 Culturability of beer-spoilage LAB

Culturability of beer-spoilage LAB strains is often changeable on QC detection media. As discussed earlier in this chapter, many beer-spoilage LAB strains are difficult to detect by conventional laboratory media. This is especially true for the primary isolation of beer-spoilage LAB strains, leading to the failure in the detection of beer-spoilage LAB by QC tests in breweries (Suzuki, 2012). The inability of beer-spoilage LAB to grow on laboratory detection media appears to be caused by the profound adaptation of these microorganisms to brewing environments (Back, 2005a; Shimokawa & Suzuki, 2021; Suzuki, 2012; Taskila, Kronlöf, & Ojamo, 2011). There appear to be several factors involved. Some beer-spoilage strains exhibit sensitivities to nutrients typically included in laboratory media, and others require beer-specific components as either growth-promoting factors or essential growth factors. Still others seem to prefer pH environments considerably lower than those found in laboratory culture media. However, in many cases, the initially hard-to-cultivate beer-spoilage strains acquire culturability on laboratory culture media when those LAB strains are gradually and stepwisely acclimatised

TABLE 7.6 The isolation of non-spoilage variants from beer-spoilage LAB and the loss of hop tolerance genes in non-spoilage variants.[a]

			Hop tolerance genes	
Species	**Strain no.**[b]	**Beer-spoilage ability**[c]	*horA*	*horC*
L. brevis	ABBC44	+	+	–
	ABBC44^{NB}	–	–	–
	ABBC45	+	+	+
	ABBC45^{CC}	–	–	–
	ABBC46	+	+	+
	ABBC46^{NB}	–	–	–
	ABBC64	+	+	+
	ABBC64^{NB}	–	–	–
	ABBC104	+	+	+
	ABBC104^{NB}	–	–	–
	ABBC400	+	+	+
	ABBC400^{NB}	–	–	–
L. paracollinoides	DSM 15502^{T}	+	+	+
	DSM 15502^{NB}	–	–	–
	LA9	+	+	+
	LA9^{NB}	–	–	–
L. lindneri	DSM 20692	+	+	+
	DSM 20692^{NB}	–	+	–
Ped. damnosus	ABBC478	+	+	+
	ABBC478^{NB}	–	–	+

[a]*The non-spoilage variants were obtained by repeatedly subculturing the wild-type strains in hop-free MRS broth at 37°C for* L. brevis, *30°C for* L. paracollinoides, *30°C for* L. lindneri, *and 35°C for* Ped. damnosus, *respectively. The superscripts NB and CC indicate the hop-sensitive variants obtained from beer-spoilage wild-type strains with the same strain number.*
[b]*ABBC and LA: Our culture collections, principally consisting of brewery isolates; DSM: Culture collections obtained from Deutsche Sammlung von Mikroorganismen und Zellkulturen.*
[c]*Beer-spoilage ability was evaluated using the degassed pilsner-type beers (pH 4.2, bitterness unit: 20 B U., ethanol content: 5.0%(v/v)).*

to the laboratory medium environments (Deng et al., 2014; Suzuki, 2012). These phenomena are observed in other fermentation industries as well. For instance, some of the initially hard-to-cultivate wine-spoilage *L. fructivorans* strains and dressing-spoilage *L. fructivorans* strains were suggested to gain culturability on laboratory media and eventually to grow well on those media (Suzuki, 2012). Conversely, the repeated subcultures in degassed beer were found to gradually reduce the culturability of initially easy-to-cultivate beer-spoilage *L. paracollinoides* and *L. lindneri* strains, eventually leading to the acquisition of hard-to-cultivate strains that mimic the state of primary isolates of these species (Suzuki, Asano, Iijima, Kuriyama, & Kitagawa, 2008). These studies indicate that the culturability of beer-spoilage LAB can change depending on the environments to which they are adapted (Shimokawa & Suzuki, 2021).

7.4.4 Subculture and preservation methods of beer-spoilage LAB

As far as the development of brewery QC tests is concerned, the most important aspects to note are the physiological/genetic traits and culturability of LAB strains to be used for the evaluation. This is because these characteristics are often changeable to a considerable extent, depending on the subculture conditions. For practical purposes, it is natural that LAB strains used for the development of a new microbiological QC test should closely mimic those actually encountered in beer products and manufacturing processes. In fact, beer-spoilage LAB rarely live in nutrient-rich environments, such as laboratory culture media, and the physiological/genetic traits and culturability of LAB strains living in the brewing environment are drastically different from those of so-called laboratory strains (Suzuki, 2009). Accordingly, if laboratory strains are to be used for the development of new microbiological QC tests, those tests may not be sufficiently suitable for practical applications. From the preceding observations, the subculture conditions have enormous impacts on the hop tolerance ability and culturability of beer-spoilage LAB strains. Considering that beer-spoilage LAB prefer the brewing environments as their habitats and that these LAB are innately adapted there, beer seems to be a natural choice for a subculture medium. The subcultures in beer or on beer agar appear to maintain the hard-to-cultivate states of beer-spoilage LAB and their hop tolerance ability/beer-spoilage ability. In contrast, the subcultures in laboratory culture media tend to change the culturability of the LAB strains and their hop tolerance ability/beer-spoilage ability, indicating that laboratory culture media are not suitable for maintaining beer-spoilage LAB strains. Therefore, the primary isolation and subsequent subcultures should be conducted using beer and beer agar as culture media, rather than traditional laboratory culture media. However, caution should be exercised upon the preparation of beer agar because the low pH of beer may dissolve agar matrix during the autoclaving and thereby hinder the solidification of agar. Accordingly, the pH of beer agar should be adjusted at around 5.0.

It is also important to prepare stock cultures as soon as the primary isolation process of beer-spoilage LAB strains is completed on beer agar. As previously stated in this chapter, the beer-spoilage ability of some LAB strains tends to decline by freeze drying, with the concomitant loss of hop tolerance genes (Suzuki, Iijima, Sakamoto, et al., 2006). In these cases, cryopreservations are more suitable. In one study, hard-to-cultivate beer-spoilage LAB strains belonging to *L. lindneri* and *L. paracollinoides* were grown in beer and concentrated by centrifugation (Suzuki, Iijima, Asano, et al., 2006). After the supernatants were discarded to remove toxic hop bitter acids, the cells were resuspended in 0.85% (w/v) NaCl solution. The hard-to-cultivate LAB strains were subsequently stored at −80°C with 10% (v/v) dimethylsulfoxide or 10% (v/v) glycerol as a cryoprotectant. After 3 months of storage, the strains were reconstituted and grown in degassed beer (pH 5.0) at 25°C. In this procedure, somewhat elevated pH of degassed beer as a recovery medium seems to improve the resuscitation rate of frozen culture stock by buffering the stress factors in beer. As a result of evaluating the reconstituted strains, no apparent changes were observed concerning the culturability and other genetic/physiological traits, suggesting this preservation method is useful for maintaining the original state of beer-spoilage LAB just as they are obtained as the primary isolates from the brewing environments. Nonetheless, the subculture and preservation method of freshly isolated beer-spoilage LAB strains has not been fully established in the brewing industry, and strain-dependent procedures may be required. It is therefore hoped that more studies will be conducted to improve these techniques.

7.5 Identification of emerging beer-spoilage LAB in the brewing industry

Since PCR was invented in 1983 by the American biochemist Kary Mullis, advancements in the field of molecular biology have made remarkable progress, creating a new era in molecular-based microbial classification. Nowadays, molecular-based (protein, DNA) microbial classification methods are widely adopted for identification of microorganisms in various fields, including for identification of beer-spoilage LAB in the brewing industry. These molecular-based methods are replacing traditional methods, such as biochemical and culture-based methods, which have been prominently used in the past for microbial classification of beer-spoilage LAB. Molecular-based methods effectively reinforce or provide alternative means for the traditional methods because they provide excellent identification accuracy, rapid analysis time and high-sensitivity detection of beer-spoilage LAB.

Nevertheless, the emergence of novel beer-spoilage LAB strains is on the rise since the 1990s. Many of these novel beer-spoilage LAB strains exhibit strong beer-spoilage ability. To date, approximately 20 LAB species belonging to several genera have been recognized as obligate and potential beer spoilers (Hutzler, Muller-Auffermann et al., 2013; Schneiderbanger, Jacob, & Hutzler, 2020). Moreover, non-traditional beverages, such as fruit-flavoured alcoholic beverages (FAB), low alcohol beers, non-alcoholic beers (NAB) etc., are soaring in popularity among drinkers who seek to reduce or eliminate their alcohol consumption for a healthier lifestyle. These non-traditional beverages are characterised by a lower alcohol content, a higher level of nutrients, elevated pH values and/or lower bitterness units. These products have lower microbiological stability than the traditional beer beverages (ale or lager). Products with low microbiological stability provide favourable environment for microbes to grow, hence are more prone to microbial contamination. Consequently, LAB that are not conventionally recognized as beer-spoilage LAB may emerge as spoilers in the non-traditional beverages, which could result in the continual emergence of spoilage LAB, leading to a significant increase of microbiological spoilage incidents. Without proper risk mitigation to prevent spoilage incidents from happening, the continual emergence of beer-spoilage LAB poses a significant threat to the brewing industry.

In this section, the current molecular-based microbiological QC methods and its development to address the problem caused by emergence of beer-spoilage LAB will be explained thoroughly. Although many different types of molecular-based methods exist, this section will focus only on the development of polymerase chain reaction (PCR)-based methods, which is the standard molecular-based method for the identification of beer-spoilage LAB.

7.5.1 Species-specific PCR for beer-spoilage LAB detection and identification method

PCR-based detection and identification methods are popularly used for routine microbiological testing in breweries, in acknowledgement of their ease of use, rapid, as well as sensitive and highly accurate identification results. They are commercially available mainly as a standalone kit or in full package with the detection system. Two of the most popular PCR-based detection methods are the conventional end-point PCR and real-time PCR methods.

The fundamental difference between these two methods is the PCR product detection mechanism. In end-point PCR, the PCR product is detected at the completion of PCR amplification and visualized by agarose or polyacrylamide gel electrophoresis for confirmation of the results. This method requires the use of short oligonucleotides, called primers. For detection of beer-spoilage LAB, the primers may be designed for genus or species-specific detection. These primers are mostly targeting the ribosomal RNA gene, for example, 16 and 23S rDNA, of LAB. In the case of species-specific detection, one primer set for each target species is required. The primers should be highly specific to the target species; otherwise, cross reactivity with nontarget species could lead to false positive results. Species-specific multiplex PCR assays, where multiple species-specific primers targeting different target species are pooled together in one reaction tube, have been developed to detect various beer-spoilage LAB, *Pectinatus* and *Megasphaera* spp. (Asano, Suzuki, et al., 2008; Iijima, Asano, Suzuki, et al., 2008). Recently, Rheonix Beer SpoilerAlert assay, a fully automated end-point PCR system capable for detection of over 47 different species of LAB in a single test, is available on the market. This system currently has the broadest detection capabilities of any end-point PCR assay on the market. According to the manufacturer, this system automatically performs cell lysis, DNA purification, end-point PCR amplification, detection and data analysis, with no user intervention required.

As opposed to the end-point PCR-based detection method, in real-time PCR-based detection method, as the name suggests, the PCR product is detected as the reaction progresses, in real time. Real-time detection of PCR products is enabled by measuring fluorescence intensity which gradually increases as the product accumulates over time. Several common real-time PCR-based detection methods are available, where probe-based real-time PCR methods become the most popular approach. In probe-based real-time PCR-based detection method, a fluorogenic probe capable of hybridizing specifically on target sequence between the two primers is used to increase detection sensitivity and specificity of the target PCR product. As a result, real-time PCR-based detection method has higher sensitivity and specificity than the end-point PCR-based detection method.

Commercial kits such as GEN-IAL QuickGEN and PIKA 4everyone Real Beer Spoiler test kit adopt the probe-based real-time PCR method for their species-specific beer-spoilage bacteria and yeast detection kit. Other commercial real-time PCR kits dedicated specifically to species-specific beer-spoilage bacteria and yeast detection include the Foodproof Beer Screening Kit. Kits are gaining popularity and have been more widely adopted for routine microbiological QC in the breweries.

Although end-point PCR and real-time PCR based species-specific detection kits are remarkably accurate and effective in detecting beer-spoilage LAB, there is a limit to the maximum number of species which could be detected by these kits. This is due to the fact that individual primers for each target species are required. As a result, only currently known beer-spoilage LAB species can be detected by the commercial kits. Moreover, detection capability of these kits is not on par with one another, as the detection scope of each kit is different, which further complicates the kit selection process. As mentioned briefly in the previous section, the emergence of novel beer-spoilage LAB species and the rising popularity of nontraditional beverages with lower microbiological stability will give rise to increasing numbers of beer-spoilage LAB at a rapid pace. Currently available commercial species-specific PCR-based detection kits will not be able to deal with as yet uncharacterized novel beer-spoilage LAB. It will be challenging and unpractical for kit manufacturers to keep up pace with the rising numbers of novel beer-spoilage LAB. Eventually, novel beer-spoilage LAB will evade detection and microbiological spoilage incidents may occur. Under these backgrounds, the applications of species-specific end-point PCR and real-time PCR based kits to detection and identification of beer-spoilage LAB will be increasingly limited. Hence, different approaches are practically necessary to cope with this emerging problem.

7.5.2 Species-independent PCR for beer-spoilage LAB detection and identification method

Since yet uncharacterised novel beer-spoilage LAB are out of detection scope for commercial species-specific detection kits, a more universal PCR-based detection method that is not restricted by species characterisation, the so-called species-independent PCR method, is necessary. Moreover, species information alone often does not give adequate information relating to the beer-spoilage potential and ability of LAB. Species-independent PCR detection method relies on detection of specific genetic markers that are independent of the species status of detected bacteria. These genetic markers are often shared among different genera and species of bacteria having common physiological traits.

As mentioned in the previous section, earlier studies suggested that hop tolerance genes e.g., *horA*, *horC*, etc., are excellent genetic markers for species-independent determination of beer-spoilage ability of LAB strains, including those that have not been or poorly characterised. In one recent study, *L. nagelii*, a previously unknown beer-spoilage LAB species, was isolated from spoiled pilsner-type beer and characterised for its beer-spoilage ability (Umegatani, Takesue, Asano, Tadami, & Uemura, 2022). It was found that the *L. nagelii* strain harbours a hop tolerance gene *horA* and caused turbidity in beer with relatively low microbiological stability (low bitterness beer) and therefore should be considered as a novel beer-spoilage LAB species. This study is just one proof of concept that hop tolerance genes are potentially useful genetic markers for determination of beer-spoilage ability of novel and uncharacterised LAB species. It also shows that depending solely on species-specific detection systems will fail to detect novel and uncharacterised beer-spoilage LAB as they will evade detection. Species-independent PCR detection kits, utilising beer-spoilage specific gene markers, for example, hop tolerance genes *horA* and *horC* and other characterised genetic markers, are available commercially.

Nevertheless, beer-spoilage LAB strains that do not harbour and non-beer-spoilage LAB strains that do harbour beer-spoilage specific gene markers have been reported, suggesting that the presence or absence of these genetic markers does not always correlate with the beer-spoilage ability of LAB. For instance, one recent study reported the existence of non-spoilage *Ped. damnosus* strains, major beer-spoilage LAB, harbouring beer-spoilage specific gene markers, e.g., hop tolerance genes *horA* and *horC* and other characterised genes (Behr et al., 2016). The study reported that beer-spoilage specific gene markers did not show a significant correlation to the beer-spoilage ability of *Ped. damnosus*. In contrast, *fabZ*, which encodes for a 3-hydroxyacyl-acyl-carrier-protein-dehydratase, shows a significant correlation, and therefore has the potential to be used as a genetic marker for the discrimination of beer-spoilage and non-spoilage *Ped. damnosus* strains. *fabZ* is a part of fatty acid biosynthetic gene cluster and does not represent a beer-spoilage specific gene. This does not mean that characterised beer-spoilage specific gene markers, e.g., hop tolerance genes *horA* and *horC*, play no roles in the beer-spoilage ability of *Ped. damnosus*. This study and other similar studies (Haakensen, Schubert, & Ziola, 2008; Bergsveinson & Ziola, 2017) suggest that the beer-spoilage ability of LAB strains is a complex multifactorial process, involving multi genes that possibly interact in a hierarchical manner. Therefore, species-independent PCR approaches alone are not sufficient to discriminate between the beer-spoilage and non-spoilage LAB. The combined use with species-specific PCR approaches will be necessary, giving the species-independent approach more emphasis on detection and identification of as yet uncharacterised novel LAB species, as it will enable brewers to conduct more reliable and comprehensive identification tests (Asano et al., 2019).

7.5.3 Third-generation DNA sequencing technology-based method for comprehensive detection and identification of beer-spoilage bacteria

Species-specific and species-independent PCR-based methods described in previous sub-sections are reliable, but nevertheless lack one essential feature in which both methods could not fully provide sufficient information on spoilage LAB strains belonging to as yet uncharacterised species. Species identification is certainly important as the number of spoilage species is increasing due to the ever-rising trend of non-traditional beverages. Therefore, a more comprehensive species identification method which allows accurate determination of broader range LAB species identity is certainly needed. Identification of these LAB species is only possible by using a sophisticated method such as DNA sequencing, a method that could provide nucleotide sequence information of a target DNA region. Moreover, the use of universal primers, primers that are not limited to species-specific detection but could universally detect a broad range of LAB species, is equally important. These universal primers, for example, primers that are targeting the 16S ribosomal RNA gene, are well established and have been shown to allow broad range microbial species identification, not limited to LAB species alone.

Conventional Sanger-based capillary DNA sequencing is one of the basic DNA sequencing technology with the highest sequencing accuracy currently available. However, this method requires substantially high initial capital investment costs and advanced operational skills, thus limiting their adoption for on-site identification tests at smaller breweries with limited capital and human resources. As an alternative, microorganisms that are unidentifiable by the species-specific PCR-based

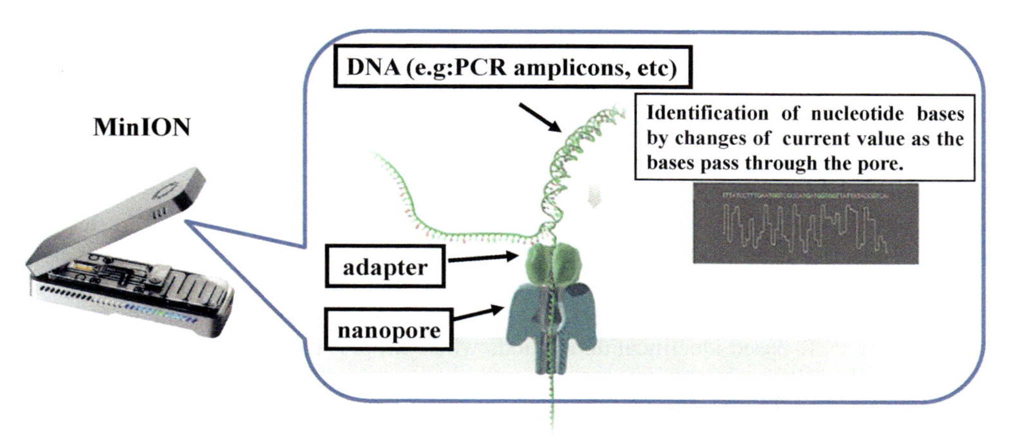

FIGURE 7.9 Schematic drawing of the MinION system and the nanopore-based sequencing. The MinION system is portable and USB-powered. The DNA molecule is translocated to the nanopore by the specially designed adapter protein that carries an enzyme motor protein. The DNA molecule then passes through the nanopore, causing changes in the electric current values that are translated into the nucleotide sequence. *Adopted from the previous literature Suzuki (2020).*

method are outsourced to third-party specialist laboratory for analysis. However, analysis by third-party takes some time, which may delay brewers from making critical decisions in a timely manner. For this reason, low-cost broad range microbial identification systems suitable for on-site identification tests in breweries, small or large, are desired.

In recent years, advancements in the field of DNA sequencing technology have made remarkable progress. One particular technology, the nanopore-based sequencing technology developed by Oxford Nanopore Technologies, stands out from all the others DNA sequencing technology. Nanopore-based sequencing is a third-generation sequencing technology that adopts an entirely different sequencing principle from a capillary sequencer or the second-generation sequencing technology. The nanopore is essentially a nano-scale biological pore made of protein. The DNA molecule is translocated to the nanopore by the specially designed adapter protein that carries an enzyme motor protein. The DNA molecule then passes through the nanopore, causing changes in the electric current values, and by measuring the changes of these current values, the DNA molecules in question are sequenced (Fig. 7.9) (Jain et al., 2016).

One of the main advantages of the nanopore-based sequencing is the ability to generate longer sequencing read lengths than the conventional capillary DNA sequencing technology and the second-generation DNA sequencing technology, for example, Illumina. While second-generation DNA sequencing technology generates short sequencing reads (limited to approximately 300 bps), nanopore-based sequencing technology generates substantially longer sequencing reads, exceeding several thousand bps. Thus, nanopore-based sequencing technology is useful for ribosomal RNA gene sequence-based bacterial identification because analysis of full length rRNA gene will provide better taxonomic resolution for species determination than the partial sequence data provided by the conventional capillary DNA sequencers and second-generation DNA sequencers (Suzuki, 2020; Yarza, Yilmaz, Pruesse, et al., 2014).

Among nanopore-based sequencing devices commercially marketed by Oxford Nanopore Technologies, the MinION system has the widest applicability for on-field test in the breweries. The MinION system is a portable 90g real-time DNA sequencing device, which has been used for environmental surveys in jungles and mountains, and so on (Maestri, Cosentino, Paterno et al., 2019; Mikheyev & Tin, 2014). Brewers do not need to spend large capital to install the MinION system in their breweries, as the initial cost for the MinION device with its sequencing kits starts from 1000 USD, substantially lower than the capital investment needed for conventional capillary DNA sequencer and other next-generation sequencers.

Even though the MinION system was not originally intended for the food and beverage industries, several studies (Kurniawan, Shinohara, Takesue, et al., 2021; Shinohara, Kurniawan, Sakai, Magarifuchi, & Suzuki, 2021) have reported the potential use of MinION microbial identification system as a new identification method for on-field tests in the breweries. When universal primers for detection of bacteria and yeast are used, conceptually, all the validly described microbial species could be identified by the MinION system. The MinION system has proven to be satisfactory for broad range identification of beer-spoilage bacteria and yeast that are frequently found in the brewery environment, including but not limited to those species belonging to the genera *Lactobacillus, Pediococcus, Pectinatus* and *Megasphaera* (Kurniawan, Shinohara, Takesue, et al., 2021). This system produces identification results that are comparable to the capillary DNA sequencer. Moreover, the MinION system was demonstrated to be useful for discrimination of closely related LAB species, for example, *L. casei-paracasei* and *L coliinoides—paracoliinoides*. These two pairs of closely related species are ordinarily difficult to discriminate by the species-specific PCR-based methods because they exhibit more than 99% nucleotide identities for 16S rRNA gene sequences. This shows the high resolution potential of the nanopore-based DNA sequencing technology.

The MinION microbial identification system also uses free and publicly available nucleotide databases, for example, NCBI nucleotide database, as a reference database for species identification. These databases are curated by experts and periodically updated so that they contain the latest information; thus, they are reliable and trusted by broad scientific disciplines. All things considered, the MinION microbial identification system can be used effectively. The rapid emergence of novel spoilage species due to the rising trend of non-traditional beverages would cause no problem because the MinION system can handle broad range identification of spoilage species, irrespective of the types of beverages, as opposed to species-specific PCR-based identification methods where kit profiles are not timely updated in response to the increasing number of novel spoilage species.

However, broad-range detection, high-accuracy identification and low initial cost alone will not be sufficient to attract brewers, particularly small or family-owned breweries, to adopt the MinION microbial identification system in their breweries. Small or family-owned breweries often have no quality control staff with working experience or background in molecular biology techniques. It will be difficult for these breweries to implement the system without extensive training and assistance from experts. Therefore, development of a simple and user-friendly protocol will be crucial, so that it can be implemented with minimal training and time.

The MinION microbial identification system protocol described in the study by Kurniawan et al. (2021) has been considerably simplified and optimised for users with no experience or skill in molecular biology. Fast analysis time is also important to allow brewers for making swift and critical judgements on-field. Total analysis time of the MinION microbial identification system protocol has been reported as *ca* 3.5h for the species identification of beer-spoilage microorganisms, significantly shorter than that originally described by the manufacturer. Because of the aforementioned developments, the MinION microbial identification system is suited for on-field microbial identification tests in the breweries, small or large. The PCR-based microbial identification methods for traditional beers and non-traditional beverages are summarised in Table 7.7.

In addition to species identification, as also mentioned in the previous section, detection of particular genetic markers, for example, beer-spoilage specific genes such as the hop tolerance genes, is equally important for intra-species discrimination between spoilers and non-spoiler strains, as this will give critical information for the microbiological QC. Since MinION system is fundamentally a DNA sequencing method, the applicability of this system could easily be expanded beyond species identification.

Another study (Kurniawan, Shinohara, Sakai, et al., 2021) reported the expansion of MinION system application to the multiplex PCR detection of hop tolerance genes, *horA* and *horC*. The PCR was done similarly with the standard multiplex PCR format to generate both *horA* and *horC* amplicons in one PCR tube. The multiplex PCR amplicons of *horA* and *horC* were then sequenced using the MinION system and further analysed by third-party genetic software for identification. The method was able to correctly detect and identify both *horA* and *horC* genes in all tested LAB species/strains.

In the case of conventional PCR systems, especially the end-point PCR system, only binary identification results are obtained because results are judged not based on the analysis of internal DNA sequences. Additionally, the results of the end-point PCR are interpreted solely on the molecular size of the PCR products. This could lead to ambiguous interpretation, e.g., if the PCR system generates non-target PCR products with similar molecular sizes or slightly different molecular sizes due to genetic polymorphisms. In these cases, further confirmatory tests are warranted in order to validate the results. Using the MinION system will avoid these kinds of problems as the MinION system will produce unambiguous results by analysing the internal DNA sequences.

The successful application of MinION system to analysis of multiplex PCR amplicons further suggests a broader applicability of the system for multiple target gene analysis. In multiplex PCR format, using any combination of target genes is virtually possible, on a condition that the thermal cycling conditions are identical. Thus, it is practically applicable as well to the concurrent identification of species and beer-spoilage specific genes. The wide applicability of the MinION system, the inexpensive initial capital investment cost, and the moderate running costs (starting from 20 to 30 US dollars per sample) will surely help this technology become more widely adopted in the breweries as an arsenal to combat the emerging spoilage species for traditional beers and non-traditional beverages.

TABLE 7.7 General guides for PCR-based microbial identification methods in the brewing industry.[a]

Methods	Initial capital investment	Running costs	Approximate time to result	Applicable categories	Operation difficulty level	Other relevant information	References
End-point PCR	Low-intermediate	Low-intermediate	3–4 h	Traditional beer	Low	Species identification; generally applicable to 20–30 species unless genus-specific approach is adopted; only binary identification results are obtained.	DiMichele and Lewis (1993), Tsuchiya et al. (1992, 1993), Yasui et al. (1997), Asano et al. (2008), Iijima et al. (2008), Rheonix Beer SpoilerAlert Assay
Real-time PCR	Intermediate	Intermediate	*ca* 2 h	Traditional beer	Low-intermediate	Species identification; generally applicable to 20–30 species unless genus-specific approach is adopted; only binary identification results are obtained.	Juvonen et al. (2008), Haakensen, Dobson, Deneer, & Ziola (2008), Haakensen, Schuber, & Ziola (2008), GEN-IALQuickGEN, Pika 4e Real Beer Spoiler
DNA sequencing	High	Intermediate-high	4–8 h	Alcoholic, low alcoholic and non-alcoholic beverages	Intermediate-high	Species identification; applicable to almost all the validly described microbial species; better suited to large brewers or outsourced to third-party specialist laboratory.	Powell and Kerruish (2017), Winand et al. (2019)
Third-generation DNA sequencer (MinION)	Low-intermediate[b]	Intermediate-high	3–4 h	Alcoholic, low alcoholic and non-alcoholic beverages	Low-intermediate	Species identification; applicable to almost all the validly described microbial species; running costs have been declining, which may make this approach more applicable to smaller breweries in the near future	Kurniawan, Shinohara, Sakai, et al. (2021), Mikheyev & Tin (2014), Jain et al. (2016), Maestri et al. (2019), Leidenfrost et al. (2020)

[a] *The table is adapted from the previous literature (Suzuki, 2020) with some modifications.*
[b] *The acquisition of the MinION system itself costs only 1000 US dollars, but an inexpensive thermal cycler for PCR amplification and a computer for data processing are also necessary.*

7.6 Other Gram-positive bacteria in brewing

7.6.1 Brewery-related LAB other than *Lactobacillus* and *Pediococcus*

Some species of LAB other than *Lactobacillus* and *Pediococcus* are occasionally isolated from the brewing environments. One of the most frequently encountered species of these LAB groups is *Lactococcus lactis*. This LAB species is a common microorganism in plants but is better known for the production of diacetyl from citrate and its role in butter manufacturing (Priest, 2003). However, the hop tolerance of *Lactococcus lactis* is rather low, and there have been no reports of *Lactococcus lactis* growing in beer except for those with microbiologically weak features, such as elevated pH values and low bitterness units (Back, 2005a). Another species isolated from brewing environments is *Leuconostoc (para)mesenteroides*. *Leuc. (para)mesenteroides* is acid tolerant and has been isolated from fruit mashes. As is the case with *Lactococcus lactis*, *Leuc. (para)mesenteroides* does not possess strong hop tolerance and is therefore unlikely to cause spoilage incidents except in beers with microbiologically weak features (Back, 2005a). The occurrence of other LAB, such as *Streptococcus* and *Enterococcus*, seems to be relatively rare in breweries.

7.6.2 Endospore-forming bacteria

Due to their strong resistance to heat treatment and disinfectants, spore-forming bacteria are difficult to eradicate from the brewing environments. Therefore, endospore-forming bacteria are sometimes isolated from work-in-process products and finished beer products, especially when QC detection media possess insufficient selectivity. Major groups of spore-forming bacteria found in brewing environments have been reported to belong to the genera *Bacillus*, *Clostridium* and *Paenibacillus* (Takeuchi, Iijima, Suzuki, Ozaki, & Yamashita, 2005). These spore-forming bacteria are generally sensitive to low pH and hop bitter acids and do not cause problems in normally hopped beer (Back, 2005a; Priest, 2003). Care may have to be taken, however, for beers with unusually high pH value and/or low bitterness units, since some *Clostridium* spp., including *Clostridium (aceto)butyricum*, might be able to grow in those beers. One worrisome feature is that the *Bacillus cereus* group was reported as one of the species isolated frequently from brewing environments (Takeuchi et al., 2005). Some of the *B. cereus* strains are known as food pathogens, causing severe nausea, vomiting and diarrhoea. However, thanks to the microbiological stability of beer described in the beginning of this chapter, no food-poisoning incidents by bacteria have been documented in beer, including those caused by *B. cereus* (Back, 2003; Bunker, 1955; Menz, Aldred, & Vriesekoop, 2011). On the other hand, spores from endospore-forming bacteria are present in malt and cereal adjuncts, and thermophilic or thermoduric spore-forming bacteria, such as *Bacillus coagulans*, are able to grow in hot sweet wort (*ca* 55−60°C) (Briggs et al., 2004). However, these bacteria are generally sensitive to hop resins and can grow only slowly in media with a pH value lower than 5.0. Therefore, they do not usually cause spoilage problems in the subsequent brewing processes and finished beer (Back, 2005a).

7.6.3 Other Gram-positive bacteria relevant in brewing

Genera belonging to *Staphylococcus*, *Kocuria* and *Micrococcus* are relatively common in breweries (Priest, 2003). These bacteria are not generally considered to be important as spoilage microorganisms, but they are known to be widely distributed in the brewing environments. Some Gram-positive cocci, such as *Staphylococcus epidermidis* and *Staphylococcus saprophyticus*, can survive in beer for long periods and are sometimes detected by QC detection media. However, these bacteria cannot grow in beer because of their hop sensitivity and their inability to grow at pH values lower than 4.5. *Kocuria kristinae*, on the other hand, is somewhat hop tolerant and acid tolerant among this group of bacteria. Unlike other *Kocuria* species that are strictly aerobic, *Kocuria kristinae* is facultatively anaerobic. Although the status of *Kocuria kristinae* as a beer-spoilage bacterium is controversial, the intensity of growth seems to be affected by the oxygen content of beer because the presence of oxygen promotes the growth of this species. When *Kocuria kristinae* grows in beer, the spoilage occurs with relatively high-pH and low-bitterness beer products. It was also reported that *Kocuria kristinae* yields a fruity aroma and an atypical taste in beer.

7.7 Concluding remarks

Beer is known as a beverage with a high microbiological stability. Most bacteria cannot grow in beer due in large part to the presence of hop bitter acids and ethanol, as well as the low pH value. Spore-forming bacteria that are the main source of concern in non-alcoholic beverages do not grow in beer and food-borne pathogens, including *Staphylococcus aureus* and *Bacillus cereus*, are also unable to grow there. These aspects are very fortunate for the brewing industries because all that brewing microbiologists have to do is to deal with very narrow subcommunities of microorganisms. In terms of Gram-positive bacteria, approximately 20 LAB species, belonging to *Lactobacillus* and *Pediococcus*, have been recognized as beer-spoilage microorganisms. Some of them, such as *L. brevis*, *L. lindneri* and *Ped. damnosus*, have been traditionally known as major beer-spoilage LAB species, whereas many of the others have become known as beer-spoilage LAB since around the year 2000. Because the culturability of many beer-spoilage LAB species is poor and the spoilage incidents are often caused by the mixed populations of LAB, the fast growers, such as *L. brevis* and *L. casei*, tend to outcompete the other hard-to-cultivate LAB species on conventional QC detection media. These phenomena make the findings of as-yet uncharacterised LAB species more difficult. Partly for these reasons, it is difficult to determine whether new beer-spoilage LAB species are constantly emerging through the horizontal transfer of the hop tolerance genes, or whether brewing microbiologists are dealing with the practically identical communities of beer-spoilage LAB species as they did in the 19th and 20th centuries. At any rate, the discoveries of novel beer-spoilage LAB species are still continuing, and brewing microbiologists are constantly trying to catch up with newly emerging opponents (Suzuki, 2020). In the face of these challenges, the recent discoveries of species-independent genetic markers, supported by the ongoing progress of hop tolerance research, have been considered significant, since at least one of those genetic markers, such as *horA* and *hor*C, have been detected in the recently recognized beer-spoilage LAB species, including *L. paracollinoides*, *L. backii*, *L. acetotolerans*, *L. paucivorans*, *L. rossiae*, *L. curtus*, *L. nagelii* and *Ped. claussenii* (Deng et al., 2014; Ehrmann, Preissler, Danne, & Vogel, 2010; Iijima et al., 2007; Pittet et al., 2012; Suzuki, 2012; Suzuki, Shinohara, & Kurniawan, 2021). These observations indicate that the species-independent genetic markers are useful for detecting as-yet uncharacterised species of beer-spoilage LAB. However, the hop tolerance mechanisms, despite the enormous progress observed in the past 30 years, have not been fully disclosed. Because of this, the intra-species determination of beer-spoilage ability is not always accurate enough to make a critical judgement when an LAB strain is detected in finished beer products; therefore, this area of research should be more vigorously conducted in future. It is also important to investigate the mechanisms as to why beer-spoilage LAB lapse into the hard-to-cultivate state. These studies not only help us to develop a new type of rapid and comprehensive QC detection medium but also reveal an entire spectrum of beer-spoilage LAB species, some of which may still remain undiscovered. Upon conducting the above research, it would become increasingly more important to use freshly isolated LAB strains. This is because all of the QC tests in breweries are actually carried out against beer-spoilage LAB strains latent in the brewing environments, rather than those subcultured with nutrient-rich laboratory culture media. The use of beer-spoilage LAB strains in the latter state often leads to an erroneous interpretation of new QC methods under development. Thus, the techniques that enable us to capture and maintain beer-spoilage LAB living in the brewing environments should be more fully established in the future. The more comprehensive and accurate QC methods for beer-spoilage LAB would emerge by overcoming the above challenges.

References

Asano, S., Shimokawa, M., & Suzuki, K. (2019). Chapter 9 Detection and identification of beer-spoilage lactic acid bacteria. In M. Kanauchi (Ed.), *Lactic acid bacteria methods and protocols-methods in molecular biology* (pp. 95–107). New York: Humana Press (Springer Protocols).

Asano, S., Suzuki, K., Iijima, K., Motoyama, H., Kuriyama, H., & Kitagawa, Y. (2007). Effect of morphological changes in beer-spoilage lactic acid bacteria on membrane filtration in breweries. *Journal of Bioscience and Bioengineering, 104*, 334–338.

Asano, S., Suzuki, K., Ozaki, K., Kuriyama, H., Yamashita, H., & Kitagawa, Y. (2008). Application of multiplex PCR to the detection of beer-spoilage bacteria. *Journal of the American Society of Brewing Chemists, 66*, 37–42.

Back, W. (1981). Bierschädliche Bakterien – Taxonomie der bierschädlichen Bakterien. *Grampositive Arten. Monatsschrift für Brauerei, 34*, 267–276.

Back, W. (1994a). Einteilung der bierschädlichen Bakterien. In W. Back (Ed.), *Farbatlas und Handbuch der Getränkebiologie* (Vol 1, pp. 62–67). Nürnberg: Verlag Hans Carl.

Back, W. (1994b). Secondary contaminations in the filling area. *Brauwelt International, 4*, 326–333.

Back, W. (2003). Biofilme in der Brauerei und Getränkeindustrie – 15 Jahre Praxiserfahrung. *Brauwelt Online, 24/25*, 1–5.

Back, W. (2005a). Brewery. In W. Back (Ed.), *Colour atlas and handbook of beverage biology* (pp. 10–112). Nürnberg: Verlag Hans Carl.

Back, W. (2005b). Winery. In W. Back (Ed.), *Colour atlas and handbook of beverage biology* (pp. 113–135). Nürnberg: Verlag Hans Carl.

Back, W., Bohak, I., Ehrmann, M., Ludwig, W., & Schleifer, K. H. (1996). Revival of the species *Lactobacillus lindneri* and the design of a species specific oligonucleotide probe. *Systematic & Applied Microbiology, 19*, 322–325.

Back, W., Leibhard, M., & Bohak, I. (1992). Flash pasteurization – membrane filtration. Comparative biological safety. *Brauwelt International, 1*, 42–49.

Barth, H. J., Klinke, C., & Schmidt, C. (1994). Hops—the brewer's gold. In C. Schmidt (Ed.), *The hop atlas* (pp. 25–29). Nürnberg: Verlag Hans Carl.

Behr, J., Gänzle, M. G., & Vogel, R. F. (2006). Characterization of a highly hop-resistant *Lactobacillus brevis* strain lacking hop transport. *Applied and Environmental Microbiology, 72*, 6483–6492.

Behr, J., Geissler, A. J., Schmid, J., Zehe, A., & Vogel, R. F. (2016). The identification of novel diagnostic marker genes for the detection of beer spoiling *Pediococcus damnosus* strains using the BlAst Diagnostic Gene findEr. *PLoS One, 11*, Article e0152747.

Behr, J., Israel, L., Gänzle, M. G., & Vogel, R. F. (2007). Proteomic approach for characterization of hop-inducible proteins in *Lactobacillus brevis*. *Applied and Environmental Microbiology, 73*, 3300–3306.

Bergsveinson, J., & Ziola, B. (2017). Investigation of beer spoilage lactic acid bacteria using omic approaches. In N. A. Bokulich, & C. W. Bamforth (Eds.), *Brewing microbiology: Current research, omics, and microbial ecology* (pp. 245–274). Norfolk, UK: Caister Academic Press.

Behr, J., & Vogel, R. F. (2010). Mechanisms of hop inhibition include the transmembrane redox reaction. *Applied and Environmental Microbiology, 76*, 142–149.

Bourdineaud, J.-P., Nehme, B., Tesse, S., & Lonvaud-Funel, A. (2004). A bacterial gene homologous to ABC transporters protect *Oenococcus oeni* from ethanol and other stress factors in wine. *International Journal of Food Microbiology, 92*, 1–14.

Briggs, D. E., Boulton, C. A., Brookes, P. A., & Stevens, R. (2004). Microbiology. In D. E. Briggs, C. A. Boulton, P. A. Brookes, & R. Stevens (Eds.), *Brewing science and practice* (pp. 606–649). Cambridge: Woodhead Publishing Limited.

Bunker, H. J. (1955). The survival of pathogenic bacteria in beer. In *Proceedings of the European brewery convention congress Baden-Baden* (pp. 330–341). Oxford: IRL Press.

Claussen, N. H. (1903). Études sur les bactéries dites sarcines et sur maladies quelles provoquent dans la bière. *Comptes Rendus des Travaux du Laboratoire Carlsberg, 6*, 64–83.

Deng, Y., Liu, J., Li, H., Li, L., Tu, J., Fang, H., et al. (2014). An improved plate culture procedure for the rapid detection of beer-spoilage lactic acid bacteria. *Journal of the Institute of Brewing, 120*, 127–132.

DiMichele, L. G., & Lewis, M. J. (1993). Rapid, species-specific detection of lactic acid bacteria from beer using the polymerase chain reaction. *Journal of the American Society of Brewing Chemists, 51*, 63–66.

Dobson, C. M., Deneer, H., Lee, S., Hemmingsen, S. Glaze, S., & Ziola, B. (2002). Phylogenetic analysis of the genus *Pediococcus*, including *Pediococcus claussenii* sp. nov., a novel lactic acid bacterium isolated from beer. *International Journal of Systematic and Evolutionary Microbiology, 52*, 2003–2010.

Dols-Lafargue, M., Lee, H. Y., Le Marrec, C., Heyraud A., Chambat, G., & Lonvaud-Funel, A. (2008). Characterization of *gtf*, a glucosyltransferase gene in the genomes of *Pediococcus parvulus* and *Oenococcus oeni*, two bacterial species commonly found in wine. *Applied and Environmental Microbiology, 74*, 4079–4090.

Ehrmann, M. A., Preissler, P., Danne, M., & Vogel, R. F. (2010). *Lactobacillus paucivorans* sp. nov., isolated from a brewery environment. *International Journal of Systematic and Evolutionary Microbiology, 60*, 2353–2357.

Ehrmann, M. A., & Vogel, R. F. (2005). Taxonomic note "*Lactobacillus pastorianus*" (Van Laer, 1892) a former synonym for *Lactobacillus paracollinoides*. *Systematic & Applied Microbiology, 28*, 54–56.

Farrow, J. A. E., Phillips, B. A., & Collins, M. D. (1988). Nucleic acid studies on some heterofermentative lactobacilli: *Lactobacillus malefermentans* sp. nov. And *Lactobacillus parabuchneri* sp. nov. *FEMS Microbiology Letters, 55*, 163–168.

Fernandez, J. L., & Simpson, W. J. (1995). Measurement and prediction of the susceptibility of lager beer to spoilage by lactic acid bacteria. *Journal of Applied Bacteriology, 78*, 419–425.

Feyereisen, M., Mahony, J., O'Sullivan, T., Boer, V., & van Sinderen, D. (2020). A plasmid-encoded putative glycosyltransferase is involved in hop tolerance and beer spoilage in *Lactobacillus brevis*. *Applied and Environmental Microbiology, 86*, Article e02268.

Foodproof Beer Screening Kit BIOTECON Diagnostics. https://www.hygiena.com/food-safety/spoilage-organism-detection/beer-spoilage-bacteria/foodproof-beer-screening-kit (accessed December 26, 2023).

Fujii, T., Nakashima, K., & Hayashi, N. (2005). Random amplified polymorphic DNA-PCR based cloning of markers to identify the beer-spoilage strains of *Lactobacillus brevis, Pediococcus damnosus, Lactobacillus collinoides* and *Lactobacillus coryniformis*. *Journal of Applied Microbiology, 98*, 1209–1220.

Gänzle, M. G. (2004). Reutericyclin: Biological activity, mode of action, and potential applications. *Applied Microbiology and Biotechnology, 64*, 326–332.

Garg, P., Park, Y., Sharma, D., & Wang, T. (2010). Antimicrobial effect of chitosan on the growth of lactic acid bacteria strains known to spoil beer. *Journal of Experimental Microbiology and Immunology, 14*, 7–12.

Geissler, A. J. (2016). Lifestyle of beer spoiling lactic acid bacteria. In *PhD Thesis, Lehrstuhl für Technische Mikrobiologie* (pp. 115–118). Technische Universität München.

Geissler, A. J., Behr, J., Schmid, J., Zehe, A., & Vogel, R. F. (2017). Beer spoilage ability of lactic acid bacteria is a plasmid-encoded trait. *Brewing Science, 70*, 57–73.

Geissler, A. J., Behr, J., von Kamp, K., & Vogel, R. F. (2016). Metabolic strategies of beer spoilage lactic acid bacteria in beer. *International Journal of Food Microbiology, 216*, 60–68.

GEN-IAL QuickGEN Beer differentiation high. https://food.r-biopharm.com/products/gen-ial-quickgen-first-beer-differentiation-pcr-kit/. (accessed December 26, 2023).

Graça da Silveira, M., Vitória San Romão, M., Loureiro-Dias, M. C., Rombouts, F. M., & Abee, T. (2002). Flow cytometric assessment of membrane integrity of ethanol-stressed *Oenococcus oeni* cells. *Applied and Environmental Microbiology, 68*, 6087–6093.

Grandvalet, C., Assad-García, J. S., Chu-Ky, S., Tollot, M., Guzzo, J., Gresti, J., et al. (2008). Changes in membrane lipid composition in ethanol- and acid-adapted *Oenococcus oeni* cells: Characterization of the *cfa* gene by heterologous complementation. *Microbiology, 154*, 2611–2619.

Groot, M. N., Klaassens, E., de Vos, W. M., Delcour, J., Hols, P., et al. (2005). Genome-based *in silico* detection of putative manganese transport systems in *Lactobacillus plantarum* and their genetic analysis. *Microbiology, 151*, 1229–1238.

Haakensen, M., Butt, L., Chaban, B., Deneer, H., Ziola, B., & Dowgiert, T. (2007). *horA*-specific real-time PCR for detection of beer-spoilage lactic acid bacteria. *Journal of the American Society of Brewing Chemists, 65*, 157–165.

Haakensen, M., Dobson, C. M., Deneer, H., & Ziola, B. (2008). Real-time PCR detection of bacteria belonging to the Firmicutes phylum. *International Journal of Food Microbiology, 125*, 236–241.

Haakensen, M., Pittet, V., Morrow, K., Schubert, A., Ferguson, J., & Ziola, B. (2009). Ability of novel ATP-binding cassette multidrug resistance genes to predict growth of *Pediococcus* isolates in beer. *Journal of the American Society of Brewing Chemists, 67*, 170–176.

Haakensen, M., Schubert, A., & Ziola, B. (2008). Multiplex PCR for putative *Lactobacillus* and *Pediococcus* beer-spoilage genes and ability of gene presence to predict growth in beer. *Journal of the American Society of Brewing Chemists, 66*, 63–70.

Hammond, J., Brennan, M., & Price, A. (1999). The control of microbial spoilage of beer. *Journal of the Institute of Brewing, 105*, 113–120.

Hayashi, N., Ito, M., Horiike, S., & Taguchi, H. (2001). Molecular cloning of a putative divalent-cation transporter gene as a new genetic marker for the identification of *Lactobacillus brevis* strains capable of growing in beer. *Applied Microbiology and Biotechnology, 55*, 596–603.

Hutzler, M., Koob, J., Grammer, M., Riedl, R., & Jacob, F. (2012). Statistische Auswertung der PCR Analysen bierschädlicher Bakterien in den Jahren 2010 und 2011. *Brauwelt, 152*, 546–547.

Hutzler, M., Müller-Auffermann, K., Koob, J., Riedl, R., & Jacob, F. (2013). Beer spoiling microorganisms – a current overview. *Brauwelt International, 31*, 23–25.

Iijima, K., Asano, S., Suzuki, K., Ogata, T., & Kitagawa, Y. (2008). Multiplex PCR method for comprehensive detection of *Pectinatus* and beer-spoilage cocci. *Bioscience, Biotechnology, and Biochemistry, 72*, 2764–2766.

Iijima, K., Suzuki, K., Asano, S., Kuriyama, H., & Kitagawa, Y. (2007). Isolation and identification of potential beer-spoilage *Pediococcus inopinatus* and beer-spoilage *Lactobacillus backi* strains carrying the *horA* and *horC* gene clusters. *Journal of the Institute of Brewing, 113*, 96–101.

Iijima, K., Suzuki, K., Asano, S., Ogata, T., & Kitagawa, Y. (2009). HorC, a hop-resistance related protein, presumably functions in homodimer form. *Bioscience, Biotechnology, and Biochemistry, 73*, 1880–1882.

Iijima, K., Suzuki, K., Ozaki, K., & Yamashita, H. (2006). *horC* confers beer-spoilage ability on hop-sensitive *Lactobacillus brevis* ABBC45^{CC}. *Journal of Applied Microbiology, 100*, 1282–1288.

Ingram, L. O. (1986). Microbial tolerance to alcohols: Role of the cell membrane. *Trends in Biotechnology, 4*, 40–44.

Jain, M., Olsen, H. E., Paten, B., & Akeson, M. (2016). The Oxford nanopore MinION: Delivery of nanopore sequencing to the genomics community. *Genome Biology, 17*.

Juvonen, R., Koivula, T., & Haikara, A. (2008). Group specific PCR-RFLP and real-time PCR methods for detection and tentative discrimination of strictly anaerobic beer-spoilage bacteria of the class *Clostridia*. *International Journal of Food Microbiology, 125*, 162–169.

Kandler, O., & Weiss, N. (1986). Genus *Lactobacillus* beijerinck 1901, 212^{AL}. In P. H. A. Sneath, N. S. Mair, M. E. Sharpe, & J. G. Holt (Eds.), *Bergey's manual of systematic bacteriology* (Vol 2, pp. 1209–1234). Baltimore: Williams and Wilkins.

Kern, C. C., Vogel, R. F., & Behr, J. (2014). Differentiation of *Lactobacillus brevis* strains using matrix-assisted-laser-desorption-ionization-time-of-flight mass spectrometry with respect to their beer spoilage potential. *Food Microbiology, 40*, 18–24.

Kitahara, K. (1974). Genus III, *Pediococcus* blacke 1884, 257. In R. E. Buchanan, & N. E. Gibbons (Eds.), *Bergey's manual of determinative bacteriology* (8th ed., pp. 513–515). Baltimore: Williams and Wilkins.

Koob, J., Jacob, F., Grammer, M., Kleucker, A., Riedl, R., & Hutzler, M. (2014). PCR Analysen bierschädlicher Bakterien 2012 und 2013. *Brauwelt, 154*, 288–290.

Kurniawan, Y. N., Shinohara, Y., Sakai, H., Magarifuchi, T., & Suzuki, K. (2021). Applications of the third-generation DNA sequencing technology to the detection of hop tolerance genes and discrimination of *Saccharomyces* yeast strains. *Journal of the American Society of Brewing Chemists, 80*, 161–168.

Kurniawan, Y. N., Shinohara, Y., Takesue, N., Sakai, H., Magarifuchi, T., & Suzuki, K. (2021). Development of a rapid and accurate Nanopore-based sequencing platform for on-field identification of beer-spoilage bacteria in the breweries. *Journal of the American Society of Brewing Chemists, 79*, 240–248.

Lawrence, D. R. (1988). Spoilage organisms in beer. In R. K. Robinson (Ed.), *Developments in food microbiology* (pp. 1–48). London: Elsevier.

Leidenfrost, R. M., Pother, D. C., Jackel, U., & Wunschiers, R. (2020). Benchmarking the MinION: Evaluating long reads for microbial profiling. *Scientific Reports, 10*, 5125.

Maestri, S., Cosentino, E., Paterno, M., Freitag, H., Garces, J. M., Marcolungo, L., et al. (2019). A rapid and accurate MinION based workflow for tracking species biodiversity in the field. *Genes, 10*, 468.

Menz, G., Aldred, P., & Vriesekoop, F. (2011). Growth and survival of foodborne pathogens in beer. *Journal of Food Protection, 74*, 1670–1675.

Mikheyev, A. S., & Tin, M. M. Y. (2014). A first look at the Oxford nanopore MinION sequencer. *Molecular Ecology Resources, 14*, 1097–1102.

Moir, M. (2000). Hops – a millennium review. *Journal of the American Society of Brewing Chemists, 58*, 131–146.

Nakagawa, A. (1978). Beer-spoilage lactic acid bacteria – principal characteristics and simple methods of their selective detection. *Bulletin of Brewing Science, 24*, 1–10.

Nakakita, Y., Maeba, H., & Takashio, M. (2003). Grouping of *Lactobacillus brevis* strains using the *gyrB* gene. *Journal of the American Society of Brewing Chemists, 61*, 157–160.

Pasteur, L. (1876). Chapitre II Recherche des causes des maladies de la bière et de celles du moût qui sert à la produire. In L. Pasteur (Ed.), *Etudes sur la bière, ses maladies, causes qui les provoquent, procédé pour la rendre inaltérable, avec une théorie nouvelle de la fermentation* (pp. 18–32). Paris: Gauthier-Villars.

Pika 4everyone *Lactobacillus* & *Pediococcus* Real Beer Spoiler PCR test kit. https://www.chaibio.com/lactobacillus-pediococcus-real-beer-spoiler-pcr-test-kit (accessed December 26, 2023).

Pittet, V., Abegunde, T., Marfleet, T., Haakensen, M., Morrow, K., Jayaprakash, T., Schroeder, K., Trost, B., Byrns, S., Bergsveinson, J., & Kusalik, A. (2012). Complete genome sequence of the beer spoilage organism *Pediococcus claussenii* ATCC BAA-344^{T}. *Journal of Bacteriology, 194*, 1271–1272.

Powell, C. D., & Kerruish, D. W. M. (2017). Beer-spoiling yeasts: Genomics, detection, and control. In N. A. Bokulich, & C. W. Bamforth (Eds.), *Brewing microbiology: Current research, omics and microbial ecology* (pp. 289–328). Norfolk, UK: Caister Academic Press.

Priest, F. G. (2003). Gram-positive brewery bacteria. In F. G. Priest, & I. Campbell (Eds.), *Brewing microbiology* (3rd ed., pp. 181–217). New York: Kluwer Academic/Plenum Publishers.

Priest, F. G. (2006). Microbiology and microbiological control in the brewery. In F. G. Priest, & G. G. Stewart (Eds.), *Handbook of brewing* (2nd ed., pp. 607–627). Boca Raton: CRC Press.

Rainbow, C. (1981). Beer spoilage microorganisms. In J. R. A. Pollock (Ed.), *Brewing science* (Vol 2, pp. 491–550). New York: Academic Press.

Rheonix Beer SpoilerAlert Assay. https://rheonix.com/food-beverage-testing/beer-spoilage-testing/ (accessed December 26, 2023).

Sakamoto, K., & Konings, W. N. (2003). Beer spoilage bacteria and hop resistance. *International Journal of Food Microbiology, 89*, 105–124.

Sakamoto, K., Margolles, A., van Veen, H. W., & Konings, W. N. (2001). Hop resistance in the beer spoilage bacterium *Lactobacillus brevis* is mediated by the ATP-binding cassette multidrug transporter HorA. *Journal of Bacteriology, 183*, 5371–5375.

Sakamoto, K., van Veen, H. W., Saito, H., Kobayashi, H., & Konings, W. N. (2002). Membrane-bound ATPase contributes to hop resistance of *Lactobacillus brevis*. *Applied and Environmental Microbiology, 68*, 5374–5378.

Sami, M., Yamashita, H., Kadokura, H., Kitamoto, K., Yoda, K., & Yamasaki, M. (1997). A new and rapid method for determination of beer-spoilage ability of lactobacilli. *Journal of the American Society of Brewing Chemists, 55*, 137–140.

Schneiderbanger, J., Grammer, M., Jacob, F., & Hutzler, M. (2018). Statistical evaluation of beer spoilage bacteria by real-time PCR analyses from 2010 to 2016. *Journal of the Institute of Brewing, 124*, 173–181.

Schneiderbanger, J., Jacob, F., & Hutzler, M. (2020). Mini-Review: The current role of lactic acid bacteria in beer spoilage. *Brewing Science, 73*, 19–28.

Segawa, S., Nakakita, Y., Takata, Y., Wakita, Y., Kaneko, T., Kaneda, H., et al. (2008). Effect of oral administration of heat-killed *Lactobacillus brevis* SBC8803 on total and ovalbumin-specific immunoglobulin E production through the improvement of Th1/Th2 balance. *International Journal of Food Microbiology, 121*, 1–10.

Segawa, S., Wakita, Y., Hirata, H., & Watari, J. (2008). Oral administration of heat-killed *Lactobacillus brevis* in ethanol-containing diet-fed C57BL/6N mice. *International Journal of Food Microbiology, 128*, 371–377.

Shimokawa, M., & Suzuki, K. (2021). Preceding subculture conditions affect growth characteristics of beer spoilage lactic acid bacteria in quality control culture media: Comparative study on hard-to-culture and culturable *Secundilactobacillus (Lactobacillus) paracollinoides* strains. *Journal of the American Society of Brewing Chemists, 79*, 340–346.

Shinohara, Y., Kurniawan, Y. N., Sakai, H., Magarifuchi, T., & Suzuki, K. (2021). Nanopore based sequencing enables easy and accurate identification of yeasts in breweries. *Journal of The Institute of Brewing, 127*(2), 160–166. https://doi.org/10.1002/jib.639

Simpson, W. J. (1993). Studies on the sensitivity of lactic acid bacteria to hop bitter acids. *Journal of the Institute of Brewing, 99*, 405–411.

Simpson, W. J., & Fernandez, J. L. (1994). Mechanism of resistance of lactic acid bacteria to *trans*-isohumulone. *Journal of the American Society of Brewing Chemists, 52*, 9–11.

Simpson, W. J., & Smith, A. R. W. (1992). Factors affecting antibacterial activity of hop compounds and their derivatives. *Journal of Applied Bacteriology, 72*, 327–334.

Storgårds, E., Pihlajamäki, O., & Haikara, A. (1997). Biofilms in the brewing process – a new approach to hygiene management. In *Proceedings of the European brewery convention congress Maastricht* (pp. 717–723). Oxford: IRL Press.

Storgårds, E., & Suihko, M.-L. (1998). Detection and identification of *Lactobacillus lindneri* from brewing environments. *Journal of the Institute of Brewing, 104*, 47–54.

Suzuki, K. (2009). Beer spoilage lactic acid bacteria. In V. R. Preedy, & R. R. Watson (Eds.), *Beer in health and disease prevention* (pp. 150–164). San Diego: Elsevier Science.

Suzuki, K. (2012). 125th anniversary review: Microbiological instability of beer caused by spoilage bacteria. *Journal of the Institute of Brewing, 117*, 131–155.

Suzuki, K. (2020). Emergence of new spoilage microorganisms in the brewing industry and development of microbiological quality control methods to cope with this phenomenon – a review. *Journal of the American Society of Brewing Chemists, 78*, 245–259.

Suzuki, K., Asano, S., Iijima, K., & Kitamoto, K. (2008). Sake and beer spoilage lactic acid bacteria – a review. *Journal of the Institute of Brewing, 114*, 209–223.

Suzuki, K., Asano, S., Iijima, K., Kuriyama, H., & Kitagawa, Y. (2008). Development of detection medium for hard-to-culture beer spoilage lactic acid bacteria. *Journal of Applied Microbiology, 104*, 1458–1470.

Suzuki, K., Funahashi, W., Koyanagi, M., & Yamashita, H. (2004). *Lactobacillus paracollinoides* sp. nov., isolated from brewery environments. *International Journal of Systematic and Evolutionary Microbiology, 54*, 115–117.

Suzuki, K., Iijima, K., Asano, S., Kuriyama, H., & Kitagawa, Y. (2006). Induction of viable but nonculturable state in beer spoilage lactic acid bacteria. *Journal of the Institute of Brewing, 112*, 295–301.

Suzuki, K., Iijima, K., Ozaki, K., & Yamashita, H. (2005a). Isolation of hop-sensitive variant from *Lactobacillus lindneri* and identification of genetic marker for beer spoilage ability of lactic acid bacteria. *Applied and Environmental Microbiology, 71*, 5089–5097.

Suzuki, K., Iijima, K., Ozaki, K., & Yamashita, H. (2005b). Study on ATP production of lactic acid bacteria in beer and development of a rapid screening method for beer-spoilage bacteria. *Journal of the Institute of Brewing, 111*, 328–335.

Suzuki, K., Iijima, K., Sakamoto, K., Sami, M., & Yamashita, H. (2006). A review of hop resistance in beer spoilage lactic acid bacteria. *Journal of the Institute of Brewing, 112*, 173–191.

Suzuki, K., Koyanagi, M., & Yamashita, H. (2004). Genetic characterization of non-spoilage variant isolated from beer-spoilage *Lactobacillus brevis* ABBC45^{C}. *Journal of Applied Microbiology, 96*, 946–953.

Suzuki, K., Sami, M., Ozaki, K., & Yamashita, H. (2004). Nucleotide sequence identities of *horA* homologues and adjacent DNA regions identified in three species of beer-spoilage lactic acid bacteria. *Journal of the Institute of Brewing, 110*, 276–283.

Suzuki, K., Shinohara, Y., & Kurniawan, Y. N. (2021). Role of plasmids in beer spoilage lactic acid bacteria: A review. *Journal of the American Society of Brewing Chemists, 79*, 1–16.

Takahashi, N., Nakakita, Y., Sugiyama, H., Shigyo, T., & Shinotsuka, K. (1999). Classification and identification of strains of *Lactobacillus brevis* based on electrophoretic characterization of D-lactate dehydrogenase: Relationship between D-lactate dehydrogenase and beer-spoilage ability. *Journal of Bioscience and Bioengineering, 88*, 500–506.

Takeuchi, A., Iijima, K., Suzuki, K., Ozaki, K., & Yamashita, H. (2005). Application of ribotyping and rDNA internal space analysis (RISA) to assessment of microflora in brewery environments. *Journal of the American Society of Brewing Chemists, 63*, 73–75.

Taskila, S., Kronlöf, J., & Ojamo, H. (2011). Enrichment cultivation of beer-spoiling lactic acid bacteria. *Journal of the Institute of Brewing, 117*, 285–294.

Tsuchiya, Y., Kaneda, H., Kano, Y., & Koshino, S. (1992). Detection of beer spoilage organisms by polymerase chain reaction technology. *Journal of the American Society of Brewing Chemists, 50*, 64–67. https://doi.org/10.1094/ASBCJ-50-0064

Tsuchiya, Y., Kaneda, H., Kano, Y., & Koshino, S. (1993). Detection of *Lactobacillus brevis* in beer using polymerase chain reaction. *Journal of the American Society of Brewing Chemists, 51*, 40–41.

Umegatani, M., Takesue, N., Asano, S., Tadami, H., & Uemura, K. (2022). Study of beer spoilage *Lactobacillus nagelii* harboring hop resistance gene *horA*. *Journal of the American Society of Brewing Chemists, 80*, 92–98.

van Laer, H. (1892). Contributions à l'histoire des ferments des hydrates de carbone. *Academie Royale de Belgique. Classe des Sciences. Memoires. Collection in Octavo, 47*, 1–37.

Vaughan, A., O'Sullivan, T., & van Sinderen, D. (2005). Enhancing the microbiological stability of malt and beer – a review. *Journal of the Institute of Brewing, 111*, 355–371.

Vogel, R. F., Preissler, P., & Behr, J. (2010). Towards an understanding of hop tolerance in beer spoiling *Lactobacillus brevis*. *Brewing Science, 63*, 23–30.

Waki, N., Yajima, N., Suganuma, H., Buddle, B. M., Luo, D., Heiser, A., et al. (2013). Oral administration of *Lactobacillus brevis* KB290 to mice alleviates clinical symptoms following influenza virus infection. *Letters in Applied Microbiology, 58*, 87–93.

Winand, R., Bogaerts, B., Hoffman, S., Lefevre, L., Delvoye, M., Van Braekel, J., et al. (2019). Targeting the 16S rRNA gene for bacterial identification in complex mixed samples: Comparative evaluation of second (Illumina) and third (Oxford nanopore Technologies) generation sequencing technologies. *International Journal of Molecular Sciences, 21*, 298.

Yarza, P., Yilmaz, P., Pruesse, E., et al. (2014). Uniting the classification of cultured and uncultured bacteria and archaea using 16S rRNA gene sequences. *Nature Reviews Microbiology, 12*, 635–645.

Yasui, T., Okamoto, T., & Taguchi, H. (1997). A specific oligonucleotide primer for the rapid detection of *Lactobacillus lindneri* by polymerase chain reaction. *Canadian Journal of Microbiology, 43*, 157–163.

Zheng, J., Wittouck, S., Salvetti, E., Franz, C. M. A. P., Harris, H. M. B., Mattarelli, P., et al. (2020). A taxonomic note on the genus *Lactobacillus*: Description of 23 novel genera, emended description of the genus *Lactobacillus* Beijerinck 1901, and union of *Lactobacillaceae* and *Leuconostocaceae*. *International Journal of Systematic and Evolutionary Microbiology, 70*, 2782–2858.

Chapter 8

Gram-negative spoilage bacteria in brewing

Ashtavinayak D. Paradh

Technical Innovation (R&D) Center, Nashik, Pernod Ricard India, Nashik, Maharashtra, India

8.1 Introduction: Gram-negative bacteria in brewing

Gram-negative beer-spoilage bacteria can be categorised into two main groups. Gram-negative obligate anaerobic bacteria belonging to genera *Pectinatus*, *Megasphaera*, *Zymophilus* and *Selenomonas* are considered as an important group of beer spoilage bacteria in low-alcohol and unpasteurised beer category. Gram-negative obligate anerobic bacteria were first described in late 1900s and were responsible for significant incidents of beer spoilage in Europe in the 1990s (Suzuki, 2011, 2020). Later, in last 2 decades, few more species were added to the genus *Pectinatus* and *Megasphaera* (Juvonen & Suihko, 2006; Schleifer et al., 1990). Recently, in 2019, novel species *Prevotella cerevisiae*, isolated from brewery wastewater exhibiting similar spoilage attributes to *Pectinatus* and *Megasphaera*, was proposed (Nakata, Kanda, Nakakita, Kaneko, & Tsuchiya, 2019). In future, further novel genera and species in the group of Gram-negative, obligate anaerobic bacteria beer spoilage bacteria are anticipated to be proposed and studied for their beer spoilage ability (Suzuki, 2020).

This chapter provides a review of the second group of Gram-negative aerobic and facultative anaerobic bacteria such as acetic acid bacteria (AAB), *Zymomonas* and selected Enterobacteriaceae species. Unlike the first group of Gram-negative, obligate anaerobes, these beer spoilage bacteria have been associated conventionally with brewing environments. This chapter deals with latest taxonomic status, occurrence in brewery environments, beer spoilage characteristics and a summary of detection methods studied for Gram-negative aerobic and facultative anaerobic brewery related bacteria.

AAB has been an extensively studied group of Gram-negative bacteria for their potential applications in the food and beverage industry (Teyssier & Hamdouche, 2016). AAB belonging to two genera, *Acetobacter* and *Gluconobacter*, are associated with haziness in beer and a vinegary off flavour due to their ability to metabolise ethanol to acetic acid (Roselli, Kerruish, Crow, Smart, & Powell, 2024; Sakamoto & Konings, 2003). AAB were conventionally an important group of beer spoilage bacteria. However, due to significant improvement in sanitation and hygiene practices in modern breweries, remarkable enhancement in sampling points and frequency of microbial analysis, rapid microbial detection methods and technological advances in filling operations resulting in effective removal of oxygen ingress in packaged product, AAB are currently not considered as an important beer spoiler. AAB bacteria can still be problematic in beer dispensing areas specially in cask conditioned ales which are dispensed through air pressure and opens kegs with oxygen ingress (Roselli et al., 2024; Teyssier & Hamdouche, 2016).

Zymomonas mobilis is a Gram-negative, facultative anaerobe and has been isolated from sugar- and alcohol-rich environments and has been widely studied as a spoilage microorganism in beers, ciders and juices (Coton, Laplace, Auffray, & Coton, 2006). *Zymomonas mobilis* has also been studied extensively as a potential ethanol-producing bacteria (Weir, 2016). In brewery environments, there has been no report of incidents of spoilage, as these microbes utilise only a narrow range of sugars (Sakamoto & Konings, 2003; Vriesekoop, Krahl, Hucker & Menz, 2012).

Certain species belonging to Enterobacteriaceae such as *Obesumbacterium*, *Hafnia*, *Klebsiella*, *Citrobacter* and *Enterobacter* are reported to be associated with spoilage of unfermented and fermenting wort (Hill & Priest, 2017; Priest, 2006). These bacteria are not normally able to grow in finished beer but are occasionally found in the initial stages of the brewing process, causing unwanted off-flavours in the final product (Priest, Hammond, & Stewart, 1994). Table 8.1 shows

Brewing Microbiology. https://doi.org/10.1016/B978-0-323-99606-8.00007-9

TABLE 8.1 Overview of gram-negative aerobic and facultative anerobic beer spoilage bacteria, beer spoilage effects and metabolic products.

Brewery related bacteria	Occurrence in brewery environments and characteristics	Off-flavour/ aroma and odour	Visual spoilage effects	Metabolic products
Acetic acid bacteria[a]				
Acetobacter	Wort, beer dispenses, and cask-conditioned ales and barrel-aged ales, brewery biofilm	Sour, vinegary	Hazy, ropiness	Acetic acid
Gluconobacter	Wort, beer dispense and cask-conditioned ales	Sour, vinegary	Hazy	Acetic acid, acetate
Zymomonas[b]	Primed beers (not found in lagers)	Fruity, rotten apple, rotten egg, sulphudic	Hazy, ropiness	Acetaldehyde and H_2S
Enterobacteriaceae[c]				
Obesumbacterium/ Shimwellia	Contaminant of Pitching yeast & fermenting wort Survive early stages of fermentation	Parsnip, sulphury	Hazy	Dimethyl sulphide (DMS), diacetyl, higher alcohols, *N*-nitrosamines, acetoin
Citrobacter	Brewing liquor, fermenting wort Early stages of fermentation	Parsnip, sulphury	_	DMS, diacetyl, lactic acid, acetaldehyde
Rahnella oxytoca	Pitching yeast, early stages of fermentation (wort)	Fruity, lactic, sulphury	_	DMS, diacetyl, methyl acetate, ethyl acetate
Klebsiella/ Raoultella	Fermenting wort, biofilm	Unp^{3}easant odou^2	_	4-vinylguaicol, DMS, diacetyl
Enterobacter	Contaminant of brewing liquor	Unpleasant odour	_	Phenolic compounds, DMS Non- volatile Nitrosoamines

[a]*Gonzalez et al. (2005), van Vuuren and Priest (2003).*
[b]*Ingledew (1979), Swings and De Ley (1977).*
[c]*Matoulková et al. (2018), Priest (2006), Smith (1994), van Vuuren and Priest (2003).*

an overview of Gram-negative aerobic and facultative anaerobic beer spoilage bacteria, their beer spoilage effects and major metabolic products.

8.2 Acetic acid bacteria

AAB are group of mesophilic, Gram-negative, strictly aerobic, non-spore-forming bacteria. AAB occur singly, in pairs or in chains, sometimes motile in nature and ellipsoidal to short rod-shaped cell morphology (Sengun & Karabiyikli, 2011; Yassunaka Hata, Surek, Sartori, Vassoler Serrato, & Aparecida Spinosa, 2023). Flagellar arrangement may vary from peritrichously to polar (Gonzalez, Hierro, Poblet, Mas, & Guillamon, 2005). AAB show a positive reaction to catalase test and a negative reaction to oxidase test (Sengun & Karabiyikli, 2011). Optimum growth of AAB occurs between pH 5 and 6.5 (Teyssier & Hamdouche, 2016), but their growth can also occur at highly acidic pH 3–4. AAB bacteria have been isolated from a variety of sources ranging from tropical fruits, rotten fruits, dried fruits, flowers, beers and wines (Sengun & Karabiyikli, 2011; Yassunaka Hata et al., 2023).

AAB bacteria are industrially important microorganisms due to their ability to oxidise sugar and ethanol into organic acid, predominantly into acetic acid. *Gluconobacter* are industrially used in the production of vinegar due to their ability to produce high concentrations of acetic acid from ethanol under aerobic conditions. AAB are primarily found in the fermented food and beverage ecosystem and have been associated with vinegar, cocoa bean, kefir, kombucha and sour beer production (Pothakos et al., 2016). In addition, AAB bacteria are used in various biotechnological applications such as production of extra cellular cellulose, sorbose and dihydroxyacetone from glycerol (Gonzalez, Hierro, et al., 2005; Sengun et al., 2022).

AAB bacteria are also important due to their spoilage effect on alcoholic beverages such as wine, beers and ciders (Bartowsky & Henschke, 2008; Kubizniaková, Kyselová, Brožová, Hanzalíková, & Matoulková, 2021). Beer spoilage AAB forms a pellicle on the surface with cloudiness in beer containing oxygen. Due to the formation of acetic acid, beer tastes sour to vinegary (Kubizniaková et al., 2021; Paradh, 2015; Storgårds, 2000, pp. 1−108). AAB associated with beer are strictly aerobic bacteria, but some of the AAB isolated from draught beer have been reported to be micro-aerotolerant (van Vuuren & Priest, 2003).

8.2.1 Taxonomic status of brewery-related AAB

AAB due to their extensive biotechnological, food and beverage applications are an extensively studied bacterial species. Taxonomic status of AAB is complex, and the taxonomic assignments of AAB have been subjected to changes on several occasions in last few decades. Based on the recent review published by Qiu et al. (2021), AAB has 19 validated genera and 92 validated species. In other work, up to 110 validated species have been reported (He et al., 2022). Out of the validated AAB genera only two genera, *Acetobacter* and *Gluconobacter*, are reported to be associated with beer spoilage (Kubizniaková et al., 2021; van Vuuren & Priest, 2003). However, due to a constantly changing taxonomic classification of AAB, the genera *Gluconacetobacter* and *Komagataeibacter* are also likely to become known as significant in the brewery environments (Kubizniaková et al., 2021).

Currently, the genus *Acetobacter* comprises 39 validated species, and the genus *Gluconobacter* has 20 validated species. Amongst the validated species of AAB, 12 species of *Acetobacter* are reported to be associated with brewing environments. *Acetobacter aceti*, *Acetobacter liquefactions*, *Acetobacter pastorianus* and *Acetobacter hansii* are most frequently occurring species (Kubizniaková et al., 2021; Priest, 2006; van Vuuren & Priest, 2003).

At least six *Gluconobacter species* namely *Gluconobacter oxydans*, *Gluconobacter cerevisiae*, *Gluconobacter cerinus*, *Gluconobacter japonicus*, *Gluconobacter wancherniae* and *Gluconobacter vitians* have been reported to be associated brewery environments as a contaminant of finished or semi-finished beer, as an indicator microorganism or as an isolate from consortia of microbes associated with sour beer production Kubizniaková et al., 2021; Spitaels et al., 2014). *Gluconobacter oxydans* has been extensively reported to be associated with brewing environments (van Vuuren & Priest, 2003). *Gluconobacter vitians* isolated from spoiled dinner beer is recently proposed *Gluconobacter species* associated with beer spoilage (Sombolestani et al., 2021). Table 8.2 describes the general characteristics of brewery related AAB.

8.2.2 Metabolic aspects of AAB

The most important metabolic characteristics of AAB are an ability to oxidise sugars, ethanol and sugar alcohols to respective organic acid (Gomes, de Fatima Borges, de Freitas Rosa, Castro-Gómez, & Spinosa, 2018). *Acetobacter* and *Gluconobacter* predominantly produce acetic acid giving vinegary and unpleasant note to the beer. Beer spoilage AAB can utilise two membrane bound enzymes *alcohol dehydrogenase* and *acetaldehyde dehydrogenase* to oxidise ethanol to acetic acid. Oxidation mainly occurs at neutral to acidic pH of around 4.5 (Lynch et al., 2019) and produced organic acid is released in the extracellular space leading to acidification of the environment. Upon depletion of other carbon sources, *Acetobacter species* further can completely oxidise acetic acid to carbon dioxide and water through the tricarboxylic acid (TCA) cycle. However, *Gluconobacter* lack a functional TCA cycle metabolic capability, which limits further oxidation of organic acids (Seo et al., 2004).

Certain AAB are capable of synthesising exopolysaccharides (EPS) such as levan and dextran resulting into high viscosity and formation of ropiness in beer (Storgårds, 2000; Paradh, 2015). *Gluconobacter* and *Komagataeibacter* can produce *levansucrase* enzyme, which is secreted through a signal peptide pathway and is accumulated in the periplasmic space and later secreted out into the surrounding medium. In the surrounding medium, *levansucrase* carry out hydrolysis of sucrose and formation of levan by transfructosylation activity (Dağbağlı & Göksungur, 2017).

8.2.3 Occurrence and beer spoilage by AAB

AAB prevail ubiquitously across brewing process especially in fermentation, filtration, storage tanks and packaging arears (Hill, 2015). Brewery-related AAB show ethanol tolerance (>10% ABV) and are acidophilic in nature. AAB found in brewery environments are generally tolerant to hop compounds (Roselli et al., 2024). AAB are found throughout the brewing operations and specially where oxygen is present as AAB in general are obligate aerobes. However, certain strains isolated from brewery environments have shown micro-aerotolerance characteristics (Jeon et al., 2015). Due to significant

TABLE 8.2 Important characteristics of brewery-related acetic acid bacteria.

	Genera			
Characteristics	**Acetobacter**	**Gluconobacter**	**Gluconoacetobacter**	**Komagataeibacter**
Motility & Flagellation	Peritrichous or non-motile	Polar or non-motile	Peritrichous or non-motile	−
Oxidation of ethanol to acetic acid	+	+	+	+
Oxidation of acetic acid to CO_2 and H_2O	+	−	+	+
Oxidation of lactase to acid to CO_2 and H_2O	+	−	+	+
Growth on 0.35% acetic acid containing medium	+	+	+	+
Growth on methanol	−	−	ND	ND
Growth on D-mannitol	+/−	+	+/−	−
Growth in presence of 30% D-glucose	−	−/+	−/+	ND
Production of cellulose	−	−	−/+	−/+
Production of water-soluble pigment	−/+	−/+	−/+	−
Fixation of molecular nitrogen	−	−	ND	ND
Ketogenesis (dihydroxyacetone) from glycerol	+/−	+	+/−	+/−
Acid production from:				
D-Mannitol	−/+	+	+/−	−
Glycerol	−/+	+	+	ND
Raffinose	−	−	−	ND
Cellular fatty acid type	$C_{18:1}$	$C_{18:1}$	ND	ND
Ubiquinone type	Q-9	Q-10	Q-10	Q-10
DNA base composition (mol % G + C)	52−60	55−63	53-65	ND

+, 90% or more of the strains positive; −, 90% or more of the strains negative; −/+- variable results; *ND*, Not determined.
Originally adapted from Gomes et al. (2018); Sengun and Karabiyikli (2011).

advancement in the beer packaging line in large breweries, significantly eliminating the oxygen from finished product, AAB are no longer considered important contaminants particularly in larger breweries (Vriesekoop et al., 2012).

AAB are also reported as an important microflora of biofilms in association with other beer spoilage microorganisms. AAB and some *enterobacteria* occur in microaerophilic niches and corners in brewery filling equipment and, at later stages of biofilm development, are protected from routine cleaning due to slime formation. AAB, due to their ability to produce exopolysaccharides provide suitable and protected environment for other microorganisms to grow and proliferate (Paradh, 2015). Yeast propagation, along with lactic acid bacteria, further provides microaerophilic and a partial anaerobic environment and substrate for growth of Gram-negative anaerobic bacteria such as *Pectinatus* and *Megasphaera* (Back, 1994; Storgårds, 2000, pp. 1−108).

Small scale breweries, micro-breweries and pub-breweries still face spoilage problems in semi-finished and finished beers as ineffective cleaning conditions and comparative lack of availability of advanced packaging technologies (Kubizniaková et al., 2021). AAB bacteria are commonly occurring microflora of beer dispensing lines mainly due to hygiene and oxygen ingress issues (Jevons & Quain, 2022; Vriesekoop et al., 2012). AAB still prevail in cask-conditioned and barrel-aged beers (Bokulich, Bamforth, & Mills, 2012). In dispensing systems, points such as taps, valves, spaces under seal, cavities, beer lines and tap stool, which are not easily accessible, are often insufficiently sanitised leading to growth of spoilage microbiota. Growth of spoilage microbes can occur singly or in association with biofilm; the latter are very difficult to remove (Back, 2005, pp. 10−112).

8.2.4 Detection of AAB

Detection of AAB can be broadly grouped into conventional methods and advanced molecular and non-molecular rapid detection methods. Conventional methods generally involve enrichment of samples on non-selective medium followed by followed by enrichment on selective or differential agar (Hill, 2009). Traditional methods are based on physiological and chemotaxonomic properties such as growth at lower pH, formation of organic acid from ethanol or glucose containing medium or growth in presence of high concentration of acetic acid (De Vuyst et al., 2008). The conventional methods are cheaper and require basic microbiological laboratory set-up and historically were widely used in detection and enumeration of AAB (Storgårds, 2000). However, the conventional media-based methods require a long incubation period and are not always accurate in detection and enumeration of target species of microorganism. There are numerous media described for isolation and identification of AAB bacteria, but no single medium has been found to be effective in supporting the growth of all AAB, and selectivity of the medium is variable for different strains (refer to Table 8.3).

For beer spoilage AAB, Frateur's differential medium containing yeast extract, ethanol and calcium carbonate and GYC medium containing glucose, yeast extract and calcium carbonate have been reported in the literature. Principally, calcium carbonate is decomposed to water, CO_2 and soluble calcium acetate resulting in clarification of the medium and appearance of 'halo effect' around the colonies (Kubizniaková et al., 2021). *Acetobacter* colonies appear clear due to formation of acid from ethanol, whereas *Gluconobacter* develop chalk-white deposit colonies (Priest, 2006).

Carr's differential medium (Carr, 1969) containing yeast extract, ethanol and bromocresol blue has also been documented. Agar (2.5% w/v) containing yeast extract (0.5% w/v) and ethanol (1.5% v/v) has been used for growth of beer spoilage AAB. Discolouration of medium is observed at bacteria growth site after incubation for 48 h. After prolong incubation the recolouration of medium is observed for *Acetobacter* and *Gluconacetobacter* species while Gluconobacter retain discoloured form (Kubizniaková et al., 2021).

Other differential media such as yeast extract, peptone and mannitol agar (YPM media) (Gullo & Giudici, 2008), AE medium (Yamada et al., 1999) and reinforced AE medium (Zahoor et al., 2006), WLN agar (Kubizniaková et al., 2021) used for isolation and identification of AAB have been reported in the literature. Use of fluorescence staining techniques such as the live/dead Baclight bacterial viability test has been studied for detection of viable but non-culturable AAB (Baena-Ruano et al., 2006).

Extensive research has been carried out on rapid detection methods for detection, enumeration and characterisation of AAB in food and beverage application. Rapid detection of AAB using polymerase chain reaction (PCR), real-time polymerase chain reaction (RT PCR) (Gammon et al., 2007; Torija, Mateo, Guillamón, & Mas, 2010), restriction fragment length polymorphism (Nanda et al., 2001; Ruiz, Poblet, Mas, & Guillamon, 2000), amplified fragment length polymorphism (Cleenwerck, de Wachter, Gonzalez, de Vuyst, & de Vos, 2009), denaturing gradient gel electrophoresis (Andorrà, Landi, Mas, Guillamón, & Esteve-Zarzoso, 2008; De Vero et al., 2006) and fluorescence in situ hybridisation (Franke et al., 1999; Franke-Whittle, O'Shea, Leonard, & Sly, 2005) are some of the examples of molecular detection methods. In recent studies, use of pyrosequencing, metagenomic DNA analysis and MALDI-TOF-MS has also been described (Bokulich et al., 2012; De Roos, Verce, Aerts, Vandamme, & De Vuyst, 2018; Snauwaert et al., 2016; Spitaels et al., 2015; Vermote et al., 2023).

8.3 Zymomonas

Zymomonas species are Gram-negative, non-endospore-forming and no capsule, intracellular lipid or glycogen former, catalase-positive, aerotolerant, facultative anaerobic bacteria. Morphologically, these bacteria are short plump rods that occur singly, in pairs, and sometimes in chains or rosettes with an average dimension of 1.0–2.0 x 4.0–5.0 μm (van Vuuren & Priest, 2003). *Zymomonas* is ethanol tolerant (below 10% ethanol v/v) and able to grow at pH more than 3.4 and temperature range of 25–30°C (Roselli et al., 2024). These bacteria metabolise glucose and fructose to produce equimolar quantities of ethanol and CO_2 using the Entner–Doudoroff pathway (refer to Fig. 8.1). Many strains can utilise sucrose mostly accompanied with levan formation but are unable to utilise maltose and maltotriose (van Vuuren & Priest, 2003; Yang et al., 2009).

Zymomonas has been primarily associated with sugar- and alcohol-rich environments (Weir, 2016). It has also been isolated as a spoilage microorganism of beers and ciders. *Zymomonas* species are reported as a spoilage microorganism from various traditional alcoholic beverages throughout the world. These bacteria are isolated from sugar-rich sap of agave, sugar cane and palm trees as a naturally occurring fauna (Coton et al., 2006; Weir, 2016). *Zymomonas* has been extensively studied for fuel ethanol application due to its fast growth, high ethanol tolerance and high productivity (Felczak, Bowers, Woyke, & TerAvest, 2021). *Zymomonas* also serves as a model microorganism for genetic modification for use of lignocellulosic biomass for ethanol production (Xia, Yang, Liu, Yang, & Bai, 2019).

TABLE 8.3 Conventional and rapid detection and characterisation methods for beer spoilage acetic acid bacteria, *Zymomonas*, and Enterobacteriaceae species.

Bacteria	Detection and characterisation methods	References
Acetic acid bacteria	Conventional media-based detection	
	1. Frateur's medium **2.** Carr's differential medium **3.** YPM medium **4.** AE medium **5.** Reinforced AE medium **6.** WLN	Carr (1969), Gullo and Giudici (2008), Kubizniaková et al. (2021), Priest (2006), Yamada et al. (1999)
	Advanced methods	
	1. DNA: DNA hybridisation **2.** PCR **3.** RT-PCR **4.** PCR TRFLP **5.** RFLP/AFLP **6.** DGGE **7.** FISH **8.** 454 Pyrosequencing **9.** Metagenomic DNA analysis **10.** MALDI-TOF-MS	Andorrà et al. (2008), Bokulich et al. (2012), Cleenwerck et al. (2009), De Roos et al. (2018), De Vero et al. (2006), Franke-Whittle et al. (2005), Gammon et al. (2007), Nanda et al. (2001), Ruiz et al. (2000), Spitaels et al. (2015)
Zymomonas	Conventional media-based detection	
	1. standard medium (SM) comprising of yeast extract and glucose. **2.** Apple juice- yeast extract medium **3.** Apple juice gelatin medium. **4.** Beer -glucose medium **5.** MGYP **6.** Beer sample with actidione **7.** Beer agar with lead acetate and Schiff reagent	Dennis and Young (1982), Goodman et al. (1982), Jespersen and Jakobsen (1996), Woodward (1982), Yanase (2014)
	Advanced methods	
	1. ARDA **2.** Duplex PCR **3.** RAPD	Coton et al. (2005), (2006)
Enterobacteriaceae	Conventional media-based detection	
	1. MacConkey agar/broth **2.** Universal Beer agar (UBA) **3.** Chromogenic agar	Finney et al. (2003), Jespersen and Jakobsen (1996); Kubizniaková et al. (2020); Priest (2006);
	Advanced methods	
	1. PCR based method. **2.** RAPD **3.** Automated ribotyping	Koivula et al. (2006), Maugueret and Walker (2002) Savard et al. (1994)

AE, acetic acid ethanol medium; *AFLP*, Amplified Fragment Length Polymorphism; *ARDA*, Amplified Ribosomal DNA; *DGGE*, Denaturing Gradient Gel Electrophoresis; *MALDI-TOF-MS*, Matrix-Assisted Laser Desorption/Ionization; *MGYP*, Malt extract Glucose Yeast extract Peptone Agar; *MS*, Mass Spectrometry; *RAPD*, Random Amplified polymorphic DNA: *RFLP*, Restriction Fragment Length Polymorphism; *RT-PCR*, real-time polymerase chain reaction; *SM*, Standard medium; *TOF*, Time Of Flight; *TRFLP*, Terminal restriction fragment-length polymorphism; *WLN*, Wallerstein Laboratory nutrient; *YPM*, yeast extract peptone mannitol medium.

8.3.1 Taxonomic status of *Zymomonas*

The genus *Zymomonas* belongs to the class *Alphaproteobacteria*, the order *Sphingomonadales* and recently proposed family *Zymomonadaceae* on the basis phylogenetic analyses of genome and 16SrRNA gene sequences (Hördt et al., 2020). *Zymomonas* to date has only one species, cited as *Z. mobilis*, which was formerly known as *Achromobacter anaerobium*,

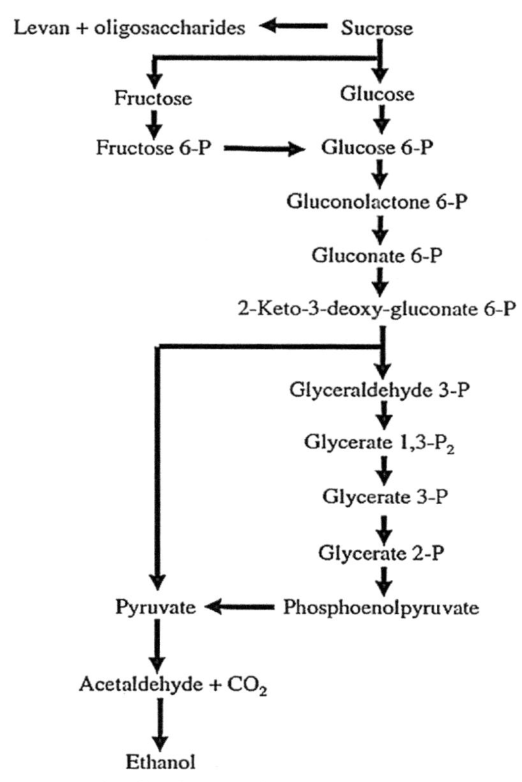

FIGURE 8.1 Schematic representation of *Entner-Doudoroff* pathway utilised by *Zymomonas mobilis* for sugar metabolism. *Originally adapted from Yanase (2014).*

originally isolated from beer (Shimwell, 1936). In the older literature, *Zymomonas* has also been cited as *Saccharomyces lindneri* and *Pseudomonas lindneri* (Hornsey, 2013).

At present, *Z. mobilis* has three validated subspecies, namely *Z. mobilis* subsp. *mobilis* (De Ley & Swings, 1976; Lindner, 1928); *Z. mobilis* subsp. *pomaceae* (De Ley & Swings, 1976; Millis, 1956) and *Z. mobilis* subsp. *francensis* (Coton et al., 2006). All three subspecies are differentiated based on phenotypic characterisation, protein and genetic characterisation and growth at 36°C (Coton et al., 2006). Out of the three validated species, only *Z. mobilis* subsp. *mobilis* is reported to be a beer spoiler (Roselli et al., 2024; van Vuuren & Priest, 2003).

8.3.2 Occurrence and beer spoilage ability of *Zymomonas*

Due to inability to metabolise maltose as a carbon source, the prevalence of beer spoilage *Zymomonas species* is limited to use of use of sucrose as an adjunct or priming sugar. *Zymomonas* are limited to ales supplemented with primed sugar such as cask conditioned ales and spoilage problems due to these bacteria have never been encountered in breweries producing lager beers as lower fermenting temperatures of lager do not favour the growth of *Zymomonas* (Bokulich & Bamforth, 2013; Dads & Martin, 1978). There have been no reported incidents of spoilage by *Zymomonas* species in beer recent times (Roselli et al., 2024).

Zymomonas occurs frequently in the brewery and cider house. However, the original source of contamination by *Zymomonas* species in the brewery and cider house is still unknown. It is suggested that soil could be the possible source of contamination in beer (Coton & Coton, 2003; Ingledew, 1979), as incidents of *Z. mobilis* contamination are linked to times of construction of new facilities and excavation in breweries (Ingledew, 1979). *Z. mobilis* can survive in the soil for several months and on certain objects contaminated with it (Weir, 2016).

Z. mobilis subsp. *mobilis* has also been reported to prevail in public houses, well-water sources, soil from brewery environments and the bottling process (Dads & Martin, 1978; Swings & De Ley, 1977). *Z. mobilis-contaminated* beer has a fruity aroma (rotten apple, due to the production of acetaldehyde), which rapidly progresses to a sulphide and rotten-egg aroma (due to the production of hydrogen sulphide) in spoiled beer (Dads & Martin, 1978).

The presence of *Zymomonas mobilis* has been reported in spoiled cider up to period of 9 weeks without losing its viability (Carr, 1983; Weir, 2016). Spoilage due to *Zymomonas* is a common problem in ciders. Rod-shaped bacteria responsible for sick cider was originally described by Barker and Hillier (1912). Later, *Z. mobilis* subsp. *pomaceae* (De Ley & Swings, 1976; Millis, 1956) was isolated from spoiled cider and described as the causal microorganism for cider sickness in English ciders. *Z. mobilis* subsp. *francensis* was proposed and characterised as the causal microorganism for 'framboisé' in French ciders (Coton et al., 2006). Spoilage due to *Zymomonas* in ciders is characterised by off-flavour typically described as like rotten banana, grassy, rotten lemon or framboisé mainly due to production of acetaldehyde in the range of 100−150 mg/L concentration. The spoilage is also associated with production of gas in bottled ciders to a high extent, decreased density and high turbidity in spoiled products (Coton & Coton, 2003).

8.3.3 Metabolic aspects of *Zymomonas*

The important physiological character of *Zymomonas* species of being naturally ethanogenic has been extensively studied for fuel ethanol production in comparison with *Saccharomyces cerevisiae* (Pérez-García, Del Rio-Arellano, Rincón, & Norma, 2022). *Saccharomyces cerevisiae* utilise Embden-Meyerhof-Parnas (EMP) pathway, a common metabolic pathway for pyruvate production. However, the EMP pathway does not operate in *Z. mobilis* (ZM4) (Fuhrer, Fischer, Sauer, & 2005). The gene for the enzyme *phosphofructokinase*, an important glycolytic enzyme, is not found in the genome, although genes for all other enzymes within the EMP pathway are present (Seo et al., 2004). *Z. mobilis* is a unique aerobic microorganism that uses the Entner Doudoroff (ED) pathway anaerobically instead of the EMP pathway for pyruvate production. Additionally, two enzymes from the tricarboxylic acid cycle, namely *two-oxoglutarate dehydrogenase complex* and *malate dehydrogenase*, are reported to be missing from the genome of *Z. mobilis* (ZM4 strain). The deficiency in these enzymes results in optimal carbon utilisation for ethanol production leading to maximum theoretical yield (Pérez-García et al., 2022). It is suggested that there may be an alternative pathway parallel to the TCA cycle for the synthesis of TCA intermediates such as oxaloacetate, malate, fumarate and succinate, as *Z. mobilis* could synthesise all necessary amino acids except lysine and methionine (Seo et al., 2004). *Z. mobilis* can ferment glucose, fructose and sucrose to ethanol and CO_2 (Seo et al., 2004; Swings & De Ley, 1977). However, *Zymomonas* is unable to utilise lactose, maltose and cellobiose due to the lack of genes responsible for production of enzymes necessary for metabolism of these sugars (Seo et al., 2004).

8.3.4 Detection of *Zymomonas*

In general, media based on fruit juice, sap and alcoholic beverages including standard medium (SM) comprising of yeast extract and glucose; Apple juice−yeast extract medium, apple juice−gelatin medium and beer-glucose medium have been recommended for isolation of *Zymomonas* species related to cider and beer associated environments (Yanase, 2014). Detection of *Zymomonas* in the brewery using malt yeast extract glucose and peptone (MYPG) agar supplemented with 50 ppm actidione and 3% ethanol or beer with 100 ppm actidione has been reported (Jespersen & Jakobsen, 1996). Synthetic medium comprising of yeast extract, essential salts, vitamins and other additives for isolation of auxotrophic Zymomonas species has also been described (Goodman et al., 1982).

In breweries, routine enrichment of the sample in primed beer supplemented with actidione for inhibition of yeast has been reported, although incubation of filtered membranes on an agar medium is not recommended as a satisfactory method (Woodward, 1982). For detection of *Zymomonas*, beer media supplemented with lead acetate (producing black colonies) and Schiff reagent (producing purple colonies) has been documented (Dennis & Young, 1982; Woodward, 1982).

Coton et al. (2005) developed a PCR-based amplified ribosomal DNA restriction analysis method for rapid detection of Zymomonas at the subspecies level. A further duplex PCR method with primers specific for 23S rRNA gene for detection of Zymomonas species has been developed. This method could detect Zymomonas species within 24 h with sensitivity of 10^2 CFU/mL. (Coton et al., 2006) characterised several strains of *Z. mobilis* with random amplified polymorphic DNA (Table 8.3).

8.4 Brewery-related *Enterobacteriaceae*

Enterobacteriaceae proposed by Lapage in (1979) is a large and diverse family of Gram-negative, facultatively anaerobic bacteria belonging to the newly proposed order *Enterobacterales* of the class Gammaproteobacteria in the phylum Proteobacteria (Adeolu, Alnajar, Naushad, & Gupta, 2016; Till, 2020, family Enterobacteriaceae comprised of 68 validated pathogenic and nonpathogenic genera and around 355 species (Janda & Abbott, 2021). Pathogenic Enterobacteriaceae such as *Salmonella*, *Serratia*, *Shigella* and *Escherichia* do not grow in beer due to a combination of intrinsic antimicrobial properties of beer along with technological and process hurdle such as wort boiling, pasteurisation and sterile filtration (Menz, Aldred, & Vriesekoop, 2009, Menz, Vriesekoop, Zarei, Zhu, & Aldred, 2010). Pathogens such as *Shigella* and *Salmonella* and *E. coli* force testing in beer showed the bacteria was inhibited within 48 h of inoculation even after inoculated at higher concentration (Boulton & Quain, 2001).

Order Enterobacterales currently has seven families and more than 40 bacterial genera amongst them Shimwellia, Obesumbacterium, Rahnella, Citrobacter, Klebsiella, Raoultella, Serratia and Enterobacter have been reported in the vicinity of fermenting wort, pitching yeast and process water and have not been identified in the finished product (Matoulková, Vontrobová, Brožová, & Kubizniaková, 2018). The brewery related Enterobacterales are facultative anaerobes are catalase positive and oxidase negative; the optimum growth temperature can range from 22–35°C. Most of the species are capable of reducing nitrates to nitrites in an anaerobic respiration mode (Matoulková et al., 2018).

8.4.1 Obesumbacterium proteus/Shimwellia pseudoproteus

The *Obesumbacterium proteus* is a Gram-negative bacterium isolated in pure culture from top fermenting yeast and was first classified as *Flavobacterium proteus* (Shimwell, 1936). Later this bacterium was assigned to the genus *Obesumbacterium* and *O. proteus* as a sole type of strain within the proposed genus (Shimwell, 1963, 1964). Genus *Obesumbacterium* was assigned to the family Enterobacteriaceae based on detailed taxonomic studies conducted (Priest, Somerville, Cole, & Hough, 1973). *O. Proteus* strains were further reclassified as biogroup-1 and biogroup-2 based on phenotypic and genetic characterisation (Priest et al., 1973). This heterogenous reclassification was later supported by data obtained from enteric repetitive intergenic consensus sequences (ERIC PCR) studies (Priest et al., 1994). In 2010, *O. proteus* biogroup-2 was assigned to new a genus *Shimwellia* and *O. proteus* was provided new taxonomic identity of *Shimwellia pseudoproteus* (Priest & Barker, 2010). *O. proteus* is commonly isolated from brewery environments, especially from fermenting wort and pitching yeast, and they have not been reported from any other source (Priest & Barker, 2010).

O. proteus in early stages of fermentation competes with yeast for nutrients, resulting a slower rate of fermentation. The growth of *O. proteus* is inhibited at alcohol concentration above 6% ABV. *O. proteus* can produce volatile components such as dimethyl sulfoxide (DMSO), acetoin, lactic acid, propanol, isobutanol and 2,3-butandiaol in initial stages of fermentation. Production of DMS imparts an undesirable parsnip flavour to the contaminated beer (Case, 1965; Priest & Hough, 1974).

Certain species of Enterobacteriaceae, especially *O. proteus* become more important due to their ability to form apparent total N-nitroso compounds (ATNCs) in fermenting wort which due to their non-volatile nature could be retained in the finished beer (refer to Fig. 8.2). ATNCs represent a possible risk to health, and consequently, their concentration is strictly monitored and limited to 20 μg/L (Maugueret & Walker, 2002). These species can utilise nitrates as an electron acceptor in anaerobic mode forming nitrites through nitrate reduction. Nitrites further could react with secondary amines present in the fermenting wort in initial stages to produce N-nitrosoamines (Matoulková et al., 2018; Smith, 1994). Apparently microbial monitoring and detection of *O. proteus* in pitching yeast and fermenting wort in initial stages became more evident due to regulatory concerns as contamination of around 0.03% of *O. proteus* in pitching yeast could result in ATNCs above the limit of 20 μg/L (Boulton & Quain, 2001; Maugueret & Walker, 2002).

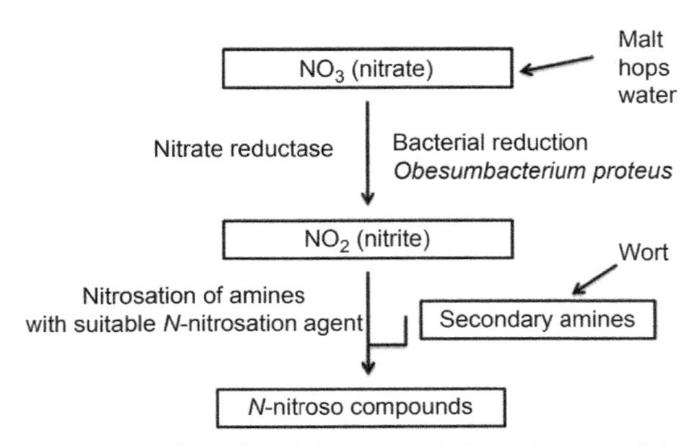

FIGURE 8.2 Role of Enterobacteriaceae in the formation of nitrosamines in fermenting wort. *Originally adapted from Priest (2006).*

8.4.2 Coliform bacteria related to brewing environments

Coliform bacteria are generally Gram-negative rod-shaped bacteria which are non-spore forming and are either motile or nonmotile in nature; they can ferment lactose with the production of acid and gas when incubated at 35−37°C within the period of 48 h (APHA, 1998). These bacteria serve as an indicator microbe reflecting the hygiene and sanitation in breweries. The presence of coliforms in water is an indication of the ineffectiveness of water treatment. These bacteria can be introduced to wort through contaminated water or seepage of external fluids through connecting pipes (Vaughan, O'Sullivan, & Sinderen, 2005). Brewery-related coliform bacteria mainly belong to genera *Citrobacter*, *Rahnella*, *Klebsiella* and *Enterobacter*. These species have been associated with unfermented and fermenting wort (Vaughan et al., 2005).

Citrobacter freundii is a facultative anaerobe, straight rod in morphology and usually occur single or in pairs and are usually motile in nature. *C. freundii* is catalase positive, and most of the strains show ability to utilise citrate as a carbon source, which is a signature characteristic of this species (Matoulková et al., 2018). *C. freundii* is occasionally associated with pitched wort and contamination of wort leads to higher fermentation rate and reduces activity of brewer's yeast. The wort spoilage is associated with production of lactate, succinate and DMS in final beer. *C. freundii* is sensitive to ethanol and only prevail in the initial stages of fermentation until the pH drops below 4.5 (Vriesekoop et al., 2012).

Genus *Rahnella*, also referred as *Enterobacter agglomerans* in older brewery literature has been isolated from soil, water, food, plant material and clinical samples (Sedlácek, 2007, p. 270; van Vuuren, Kersters, Ley, Toerien, Meisel, 1978). *Rahnella* has been reported to influence the main fermentation top fermenting yeast and can survive up to 40−60 h from the onset of fermentation (Matoulková et al., 2018). *Rahnella aquaticus* is considered as a potential beer spoiler due to its ability to survive through fermentation and can show carry over contamination by accumulating on the surface of the yeast (Hamze et al., 1991; van Vuuren et al., 1978). *Rahnella aquaticus* causes fruity, lactate, sulphury aroma due to production of DMS, acetaldehyde and methyl acetate in the fermenting wort (Sakamoto & Konings, 2003).

Genus *Klebsiella* in general are non-motile rod-shaped bacteria and the cells occur singly, in pairs or short chains. Cells often show presence of capsule and colonies on growth media show slimy colonies. The stickiness varies significantly based on strains and composition of the growth media (Matoulková et al., 2018). Two species of Genus *Klebsiella* namely *K. oxytoca* and *K. terragena* have been reported in the brewery environment (Vaughan et al., 2005). During phylogenetic analyses of existing *Klebsiella species*, a new genus *Raoultella* was proposed and the species *K. terragena* was given new taxonomic identity of *Raoultella terragena* (Drancourt et al., 2001). The *Klebsiella* species are associated with fermented wort and reported for production of phenolic off flavour due to production of 4- vinyl-guaiacol, which is formed by decarboxylation of ferulic acid present in the wort (van Vuuren & Priest, 2003). *R. terragena* (formerly known as *K. terragena*) is positive to Voges-Proskauer test and reported to produce a high concentration of acetoin and 2,3 butanediol through the 2,3 butanediol pathway.

Other Enterobacteriaceae species which have been reported in brewery environments are *Serratia* and *Enterobacter* species mainly as a contamination of source water and their activity is limited to the beginning of fermenting wort. *E. cloacae* has been cited for production of phenolic compound and DMS and affect the sensorial attributes of finished beer (Matoulková et al., 2018).

8.4.3 Detection of Enterobacteriaceae in brewery environments

Detection of Enterobacteriaceae in brewing samples such as fermenting wort, yeast slurry and process water are conducted using McConkey agar and chromogenic agar media by incubation at 37°C (Kubizniaková et al., 2020). McConkey agar is a selective medium for detection of coliform and Enterobacteriaceae. It contains bile salt and crystal violet for inhibition of Gram-positive bacteria (Finney, Smullen, Foster, Brokx, & Storey, 2003), and lactose is utilised as a sole source of carbon. McConkey agar supplemented with actidione (10 ppm) for suppression of brewer's yeast is recommended by the European Brewing convention (Jespersen & Jakobsen, 1996). However, lactose genitive bacteria such as *O. proteus* and *Shimwellia* has shown slower growth on McConkey agar and formed creamy coloured colonies with discolouration of media to light yellow was observed (Kubizniaková et al., 2020; Priest, 2006). All brewery-related *Enterobacterales* can grow on nutrient agar and plate count agar. Endo agar on the other hand is limited for detection of coliform bacteria (Kubizniaková et al., 2020). For enrichment, use of Universal beer medium (UBA) for wort samples and Wallerstein Laboratory Nutrient medium (WLN) for enrichment of beer has been recommended (Priest, 2006).

Chromogenic agar has been formulated for simultaneous detection and enumeration of total coliform bacteria and *E. coli* in single medium. The medium consists of chromogenic substrates salmon-GAL and X-glucuronide. Salmon-GAL detects the presence of enzyme ß-D- *galactosidase*, which facilitates the breakdown of lactose into glucose and galactose, and this enzyme is generally present in all coliform bacteria. *X-glucuronide* is utilised for presence of β -glucuronidase produced by *E. coli*. For inhibition of Gram-positive and other bacteria tergitol-7 is utilised (Finney et al., 2003). Coliform bacteria form pink- to red-coloured colonies, while *E. coli* form blue-purple colonies on chromogenic agar Kubizniaková et al., 2020).

Enterobacteriaceae have been extensively studied for advanced detection in the field of clinical and food application. However, the application of advance detection methods for rapid detection and characterisation of brewery related Enterobacteriaceae has been not documented extensively. The major reason could be as these microorganisms are indicators of hygienic condition and do not show direct threat of beer contamination. Random amplified polymorphic DNA has been cited for detection and characterisation of *O. proteus* both biotype-1 and biotype-2 strains (Savard, Hutchinson, & Dowhanick, 1994). A PCR-based method for specific detection and discrimination of *O. proteus* biogroup-2 strains from *O. proteus* biogroup-1 and other related microorganisms has been documented (Maugueret & Walker, 2002). Automated ribotyping and PCR-based methods have also been reported for characterisation of *O. proteus* biogroup-1 strains (Koivula, Juvonen, Haikara, Suihko, 2006).

8.5 Conclusions

Gram-negative aerobic and facultative anaerobes such as AAB, *Zymomonas* and certain species of Enterobacteriaceae were of serious concern to brewers a few decades ago. As a result of technological improvements in packaging aspects, advancement in brewery hygiene levels and enhancement in rapid detection methods, these microorganisms are less likely to be the focus of brewing microbiology research in the near future. However, due to the extent of consumer awareness about food and beverage safety and the concern with maintaining corporate brand image, beer spoilage microorganisms will remain a serious concern to breweries worldwide. Brewery-related Enterobacteriaceae need to be monitored in breweries due to the high level of hygiene required and due to aspects of controlling ATNC below regulatory limits in finished beer. *Zymomonas* seems to be limited to primed beer in the brewing industry, but strains of *Zymomonas* are still a problem in the cider industry. Also, AAB still prevail as a common contaminant in beer dispense and cask conditioned ales. In addition there is the emergence of new species in brewery environments, especially in AAB, and there is need to extensively study these species for beer spoilage ability and further characterisation in brewery environments. It would also be interesting to review prevalence of these microorganisms in no and low alcoholic beverages and small craft breweries, where varieties of creative beer styles and use of diverse raw materials is implemented in basic brewing infrastructure and laboratory set-up.

Selective Enterobacteriaceae play a significant role in flavour development in Lambic and American coolship ales, and the role of AAB will be an important topic of research for their controlled use in sour beer production. Biotechnological applications of *Zymomonas* in fuel ethanol fermentation will also still be an important topic of research.

Acknowledgements

The author would like to acknowledge Professor A. E. Hill, International Centre for Brewing and Distilling (ICBD), Heriot Watt University, Edinburgh, UK, for her valuable advice and information during the preparation of this manuscript. The author would also like to thank Ruchika Malhotra, Head, Technical Innovation, Pernod Ricard India and Mahesh Patil, Master Blender, Pernod Ricard India for their support and technical input during completion of this manuscript.

References

Adeolu, M., Alnajar, S., Naushad, S., & Gupta, R. (2016). Genome-based phylogeny and taxonomy of the 'enterobacteriales': Proposal for *Enterobacterales ord. nov.* Divided into the families *Enterobacteriaceae, Erwiniaceae fam. nov., Pectobacteriaceae fam. nov., Yersiniaceae fam. nov., Hafniaceae fam. nov., Morganellaceae fam. nov., and Budviciaceae fam. nov. International Journal of Systematic and Evolutionary Microbiology, 66*(12), 5575–5599.

Andorrà, I., Landi, S., Mas, A., Guillamón, J. M., & Zarzosa, B. (2008). Effect of oenological practices on microbial populations using culture-independent techniques. *Food Microbiology, 25*(7), 849–856.

APHA. (1998). *Standard methods for the examination of water and wastewater*. Washington, DC, USA: American Public Health Association.

Back, W. (1994). Secondary contamination in the filling area. *Brauwelt International, 4*, 326–328.

Back, W. (2005). *Brewery. Colour atlas and handbook of beverage biology*. Nürnberg: Fachverlag Hans Carl.

Baena-Ruano, S., Jiménez-Ot, C., Santos-Dueñas, I. M., Cantero-Moreno, D., Barja, F., & García-García, I. (2006). Rapid method for total, viable and non-viable acetic acid bacteria determination during acetification process. *Process Biochemistry, 4*(5), 1160–1164.

Barker, B. T. P., & Hillier, V. F. (1912). Cider sickness. In *Annual report of the agricultural and horticultural research station of long ashton, bristol, UK* (pp. 174–181).

Bartowsky, E. J., & Henschke, P. A. (2008). Acetic acid bacteria spoilage of bottled red wine—a review. *International Journal of Food Microbiology, 125*(1), 60–70.

Bokulich, N. A., & Bamforth, C. W. (2013). The microbiology of malting and brewing. *Microbiology and Molecular Biology Reviews, 77*(2), 157–172.

Bokulich, N. A., Bamforth, C. W., & Mills, D. A. (2012). Brewhouse-resident microbiota are responsible for multi-stage fermentation of American coolship ale. *PLoS One, 7*(4), Article e35507.

Boulton, C., & Quain, D. (2001). *Brewing yeast & fermentation* (1st ed.). Oxford: Blackwell Science.

Carr, J. G. (1969). Identification of acetic acid bacteria. In B. M. D. A. Shapton (Ed.), *Identification methods for microbiologists. Part B* (pp. 1–8). London: Academic Press.

Carr, J. G. (1983). Microbes I have known: A study of those associated with fermented products. *Journal of Applied Bacteriology, 55*(3), 383–401.

Case, A. C. (1965). Conditions controlling flavobacterium proteus in brewery fermentations. *Journal of the Institute of Brewing, 71*(3), 250–256.

Cleenwerck, I., de Wachter, M., Gonzalez, A., de Vuyst, L., & de Vos, P. (2009). Differentiation of species of the family *acetobacteraceae* by AFLP DNA fingerprinting: *Gluconoacetobacter kombuchae* is a later heterotypic synonym of *Gluconoacetobacter hansenii*. *International Journal of Systematic and Evolutionary Microbiology, 59*, 1771–1786.

Coton, E., & Coton, M. (2003). Microbiological origin of "Framboisé" in French ciders. *Journal of the Institute of Brewing, 109*(4), 299–304.

Coton, M., Laplace, J. M., Auffray, Y., & Coton, E. (2005). Duplex PCR method for rapid detection of Zymomonas mobilis in cider. *Journal of the Institute of Brewing, 111*(3), 299–303.

Coton, M., Laplace, J. M., Auffray, Y., & Coton, E. (2006). Polyphasic study of *Zymomonas mobilis* strains revealing the existence of a novel subspecies *Z. mobilis* subsp. *francensis subsp. nov.*, isolated from French cider. *International Journal of Systematic and Evolutionary Microbiology, 56*(1), 121–125.

Dağbağlı, S., & Göksungur, Y. (2017). Exopolysaccharide production of acetic acid bacteria. In *Acetic acid bacteria* (pp. 120–141). CRC Press.

Dads, M. J. S., & Martin, P. A. (1978). The genus *Zymomonas*-a review. *Journal of the Institute of Brewing, 79*(5), 386–391.

De Ley, J., & Swings, J. (1976). Phenotypic description, numerical analysis, and proposal for an improved taxonomy and nomenclature of the genus *Zymomonas* Kluyver and van Niel 1936. *International Journal of Systematic Bacteriology, 26*(2), 146–157.

De Roos, J., Verce, M., Aerts, M., Vandamme, P., & De Vuyst, L. (2018). Temporal and spatial distribution of the acetic acid bacterium communities throughout the wooden casks used for the fermentation and maturation of lambic beer underlines their functional role. *Applied and Environmental Microbiology, 84*(7), Article e02846-17.

De Vero, L., Gala, E., Gullo, M., Solieri, L., Landi, S., & Giudici, P. (2006). Application of denaturing gradient gel electrophoresis (DGGE) analysis to evaluate acetic acid bacteria in traditional balsamic vinegar. *Food Microbiology, 23*(8), 809–813.

De Vuyst, L., Camu, N., De Winter, T., Vandemeulebroecke, K., Van de Perre, V., Vancanneyt, M., et al. (2008). Validation of the (GTG) 5-rep-PCR fingerprinting technique for rapid classification and identification of acetic acid bacteria, with a focus on isolates from Ghanaian fermented cocoa beans. *International Journal of Food Microbiology, 125*(1), 79–90.

Dennis, R. T., & Young, T. W. (1982). A simple, rapid method for the detection of subspecies of Zymomonas mobilis. *Journal of the Institute of Brewing, 88*(1), 25–29.

Drancourt, M., Bollet, C., Carta, A., & Rousselier, P. (2001). Phylogenetic analyses of *Klebsiella* species delineate *Klebsiella* and *Raoultella gen. nov.*, with description of *Raoultella ornithinolytica comb. nov.*, *Raoultella terrigena comb. nov.* and *Raoultella planticola comb. nov. International Journal of Systematic and Evolutionary Microbiology, 51*(3), 925–932.

Felczak, M. M., Bowers, R. M., Woyke, T., & TerAvest, M. A. (2021). *Zymomonas* diversity and potential for biofuel production. *Biotechnology for Biofuels, 14*(1), 112.

Finney, M., Smullen, J., Foster, H. A., Brokx, S., & Storey, D. M. (2003). Evaluation of chromocult coliform agar for the detection and enumeration of *Enterobacteriaceae* from faecal samples from healthy subjects. *Journal of Microbiological Methods, 54*(3), 353–358.

Franke, I. H., Fegan, M., Hayward, C., Leonard, G., Stackebrandt, E., & Sly, L. I. (1999). Description of *Gluconacetobacter sacchari* sp. *nov.*, a new species of acetic acid bacterium isolated from the leaf sheath of sugar cane and from the pink sugar-cane mealy bug. *International Journal of Systematic Bacteriology, 49*, 1681–1693.

Franke-Whittle, I. H., O'Shea, M. G., Leonard, G. J., & Sly, L. I. (2005). Design, development, and use of molecular primers and probes for the detection of *Gluconacetobacter species* in the pink sugarcane mealy bug. *Microbial Ecology, 50*, 128–139.

Fuhrer, T., Fischer, E., & Sauer, U. (2005). Experimental identification and quantification of glucose metabolism in seven bacterial species. *Journal of Bacteriology, 187*(5), 1581–1590.

Gammon, K. S., Livens, S., Pawlowsky, K., Rawling, S. J., Chandra, S., & Middleton, A. M. (2007). Development of real-time PCR methods for the rapid detection of low concentrations of *Gluconobacter* and *Gluconacetobacter* species in an electrolyte replacement drink. *Letters in Applied Microbiology, 44*(3), 262–267.

Gomes, R. J., de Fatima Borges, M., de Freitas Rosa, M., Castro-Gómez, R. J. H., & Spinosa, W. A. (2018). Acetic acid bacteria in the food industry: Systematics, characteristics, and applications. *Food Technology and Biotechnology, 56*(2), 139.

Gonzalez, A., Hierro, N., Poblet, M., Mas, A., & Guillamon, J. M. (2005). Application of molecular methods to demonstrate species and strain evolution of acetic acid bacteria population during wine production. *International Journal of Food Microbiology, 102*, 295–304.

Goodman, A. E., Rogers, P. L., & Skotnicki, M. L. (1982). Minimal medium for isolation of auxotrophic zymomonas mutants. *Applied and Environmental Microbiology, 44*(2), 496–498.

Gullo, M., & Giudici, P. (2008). Acetic acid bacteria in traditional balsamic vinegar: Phenotypic traits relevant for starter cultures selection. *International Journal of Food Microbiology, 125*, 46–53.

Hördt, A., López, M. G., Meier-Kolthoff, J. P., Schleuning, M., Weinhold, L. M., Tindall, B. J., et al. (2020). Analysis of 1,000+ type-strain genomes substantially improve taxonomic classification of Alphaproteobacteria. *Frontiers in Microbiology, 11*, Article 493139.

Hamze, M., Mergaert, J., Van Vuuren, H. J. J., Gavini, F., Beji, A., Izard, D., et al. (1991). Rahnella aquatilis, a potential contaminant in lager beer breweries. *International Journal of Food Microbiology, 13*(1), 63–68.

He, Y., Xie, Z., Zhang, H., Liebl, W., Toyama, H., & Chen, F. (2022). Oxidative fermentation of acetic acid bacteria and its products. *Frontiers in Microbiology, 13*, Article 879246.

Hill, A. E. (2009). Microbiological stability of beer. In C. Bamforth (Ed.), *Beer: A quality perspective* (pp. 163–184). Academic Press.

Hill, A. (Ed.). (2015). *Brewing microbiology: Managing microbes, ensuring quality and valorising waste*. Woodhead Publishing.

Hill, A. E., & Priest, F. G. (2017). Microbiology and microbiological control in the brewery. *Handbook of brewing* (pp. 529–546).

Hornsey, I. (2013). *Brewing*. London: Royal Society of Chemistry.

Ingledew, M. W. (1979). Effect of bacterial contaminants on beer. A review. *Journal of the American Society of Brewing Chemists, 37*, 145–150.

Janda, J. M., & Abbott, S. L. (2021). The changing face of the family *Enterobacteriaceae* (order: "*Enterobacterales*"): New members, taxonomic issues, geographic expansion, and new diseases and disease syndromes. *Clinical Microbiology Reviews, 34*(2), 10–1128.

Jeon, S. H., Kim, N. H., Shim, M. B., Jeon, Y. W., Ahn, J. H., Lee, S. H., ... Rhee, M. S. (2015). Microbiological diversity and prevalence of spoilage and pathogenic bacteria in commercial fermented alcoholic beverages (beer, fruit wine, refined rice wine, and yakju). *Journal of Food Protection, 78*(4), 812–818.

Jespersen, L., & Jakobsen, M. (1996). Specific spoilage organisms in breweries and laboratory media for their detection. *International Journal of Food Microbiology, 33*, 139–155.

Jevons, A. L., & Quain, D. E. (2022). Identification of spoilage microflora in draught beer using culture-dependent methods. *Journal of Applied Microbiology, 133*(6), 3728–3740.

Juvonen, R., & Suihko, M. L. (2006). *Megasphaera paucivorans* sp. *nov., Megasphaera sueciensis* sp. *nov. and Pectinatus haikarae* sp. *nov.*, isolated from brewery samples, and emended description of the genus Pectinatus. *International Journal of Systematic and Evolutionary Microbiology, 56*(4), 695–702.

Koivula, T. T., Juvonen, R., Haikara, A., & Suihko, M. L. (2006). Characterization of the brewery spoilage bacterium *Obesumbacterium proteus* by automated ribotyping and development of PCR methods for its biotype 1. *Journal of Applied Microbiology, 100*(2), 398–406.

Kubizniaková, P., Brožová, M., Štulíková, K., Vontrobová, E., Hanzalíková, K., & Matoulková, D. (2020). Microbiology of brewing production-bacteria of the order *Enterobacterales* and culture methods for their detection. *Kvasny prumysl, 66*(5), 345–350.

Kubizniaková, P., Kyselová, L., Brožová, M., Hanzalíková, K., & Matoulková, D. (2021). The role of acetic acid bacteria in brewing and their detection in operation. *Kvasny prumysl, 67*(5), 511–522.

Lapage, S. P. (1979). Proposal of *Enterobacteraceae nom. nov*. As a substitute for the illegitimate but conserved name *Enterobacteriaceae* rahn 1937: Request for an opinion. *International Journal of Systematic Bacteriology, 29*(3), 265–266.

Lindner, P. (1928). *Atlas mikrosk. Grundl. Gdrungsk. 2 3 aufl.*

Lynch, K. M., Zannini, E., Wilkinson, S., Daenen, L., & Arendt, E. K. (2019). Physiology of acetic acid bacteria and their role in vinegar and fermented beverages. *Comprehensive Reviews in Food Science and Food Safety, 18*(3), 587–625.

Matoulková, D., Vontrobová, E., Brožová, M., & Kubizniaková, P. (2018). Microbiology of brewery production—bacteria of the order Enterobacterales. *Kvasny Prumysl, 64*(4), 161–166.

Mauguerct, T. J., & Walker, S. L. (2002). Rapid detection of obesumbacterium proteus from yeast and wort using polymerase chain reaction. *Letters in Applied Microbiology, 35*(4), 281–284.

Menz, G., Aldred, P., & Vriesekoop, F. (2009). Pathogens in beer. In V. R. Preedy (Ed.), *In beer in health and disease prevention* (pp. 403–413). Amsterdam: Academic Press.

Menz, G., Vriesekoop, F., Zarei, M., Zhu, B., & Aldred, P. (2010). The growth and survival of food-borne pathogens in sweet and fermenting brewers' wort. *International Journal of Food Microbiology, 140*(1), 19–25.

Millis, N. F. (1956). A study of the cider-sickness bacillus—a new variety of Zymomonas anaerobia. *Journal of General Microbiology, 15*(3), 521–528.

Nakata, H., Kanda, H., Nakakita, Y., Kaneko, T., & Tsuchiya, Y. (2019). *Prevotella cerevisiae* sp. nov., beer-spoilage obligate anaerobic bacteria isolated from brewery wastewater. *International Journal of Systematic and Evolutionary Microbiology, 69*(6), 1789–1793.

Nanda, K., Taniguchi, M., Ujike, S., Ishihara, N., Mori, H., Ono, H., et al. (2001). Characterization of acetic acid bacteria in traditional acetic acid fermentation of rice vinegar (komesu) and unpolished rice vinegar (kurosu) produced in Japan. *Applied and Environmental Microbiology, 67*(2), 986–990.

Pérez-García, L. A., Del Rio-Arellano, C. N., Rincón, D. F. L., & Norma, M. (2022). Physiology of ethanol production by *Zymomonas mobilis*. In *Bioethanology* (pp. 21–42). Apple Academic Press.

Paradh, A. D. (2015). Gram-negative spoilage bacteria in brewing. In *Brewing microbiology* (pp. 175–194). Woodhead Publishing.

Pothakos, V., Illeghems, K., Laureys, D., Spitaels, F., Vandamme, P., & De Vuyst, L. (2016). Acetic acid bacteria in fermented food and beverage ecosystems. Acetic acid bacteria. *Ecology and physiology*, 73–99.

Priest, F. G. (2006). Microbiology and microbial control methods in the brewery. In F. G. Priest, & G. G. Stewart (Eds.), *Handbook of brewing*. CRC Press.

Priest, F. G., & Barker, M. (2010). Gram-negative bacteria associated with brewery yeasts: Reclassification of *Obesumbacterium proteus* biogroup 2 as *Shimwellia pseudoproteus gen. Nov., sp. nov.*, and transfer of *Escherichia blattae* to *Shimwellia blattae comb. Nov. International Journal of Systematic and Evolutionary Microbiology, 60*(4), 828–833.

Priest, F. G., Hammond, J. R., & Stewart, G. S. (1994). Biochemical and molecular characterization of Obesumbacterium proteus, a common contaminant of brewing yeasts. *Applied and Environmental Microbiology, 60*(5), 1635–1640.

Priest, F. G., & Hough, J. S. (1974). The influence of Hafnia protea (*Obesumbacterium proteus*) on beer flavour. *Journal of the Institute of Brewing, 80*(4), 370–376.

Priest, F. G., Somerville, H. J., Cole, J. A., & Hough, J. S. (1973). The taxonomic position of *Obesumbacterium proteus*, a common brewery contaminant. *Journal of General Microbiology, 75*(2), 295–307.

Qiu, X., Zhang, Y., & Hong, H. (2021). Classification of acetic acid bacteria and their acid resistant mechanism. *AMB Express, 11*(1), 29.

Roselli, G. E., Kerruish, D. W., Crow, M., Smart, K. A., & Powell, C. D. (2024). The two faces of microorganisms in traditional brewing and the implications for no-and low-alcohol beers. *Frontiers in Microbiology, 15*, Article 1346724.

Ruiz, A., Poblet, M., Mas, A., & Guillamon, J. M. (2000). Identification of acetic acid bacteria by RFLP of PCR-amplified 16S rDNA and 16S-23S rDNA intergenic spacer. *International Journal of Systematic and Evolutionary Microbiology, 50*(6), 1981–1987.

Sakamoto, K., & Konings, W. N. (2003). Beer spoilage bacteria and hop resistance. *International Journal of Food Microbiology, 89*(2), 105–124.

Savard, L., Hutchinson, J. N., & Dowhanick, T. M. (1994). Characterization of different isolates of Obesumbacterium proteus using random amplified polymorphic DNA. *Journal of the American Society of Brewing Chemists, 52*(2), 62–65.

Schleifer, K. H., Leuteritz, M., Weiss, N., Ludwig, W., Kirchhof, G., & Seidel-Rüfer, H. E. L. G. A. (1990). Taxonomic study of anaerobic, gram-negative, rod-shaped bacteria from breweries: Emended description of *Pectinatus cerevisiiphilus* and description of *Pectinatus frisingensis* sp. *nov., Selenomonas lacticifex* sp. *nov., Zymophilus raffinosivorans gen. Nov., sp. nov., and Zymophilus paucivorans* sp. *nov. International Journal of Systematic Bacteriology, 40*(1), 19–27.

Sedlácek, I. (2007). *Taxonomie prokaryot*. Brno, Czech Republic: Masarykova Univerzita.

Sengun, I. K., & Karabiyikli, S. (2011). Review: Importance of acetic acid bacteria in food industry. *Food Control, 22*, 647–656.

Sengun, I. Y., Kilic, G., Charoenyingcharoen, P., Yukphan, P., & Yamada, Y. (2022). Investigation of the microbiota associated with traditionally produced fruit vinegars with focus on acetic acid bacteria and lactic acid bacteria. *Food Bioscience, 47*, Article 101636.

Seo, J. S., Chong, H., Park, H. S., Yoon, K. O., Jung, C., Kim, J. J., et al. (2004). The genome sequence of the ethanologenic bacterium Zymomonas mobilis ZM4. *Nature Biotechnology, 23*(1), 63–68.

Shimwell, J. L. (1936). Study of a new species of Acetobacter (*A. capsulatum*) producing ropiness in beer and beer-wort. *Journal of the Institute of Brewing, 42*(6), 585–595.

Shimwell, J. L. (1963). Obesumbacterium gen. nov. *Brewers' Journal., 99*, 759–760.

Shimwell, J. L. (1964). Obesumbacterium, a new genus for the inclusion of "*Flavobacterium proteus*". *Journal of the Institute of Brewing, 70*(3), 247–248.

Smith, N. A. (1994). Cambridge prize lecture nitrate reduction and n-nitrosation in brewing. *Journal of the Institute of Brewing, 100*(5), 347–355.

Snauwaert, I., Roels, S. P., Van Nieuwerburgh, F., Van Landschoot, A., De Vuyst, L., & Vandamme, P. (2016). Microbial diversity and metabolite composition of Belgian red-brown acidic ales. *International Journal of Food Microbiology, 221*, 1–11.

Sombolestani, A. S., Cleenwerck, I., Cnockaert, M., Borremans, W., Wieme, A. D., De Vuyst, L., et al. (2021). Characterization of novel *gluconobacter* species from fruits and fermented food products: *Gluconobacter cadivus* sp. *nov.*, *Gluconobacter vitians* sp. *nov.* and *Gluconobacter potus* sp. *nov. International Journal of Systematic and Evolutionary Microbiology, 71*(3), Article 004751.

Spitaels, F., Wieme, A., Balzarini, T., Cleenwerck, I., Van Landschoot, A., De Vuyst, L., et al. (2014). *Gluconobacter cerevisiae* sp. *nov.*, isolated from the brewery environment. *International Journal of Systematic and Evolutionary Microbiology, 64*(Pt_4), 1134–1141.

Spitaels, F., Wieme, A. D., Janssens, M., Aerts, M., Van Landschoot, A., De Vuyst, L., et al. (2015). The microbial diversity of an industrially produced lambic beer shares members of a traditionally produced one and reveals a core microbiota for lambic beer fermentation. *Food Microbiology, 49*, 23–32.

Storgårds, E. (2000). *Process hygiene control in beer production and dispensing* (Academic Dissertation), VTT Publication 410. Helsinki: Finland.

Suzuki, K. (2011). 125th anniversary review: Microbiological instability of beer caused by spoilage bacteria. *Journal of the Institute of Brewing, 117*(2), 131–155.

Suzuki, K. (2020). Emergence of new spoilage microorganisms in the brewing industry and development of microbiological quality control methods to cope with this phenomenon: A review. *Journal of the American Society of Brewing Chemists, 78*(4), 245–259.

Swings, J., & De Ley, J. (1977). The biology of Zymomonas. *Bacteriological Reviews, 4*(1), 1.

Teyssier, C., & Hamdouche, Y. (2016). Acetic acid bacteria. In , *Vol 97. Fermented foods. Part I: Biochemistry and biotechnology*. Boca Raton, FL, USA: CRC Press.

van Vuuren, H. J. J., Kersters, K., Ley, J. D., Toerien, D. F., & Meisel, R. (1978). *Enterobacter agglomerans*—a new bacterial contaminant isolated from lager beer breweries. *Journal of the Institute of Brewing, 84*(6), 315–317.

van Vuuren, H. J. J., & Priest, F. G. (2003). Gram-negative brewery bacteria. In F. G. Priest, & Campbell (Eds.), *Brewing microbiology* (pp. 219–245). New York: Kluwer Academic/Plenum Publishers.

Vaughan, A., O'Sullivan, T., & Sinderen, D. (2005). Enhancing the microbiological stability of malt and beer—a review. *Journal of the Institute of Brewing, 111*(4), 355–371.

Vermote, L., De Roos, J., Cnockaert, M., Vandamme, P., Weckx, S., & De Vuyst, L. (2023). New insights into the role of key microorganisms and wooden barrels during lambic beer fermentation and maturation. *International Journal of Food Microbiology, 394*, Article 110163.

Vriesekoop, F., Krahl, M., Hucker, B., & Menz, G. (2012). 125th Anniversary Review: Bacteria in brewing: The good, the bad and the ugly. *Journal of the Institute of Brewing, 118*(4), 335–345.

Weir, P. M. (2016). The ecology of Zymomonas: A review. *Folia Microbiologica, 61*, 385–392.

Woodward, J. D. (1982). Detection of Zymomonas. *Journal of the Institute of Brewing, 88*(2), 84–85.

Xia, J., Yang, Y., Liu, C. G., Yang, S., & Bai, F. W. (2019). Engineering Zymomonas mobilis for robust cellulosic ethanol production. *Trends in Biotechnology, 37*(9), 960–972.

Yamada, Y., Hosono, R., Lisdyanti, P., Widyastuti, Y., Saono, S., & Uchimura, T. (1999). Identification of acetic acid bacteria isolated from Indonesian sources, especially of isolates classified in the genus *Gluconobacter*. *Journal of General and Applied Microbiology, 45*(1), 23–28.

Yanase, H. (2014). Zymomonas. In C. A. Batt (Ed.), *Encylopedia of food microbiology* (2nd ed.). London: Elsevier.

Yang, S., Pappas, K. M., Hauser, L. J., Land, M. L., Chen, G. L., & Hurst, G. B. (2009). Improved genome annotation for *Zymomonas mobilis*. *Nature Biotechnology, 27*(10), 893–894.

Yassunaka Hata, N. N., Surek, M., Sartori, D., Vassoler Serrato, R., & Aparecida Spinosa, W. (2023). Role of acetic acid bacteria in food and beverages. *Food Technology and Biotechnology, 61*(1), 85–103.

Zahoor, T., Siddique, F., & Farooq, U. (2006). Isolation and characterization of vinegar culture (Acetobacter aceti) from indigenous sources. *British Food Journal, 108*, Article 429439.

Further reading

Comprehensive review articles on the Gram negative bacteria recently published on the topic of role of acetic acid bacteria in brewing and their detection by Kubizniaková, Kyselová, Brožová, Hanzalíková, & Matoulková, 2021, comprehensive review on brewery related *Enterobacterales* by Matoulková, Vontrobová, Brožová, & Kubizniaková, 2018 and beneficial and spoilage aspects of microorganisms in traditional brewing and their implications of No and low alcohol beers by Roselli, Kerruish, Crow, Smart, & Powell, 2024 are recommended for further reading on the topic.

Chapter 9

Strictly anaerobic beer-spoilage bacteria

Riikka Juvonen
VTT Technical Research Centre of Finland, Espoo, Finland

9.1 Introduction

Strictly anaerobic beer-spoilage bacteria are a group of evolutionarily and physiologically related microorganisms. Unlike other brewery-related spoilage microbes, they require a nearly oxygen-free environment to grow in beer. The first species were discovered only in the late 1970s. However, there is indirect evidence of their occurrence in breweries as early as 1946 (Haikara, 1984, pp. 1—47). It has been postulated that the improvements in the filling technology to reduce oxygen levels in the final beer, coupled with increased production of unpasteurised products, made the growth of the strictly anaerobic bacteria in beer eventually possible (Haikara & Helander, 2006). These bacteria are amongst the most detrimental organisms in the beer production chain and show a worldwide distribution. They mainly spoil non- or flash-pasteurised beers produced in modern breweries with effective filling technology. Spoilage is evidenced by development of turbidity and unpleasant off-odours described as rotten-egg-, rancid- or faeces-like, which render the product undrinkable. High economic losses usually ensue from spoilage due to damaging of the corporate brand and high costs of disposing contaminated batches and/or keeping the produced beer in quarantine. As the production of no- and low-strength products expands, and a broader range of raw materials is utilized in breweries, the risk of spoilage could increase in the future.

This chapter will give the reader an overview of the types, evolution, occurrence and properties of the strictly anaerobic beer-spoilage bacteria. Moreover, the prevention and elimination of contaminations will be discussed, and finally an outlook of the future importance of the strictly anaerobic beer-spoilage bacteria will be given. The detection and identification methods for the strictly anaerobic beer-spoilage bacteria are presented in Chapters 14 and 15. *Zymomonas mobilis*—a specialised spoilage organism of primed beers (sugar added after bottling) and ciders produced in the UK (Van Vuuren & Priest, 2003)—is excluded, since it differs in many ways from the other strictly anaerobic beer spoilers.

9.2 The types of strictly anaerobic beer-spoilage bacteria

The strictly anaerobic beer-spoilage bacteria comprise currently 10 species that are distributed between the genera *Megasphaera* (Engelmann & Weiss, 1985; emended by Marchandin Haikara, & Juvonen, 2009), *Pectinatus* (Lee, Mabee, & Jangaard, 1978; emended by Caldwell, Juvonen, Brown, & Breidt, 2013, Schleifer et al., 1990), *Selenomonas* (Von Prowazek, 1913 as quoted by Shouche, Dighe, Dhotre, Patole, & Ranade, 2009), *Propionispira* (Ueki, Watanabe, Ohtaki, Kaku, & Ueki, 2014) and *Prevotella* (Nakata, Kanda, Nakakita, Kaneko, & Tsuchiya, 2019). The analysis of brewery samples with culture-independent techniques has suggested that there are still new anaerobic spoilage bacteria to be discovered in the beer production chain (Nakakita, Takahashi, Sugiyama, Shigyo, & Shinotsuka, 1998; Suzuki, 2020; Timke, Wang-Lieu, Altendorf, & Lipski, 2005).

The science of classification of microorganisms is undergoing constant changes. DNA techniques, especially the sequence analysis of molecular chronometers and 16S rRNA gene in particular, have allowed the scientist to construct a classification system reflecting evolutionary (phylogenetic) relationships between the organisms. 16S rRNA gene sequence analysis has shown that despite the Gram-negative cell envelope the strictly anaerobic beer-spoilage species of the genera *Megasphaera, Pectinatus*, *Propionispira* and *Selenomonas* originate from Gram-positive bacteria (Fig. 9.1). They are currently classified in the phylum *Bacillota* (Oren & Garrity, 2021), which contains otherwise mainly Gram-positive bacteria, such as *Clostridium* and *Bacillus* species (Doyle et al., 1995; Willems & Collins, 1995), and within the class

Brewing Microbiology. https://doi.org/10.1016/B978-0-323-99606-8.00014-6

Negativicutes, which was established for bacteria with a Gram-negative cell envelope (Marchandin et al., 2010). It has been suggested that these bacteria represent a line of evolution for developing a protective barrier for escaping from the lethal effects of antibiotics produced by other microbes (Gupta, 2011). The genus *Prevotella* is phylogenetically distinct from the other anaerobic beer-spoilage species.

9.2.1 Pectinatus

The genus *Pectinatus* ('combed bacteria') currently includes three recognised beer spoilers: *Pectinatus cerevisiiphilus* ('beer lover') (Lee et al., 1978; emended by Schleifer et al., 1990), *Pectinatus frisingensis* ('from Freising') (Schleifer et al., 1990) and *Pectinatus haikarae* ('named after Dr Auli Haikara for her many contributions to the study of *Pectinatus* bacteria') (Juvonen & Suihko, 2006). The first species, *P. cerevisiiphilus*, was named to describe an unusual strictly anaerobic bacterium isolated from spoiled beer in the USA in the late 1970s (Lee et al., 1978). A similar organism, first misidentified as *P. cerevisiiphilus*, was found shortly after from spoiled beers in Finland (Haikara, Enari, & Lounatmaa, 1981). It was eventually described in 1990 as a new species, *P. frisingensis* (Schleifer et al., 1990). In the late 1980s, a *Pectinatus*-like beer spoiler genetically different from the already recognised species was deposited to a German culture collection (www.dsmz.de/). The finding of similar bacteria in Finland a few years later led (Suihko & Haikara, 2001) to the description of the third beer-spoilage species, *P. haikarae* (Juvonen & Suihko, 2006).

Subsequent years have seen the discovery of the first *Pectinatus* species not associated with beer production. Gonzalez et al. (2005) found a new species, *Pectinatus portalensis* ('of El Portal'), from winery wastewater. However, the cultures cited as the type strain of the species do not conform to the original species description, and the International Committee on Systematics of Prokaryotes officially endorsed the rejection of *P. portalensis* as a validly described species in 2020 (Arahal, 2020; Vereecke & Arahal, 2008). Another new species, *Pectinatus brassicae* ('of cabbage'), was isolated from salty pickle wastewater of cabbage production (Zhang et al., 2012). Shortly after this, *Pectinatus sottacetonis* ('of pickle') was discovered from a commercial pickle spoilage tank of cucumbers in the USA (Caldwell et al., 2013; Parks, Chuvochina, & Waite, 2018).

Based on 16S rRNA gene sequence comparisons, the closest relative of *P. frisingensis* is *P. portalensis* (Gonzalez et al., 2005), whereas *P. cerevisiiphilus* is most closely related to *P. haikarae* (Juvonen & Suihko, 2006) (Fig. 9.1). The closest known relatives of *P. sottacetonis* are *P. haikarae* and *P. brassicae* (Caldwell et al., 2013). The analysis of different molecular clocks has suggested that *P. frisingensis* is an older and more diverse species than *P. cerevisiiphilus* (Chaban et al., 2005; Motoyama, Ogata, & Sakai, 1998; Suihko & Haikara, 2001). The most closely related other bacteria to *Pectinatus* are anaerobic *Megamonas* species mainly found in caecal contents of birds (Fig. 9.1).

9.2.2 Megasphaera

The genus *Megasphaera* ('a big sphere') was created in 1971 (Rogosa, 1971; emended by Marchandin et al., 2003) and is currently assigned to a family *Veillonellaceae* (Marchandin et al., 2010). It includes three beer-spoilage species: *Megasphaera cerevisiae* ('of beer'), *Megasphaera paucivorans* ('user of a few substrates') and *Megasphaera sueciensis* ('of Swedish origin'). Twelve other species in this genus have been discovered from human gut and vaginal microbiota, clinical samples, rumen, and chicken (https://lpsn.dsmz.de/search?word=Megasphaera).

M. cerevisiae shares 93.9% 16S rRNA sequence similarity with *M. sueciensis* and *M. paucivorans* (Marchandin et al., 2003). *M. sueciensis* and *M. paucivorans* have nearly identical 16S rRNA gene sequences but can be differentiated from each other using DNA—DNA reassociation and few physiological tests (Juvonen, 2009). *Anaeroglobus geminatus*, *Allisonella histaminiformans* and *Dialister* species are among the nearest relatives of *Megasphaera* species (Carlier et al., 2002; Marchandin et al., 2003) (Fig. 9.1).

9.2.3 Selenomonas and Propionispira (Zymophilus)

The genus *Zymophilus* with two brewery-associated species, that is, *Zymophilus paucivorans* ('user of a few substrates') and *Zymophilus raffinosivorans* ('raffinose devouring') (Schleifer et al., 1990) has been combined with the genus *Propionispira* as they were shown to have a common ancestor (Ueki et al., 2014). In addition to *Propionispira paucivorans* and *Propionispira raffinosivorans*, this genus includes *Propionispira arboris* (Schink, Thompson, & Zeikus, 1982) from wetwoods of living trees (Schink et al., 1982) and *Propionispira arcuata* (Ueki et al., 2014) from methanogenic cattle waste. The brewery-related species are evolutionarily closest to each other (Motoyama & Ogata, 2000; Schleifer et al., 1990) (Fig. 9.1).

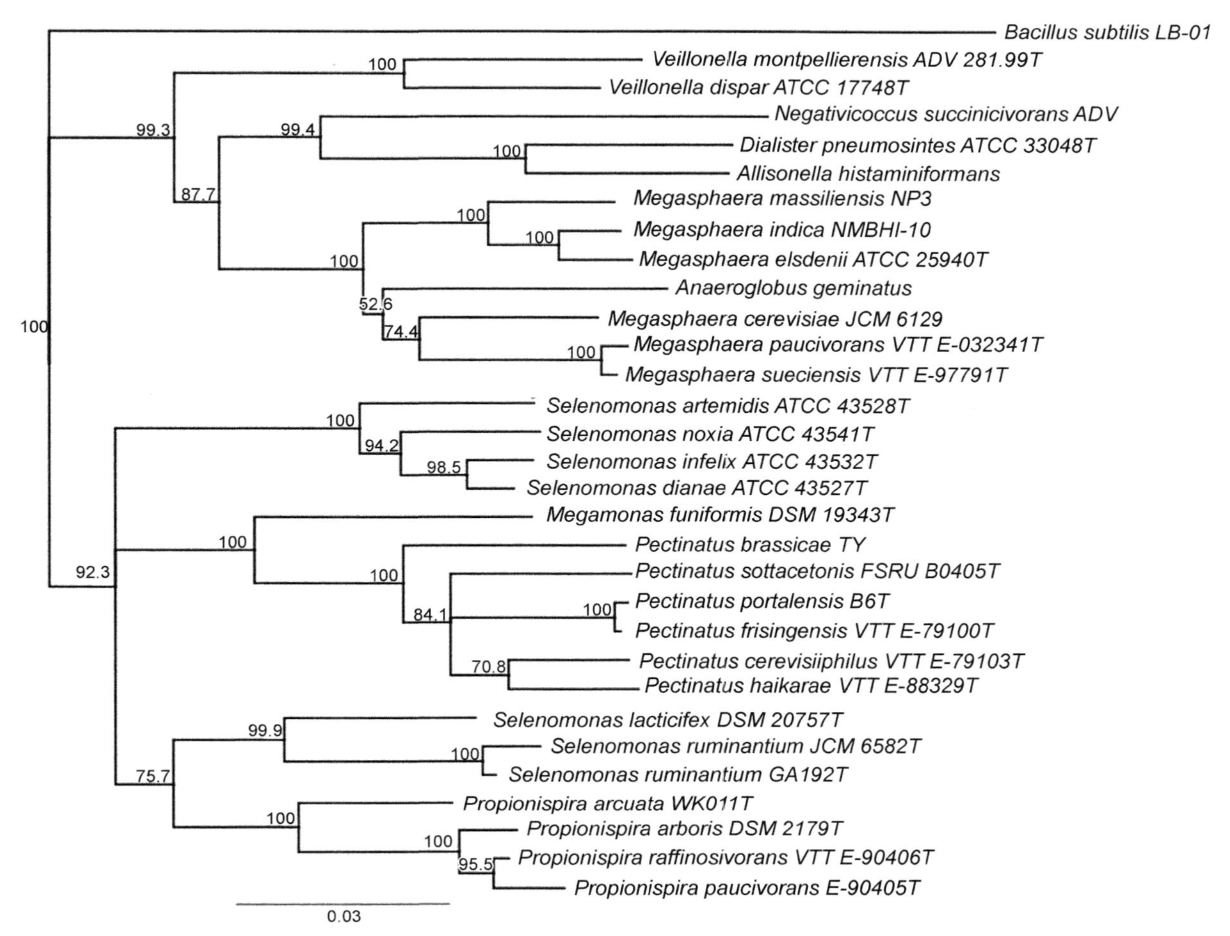

FIGURE 9.1 16S rRNA gene sequence-based tree describing evolutionary relationships of the strictly anaerobic beer-spoilage bacteria of the class *Negativicutes*.

Selenomonas ('crescent-shaped') *lacticifex* ('a maker of lactic acid') is the only brewery-related species in its genus (Schleifer et al., 1990) that comprises 13 other species mainly found from oral and ruminal habitats (Poothong, Tanasupawat, Chanpongsang, Kingkaew, & Nuengjamnong, 2024; Shouche et al., 2009; Zhang & Dong, 2009). The genus *Selenomonas* appears to originate from several ancestors and should be split and reclassified in the future (Juvonen, 2009) (Fig. 9.1).

9.2.4 Prevotella

The newest addition to the strictly anaerobic bacteria, *Prevotella cerevisiae*, was isolated from brewery wastewater in Japan (Nakata et al., 2019). The type strain is most closely related to *Prevotella bryantii* and *P. multisaccharivorax* but differs in fermentation products and sugar assimilation. The genus *Prevotella* contains more than 50 anaerobic, non-spore-forming, Gram-negative species with *P. cerevisiae* is the first reported beer-spoiling member of the family Prevotellaceae (Betancur-Murillo, Aguilar-Marin, & Jovel, 2022; Nakata et al., 2019).

9.3 Occurrence in artificial and natural environments

The beer-spoilage *Megasphaera*, *Pectinatus*, *Propionispira*, *Prevotella* and *Selenomonas* species have mainly or solely been detected from the beer production process, spoiled beers, or brewery environment, and their habitats outside breweries are poorly characterised.

9.3.1 Pectinatus

The beer-spoilage *P. cerevisiiphilus*, *P. frisingensis* and *P. haikarae* species have mainly been isolated from spoiled unpasteurised and flash-pasteurised beers and from the brewing process, and their natural sources and mode of transmission to breweries are still not entirely clear. *P. cerevisiiphilus* and *P. frisingensis* have been found in the beer production chain worldwide (Hage & Wold, 2003; Haikara & Helander, 2006; Lee et al., 1978; Matoulková, Kosar, & Slabý, 2012; Paradh, Mitchell, & Hill, 2011; Schleifer et al., 1990), whereas *P. haikarae* findings have been restricted to the Nordic countries and Germany (Juvonen, 2009; Voetz, Pahl, & Folz, 2010).

The findings of *Pectinatus* bacteria in breweries have concentrated in filling halls (Haikara & Helander, 2006; Hakalehto, 2000; Juvonen, 2009; Matoulková et al., 2012; Paradh et al., 2011). Brewery filling halls provide a good growth environment for microbes due to their relatively high temperature and humidity and the presence of nutrients from product residues (Henriksson & Haikara, 1991). Matoulková et al. (2012) studied the occurrence of *Pectinatus* in 11 filling lines in 10 different brewery plants in the Czech Republic. *Pectinatus* could be isolated from all breweries regardless of their size, output and type of beer produced, the filling line capacity, rate, design, age or the method of cleaning. The most frequently contaminated areas (*Pectinatus* in more than 50% of samples) were the difficult to clean parts inside and underneath the conveyor belts and the various monoblock constructions, such as the surface of piping below the bench. Overall the highest percentage of positive samples was found in the floor sampling area, including drainage systems and cracks and crumbling joints in the floor. The occurrence in drainage systems has also been noted in many previous studies, and water has been suspected as one primary source of contamination (Haikara & Helander, 2006; Juvonen, 2009). Other reported sources within the filling halls include the air and ceiling of the filling halls and chain lubricants (for a review, see Haikara & Helander, 2006; Juvonen & Suihko, 2006). *Pectinatus* has occasionally been detected in biofilms in the filling machines but also in places with no usual biofilm occurrence (Matoulková et al., 2012; Timke et al., 2005; Voetz et al., 2010). Genes related to biofilm formation have been recently detected in the genomes of *Pectinatus* species (Kramer et al., 2020).

Unlike previously thought, *Pectinatus* bacteria may contaminate all stages of the beer production process. *Pectinatus* bacteria have occasionally been detected in the fermentation area, such as in carbon dioxide collection pods of fermenters, maturing beer in the cellar and in finished beer as well as in bright beer tanks (Juvonen, unpublished data; Matoulková et al., 2012; Paradh et al., 2011). However, to our knowledge, viable cells have only been recovered from the filling halls and spoiled beers. The role of brewing raw materials as contamination sources is still unresolved. Some of the early reports about the occurrence of *Pectinatus* in malt steeping water and pitching yeast have later proven to be misidentifications (Haikara & Helander, 2006).

There appears to be a seasonal variation in the occurrence of *Pectinatus* in breweries. Spoilage incidences and *Pectinatus* findings tend to peak during the warm months of the year (Paradh et al., 2011). Despite the seasonal variation, *Pectinatus* species are considered permanent rather than occasional invaders in breweries (Hakalehto, 2000). They are typically detected from several sources in a single brewery (Hakalehto, 2000; Matoulková et al., 2012; Paradh et al., 2011; Suiker, O'Sullivan, & Vaughan, 2007). The communities can be rather complex, comprising several different genotypes (Juvonen, unpublished data). As a result, several strains may be involved in spoilage incidents, which makes tracing of contaminations challenging. Suzuki (2011) recently postulated that the beer-spoilage *Pectinatus* bacteria have adapted to live in a mutually beneficial association with brewer's yeast and lactic acid bacteria since the early times of brewing.

Since the peak in the late 1980s and early 1990s, a decreasing trend in documented spoilage incidents has been observed (Back, 2005). However, it needs to be taken into consideration that most spoilage cases are not reported, and information has systematically been gathered only in Germany (Back, 2005). Suzuki (2011) estimated that *Pectinatus* bacteria are responsible for 20%−30% of beer-spoilage incidents. *P. frisingensis* is the dominant species in spoilage incidents and is also the most frequently reported species in breweries (Matoulková et al., 2012; Motoyama et al., 1998; Paradh et al., 2011; Suihko & Haikara, 2001; Suiker et al., 2007).

During the past few years the known habitats of *Pectinatus* bacteria have widened to fermentation processes other than brewing, such as waste streams of wineries and distilleries and pickle production (Caldwell et al., 2013; Castelló et al., 2009; Gonzalez et al., 2005; Mota, Delforno, Ribeiro, Zaiat, & Oliveira, 2024; Temudo, Muyezer, Kleerebezem, & van Loodsrecht, 2008; Zhang et al., 2012). Hence, the *Pectinatus* bacteria appear to be typically associated with lactic or ethanol fermentation processes of plant raw materials. In 2013, a beer-spoilage *Pectinatus* species was, for the first time, isolated outside the beer production chain when *P. cerevisiiphilus* strains were found in mangrove sediments in Thailand (http://www.ncbi.nlm.nih.gov/genbank/). The available evidence suggests that *Pectinatus* species are plant-originating bacteria, which is also supported by their cell wall structures and their ability to grow with common plant sugars and in the presence of high concentrations of plant phenolic compounds (Caldwell et al., 2013; Helander, Haikara, Sadovskaya,

Vinogradov, & Salkinoja-Salonen, 2004; Juvonen, 2009; Kramer et al., 2020). On the other hand, their close evolutionary relations to intestinal species (Chevrot et al., 2008) pinpoints to an animal or human source. Indeed, reports on the discovery of *Pectinatus* species, including the beers-spoilage organisms, in poultry, bees, pigs and rats, as well as humans are increasing (Anisha, 2018; Callaway et al., 2009; Kim, Nguyen, Guevarra, Lee, & Unno, 2015; Kwong et al., 2017; Liu et al., 2017).

9.3.2 Megasphaera

M. cerevisiae appears to be geographically more restricted compared to the beer-spoilage *Pectinatus* species. Contaminations have been reported in Australia, Finland, Germany, Norway, Sweden and the UK (Hage & Wold, 2003; Haikara & Helander, 2006; Paradh et al., 2011). *M. cerevisiae* shares its ecological niche with the beer-spoilage *Pectinatus* bacteria. It has been detected in spoiled beers and in brewery filling halls. Sporadic findings from pitching yeast and a brewery carbon dioxide line have also been reported (Haikara & Helander, 2006).

M. cerevisiae is a less frequent brewery contaminant and beer spoiler than *Pectinatus* bacteria. In Germany, *M. cerevisiae* caused 2%–7% of the documented beer-spoilage cases during the time period 1990–2002 (Suzuki, 2011). *M. paucivorans* was originally isolated from spoiled beer produced in Italy, whereas *M. sueciensis* was discovered from a spoiled Swedish beer. Paradh et al. (2011) detected *M. sueciensis/M. paucivorans* (not separable by the DNA analysis used in the study) also in a brewery filling hall in the UK (conveyer belt of canning line). Moreover *M. sueciensis/M. paucivorans* have been found in anaerobic biohydrogen production systems using cheese whey and anaerobic sludge as raw materials (Castelló et al., 2009; Jin, Sun, & Shi, 2010; Ning, Jin, Sheng, Harada, & Shi, 2012). Many *Megasphaera* species produce hydrogen and can be beneficial organisms in a biohydrogen production process.

9.3.3 Selenomonas and Propionispira

Propionispira paucivorans and *P. raffinosivorans* as well as *S. lacticifex* have been isolated from pitching yeast samples in Germany and Finland. In the late 1980s, Seidel-Rüfer (1990) examined more than 3000 yeast samples from German breweries. Of these, 0%–0.03% were contaminated with *S. lacticifex* and 0.12%–0.7% with *Propionispira* species. Moreover, brewery waste streams and drainage systems have been mentioned as possible sources of *P. raffinosivorans* (Haikara, 1989; Schleifer et al., 1990; Seidel-Rüfer, 1990).

Culture-independent analysis of microbial diversity in industrial and natural ecosystems has revealed new habitats for *S. lacticifex*. DNA sequences from *S. lacticifex* have been detected from a biomass of a continuous stirred tank reactor, sewage from a wastewater treatment plant and from the food waste hydrolysate (http://www.ncbi.nlm.nih.gov/genbank/). The close evolutionary relation of the beer-spoilage *P. raffinosivorans* to a tree pathogen *P. arboris* suggests that it may be carried to breweries with plant raw materials (Juvonen, 2009).

9.3.4 Prevotella

Prevotella cerevisiae has only been isolated from brewery wastewater, with two strains isolated during routine surveillance between 1995 and 2001 (Nakata et al., 2019). The genus is ubiquitous with species found in various body environments and across multiple species of animals including humans, livestock, rodents and insects, making transmission to the brewing environment difficult to trace (Betancur-Murillo et al., 2022).

9.4 Appearance of cells and laboratory cultures

Properties, such as the cellular shape, motility and structures and appearance of laboratory cultures in solid and liquid media have traditionally been used as first steps to identify unknown microbes. The double staining method of Hans Christian Gram can reveal differences in the cell wall structures, and it has been used to classify bacteria to Gram-positive and Gram-negative ones. Gram-positive bacteria retain crystal violet staining purple, whereas Gram-negative bacteria decolorise and can be counterstained red with safranin. *Pectinatus*, *Megasphaera*, *Selenomonas* and *Propionispira* species and related organisms are special in that they stain Gram-negative but are evolutionarily related with Gram-positive bacteria. The cells possess features of both Gram-negative and Gram-positive bacteria being surrounded by a thick peptidoglycan layer typical of Gram-positive bacteria and an outer membrane typical of Gram-negative bacteria. The cellular and cultural characteristics of the strictly anaerobic beer spoilers are discussed below.

9.4.1 Pectinatus

Pectinatus cells are non-spore-forming, slightly curved helical rods, 0.4–1.0 by 2–50 μm or more, with rounded ends (Fig. 9.2). They typically occur singly, in pairs or rarely in short chains. In older cultures, elongated snake-like cells and various round cell formations can be seen (Haikara & Juvonen, 2009). A distinctive feature of the *Pectinatus* cells is their comb-like flagellar arrangement in which flagella only emanate from one side of a cell which leads to the formation of an X-pattern during movement.

The cell surface structures of *P. cerevisiiphilus* and *P. frisingensis* strains have been studied in detail (for a review see Helander, Haikara, Sadovskaya, Vinogradov, & Salkinoja-Salonen, 2004). Structures and composition of their lipopolysaccharides (LPS) are exceptional in many ways. LPS are unique functional components of the outer membranes of Gram-negative bacteria, consisting of a lipid and a polysaccharide part (Helander et al., 2004). Each *Pectinatus* strain appears to be capable of producing at least two types of LPS with distinct carbohydrate structures. Moreover, their LPS contain 3-deoxy-D-manno-oct-2-ulapyranosonic acid and the lipid A linkage with polysaccharide is very acid-stable (Haikara & Helander, 2006). The outer membranes of classic Gram-negative bacteria are typically an efficient permeability barrier for cationic substances and large molecules. However, the outer membranes of *Pectinatus* cells fulfil this function variably (Caldwell et al., 2013; Haikara & Juvonen, 2009). Peptidoglycan of *P. frisingensis* and *P. cerevisiiphilus* contains cross-linked meso-diaminopimelic with putrescine or cadaverine in the peptide subunit (Schleifer et al., 1990). The fatty acid composition of cells is similar between different species being dominated by odd-numbered fatty acids, that is, $C_{11:0}$, $C_{13:0}$, $C_{15:0}$, $C_{13:0\ (3OH)}$ (most probably misidentified as $C_{14:0}$ in MIDI), $C_{17:1}$ and $C_{18:1trans11}$ (Caldwell et al., 2013; Haikara & Helander, 2006).

Cultural characteristics of *Pectinatus* species have been studied in peptone yeast extract fructose (PYF) medium (http://culturecollection.vtt.fi) and deMan Rogosa Sharpe (MRS) medium. The beer-spoilage species form circular, entire, glistening and opaque colonies on PYF medium. The colour of *P. haikarae* colonies varies from cream to greyish and in the case of the other two species from beige to white (Haikara & Juvonen, 2009).

9.4.2 Megasphaera

Unlike the other strictly anaerobic beer-spoilage species, *Megasphaera* cells are non-motile cocci (Fig. 9.2). They are normally arranged singly, in pairs and sometimes in short chains. The three beer-spoilage species can be discriminated from each other by their cell size; *M. cerevisiae* is the biggest and *M. sueciensis* is the smallest (Juvonen & Suihko, 2006; Marchandin et al., 2009). The peptidoglycan of *M. cerevisiae* is of the meso-diamonopimelic acid direct type with putrescine residues (Engelmann & Weiss, 1985). The main fatty acid components are $C_{12:0}$, $C_{16:0}$, $C_{16:1}$, $C_{18:1}$, $C_{17cyclo}$, $C_{19\ cyclo}$ and $C_{14:0\ 3OH}$ (Marchandin et al., 2009). The cell surface structures of the other beer-spoilage species have not been studied.

All the beer-spoilage species grow on PYF agar medium. However, the growth of *M. sueciensis* and *M. paucivorans* is greatly improved when pyruvic acid or gluconic acid is used instead of fructose (Juvonen & Suihko, 2006). The three species differ in their growth rate on PYF agar medium. *M. cerevisiae* is normally detected within 1–2 days, whereas *M. paucivorans* and *M. sueciensis* require 3–4 days to form visible colonies (Juvonen, 2009). The colonies are circular, glossy and opaque. The colonies of *M. cerevisiae* are whitish in colour, whereas those of *M. paucivorans* and *M. sueciensis* are yellowish (Juvonen & Suihko, 2006; Marchandin et al., 2009).

9.4.3 Selenomonas and Propionispira

S. lacticifex and the beer-spoilage *Propionispira* species are motile rod-shaped bacteria. They may lose their mobility upon repeated cultivations. None of these species forms endospores. The cells of *S. lacticifex* are curved crescent-shaped rods, 0.6–0.9 to 5–15 μm in size (Schleifer et al., 1990). The cells of *P. paucivorans* are curved, helical or crescent shaped and up to 15 μm long. They occur singly, in pairs or in short chains. The cells of *P. raffinosivorans* are straight to slightly curved 'sausage-shaped' rods (0.8 × 6 μm). Cells in various helical arrangements may be seen even in young cultures (Seidel-Rüfer, 1990; Ueki et al., 2014). The three species share a similar peptidoglycan structure with *Pectinatus* bacteria (Schleifer et al., 1990; Ziola et al., 1999).

Cultural characteristics have been determined on modified MRS agar medium in which the beer-spoilage *Propionispira* species form circular, smooth, opaque and slightly yellow colonies with a diameter of 1–2 mm after 3 days of incubation (Schleifer et al., 1990). The colonies of *S. lacticifex* are also smooth, circular, opaque and yellowish with a diameter of 2–3 mm after 3 days of incubation.

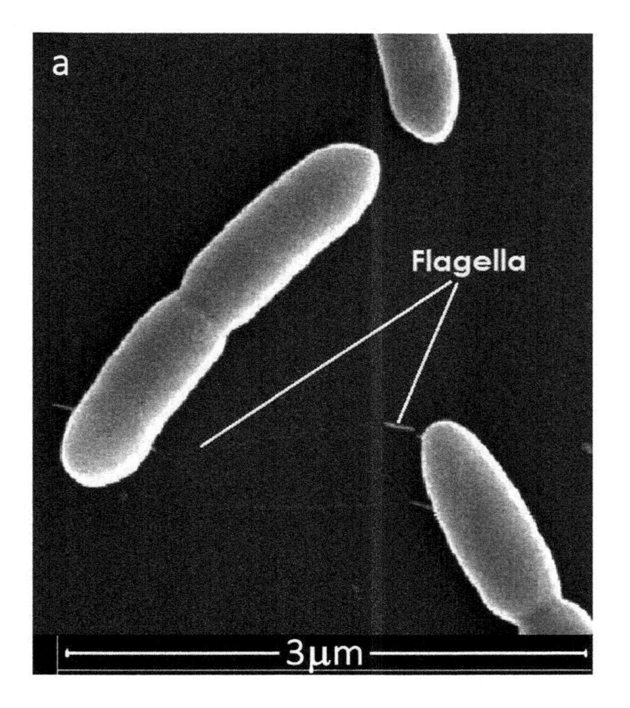

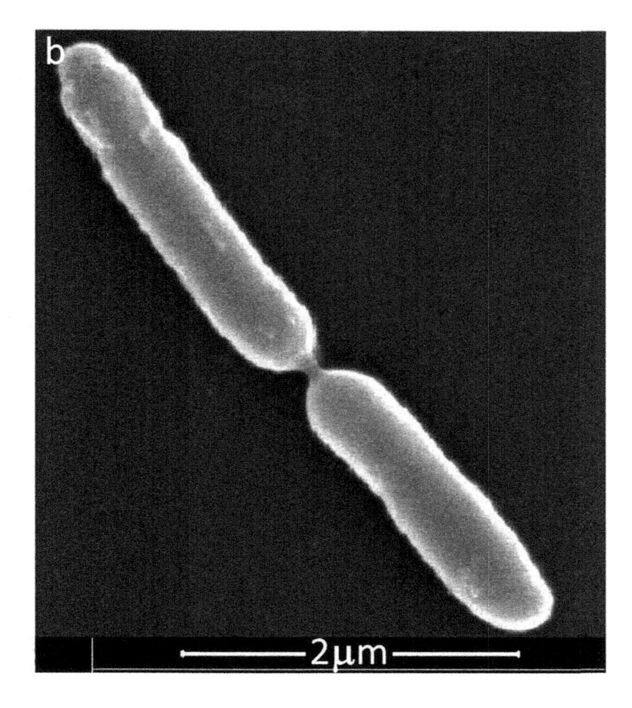

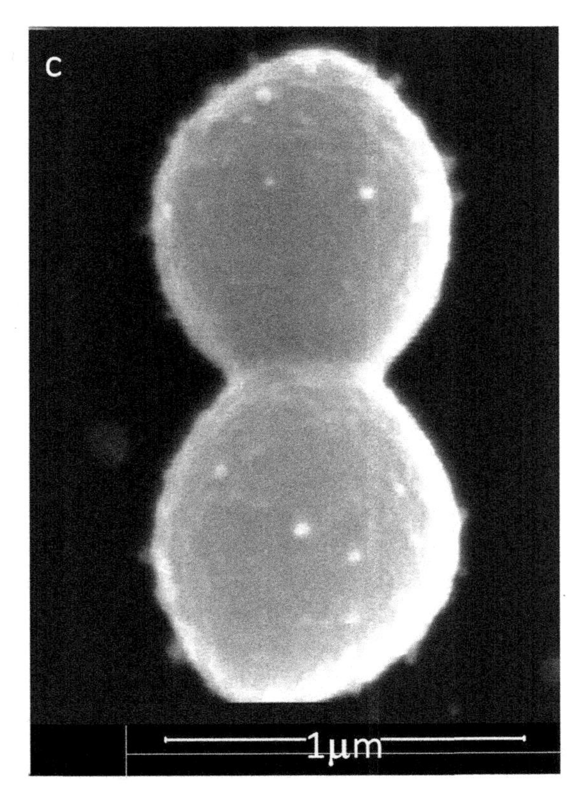

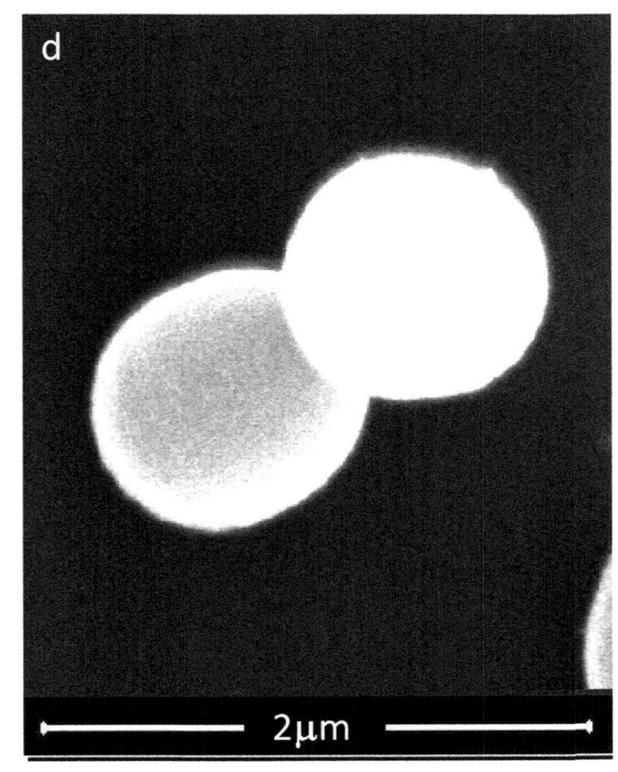

FIGURE 9.2 SEM images of (a) *Pectinatus cerevisiiphilus*, (b) *Pectinatus frisingensis* and (c) & (d) *Megasphaera cerevisiae* brewery isolates. *Courtesy of A. Paradh.*

9.4.4 Prevotella

P. cerevisiae are non-motile rods, 2–5 μm in size. Isolates are typically maintained on modified MRS agar using beer instead of water under anaerobic conditions. They form white to off-white brown colonies with a diameter of 0.5–1 mm after 5 days at 30°C (Hitch et al., 2022).

9.5 General physiology and metabolism

The strictly anaerobic beer-spoilage bacteria share a few metabolic and physiological features. However, clear differences exist in their nutrition and physiology, which is reflected in their beer-spoilage ability and can be exploited for their detection and identification. The key characteristics discriminating the beer-spoilage species from each other are shown in Tables 9.1 and 9.2.

9.5.1 Pectinatus

The *Pectinatus* bacteria are fermentative organisms growing well with glucose and fructose. The different species vary to some extent in their carbohydrate utilisation, which can be exploited in the phenotypic identification (Table 9.1). *P. frisingensis* grows with a wider range of carbohydrates compared to *P. cerevisiiphilus* (Schleifer et al., 1990). *P. haikarae* is the only species using lactose but not salicin. Interestingly only a few beer-spoilage strains utilise the main carbohydrate of malt, maltose (Juvonen & Suihko, 2006; Motoyama et al., 1998; Schleifer et al., 1990). Citric acid and lactic acid, which can be found in small quantities in beer, are metabolised (Haikara & Juvonen, 2009; Watier, Chowdhury, Leguerinel, & Hornez, 1996b). Due to the ability of *Pectinatus* to utilise lactic acid, contamination of beer by lactic acid bacteria may promote their growth. Ethanol or the main amino acids of beer are not metabolised by *P. frisingensis* or *P. cerevisiiphilus* (Schleifer et al., 1990; Tholozan, Membré, & Kubaczka, 1996).

The fermentation end products from simple sugars and some organic acids have been determined. Glucose is mainly fermented to propionic and acetic acids and acetoin, but succinic and lactic acids may also be produced (Haikara, Penttilä, Enari, & Lounatmaa, 1981; Juvonen & Suihko, 2006; Tholozan, Membré, & Grivet, 1997; Watier et al., 1996b). The relative proportions of the end products depend on the substrate. *Pectinatus* spp. use the same metabolic pathway for propionic acid synthesis as propionibacteria (Haikara, Penttilä et al., 1981; Tholozan et al., 1994). In this pathway, succinate oxidoreductase reduces fumaric acid to succinic acid. Biomass and volatile fatty acid concentrations have been found to be proportional to glucose and lactate concentrations in the medium (Tholozan et al., 1996). Propionic acid synthesis is not directly linked with growth (Watier et al., 1996b).

The *Pectinatus* bacteria are relatively acid and ethanol tolerant. In a laboratory medium, Tholozan et al. (1996, 1997) found that the optimal pH value for the growth of *P. cerevisiiphilus* and *P. frisingensis* was 6.0–6.2 and 4.5–4.9. Somewhat broader optimum pH ranges have been more recently reported for other strains (Kramer et al., 2020). The minimum pH for the growth was in the range of 3.5–4.4.5 (Caldwell et al., 2013; Haikara, Penttilä, et al., 1981; Kramer et al., 2020; Watier et al., 1996b). The maximum ethanol concentrations for growth varied from 5.8% to 8% (Tholozan et al., 1996, 1997; Watier et al., 1996b).

The *Pectinatus* species are mesophiles in their temperature preferences. Depending on the strain, they can grow at 8–15°C, but the optimum is 30–37°C. *P. cerevisiiphilus* still grows at 40–45°C and *P. frisingensis* 37–45°C (Haikara, 1989; Kramer et al., 2020; Lee et al., 1978; Schleifer et al., 1990). *P. frisingensis* has been shown to be able to survive rapid temperature changes and recover quickly at suitable growth conditions (Chihib & Tholozan, 1999).

Despite being anaerobes, *P. cerevisiiphilus* and *P. frisingensis* are relatively oxygen tolerant. The decimal reduction times (D_{oxy}) at the dissolved oxygen content of 4.78 mg/L (32°C) varied from 3.3 to 55 h (Chowdhury, Watier, & Hornez, 1995). The dissolved oxygen content in the medium affected the inactivation rates (Chowdhury et al., 1995). For instance, D_{oxy} values for *P. cerevisiiphilus* increased from 4.8 to 13.3 h when the oxygen content of wort decreased from 5.74 to 3.34 mg/L. The oxygen tolerance is also influenced by temperature, increasing with the temperature decrease. *P. haikarae* appears to also be oxygen tolerant, as it has been isolated alive from the air of brewery bottling halls (Juvonen & Suihko, 2006).

9.5.2 Megasphaera

M. cerevisiae strains form a uniform group in terms of utilised carbon sources that include arabinose, fructose, lactic acid and pyruvic acid (Engelmann & Weiss, 1985). However, the only known carbon sources used by *M. sueciensis* and *M.*

TABLE 9.1 Phenotypic characteristics discriminating strictly anaerobic rod-shaped beer-spoilage bacteria.

Characteristic	*Pectinatus cerevisiiphilus*	*Pectinatus frisingensis*	*Pectinatus haikarae*	*Selenomonas lacticifex*	*Propionispira paucivorans*	*Propionispira raffinosivorans*
Catalase	−	−	+	−	−	−
Growth at 37°C	+	+	−	+	−	−
Acid from:						
N-acetyl-glucosamine	−	+	−	nd	−	+
Cellobiose	−	+	−	+	+	+
Inositol	−	+	+	−	−	+
Lactose	−	−	+	+	−	+
Maltose	−	−	−	d	+	+
Mannitol	+	+	+	−	+	+
Melibiose	+	−	+	+	−	+
Raffinose	−	−	−	+	−	+
Rhamnose	+	+	+	−	−	+
Salicin	+	+	−	−	−	v
Sorbitol	+	+	nd	−	+	+
Sucrose	−	−	nd	+	+	+
Xylitol	−	d	+	−	−	+
Xylose	+	−	+	+	−	+
Acetoin production	+	+	+	nd	−	−
Succinic acid production	+	+	nd	−	−	−
Lactic acid as the main metabolite	−	−	−	+	−	−

+, 75% or more of the strains are positive; −, 75% or more of the strains are negative; *d*, delayed; *nd*, not determined.
Modified from Haikara and Juvonen (2009) and Juvonen (2009).

TABLE 9.2 Phenotypic characteristics discriminating beer-spoilage *Megasphaera* species.

Characteristic	*M. cerevisiae*	*M. paucivorans*	*M. sueciensis*
Cell size (μm)	1.5–2.1	1.2–1.9 × 1.0–1.4	1.0–1.4 × 0.8–1.2
Colonies visible on solid media	1–2 days	3 days	4 days
Acid production from fructose	+	–	–
Growth with lactic acid	+	–	–
Major metabolites[a]	**C**, iV, B	**iV, C**	**iV**, B, C, V

+, 75% or more of the strains are positive; –, 75% or more of the strains are negative.
[a]*B, butyric acid; C, caproic acid; iV, isovaleric acid; V, valeric acid. The products in bold constitute 40%–60% of the total amount.*
Modified from Juvonen (2009) and Juvonen and Suihko (2006).

paucivorans strains are pyruvic, gluconic and glucuronic acids (Juvonen & Suihko, 2006). *Megasphaera* bacteria have a fermentative type of metabolism. The exact composition of the fermentation products depends on the energy source in the medium and may include acetic, propionic, iso- and n-butyric, iso- and n-valeric and caproic acids. Moreover, hydrogen sulphide, hydrogen and carbon dioxide are produced (Engelmann & Weiss, 1985; Juvonen & Suihko, 2006). The three beer-spoilage species can be discriminated from each other (Table 9.2) based on their volatile fatty acid metabolites, which can be determined by using gas chromatography. Metabolite analysis is a useful method to identify the growth of *Megasphaera* and *Pectinatus* in beer, especially when the cells are no longer cultivable (Table 9.3).

M. cerevisiae is moderately acid tolerant. It still grows weakly in laboratory media at pH 4.1–4.2 but not anymore at pH 4.0 (Haikara, 1989). The acid tolerance of the other beer-related species has not been studied. The growth temperature of the beer-spoilage species ranges from 15 to 37 °C (Haikara, 1989; Juvonen & Suihko, 2006). *M. cerevisiae* appears to tolerate oxygen at least at low temperatures (Juvonen, 2009). The antibiotic sensitivity of the type strains has been studied. All the type strains are resistant to vancomycin. *M. sueciensis* and *M. paucivorans* are also resistant to colistin, whereas *M. cerevisiae* is sensitive to this compound (Juvonen & Suihko, 2006).

9.5.3 Selenomonas and Propionispira

The physiological properties of *S. lacticifex* and *Propionispira* species have been little studied. *S. lacticifex* utilises a wide range of carbon sources, including arabinose, cellobiose, glucose, lactic acid and maltose (Schleifer et al., 1990). It differs from the other strictly anaerobic beer spoilers by producing lactic acid as the major fermentation end product (Schleifer et al., 1990). *P. raffinosivorans* uses a greater variety of carbon substrates compared to *P. paucivorans* (Schleifer et al., 1990; Ueki et al., 2014). Both species ferment glucose, pyruvic, lactic and fumaric acids to acetic and propionic acids. Moreover, propionic acid is produced from succinic acid.

S. lacticifex strains from yeast samples have been shown to grow well at pH 4.3 but not at pH 4.2. The *Propionispira* strains were less acid tolerant (Seidel-Rüfer, 1990). The optimum growth temperature for all three species is close to 30°C. *P. raffinosivorans* and *P. paucivorans* do not grow at 37°C, whereas *S. lacticifex* strains vary in this respect (Haikara, 1989; Schleifer et al., 1990; Seidel-Rüfer, 1990). *S. lacticifex* shows a lower minimum temperature for growth compared to *Megasphaera* and *Pectinatus* bacteria and can even grow at 10 °C.

9.5.4 Prevotella

The optimum pH range for *Prevotella cerevisiae* is pH 5.7 in modified De Man Rogosa Sharpe (MRS) medium. Cells can ferment glucose, lactose, maltose, mannose, raffinose, salicin, arabinose and cellobiose, but not mannitol, melezitose, rhamnose, sorbitol, trehalose or xylose (Nakata et al., 2019). The major organic acids of fermentation in pre-reduced anaerobically sterilised peptone–yeast broth were acetic acid and succinic acid.

9.6 Growth and effects in beer

The strictly anaerobic beer-spoilage bacteria can be divided into absolute and potential beer-spoilage organisms. By definition, an absolute spoiler is able to grow in beer without a long adaptation period and to cause obvious quality defects.

TABLE 9.3 Effects of strictly anaerobic beer-spoilage bacteria on beer quality.

		Effects in finished beer		
Genus	Effects on fermentation	Metabolites[a]	Off-flavour	Turbidity
Megasphaera	Not known	Acetic, **butyric**, caproic, isobutyric, isovaleric, propionic, and valeric acids, $\mathbf{H_2S}$	Rancid, rotten egg	+
Pectinatus[b]	Possible inhibition	Acetic, **propionic**, succinic and lactic acids, acetoin, $\mathbf{H_2S}$, organic sulphur compounds	Rotten egg	+
Selenomonas[c]	Not known	Acetic, **lactic**, and propionic acids	Not known	+
Propionispira[d]	Not known	Acetic and propionic acids (H_2S by *P. raffinosivorans*)	Not known	+

[a] *The major metabolites are in bold.*
[b] Pectinatus haikarae *may also cause foaming but are slight, off-flavour and turbidity.*
[c] Selenomonas lacticifex.
[d] P. paucivorans *and* P. raffinosivorans *grow at an elevated pH value of five to six.*

A potential spoiler does not grow in standard beers under normal conditions and does not always cause obvious quality defects or requires a long adaptation time (Back, 2005). It is currently considered that even a few viable cells of the strictly anaerobic bacteria in a package of beer may eventually lead to spoilage.

9.6.1 Pectinatus

P. cerevisiiphilus, *P. frisingensis* and *P. haikarae* are absolute beer spoilers. They spoil mainly unpasteurised and flash-pasteurised beers in a package. All strains are regarded as potentially harmful, although strain-specific differences in the ability to survive and grow in beer have been noted (Suiker, Vaughan, & O'Sullivan, 2009; Rodríguez-Saavedra et al., 2021). The spoilage results from the production of large quantities of propionic acid (up to >1000 mg/L), some acetic acid, hydrogen sulphide (20–300 μg/L) and turbidity (Fig. 9.3; Table 9.3), which is evident at cell concentrations of approximately 10^5 cfu/mL (Haikara, Enari, et al., 1981). Organic sulphur compounds, dimethyl trisulphide and methyl mercaptan, may also be produced above their taste threshold levels. The growth of *Pectinatus* too low to cause turbidity may produce metabolites in concentrations high enough to cause spoilage (Haikara, Enari, et al., 1981). The spoiled beer has an odour of rotten eggs that makes it fully unfit for consumption (Haikara & Helander, 2006). The off-flavour and turbidity of beer spoiled by *P. haikarae* appear to be less noticeable compared to the defects caused by the other species (Voetz et al., 2010).

The pH value, ethanol concentration and dissolved oxygen content are among the key factors controlling in concert the growth of *P. cerevisiiphilus* and *P. frisingensis* in beer. However, there are other unidentified factors that affect the susceptibility of a beer to *Pectinatus* spoilage. *P. frisingensis* and *P. cerevisiiphilus* appear to be the most acid-tolerant species among the strictly anaerobic beer spoilers. *P. frisingensis* grows well at pH values of typical lager beers. Some growth retardation was observed in beer at pH 4.1 (Haikara, 1984, pp. 1–47). *Pectinatus* species tolerate ethanol better than classical Gram-negative bacteria. *P. frisingensis* grew well in commercial beers with 3.7%–4.5% (w/v) ethanol, although the growth was slower than in low-alcohol products. Strong beers with ≥5.2% (w/v) ethanol were not spoiled (Haikara, 1984, pp. 1–47; Haikara, Enari, et al., 1981; Seidel-Rüfer, 1990). Good growth of all known species in 5% (v/v) has been recently reported (Kramer et al., 2020). *Pectinatus* bacteria are more hop tolerant compared to lactic acid bacteria and resist well the levels of hop bitter acids normally found in beers (EBU 33–38) (Haikara & Helander, 2006; Kramer et al., 2020; Matoulková et al., 2012). The available information suggests that the hop tolerance may not be related to the properties of their outer membranes (Helander et al., 2004).

Despite their anaerobic nature, *Pectinatus* bacteria tolerate oxygen relatively well, especially at low temperatures, and viable bacteria have been isolated from various aerobic niches in breweries. However, low dissolved oxygen content is necessary for the growth in beer or wort. With the modern filling techniques the dissolved oxygen content of beer typically ranges from 0.3 to 0.8 mg/L. The growth of *Pectinatus* has been reported in beers with up to 1.9 mg/L of dissolved oxygen (Soberka Sciazko, Warzecha, 1988). The laboratory studies suggest that *P. cerevisiiphilus* may also grow slowly in oxygenated wort in the presence of brewer's yeast. In the study of Chowdhury, Watier, Leguerine, Hornez, and (1997) *P. cerevisiiphilus* started to inhibit the yeast activity at fermentation temperatures above 15°C. Hence, *Pectinatus* bacteria could also cause fermentation problems.

FIGURE 9.3 Beer spoiled by *Megasphaera cerevisiae* (*left*) and *Pectinatus frisingensis* (*right*).

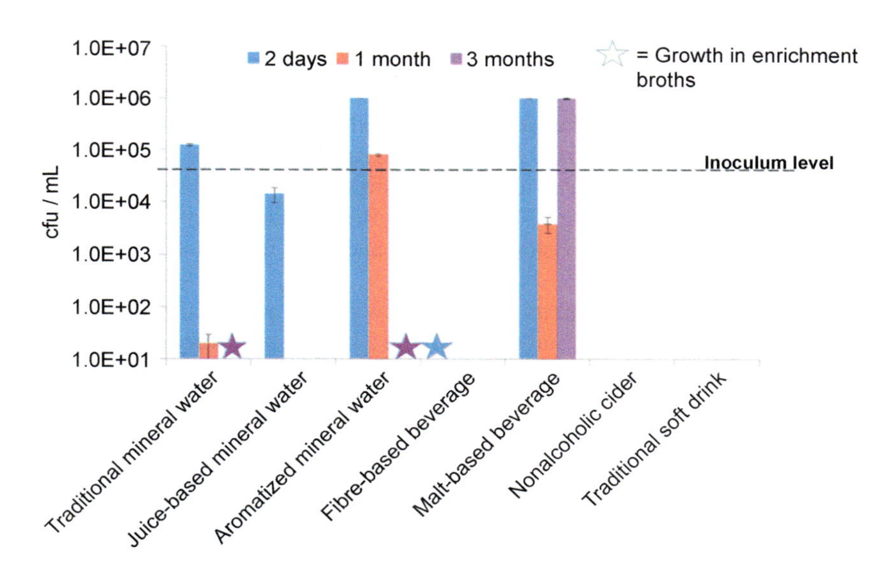

FIGURE 9.4 The growth of *Pectinatus frisingensis* in traditional and novel types of nonbeer beverages. Commercial beverage products were artificially contaminated with approximately 5×10^4 cfu/mL. Viable counts were determined after 2 days, 1 month and 3 months of ambient storage using the plate count technique.

Reported beer-spoilage incidents by *P. haikarae* have concerned low-alcohol products (Juvonen, 2009; Voetz et al., 2010). However, artificial inoculation experiments have shown that it has similar beer spoilage characteristics with other species (Kramer et al., 2020). It grows in typical pH values and ethanol concentrations of beer. *P. haikarae* appears to be better adapted to the brewing environment compared to its closest relative, *P. cerevisiiphilus*. It grows at lower temperatures and produces a catalase enzyme protecting the cells from toxic oxygen radicals. The isolation of *P. haikarae* from air samples indicates that it can survive in the air at least for short periods of time (Henriksson & Haikara, 1991).

P. brassicae and P. sottacetonis species have not yet been found in the beer production chain. However, experiments with artificially contaminated beers have shown that they can grow in beer containing 5 vol-% alcohol (Caldwell et al., 2013; Kramer et al., 2020).

Nowadays breweries produce increasingly new types of non-beer beverages that are expected to be more sensitive to microbiological spoilage compared to traditional soft drinks due to their higher pH value or nutrient content or milder carbonation level (Juvonen et al., 2011). We have evaluated the ability of various emerging and established beverage spoilage organisms and food pathogens to survive and grow in a range of functional drinks and modified waters (Juvonen, unpublished results). A strain of *P. frisingensis* was able to spoil a flavoured mineral water and a malt-based drink (Fig. 9.4), and it could be considered a potential threat to the quality of some non-beer beverages. It also tolerated relatively well organic acid preservatives used in soft drinks (Juvonen, unpublished data).

9.6.2 Megasphaera

M. cerevisiae, *M. paucivorans* and *M. sueciensis* are regarded as absolute beer spoilers. *M. cerevisiae* spoils beer by producing copious amounts of butyric acid with minor amounts of C-5 and C-6 fatty acids and H_2S, which cause a particularly unpleasant off-flavour (Haikara & Lounatmaa, 1987). The contaminated beer normally becomes turbid in 4–6 weeks (Fig. 9.3). The major organic acid produced by *M. paucivorans* in beer is also butyric acid, possibly deriving from the metabolism of pyruvic acid (Juvonen, unpublished data).

M. cerevisiae mainly spoils low-alcohol products due to its poor ethanol tolerance. In the study of Haikara and Lounatmaa (1987), the growth of *M. cerevisiae* strains was reduced in beers above 2.1% (w/v) ethanol concentration and ceased at 4.2% ethanol (w/v) concentration. The rate of spoilage was inversely proportional to the concentration of ethanol. *M. cerevisiae* is sensitive to the normal low pH of beer. In a commercial beer, no growth occurred at pH 4.0, and the spoilage rate was reduced at the pH of normal beer (Haikara & Lounatmaa, 1987). The results of our survival experiments suggest that *M. cerevisiae* can persist in nutrient poor and hostile conditions fully viable and active for long periods and initiate rapid growth when conditions improve (Juvonen, 2009).

There is a lack of data about beer-spoilage properties of the other species. *M. sueciensis* has been found as a spoilage microbe in low-alcohol beer, whereas *M. paucivorans* was found in spoiled product with an ethanol content of 3.9% (w/v). The pH of the beers spoiled by these species varied from 4.3 to 4.9 (Juvonen & Suihko, 2006).

9.6.3 Selenomonas and Propionispira

S. lacticifex is considered an absolute beer spoiler owing to its ability to grow in beer at pH 4.3−4.6 (Seidel-Rüfer, 1990). Since no beer-spoilage incidents caused by this species have been reported, it could be regarded as a potential threat to beer quality. *S. lacticifex* is more acid tolerant compared to brewery-related *Propionispira* species and less acid tolerant compared to *M. cerevisiae* or *P. frisingensis*. *S. lacticifex* is relatively alcohol tolerant. Growth has been observed in beer with an alcohol content of 4.2% (w/v) and in PYF medium containing 5%−6% ethanol (w/v). Laboratory studies also suggest that *S. lacticifex* could still grow at the low temperature of yeast storage (10 °C) and lager fermentation (Haikara, 1989).

P. raffinosivorans is considered to be a potential beer spoiler owing to its ability to grow in beer at pH 5.0 but not at pH 4.6. *P. paucivorans* was able to grow in beer at pH 6.0 but not at pH 5.0 and appears to be a harmless brewery contaminant (Seidel-Rüfer, 1990). There is a lack of data about other beer-spoilage properties of the *Propionispira* species.

9.6.4 Prevotella

Although beer spoilage ability has been demonstrated, further information on spoilage characteristics of *P. cerevisiae* has not yet been published.

9.7 Management of contaminations

9.7.1 Prevention

The fact that natural sources of the strictly anaerobic beer spoilers in breweries are largely unknown complicates the prevention of contaminations. Potential primary sources include plant raw materials, such as hops and malt, bird droppings and water. It is, however, apparent that once *Megasphaera* and *Pectinatus* bacteria have found their way into breweries, they can establish themselves and persist in suitable niches for years. Therefore, maintenance of good factory and process hygiene, regular monitoring of critical points and rapid countermeasures in case of positive findings are key factors in preventing contaminations.

Iron has been found to enhance the growth of *Pectinatus* species in beer (Takesue, Suzuki, Mizutani, & Nakamura, 2023), and one potential source is diatomaceous earth (DE) used in filtration. Although use of DE in brewing is declining, for those companies using this aid, it is advised that a source low in iron is selected.

Dirty return bottles are one possible mode of transmission between and within the breweries (Haikara & Helander, 2006; Matoulková et al., 2012). Disinfection of the bottle racks and empty bottles before their entry into the filling halls, physical separation of bottle washing from the filling operations and the configuration of bottle washers so that bottle inlet and outlet are on the opposite sides have been suggested as preventive measures to reduce spreading of contaminations. Drainage systems and other anaerobic niches in the filling halls, such as broken floor structures, are often permanently inhabited by *Pectinatus* and *Megasphaera* bacteria, from where they easily spread via aerosols and human activities. Hygienic factory design and maintenance of good hygienic conditions, not only in the filling machines but also in the filling hall environment, is important to minimise the colonisation of strictly anaerobic beer-spoilage bacteria in the breweries.

Filling lines are often structurally complex and contain difficult to clean areas prone to accumulation of biofilm (Storgårds & Priha, 2009). Biofilm is formed when microbial cells attach to surfaces and form complex communities that are protected by the self-produced slime. Avoiding complex constructs and regular sanitation of the complete filling lines, including dismantling and mechanical cleaning of difficult to access parts, is advised. In the study of Matoulková et al. (2012), the side ledge of the conveyor belt cover, cable line bundles beneath the conveyor belt and structural elements of the belt and monoblock parts were the critical areas, spreading contamination to the whole filling machine.

Investigation of microbial adhesion to stainless steel has shown that *Pectinatus* species may be found in all brewery plants regardless of machinery construction and cleaning methods (Bittner et al., 2016). Adhesion to stainless steel by both *Pectinatus* and *Megasphaera* could however be reduced by creating an environment of alkaline pH with low ionic strength. An alkaline rinse at the end of a cleaning cycle could be effective in reducing adhesion, as could use of functional coatings on critical parts of the filling machinery (Bittner et al., 2016).

9.7.2 Elimination

Pectinatus and *Megasphaera* bacteria are able to find suitable niches in breweries where they may survive for years without causing any obvious defects (Hakalehto, 2000). Then due to some technological faults or inadequate cleaning, they may cause beer contamination and spoilage. Finding of contamination sources is the first step for their elimination. High contamination frequency of packaged products indicates that the contamination is affecting the whole production batch and sources should be looked at throughout the production process. Sporadic incidences refer to a secondary contamination in the filling stage. Frequently several contamination sites can be found in the process, and it may be difficult to relate a specific source with spoilage incidents. However, occurrence of the strictly anaerobic beer spoilers in the areas where open product is handled is always a risk for the product quality and should lead to counteractions.

As counteractions, proper mechanical cleaning followed by disinfection and replacement of worn-out parts and surfaces should be promptly undertaken to eliminate contaminations. *M. cerevisiae* and *Pectinatus* are relatively sensitive, especially to oxidative biocides such as peracetic acid, and their use is primarily recommended (Haikara, 1984, pp. 1–47). Addition of biocides to lubricants used in conveyer lines, filler heads and CO_2 recovery systems is also pragmatic. It should be borne in mind that microbes aggregated in a biofilm may have a much higher resistance towards biocides (up to 10–100 times) in comparison to planktonic cells (Storgårds & Priha, 2009).

Heat resistance studies in laboratory conditions have indicated that flash pasteurisation treatments applied in the brewing process are normally sufficient to inactivate *Pectinatus* and *Megasphaera* cells (Watier, Chowdhury, Leguerinel, & Hornez, 1996a; Watier, Leguerinel, Hornez, Chowdhury, & Dubourguier, 1995). These bacteria cannot survive wort boiling. Decimal reduction time of *Pectinatus* strains at 60 °C (D_{60}) was reported to be close to 0.4 min (Watier et al., 1996b). However, *Pectinatus* cells may adapt to heat, which increases their heat tolerance (Flahaut, Tierny, Watier, Hornez, & Jeanfils, 2000). *M. cerevisiae* appears to tolerate heating better than *Pectinatus* species. The D_{60} value for this organism in wort and beer was determined to be 0.55 min. It should be remembered that any preservative method is only effective when the initial contamination level is low. In case of heavy primary contamination, flash pasteurisation might not eliminate the risk of spoilage by the strictly anaerobic beer spoilers (Watier et al., 1996a).

9.8 Future outlook and research needs

Naturalness and healthiness remain megatrends in the food and beverage industry. Functional non-beer beverages and no/low-alcohol beer products are increasing. It is also increasingly popular to mix various types of drinks together to create new flavours. Low-alcohol beers are particularly susceptible to spoilage by *Pectinatus* and *Megasphaera* bacteria. Our studies have indicated that *Pectinatus* bacteria may also be capable of spoiling various new types of non-beer beverages. Hence, it may be expected that the importance of the strictly anaerobic beer-spoilage bacteria will increase. Moreover, the use of fermented vegetable juices as ingredients in the beverages could introduce new *Pectinatus* species into the brewery environment and create new spoilage risks.

The strictly anaerobic beer spoilers are still a relatively little-studied group of microbes. The spoilage properties of the latest described *Pectinatus*, *Megasphaera* and *Prevotella* species should be further studied to understand the real risks they pose to beer and beverage production. Moreover, there appear to be many factors apart from the basic variables (pH, bitterness, alcohol, oxygen content) which affect the growth of the anaerobic bacteria in beer. Understanding of the molecular basis of beer adaptation could help in developing increasingly effective control measures. Recent studies have indicated that some of the strictly anaerobic beer-spoilage species may also be beneficial to mankind, playing a role in biohydrogen production and anaerobic wastewater treatment processes. Beneficial aspects of these intriguing organisms certainly warrant further investigation.

9.9 Sources of further information and advice

Detailed information regarding strictly anaerobic beer-spoilage bacteria can be found from the reviews of Haikara and Helander (2006), Haikara and Juvonen (2009), Marchandin et al. (2009) and Suzuki (2011, 2020). Comprehensive information regarding lipopolysaccharides of *Pectinatus* can be found from the review of Helander et al. (2004). Detection and identification methods for these organisms have been described by Juvonen (2009) and are also found in Chapters 14 and 15.

References

Anisha, D. J. (2018). *Identification of subgingival microbiome in periodontal health and gingival recession using next generation sequencing technology.* Doctoral dissertation. Chennai: Ragas Dental College and Hospital.

Arahal, D. R. (2020). Opinions 97, 98 and 99. *International Journal of Systematic and Evolutionary Microbiology, 70*, 1439–1440.

Back, W. (2005). Brewery. In W. Back (Ed.), *Colour atlas and handbook of beverage biology* (pp. 10–112). Nürnberg, Germany: Fachverlag Hans Carl.

Betancur-Murillo, C. L., Aguilar-Marín, S. B., & Jovel, J. (2022). Prevotella: A key player in ruminal metabolism. *Microorganisms, 11*(1), 1.

Bittner, M., de Souza, A. C., Brozova, M., Matoulkova, D., Dias, D. R., & Branyik, T. (2016). Adhesion of anaerobic beer spoilage bacteria *Megasphaera cerevisiae* and *Pectinatus frisingensis* to stainless steel. *Food Science and Technology, 70*, 148–154.

Caldwell, J. M., Juvonen, R., Brown, J., & Breidt, F. (2013). *Pectinatus sottacetonis* sp. nov., isolated from a commercial pickle spoilage tank. *International Journal of Systematic and Evolutionary Microbiology, 63*, 3609–3616.

Callaway, T. R., Dowd, S. E., Wolcott, R. D., Sun, Y., McReynolds, J. L., Edrington, T. S., et al. (2009). Evaluation of the bacterial diversity in cecal contents of laying hens fed various molting diets by using bacterial tag-encoded FLX amplicon pyrosequencing. *Poultry Science, 88*(2), 298–302.

Carlier, J.-P., Marchandin, H., Jumas-Bilak, E., Lorin, V., Henry, C., Carrière, C., et al. (2002). *Anaeroglobus geminatus* gen. nov., sp. nov., a novel member of the family *Veillonellaceae*. *International Journal of Systematic and Evolutionary Microbiology, 52*, 983–986.

Castelló, E., y Santos, C. G., Igleasias, T., Paulino, G., Wenzel, J., Borzacconi, L., et al. (2009). Feasibility of biohydrogen production form cheese whey using a UASB reactor: Links between microbial community and reactor performance. *International Journal of Hydrogen Energy, 34*, 5674–5682.

Chaban, B., Deneer, H., Dowgiert, T., Hymers, J., & Ziola, B. (2005). The flagellin gene and protein from the brewing spoilage bacteria *Pectinatus cerevisiiphilus* and *Pectinatus frisingensis*. *Canadian Journal of Microbiology, 51*, 863–874.

Chevrot, R., Carlotti, A., Sopena, V., Marchand, P., & Rosenfeld, E. (2008). *Megamonas rupellensis* sp. nov., an anaerobe isolated from the caecum of a duck. *International Journal of Systematic and Evolutionary Microbiology, 58*, 2921–2924.

Chihib, N.-E., & Tholozan, J.-L. (1999). Effect of rapid cooling and acidic pH on cellular homeostasis *Pectinatus frisingenis*, a strictly anaerobic beer-spoilage bacterium. *International Journal of Food Microbiology, 48*, 191–202.

Chowdhury, I., Watier, D., & Hornez, J. P. (1995). Variability in survival of *Pectinatus cerevisiiphilus*, strictly anaerobic bacteria, under different oxygen conditions. *Anaerobe, 1*(3), 151–156.

Chowdhury, I., Watier, D., Leguerine, I., & Hornez, J.-P. (1997). Effect of *Pectinatus cerevisiiphilus* on *Saccharomyces cerevisiae* concerning its growth and alcohol production in wort medium. *Food Microbiology, 14*, 265–272.

Doyle, L. M., McInerney, J. O., Mooney, J., Powell, R., Haikara, A., & Moran, A. P. (1995). Sequence of the gene encoding the 16S rRNA of the beer spoilage organism *Megasphaera cerevisiae*. *Journal of Industrial Microbiology, 15*, 67–70.

Engelmann, U., & Weiss, N. (1985). *Megasphaera cerevisiae* sp. nov.: A new gram-negative obligately anaerobic coccus isolated from spoiled beer. *Systematic and Applied Microbiology, 6*, 287–290.

Flahaut, S., Tierny, Y., Watier, D., Hornez, J. P., & Jeanfils, J. (2000). Impact of thermal variations on biochemical and physiological traits in *Pectinatus* sp. *International Journal of Food Micorbiology, 55*, 53–61.

Gonzalez, J. M., Jurado, V., Laiz, L., Zimmermann, J., Hermosin, B., & Sainz-Jimenez, C. (2005). *Pectinatus portalensis* nov. sp., a relatively fast-growing, coccoidal, novel *Pectinatus* species isolated from a wastewater treatment plant. Validation List N° 102. International Journal of Systematic and Evolutionary Microbiology 55, 547–549 *Antonie Van Leeuwenhoek, 86*.

Gupta, R. S. (2011). Origin of diderm (gram-negative) bacteria: Antibiotic selection pressure rather than endosymbiosis likely led to the evolution of bacterial cells with two membranes. *Antonie Van Leeuwenhoek, 100*, 171–182.

Hage, T., & Wold, K. (2003). Practical experiences on the combat of a major *Pectinatus* and *Megasphaera* infection with the help of TaqMan realtime-PCR. CD-ROM. In *Proc. 29th EBC congr., Dublin. nürnberg, Germany* (pp. 1145–1148). Fachverlag Hans Carl. CD-ROM.

Haikara, A. (1984). *Beer spoilage organisms. Occurrence and detection with particular reference to a new genus Pectinatus* (Ph.D. thesis). Espoo, Finland: Technical Research Centre of Finland Publications 14.

Haikara, A. (1989). Invasion of anaerobic bacteria into pitching yeast. In *Proceedings of 22nd congress* (pp. 537–544). Zürich: Eur. Brew. Conv.

Haikara, A., Enari, T.-M., & Lounatmaa, K. (1981). The genus *Pectinatus*, a new group of anaerobic beer spoilage bacteria. In *Proceedings of 18th congress Europian brewing conversation, copenhagen* (pp. 229–240).

Haikara, A., & Helander, I. (2006). *Pectinatus, Megasphaera* and *Zymophilus*. In M. Dworkin, S. Falkow, E. Rosenberg, K.-H. Schleifer, & E. Stackebrandt (Eds.), *The prokaryotes: A handbook on the biology of bacteria* (pp. 965–981). New York: Springer Science + Media, LLC.

Haikara, A., & Juvonen, R. (2009). Genus XV. *Pectinatus*. In P. De Vos, G. Garrity, D. Jones, N. R. Krieg, W. Ludwig, F. A. Rainey, et al. (Eds.), *Bergey's manual of systematic bacteriology* (2nd ed., pp. 1094–1099). New York: Springer.

Haikara, A., & Lounatmaa, K. (1987). Characterization of *Megasphaera* sp., a new anaerobic beer spoilage coccus. In *proceedings of the congress - European brewery convention, madrid* (pp. 473–480).

Haikara, A., Penttilä, L., Enari, T.-M., & Lounatmaa, K. (1981). Microbiological, biochemical, and electron microscopic characterization of a *Pectinatus* strain. *Applied and Environmental Microbiology, 41*, 511–517.

Hakalehto, E. (2000). *Characterization of* Pectinatus cerevisiiphilus *and* P. frisingiensis *surface components. Use of synthetic peptides in the detection of some gram-negative bacteria.* Ph.D. thesis. Kuopio, Finland: Kuopio University Publications. Natural and Environmental Sciences, 112. 70 pp..

Helander, I., Haikara, A., Sadovskaya, I., Vinogradov, E., & Salkinoja-Salonen, M. (2004). Lipopolysaccharides of anaerobic beer spoilage bacteria of the genus *Pectinatus* – lipopolysaccharides of a gram-positive genus. *FEMS Microbiology Reviews, 28*, 543–552.

Henriksson, E., & Haikara, A. (1991). Airborne microorganisms in the brewery filling area and their effect on microbiological stability of beer. *Monatsschrift für Brauwissenschaft, 44*, 4–8.

Hitch, T. C., Bisdorf, K., Afrizal, A., Riedel, T., Overmann, J., Strowig, T., et al. (2022). A taxonomic note on the genus *Prevotella*: Description of four novel genera and emended description of the genera Hallella and Xylanibacter. *Systematic and Applied Microbiology, 45*(6), Article 126354.

Jin, D.-W., Sun, Q.-Y., & Shi, X.-Y. (2010). Microbial diversity analysis of anaerobic sludge with chloroform treatment for fermentative H_2 production. *Microbiology China, 37*, 811–816.

Juvonen, R. (2009). *DNA-based detection and characterisation of strictly anaerobic beer-spoilage bacteria. Espoo 2009*. Espoo, Finland: VTT Publications. p. + app. 50 pp. VTT Publications 723 134. p. + app. 50 pp.

Juvonen, R., & Suihko, M.-L. (2006). *Megasphaera paucivorans* sp. nov., *Megasphaera sueciensis* sp. nov. and *Pectinatus haikarae* sp. nov., isolated from brewery samples, and emended description of the genus *Pectinatus*. *International Journal of Systematic and Evolutionary Microbiology, 56*(Pt 4), 695–702. https://doi.org/10.1099/ijs.0.63699-0. PMID: 16585679.

Juvonen, R., Virkajärvi, V., Priha, O., & Laitila, A. (2011). *Microbiologial spoilage and safety risks in non-beer beverages produced in a brewery environment*. Espoo, Finland: VTT. Research Notes 2599.107 pp. + app. 4 pp. VTT Tiedotteita: Espoo.

Kim, J., Nguyen, S. G., Guevarra, R. B., Lee, I., & Unno, T. (2015). Analysis of swine fecal microbiota at various growth stages. *Archives of Microbiology, 197*, 753–759.

Kramer, T., Kelleher, P., van der Meer, J., O'Sullivan, T., Geertman, J. A., Duncan, S. H., … Louis, P. (2020 Sep). Comparative genetic and physiological characterisation of Pectinatus species reveals shared tolerance to beer-associated stressors but halotolerance specific to pickle-associated strains. *Food Microbiology, 90*, 103462. https://doi.org/10.1016/j.fm.2020.103462. Epub 2020 Feb 21. PMID: 32336380.

Kwong, W. K., Medina, L. A., Koch, H., Sing, K. W., Soh, E. J. Y., Ascher, J. S., et al. (2017). Dynamic microbiome evolution in social bees. *Science Advances, 3*(3).

Lee, S. Y., Mabee, M. S., & Jangaard, N. O. (1978). *Pectinatus*, a new genus of the family *Bacteroidaceae*. *International Journal of Systematic Bacteriology, 28*, 582–594.

Liu, D., Li, T., Zheng, H., Yin, X., Chen, M., Liao, Z., et al. (2017). Study on alterations of physiological functions in aged constipation rats with fluid-deficiency based on metabonomic and microbiology analysis. *RSC Advances, 7*(76), 48136–48150.

Marchandin, H., Haikara, A., & Juvonen, R. (2009). Genus XII. *Megasphaera*. In P. De Vos, G. Garrity, D. Jones, N. R. Krieg, W. Ludwig, F. A. Rainey, et al. (Eds.), *Bergey's manual of systematic bacteriology* (2nd ed., pp. 1082–1090). New York: Springer.

Marchandin, H., Jumas-Bilak, E., Gay, B., Teyssier, C., Jean-Pierre, H., Siméon de Buochberg, M., et al. (2003). Phylogenetic analysis of some *Sporomusa* sub-branch members isolated from human clinical specimens: Description of *Megasphaera micronuciformis* sp. nov. *International Journal of Systematic and Evolutionary Microbiology, 53*, 547–553.

Marchandin, H., Teyssier, C., Campos, J., Jean-Pierre, H., Roger, F., Gay, B., et al. (2010). *Negativicoccus succinivorans* gen. nov., sp. nov., isolated from human clinical samples, emended description of the family *Veillonellaceae* and description of *Negativicutes* classis nov., *Selenomonadales* ord. nov. and *Acidaminococcaceae* fam. nov. in the bacterial phylum *Firmicutes*. *International Journal of Systematic and Evolutionary Microbiology, 60*, 1271–1279.

Matoulková, D., Kosar, K., & Slabý, M. (2012). Occurrence and species distribution of strictly anaerobic bacterium *Pectinatus* in brewery bottling halls. *Journal of the American Society of Brewing Chemists, 70*, 262–267.

Mota, V. T., Delforno, T. P., Ribeiro, J. C., Zaiat, M., & Oliveira, V. M. (2024). Understanding microbiome dynamics and functional responses during acidogenic fermentation of sucrose and sugarcane vinasse through metatranscriptomic analysis. *Environmental Research, 246*, 118150. https://doi.org/10.1016/j.envres.2024.118150. Epub 2024 Jan 11. PMID: 38218518.

Motoyama, Y., & Ogata, T. (2000). 16S-23S rDNA spacer of *Pectinatus, Selenomonas* and *Zymophilus* reveal new phylogenetic relationships between these genera. *International Journal of Systematic and Evolutionary Microbiology, 50*, 883–886.

Motoyama, Y., Ogata, T., & Sakai, K. (1998). Characterization of *Pectinatus cerevisiiphilus* and *P. frisingensis* by ribotyping. *Journal of the American Society of Brewing Chemists, 56*, 19–23.

Nakakita, Y., Takahashi, T., Sugiyama, H., Shigyo, T., & Shinotsuka, K. (1998). Isolation of novel beer-spoilage bacteria from brewery environment. *Journal of the American Society of Brewing Chemists, 56*, 114–117.

Nakata, H., Kanda, H., Nakakita, Y., Kaneko, T., & Tsuchiya, Y. (2019). *Prevotella cerevisiae* sp. nov., beer-spoilage obligate anaerobic bacteria isolated from brewery wastewater. *International Journal of Systematic and Evolutionary Microbiology, 69*(6), 1789–1793.

Ning, Y.-Y., Jin, D.-W., Sheng, G.-P., Harada, H., & Shi, X.-Y. (2012). Evaluation of the stability of hydrogen production and microbial diversity by anaerobic sludge with chloroform treatment. *Renewable Energy, 38*, 253–257.

Oren, A., & Garrity, G. M. (2021). Valid publication of the names of forty-two phyla of prokaryotes. *International Journal of Systematic and Evolutionary Microbiology, 71*, 5056.

Paradh, A. D., Mitchell, W. J., & Hill, A. E. (2011). Occurrence of *Pectinatus* and *Megasphaera* in the major UK breweries. *Journal of the Institute of Brewing, 117*, 498–506.

Parks, D. H., Chuvochina, M., Waite, D. W., et al. (2018). A standardized bacterial taxonomy based on genome phylogeny substantially revises the tree of life. *Nature Biotechnology, 36*, 996–1004.

Poothong, S., Tanasupawat, S., Chanpongsang, S., Kingkaew, E., & Nuengjamnong, C. (2024). Anaerobic flora, *Selenomonas ruminis* sp. nov., and the bacteriocinogenic *Ligilactobacillus salivarius* strain MP3 from crossbred-lactating goats. *Scientific Reports, 14*, 4838.

Rodríguez-Saavedra, M., de Llano, D. G., Beltran, G., Torija, M. J., & Moreno-Arribas, M. V. (2021). *Pectinatus* spp.—Unpleasant and recurrent brewing spoilage bacteria. *International Journal of Food Microbiology, 336*, Article 108900.

Rogosa, M. (1971). Transfer of *Peptostreptococcus elsdenii* Gutierrez et al. to a new genus, *Megasphaera* [*M. elsdenii* (Gutierrez et al. comb. nov.]. *International Journal of Systematic Bacteriology, 21*, 187—189.

Schink, B., Thompson, T. E., & Zeikus, J. G. (1982). Characterization of *Propionispira arboris* gen. nov. sp. nov., a nitrogen-fixing anaerobe common to wetwoods of living trees. *Journal of General Microbiology, 128*, 2771—2779.

Schleifer, K. H., Leuteritz, M., Weiss, N., Ludwig, W., Kirchhof, G., & Seidel-Rüfer, H. (1990). Taxonomic study of anaerobic, gram-negative, rod-shaped bacteria from breweries: Emended description of *Pectinatus cerevisiiphilus* and description of *Pectinatus frisingensis* sp. nov., *Selenomonas lacticifex* sp. nov., *Zymophilus raffinosivorans* gen. Nov., sp. nov. and *Zymophilus paucivorans* sp. nov. *International Journal of Systematic Bacteriology, 40*, 19—27.

Seidel-Rüfer, H. (1990). *Pectinatus* und andere morphologisch ähnliche gram-negative, anaerobe Stäbchen aus dem Brauereibereich. *Monatsschrift für Brauwissenschaft, 3*, 101—105.

Shouche, Y. S., Dighe, A. S., Dhotre, D. P., Patole, M. S., & Ranade, D. R. (2009). Genus XXI. *Selenomonas* von Prowazek 1913, 36[AL]. In P. De Vos, G. Garrity, D. Jones, N. R. Krieg, W. Ludwig, F. A. Rainey, et al. (Eds.), *Bergey's manual of systematic bacteriology* (2nd ed., pp. 1086—1092). New York: Springer.

Soberka, R., Sciazko, D., & Warzecha, A. (1988). *Pectinatus*: nouvelle bactérie pouvant affecter la stabilité biologique du moût et de la bière. *Bios, 19*, 31—37.

Storgårds, E., & Priha, O. (2009). In P. M. Fratamico, A. A. Bassam, & N. W. Gunther, IV. (Eds.), *Biofilms in the food and beverage industries* (pp. 432—454). Cambridge, UK: Woodhead Publishing Limited.

Suihko, M.-L., & Haikara, A. (2001). Characterization of *Pectinatus* and *Megasphaera* strains by automated ribotyping. *Journal of the Institute of Brewing, 107*, 175—184.

Suiker, I., O'Sullivan, T., & Vaughan, A. (2007). Diversity analysis of beer-spoiling gram-negative isolates using PCR fingerprinting and computer-assisted analysis. CD-ROM. In *Proceedings of 31st EBC congress of venice. Nürnberg, Germany* (pp. 1099—1105). Fachverlag Hans Carl. CD-ROM.

Suiker, I., Vaughan, A., & O'Sullivan, T. (2009). Differences in growth behaviour of *Pectinatus frisingensis* isolates in beer. Poster presentation. In *Proc. 32nd EBC congress*. Hamburg: Poster presentation.

Suzuki, K. (2011). 125th anniversary review: Microbiological instability of beer caused by spoilage bacteria. *Journal of the Institute of Brewing, 117*, 131—155.

Suzuki, K. (2020). Emergence of new spoilage microorganisms in the brewing industry and development of microbiological quality control methods to cope with this phenomenon: A review. *Journal of the American Society of Brewing Chemists, 78*(4), 245—259.

Takesue, N., Suzuki, K., Mizutani, M., & Nakamura, Y. (2023). Iron enhances the growth of the genus *Pectinatus* in beer. *Journal of the American Society of Brewing Chemists, 81*(1), 162—170.

Temudo, M. F., Muyezer, G., Kleerebezem, R., & van Loodsrecht, M. C. M. (2008). Diversity of microbial communities in open mixed culture fermentations: Impact of the pH and carbon source. *Applied Microbiology and Biotechnology, 80*, 1121—1130.

Tholozan, J. L., Grivet, J. P., & Vallet, C. (1994). Metabolic pathway to propionate of *Pectinatus frisingensis*, astrictly anaerobic beer-spoilage bacterium. *Archives of Microbiology, 162*, 401—408.

Tholozan, J.-L., Membré, J.-M., & Grivet, J.-P. (1997). Physiology and development of *Pectinatus cerevisiiphilus* and *Pectinatus frisingensis*, two strict anaerobic beer spoilage bacteria. *International Journal of Food Microbiology, 35*, 29—33.

Tholozan, J.-L., Membré, J.-M., & Kubaczka, M. (1996). Effects of culture conditions on *Pectinatus cerevisiiphilus* and *Pectinatus frisingensis* metabolism: A physiological and statistical approach. *Journal of Applied Bacteriology, 80*, 418—424.

Timke, M., Wang-Lieu, N. Q., Altendorf, K., & Lipski, A. (2005). Fatty acid analysis and spoilage potential of biofilms from two breweries. *Journal of Applied Microbiology, 99*, 1108—1122.

Ueki, A., Watanabe, M., Ohtaki, Y., Kaku, N., & Ueki, K. (July 24, 2014). *Propionispira arcuata* sp. nov., isolated from a methanogenic reactor of cattle waste and reclassification of *Zymophilus raffinosivorans* and *Zymophilus paucivorans* as *Propionispira raffinosivorans* comb. Nov. and *Propionispira paucivorans* comb. Nov., and emended description of the genus *Propionispira*. *International Journal of Systematic and Evolutionary Microbiology*. https://doi.org/10.1099/ijs.0.063875-0

Van Vuuren, H. J. J., & Priest, F. G. (2003). In F. G. Priest, & I. Campbell (Eds.), *Brewing microbiology* (3rd ed., pp. 219—246). New York: Kluwer Academic/Plenum Publishers.

Vereecke, C., & Arahal, D. R. (2008). The status of the species *Pectinatus portalensis* Gonzalez et al. Request for an opinion. *International Journal of Systematic and Evolutionary Microbiology, 58*, 1507.

Voetz, M., Pahl, R., & Folz, R. (2010). *Der etwas andere* Pectinatus. Brauwelt, 9—10 261—263.

Watier, D., Chowdhury, I., Leguerinel, I., & Hornez, J. P. (1996a). Survival of *Megasphaera cerevisiae* heated in laboratory media, wort and beer. *Food Microbiology, 13*, 205—212.

Watier, D., Chowdhury, I., Leguerinel, I., & Hornez, J. P. (1996b). Response surface models to describe the effects of temperature, pH, and ethanol concentration on growth kinetics and fermentation end products of a *Pectinatus* sp. *Applied and Envrionmental Microbiology, 62*, 1233—1237.

Watier, D., Leguerinel, I., Hornez, J. P., Chowdhury, I., & Dubourguier, H. C. (1995). Heat resistance of *Pectinatus* sp., a beer spoilage anaerobic bacterium. *Journal of Applied Microbiology, 78*, 164—168.

Willems, A., & Collins, M. D. (1995). Phylogentic placement of *Dialister pneumosintes* (formerly *Bacteroides pneumosintes*) within the *Sporomusa* subbranch of the *Clostridium* subphylum of the gram-positive bacteria. *International Journal of Systematic Bacteriology, 45*, 403—405.

Zhang, K., & Dong, X. (2009). *Selenomonas bovis* sp. nov., isolated from yak rumen contents. *International Journal of Systematic and Evolutionary Microbiology, 59*, 2080–2083.

Zhang, W. W., Fang, M. X., Tan, H. Q., Zhang, X. Q., Wu, M., & Zhu, X. F. (2012). *Pectinatus brassicae* sp. nov., a Gram-negative, anaerobic bacterium isolated from salty wastewater. *International Journal of Systematic and Evolutionary Microbiology, 62*, 2145–2149.

Ziola, B., Gares, S. L., Lorrain, B., Gee, L., Ingledew, W. M., & Lee, S. Y. (1999). Epitope mapping of monoclonal antibodies specific for the directly cross-linked mesodiaminopimelic acid peptidoglycan found in the anaerobic beer spoilage bacterium *Pectinatus cerevisiiphilus*. *Canadian Journal of Microbiology, 45*, 779–785.

Part III

Reducing microbial spoilage: Design and technology

Chapter 10

Hygienic design and cleaning-in-place (CIP) systems in breweries

Ben Connolly[1], **Scott Davies**[1], **Trevor Sykes**[1], **Mark Phillips**[1], **John Hancock**[1], **Nicholas Watson**[2] **and Alex Bowler**[2]

[1]*Briggs of Burton Plc, Staffordshire, United Kingdom;* [2]*Faculty of Environment, School of Food Science and Nutrition, University of Leeds, Leeds, United Kingdom*

10.1 Introduction

Brewers would be extremely disappointed to find that the beer leaving their brewery was compromised in flavour and quality as a result of contaminating microorganisms. A brewer may also be disappointed to find that their brewery was unclean with respect to equipment fouling. Fouling of key processes required for heat transfer would directly affect the heating and cooling medium temperature required and extend the process time. This has a large impact on the brewery operation, energy requirements and cycle time. Therefore, to minimise the risk of contamination from spoilage microorganisms and reduce the extent of vessel and equipment fouling, it is important that the brewery has been designed and engineered with hygiene in mind. Reviewed here are the fundamentals of hygienic process design and the implementation of a cleaning-in-place (CIP) system as applied in the brewery brewhouse. It should be noted that the actual design for an effective CIP system depends on the appropriate implementation of hygienic plant design.

The practice of CIP is thought to have been originally developed for the dairy industry as a method to effectively clean vessels and pipework without the requirement to dismantle the process equipment (Meyers, 1959). CIP technology was adopted by brewers as a method to eliminate the need for manual cleaning. This as a consequence reduced the requirement for manual labour and its associated cost. Modern health and safety regimes seek to minimise the involvement of manual labour operations and therefore reduce the risk to the plant operators. Automatically cleaning large-scale breweries using a CIP system through automation is essential today to achieve the brewery throughput and required process turnaround time (TAT) to meet the market demand. Fundamentally, the CIP system removes residual soil that could lead to the introduction, growth and establishment of microorganisms.

Sterilisation-in-place (SIP) is a technology that is used in conjunction with a CIP system to provide a sterile environment. SIP is only briefly mentioned here as another process that is used to ensure a hygienic environment. An SIP system as the name suggests uses (sterile) steam to create an appropriately 'sterile' environment. This is still, however, reliant on the environment actually being clean. For example, applying steam to an unclean vessel containing residual soil would cause further physical bake on to the equipment. In the brewery, SIP systems are only found on low-temperature processes such as the yeast propagation system where microbial contamination is most likely. It is very difficult to steam large vessels such as fermentation vessels (FVs). The total steam required for this process is high and often more expensive than performing a CIP cycle. Cooling the large vessels takes a long time, and many of the brewery vessels do not have the pressure and temperature rating to undergo the sterilisation process conditions. Vessels must also have adequate venting systems to cope with filling with cold product. Failure to do so may cause the vessel to collapse. CIP systems are the most prevalent cleaning process in large breweries and are used to clean all major brewhouse processes and vessels including the mash tun/lauter tun, wort kettle and FVs. CIP also has uses in keg handling; however, this review is focused on the brewhouse.

Brewing Microbiology. https://doi.org/10.1016/B978-0-323-99606-8.00015-8

From a process engineering perspective, the CIP system is often more intricate than the actual main brewery process. This is predominately due to the tight integration of the CIP system around the main brewing process, and also the organisation of the pipework and number of valves required to control the flow of wort and beer separately from the cleaning fluid. The implementation of a CIP system should not be simply an afterthought to the brewery process but recognised as an integral design consideration to ensure hygiene. The overall brewery process should therefore be designed for cleanability in the first instance. The presence of a CIP system in a brewery that has not been designed appropriately may still lead to equipment fouling and poor hygiene. For example, CIP systems that feature inadequate drainage, dead legs in pipework and unhygienic valve designs are each discussed here as examples of poor hygienic design practice and are recognised as likely factors that can contribute to equipment fouling and microbial contamination.

10.2 Brewery contamination

The introduction of contaminating microorganisms can occur from raw materials such as malt or hops, through airborne transmission, but also through the brewery process pipework or vessels if they have not been appropriately designed for hygiene. The brewery is not a sterile environment. However, the presence and prevalence of foreign microorganisms in the brewery should be minimised through appropriate brewery design and cleaning practices, as the presence of contaminating microorganisms can cause stuck fermentations and affect product yield and beer flavour, and consequently brewery profitability (Hill, 2009, pp. 163–184). The actual brewing process and final product (beer) are actually quite inhospitable environments to many microorganisms. However, as is recognised throughout nature, there are a select few microorganisms that have the capacity to withstand this environment. Unfortunately for the brewmaster, these undesirable, contaminating microorganisms may potentially cause undesirable off-flavours and affect the beer quality.

10.2.1 Beer is a hostile environment

From the perspective of a microorganism, the chemical composition of beer makes this product quite a hostile environment and poor growth medium. Beer typically contains ethanol in the range of 0.5%–10% w/w, hop bitter compounds (approximately 17–55 ppm of iso α-acids), a high content of carbon dioxide (approximately 0.5% w/w) and a reduced oxygen content (<0.1 ppm), a low pH (3.8–4.7), and only traces of nutritive substances such as glucose, maltose and maltotriose (Sakamoto & Konings, 2003).

In comparison, wort is a far more favourable environment for microorganisms to grow. Therefore, ensuring that wort remains free from contaminating microorganisms is an important process consideration, especially when the presence of competing microorganisms can affect ethanol and product yields. Wort is rich in free amino nitrogen (FAN) and fermentable sugars (Briggs et al., 2004; Lekkas, Stewart, Hill, Taidi, & Hodgson, 2005), which are as essential to the Brewers' yeast as they are to other undesirable but opportunistic fermentative microorganisms. The contamination of wort is largely minimised by the brewing process itself, along with the introduction of compounds such as iso α-acids from hops. For example, boiling wort in the wort kettle serves as a method of sterilisation. After the kettle, the wort is cooled and pitched with yeast, and then transferred to the fermentation vessels. Boiling wort is unique to the brewing industry. The production of Scotch Malt whisky, which has a very similar process in the preparation of wort/wash from cereal grains, does not involve wort boiling. This difference is largely due to the requirement and presence of microorganisms such as lactic acid bacteria (LAB) in the FVs (washbacks) in the production of Scotch Malt whisky, which are recognised to contribute and influence the spirit flavour. Furthermore, the high alcohol content and distillation step in Scotch Malt whisky make the final product a largely unfavourable environment for microbial growth. Whilst wort boiling in the brewery is known to improve the sterility of the wort, all the downstream interconnecting pipework and process equipment should be clean. This is especially important because the wort is cooled to temperatures that are favourable for both the brewers yeast and other potentially contaminating microorganisms.

10.2.2 Types of contamination recognised in the brewery

From a brewers perspective, the presence of spoilage microorganisms can be detrimental to the production of beer, affecting its flavour through the production of unfavourable smells/off-flavours including diacetyl (Chuang & Collins, 1968) or hydrogen sulphide (Sakamoto & Konings, 2003). Spoilage bacteria are also known to affect beer turbidity and acidity

(Sakamoto & Konings, 2003). Microbial infections present in breweries with less stringent hygiene and cleaning regimes often find gram-positive anaerobic bacilli such as *Lactobacillus* spp. growing (Ault, 1965; Sakamoto & Konings, 2003; Suzuki, Funahashi, Koyanagi, & Yamashita, 2004). The predominance of this particular bacterial genus in breweries and other fermentation-based industries such as first generation bioethanol facilities and distilleries is due to their similarity to yeast with tolerances to an acidic pH and ethanol. A more in-depth review of these microorganisms is reviewed elsewhere in this book.

In addition to bacterial-based spoilage microorganisms, the brewery is also susceptible to contamination by wild yeast. Wild yeast refers to yeast that were not intentionally pitched into the fermentation vessel by the brewer. To a brewer, the yeast strain used in brewing fermentation is one of the key factors that contribute to beer flavour, in conjunction with the raw materials malt and hops. Therefore, it is important that the brewer has confidence that the yeast strain pitched into the fermentation vessel is the desired strain and that this can be achieved consistency.

Hygienic design and automation of the brewers' yeast propagation, storage, and pitching systems are therefore an important consideration to ensuring only the desired yeast strain is grown. Large breweries feature on-site propagation systems to specifically manage the growth and handling of their own specific yeast strain. To reduce the potential risk of contamination from wild yeast on-site, the yeast propagation systems usually feature high levels of automation and control to ensure contamination is minimised and hygiene maintained. A dedicated single-use CIP system is often used to minimise cross-contamination across the brewery. The application and details of the single CIP system for yeast propagation are discussed in the overview of CIP systems later.

10.2.3 The prevalence of microorganisms in the brewery

If microorganisms have successfully infiltrated the brewery through the raw materials or poor hygienic process design, they can remain prevalent in pipework and crevices through their capacity to develop biofilms. Biofilms are essentially a community of cells that exist in a polymer network comprising of proteins, lipids, and polysaccharides (Costerton, Stewart, & Greenberg, 1999; Sutherland, 2001). The establishment of a biofilm causes the contained microbial cells to undergo both morphological and genetic alterations that make it distinct from the planktonic state, where the microbes exist in a free floating environment. The development of a biofilm provides the microbial community with greater resistance to mechanical and chemical treatment. However, microorganisms can only synthesise exo-polysaccharides required for the development of biofilms if there is an available carbon and nutrient source (Sutherland, 2001). Microbially synthesised exo-polysaccharides present in the biofilm are typically structurally long ($0.5-2 \times 10^6$ Da), thin and polyanionic. The structural and chemical heterogeneity of the exo-polysaccharide allows various associations through electrostatic, hydrogen bonding and ionic interactions (Sutherland, 2001). With respect to cleaning, it should be noted that the biofilm exo-polysaccharides feature the greatest ordered state at low temperatures and in the presence of salts (Sutherland, 2001). Therefore, the application of high temperatures during the CIP operation is necessary to disrupt the exo-polysaccharides native state. The requirement for hygienic design is therefore based on minimising the opportunity for microorganisms to adhere and proliferate on surfaces, in crevices and in key process equipment.

10.3 The main principles of hygienic design as applied in the brewery

The concept of hygienic plant design evolved in the food, beverage, and pharmaceutical industries. All of these industries require processes that form products that are free from contamination and are safe for human consumption or use. As mentioned previously, the implication of microbial contamination can have detrimental effects on the quality of beer. Therefore, designing and engineering a brewery that is cleanable requires an appreciation of several hygienic plant design concepts. The European Hygienic Engineering and Design Group (EHEDG) and The American Society of Mechanical Engineers (ASME) provide a set of guiding principles that have been developed over time from the contributions of their members. The EHEDG guidelines and ASME BPE 2009 Bioprocess document are both extremely comprehensive in specifying and justifying best practices of hygienic process design. Discussed here are the main hygienic design principles that are most appropriate in the context of the brewery and include the following:

- Pipe layout, design and overall process flow, which focuses on design considerations such as pipework deadlegs
- The presence of crevices and imperfections in material surfaces, which arise through fabrication and material selection and are known to promote the formation of biofilms

10.3.1 Brewery pipework design and layout to minimise contamination

Designing pipelines hygienically maximises cleanability while minimising the prevalence of soil or potentially contaminating microorganisms. Cleaning process pipework using the CIP system requires a combination of valves and pumps to control and direct the cleaning fluid through the brewery process. The shape of pipework fixtures such as dead legs and T-shaped junctions naturally lend themselves as sinks to organic material deposition that could harbour and support the growth of undesirable microorganisms (Fig. 10.1). In addition to the pipework design, it is also important that the CIP fluid has a turbulent flow in the process pipework to scour the surface and remove the soil. Both of these aspects are reviewed next.

10.3.2 Operating conditions required to achieve a cleaning action in pipework

The effect of surface fouling has a large impact on the heat transfer coefficient. Protein fouling creates a thermal barrier, which during wort boiling for example will increase the temperatures required by the heating medium (steam) to heat the product. Ineffective cleaning will therefore have an impact on steam usage.

A turbulent flow is required in the pipework of the CIP system to remove any residual soil. The velocity of the CIP cleaning fluid in the process pipe should be between 1.5 and 2.0 m/s. Practically obtaining this velocity range is dependent on the cleaning fluid flow rate (m^3/h) and the pipe diameter. A fluid is recognised as turbulent when its Reynolds number is greater than 4000 (Equation 10.1). Fluids with a Reynolds number of 2100 and below have a laminar flow, which is not effective at scouring the pipework surface (Chisti & Moo-Young, 1994). During laminar flow, the fluid has the greatest

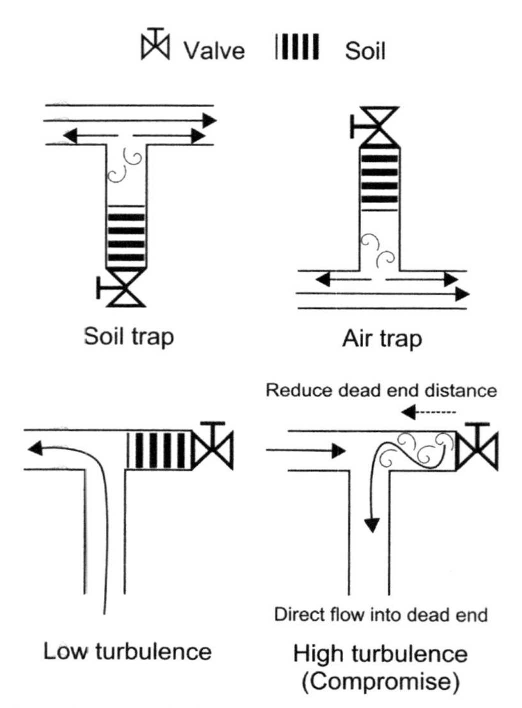

FIGURE 10.1 Examples of best practice in hygienic process design. Dead ends and T-shaped bends can potentially harbour soil. *Reproduced with permission from Brian Eaton from the lecture material used in IBD Diploma in Distilling Module 2 on Plant Cleaning.*

EQUATION 10.1 Calculation of Reynolds number to determine the whether the flow is laminar, transient or turbulent.

$Re = \frac{\rho u d_h}{\mu}$, where Re is the Reynolds number (non-dimensional), ρ is the density (kg/m^3), u is the velocity (m/s), d_h is the hydraulic diameter (m), and μ the dynamic viscosity (m^2/s) from The Engineering toolbox (www.theengineeringtoolbox.com).

EQUATION 10.2 Pressure loss in a pipe

$\Delta P = kV^2$, where ΔP is the Pressure difference, k a coefficient, V is the velocity (m/s) of the fluid travelling through the pipe from The Engineering toolbox (www.theengineeringtoolbox.com).

velocity (V_{max}) at the centre and lowest (zero) velocity from the pipe wall surface. Therefore, there is no movement/mechanical action at the surface that you want to clean. The CIP cleaning fluids, wort and beer are all turbulent at velocities of less than 1.5 m/s due to their density (Equation 10.1). For example, a CIP cleaning fluid at 0.3 m/s would have a Re of approximately 10,000. It should be noted that even if the fluid has a turbulent flow, if their velocity is too low they will have a thick boundary layer, which in the pipework will still cause potential fouling and deposition. Cleaning fluid velocities greater than 2.0 m/s in the pipework are not recognised to have any additional effect on cleaning. Therefore, increasing the fluid velocity beyond this value only increases the energy used to pump the fluid due to an increased pressure drop. For example, doubling the fluid velocity quadruples the pressure drop (Equation 10.2).

10.3.3 Hygienic design and operating practice of valves, fixtures and fittings

Pipework in a brewery is fundamental to moving products such as wort and beer. Therefore, ensuring the pipes are arranged in a manner to promote cleanability is an important consideration for hygienic design. Deadlegs in pipework should be avoided (Fig. 10.1). Fig. 10.1 shows several T-junctions that create environments that lead to poor cleanability and should be avoided in the brewery process design. If T-junctions are present, the cleaning fluid should be pumped in the direction of the dead leg so that sufficient turbulence action can be achieved in the dead leg space (Fig. 10.1). Fig. 10.1 shows how T-junctions may lead to the development of air and soil traps, which are both undesirable from a hygiene perspective. It is also important that the CIP supply routes are appropriately considered and do not split across the process pipework. Instead a single route that systematically works through the brewery pipework should be considered. However, additional pipework may be necessary to route the CIP cleaning fluids around the process through dedicated, separate pipework. Controlling the direction of cleaning fluids is achieved using double-seated mix-proof valves (Fig. 10.2). These have become an integral part of a CIP system for routing both product and cleaning fluid. An example of a valve manifold containing an array of double-seated valves is shown in Fig. 10.2 and 10.3.

Instrumentation such as pH probes can be fitted to the vessels either directly or where greater hygiene is required using hygienic housing that can retract the probe (Chisti & Moo-Young, 1994). Probes are usually directly fitted to the vessel for cost purposes. The benefit of the retractable housing is that it prevents the probe from being damaged and allows periodic cleaning, independently from the CIP system. This is important, for example, in yeast systems where the probe will require more periodic cleaning to reduce fouling and ensure accurate readings.

10.3.4 The effect of material surface finish on microbial surface adhesion

The surface finish of metals has a large impact on the capacity of microorganisms to adhere to pipes and vessels (Milledge, 2010). The surface characteristic of metal can be changed through processes such as welding and polishing. Welds, for

FIGURE 10.2 Mix-proof valves and valve array commonly used next to a chain of fermenters to hygienically control the transfer of beer and the CIP cleaning fluid.

FIGURE 10.3 Valve manifold of the cleaning-in-place system with caustic tanks in the background.

example, introduce both physical and chemical changes to the metal surface from both the metal filler composition (steel) and solidified slag. Surface defects and the material topography are both known to influence the cleanability of stainless steel. The changes to the metal surface through processes such as welding are thought to facilitate the accumulation of organic material that can lead to the growth of microorganisms. The preferential colonisation of welds by microorganisms has been correlated with the material surface roughness (Sreekumari, Ozawa, & Kikuchi, 2000). As a material for vessels and pipework stainless steel benefits from the development of a passive layer when exposed to air (chromium oxides). This effectively serves as a barrier between the fluid and pipe wall itself. Periodically using acid detergents such as citric or nitric acid in the CIP system is important to re-establish this passivation layer (oxidation) and helps to ensure that the stainless steel remains rust free ("BSSA - Passivation of stainless steels," 2014).

One method of evaluating the surface finishes of a metal is the roughness average (Ra) value or the root mean square average (RMS). The development of several surface characterisation methods has arisen due to the different possibilities of representing a material's surface using an average and single digit metric. Both methods are recognised and included as part of the ASME B46.1 standard in determining the surface properties of materials. From a hygiene perspective, a lower roughness average (Ra) value or the root mean square average (RMS) value indicates a reduction in the depth of crevices (peaks/troughs) across the metal surface and therefore minimises the amount of organic material that may reside on the surface. A reduction in the metal surface roughness can be achieved through more extensive polishing operations. Surface variations at the macroscopic level can be reduced using mechanical polishing and at the microscopy level using electropolishing. The pharmaceutical industry has long required highly polished vessels and process equipment to improve the levels of hygiene of their plant equipment. Similarly, the food industry demands roughness average (Ra) values of less than 0.8 μm (Milledge, 2010). However, increasing the extent of material polishing from a manufacturing perspective increases the cost of the material. The brewing industry has never implemented the same stringent control over the material surface finish as the pharmaceutical and food industries, which will largely be due to additional cost and the potential to damage the material surface. The smoothest surfaces for steel are achieved through a cold rolling process, followed by chemical descaling. This process produces a material with crevices. This is in contrast to polishing, which effectively only scratches the surface.

10.4 An overview of cleaning-in-place systems used in the brewery

Brewery maintenance is an important aspect from a hygiene perspective to prevent contamination and fouling, both of which can affect the brewery yield and process efficiency. Small breweries (<50 UK Barrel brewlength) will typically clean the process equipment by hand using brushes and spray hoses, or have simple CIP systems involving a detergent tank and pump. Larger breweries would be expected to automate the cleaning process using a fully integrated CIP system. A complete CIP system used in large breweries features detergent makeup tanks, interconnecting pipework, pumps, valves and heat exchangers (Fig. 10.4). The whole CIP system is usually automated, relying on flow metres, temperature probes, and conductivity metres to monitor the process. The complexity and functionality of the overall CIP system is highly dependent on the brewers requirements and the brewery operation. The cleaning fluid from the CIP system can be pumped in either the same or opposing direction to the process flow. Pumping the cleaning fluid in the reverse direction is

FIGURE 10.4 Tanks used in a brewery clean-in-place system. Left to right are the pre-rinse storage tank, the caustic tank and the CIP return tank.

sometimes necessary to remove soil. The number of vessels and level of automation that is required is dependent on the application of the CIP system. The insides of vessels are cleaned using cleaning machines and spray heads. The major difference between the operations of these two types of vessel cleaning equipment relates to the cleaning fluid flow rate and pressure. Cleaning machines are typically high pressure with a low flow rate. This provides greater mechanical impact and does not simply rely on the chemical action of the CIP detergent. Spray heads in comparison operate at lower pressure with a higher flow rate. Fig. 10.5 shows examples of cleaning machines and spray heads that are used on brewery vessels. Cleaning machines are typically used in brewery equipment such as the mash tun and wort kettle. The application and suitability of the cleaning system is highly dependent on the vessel scale. The number and position of spray heads in the vessel is also an important consideration. Vessel equipment such as agitators can obstruct the spray ('shadowing'), which impacts the effectiveness of the cleaning fluid. During the CIP cycle the agitators should be activated to prevent 'shadowing' from occurring. Spray heads are usually situated at the top of the vessel to allow cleaning fluid to be sprayed across the body of the tank, which then runs down the sides of the vessel. Vessels may feature several spray devices to ensure the whole surface is covered and no shadowing occurs. An internal kettle fountain, for example, required several spray balls to reach all the crevices in its design. To ensure sufficient mechanical action during cleaning a high pressure is required to remove soil material. Static spray heads feature no moving parts and are low cost. Spray heads use more water and energy than cleaning machines due to the higher flow rate. Cleaning fluid exiting the spray head atomises, which increases the adsorption of CO_2 by caustic resulting in the formation of carbonates. Cleaning machines are the most effective and aggressive cleaning strategy, benefitting from the lowest energy and water usage. Both spray heads and cleaning machines can become blocked with soil from pre-rinse and caustic featuring contaminant material from previous CIP batches. This problem can be overcome using a strainer in the CIP supply. Ideally the solids are completely removed from the system during the initial pre-rinse stage, which is discharged to drain and discussed in more detail later. As an alternative, a strainer may be fitted to the CIP return, which will reduce the problem and reduce any sediment reaching the CIP chemical storage tanks.

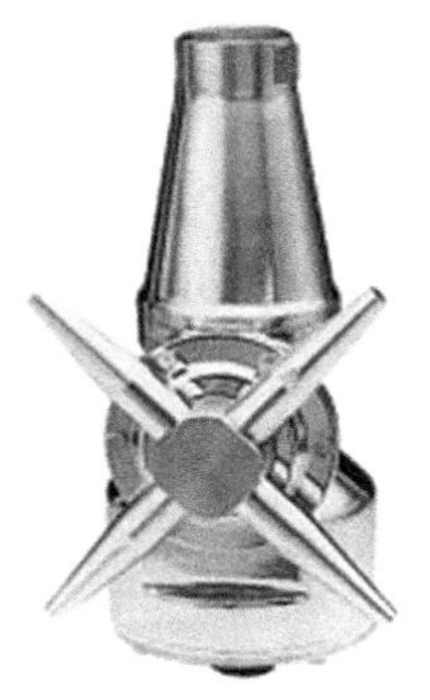

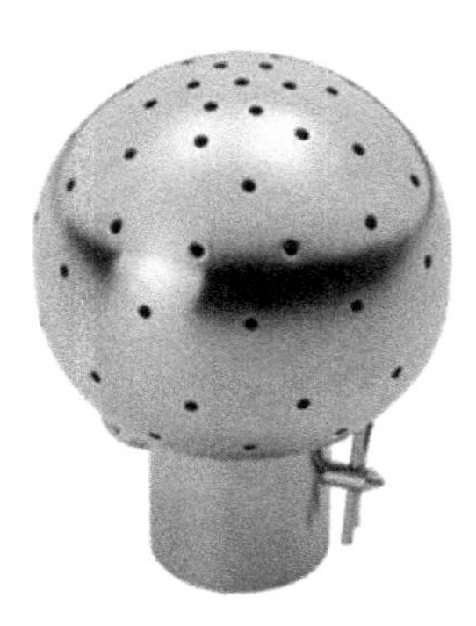

FIGURE 10.5 Example of vessel cleaning machines (GamJet TZ-74; left), static spray ball (middle) and rotary spray ball (left) as applied in brewery vessels.

The main terminologies used to describe the operation of the CIP system relate to the direction of the CIP cleaning fluid. The CIP fluid leaving the CIP storage tank is referred to as the CIP supply. The CIP supply is usually heated in a heat exchanger and pumped through the process pipework, reaching spray heads or cleaning machines inside the vessels. The CIP cleaning fluid that is recovered from vessels is referred to as the CIP return and is collected using a CIP scavenge pump. There are three main types of CIP system that are installed in a brewery. These include a single use, partial recovery, and full recovery system. The variations of these different CIP systems relate to the extent that the cleaning fluids are recovered. The selection of the appropriate CIP system, its operation temperature and the extent of the number of CIP channels used in the recovery CIP system are based on the specific brewery operation and brewhouse process.

10.4.1 Operating conditions of a Cleaning-In-Place system in the brewery

An effective CIP system involves three types of processes: mechanical, chemical, and sanitisation. An effective CIP system is a balance between an optimum temperature, residence time, mechanical and chemical treatment. Mechanical processes physically remove materials that soil process equipment through turbulence or a scouring action. Spray balls used inside vessels and CIP pumps are necessary to remove residual particulate such as proteinaceous materials. Ineffective removal of the residual soil reduces the effectiveness of brewery process equipment including plate heat exchangers and vessels. As mentioned previously, the residual soil may also provide an adherence and nutrient rich site for biofilm development. Additional energy is required as the fouling of process equipment negatively affects heat transfer. In combination with the turbulent flow generated by mechanical action, the CIP system will use chemical reagents including both base and/or acid to clean. CIP systems are a relatively large user of water in the brewhouse ('Reducing water use through Cleaning-In-Place (CIP) envirowise - EN894', 2008). Efforts to minimise the use of water use in the brewery have been achieved through modifications of the CIP program ('Reducing water use through Cleaning-In-Place (CIP) envirowise - EN894', 2008).

A typical CIP cycle would include the following stages below:

- A pre-rinse with water to remove any loose material. The wash water and any soluble material are discharged directly to the drain to eliminate any material carryover. An effective pre-rinse will also remove solids that cause equipment blockages.
- A hot caustic wash to chemically remove material that has soiled and fouled the equipment is used. A 2%–3% caustic wash at 75–80°C is used in the brewhouse and for processes involving wort. Lower strength caustic (1%) is used on lower soil environments such as bright beer. The hot caustic should digest and dissolve any soiled material. During the CIP cycle the caustic solution is recirculated several times. Heating the caustic CIP cleaning fluid is achieved using a heat exchanger, which can use waste heat to prewarm the CIP cleaning fluid.
- A further washing stage to remove the caustic.
- Finally an acid wash can be applied on a periodic operation of the CIP cycle. This minimises the CIP TAT and operating cost. The use of acid has several benefits in the brewhouse. It is used in cold processes such as FVs where the acid serves to eliminate bacteria, effectively serving as the sanitising agent. The acid detergent does not suffer from degradation by CO_2 as recognised with caustic, which is known to reduce its effectiveness (forming carbonates). The acid wash can also serve to re-establish the passivation layer at the stainless steel surface.
- A final wash using either reverse osmosis or deionised water is applied to remove any residual detergent.

Indicative timings for each of these CIP stages for cleaning brewery vessels and process pipework are in Table 10.1. The operating cost of the CIP system is influenced by the amount and concentration of detergent used, and whether it is recovered or not. Therefore, the specific recipe and conditions of the CIP system are important to the brewer.

TABLE 10.1 Timings of the main stages of a CIP system for both vessels and the main process systems pipework.

Unit operation	Function	Vessel CIP (min)	Mains CIP (min)
Prerinse	Mechanical removal of soil	10–20	5–10
Caustic detergent	Cleaning of remaining soil	30–40	20–30
Rinse	Wash any residual detergent	10–15	5–10
Acid detergent		20–30	15–20
Final rinse	Wash any residual detergent	15–20	10–15
Steriliant		10–15	5–10

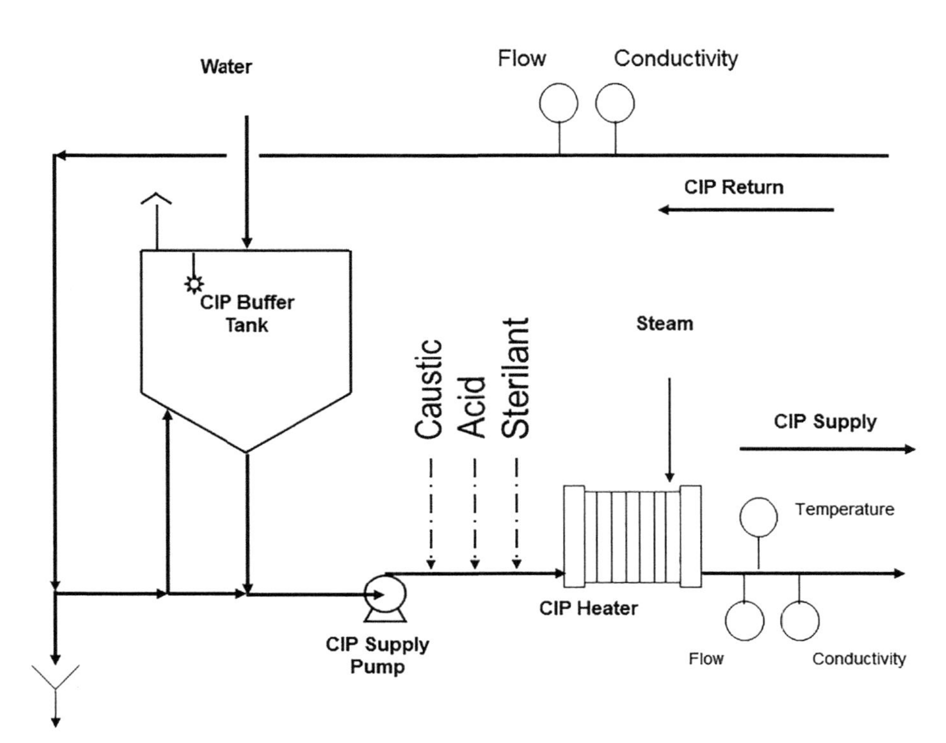

FIGURE 10.6 Example of a single-use CIP system. In-line heating and chemical dosing.

10.4.2 Types of Cleaning-In-Place systems recognised across the brewery

10.4.2.1 Single use

A single-use CIP system pumps the cleaning fluid around the process pipework and vessel, and on its return sends the water phase directly to the drain (Fig. 10.6). The single-use CIP system is the simplest cleaning system and is important for processes such as yeast handling and propagation that require the highest levels of hygiene. Brewery yeast systems are often single-use CIP systems. Both the pre-rinse and detergent used to clean the process vessels are sent to the drain. This essentially serves to reduce the risk of cross-contamination by CIP systems that are less than optimal in recovering and recycling the pre-rinse water and detergent.

10.4.2.2 Partial and full recovery cleaning system

Partial recovery CIP systems recover the detergent for use in the next detergent step or as a pre-rinse. A full recovery CIP system is designed to recover the final rinse for the next pre-rinse and return the cleaning fluid streams back to their chemical supply tanks. Examples of recovery CIP systems are shown in Figs 10.7 and 10.8. Fig. 10.8 shows a CIP system that features both caustic and acid tanks as a detergent. This is a more complicated system with respect to the number of chemical detergent tanks, valves and extent of pipework routing.

In summary, single-use CIP systems are less capital intensive, require less space and reduce the risk of cross-contamination compared with recovery CIP systems. The single-use CIP system is important for specific applications such as yeast handling. In comparison, recovery CIP systems have lower chemical losses and therefore costs, energy demand and volumes of water and effluent. The recovery CIP system also benefits from a shorter turnaround time and typically setup to be ready with no delay. For instance, the chemical detergents tanks already contain the required detergent concentration at the necessary temperature.

10.4.3 Cleaning-In-Place fluid composition

CIP systems can pump both pre-rinse material, detergents and final rinse water to remove soil. The pre-rinse material (potentially dilute caustic) is used to remove any loose debris, whereas the detergents chemically remove the soil.

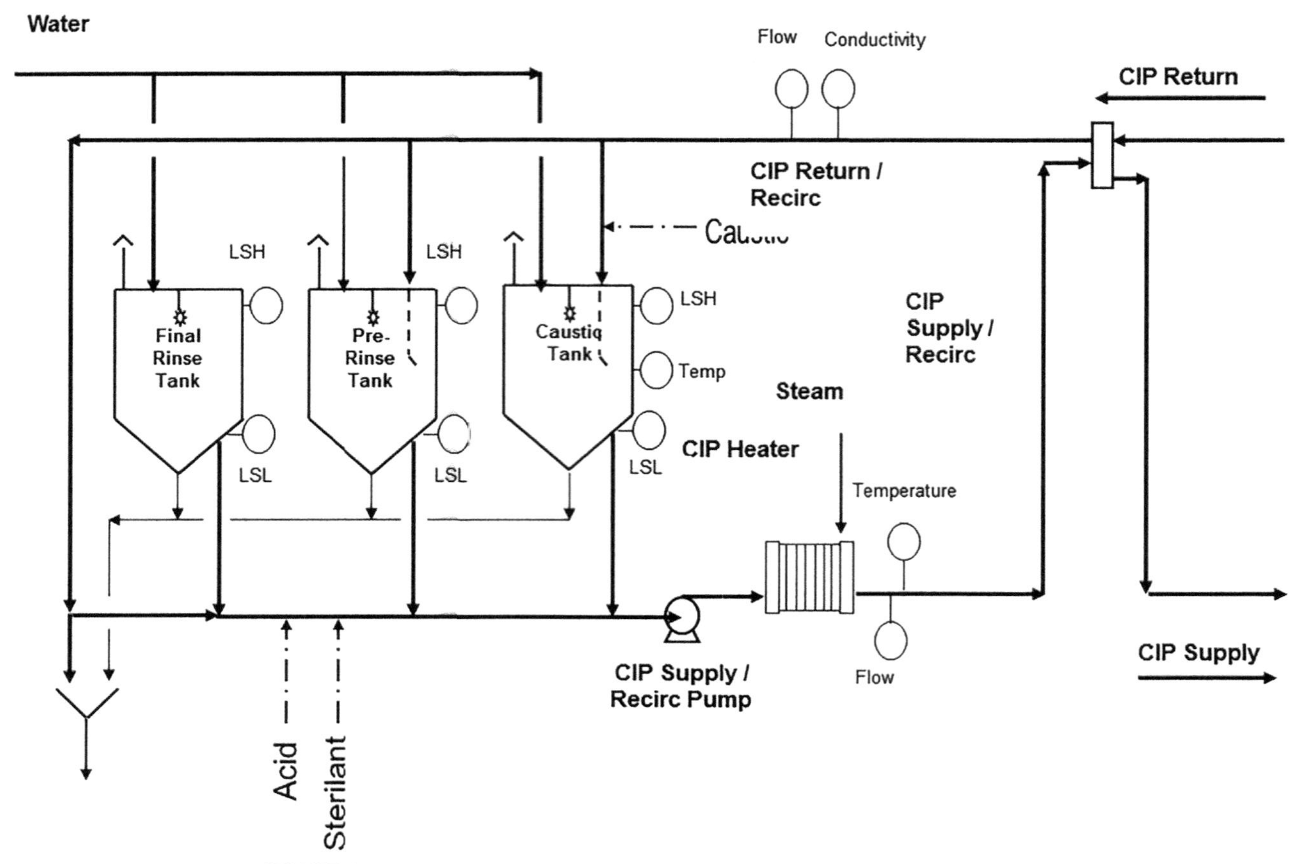

FIGURE 10.7 Example of a recovery CIP system with one CIP supply and three tanks.

Described in the following are the different detergents that can be used in CIP systems and also chemical additives such as sequestrates (chelating agents) and surfactants that improve detergent penetration.

10.4.3.1 Detergents

Detergents are used to chemically remove soil. Ideally, the detergents are non-foaming or include anti-foam, free rinsing/non-tainting, non-corrosive and have minimal environmental impact. Appropriately formulated detergents are effective at removing soil. Caustic-based detergents are more effective than acid detergents on high soil environments. Sodium hydroxide reacts with proteins and oils converting them into their respect salts, which increases their solubility and therefore their removal from stainless steel surfaces. Applying detergents at elevated temperatures using a hot water medium provides a level of disinfection. Detergents are effective at removing protein soil, which are especially important as fouling can reduce the effectiveness of heat transfer surfaces. This is especially important in brewhouse processes such as wort kettle or equipment such as heat exchangers where the protein can become baked onto the surface at high temperatures.

Controlling the strength of detergents can be achieved using conductivity metres. Disadvantages to caustic-based detergents are their degradation by CO_2, forming carbonates. As a result of this, the brewer would require more caustic to achieve the required working concentration. This is most important in the FVs where the CO_2 levels are greatest. Pre-rinse water can absorb some of the CO_2 present in the FVs and therefore reduce the chance of producing carbonates. It should be noted that the absorption of CO_2 by caustic creates a risk of forming a vacuum that cause the FV to collapse (Manzano et al., 2011).

Acid-based detergents are more frequently used to clean and sterilise fermentation vessels, whereas as caustic detergents are used to remove soil from the main brewery operations. The activity of caustic detergents is also affected by water hardness. Caustic detergents have poor rinsability compared with acid detergents and therefore require more water to

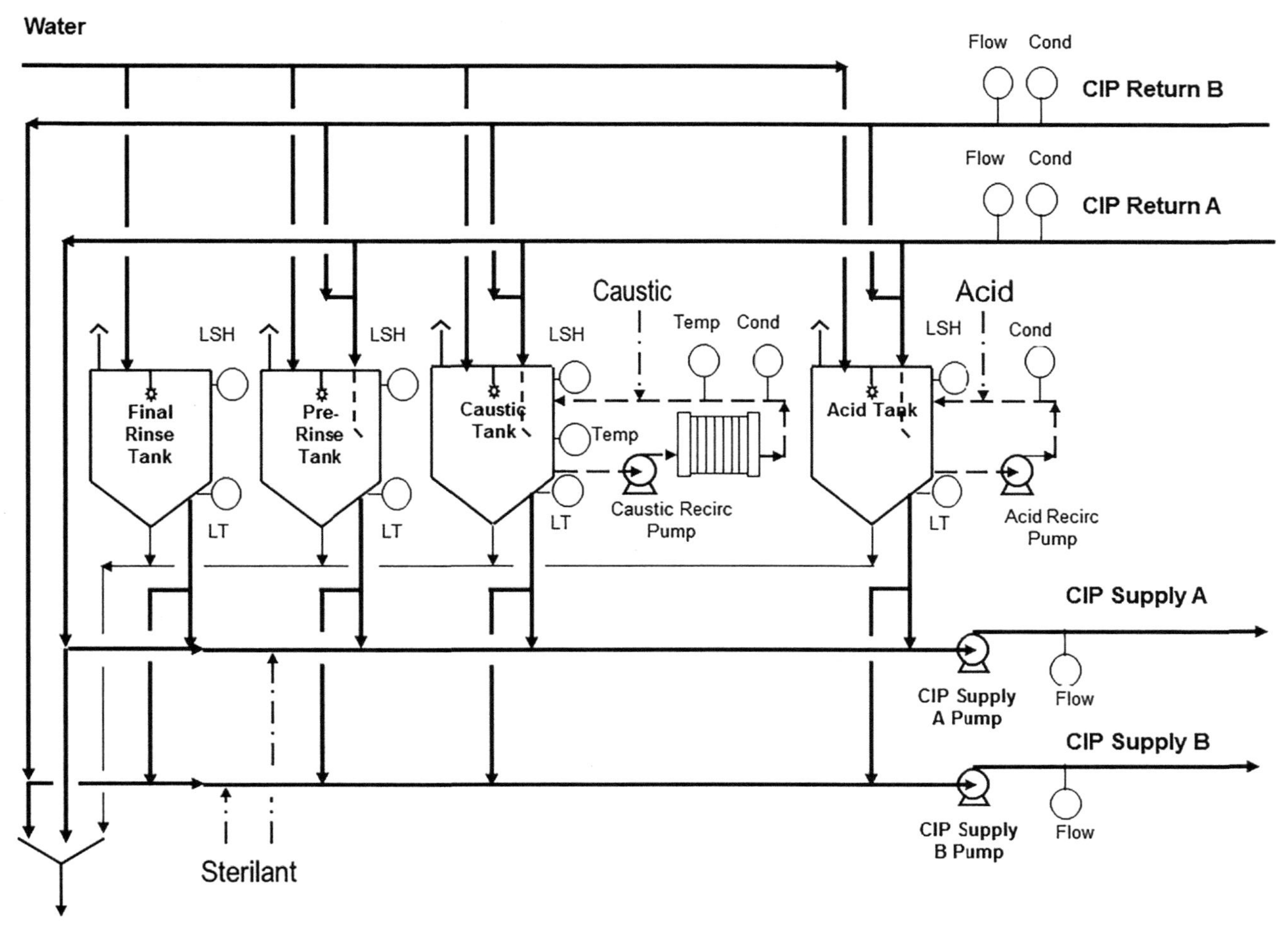

FIGURE 10.8 Example of a recovery CIP system with two CIP supply and four tanks. Chemical tanks feature a recirculation loop with an independent heat exchange, which allows the tanks to be heated during a CIP cycle for immediate use. Important on larger breweries to minimise turnaround time (TAT).

remove. Caustic detergents are ineffective at removing inorganic scale such as the gypsum and beerstone. Phosphoric acid and nitric acid are often used to remove inorganic scale. It should, however, be noted that there are potential environmental issues of releasing phosphates and nitrates into the effluent. Selecting the appropriate detergent is based on the unit operation in the brewery, the type of equipment fouling (organic or inorganic) and also the environmental impact on discharging detergents to effluent.

Detergent additives including sequestrate (chelating agents) and surfactants can also be added to the CIP cleaning fluid. Sequestrates such as ethylenediaminetetraacetic acid (EDTA), nitrilotriacetic acid (NTA), gluconates and phosphonates complex with metal ions in solution and prevent the precipitation of the insoluble salts of the metal ions. The main rationale for adding the sequestrate is to inhibit the formation of scaling with the design philosophy that prevention is better than cure. Surfactants (wetting agents) added to detergents reduce the cleaning fluids surface tension, which allows the detergent to penetrate the metal surface more effectively.

10.4.3.2 Steriliant

A steriliant can be applied following the detergent-based cleaning process to remove any residual microorganisms. Steriliants include chlorine, ionophores and peracetic acid (PAA). PAA degrades into acetic acid and hydrogen peroxide. Hydrogen peroxide is a strong oxidising agent, which can be used to enhance the CIP detergent (caustic or acid). The formation of acetic acid from the decomposition of PAA increases the organic load in the effluent waste.

10.4.4 Evaluation of the effectiveness of Cleaning-In-Place systems

Evaluation of the brewery CIP system and its effectiveness is usually determined using off-line laboratory analysis or portable measuring devices. This typically involves dyes such as riboflavin or ATPase activity assays used to determine the presence of living microorganisms.

Online sensors are used as part of the CIP system to check aspects such as the concentration and quality of the caustic cleaning fluid. Conductivity metres are fitted within the CIP set itself to evaluate the detergent concentration and control a dosing pump to top up the detergent as required. From an overall automation perspective, all the basic parameters for flow, temperature and time are each controlled and monitored as part of the CIP system.

10.5 Conclusions

Hygienic design is an important factor in brewery design and engineering. The brewer and beer drinker expect that resultant beer is safe to drink and consistent in quality and flavour. Reviewed here were the main design considerations that a brewery must employ to minimise the risk of contamination and fouling. A real appreciation of the intricacies of the CIP system and the requirement for hygienic process design become apparent with larger breweries. Larger breweries typically require more extensive levels of automation and control to coordinate the vast array of mix proof valves to correctly direct and route the CIP cleaning fluid. Furthermore, the amount of water and cleaning fluid reagent used in these large-scale breweries will have a large financial and environmental impact.

Efforts to minimise water and energy usage during the CIP process are an important aspect of the overall process, particularly when considering the frequency of cleaning between batches. High-pressure, low-flow cleaning machines are employed to remove soiled material that has fouled equipment using less water. An effective prerinse and sufficient mechanical action are required to disrupt materials that cause fouling. Ineffective pre-rinsing requires more water and chemicals during the CIP cleaning cycle. Pumping cleaning fluids at lower flow rates or adopting lower cleaning temperatures are potential strategies to minimise water and energy usage during CIP operations. However, to effectively reduce water and energy use while maintaining an effective cleaning regime requires careful consideration of the whole CIP operation. For example, current brewery operations could simply benefit from optimisation of the CIP system schedule and cleaning detergent recipe.

The demand for greater flexibility in the brewery increases the requirement for cleanability. There appears to be a shift from operating vessels with a single function to adopting a more flexible approach, where the brewer can use equipment for a multitude of purposes to satisfy changes in market demand. For example, a contract brewer may require several yeast strains to produce different beers. Propagating and cropping several different yeast strains from a single on-site yeast propagation system is only practically achievable with hygienic process engineering and cleaning systems. As outlined in this review, a single-use CIP system would be most appropriate for this application. Expanding on the brewer's requirement to demand more from their current brewery equipment could see storage vessels and tanks re-purposed to hold scrap or waste yeast for example. Again, changing the operation of the brewery vessels without changing the equipment outright is only practical if it has been appropriately designed.

As mentioned in the introduction, the implementation and execution of the CIP used to be an afterthought to the design of a brewery. However, as hopefully highlighted here, the integration of the CIP system with the main brewery operations is fundamental to its effectiveness. The cost of contamination and equipment with respect to downtime and product loss in fermentation-based processes such as brewing industry is not always considered. However, with increasing raw material costs and utility costs, the brewer cannot afford to suffer from contamination issues that compromise their fermentations or beer, or a brewery that is unclean demands more steam.

10.6 Commentary/future trends

10.6.1 Future brewery designs and the impact on water and energy recovery

Future megabrewery designs will put additional technological pressure on cleanability. It would be expected that the larger breweries require larger diameter pipework to satisfy the greater volume capacity. Breweries featuring larger diameter pipework will require larger pumps to achieve the same flow velocities to obtain the necessary turbulent and scouring action during the CIP process. An impact of the larger pump size will be the energy required to achieve the necessary velocity of the cleaning fluid. Heat exchangers are already used to recover heat from heat intensive processes such as the wort kettle to pre-heat the CIP cleaning fluid. New technologies such as electrochemically activated water generated from the electrolysis of a saline solution could replace the requirement for the delivery of bulk caustic to site (CIP and Sanitation of Process Plant - SPX, n.d.). In the short term, the amount of water that would be required to clean these processes will put additional stress on freshwater aquifers. The future for CIP is therefore expected to feature benchmarking similar to the brewery benchmarking, which compares the number of hectolitres of water per hectolitre of beer. One question would be that is there still a requirement for water as part of the cleaning process, or could self-cleaning materials be the future?

10.6.2 Developments in nanotechnology to provide antimicrobial surfaces and materials

The application of silver nanoparticle technology as a future antimicrobial material is an interesting area of research. Silver nanoparticles have been shown to be prevent the development and establishment of biofilms (Palanisamy et al., 2014). This in principle would be an effective strategy to prevent the growth of undesirable microorganisms. The application of silver nanoparticle technology seems ideal for medical equipment; however, due to its non-selective mechanism and its detrimental effect towards yeast, it has less use in the brewery. The application of the silver nanoparticle technology in pipework would not be expected to directly replace a CIP system, owing to its role in removing both contaminating microorganisms and soil. Extensive trials would be expected to be undertaken in adding this technology to a brewery. Particularly as the size of the silver nanoparticles could be a potential health risk to both the operator and consumer, it could, however, find more suitable applications on discharge pipework. If self-cleaning materials add a high-cost premium to the materials of construction used for brewery vessels and pipework, it could be more practical to explore advances in software and sensors. An added benefit to the brewery is that these systems could improve over time through machine learning.

10.6.3 Developments in data-driven approaches to cleaning

CIP processes in breweries are resource-intensive, consuming significant amounts of water, energy, chemicals, and time. These processes tend to over-clean, as they are typically conducted for a pre-determined period chosen to eliminate cleaning failure. As a result, breweries can focus on improving the efficiency of their cleaning processes in the future. A 2022 Engineering and Physical Sciences Research Council (EPSRC) report highlighted the need for optimization of cleaning processes to enhance productivity, reduce environmental impact, and improve safety (Wilson et al., 2022). Specifically, the report identified the opportunity for digital technologies and the combination of data-driven approaches, quantitative modelling for simulation of cleaning processes, and sensing techniques.

Regarding data-driven approaches, Design of Experiment (DoE) techniques or Bayesian optimization can be used to minimize the number of trials required to optimize objectives such as energy consumption, wastewater, carbon footprint, or time whilst ensuring cleanliness (Bowler et al., 2023). Several parameters can be evaluated through data collection, such as temperature and pressure of the cleaning fluid, flow rate, cleaning time, and cleaning mechanisms (Brooks & Roy, 2022; Palabiyik et al., 2015; Piepiórka-Stepuk et al., 2016; Piepiórka-Stepuk et al., 2021). Data-driven models are developed to correlate cleaning process parameters with the objective functions, and analyzing the response of the models to varying inputs can be used to find optimal solutions (Fig. 10.9). Multi-objective optimisation produces a set of non-dominated solutions, or Pareto-optimal solutions, where no single objective can be improved without sacrificing another objective. Ultimately, the selection of a single non-dominated solution to implement depends on the priorities of the stakeholders involved in the decision-making process, which could include brewery managers, production engineers, environmental consultants, or regulatory authorities.

Methods to simulate CIP processes have modelled acidity, temperature, turbidity, conductivity, and sugar content during cleaning (Pettigrew et al., 2015) or used parameters such as the Reynolds number, the ratio of the density of the fouling to the density of the cleaning agent, and properties of the fouling material (Deponte et al., 2020). Simulations can be used to suggest optimal cleaning process parameters, evaluate proposed cleaning solution recipes, or predict fouling degrees of different product formulations.

Sensor measurements can monitor the variability in fouling instances for each CIP process conducted. Turbidity, conductivity, and pH sensors may be used to monitor residue in cleaning fluid downstream of the fouling and estimate the time until cleaning completion, with commercial technologies available for this purpose. Other research has focused on real-time monitoring of flow rate, pressure, conductivity, and temperature to ensure that the cleaning equipment is working as expected (Yang et al., 2018). However, to ensure that all fouling has been removed, direct sensing of the fouled area is

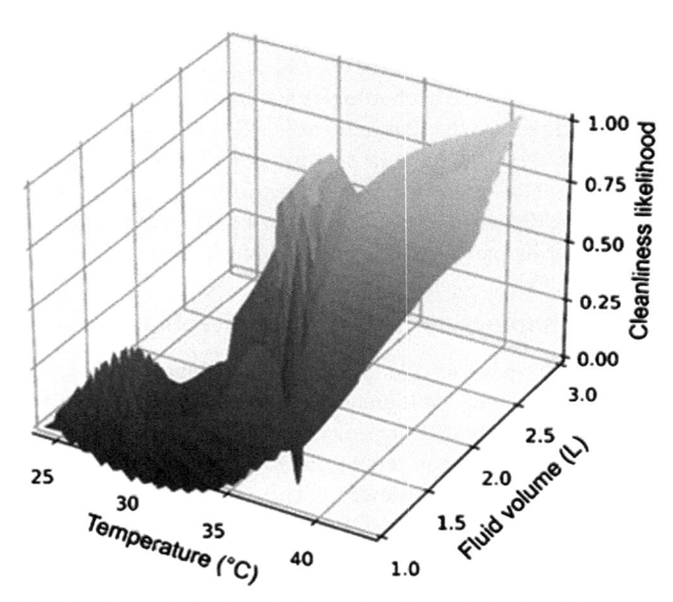

FIGURE 10.9 A response surface of the cleanliness obtained by the interactive effects of cleaning fluid temperature and volume. This can be used to optimise cleaning process parameters for multiple objectives such as energy consumption, wastewater, carbon footprint, or time whilst achieving the required level of cleanliness.

required. Low-power (intensities below 1 Wcm^2) and high-frequency (higher than 100 kHz) ultrasonic (US) sensors monitor the interaction of materials with mechanical sound waves. US sensors have previously been used to monitor fouling in heat exchangers (Úbeda et al., 2016), pipes (Escrig, Woolley et al., 2020), and duct sections (Chen et al., 2019). Other methods for in-line, direct monitoring of fouling removal consist of vibrational (Ikonen et al., 2023), electrical (Ren et al., 2019), and optical (Simeone et al., 2018) methods. These methods may be used to benchmark processes, trial cleaning parameters (e.g., cleaning mechanisms such as direct, pulsed, or intermittent flows), or to be permanent installations in breweries. Optical (visible and near-infrared), electrical, ultrasonic, and fluorescence methods can also be used for detection of biofilms (Achinas et al., 2020).

Sensors may produce a sigmoidal response during fouling removal by dissolution or a step-change response where fouling is removed by mechanical action (Escrig, Simeone et al., 2020). Furthermore, Machine Learning (ML) approaches can be used to monitor cleaning when multi-dimensional sensor responses are obtained (i.e., multiple data points received with every measurement, such as a full US waveform) or to monitor single-dimensional sensor readings over time (e.g., turbidity measurements) (Bowler et al., 2022). The supervised ML process involves collecting and labelling (i.e., assigning an output, such as whether the equipment was clean) sensor data, performing feature engineering and selection from the sensor measurements, selecting ML algorithms and hyperparameters, model training, validating and optimizing the models, and finally testing the models on unseen data for unbiased evaluation (Watson et al., 2021). Additional ML techniques can be considered, such as explainability methods to provide insights into the underlying mechanisms of the model, uncertainty quantification to enhance trust, or feature importance to enable operators to focus on relevant factors affecting the cleaning outcome (Molnar et al., 2020).

Data-driven approaches, such as ML, can be combined with first principles simulations to leverage the strengths of each. This is achieved by constraining the patterns identified in the empirical data to known mechanisms while incorporating the complexity not captured in physical laws (Bradley et al., 2022). This leads to more accurate and generalizable models. Furthermore, real-time sensing can automatically collect the data required to inform these hybrid models. The combination of data-driven, simulation, and sensing techniques could allow for continuous optimisation and control of CIP processes through the creation of digital twins. Ownership of the digital twin model should be considered among the stakeholders involved in its development and utilization.

The scale of operations in the brewery industry varies widely, ranging from financially constrained craft brewers to market-leading multinational companies. When considering the adoption and application of the discussed approaches, it may be simpler to validate and implement them on a smaller scale or in non-production environments. In the near term, smaller breweries or those with more flexible operations may be more able to implement these advancements. However, in the longer term, as the benefits become more evident and proven, larger breweries and industry leaders may begin to adopt these practices. The interplay between producers, Original Equipment Manufacturers (OEMs), and industrial and academic research and development plays a crucial role in driving advancements in cleaning processes. Collaboration and knowledge sharing among these stakeholders can facilitate the development of best practices, access to data, and improved equipment design. Progress will not only depend on technological advancements but also on the education and mindset of operators and plant owners. Embracing a culture of continuous improvement and a willingness to adopt new approaches will be essential for maximizing the benefits of evolving technologies.

10.7 Further information and advice

Further information around the guiding hygienic design standards and frameworks are included in the reference section and stated in the following for convenience: European Hygienic Engineering and Design Group (EHEDG) and The American Society of Mechanical Engineers (ASME) BPE 2009 Bioprocess. WRAPs Envirowise EN894 provides a useful overview to minimising water in CIP systems. ('Reducing water use through Cleaning-In-Place (CIP) envirowise - EN894', 2008). Chisti and Moo-Young (1994) also offer a comprehensive review of CIP in bioprocessing and fermentation-based systems.

References

Achinas, S., Yska, S. K., Charalampogiannis, N., Krooneman, J., & Euverink, G. J. W. (2020). A technological understanding of biofilm detection techniques: A review. *Materials, 13*, 3147. https://doi.org/10.3390/ma13143147

Ault, R. G. (1965). Spoilage bacteria in brewing—A review. *Journal of the Institute of Brewing, 71*, 376–391. https://doi.org/10.1002/j.2050-0416.1965.tb06362.x

Briggs, D., Boulton, C., Brookes, P., & Stevens, R. (2004). *Brewing: Science and Practice*. CRC Press. https://doi.org/10.1201/9780203024195

Bradley, W., Kim, J., Kilwein, Z., Blakely, L., Eydenberg, M., Jalvin, J., Laird, C., & Boukouvala, F. (2022). Perspectives on the integration between first-principles and data-driven modeling. *Computers & Chemical Engineering, 166*, Article 107898. https://doi.org/10.1016/j.compchemeng.2022.107898

Brooks, S., & Roy, R. (2022). Design and complexity evaluation of a self-cleaning heat exchanger. *International Journal of Heat and Mass Transfer, 191*, Article 122725. https://doi.org/10.1016/j.ijheatmasstransfer.2022.122725

BSSA - Passivation of stainless steels. (2014). http://www.bssa.org.uk/topics.php?article=68.

Bowler, A. L., Pound, M. P., & Watson, N. J. (2022). A review of ultrasonic sensing and machine learning methods to monitor industrial processes. *Ultrasonics, 124*, Article 106776. https://doi.org/10.1016/j.ultras.2022.106776

Bowler, A. L., Rodgers, S., Cook, D. J., & Watson, N. J. (2023). Bayesian and ultrasonic sensor aided multi-objective optimisation for sustainable clean-in-place processes. *Food and Bioproducts Processing, 141*, 23–35. https://doi.org/10.1016/j.fbp.2023.06.010

Chen, B., Callens, D., Campistron, P., Moulin, E., Debreyne, P., & Delaplace, G. (2019). Monitoring cleaning cycles of fouled ducts using ultrasonic coda wave interferometry (CWI). *Ultrasonics, 96*, 253–260. https://doi.org/10.1016/j.ultras.2018.12.011

Chisti, Y., & Moo-Young, M. (1994). Clean-in-place systems for industrial bioreactors: Design, validation and operation. *Journal of Industrial Microbiology, 13*, 201–207. https://doi.org/10.1007/BF01569748

Chuang, L. F., & Collins, E. B. (1968). Biosynthesis of diacetyl in bacteria and yeast. *Journal of Bacteriology, 95*, 2083–2089. https://doi.org/10.1128/jb.95.6.2083-2089.1968

CIP and Sanitation of Process Plant - SPX. (n.d.).

Costerton, J. W., Stewart, P. S., & Greenberg, E. P. (1999). Bacterial biofilms: A common cause of persistent infections. *Science, 284*, 1318–1322. https://doi.org/10.1126/science.284.5418.1318

Deponte, H., Tonda, A., Gottschalk, N., Bouvier, L., Delaplace, G., Augustin, W., & Scholl, S. (2020). Two complementary methods for the computational modeling of cleaning processes in food industry. *Computers & Chemical Engineering, 135*, Article 106733. https://doi.org/10.1016/j.compchemeng.2020.106733

Escrig, J., Woolley, E., Simeone, A., & Watson, N. J. (2020). Monitoring the cleaning of food fouling in pipes using ultrasonic measurements and machine learning. *Food Control, 116*, Article 107309. https://doi.org/10.1016/j.foodcont.2020.107309

Escrig, J. E., Simeone, A., Woolley, E., Rangappa, S., Rady, A., & Watson, N. J. (2020). Ultrasonic measurements and machine learning for monitoring the removal of surface fouling during clean-in-place processes. *Food and Bioproducts Processing, 123*, 1–13. https://doi.org/10.1016/j.fbp.2020.05.003

Hill, A. E. (2009). Microbiological stability of beer. In C. Bamforth (Ed.), *Beer: A quality perspective*. Academic Press.

Ikonen, E., Liukkonen, M., Hansen, A. H., Edelborg, M., Kjos, O., Selek, I., & Kettunen, A. (2023). Fouling monitoring in a circulating fluidized bed boiler using direct and indirect model-based analytics. *Fuel, 346*, Article 128341. https://doi.org/10.1016/j.fuel.2023.128341

Lekkas, C., Stewart, G. G., Hill, A., Taidi, B., & Hodgson, J. (2005). The importance of free amino nitrogen in wort and beer. *Technical Quarterly, 42*, 113.

Manzano, M., Iacumin, L., Vendrames, M., Cecchini, F., Comi, G., & Buiatti, S. (2011). Craft beer Microflora identification before and after a cleaning process. *Journal of the Institute of Brewing, 117*, 343–351. https://doi.org/10.1002/j.2050-0416.2011.tb00478.x

Meyers, V. E. (1959). Recent developments in automatic cleaning of storage tanks. *Journal of Dairy Science, 42*, 1730–1733. https://doi.org/10.3168/jds.S0022-0302(59)90794-5

Milledge, J. J. (2010). The cleanability of stainless steel used as a food contact surface: An updated short review. *Food Science and Technology Journal, 24*, 27–28.

Molnar, C., Casalicchio, G., & Bischl, B. (2020). Interpretable machine learning — A brief history, state-of-the-art and challenges. In *ECML PKDD 2020 Workshops* (pp. 417–431). Cham: Springer International Publishing. https://doi.org/10.1007/978-3-030-65965-3_28

Palabiyik, I., Yilmaz, M. T., Fryer, P. J., Robbins, P. T., & Toker, O. S. (2015). Minimising the environmental footprint of industrial-scaled cleaning processes by optimisation of a novel clean-in-place system protocol. *Journal of Cleaner Production, 108*, 1009–1018. https://doi.org/10.1016/j.jclepro.2015.07.114

Palanisamy, N. K., Ferina, N., Amirulhusni, A. N., Mohd-Zain, Z., Hussaini, J., Ping, L. J., et al. (2014). Antibiofilm properties of chemically synthesized silver nanoparticles found against Pseudomonas aeruginosa. *Journal of Nanobiotechnology, 12*, 2. https://doi.org/10.1186/1477-3155-12-2

Pettigrew, L., Blomenhofer, V., Hubert, S., Groß, F., & Delgado, A. (2015). Optimisation of water usage in a brewery clean-in-place system using reference nets. *Journal of Cleaner Production, 87*, 583–593. https://doi.org/10.1016/j.jclepro.2014.10.072

Piepiórka-Stepuk, J., Diakun, J., & Mierzejewska, S. (2016). Poly-optimization of cleaning conditions for pipe systems and plate heat exchangers contaminated with hot milk using the cleaning in place method. *Journal of Cleaner Production, 112*, 946–952. https://doi.org/10.1016/j.jclepro.2015.09.018

Piepiórka-Stepuk, J., Diakun, J., Sterczyńska, M., Kalak, T., & Jakubowski, M. (2021). Mathematical modeling and analysis of the interaction of parameters in the clean-in-place procedure during the pre-rinsing stage. *Journal of Cleaner Production, 297*, Article 126484. https://doi.org/10.1016/j.jclepro.2021.126484

Reducing water use through Cleaning-In-Place (CIP) envirowise - EN894, 2008.

Ren, Z., Trinh, L., Cooke, M., De Hert, S. C., Silvaluengo, J., Ashley, J., Tothill, I. E., & Rodgers, T. L. (2019). Development of a novel linear ERT sensor to measure surface deposits. *IEEE Transactions on Instrumentation and Measurement, 68*, 754–761. https://doi.org/10.1109/TIM.2018.2853380

Sakamoto, K., & Konings, W. N. (2003). Beer spoilage bacteria and hop resistance. *International Journal of Food Microbiology, 89*, 105–124. https://doi.org/10.1016/S0168-1605(03)00153-3

Simeone, A., Deng, B., Watson, N., & Woolley, E. (2018). Enhanced clean-in-place monitoring using ultraviolet induced fluorescence and neural networks. *Sensors, 18*, 3742. https://doi.org/10.3390/s18113742

Sreekumari, K. R., Ozawa, M., & Kikuchi, Y. (2000). Effect of surface condition on attachment of bacteria to stainless steel welds (materials, metallurgy & weldability). *Transactions of JWRI, 29*, 45–51.

Sutherland, I. W. (2001). Biofilm exopolysaccharides: A strong and sticky framework. *Microbiology, 147*, 3–9.

Suzuki, K., Funahashi, W., Koyanagi, M., & Yamashita, H. (2004). Lactobacillus paracollinoides sp. nov., isolated from brewery environments. *International Journal of Systematic and Evolutionary Microbiology, 54*, 115–117. https://doi.org/10.1099/ijs.0.02722-0

Úbeda, M. A., Hussein, W. B., Hussein, M. A., Hinrichs, J., & Becker, T. M. (2016). Acoustic sensing and signal processing techniques for monitoring milk fouling cleaning operations. *Engineering in Life Sciences, 16*, 67–77. https://doi.org/10.1002/elsc.201400235

Watson, N. J., Bowler, A. L., Rady, A., Fisher, O. J., Simeone, A., Escrig, J., Woolley, E., & Adedeji, A. A. (2021). Intelligent sensors for sustainable food and drink manufacturing. *Frontiers in Sustainable Food Systems, 5*, Article 642786. https://doi.org/10.3389/fsufs.2021.642786

Wilson, D., Landel, J., Christie, G., Fryer, P., Hall, I., & Whitehead, K. (2022). *A roadmap for quantitative modelling of cleaning and decontamination.* https://doi.org/10.17863/CAM.83082

Yang, J., Jensen, B. B. B., Nordkvist, M., Rasmussen, P., Pedersen, B., Kokholm, A., Jensen, L., Gernaey, K. V., & Krühne, U. (2018). Anomaly analysis in cleaning-in-place operations of an industrial brewery fermenter. *Industrial and Engineering Chemistry Research, 57*, 12871–12883. https://doi.org/10.1021/acs.iecr.8b02417

Chapter 11

Reducing microbial spoilage of beer using filtration

Gary J. Freeman
baftec, Surrey, United Kingdom

11.1 Introduction

Filtration of beer is a challenging operation. Prefilter beer contains a significant concentration of suspended particles. Most often, the volumetric bulk of these particles comprises yeast cells, many of which are joined together by the natural flocculation process. These particles, therefore, are at least several microns in size. This makes their removal relatively simple, although the volumetric bulk will add to filtration costs. Indeed, a single filtration stage can result in effectively zero yeast cells in the filtered product. This is significant because in most beer as served to the consumer, the presence of brewing yeast must be regarded as contamination. However, also present in prefilter beer are much smaller particles, many of which are as small as less than a micron. These particles mostly comprise protein—polyphenol and are known as chill haze. Beer will not be visually clear unless we remove most of these particles to as small a size as below one micron. This greatly increases the difficulty of beer filtration and limits the technologies that are suitable. However, this also means that all beer filtration technologies will reduce the count of any bacterial contamination.

11.2 Filtration technologies in brewing

Filtration processes may be classified as either depth filtration or surface filtration. Depth filtration relies on a layer of porous media in which suspended particles in the beer are trapped within the media. Examples in brewing include filter aid filtration, sheet filtration and some forms of filter cartridge. Surface filtration normally refers to membrane technology. A thin layer of membrane has pores throughout the structure. This means that it is possible to achieve very exact filtration, perhaps enabling sterilisation, but typically the quantity of suspended beer particles that may be removed is less than for depth filtration.

Fine filtration processes imply increased energy usage, most obviously increased pressure, and likely a reduced capacity for suspended beer particles. This means that there is scope to perform the filtration with more than one technology in series. As the beer progresses through the series, each filtration step becomes progressively finer. For example, a filter aid filter may be followed by a finer depth filter such as a sheet filter. A membrane filter, most commonly in the form of a filter cartridge, would normally be at least the third in a series. A filter cartridge enables the possibility of guaranteed sterile filtration. A filter sheet of suitable grade, although a depth filter, may enable effective sterile filtration to the satisfaction of the brewer.

There is a need to design a sequence of filtration operations to maximise the throughput. For example, a relatively coarse initial stage will increase the duty on the second stage and therefore may not increase the total throughput. There is scope to perform this optimisation on the pilot or laboratory scale (Freeman, 1996).

The process sequence may be designed to enable sterile filtration. Other technologies to achieve microbiological stability in product include pasteurisation or maintenance of a yeast culture in product to prevent the growth of damaging contaminants. The relative advantages of sterile filtration and pasteurisation pertain to both process costs and product quality. Pasteurisers, either in-line (flash) or for small package (tunnel), are expensive capital items compared to a sterile

Brewing Microbiology. https://doi.org/10.1016/B978-0-323-99606-8.00009-2

filtration unit. However, the ongoing need to replace the sterile filtration media means that, typically, the operating costs for sterile filtration are higher. It may be, therefore, that sterile filtration is more viable for a small brewery and pasteurisation for a large brewery. The relative merits for product quality are a matter of some dispute (White, 2008). Thermal treatment of beer accelerates chemical reactions and therefore reduces flavour stability and produces 'cooked' off-flavours, although good operations in a modern brewery, resulting in reduced microbial contamination, have reduced the extent of thermal treatment required. Sterile filtration of course does not include thermal treatment at all. However, all filtration processes, especially fine filtration processes, result in some removal of positive beer characteristics such as colour, foam stability bitterness and 'mouthfeel'.

11.3 Filter aid filtration

The majority of the volume of beer in the world is processed by filter aid filtration. This technology relies on adding powders to the beer that form a very porous bed when they impinge on the filter surface. This is achieved by slurrying the filter aid in water, deaeration and pumping the slurry into the beer. Thus, the prefilter beer on reaching the filter surface encounters a fresh layer of filter aid as the bed develops. This prolongs the filtration run.

The most common filter aid employed is kieselguhr (diatomaceous earth). This comprises the fossils or skeletons of fresh or saltwater algae known as diatoms (Fig. 11.1). The highly porous nature of the particles enables effective liquid flow but also entrapment of particles. Other types of filter aid that are commonly used include perlite (volcanic glass) and cellulose fibres. However, the porous internal structure of the diatoms makes them more effective than the alternatives.

Filter vessel technologies include plate and frame, leaf and candle. The latter two are preferred today because they are easier and faster to clean and restart than the plate and frame. Candle filters are simpler constructions, but leaf filters are probably more flexible (Hermia & Brocheton, 1994).

The filtration operation is preceded by precoating of the filter. This comprises recirculation of deaerated water around the vessel while adding filter aid slurry. Thus, an initial bed of filter aid exists at the start of beer filtration. This means that the brewer has options to optimise the process other than simply the selection of the main filtration grade. For example, if the precoat is a finer (smaller) grade than the bodyfeed (admixed to the beer) grade, then we have a genuine two-stage process. Beer clarity would be improved at the expense of some run time.

An optimised filter aid filtration process is capable of removing the vast majority of particles down to as small as half a micron. This means that contaminating bacteria that typically have a minimum dimension of half a micron are removed in significant quantities by a filter aid filter. Indeed, it is likely that a well-operated operation will reduce bacteria by a factor of 1000 (log reduction value of three). This has several beneficial implications. It reduces the requirement for microbiological stabilisation. For example, the brewer can pasteurise with less intensity. If sterile filtration is employed, excellent beer clarity post filter aid filter increases the run time of the sterile filtration process. To this end, there is generally an advantage to using kieselguhr as opposed to other filter aids because of the superior beer clarity obtained.

FIGURE 11.1 An electronmicrograph of a kieselguhr filter aid.

11.4 Crossflow microfiltration

Filter aid filtration was in many ways the only viable option for bulk filtration of large volumes of beer for many years. However, there are significant problems with the technology. In particular, the material in kieselguhr (the most efficient filter aid) comprises crystalline silica (cristabolite). This is carcinogenic if inhaled, causing the disease known as silicosis. Alternative filter aids such as perlite are not crystalline silica but as mentioned are not as effective. Thus, it may be a hazard to brewery personnel if packaging fails or during transfer from the package to the slurry tank. Also, disposal of the spent filter aid cake usually has landfilling as the only viable option. Depending on location, landfilling is becoming increasingly expensive and restricted.

Another significant issue pertains to beer quality. Although the kieselguhr manufacturing process includes a calcination (furnacing) step, which is designed to remove metal ions other than silicon, these are still present in the kieselguhr. Transition metal ions such as iron, copper and manganese instigate oxidative damage to the beer that causes turbidity and stale flavours. Kieselguhr filtration commonly results in a doubling of the concentration of these ions. Thus, especially in these times of increasing corporate social responsibility, there is real incentive to use technologies that do not employ filter aids.

At the latter end of the twentieth century, a technology emerged that was competitive in terms of cost with filter aid filtration. Crossflow microfiltration, alternatively known as tangential flow microfiltration, utilises a membrane filtration process in a single stage. Deposition of particles as a 'cake' on the membrane surface would normally cause a membrane to foul very quickly and make the process impractical. However, this effect is minimised by causing the prefilter beer to flow across the membrane surface. This causes the 'cake' to be re-suspended back into the prefilter beer and helps to maintain a satisfactory flow of filtered beer. The downside is that the operation requires a lot of pumping energy and, as a consequence, requires a lot of refrigeration energy also.

The membrane format currently employed in brewing is a tube, with the beer flowing longitudinally down the inside so that filtrate passes through the tube wall to the tube set housing. The membrane composition may be either polymeric or ceramic. Polymeric membranes underwent a step-change improvement with the development of the ability to manufacture membranes in polyether sulphone (PES). This material demonstrably improved the flow rate performance of the membranes because PES is less inclined to adsorb beer components such as colour and proteins. Ceramic membranes currently achieve slower rates of filtration per unit area of membrane than polymerics. However, they have a key advantage in that if operated responsibly, they are extremely long-lasting, perhaps for well over 10 years. Whereas polymeric membranes have to be replaced more frequently, currently every 1 to 2 years (a significant operating cost) typically, the ceramics are more rugged. It is unclear as to which type will evolve to become more attractive in the future. At the moment, the technology mostly used in bulk beer filtration is polymeric membranes. However, it is clear that membrane efficiency and reduced effective cost will continue to improve for both ceramic and polymeric membranes. Compare this with filter aid filtration where costs will increase, in particular for disposal of the used filter aid.

The polymeric membranes are installed, several hundred at a time, in housings of suitable hygienic material. Fig. 11.2 shows the top of an example of such housings. The pipe and valve arrangements mean that flow through the individual

FIGURE 11.2 The upper part and pipe and valve arrangements of a polymeric crossflow filtration plant.

housings can be separated from the other housings. Thus, it is possible to clean one individual housing while continuing to filter beer through the others. Therefore, crossflow filtration plants are compatible with continuous processing.

The pore-size ratings of the crossflow membranes are most commonly in the range of 0.4–0.8 microns. It should be noted that a 0.45 micron membrane is regarded as being capable of removing all beer spoilage organisms. Thus, at first sight, the brewer has the potential to perform sterile filtration in a single stage. However, there are problems with this approach. Sterile filtration necessitates integrity testing (see later) of filter modules before the run. Although this is feasible, the complexity of the highly modularised (Fig. 11.2) and large filtration plant makes it an engineering challenge. Also, many brewers have further processes between the filtration and packaging lines, for example, stabilisation processes comprising adsorbents such as polyvinyl polypyrollidone or agarose gels. However, there are options for stabilisation upstream or even employment of the additions in the prefilter beer that is recirculating around the crossflow plant. Furthermore, conventional brewing employs a process step of holding the beer immediately after filtration in a 'bright beer tank' before microbiological stabilisation and packaging. This step is mainly to perform the last quality control checks on the product, enabling adjustments or blending as required and is not normally maintained to a high level of sterility. Therefore, improvements to product consistency and process hygiene would need to be achieved to employ crossflow microfiltration as the sterilising process. In the author's opinion these difficulties could be overcome in the future with potential advantages in process simplicity and costs.

11.5 Sterile filtration

11.5.1 Cartridge filtration

At the time of writing, the most important technique to achieve sterile filtration in a brewery is through the use of cartridge filters. These comprise a membrane or depth filter that is a relatively thin layer and is pleated within the support structure to provide a high filtration area in a compact volume (Fig. 11.3).

The unit needs to be readily cleanable and disinfectable. Ideally the unit may also be backflushed. This means that cleaning fluids or rinses can be flowing in the reverse direction to normal filtration. This enhances the removal of filtered particles from the cartridge. The lifetime of the cartridge is affected by both the filtration duty that is placed upon it and the consequent number of aggressive cleans. The cartridges are relatively expensive, and if the lifetime is short, then costs become significant.

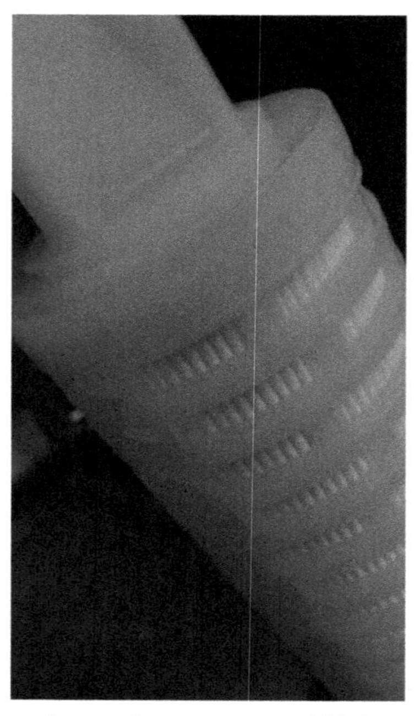

FIGURE 11.3 A cartridge filter showing the pleated membrane and support structure. This is contained in a stainless steel housing and filtration is from outside to inside this unit.

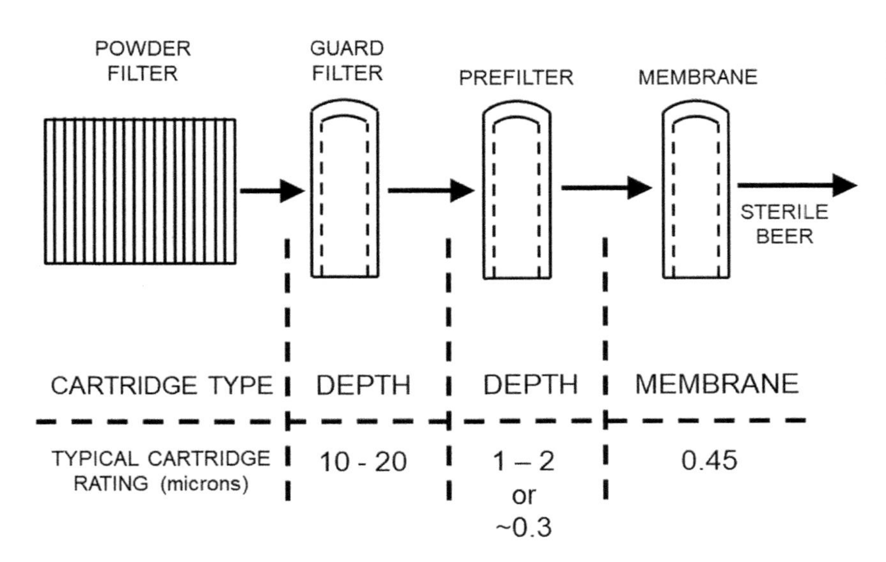

FIGURE 11.4 A typical sterile filtration process utilising cartridge filtration.

As described earlier, to manage the filtration of turbid beer from maturation and conditioning to sterile product requires several stages. Typically, these stages will resemble something akin to that described in Fig. 11.4. The main beer filter in this process is a powder (filter aid) filter. This filter will remove the vast majority of the volume of suspended material, but some very fine particles including some bacteria may remain. Most often the filter aid filter is followed by a relatively coarse guard or trap filter cartridge. The purpose of this is to catch any filter aid that leaks through the main filter. The next filter in the sequence is a cartridge filter that operates as a depth filter, so that the active layer in the cartridge is a fibrous mass, perhaps of polypropylene. The rating as shown may be in the range of one to two microns, although more recently many brewers will use a smaller rating, perhaps below half a micron, to ensure that the relatively expensive final filter performs negligible filtration duty. For a depth filter that does not exhibit an absolute cutoff size, this is known as a nominal cutoff or nominal pore size. It represents the size above which effectively all particles are removed. However, it is a feature of depth filters that they will also remove a lot of particles below the nominal cutoff. Hence, the beer that leaves this cartridge will be of 'sparkling' clarity with few particles that will require removal by the final membrane (surface filter). This is essential because the particle removal capacity of the surface filter is much less than that of the depth filter. The final 0.45 micron membrane cartridge acts only as a guarantee of total removal of bacteria (effectively sterility for the brewer) prior to packaging.

To maximise the efficiency of the process, it is necessary for the operators to monitor the whole system performance. Maximising value is largely an issue of maximising the run times through the cartridges. There is a risk that different beer products will affect the filtration sequence in different ways. For example, if a batch contains a great deal of very small particles, it may increase the loading on the final two cartridges, especially damaging if it is the final membrane, and the run time may become very suboptimal. The most obvious monitoring, which is often overlooked, is to have pressure gauges or transmitters on either side of each of the filtration steps. It is then simple for the operator to identify the step that is taking more of the filtration load than is optimal for that step. Actions that may optimise the process, depending on the pressure drop characteristics that are achieved, include.

- changing the filter aid specifications in the filter aid filter,
- preventative maintenance on the filter aid filter to prevent leakage of filter aid and blinding of the trap filter,
- changing the rating of the cartridge(s) that protect the final membrane, and
- simply adding or removing filtration area from the step as appropriate.

11.5.2 Integrity testing

One requirement of sterile filtration is that the final membrane filter must be integrity tested. This refers to ensuring that the membrane pore structure is still intact and will remove the microorganisms as required. In the food and beverage industry sectors where pathogens are an issue, it is likely that integrity testing is a legal requirement, as required by the Food and Drug Administration (FDA) in the United States. However, in brewing, there are no pathogens in conventional beer, and the concern is about spoilage. Nevertheless, the threat is still significant enough to make integrity testing an important business requirement. Integrity testing of a membrane in process involves wetting the membrane and then draining it. The housing is full of air or gas, but the microscopic pores of the membrane are still full of water. The most common integrity test is known as bubble point and involves increasing the gas pressure upstream of the membrane until the pressure pushes the water out of the pores and gas flow occurs. The pressure difference must exceed a certain value, or the integrity of the membrane has been compromised. Similar tests involve measuring the small flow rate of gas at a small pressure drop caused by diffusion, known as the diffusional flow technique, or a similar test that measures pressure difference decay, known as the pressure decay test. These tests can be performed manually although many brewers employ automated systems that are compatible with modern brewery automation.

11.5.3 Other sterile filters

Sterile filtration can potentially be achieved by fine depth filtration. As mentioned previously, sheet filters have been used to filter beer to a satisfactory microbiological stability. Sheet filters are essentially compressed pads of (normally) cellulose fibres that are arranged into a plate and frame filter press. Performance is sometimes enhanced by incorporating kieselguhr into the structure. Colloidal stabilisation may also be achieved by the presence of polyvinyl polypyrollidone. Process efficiency can be improved by making use of the fact that the plate and frame pack may be arranged so that the beer passes through relatively coarse grades of sheet before a finer grade. Such a two-stage process enables the finer of the two grades to be effectively a sterilising grade. Sheet filters operate less effectively if a maximum flow rating is exceeded. Also, best performance is gained by minimising interruptions to the flow of beer that cause 'dislodging' of previously trapped particles.

11.5.4 Downstream process

One issue that arises with sterile filtration is the requirement to fill into package in a sterile manner. Sterile filling of large containers (kegs) has been performed for many years. In this case, the high beer flow rates in the filling machinery greatly reduce, but do not eliminate, risks. Filling small pack such as plastic or glass bottles and cans requires many more filling heads and a greater opportunity for contaminants to get into product if operation is poor. Sterile filling lines rely on techniques such as positive air pressure in the environment around the sensitive areas to prevent airborne bacteria approaching and also tightly controlled hygienic practices by the operators. It should be noticed, however, that there is a distinct trend away from tunnel pasteurisation of small pack products. Tunnel pasteurisers are much more energy expensive and also water expensive than in-line flash pasteurisers. The latter of course will also require sterile filling.

11.6 Improving filtration performance

As discussed above, sterile filtration is capable of producing beer of high quality. However, because the process is multistage, the costs become significant. Some opportunities for optimisation have already been outlined. However, ultimately, the limiting factor for performance of a filtration step is the characteristics of the beer requiring filtration. Therefore, there is scope for the introduction of technologies upstream of the filtration steps that improve the 'filterability' of the beer.

11.6.1 Centrifugation

Disc stack centrifuges are powerful solid–liquid separators. In years past, they were often problematic. They often gave very undesirable effects such as warming the beer up and drawing in oxygen. Today, however, superior engineering design features such as hermetic sealing have largely eliminated these problems. In addition, disc stack centrifuges are available that are more powerful than previous versions. Some may remove some of the colloidal particles as small as one micron. In the context of filtration, it may be viable to use the centrifuge upstream of the bulk beer filter. In the case of a filter aid filter, this will allow increased run time because of the reduced solids loading. However, an often-overlooked benefit is that the brewer could employ finer grades of filter aid. This improves the filtered beer clarity and in the case of a sterile filtration sequence will reduce the loading onto the subsequent stages, prolonging run time.

11.6.2 Flocculants (finings)

There are several types of flocculants employed in brewing. One example is isinglass, which is a suspension of collagen, derived from the swim bladders of fish, in weak mineral acids. The collagen macromolecules unusually form a net positive charge in the mildly acidic beer. Thus, they can interact electrostatically with the suspended beer particles that almost exclusively exhibit a net negative charge. This results in coagulation and flocculation processes that cause the particles to group into large flocs, making sedimentation and removal more simple and rapid. Indeed, this process can be so effective that isinglass can be used to produce acceptable beer clarity on its own. An example is traditional UK cask ale, which is a clear product that is not filtered.

However, the real opportunity for the employment of flocculants such as isinglass in the context of sterile filtration is that flocculants are very effective at the removal of small, colloidal particles. If compared with a centrifuge, which has a mode of operation that makes it more effective at removing relatively large particles, flocculants can be seen to be effective at all particle sizes. Hence, it is clear that application of flocculants will reduce the loading of colloidal material that will in particular curtail the run length of the sterilising filters (such as the membrane) at the end of the process. Isinglass may be employed in the cold storage stage that precedes beer filtration. Improved colloidal (clarity) stability of the final product in package is a benefit as well as improved performance of the filters.

Other flocculants are often employed in the brewhouse. Copper finings are employed in the wort boiling stage. They comprise carrageenan derived from certain seaweeds. Their main mode of operation is that they very significantly increase the precipitation of protein–polyphenol material when the wort is subsequently cooled, ready for fermentation. Removal of this material greatly reduces the potential for 'chill haze' formation in beer processing and package. Hence, copper finings have a much greater effect on both colloidal stability and beer filter performance than is often realised. An alternative to copper finings is silica sol, which is an aqueous suspension of colloidal silica. There are similar stabilising and precipitating effects, although usually silica sol is only preferred in beers that aim to obey Reinheitsgebot, the German Beer Purity Law.

11.6.3 Enzyme treatments

Exogenous enzymes are those that are added by the brewer rather than those that occur naturally from the malted cereals and yeast. Enzymes are available that exhibit proteolytic, cytolytic (plant cell wall material) and carbohydrase activity. Cost-effective enzyme preparations are not in general pure enzymes; however, a blend is often an advantage, enabling more than one activity to occur.

Enzymes may be employed throughout the process from brewhouse through to cold storage to eliminate problematic components or in some cases significantly change beer flavour and quality. A common example is beta-glucanase. This originates from the cell walls of the raw materials most obviously from barley malt. Its substrate beta-glucan is capable of causing problems throughout the process. In particular, late in the process, it is likely to precipitate from the beer as the wort sugars, which stabilise it in solution, are reduced by fermentation. Worse still in some cases, it will form very high-molecular-weight gels that will seriously impede the beer filters. Application of beta-glucanase at a judicious point in the process may eliminate the detrimental effects. Similar applications may be found on occasions for proteases, pentosanases (xylanases) and other carbohydrases such as amylases.

Membrane filters, including both final sterilising filters and crossflow filters, are often found to be difficult to clean to a suitable standard for effective processing. In some cases, it is viable to use specialised enzymatic cleaners even though they may be relatively expensive compared to 'non-biological' detergents. It should be noted that different beer compositions may call for different enzyme activities, depending upon what is fouling the membrane (Taylor et al., 2001). Several suppliers of membrane technologies also supply their own propriety enzymatic cleaners.

11.7 Future trends

The pressure to eliminate kieselguhr from the brewing process on the grounds of health and safety and environmental friendliness will continue to grow. The latter will directly increase the costs of usage. Zuber (2009) describes an example of some developments in man-made, regenerable filter aids, thus reducing landfill disposal requirements. Often these may be readily retrofitted to existing plants. However, in the medium-long term, it seems likely that crossflow microfiltration will become the bulk filtration process of choice. This is because membranes are likely to come down in cost (in real terms) and improve in efficiency. At this time, there is a need for membrane suppliers to increase the ruggedness of, in particular polymeric, membrane modules. This will reduce repair or replacement costs and also reduce the risk to potential purchasers of crossflow plants.

At the time of this writing, it is clear that there has been a divergence in the brewing industry. On the one hand, there are multinational brewing companies with large, global brands. On the other hand, there are smaller brewing companies that produce so-called 'craft' beers. As described earlier, the economics of sterile filtration over pasteurisation are favourable for smaller brewers. Sterile filtration may give them improved options to produce beer in conventional bottles and cans on their own site. It could be that sterile filtered beer brands will become more common in the future. Companies that supply and maintain mobile canning and bottling lines that service multiple brewing sites enable small brewers to avoid the significant capital expenditure on such items.

11.8 Sources of further information and advice

An excellent practical guide to filtration processing is available from the European Brewery Convention (EBC manual of good practice, 1999). Further learning materials are available from the Institute of Brewing and Distilling (IBD). The IBD is a UK-based international professional body for personal development and learning. It has a free-to-access search facility for material that is then free to IBD members but can be purchased by nonmembers (www.ibd.org.uk).

Another body that readers may be interested in is the Filtration Society (a mixture of academics, commercial suppliers and users) at www.filtsoc.org.

Campden BRI operates a commercial database on all aspects of brewing and can be contacted for information support at www.campdenbri.co.uk.

References

EBC manual of good practice. (1999). In *Beer filtration, stabilisation and sterilisation*. Brussels: European Brewery Convention and Fachverlag Hans Carl.

Freeman, G. (1996). Liquid and solid separation—in beer. *Brewers' Guardian, 125*(12), 27–32.

Hermia, J., & Brocheton, S. (1994). Comparison of modern beer filters. *Filtration & Separation, 31*(7), 720–725.

Taylor, M., Faraday, D. B. F., O'Shaughnessy, C. L., Underwood, B. O., & Reed, R. J. R. (2001). Quantitative determination of fouling layer composition in the microfiltration of beer. *Separation and Purification Technology, 22/23*(1/3), 133–142 [Special issue].

White, R. (2008). Some like it hot. *Brewers' Guardian, 137*(3), 30–34.

Zuber, J. P. (2009). Industrial results of precoat filtration on a candle filter with regenerable filter aid (abstract published online). In *Proceedings of the European brewing convention, hamburg* (abstract published online).

Chapter 12

Reducing microbial spoilage of beer using pasteurisation

Edward Wray
Hepworth and Co. Ltd, West Sussex, United Kingdom

12.1 Introduction

Pasteurisation has long been used for the preservation of beer, following on from the work of Louis Pasteur. It is a process of applying heat to preserve food and drinks. It is effective in stabilising beer with regard to microbial contaminants and is in widespread use across the brewing industry. A number of concepts have been developed to measure and analyse the degree of pasteurisation, which can show wide variation depending on time, temperature, drink composition and organisms present. These concepts include the D value, the z value and pasteurisation units (PUs).

The composition of beer makes it an inherently stable product and factors such as alcohol content, low pH, low nutrients and anaerobic conditions mean only a low degree of pasteurisation needs to be applied to achieve microbial stability, though differences in beer composition mean these factors vary across brands. Different microorganisms also have different degrees of heat resistance, so selecting the level of pasteurisation to apply to beer is not always a straightforward matter.

Two main methods are used to pasteurise beer: Tunnel pasteurisation, where bottles or cans are passed through a series of water jets applying heat and flash pasteurisation, where the beer is heated rapidly in a plate heat exchanger and holding tube before packaging.

The application of heat can affect the flavour of beer in a number of ways, particularly if oxygen levels are high. Nevertheless, with good practice and attention to quality control, a high-quality product can be produced with minimal flavour changes and a high degree of microbial stability.

12.2 History

The term 'pasteurisation' takes its name from the great French scientist Louis Pasteur. Prior to his work, heat preservation of some foods and drinks was already employed and had been for some time. However, this was carried out on an empirical basis, and it was Pasteur who was able to elucidate the scientific reasoning behind how it works.

In 1865, he patented a heat treatment for wine preservation, and in 1866, he published his *Studies on Wine* where he stated that heating to as low as 50°C could preserve wine by killing the microorganisms that caused spoilage. The potential for preserving beer in a similar way was of immediate interest to brewers, though it was not until after the Franco-Prussian war of 1870—71 that Pasteur himself turned his attention to beer. He hoped to gain revenge against the German victors by improving beer production in other countries to such an extent that it would undermine German beer exports. Working with other scientists and prominent French brewers, he developed means of producing beer with much less risk of infection, which were published as *Studies on* Beer (1876) and included details on pasteurising beer, though not without reservations.

A number of large breweries rapidly adopted pasteurisation of their bottled beers, and various methods were developed for carrying it out in cabinets using steam and hot water. These had high energy use so ways of recovering heat were developed. Methods of moving crates through zones of water at different temperatures were developed in the early twentieth century, as were using water sprays (European Brewery Convention, 1995).

Brewing Microbiology. https://doi.org/10.1016/B978-0-323-99606-8.00017-1

The development of walking beam technology and improvements in mechanisation improved the technology further and led to modern tunnel pasteurisers. Walking beams are stainless steel strips that run the length of the pasteuriser. Alternate beams can be lifted and moved forward a short distance before lowering and returning, having moved the can or bottle slowly forward in the process (Wilson, 1981).

The introduction of keg beer leads to the next main method used in beer pasteurisation as in-package pasteurisation of such large containers is impractical. By passing beer though a plate heat exchanger, it can be rapidly heated and held in a holding tube at the required temperature before cooling for packaging (flash pasteurisation). As the beer is cooled before packaging, scrupulous hygiene must be maintained to ensure the pasteurised beer does not become re-infected. Automation and control improvements from the 1970s onwards improved this process, and it is now also used for bottles and cans.

12.3 Principles of pasteurisation

Pasteurisation is a means of achieving microbial stability in food and drinks by applying enough heat to destroy organisms capable of growing during the subsequent storage period. It works in conjunction with other parameters of the product to ensure microbial stability. The other parameters which contribute to beer stability include low pH, alcohol content and the presence of hop compounds. These will be discussed further in the section on hurdle technology (Section 12.5.1).

Pasteurisation is not full sterilisation; it is what has been termed 'practical sterility' (European Brewery Convention, 1995). Microorganisms that are able to grow in beer need to be killed in order to achieve stability, but the heat treatment does not have to be to such a level to kill heat-resistant spore forming bacteria, as they are unable to grow in beer. Applying the minimal amount of heat necessary to achieve stability ensures that both the effect on flavour and energy use is minimised. Pasteurisation will not, however, solve problems in beer quality that poor hygiene can cause before pasteurisation is carried out!

The temperature applied during pasteurisation and the length of time it is applied for are of critical importance in pasteurisation, as is which microorganisms are present in the beer and the quantity in which they are found.

12.4 D value, z value, *P* value, process time, pasteurisation units and L value

A number of concepts have been developed to explain and quantify the effects of pasteurisation on a product.

The D value is the time required at a set temperature for a decimal reduction (i.e., one log or 90%) in the population numbers of a known organism. The size of the D value depends on the temperature, the microorganism and the other parameters in the beer that affect microbial grown (see Section 12.5.1 on hurdle technology). A higher temperature or lower pH will lead to a lower D value. A D value can be expressed in minutes or seconds.

The z value is the change in temperature required to bring about a 10-fold change (i.e., one log) in the D value. The z value is expressed in degrees. In the brewing industry, a z value of 6.94°C (often rounded to 7°C) is generally used after the work of Del Vecchio, Dayharsh, and Baselt (1951).

The *P* value is the time required to achieve a stated reduction in numbers of a microbial population at a given temperature and z value. It is a function of the D value and the z value to give a total pasteurisation value. A *P* value must therefore have the temperature and the z value specified for it to have meaning. A *P* value expressing the pasteurisation value at 60°C for a z value of 6°C is written as $P^{6}{}_{60}$. *P* value is used to give the time for a specified log reduction in organism numbers. For example, when the D value is 3 min at 60°C, with a z value of 6°C, to achieve a 6 log reduction will require 18 min, and the *P* value will be written as = 18.

In the brewing industry, things have been simplified somewhat, and the PU is routinely used. The PU is based on the temperature of 60°C, and a z value of 6.94°C is used. One PU = 1 min at 60°C.

As the effect of pasteurisation is highly temperature dependent and process temperatures will not be fixed, the Lethality rate at a given temperature (L_T) can be related to the D value at 60°C (as in PU) using the equation:

$$L_T = D_{60}/D_T \tag{12.1}$$

The lethality rate is usually expressed as the time in minutes which will give one PU (i.e., is equivalent to the 1 min at 60°C).

To calculate the PU of 1 min at any given temperature (*T*), the following equation can be used:

$$PU = 1.393^{(T-60)} \tag{12.2}$$

Total PU can be obtained by multiplying the result by the number of minutes at that temperature.

TABLE 12.1 The effects of different temperatures on pasteurisation units.

Temperature °C	PU/minute
60.0	1.0
62.1	2
67.0	10
70.0	28
73.9	100

As a useful approximation, a temperature increase of 2°C doubles the PU and an increase of 7°C increases the PU 10-fold. Lethality tables can be produced showing the PUs for a range of given times and temperatures and can be used to calculate the total effect of a pasteurisation process. Table 12.1 shows the effects of different temperatures on PU at certain key temperatures.

A more detailed look at how PUs alter with temperature is shown in Table 12.2.

12.5 Spoilage hurdles

Pasteurisation is only effective in conjunction with other factors that make a product inhospitable to contaminating organisms. The numerous hurdles present in beer have a synergistic effect and mean that only mild pasteurisation is required for practical sterility and the effects of heat treatment on flavour are minimised.

Mashing and boiling during wort production produce the required degree of sterility prior to yeast pitching, and the fermented beer has a number of hurdles (Vriesekoop, Krahl, Hucker, & Menz, 2012, Table 12.3).

12.5.1 Ethanol

The ethanol content of beer (generally between 3.5% and 5% by volume) is a large hurdle to microbial growth. Ethanol inhibits cell membrane functions and induces cell membrane leakage. Increased cell membrane permeability increases the effects of low pH on the cell by increasing proton passage into the cytoplasm and reducing the ability of the cell to maintain pH homoeostasis.

12.5.2 Low pH

The low pH of beer (generally 3.7−4.1) inhibits the growth of many microorganisms. At low pH entry of organic acids into the cell is enhanced, causing intracellular acidification. This leads to the destruction of enzyme systems and the reduction in nutrient uptake.

The microorganism will attempt to maintain pH homoeostasis by using energy to pump cations across the cell membrane. When the ability of the cell to do this is overwhelmed starvation and cell death follow.

The low pH of beer also has a synergistic effect with the antimicrobial properties of hop resins. Low pH is an important reason why beer is generally considered to be unable to support pathogen growth. For example, *Clostridia* are unable to grow below pH 4.5 and *Salmonella* are unable to grow below pH 4.

12.5.3 Hop resins

The antimicrobial effect of hop resins is mainly derived from the isomerised alpha-acids. Though beta-acids also have antimicrobial effects, their low solubility means they contribute little to this in beer. The isomerised alpha-acids cause the cell membranes of many Gram-positive bacteria to leak, dissipate the transmembrane pH gradient, deplete the proton motive force, inhibit the uptake of nutrients, deplete divalent cations and cause oxidative stress. However, beer spoilage bacteria have a number of mechanisms to resist these effects.

12.5.4 Carbon dioxide

Carbon dioxide is one of the main products of wort fermentation, and extraneous CO_2 is commonly added to beer to increase carbonation. It creates an anaerobic environment, lowers the pH, affects reactions in the cell and inhibits cell growth.

TABLE 12.2 Lethality table showing pasteurisation units for 1 min at the given temperature (degrees given in column one, decimals in row one).

T °C	0.0	0.1	0.2	0.3	0.4	0.5	0.6	0.7	0.8	0.9
50	0.04	0.04	0.04	0.04	0.04	0.04	0.04	0.05	0.05	0.05
51	0.05	0.05	0.05	0.06	0.06	0.06	0.06	0.06	0.07	0.07
52	0.07	0.07	0.08	0.08	0.08	0.08	0.09	0.09	0.09	0.10
53	0.10	0.10	0.10	0.11	0.11	0.12	0.12	0.12	0.13	0.13
54	0.14	0.14	0.15	0.15	0.16	0.16	0.17	0.17	0.18	0.18
55	0.19	0.20	0.20	0.21	0.22	0.23	0.23	0.24	0.25	0.26
56	0.27	0.27	0.28	0.29	0.30	0.31	0.32	0.33	0.35	0.36
57	0.37	0.38	0.40	0.41	0.42	0.44	0.45	0.47	0.48	0.50
58	0.52	0.53	0.55	0.57	0.59	0.61	0.63	0.65	0.67	0.69
59	0.72	0.74	0.77	0.79	0.82	0.85	0.88	0.91	0.94	0.97
60	1.00	1.03	1.07	1.10	1.14	1.18	1.22	1.26	1.30	1.35
61	1.39	1.44	1.49	1.54	1.59	1.64	1.70	1.76	1.82	1.88
62	1.94	2.01	2.07	2.14	2.22	2.29	2.37	2.45	2.53	2.61
63	2.70	2.79	2.89	2.99	3.09	3.19	3.30	3.41	3.52	3.64
64	3.77	3.89	4.02	4.16	4.30	4.44	4.59	4.75	4.91	5.07
65	5.25	5.42	5.60	5.79	5.99	6.19	6.40	6.61	6.84	7.07
66	7.31	7.55	7.81	8.07	8.34	8.62	8.91	9.21	9.53	9.85
67	10.18	10.52	10.88	11.24	11.62	12.01	12.42	12.84	13.27	13.72
68	14.18	14.66	15.15	15.66	16.19	16.73	17.30	17.88	18.48	19.11
69	19.75	20.42	21.10	21.81	22.55	23.31	24.10	24.91	25.75	26.61
70	27.51	28.44	29.40	30.39	31.41	32.47	33.56	34.70	35.87	37.07
71	38.32	39.61	40.95	42.33	43.76	45.23	46.76	48.33	49.96	51.64
72	53.38	55.18	57.04	58.97	60.95	63.01	65.13	67.33	69.59	71.94
73	74.36	76.87	79.46	82.14	84.91	87.77	90.73	93.78	96.94	100.21
74	103.59	107.08	110.69	114.42	118.28	122.26	126.38	130.64	135.04	139.60
75	144.30	149.16	154.19	159.39	164.76	170.31	176.05	181.98	188.12	194.46
76	201.01	207.78	214.79	222.03	229.51	237.24	245.24	253.50	262.05	270.88
77	280.01	289.44	299.20	309.28	319.70	330.48	341.62	353.13	365.03	377.33
78	390.05	403.20	416.78	430.83	445.35	460.36	475.87	491.91	508.49	525.63
79	543.34	561.65	580.58	600.15	620.37	641.28	662.89	685.23	708.32	732.20
80	756.87	782.38	808.75	836.00	864.18	893.30	923.41	954.53	986.70	1019.95

12.5.5 Low oxygen level

The low oxygen levels found in beer inhibit the growth of many microorganisms, preventing some growing entirely and slowing the growth rate of others.

12.5.6 Low nutrient content

As beer is a fermented beverage, many of the nutrients present in wort are utilised by yeast during the production process. This leaves a nutrient depleted environment for spoilage organisms; the more attenuated the beer, the more the nutrients are depleted and the less are available for other organisms to utilise.

TABLE 12.3 Hurdles to microbial growth.

Hurdle	Mode of action
Ethanol	Inhibits cell membrane function
Low pH	Affects enzyme activity Enhances inhibitory effect of hop resins
Hop resins	Inhibits cell membrane function in Gram-positive bacteria
CO_2	Creates anaerobic conditions Lowers the pH Affects enzyme activity Affects cell membranes
Low O_2	Anaerobic conditions inhibit the growth of obligate aerobes
Low nutrient content	Starves cells

The degree of protection from microbial growth that hurdles will provide in beer will vary with the composition of the beer. High pH, low attenuation, low CO_2, low ethanol and low hopping rates will all make beer less inherently microbiologically stable. Low alcohol and nonalcoholic beers in particular will require more PUs to be applied for practical sterility to be obtained, although even in low-alcohol beers, it has been show that a high hopping rate will reduce PU requirements (Rachon, Raleigh, & Betts, 2022).

12.6 Microorganism heat resistance

Pasteurisation must be effective against the most heat-resistant organism likely to be present as a contaminant. As has been stated previously practical sterility, not absolute sterility, is required and the hurdles to microbial growth mean the pasteurisation of beer is a mild process compared to many other beverages (Table 12.4).

Del Vecchio et al. (1951) based the z value used in PUs as 6.94°C because the most heat-resistant organism they found was an abnormal yeast with this z value. It has since been found that some beer contaminants have a higher z value than this, though in practice for most common contaminants, it is less (O'Connor-Cox, Yiu, & Ingledew, 1991a; Rachon, Rice, Pawlowsky, & Raleigh, 2018).

D values similarly show variation between organisms at the same temperature, and the changes to the D value with temperature changes also vary greatly (Boulton & Quain, 2006). As is often the case caution is required when utilising the calculations in practice. The different compositions of each beer brand will further complicate matters, as the level of protection provided by hurdle technology will vary. It has been suggested that pasteurisation regimes will need to be

TABLE 12.4 D and z values for a number of microorganisms.

Organism	D_{60} value (min)	z value (°C)
Saccharomyces cerevisiae	0.01	4.6
Saccharomyces pastorianus	0.004	4.4
Saccharomyces diastaticus	0.06	7.8
Lacticaseibacillus paracasei	0.02	6.5
Aspergillus niger	0.04	3.7
Pediococcus sp.	0.00073	4.0
Hansenula anomala	0.0039	4.6
Pichia membranaefaciens	0.00025	2.8
Lentilactobacillus parabuchneri	0.44	15
Lactobacillus delbrueckii	0.091	12

European Brewery Convention (1995), Gaze (2006), Kilgour and Smith (1985).

established for each beer according to its composition (Garrick & Mc Neil, 1984) and common contaminants found in a plant (O'Connor-Cox et al., 1991a; O'Connor-Cox, Yiu, & Ingledew, 1991b).

Despite these complications, PUs remain in widespread use in the brewing industry and in practice reflect well actual microbial destruction at the temperature range (60—72°C) used in beer pasteurisation (Zufall & Wackerbauer, 2000). To relate these figures to PU, it has been reported that one PU is sufficient to achieve practical sterility with regards to brewers' yeast and *Pediococcus* sp., 5 PUs are required for *Lactobacillus* sp. and 10 PUs are required for wild yeast (O'Connor-Cox et al., 1991b). In practice, more are often used for security reasons, though the general trend has been for the amount of PU applied to come down over time. A laboratory method has been developed for validating pasteurisation processes in different beers using heat resistant ascospores of an *S. cerevisiae* strain as the test organism (Rachon, Raleigh, & Pawlowsky, 2021).

Though 10—12 PUs should be adequate and has been confirmed as satisfactory in some breweries (Wackerbauer & Zufall, 1997) the EBC Manual of Good Practice (1995) on beer pasteurisation makes the following recommendations (Table 12.5).

The inclusion of low- and non-alcoholic beers as well as lemonade and fruit juices shows how much the properties of beer minimised the amount of PU that are required for beer pasteurisation. The growth in popularity of low- and non-alcoholic beers, as well as the increase in the use of novel ingredients, presents a challenge for brewers seeking to stabilise their product without over pasteurisation. Hard Seltzers have also become significant products for some breweries, and although they lack the protection of hops, typically 10—15 PUs have proved sufficient for microbiological stabilisation.

12.7 Tunnel pasteurisation

With tunnel pasteurisation, the beer is filled into a container (bottle or can), which is sealed before pasteurisation. The container is then transported through the tunnel pasteuriser. Water is sprayed on the containers in stages as they pass through the tunnel, at first heating them until the desired holding temperature is reached and then cooling them to the required discharge temperature (Fig. 12.1).

TABLE 12.5 Typical PU values for different brewery products.

Product	Typical minimum PU	Typical maximum PU
Pilsner and lager beer	15	25
Ale and stout	20	35
Low alcohol beer	40	60
Nonalcoholic beer	80	120
Lemonade	300	500
Fruit juices	3000	5000

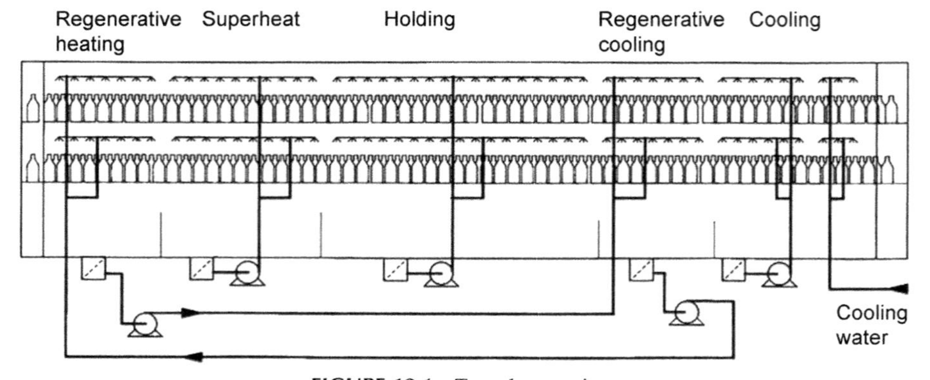

FIGURE 12.1 Tunnel pasteuriser.

Tunnel pasteurisers are divided into zones where water is sprayed at different temperatures, with the total transit time and temperature profile calculated to provide the required degree of pasteurisation. It is common in practice to have a 'superheat' zone before the holding zone to ensure the required temperature has been reached. Heat recovery is important to ensure maximum energy efficiency and minimise costs. Each zone of the tunnel pasteuriser will contain a water tank, a pump and a water distribution system. Heat recovery is achieved by water being moved to different zones where its temperature is appropriate.

Heat transfer to beer inside bottles or cans takes place through the walls of the beer container, which causes a lag in the heating process, bottles having a longer lag than cans. Convection currents are also generated in the beer being heated. Because of this there will be a point near the base of the container, known as the cold spot, where the lowest rate of heat transfer occurs. It is this point that calculations of PU applied to the beer must be made. The temperature rise cannot be too rapid or there is a risk of glass bottles breaking or the pressure rise in the carbonated beer causing the container to burst. The amount of headspace has a large effect on the internal pressure generated during pasteurisation and must be carefully controlled. Pasteurisation in a tunnel pasteuriser can take up to an hour in total.

The first heating stage will gently warm the container approximately 10°C and subsequent stages will steadily raise the temperature to at least 60°C. The most important stage is the superheat zone, which must be accurately controlled at 61−65°C to ensure the container has reached 60°C as it enters the holding zone.

To prevent over pasteurisation, and associated deterioration in product flavour and risk of producing haze, it is important that controls are in place to adjust the heat delivered in the event of a stoppage. Modern tunnel pasteurisers can calculate the total PUs delivered to containers as they travel though the pasteuriser.

Travelling recorders are also routinely used, consisting of a dummy bottle or can on a base plate with sensors and recorders. This can be passed through the pasteuriser to measure the temperature profile that containers are exposed to both as they pass through different zones of the pasteuriser and in different positions that containers may occupy. This means 'cold spots' due to blocked spray jets or other problems can be detected.

Each section of the pasteuriser will have a water tank, water pump and spraying system (spray nozzles or spray pans). A float valve and overflow maintain a constant water level in the tanks. Heaters bring the water to the required temperatures on startup and for adjustment, though during normal operation are only needed in the superheating and holding sections.

The different heating and cooling sections are interconnected and therefore make efficient use of the water at different temperatures. For example, cooling zones towards the end of the pasteuriser will pick up heat from warm containers leaving the holding zone. This warmed water will be pumped to the front of the pasteuriser where it can start to warm cold containers entering the pasteuriser. This will cause it to lose heat so it can be pumped back to carry out more cooling duty. This can be repeated several times as the containers pass through the various stages. Excess heat in cooling sections is reduced by letting hot water overflow drain and bringing in more cold water. Heat recovery in tunnel pasteurisers is around 50% (European Brewery Convention, 1995, Table 12.6).

Movement of the containers through the pasteuriser can be by a 'walking beam' system or on a conveyor belt through a flat-bed system. The former copes well with broken fragments from glass bottles and the latter was originally developed for cans, though has now been found to cope well with glass bottles, too. Double-deck tunnel pasteurisers are available offering greater capacity and/or saving space.

Products passed through a tunnel pasteuriser can have a shelf life of up to a year (Boulton & Quain, 2006).

TABLE 12.6 Typical temperature time profile for a seven zone pasteuriser.

Stage number	Function	Spray temp (°C)	Spray time (minutes)	Container temperature in (°C)	Container temperature out (°C)
1	Preheat	22	6	2	9
2	Preheat	32	7	9	21
3	Superheat	65	14	21	60
4	Holding	60	6	60	60
5	Cooling	40	10	60	43
6	Cooling	32	7	43	36
7	Cooling	22	6	36	28

Dunn (2006).

12.7.1 Cleaning

The warm and wet conditions inside a tunnel pasteuriser make it a good environment for slime to grow and corrosion problems to develop. The water used for spraying may be treated with biocides, softened to prevent scaling and treated to minimise corrosion. Care must be taken with addition rates, however, to ensure compliance with local regulations, the effects on the equipment and personnel and even possible effects on cans.

A 'strainer box' will be present at the outlet from water tanks to trap any material that may cause nozzle blockages. These will require frequent cleaning. Water in the tanks will also need to be changed frequently due to build up of broken glass and other solid materials and spilt beer from damaged containers.

A key design parameter for tunnel pasteurisers is ease of cleaning. Water tanks, spray jets and strainers must have easy access so scale, broken glass and other foreign bodies can be easily removed. Spray nozzles must have large bores and be easily removed and replaced.

Spraying all sections of the pasteuriser with water at >80°C can be used as part of the cleaning process to limit biological growth (European Brewery Convention, 1995).

12.8 Flash pasteurisation

In flash pasteurisation, the beer is rapidly heated in a plate heat exchanger and held in holding tubes where the required number of PUs is applied in a matter of seconds. Typically the beer will be heated to around 72.5°C and held for 20 s, which will give 20 PUs (Fig. 12.2).

The plate heat exchanger consists of a series of sealed metal plates linked by connections in the corners and clamped together at the ends. Beer will be allowed to flow through one side of the plates and pass though the corner ports to the next and heating liquid will flow in a similar manner through the other side of the plate. When the beer is at the required temperature, it will enter a holding tube in order to give the time at the appropriate temperature to give the desired degree of pasteurisation. The flow through the pasteuriser is highly turbulent which aids rapid heat exchange and little temperature difference across the diameter of the holding tube, though beer in the centre will flow slightly faster than that close to the wall of the tube. The rapid passage of beer through a flash pasteuriser means that most of the PU is applied in the holding tube section. A regeneration section with flash pasteurisers allows for a high degree of heat recovery (over 90%) as beer is cooled from the pasteurisation temperature (Dunn, 2006, Table 12.7).

As flash pasteurised beer is not pasteurised inside a sealed container, a higher degree of hygiene and sterility is required than for tunnel pasteurisation. It is particularly critical once the beer has left the pasteuriser, and the buffer tank, filling equipment and all associated pipework are all areas of particular risk.

A thorough microbiological monitoring system should be in place to ensure that hygiene is maintained with samples taken before and after pasteurisation. According to Boulton and Quain (2006), typically a shelf life of six to eight weeks is given to flash pasteurised beer, though in trade keg beer with a shelf life of up to six months is commonly seen.

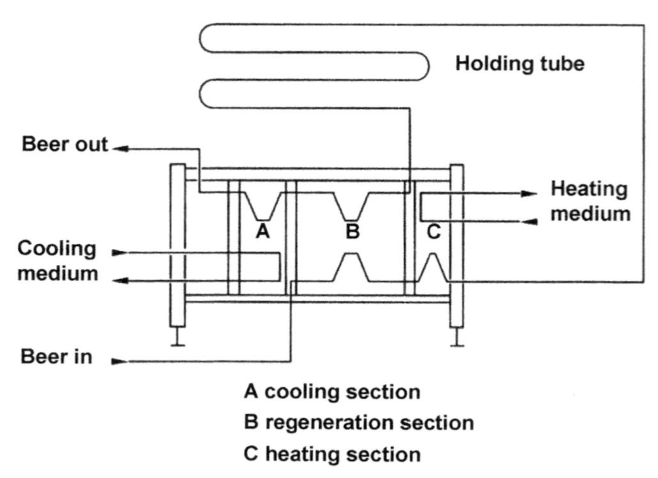

FIGURE 12.2 Flash pasteuriser: heating section, cooling section, regenerative section.

TABLE 12.7 A typical time–temperature programme.

Beer inlet temperature	3.0°C
Outlet from regenerative heating section	65.6°C
Outlet from heating section/entry into holding tube	70.3°C
Outlet from regenerative cooling section	7.7°C
Outlet from cooling section	3.0°C
Holding time	30 s

European Brewery Convention (1995).

12.8.1 Process control

The short time that flash pasteurisation takes means that maintaining the correct temperature and holding time is critical.

PUs can be calculated using the following equation (European Brewery Convention, 1995):

$$PU = (V/Q) \cdot 1.393^{(T-60)}$$

$$V = \text{Volume of holding tube}$$

$$Q = \text{Flow rate in m}^3/\text{minute}$$

$$T = \text{Temperature} \tag{12.3}$$

Even though the flow through the holding tube is turbulent, the beer in the centre will still flow faster than the beer adjacent to the pipe. To ensure that even beer flowing at the fastest rate obtains the correct number of PU, the calculated holding time needs to be increased by a factor of 1.25 (European Brewery Convention, 1995).

Flow rate is controlled by a valve on the outlet, which can be either manually or automatically controlled. In the case of automatic control, it can adjust in response to the level in the buffer tank. The temperature of pasteurisation is controlled by adjusting the temperature of the heating medium. This can be linked to a PU controller, which calculates the temperature required in relation to the flow rate. The outlet temperature of the beer needs to be at a level suitable for packaging and is controlled by adjusting the flow rate of coolant.

Flash pasteurisers work best at a constant flow rate but packaging operations mean that this is not always possible and pasteurisers will need to be able to work at variable speeds. It is usual to have an outlet buffer tank to smooth out the flow rate prior to filling (Dunn, 2006).

If the temperature at the pasteuriser outlet is low, the system should be designed to ensure no under processed product leaves the pasteuriser by diverting the flow back to the inlet until the correct operating temperature is restored. In the case of serious faults, prolonged recirculation should be avoided by shutting down the system.

12.8.2 Gas breakout

At the temperatures used in flash pasteurisation, CO_2 has very low solubility, so the beer needs to be kept under high pressure to prevent gas breakout and fobbing. If fobbing does occur, it increases the flow rate of the beer leading to under pasteurisation, a problem compounded by the possibility of microorganisms avoiding heating inside gas bubbles. Collapsing foam can lead to haze formation and can also bake onto the holding tube, increasing the risk of infection. A pressure monitor needs to be installed to detect when problems have occurred so the process can be stopped. The pasteuriser will need to be cleaned and sterilised before being restarted (Dunn, 2006).

12.8.3 Plate failure

Another potential problem with flash pasteurisers is that of plate failure, when corrosion causes a hole to develop. This can allow unpasteurised beer or coolant to leak into the pasteurised beer. A good maintenance programme will minimise occurrence. Also a booster pump is typically used to ensure pasteurised beer is maintained at the highest pressure in the system (at least 0.5 bar higher than the product) so that in the event of any leakage, it will be of pasteurised beer into unpasteurised beer or coolant, ensuring that product is not contaminated (Hyde, 2001).

12.9 Flavour change

Pasteurisation can affect the flavour of beer, and Pasteur himself had concerns about the effects of pasteurisation on beer (Pasteur, 1876):

> *To preserve bottled beer from deterioration, some bottlers employ, at the moment of filling, a small quantity of bisulphite of lime [calcium bisulphite]. Others heat the bottles to a temperature of 55°C (131°F) in the north of Germany and in Bavaria, this practice has been widely adopted since the publication of the author's 'Studies on Wine', and some of M. Velten's writings. The process has been termed pasteurisation in recognition of the author's discovery of the causes of deterioration in fermented liquors, and of the means of preserving such liquors by the application of heat. Unfortunately this process is less successful in the case of beer than in that of wine, for the delicacy of flavour which distinguishes beer is affected by heat, especially when the beer has been manufactured by the ordinary process.*

A more recent study (O'Connor-Cox et al., 1991a) has described the detrimental effects of pasteurisation:

> *Perhaps the worst effect may be the off-flavours accompanying the processing. Pasteurisation flavour has been described as oxidised, bread crust-like, or possessing a cooked quality. These off-flavours have been shown to be associated with a wide range of carbonyl compounds including unsaturated aldehydes. Prolonged pasteurisation and/or exposure to oxygen have both been shown to have a significant effect on the development of carbonyl compounds. If oxygen is present, pasteurisation generally also results in darkening of beer colour.*

The effects of pasteurisation are more apparent in lighter flavoured beers, which is why less PUs are typically applied to lagers than to ales and stouts.

The negative effects of pasteurisation on beer flavour are a particular problem when oxygen levels are high or excessive pasteurisation is employed. The presence of metal ions in beer, particularly copper, also increases the staling effects of pasteurisation. Keeping oxygen levels in the beer below 0.3 ppm (O'Connor-Cox et al., 1991a), not using copper brewing vessels and only applying the level of pasteurisation required minimises these effects. It has been found with flash pasteurisation that applying the required PU with shorter time and higher temperature has the least effect on flavour (Meilgaard, 2001).

Pasteurisation can also cause hazes to form in beer. It is mainly due to prolonged heat treatment causing proteins to denature (O'Connor-Cox et al., 1991a).

12.10 Good practice and quality control

Once a pasteurisation process has been established resulting in a stable product of the desired quality, the time and temperature of the process will need to be routinely monitored (Gaze, 2006). It should be included in the HACCP plan as a Critical Control Point. Monitoring devices will need to be regularly calibrated and calibration records must be maintained.

Verification of the process will need to be based on likely errors that may occur and the organisms that are most likely to cause problems. The different processes employed in tunnel pasteurisation and flash pasteurisation mean they have different potential problems and monitoring must take this into account. For example, blocked jets or gas breakout are specific to particular methods though poor maintenance or calibration can apply to both.

For tunnel pasteurisers, time—temperature indicators that travel through the tunnel recording the temperatures that containers are subjected to, and for what time, can be used to monitor the process.

Fig. 12.3 illustrates data from a time temperature indicator showing the temperature profiles during low (10.8 PU) and high (36.8 PU) levels of pasteurisation.

For flash pasteurisers, the pasteurisation process needs very close control as the potential risks of both errors and infection are greater. The recirculating system must be assessed to ensure that the correct level of pasteurisation has been reached before any product leaves the system. If microbiological monitoring indicates under pasteurisation is taking place the calibration of the monitoring instruments must be checked to ensure that correct temperatures, timings and flow rates for the required PU are being achieved. Similar checks must be made if over pasteurisation is suspected.

It is good practice to ensure pasteurisers run steadily and process interruptions are as limited as possible. The correct sizing of equipment and buffer tanks can be used to help ensure this (European Brewery Convention, 1995).

12.10.1 Microbiological problems

To ensure the effectiveness of pasteurisation, a sampling plan will need to be developed for the process, taking into account likely places where microorganisms will be able to grow and using specific culture media and conditions for the

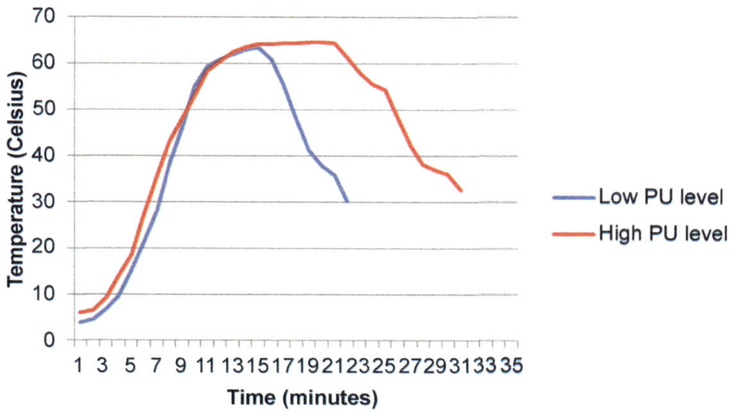

FIGURE 12.3 Graph to show temperature during travel through a tunnel pasteuriser.

microorganisms in question. Efforts must be taken to maximise recovery of microorganisms, even if the cells are damaged, otherwise contaminated product may be undetected. This is particularly important for flash pasteurisation as the product is at risk of infection once it has left the pasteuriser.

Enzyme activity has also been used to monitor pasteurisation in beer (European Brewery Convention, 1995). Yeast cells excrete cell materials, including enzymes, into beer and as enzymes are highly temperature sensitive measuring their activity can be used to determine the degree of pasteurisation. The ability of invertase to produce glucose from sucrose has been used in this way; though as the enzyme is quite heat labile, this method is only appropriate up to about 5 PUs. The ability of enzyme melibiase, found in the cell walls of lager yeast, to produce glucose from melibiose can be used to determine the degree of pasteurisation and this method works up to about 80 PUs.

12.11 Future trends

As breweries strive to reduce their water and energy usage, the trend has been to move away from tunnel pasteurisation to flash pasteurisation. It has been estimated that the costs of flash pasteurisation are only 15% of that of tunnel pasteurisation (Hyde, 2001). The capital expenditure required for replacement and the durability of tunnel pasteurisers means their replacement will proceed slowly, but it has been predicted that they will be gone by 2030 (Nelson, 2009). There has also been an increase in using sterile filtration as an alternative to pasteurisation. However, as larger brewers have moved away from tunnel pasteurisation, there has been increased demand for it from craft brewers producing sweet fruit beers that are particularly prone to microbial instability.

A number of novel methods, many non-thermal, for pasteurisation of beer have also been investigated (Hill, 2009; Milani & Silva, 2022).

Ultrasound was used in conjunction with heat (thermosonication) by Deng et al. (2018). They found 2.7 W/mL volumetric power at 24 kHz frequency on lager at 50°C for 2 minutes gave 12 months microbial stability with good flavour and haze stability.

Pulsed electrical fields have been used to inactivate microorganisms by electroporation. This has the benefit of causing little or no change to the organoleptic properties of the product. High levels of hydrostatic pressure (100–1000 MPa) have also been used successfully to enhance microbial stability in beer to a level similar to heat treatment. This works by increasing the permeability of the cytoplasmic membrane and inactivating hop resistance mechanisms. No chemical changes to the beer were found after this treatment in filtered beer, but when used for unfiltered beer, an increase in carbonyl compounds was found (Štulíková et al., 2020).

High-pressure homogenisation as a means of inactivating spoilage microorganisms in beer has also been investigated (Franchi, Tribst, & Christianini, 2013). This is a continuous process where the fluid is forced under pressure through a narrow gap where it undergoes rapid acceleration (200 m/s at 340 MPa) followed by an extreme drop in pressure. This can lead to microbial inactivation by causing cell permeability changes and a reduction in fluid viscosity. However, a number of chemical changes also occur, which can alter the colour of the beer and increase haze (Franchi et al., 2011).

Whether any of these novel processes make it into production only time will tell.

12.12 Sources of further information and advice

The European Brewery Convention Beer Pasteurisation (Manual of Good Practice, 1995) remains the best source of detailed information on the pasteurisation of beer. *Excellence in Packaging of* Beverages (2001) also contains a wealth of information. Wilson's chapter on microbial stabilisation in the Master Brewers' Association of the Americas book *Beer* Packaging (1981) provides a good overview of the process, as does Dunn's (2006) section on pasteurisation in *The Handbook of* Brewing (2006).

The pair of articles by O'Connor-Cox et al. (1991a, 1991b) provides useful information on what happens during pasteurisation and actual industrial practice.

The Lemgo D and z value Database for Food provides an excellent resource for finding these values for a number of beer contaminant microorganisms:

http://www.hs-owl.de/fb4/ldzbase/index.pl.

References

Boulton, C., & Quain, D. (Eds.). (2006). *Brewing yeast and fermentation*. Oxford: Blackwell.

Del Vecchio, H. W., Dayharsh, C. A., & Baselt, F. C. (1951). Thermal death time studies on beer spoilage organisms. *ASBC Proceedings*, 45–50.

Deng, Y., Bi, H., Yin, H., Yu, J., Dong, J., Yang, M., et al. (2018). Influence of ultrasound assisted thermal processing on the physicochemical and sensorial properties of beer. *Ultrasonics Sonochemistry, 40*(Part A), 166–173.

Dunn, A. R. (2006). Packaging technology. In F. G. Priest, & G. G. Stewart (Eds.), *Handbook of brewing* (pp. 596–603). Boca Raton, FL: CRC Press.

European Brewery Convention. (1995). *Beer pasteurisation (Manual of good practice)*. Nuremberg, Germany: Fachverlag Hans-Carl.

Franchi, M. A., Tribst, A. A. L., & Christianini, M. (2011). Effects of high pressure homogenization on beer quality. *Journal of the Institute of Brewing, 117*(2), 195–198.

Franchi, M. A., Tribst, A. A. L., & Christianini, M. (2013). High-pressure homogenization: A non-thermal process applied for inactivation of spoilage microorganisms in beer. *Journal of the Institute of Brewing, 119*, 237–241.

Garrick, C. C., & Mc Neil, K. E. (1984). Influence of product composition on pasteurisation efficiency. In *Proceedings of the 18th institute of brewing convention (Australia and New Zealand section)* (pp. 244–251). Adelaide.

Gaze, J. E. (2006). *Pasteurisation: A food industry practical guide*. (Chipping Campden, CCFRA).

Hill, A. E. (2009). Microbial stability of beer. In C. Bamforth (Ed.), *Beer: A quality perspective* (pp. 163–184). USA: Academic Press.

Hyde, A. (2001). Plate pasteurisation. In J. Browne, & E. Candy (Eds.), *Excellence in packaging of beverages* (pp. 109–119). Andover: Binstead Group.

Kilgour, W., & Smith, P. (1985). The determination of pasteurisation regimes for alcoholic and alcohol-free beer. In *Proceedings of the European brewing convention congress* (pp. 435–442).

Meilgaard, M. (2001). Effects on flavour of innovations in brewery equipment and processing: A review. *Journal of the Institute of Brewing, 107*(5), 271–286.

Milani, E. A., & Silva, F. V. M. (2022). Pasteurization of beer by non-thermal technologies. *Frontiers in Food Science and Technology, 1*, Article 798676.

Nelson, L. (Ed.). (2009). *138* (p. 26). Brew. Guardian.

O'Connor-Cox, E. S. C., Yiu, P. M., & Ingledew, W. M. (1991). Pasteurization: Thermal death of microbes in brewing. *MBBA Technical Quarterly, 28*, 67–77.

O'Connor-Cox, E. S. C., Yiu, P. M., & Ingledew, W. M. (1991). Pasteurization: Industrial practice and evaluation. *MBBA Technical Quarterly, 28*, 99–107.

Pasteur, L. (1876). *Études Sur La Bière*. UK: translated as Studies on fermentation Macmillan and Company.

Rachon, G., Raleigh, C. P., & Betts, G. (2022). The impact of isomerised hop extract on the heat resistance of yeast ascospores and *Lactobacillus brevis* in premium and alcohol-free beers. *Journal of the Institute of Brewing, 128*(1), 22–27.

Rachon, G., Raleigh, C. P., & Pawlowsky, K. (2021). Heat resistance of yeast ascospores and their utilisation for the validation of pasteurization processes for beers. *Journal of the Institute of Brewing, 127*(2), 149–159.

Rachon, G., Rice, C. J., Pawlowsky, K., & Raleigh, C. P. (2018). Challenging the assumptions around the pasteurisation requirements of beer spoilage bacteria. *Journal of the Institute of Brewing, 124*(4), 443–449.

Štulíková, K., Bulíř, T., Nešpor, J., Jelínek, L., Karabín, M., & Dostálek, P. (2020 May 21). Application of high-pressure processing to assure the storage stability of unfiltered lager beer. *Molecules, 25*(10), 2414.

Vriesekoop, F., Krahl, M., Hucker, B., & Menz, G. (2012). 125th anniversary review: Bacteria in brewing: The good, the dab and the ugly. *Journal of the Institute of Brewing, 118*(4), 335–345.

Wackerbauer, K., & Zufall, C. (1997). Flash pasteurisation and its impact on beer quality. *Brauwelt International, 15*(2), 94–96.

Wilson, J. R. (1981). Microbial stabilization. In *Beer packaging*. Madison, WI: MBBA.

Zufall, C., & Wackerbauer, K. (2000). The biological impact of flash pasteurization over a wide temperature interval. *Journal of the Institute of Brewing, 106*(3), 163–168.

Chapter 13

Maintaining microbiological quality control in NOLO beverages and hard seltzers

Scott J. Britton[1,2], Frank Vriesekoop[3] and Annie E. Hill[2]

[1]*Brouwerij Duvel Moortgat, Puurs-Sint-Amands, Belgium;* [2]*The International Centre for Brewing & Distilling, Heriot-Watt University, Edinburgh, United Kingdom;* [3]*Harper Food Innovation, Harper Adams University, Newport, United Kingdom*

With revenue reaching nearly $37bn USD in 2024 and compound annual growth of 9%, the global no- and low-alcoholic beer market continues to grow (Statistica.com). This trend reflects the growing desire for a wholesome lifestyle and improved production techniques that do not compromise the experience of consuming beverages traditionally associated with alcohol (Mellor, Hanna-Khalil, & Carson, 2020). It enables the consumer to savour novel beverage tastes and relish social interactions without the need for alcohol. The consumption of no- and low-alcohol beverages has become firmly established and is currently on an upward trajectory, especially in Western Europe, which held 41% of the market value in 2017 (Quain, 2021; Anderson, O'Donnell, Kokole, Llopis, & Kaner, 2021; Salanță et al., 2020).

The definition of low or non-alcoholic beverages varies, as does the terminology (Quain, 2021; Labrado et al., 2020; Salanță et al., 2020; De Fusco et al., 2019; Bellut et al., 2019; Brányik, Silva, Baszczyňski, Lehnert, & Almeida e Silva, 2012). In most of the EU, regulations stipulate that low-alcohol beers (LABs) must have an alcohol by volume (ABV) content of less than 1.2%, while alcohol-free beers (AFBs) must contain less than 0.5% ABV (Montanari, Marconi, Mayer, & Fantozzi, 2009; Quain, 2021). In the UK, AFBs must contain less than 0.05% ABV. In the United States, malt beverages labelled as low or reduced alcohol must not exceed 2.5% ABV with non-alcoholic beverages less than 0.5% ABV, and those labelled as 'alcohol-free' must contain no alcohol (Simon, Vuylsteke, & Collin, 2022).

In recent years, the hard seltzer category has also experienced substantial growth, propelled primarily by its popularity as a ready-to-drink (RTD) beverage with lower calorie content. The EU and TTB (Alcohol and Tobacco Tax and Trade Bureau) regulations do not explicitly categorise hard seltzers. However, they are commonly designated as beer for labelling and taxation considerations. Typically, they comprise carbonated water, flavouring and alcohol (2.5%–8% ABV). An increasing number of breweries are now crafting hard seltzers using malted barley or other fermentable substrates, and as such, the considerations of incorporating this product type into a brewery's portfolio deserve careful consideration.

Ensuring the safety and stability of brewery products remains a consistent and significant challenge for brewers. While traditional beers have inherent characteristics that minimise the risk of microbial spoilage (Menz, Aldred, & Vriesekoop, 2009), low- or non-alcoholic beverages and complex RTD options are more vulnerable to compromise. In this chapter, we delve into the production processes of low- and non-alcoholic beer and hard seltzers and examine the intrinsic properties of these beverages. Additionally, we evaluate how production methods and the ultimate product characteristics impact microbiological stability and explore effective strategies to protect against spoilage.

13.1 Production of LAB, AFB and hard seltzers

13.1.1 LAB and AFB

The manufacturing of 'non-intoxicating' beer has a longstanding history. As early as the 1830s, several companies in Scotland were already promoting LAB and stout (Lawrie, 2021, pp. 48–52), with such beers likely to have been made by

Brewing Microbiology. https://doi.org/10.1016/B978-0-323-99606-8.00011-0

halting fermentation prematurely. Low- and non-alcoholic beers are crafted through adaptations to biological or physical processes (Fig. 13.1). In biological approaches, ethanol formation is restricted by employing yeast strains with reduced fermentation capabilities, employing cold fermentation, or prematurely halting fermentation. On the other hand, physical methods involve removing alcohol from beer using techniques such as thermal methods, extraction, or membrane-mediated processes.

Traditional beer production relies on *Saccharomyces cerevisiae* (for ale, employing top fermentation at temperatures of 18–24°C) or *Saccharomyces pastorianus* (for lager, using bottom fermentation at temperatures of 8–14°C). There is the potential to utilise standard production strains to create beverages with lower alcohol content by decreasing the initial concentration of fermentable sugars present in the wort, lowering the fermentation temperature or stopping fermentation prematurely (Dalberto et al., 2021; Ivanov et al., 2016). Another biological method is to employ alternative yeast strains that are restricted in their ability to metabolise all fermentable sugars present in wort (Johansson et al., 2021). For example, *Saccharomycodes ludwigii* and specific varieties of *Saccharomyces cerevisiae* cannot metabolise maltose (De Francesco et al., 2018). *Torulaspora delbrueckii* has also demonstrated effectiveness in cold-contact fermentation (1.0 ± 0.5°C; Nikulin, Aisala, & Gibson, 2022), and several *Lachancea* species isolated from kombucha have shown promise (Bellut, Krogerus, & Arendt, 2020). A range of brewery contaminant yeast have been investigated as potential production strains for LABs, with two strains, *Trigonopsis cantarelli* and *Candida sojae,* performing comparably with a commercial reference strain (Krogerus, Eerikäinen, Aisala, & Gibson, 2022). In particular, *T. cantarelli* was discovered to yield minimal off-flavours while offering a spectrum of appealing flavour components.

An alternative to limiting alcohol formation involves physically extracting alcohol from full-strength beer. Various 'thermal' techniques can be employed to vapourise or distil ethanol. With vacuum distillation, which emerged in some brewing operations during the 1980s, evaporation of alcohol can be achieved at a relatively low temperature, thus avoiding the development of burnt, caramel or staling characters (Jiang et al., 2017; Liguori et al., 2015; Loredana et al., 2018).

Most extraction and membrane processes to remove ethanol may occur at low temperatures, even as low as 10°C, presenting no noticeable thermal stress on the sensory characteristics. Taking reverse osmosis as an illustration, it employs a semipermeable membrane to separate very small molecules like ethanol and water from the more substantial ones found in beer. These larger molecules include those that contribute to the beer's flavour and colour, as well as proteins maintaining many of the beer's sensory characteristics without the alcohol (Peyravi, Jahanshahi, & Banafti, 2020). Supercritical fluid extraction techniques exhibit a notable preference for extracting ethanol, thereby reducing the influence of the removal process on the flavour and aroma of the beer (Muller, Neves, Gomes, Guimarães, & Ghesti, 2019).

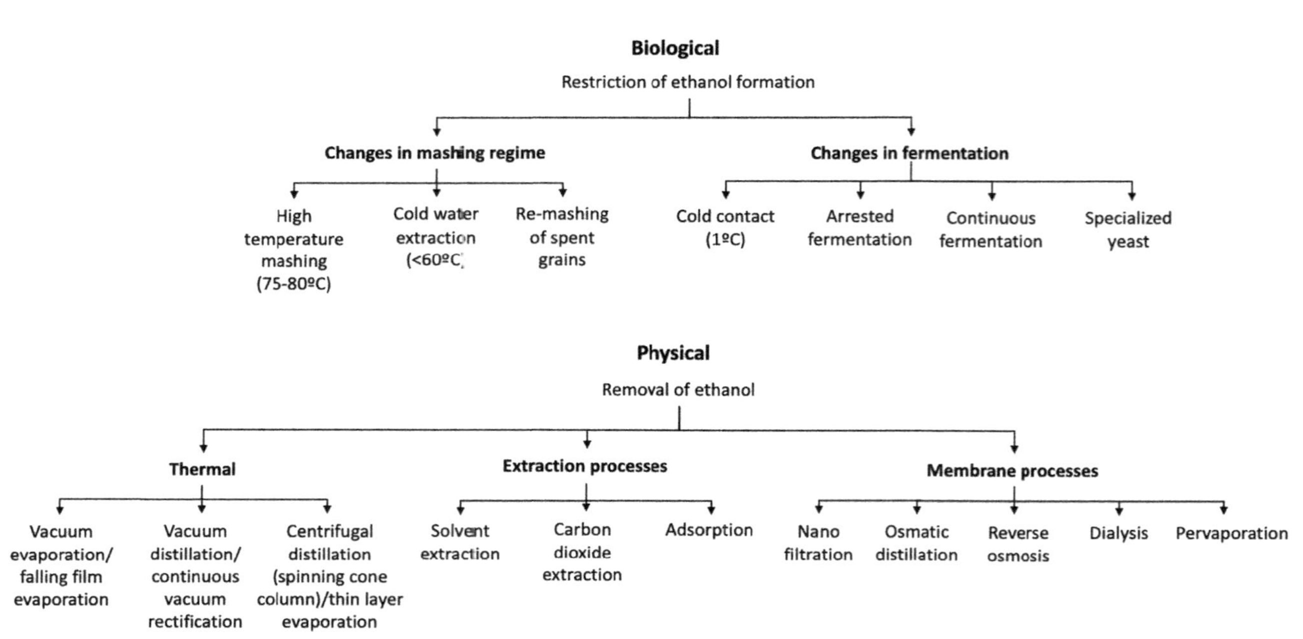

FIGURE 13.1 Strategies to produce low- and non-alcoholic beer.

It is important to note that if the production process includes the removal of alcohol from beer through reverse osmosis, or other processes that separate alcohol from the remaining components of a beverage, the process may be considered for tax or regulatory purposes as a distilling operation, and therefore may require additional licences/permits.

13.1.2 Production of hard seltzers

In contrast to LABs and AFBs, the production of hard seltzers from sugar or grain is typically more straightforward (Fig. 13.2).

The initial phase entails preparing an alcoholic base, commonly derived from the fermentation of dextrose, sucrose, cane sugar, liquid invert sugar, agave syrup, molasses and honey, among others. Nevertheless, dextrose is frequently preferred due to its cost-effectiveness, widespread accessibility and its ability and capacity to yield a complete fermentation. In contrast, fermentables based on sucrose may hinder fermentation efficiency, potentially leaving residual fructose after the fermentation process concludes. The fermented sugar base (FSB) undergoes a scrubbing process to diminish flavour compounds, resulting in a flavour neutral base suitable for subsequent flavouring. A neutral base is typically achieved through filtration using plate or charcoal filters. After attaining the desired alcohol content (2.5%–8% ABV), sweeteners and flavourings are incorporated prior to the carbonation and packaging stages.

13.2 Characteristics of LABs, AFBs and hard seltzers

Aside from the apparent decrease or absence of alcohol in LABs and AFBs, the broader composition of the final beverage compared to typical or original beer will vary depending on the production method.

Typically, beer has an ethanol content of up to 8% ABV, pH of 3.8–4.7, hop bitters (17–55 mg α-acid/L), CO_2 (approximately 0.5% w/v), SO_2 (approximately 5–30 mg/L), organic acids, acetaldehyde and a range of other residual nutrients. AFBs will have relatively higher or lower levels of components depending on whether they are produced by shortened fermentation or by removal of alcohol from fully attenuated beer.

Several research investigations have examined the characteristics of commercially accessible AFBs and LABs (Bellut et al., 2018; Riu-Aumatell, Miró, Serra-Cayuela, Buxaderas, & López-Tamames, 2014; Sohrabvandi et al., 2010). For instance, Bellut et al. (2018) conducted studies on factors such as pH level and the influence of yeast strains on volatile aromatic compounds. The average pH of AFB was found to be 4.59, reaching a maximum of 5.10. This upper limit is towards the higher end of the customary pH range for beer (3.8–4.7; Bellut et al., 2018). Comparison of the production of volatile compounds identified significant differences between the *Saccharomyces cerevisiae* control in contrast to non-*Saccharomyces* yeasts *Hanseniaspora valbyensis, Hanseniaspora vineae, Torulaspora delbrueckii, Zygosaccharomyces bailii* and *Zygosaccharomyces kombuchaensis* (Table 13.1), with non-*Saccharomyces* yeast producing 3–5 times less of the higher alcohols n-propanol, iso-butanol and isoamyl alcohol (Bellut et al., 2018; Gernat, Brouwer, & Ottens, 2020).

A similar study by De Francesco, Freeman, Lee, Marconi, and Perretti (2014), and later by De Francesco et al. (2021), demonstrated a direct relationship between ethanol removal and substantial loss of volatile compounds (Table 13.2).

Analysis of a selection of commercially available European AFBs and LABs by Quain (2021) found a greater quantity of remaining fermentable substrates than control beers. The fermentable sugar concentration was six times higher than that typically found in standard beer (Table 13.3).

For hard seltzers, the typical alcohol content ranges from 2.5% to 8% ABV. They are also typically nutrient-poor, with low sugar content (0–2 g/L) and carbonated (2.5 g/L-6.5 g/L CO_2) and have a low pH (pH 2.8–4.4).

13.3 Spoilage risk in LABs, AFBs and hard seltzers

Elevated levels of ethanol have long been linked to enhanced microbial stability and have been noted as an effective antimicrobial at concentrations as low as 2.5% (Chatterjee et al., 2006; Menz et al., 2009, Menz, Aldred, & Vriesekoop, 2011; Shimwell, 1935). However, it is also acknowledged that the broader composition of the beverage influences alcohol tolerance. In a study by Quain (2021) involving challenge testing LAB and AFB, there was a modest reduction in microbial growth as ethanol concentration increased, but instances of spoilage were estimated to be 2–5 times higher than in

FIGURE 13.2 Typical process flow for hard seltzer production.

TABLE 13.1 Volatile compounds found in fermentation with non-*Saccharomyces* yeasts in LAB and AFBs.

Volatile compound (mg/L)	*Saccharomyces cerevisiae*	*Saccharomycodes ludwigii*	*Hanseniaspora valbyensis*	*Hanseniaspora vineae*	*Torulaspora delbrueckii*	*Zygosaccharomyces bailii*	*Zygosaccharomyces kombuchaensis*
Alcohols							
n-propanol	13.7	2.6	2.1	2.2	2.9	2.7	2.1
Iso-butanol	17.9	6.4	4.8	4.6	4.9	5.7	7.1
Isoamyl alcohols	50.8	12.1	16.5	13.4	10.4	14.8	12.9
Esters							
Ethyl acetate	4.05	0.8	0.9	6.0	0.77	1.0	1.0
Ethyl formate	1.05	1.01	0.78	0.76	0.9	0.56	0.72
Isoamyl acetate	0.2	<0.1	<0.1	<0.1	<0.1	<0.1	<0.1
Aldehydes							
Acetaldehyde	7.8	8.5	3.3	4.1	9.1	4.9	6.8
Diketones							
Diacetyl total	0.04	0.03	0.21	0.05	0.06	0.03	0.15

From Bellut et al. (2018).

TABLE 13.2 Loss of aroma compounds during the production of low and non-alcoholic beer using physical methods.

Volatile compound	Percentage (%) loss of volatile compounds				
	Dialysis	Falling film evaporation	Vacuum distillation	Reverse osmosis	Osmotic distillation
Alcohols					
Ethanol	90	89	99	91	87
n-propanol	94	92	100	70	74
Iso-butanol	95	95	100	66	64
Esters					
Ethyl acetate	99	N/A	100	88	80
Isoamyl acetate	95	25	100	89	93
Aldehydes					
Acetaldehyde	31	N/A	60	N/A	54

From De Francesco et al. (2014).

standard beers. It was demonstrated that high concentrations of free amino nitrogen, maltotriose, and other fermentable compounds bolster microbial growth (Quain, 2021; Fernandez & Simpson, 1995) and can partially counteract the antimicrobial properties inherent to beer. Other innate qualities of beer that limit the risk of infection include acidic pH and competition for nutrients in the wort (Sakamoto & Konings, 2003). Typically, LAB and AFB produced through arrested fermentation have a higher pH and an elevated level of residual nutrients compared to those where fermentation ran to completion before the removal of alcohol. The availability of micronutrients such as nitrogen, vitamins, mineral ions and sulphur combined with the high fermentable sugar concentration makes these beers more susceptible to spoilage (Quain, 2021).

13.4 Potential spoilers of LABs, AFBs and hard seltzers

Until recently, the spectrum of wild yeast and bacteria recognised as potential adversaries to beer was confined to a limited number of genera (Suzuki, 2020). Primarily, the resilience against spoilage by most microorganisms in traditional beer

TABLE 13.3 Composition of low- and non-alcoholic beer (*ABV*, alcohol by volume; gravity, current gravity; *IBC*, colour; sugars, fermentable sugars; *FAN*, free amino nitrogen; *IBU*, International Bitterness Units; ingredient 1, malted barley; 2, maize; 3, rice; 4, sugar; 5, hops/hop products; 6, natural flavourings).

Origin	ABV (%)	Gravity (g/mL)	pH	IBC	Sugars (g/L)	FAN (g/L)	IBU	Glycerol (g/L)	Ingredient
UK	0.01	1.0259	4.06	5.5	43.7	141	11.5	0.07	1,5
Italy	0.03	1.0215	4.28	2.9	42.5	91	8.1	0.07	1,2,5,6
Belgium	0.00	1.0344	3.90	4.1	27.3	19	18.9	1.55	1,2,4,5,6
Spain	0.02	1.0233	4.04	4.6	41	152	11.7	0.06	1,5,6
Netherlands	0.03	1.0206	4.44	3.8	21	102	13.9	1.03	1,5,6
EU	0.00	1.0343	4.38	3.9	30.5	26	<5	1.01	1,3,5,6
Sweden	0.40	1.0269	4.15	14.0	29	125	40.3	0.29	1,5
Germany	0.38	1.0255	4.61	4.6	39	89	22.5	0.25	1,4

From Quain (2021).

varieties arises from the intrinsic chemical attributes characterising them, including appreciable concentrations of ethanol and antimicrobial hop bittering compounds; low pH; restricted oxygen availability; heightened carbon dioxide levels, and comparatively diminished residual nutritional value due to the fermentative activities of brewing yeast (Bergsveinson, Redekop, Zoerb, & Ziola, 2015; Menz et al., 2011; Michel et al., 2020; Sakamoto & Konings, 2003; Vaughan, O'Sullivan, & Van Sinderen, 2005; Vriesekoop, Krahl, Hucker, & Menz, 2012). These inherent chemical characteristics, in conjunction with external processing hurdles such as boiling, pasteurisation, sterile filtration, and cold storage, further elevate the defence against a broad spectrum of microorganisms, including most pathogens (Menz et al., 2011).

However, due to the dynamic landscape of consumer preferences, there has been a significant increase in the popularity and consumption of LABs, AFBs and hard seltzers, and the integration of avant-garde ingredients (Vriesekoop, 2021), accompanied by the embrace of pioneering processing techniques (Bellut & Arendt, 2019; Simon et al., 2022). Yet, despite their popularity, an inherent challenge persists: the formulation of many of these libations falls short of the time-honoured protective chemical profiles of their predecessors. Consequently, they find themselves increasingly vulnerable to spoilage by both conventional and unconventional biological agents, heralding a call for refined craftsmanship in the face of evolving tastes.

The preceding chapters extensively covered the conventional beer spoilers, as detailed in Chapters 6–9. In contrast, this section shifts focus to the exploration of bacterial spoilage agents, which, although less prevalent and concerning in traditional finished beer, pose a heightened risk in the production of commercial no- and-low alcohol beers and hard seltzers. Notably, we will refrain from discussing potential spoilers related to moulds and yeast. Molds typically flourish and thrive in aerobic environments and are deterred by atmospheres with restricted oxygen and elevated concentrations of carbon dioxide. Similarly, the presence of yeast, by and large, is deemed undesirable in the later stages of no- and-low alcoholic beverage and hard seltzer production.

Before we embark on the discussion surrounding spoilage bacteria as they pertain to no- and low-alcohol beverages and hard seltzers, it is imperative to recognise that the metabolites serve as a reflection of the physiological functions of the cell, and their expression is subject to a myriad of environmental influences, including, but not limited, to oxygen levels, pH, and nutrient availability. Therefore, as the investigation into spoilers of no- and low-alcoholic beverages remains underexplored, the consequential impact of any microorganisms discussed herein may yield unforeseen and divergent outcomes, contingent upon the unique circumstances of each case.

13.4.1 *Enterobacteriaceae*

The *Enterobacteriaceae* taxonomic family contains more than 60 genera of straight rod-shaped, facultatively anaerobic Gram-negative bacteria that are predominantly hop-insensitive, non-sporulating, nitrate-to-nitrite reducing, catalase-positive and oxidase-negative (Brenner & Farmer III, 2015; Janda & Abbott, 2021). The lesser-known aspect of the *Enterobacteriaceae* family is the wide environmental diversity within which its members reside, ranging from vertebrate, plant, insect, aquatic and other bio-based systems to aquatic environments (Brenner & Farmer III, 2015; Janda & Abbott, 2021; Smith & Fratamico, 2015). Therefore, it should not be surprising that *Enterobacteriaceae* occasionally enter the brewing process through co-entry with tainted dry raw materials (such as malt, spices, fruit, etc.) or by using contaminated water sources (Al-Kharousi, Guizani, Al-Sadi, Al-Bulushi, & Shaharoona, 2016; Figueras & Borrego, 2010; Iwu & Okoh, 2019; Santamaría & Toranzos, 2003; Van Doren et al., 2013).

Prominent among this taxonomic cohort are the foodborne pathogens originating from the genera *Escherichia*, *Salmonella*, *Shigella*, *Yersinia* and *Cronobacter*, with the first four accounting for approximately 40% of reported bacterial-induced foodborne maladies annually in the United States alone (Brenner & Farmer III, 2015; Smith, 2012).

Among the aforementioned genera, two were previously scrutinised by Menz and colleagues within various alcoholic beers. Their findings indicated that *Escherichia coli* O157:H7, a notorious pathogenic strain infamous for triggering severe haemorrhagic diarrhoea in humans, and *Salmonella typhimurium* demonstrated an inability to flourish and ultimately succumbed in full-strength beer containing 5% alcohol-by-volume, or a range of mid-strength beers, ranging from 2.5% to 3.6% alcohol-by-volume (Menz et al., 2011). Menz et al. concluded this restraint was ascribed to the concurrent inhibitory effects of the low pH ($\leq$4.3) and ethanol concentration (2.3%–5.0% ABV) inherent to these beers. Intriguingly, however, both bacterial species managed to maintain viability for over 10 days in full-strength beer and over 30 days in mid-strength beer when stored at 4°C; however, when exposed to storage at ambient temperature (25°C) or warmer conditions (37°C) post-inoculation, the survival of the bacteria was significantly compromised, with viability lasting less than 2 days (Menz et al., 2011). The prolonged survival and heightened resilience in colder temperatures are consistent with previous findings regarding the inactivation dynamics of *Enterobacteriaceae* in growth-inhibiting environments (Escartin, Castillo, Hinojosa-Puga, & Saldaña-Lozano, 1999; Keerthirathne, Ross, Fallowfield, & Whiley, 2019; Massa, Altieri, Quaranta, &

De Pace, 1997; Rocelle, Clavero, & Beuchat, 1996; Tsai & Ingham, 1997; Zaika, 2002). Largely, this phenomenon is attributed to the discovery of adaptive response mechanisms, which modulate the composition of fatty acids and upregulate protective cold-shock proteins in colder environmental conditions (Berry & Foegeding, 1997; Ingram & Buttke, 1985; Jones, Inouye, Wood, & Medical, 1994; Ross, Zhang, & McQuestin, 2008; Weber & Marahiel, 2003).

In stark contrast, experiments conducted in AFBs containing a mere 0.5% ABV revealed a concerning reality. *Escherichia coli* O157:H7, *Salmonella typhimurium*, *Shigella flexneri*, *Klebsiella pneumoniae*, *Yersinia enterocolitica* and *Enterococcus faecium* were able to thrive and rapidly multiply possessing a pH equal to or above 4.3 (L'Anthoën & Ingledew, 1996; Menz et al., 2011). The investigations conducted by Menz et al. (2011), which largely focused on *Escherichia coli* O157:H7 and *Salmonella typhimurium* further discovered that the adjustment of pH levels in an existing AFB from 4.3 to 4.5 and 5.0 expedited their growth rates and curtailed lag phases, conversely, lowering the pH to 4.0 impeded growth, ultimately leading to their deactivation over time (Menz et al., 2011).

Despite the wealth of evidence supporting these assertions, a persistent misconception lingers, the erroneous belief that beer serves as an immediate deterrent to foodborne pathogens, swiftly neutralising them upon exposure (Jeon et al., 2015; Kim, Kim, Lee, Hwang, & Rhee, 2014; L'Anthoën & Ingledew, 1996; Medina, Romero, Brenes, & De Castro, 2007; Menz et al., 2011). Consequently, from a standpoint of food safety, meticulous scrutiny of the manufacturing and formulation processes governing no- and low-alcohol beer and hard seltzers becomes imperative. The absence of stringent preventive measures holds the potential to exacerbate the risk of contamination and foster the unchecked proliferation of potentially hazardous *Enterobacteriaceae*, capable of producing toxins and subsequent illness. The ramifications of pathogenic *Enterobacteriaceae* on the production of microbiologically non-stabilised no- and low-alcohol beers and hard seltzers cannot be overstated. Their manufacture inherently harbours risks and demands a prudent approach underscored by an abundance of caution.

Apart from the pathogenic *Enterobacteriaceae*, other species like *Shimwellia pseudoproteus; Obesumbacterium proteus* (formerly known as *Flavobacterium proteus*), and *Termobacterium lutescens*; *Rahnella aquatilis*; *Citrobacter freundii*; *Klebsiella oxytoca* and *Enterobacter cloacae*, traditionally regarded as agents of wort spoilage, present an additional threat to no- and low-alcohol beer, especially those produced from arrested fermentations. While Chapter 8 extensively explores this topic, here we offer a concise overview of the potential spoilage implications posed by these spoilers on no-alcohol beers and LABs, as well as low-alcohol seltzers. Still, it is important to note that an even broader spectrum of uncommon *Enterobacteriaceae* bacteria may contaminate non-alcoholic and LAB, as well as low-alcohol seltzers.

Shimwellia pseudoproteus and *Obesumbacterium proteus*, often found in conjunction with yeast, possess the ability to thrive in wort and exhibit a notable resistance to ethanol levels below 6% v/v.; however, their ethanol tolerance is significantly influenced by the pH of the surrounding milieu (Case, 1965; Farmer & Brenner, 2015; Van Vuuren & Priest, 2003). Given the similarities of no- and low-alcohol beers to wort, it is reasonable to anticipate these bacteria will behave similarly in such settings, potentially introducing undesirable haze and flavour attributes through the production of organosulphur compounds, such as dimethyl sulfide and dimethyl disulfide; ethanol, n-propanol; acetoin; lactic acid; and 2,3-butanediol; however, a broader array of metabolites has been documented in tainted beer also cohabitated with yeast (Thomas, Cole, & Hough, 1972, Priest, Cowbourne, & Hough, 1974). This blend of components imbues a nuanced vegetal flavour, often described as reminiscent of parsnip (Thomas et al., 1972). Although, much like other potential spoiling agents that may challenge the landscape of non-alcoholic beverages, such as no-alcohol beer and LAB or hard seltzers, these species are inhibited below a pH threshold of 4.0 (Thomas et al., 1972). *Citrobacter freundii* is also a notable spoiler of wort and a formidable threat even in minimal quantities and brief exposures (Keevil, Hough, & Cole, 1979). Originating predominately from the intestinal tracts of humans and animals, its presence in wort arises from cross-contamination with tainted sewage, water or soil, subsequently released into the brewing environment (Frederiksen, 2015). If introduced into non-alcoholic beers or LABs, these bacteria could elevate levels of acetic acid, lactic acid, diacetyl, acetaldehyde, and dimethyl sulfide (Keevil et al., 1979; Martens, Dawoud, & Verachtert, 1991; Priest et al., 1974; Van Vuuren & Priest, 2003). *Citrobacter freundii* exhibits vigorous growth between pH 5.0 and 10.0; however, its proliferation is notably impeded at pH 4.0, and it succumbs to adverse conditions at pH 3.0, unable to survive (Martens et al., 1991; Wang, Elzenga, & van Elsas, 2021; Zhao, Wen, Huang, Weng, & He, 2022).

Rahnella aquatilis, a resilient bacterium thriving in both cold temperatures (<4°C) and acidic conditions (pH 4.0), displays remarkable tolerance to ethanol levels of 11%–12% v/v, which are commonly found in freshwater environments and other diverse habitats, including humans, mountain soils, and various others (Kämpfer, 2015; Liu et al., 2008; Magnus, Ingledew, & Casey, 1986; Peng, Xu, Li, Zhao, & Guo, 2023; Van Vuuren & Priest, 2003). Like *Shimwellia* and *Obesumbacterium*, *Rahnella* is known to accumulate in pitching yeast, posing a persistent threat to subsequent batches and leading to chronic spoilage. Alcoholic beers tainted by *R. aquatilis* often exhibit detrimental sensory qualities, characterised by fruity, milky, and sulphurous notes, attributed mainly to elevated levels of dimethyl sulfide, acetaldehyde, methyl acetate, 2,3-pentanedione, and diacetyl (McCaig & Morrison, 1984; Van Vuuren, Cosser, & Prior, 1980). Therefore, considering their adverse impact within the challenging milieu of more traditional alcoholic beverages, it is reasonable to extrapolate that these organisms would similarly precipitate flaws in across the broad range of LABs and non-alcoholic beers as well as hard seltzers. *Klebsiella* spp. are commonly encountered in diverse aquatic settings, particularly those receiving industrial wastewater; plants and plant-derived products; mammalian intestinal tracts; as well as sugary and acidic foods and various other ecological niches (Grimont & Grimont, 2005b). In the context of brewery environments, notable species include *Klebsiella aerogenes, Klebsiella terrigena, Klebsiella variicola* and *Klebsiella oxytoca* (De Roos & De Vuyst, 2019; De Roos, Vandamme, & De Vuyst, 2018; Martens et al., 1991; Van Vuuren & Priest, 2003; Vaughan et al., 2005). In non-alcoholic beers and LABs, indole-negative *Klebsiella* spp. may lead to the development of undesirable phenolic off-flavours occurring from the decarboxylation of *p*-coumaric and ferulic acids to 4-vinylphenol and 4-vinylguaiacol, along with the production of volatile organo-sulphur compounds, like dimethyl sulfide (Blomqvist et al., 1993; Hunter, Manter, & Van Der Lelie, 2012; Uchiyama, Hashidoko, Kuriyama, & Tahara, 2008; Van Vuuren & Priest, 2003; Virkajärvi, Vauhkonen, & Storgårds, 2001). One species within this group, *K. terrigena*, thrives within the pH range of 6.0−10.0 yet exhibits significantly lower growth at pH levels of 4.0 and 10.0, failing to proliferate under strongly acidic conditions (pH $\leq$ 3.0) (Tantasuttikul & Mahakarnchanakul, 2019).

Enterobacter spp. are pervasive inhabitants of diverse natural ecosystems, spanning water bodies, sewage systems, soil, meat, and the epidermal landscapes of both humans and animals, as well as within their gastrointestinal tracts (Grimont & Grimont, 2005a, pp. 1−17). Notably, within the realm of brewing, *Enterobacter cloacae*, *Enterobacter hormaechei* and *Enterobacter kobei* emerge as the most prevalent (De Roos & De Vuyst, 2019; De Roos et al., 2018; Spitaels et al., 2015). Aligned with their counterparts in the *Enterobacteriaceae* family, contamination by *Enterobacter* spp. poses a notable risk, capable of instilling unwanted phenolic off-flavours and releasing volatile dimethyl sulfide into non-alcoholic beer and LABs, including low-alcoholic seltzers (Anness, 1980; Hunter et al., 2012). Across different matrices, the inhibitory effect on *E. cloacae* is complete at a pH of 4, with substantial inhibition observed at a pH of 4.5 (Azis et al., 2019). However, interestingly, certain strains of *Enterobacter* have exhibited a wider growth ambit, thriving within a pH range of 4.0−10.2 (Bevilacqua, Cannarsi, Gallo, Sinigaglia, & Corbo, 2010). This poses a pertinent concern as this pH spectrum coincides with the range commonly observed in LABs and non-alcoholic beverages, accentuating the need for stringent vigilance.

13.4.2 *Clostridium*

The genus *Clostridium* is one of the largest bacterial genera, boasting more than 160 validly recognised species. Its hallmark characteristic lies in the remarkable diversity exhibited by its member species; however, broadly speaking, organisms belonging to this taxonomic group are obligately anaerobic to aerotolerant, characterised by their catalase-negative and oxidase-negative nature, and are Gram-positive pleiomorphic rod-shaped cells that typically form oval or spherical endospores that often distend the vegetative cell due to their size (Dürre, 2016; Rainey, Hollen, & Small, 2015). These microorganisms are ubiquitous in nature and have been isolated from various ecological niches, including soil, river and marine sediments, the intestinal tracts of animals and various other biotopes, to name a few (Dürre, 2016; Rainey et al., 2015). Therefore, it should not be surprising to learn that bacteria belonging to the *Clostridium* genus frequently find their way into the brewing process alongside the primary raw materials, including hops, malt, malt substitutes (such as unmalted grains, extracts, syrups) and various flavour enhancers (such as juices and concentrates) and therefore can be isolated in their vegetative form from intermediate brewing stages, like boiled wort (Bokulich & Bamforth, 2013; Brožová, Kubizniaková, & Matoulková, 2018; Feng, Churey, & Worobo, 2010; Hawthorne et al., 1991).

In traditional beer varieties, the combined suppressive effects of ethanol, hop bitter acids and acidic pH present sufficient hurdles to inhibit their development; however, the resilience as endospores permits them to persist throughout mashing, wort boiling and pasteurisation (Bortoluzzi, Menten, Romano, Pereira, & Napty, 2014; Daifas et al., 2003; Doyle et al., 2015; Hawthorne et al., 1991; Munford et al., 2017; Sleha et al., 2021; Vriesekoop et al., 2012). In more vulnerable mediums, such as an unhopped wort or a fruit-infused non-alcoholic beer with an elevated pH and minimal bitterness, *Clostridium* bacteria may thrive. As they multiply, their metabolic processes could yield an array of undesirable compounds, profoundly impacting the sensory experience of the final product. These metabolites include acetic, butyric, propionic, valeric, and caproic acids, alongside sulphur compounds, which often evoke descriptors like cheesy, buttery, putrid and rancid in both taste and aroma (Atasoy & Cetecioglu, 2020; Rainey et al., 2015; Stadtman, Stadtman, & Barker, 1949; Storgårds, 2000). Moreover, certain members of the *Clostridium* genus, notably *Clostridium botulinum* and *Clostridium perfringens*, possess the ability to synthesise toxins that pose a significant threat to human health when consumed (Chaidoutis et al., 2022; Dürre, 2016; Grenda et al., 2023). While standard-strength traditional-style beers typically thwart the proliferation of both non-toxigenic and toxigenic clostridia, these bacteria present a potential hazard in more susceptible environments such as those of non-alcoholic beer and LABs. However, as the ethanol concentration necessary to inhibit the growth of certain clostridial strains (6% v/v) contradicts the formulation of no-alcoholic beer or LABs and many hard seltzers, and the available data on alpha-acid sensitivities from hops is limited, stringent pH control (<4.0 pH) throughout the entire process remains the most practical measure to prevent potential product spoilage and intoxication by commonly encountered *Clostridium* spp. (Daifas et al., 2003; Juneja, Baker, Thippareddi, Snyder, & Mohr, 2013; Lund, Graham, & Franklin, 1987; Rainey et al., 2015; Valero et al., 2020).

13.4.3 *Bacillus, Brevibacillus* and *Paenibacillus*

Given their close taxonomic proximity, the genera *Bacillus*, *Brevibacillus*, *Paenibacillus* and other previously classified *Bacillus* species will be collectively examined. These predominately catalase-positive bacteria, characterised predominantly by their rod-shaped morphology, saprophytic nature and capacity for endospore formation, exhibit a remarkable array of physiological traits. Among this taxonomic group, Gram-staining may vary, as can motility, ranging from peritrichous flagella to non-motile forms. Additionally, they display diverse respiratory requirements, spanning from strictly aerobic to strictly anaerobic (Logan & DeVos, 2015a, 2015b, pp. 1–22; Priest, 2015). Moreover, they thrive in various environmental conditions, from psychrophilic to thermophilic, halophilic, acidophilic and alkaliphilic habitats (Logan & DeVos, 2015a, 2015b, pp. 1–22; Priest, 2015). Microbial species within these genera are ubiquitous in nature, often discovered in soils, soil-contaminated environments, water bodies, foodstuffs and even beverages such as beer (Logan & DeVos, 2015a, 2015b, pp. 1–22; Munford et al., 2017; Priest, 2015). Their endurance against harsh conditions including heat, radiation, disinfectants and desiccation, owing to the remarkable resilience of their endospores, renders their eradication from food and beverages a formidable challenge (Wells-Bennik et al., 2016). Consequently, due to their ubiquitousness in nature and the resistant nature of their endospores, these bacteria readily infiltrate entry into the brewing process with raw materials like water, malt, hops and spices (Baik et al., 2010; Cufaoglu & Ayaz, 2022; Noots, Delcour, & Michiels, 1999; Vaughan et al., 2005; Wells-Bennik et al., 2016). Nevertheless, in most scenarios, the growth of these genera is impeded by the intrinsic antimicrobial defences inherent to traditional beer, notably the presence of hop acids and the acidic pH (Bokulich, Bamforth, & Mills, 2012; Haas & Barsoumian, 1994; Teuber & Schmalreck, 1973; Vriesekoop et al., 2012). A study led by Munford et al. (2017) explored the spoilage potential of these endospore-producing genera, scrutinising 248 strains for the *horA* and *horC* genes associated with hop resistance. Among them, only 14 strains were found to harbour these hop-resistant genes. Further examination involving five strains revealed no capacity to spoil both lager-type (~5% ABV) and AFBs. The researchers postulated that the relatively low pH (<4.5) of the experimental beers played a pivotal role in inhibiting the germination and proliferation of spore-forming bacteria.

In certain instances, well-documented occurrences have revealed that spore-forming bacteria such as *Bacillus cereus*, *Bacillus licheniformis* and *Paenibacillus humicus*, each harbouring the hop-resistance *horA* gene, can precipitate undesired aesthetic characteristics in beer with alcohol volumes of 4% v/v and 5% v/v (Haakensen & Ziola, 2008); however, consistent with the findings of Munford et al. (2017) the pH measurements recorded for these contaminated beers, which harbour spore-forming microorganisms, ranged between 4.8 and 5.2. However, a solitary *Bacillus cereus* strain has been documented for its capability to spoil commercially available beer with a modest alcohol-by-volume content (3.2%) and a pH level of 4.3 (Wang, Liu, Sun, Du, & Li, 2017). This contamination led to an increase in glyoxylic acid, pyruvic acid, lactic acid, ethyl formate, ethyl acetate, isobutyl acetate, ethyl caprylate and biogenic amines in comparison to uncontaminated samples (Wang et al., 2017). Given the diverse pH growth spectrum observed among spore-forming organisms within these genera, it is conceivable that certain species may possess an inherent propensity to flourish in no- and low-alcoholic beers, even at traditional pH ranges.

Most species within these genera are typically deemed non-pathogenic or exhibit low pathogenicity, rarely associated with human ailments. However, specific strains possess the capacity to induce foodborne illnesses in humans, with *Bacillus cereus* standing out as the most notable and prevalent culprit. *Bacillus cereus* can prompt food poisoning in humans through two distinct mechanisms. Diarrhoeal illness ensues from the synthesis of enterotoxins concurrent with *Bacillus cereus* proliferation in the small intestine post-consumption, whereas the emetic toxin, known as cereulide, induces symptoms such as nausea, vomiting and malaise through the proliferation of cells present within food under aerobic conditions (Granum & Lund, 1997; Jääskeläinen, Häggblom, Andersson, & Salkinoja-Salonen, 2004). Given the remarkable adaptability of certain strains of *B. cereus* to thrive across a broad temperature spectrum spanning from 4 to 48°C, and within a pH range extending from 3.5 to 9.3, albeit with inhibition occurring below a pH threshold of 4.8, it is conceivable that *B. cereus* presents both spoilage and food safety concerns within the context of non-alcoholic beer and LAB as well as hard seltzers production (Carlin et al., 2013; Lanciotti, Sinigaglia, Gardini, Vannini, & Guerzoni, 2001;

Schneider et al., 2017; Wang et al., 2017). The heightened risk becomes particularly pronounced when beverage products lack hop acids and exhibit an elevated pH. Nonetheless, any potential health concerns are more apt to arise from enterotoxins, as the aerobic conditions needed for emetic toxin generation are generally scarce throughout the brewing procedure.

13.4.4 *Listeria* and *Staphylococcus*

The *Listeria* genus comprises a group of Gram-positive, short rod-shaped bacteria that thrive aerobically or under facultative anaerobic conditions. They are characterised by their catalase-positive, oxidase-negative nature, absence of sporulation and consistent motility through peritrichous flagella (Liu & Zhang, 2011). Members commonly inhabit diverse natural settings, including soil, sewage, vegetation and the faecal matter of both animals and humans (Liu & Zhang, 2011). One of the most prevalent species in this genus is *Listeria monocytogenes*, renowned for causing listeriosis, a foodborne ailment marked by symptoms such as fever, muscle discomfort, nausea, vomiting, diarrhoea, and in severe instances cognitive confusion, loss of balance, convulsions and even death. Previous studies by Menz et al. (2010) examining the viability of foodborne pathogens in sweet wort and beer revealed that even at low to moderate concentrations, hop acids exceeding 5 International Bitterness Units (IBU) effectively suppressed the proliferation of *Listeria monocytogenes* in unfermented sweetened wort. Considering the widespread utilisation of hops in the beer brewing process, brewers commonly overlook *L. monocytogenes* as a substantial concern, even amidst the crafting of LABs or AFBs. However, a notable escalation in risk emerges within the production of hard seltzers. This concern arises due to the non-essential incorporation of hops into seltzer production. Notably, *L. monocytogenes* exhibits remarkable resilience to acidity, thriving within a pH range of 4.0–9.5 in certain environments, and demonstrates significant tolerance to ethanol (~5% v/v) (Conner, Scott, & Bernard, 1990; Oh & Marshall, 1993; Schulz, Konrath, Rismondo, 2023).

Staphylococcus spp. are characterised as non-motile, spherical Gram-positive facultative anaerobes, with very few exceptions, which are predominately catalase-positive and oxidase-negative, accelerated growth under aerobic environments (Schleifer & Bell, 2015). In nature, these inhabit the skin, skin glands and mucous membranes of warm-blooded creatures; however, their presence is not confined solely to such domains; one may also encounter them in connection with various animal byproducts and other environmental reservoirs (Schleifer & Bell, 2015). Regarding foodborne illness, *Staphylococcus aureus* demands scrutiny due to its ability to produce enterotoxins, which can lead to staphylococcal food poisoning (Argudín, Mendoza, & Rodicio, 2010). This intoxication arises subsequent to the consumption of food or beverages imbued with quantities of one or more pre-formed enterotoxins, leading to the rapid onset, typically within 2–8 h, of symptoms encompassing nausea, intense vomiting, abdominal distress and diarrhoea (Argudín et al., 2010). Typically, the contamination of food and beverages originates from cross-contamination facilitated by food handlers harbouring enterotoxin-producing *S. aureus*; however, prior research demonstrated *S. aureus* is unable to endure prolonged periods in sweet, unhopped wort at a pH of 4.3 (Menz et al., 2010). While the findings of Menz and colleagues did not explicitly demonstrate this phenomenon, supplementary evidence suggests that *S. aureus* may indeed thrive in an environment with a pH of 4.0, given particular environmental circumstances (Lund & Gould, 2000; Scientific Committee on Veterinary Measures Relating to Public Health, 2003).

Moreover, akin to the response observed in *Listeria monocytogenes*, *Staphylococcus aureus* demonstrated heightened inhibition to increasing concentrations of hop iso-α-acids (Menz et al., 2010). While the impact of *S. aureus* on nonalcoholic beer and LABs might be insignificant, this might not be the case for hard seltzers. In a study conducted by Pando et al. (2017), an analysis of over 100 strains of *S. aureus* revealed varying minimum inhibitory concentrations (MICs) of ethanol, ranging from 6% to 11% v/v, and minimum bactericidal concentrations (MBCs) between 8% and 15% v/v, as determined in Mueller-Hinton broth (Pando et al., 2017). Considering the typical alcohol content of hard seltzers falls between 2.5% and 8.0% ABV, *S. aureus* presents a potential food safety hazard.

13.4.5 *Leuconostoc* and *Oenococcus*

Leuconostoc, a genus within the order *Lactobacillales*, are characterised as Gram-positive bacteria that are obligate fermenters, relying on organic compounds for energy (Tieghem, 2015). These organisms are catalase-negative and manifest as immotile, ellipsoidal-to-spherical cells, frequently observed in paired or abbreviated chains (Tieghem, 2015). Characterised as psychrotrophic mesophiles, optimal growth occurs between 14 and 30°C; however, the range differs between distinctive species and strains span 1–10 to 30–40°C (Tieghem, 2015). Leuconostocs thrive within both natural and manmade settings, often in the company of other lactic acid bacteria, including lactobacilli, pediococci and carnobacteria (Tieghem, 2015). They frequently inhabit environments rich in decomposing plant matter and its derivatives and are also regularly found in many animal-derived products like raw milk, dairy goods, meat, poultry and fish, although documentation of their occurrence in the gastrointestinal tracts and mucous membranes of warm-blooded animals is infrequent (Tieghem, 2015). *Leuconostoc* spp. are nutritionally fastidious organisms, having limited biosynthetic capacity and their requirements for growth factors, pre-existing amino acids, purine and pyrimidine bases and other nutrients, like D-pantothenate, to thrive, and unlike other lactic acid bacteria, *Leuconostoc* spp. lack the Embden-Meyerhof-Parnas pathway, the prevailing form of glycolysis. Instead, they function as obligate heterofermentatives and ferment glucose via the phosphoketolase pathway, yielding equivalent quantities of lactic acid, carbon dioxide and ethanol (Tieghem, 2015). Consequently, the distinctive sour note linked with *Leuconostoc* spoilage originates from the accumulation of D(−) lactic and acetic acid; nevertheless, the formation of acids is contingent upon the particular species and strain, the availability of oxygen and the substrates furnished (Borch & Molin, 1989; Zaunmüller, Eichert, Richter, Unden, 2006). Leuconostocs are distinguished for their capacity to produce diacetyl, acetoin and 2,3-butanediol in environments containing citrate or fructose, as well as for their ability to generate a viscous, slime-like texture through the excretion of long-chain, extracellular polysaccharides (Sarwat, Qader, Aman, & Ahmed, 2008). As acid-sensitive microorganisms, this particular group's capacity to flourish in and compromise the quality of non-alcoholic beers and LABs, as well as hard seltzers, parallels that of numerous other organisms, primarily contingent upon the beverage's acidity level and substrate accessibility. Given the recognised sensitivity of leuconostocs to acidity, with notable growth suppression observed below a pH threshold of 4.8, a prudent strategy in crafting novel products might entail precise pH regulation and the omission of ingredients containing D-pantothenate or its derivatives (Hemme & Foucaud-Scheunemann, 2004). Moreover, it is critical to mention that ethanol poses a minimal hindrance to leuconostocs, with experimental evidence revealing that ethanol inhibition was observed only at 8% (v/v) and 10% (v/v) is needed for complete inhibition (Hemme & Foucaud-Scheunemann, 2004; Pittet, Morrow, Ziola, 2011; Säde, 2011; Tieghem, 2015).

The genus *Oenococcus*, in close kinship with *Leuconostoc*, mirrors its counterpart in several respects. Much like *Leuconostoc*, species within this genus exhibit facultative anaerobiosis, stain Gram-positive, lack spore formation, are catalase-negative, lack motility and commonly manifest an ellipsoidal to spherical morphology, often arranged in pairs or chains (Dicks & Holzapfel, 2015). Likewise, *Oenococcus* metabolises glucose to produce equal quantities of D(−)lactic acid, carbon dioxide and either ethanol or acetate via fermentation (Dicks & Holzapfel, 2015). However, in contrast to their close relatives, *Oenococcus* members exhibit diverse pH preferences and predominantly inhabit grape must and wine environments (Dicks & Holzapfel, 2015). Select oenococci excel in acidic environments, thriving at an initial growth pH of 4.8, while others thrive in neutral to slightly acidic conditions, demonstrating optimal growth within a pH range of 6.0−6.8 (Dicks & Holzapfel, 2015). Certain strains of *O. oeni* have demonstrated a remarkable ability to thrive even in environments with a pH below 3.5 (G-Alegría et al., 2004; Mota et al., 2018). Furthermore, research has revealed their remarkable resilience in adverse environmental conditions. Certain species demonstrate an impressive capacity to thrive even in ethanol concentrations reaching up to 13% v/v (Dicks & Holzapfel, 2015; G-Alegría et al., 2004; Pittet et al., 2011).

13.5 Mitigating against spoilage of LABs, AFBs and hard seltzers

The method of production and properties of the final product have been shown to influence the susceptibility of LABs, AFBs and hard seltzers to microbial infection. Similarly, extrinsic and intrinsic methods may be used to reduce the likelihood of contamination and subsequent spoilage.

Operational-based (extrinsic) methods include the process and filling equipment design, cleaning and disinfection procedures (type, concentration, contact time, flow rate, etc), raw material and water quality, as well as product treatment (heat, filtration). Mitigation strategies to reduce spoilage incidents should be in place regardless of whether the product is low, or free of, alcohol, as discussed within various chapters within this text. However, particular attention should be paid to pasteurisation regime, if used.

Pasteurisation regimes must be effective against heat-resistant microorganisms potentially present as a spoilage contaminant. The brewing industry routinely uses a simplified Pasteurisation Unit (PU) metric, where **P** is the required time to reduce the microbial population by the **z** required temperature changes. The **D** value means the required time over a temperature defined for a microbial decimal reduction (Table 13.4; Wray, 2015).

Microbial thermal resistance to heat is dependent on ethanol concentration (Kilgour & Smith, 1985). Consequently, achieving optimal sterility levels in LAB and AFB necessitates more PUs. Specifically, LAB demands twice the standard amount typically needed for a standard ale or lager, while AFB requires four times as much (Table 13.5; Wray, 2015; Rachon, Rice, Pawlowsky, & Raleigh, 2018).

TABLE 13.4 Microorganism decimal-reduction stated by D and z values (Wray, 2015).

Microorganism	D_{60} value (min)	z value (°C)
Saccharomyces cerevisiae	0.01	4.6
Saccharomyces pastorianus	0.004	4.4
Saccharomyces diastaticus	0.06	7.8
Pichia membranaefaciens	0.00025	2.8
Hansenula anomala	0.0039	4.6
Aspergillus niger	0.04	3.7
Lacticaseibacillus paracasei	0.02	6.5
Lentilactobacillus parabuchneri	0.44	15
Lactobacillus delbrueckii	0.091	12
Pediococcus species	0.00073	4

Other operational-based methods of significance include those relating to draft dispense. Given the enhanced susceptibility to spoilage of LAB and AFB, robust quality assurance protocols should be implemented to ensure that cellars are kept at an appropriate temperature and that dispense lines are sufficiently clean (Quain, 2015, 2021). Silver impregnation of lines has been suggested as a method to produce an antimicrobial effect and to aid in preventing biofilm formation (Mohanta et al., 2020). Pulsed electric fields could also be used to inactivate any microbes present (Puligundla, Pyun, & Mok, 2018).

Intrinsic methods relate to the attributes of the product and to changes that may be made to the composition to improve microbial stability. As we have found, beers have a range of pH values that span between 3.4 and 4.8 (Vriesekoop et al., 2012). This acidic environment presents a challenge to most microorganisms, as weak organic acids, such as lactic or acetic acid, are protonated at low pH and can permeate the lipid bilayer, causing cytoplasmic acidification. This collapse in the proton gradient across the cell membrane disrupts cell enzyme function and nutrient transport (Mira & Teixeira, 2013). Hop iso-α-acids are also characterised as weak acids and act as ionophores, stimulated at low pH (Behr & Vogel, 2009). Lowering the pH of LAB and AFB would improve resistance to infection, limited only by how acidic the beer can be made without compromising sensory characteristics.

Preservatives are commonly used in various food products, including soft drinks. They encompass a variety of substances, such as sorbates, benzoates, sulphites and nitrites. When it comes to beer production, nitrite addition should be avoided, and regulations in some countries restrict the other preservative options available. Among the approved choices, sodium benzoate and potassium sorbate are two of the limited options, with a maximum allowable limit of 200 parts per million (ppm) specifically for kegged beer in Europe (EU Regulations, 2016). The regulations regarding the use of preservatives vary between countries (Table 13.6), and care should be taken in case of export.

Potassium sorbate is commonly used in apple cider, wine and soft drinks, as is sulphur dioxide. The addition of preservatives to LAB and AFB was investigated by Bartlett (2023). Experiments on spoilage revealed that, within

TABLE 13.5 Common Pasteurisation Unit (PU) values for beverages (Wray, 2015).

Product	Usual minimum PU	Usual maximum PU
Pilsner and lager beer	15	25
Ale and stout	20	35
Low-alcohol beer	40	60
Non-alcohol beer	80	120
Lemonade	300	500
Fruit juices	3000	5000

TABLE 13.6 Allowable application of preservative in non-alcoholic beers.

Country/Region	Allowable preservative	Concentration
EU	Benzoic acid/benzoates Sorbic acid/sorbates	200 mg/L in kegged beer 200 mg/L in kegged beers
UK	Benzoic acid/benzoates	70 mg/L in small pack 200 mg/L in kegged beer
USA	Benzoic acid/benzoates Sorbic acid/sorbates	0.1% (w/v) = 1000 mg/L 0.1% (w/v) = 1000 mg/L
Australia & New Zealand	Benzoic acid/benzoates Sorbic acid/sorbates	70 mg/L 70 mg/L
Russia	Benzoic acid/benzoates	200 mg/L kegged only

allowable concentrations in various food items, potassium sorbate and sodium benzoate could suppress the proliferation of *Pichia membranifaciens* and *Levilactobacillus brevis* to a degree comparable to a complete alcohol-based alternative. Sulphur dioxide displayed effectiveness against Gram-negative bacteria, while its impact on Gram-positive bacteria and yeasts was much more limited. Sulphur dioxide has a limited antimicrobial role in beers because at the pH of typical beers, sulphur dioxide is predominantly present in bisulphite form. The undissociated form (SO_2 H_2O) is by far the most-effective bacteriostatic form, while the bisulphite form (H SO_3^-) has only a limited bacteriostatic effect and only when unbound (Vriesekoop, 2021). At the pH of beer, the bisulphite ions are mostly bound to carbonyl compounds and hence not readily available to act as preservatives. However, regardless of the potential for sulphur dioxide to play a preservative role in beers, it is a reportable allergen in all countries when present over 10 ppm with variable upper limits (15–50 ppm) in different countries (Vriesekoop, 2021).

Another internal technique, primarily designed for preserving fruit juice, is Velcorin. This is the commercial term for dimethyl decarbonate (DMDC), a cold sterilising agent that demonstrates effectiveness against a broad spectrum of yeast, bacteria and moulds (Threlfall & Morris, 1995). DMDC penetrates the cell and deactivates enzymes involved in microbial metabolism, rapidly killing cells and breaking down into methanol and carbon dioxide. It was approved for use in wines in 1988 by the FDA and has also found application in hard seltzers and beer-based products.

The potential application of naturally occurring antimicrobial peptides as bio-preservatives has been determined for both nisin and human defensin. Nisin is a polycyclic antibacterial peptide produced by *Lactococcus lactis* that is effective against a wide range of Gram-positive bacteria (Delves-Broughton, Blackburn, Evans, & Hugenholtz, 1996). Nisin could be applied at any stage during the brewing process to improve the biological stability of beers since it is extremely heat stable (withstands pasteurisation and wort boil) and is retained post kieselguhr filtration (Müller-Auffermann, Grijalva, Jacob, & Hutzler, 2015). A bio-engineered strain of *Saccharomyces pastorianus* has been created to excrete human defensin during fermentation, which is effective against a range of common beer spoilage organisms (James et al., 2014).

Biopreservation through the application of bacteriophages, viruses that infect bacteria, has been investigated in a range of food products with commercially available 'phage cocktails' in use for ready-to-eat meat products and cheese (Kawacka, Olejnik-Schmidt, Schmidt, & Sip, 2020). Several phages capable of infecting *Levilactobacillus brevis* beer-spoiling strains have been isolated and shown to demonstrate lytic activity (Deasy, Mahony, Neve, Heller, & Van Sinderen, 2011; Feyereisen et al., 2019), but further work is needed in this area to demonstrate phage propagation beyond lysis (Feyereisen et al., 2019). Further methods of biopreservation have recently been reviewed by Kordialik-Bogacka (2022).

13.6 Conclusions and future considerations

For ages, the brewing world has acknowledged the varying susceptibilities of beers, whether alcoholic or non-alcoholic, to adulteration and spoilage by microorganisms. Nevertheless, the recent unprecedented rise in demand for innovative non-alcoholic beers and LABs and hard seltzers poses an added challenge to brewers, as these beverages with their unique composition, lack the inhibitory hurdles inherent to traditional, standard strength beers and are particularly susceptible not only to contamination by traditional beer spoilers but also to a wider array of opportunistic microorganisms. As a result, a growing number of brewers feel compelled to venture into the realm of non-alcoholic and low-alcoholic beverages and hard seltzers, driven by mounting market competition. However, this expansion presents numerous challenges, as they often lack familiarity with the expanded set of potential spoilage organisms, lack validated methods for their precise detection and enumeration and face heightened risks associated with inadequate protective food safety measures. Underestimating these inadequacies can result in substantial financial losses due to compromised product quality.

In our collective perspective, drawing upon extensive literature, we encourage non-alcoholic beers and LABs, as well as hard seltzers, to maintain a strict pH below 4.0, restrict excess fermentable extract and free amino nitrogen content and undergo suitable stabilisation. This may entail a combination of processing methods such as sterile filtration, flash and tunnel pasteurisation, and/or the incorporation of preservatives or anti-microbial agents to assist in mitigating potential microbiological hazards. It is imperative to acknowledge that stabilisation techniques implemented post-packaging, such as tunnel pasteurisation, preservatives and antimicrobial agents, offer the highest degree of security and efficacy. When considering non-alcoholic beers and LABs, or seltzers infused with hops, we additionally advise ensuring a bitterness level exceeding 5 International Bitterness Units (IBU) where feasible as a preventive measure against some notable pathogens. These measures, in combination with the meticulous application of stringent hygienic protocols coupled with the strategic implementation of chilled temperature storage, can greatly mitigate against the risk of spoilage, ensuring product integrity and safety. It is imperative to bear in mind that for draught products, interaction with tainted nozzles, sparklers and couplers or inadequately upheld long-draw draught systems could usher in organisms capable of spoiling the product or posing potential health hazards to consumers. While it is generally not advisable to serve non-alcoholic beers and LABs on draught, if done, it is recommended that draught non-alcoholic beers and LABs and seltzers be dispensed through meticulously crafted standalone systems designed to mitigate the risk of microbiological contamination, for instance through the incorporation of a short-draw disposable dispense line.

In essence, the burgeoning interest in non-alcoholic beers and LABs and hard seltzers presents significant hurdles for the brewing industry, challenges that may prove daunting for brewers lacking sufficient resources. Thus, it is prudent for aspiring entrants to this market segment to engage with seasoned experts in food safety and microbiology before embarking. Moreover, given the escalating demand for these alternative beverages, further research is needed to thoroughly explore and fully understand any potential hazards inherent to their production processes or consumption.

References

Al-Kharousi, Z. S., Guizani, N., Al-Sadi, A. M., Al-Bulushi, I. M., & Shaharoona, B. (2016). Hiding in fresh fruits and vegetables: Opportunistic pathogens may cross geographical barriers. *International Journal of Microbiology, 2016*. https://doi.org/10.1155/2016/4292417

Anderson, P., O'Donnell, A., Kokole, D., Llopis, E. J., & Kaner, E. (2021). Is buying and drinking zero and low alcohol beer a higher socio-economic phenomenon? Analysis of British survey data, 2015–2018 and household purchase data 2015–2020. *International Journal of Environmental Research and Public Health, 18*(19), Article 10347. https://doi.org/10.3390/ijerph181910347

Anness, B. J. (1980). The reduction of dimethyl sulphoxide to dimethyl sulphide during fermentation. *Journal of the Institute of Brewing, 86*(3), 134–137.

Argudín, M.Á., Mendoza, M. C., & Rodicio, M. R. (2010). Food poisoning and *Staphylococcus aureus* enterotoxins. *Toxins, 2*(7), 1751–1773. https://doi.org/10.3390/toxins2071751

Atasoy, M., & Cetecioglu, Z. (2020). Butyric acid dominant volatile fatty acids production: Bio-Augmentation of mixed culture fermentation by *Clostridium butyricum*. *Journal of Environmental Chemical Engineering, 8*(6), Article 104496. https://doi.org/10.1016/j.jece.2020.104496

Azis, K., Zerva, I., Melidis, P., Caceres, C., Bourtzis, K., & Ntougias, S. (2019). Biochemical and nutritional characterization of the medfly gut symbiont *Enterobacter* sp. AA26 for its use as probiotics in sterile insect technique applications. *BMC Biotechnology, 19*(Suppl. 2). https://doi.org/10.1186/s12896-019-0584-9

Baik, K. S., Lim, C. H., Park, S. C., Kim, E. M., Rhee, M. S., & Seong, C. N. (2010). *Bacillus rigui* sp. nov., isolated from wetland fresh water. *International Journal of Systematic and Evolutionary Microbiology, 60*(9), 2204–2209. https://doi.org/10.1099/ijs.0.018184-0

Bartlett, C. (2023). The growth of pathogens and beer spoilage organisms in No and low alcohol beers and strategies to prevent their growth. In *MRes brewing science*. UK: University of Nottingham.

Behr, J., & Vogel, R. F. (2009 Jul 22). Mechanisms of hop inhibition: Hop ionophores. *Journal of Agricultural and Food Chemistry, 57*(14), 6074–6081. https://doi.org/10.1021/jf900847y

Bellut, K., & Arendt, E. K. (2019). Chance and challenge: Non-*Saccharomyces* yeasts in nonalcoholic and low alcohol beer brewing—A review. *Journal of the American Society of Brewing Chemists, 77*(2), 77–91. https://doi.org/10.1080/03610470.2019.1569452

Bellut, K., Krogerus, K., & Arendt, E. K. (2020). *Lachancea fermentati* strains isolated from kombucha: Fundamental insights, and practical application in low alcohol beer brewing. *Frontiers in Microbiology*, 764.

Bellut, K., Michel, M., Hutzler, M., Zarnkow, M., Jacob, F., De Schutter, D. P., et al. (2019). Investigation into the potential of *Lachancea fermentati* strain KBI 12.1 for low alcohol beer brewing. *Journal of the American Society of Brewing Chemists, 77*(3), 157–169.

Bellut, K., Michel, M., Zarnkow, M., Hutzler, M., Jacob, F., De Schutter, D. P., et al. (2018). Application of non-*Saccharomyces* yeasts isolated from kombucha in the production of alcohol-free beer. *Fermentation, 4*(3), 66.

Bergsveinson, J., Redekop, A., Zoerb, S., & Ziola, B. (2015). Dissolved carbon dioxide selects for lactic acid bacteria able to grow in and spoil packaged beer. *Journal of the American Society of Brewing Chemists, 73*(4), 331–338. https://doi.org/10.1094/ASBCJ-2015-0726-01

Berry, E. D., & Foegeding, P. M. (1997). Cold temperature adaptation and growth of microorganisms. *Journal of Food Protection, 60*(12), 1583–1594. https://doi.org/10.4315/0362-028X-60.12.1583

Bevilacqua, A., Cannarsi, M., Gallo, M., Sinigaglia, M., & Corbo, M. R. (2010). Characterization and implications of *Enterobacter cloacae* strains, Isolated from Italian table olives "bella di cerignola." *Journal of Food Science, 75*(1), 53–60. https://doi.org/10.1111/j.1750-3841.2009.01445.x

Blomqvist, K., Nikkola, M., Lehtovaara, P., Suihko, M. L., Airaksinen, U., Straby, K. B., et al. (1993). Characterization of the genes of the 2,3-butanediol operons from *Klebsiella terrigena* and *Enterobacter aerogenes*. *Journal of Bacteriology, 175*(5), 1392–1404. https://doi.org/10.1128/jb.175.5.1392-1404.1993

Bokulich, N. A., & Bamforth, C. W. (2013). The microbiology of malting and brewing. *Microbiology and Molecular Biology Reviews, 77*(2), 157–172. https://doi.org/10.1128/mmbr.00060-12

Bokulich, N. A., Bamforth, C. W., & Mills, D. A. (2012). Brewhouse-resident microbiota are responsible for multi-stage fermentation of American coolship ale. *PLoS One, 7*(4). https://doi.org/10.1371/journal.pone.0035507

Borch, E., & Molin, G. (1989). The aerobic growth and product formation of *Lactobacillus*, *Leuconostoc*, *Brochothrix*, and *Carnobacterium* in batch cultures. *Applied Microbiology and Biotechnology, 30*(1), 81–88. https://doi.org/10.1007/BF00256001

Bortoluzzi, C., Menten, J. F. M., Romano, G. G., Pereira, R., & Napty, G. S. (2014). Effect of hops β-acids (*Humulus lupulus*) on performance and intestinal health of broiler chickens. *The Journal of Applied Poultry Research, 23*(3), 437–443. https://doi.org/10.3382/japr.2013-00926

Brányik, T., Silva, D. P., Baszczyňski, M., Lehnert, R., & Almeida e Silva, J. B. (2012). A review of methods of low alcohol and alcohol-free beer production. *Journal of Food Engineering, 108*(4), 493–506.

Brenner, D. J., & Farmer III, J. J. (2015). Bergey's manual of systematics of archaea and bacteria. In *Bergey's manual of systematics of archaea and bacteria* (2015th ed.). Springer. https://doi.org/10.1002/9781118960608

Brožová, M., Kubizniaková, P., & Matoulková, D. (2018). Brewing microbiology - bacteria of the genus *Clostridium*. *Kvasny Prumysl, 64*(5), 242–247. https://doi.org/10.18832/kp201830

Carlin, F., Albagnac, C., Rida, A., Guinebretière, M. H., Couvert, O., & Nguyen-the, C. (2013). Variation of cardinal growth parameters and growth limits according to phylogenetic affiliation in the *Bacillus cereus* Group. Consequences for risk assessment. *Food Microbiology, 33*(1), 69–76. https://doi.org/10.1016/j.fm.2012.08.014

Case, A. C. (1965). Conditions controlling *Flavobacterium proteus* in brewery fermentations. *Journal of the Institute of Brewing, 71*, 250–256.

Chaidoutis, E., Keramydas, D., Papalexis, P., Migdanis, A., Migdanis, I., Lazaris, A. C., et al. (2022). Foodborne botulism: A brief review of cases transmitted by cheese products (review). *Biomedical Reports, 16*(5), 1–7. https://doi.org/10.3892/BR.2022.1524

Chatterjee, I., Somerville, G. A., Heilmann, C., Sahl, H. G., Maurer, H. H., & Herrmann, M. (2006). Very low ethanol concentrations affect the viability and growth recovery in post-stationary-phase *Staphylococcus aureus* populations. *Applied and Environmental Microbiology, 72*(4), 2627–2636.

Conner, D. E., Scott, V. N., & Bernard, D. T. (1990). Growth, inhibition, and survival of *Listeria monocytogenes* as affected by acidic conditions. *Journal of Food Protection, 53*(8), 652–655.

Cufaoglu, G., & Ayaz, N. D. (2022). Potential risk of *Bacillus cereus* in spices in Turkey. *Food Control, 132*(September 2021), Article 108570. https://doi.org/10.1016/j.foodcont.2021.108570

Dürre, P. (2016). Physiology and sporulation in clostridium. *The Bacterial Spore: From Molecules to Systems*, 313–329. https://doi.org/10.1128/9781555819323.ch15

Daifas, D. P., Smith, J. P., Blanchfield, B., Cadieux, B., Sanders, G., & Austin, J. W. (2003). Effect of ethanol on the growth of *Clostridium botulinum*. *Journal of Food Protection, 66*(4), 610–617. https://doi.org/10.4315/0362-028X-66.4.610

Dalberto, G., Da Rosa, M. R., Niemes, J. P., Leite, K., Kutkoski, R. F., & Da Rosa, E. A. (2021). Cold mash in brewing process: Optimization of innovative method for low-alcohol beer production. *ACS Food Science & Technology, 1*(3), 374–381.

De Francesco, G., Freeman, G., Lee, E., Marconi, O., & Perretti, G. (2014). Effects of operating conditions during low-alcohol beer production by osmotic distillation. *Journal of Agricultural and Food Chemistry, 62*(14), 3279–3286.

De Francesco, G., Marconi, O., Sileoni, V., Freeman, G., Lee, E. G., Floridi, S., et al. (2021). Influence of the dealcoholisation by osmotic distillation on the sensory properties of different beer types. *Journal of Food Science and Technology, 58*(4), 1488–1498.

De Francesco, G., Sannino, C., Sileoni, V., Marconi, O., Filippucci, S., Tasselli, G., et al. (2018). Mrakia gelida in brewing process: An innovative production of low alcohol beer using a psychrophilic yeast strain. *Food Microbiology, 76*, 354–362.

De Fusco, D. O., Madaleno, L. L., Del Bianchi, V. L., Bernardo, A. D. S., Assis, R. R., & de Almeida Teixeira, G. H. (2019). Development of low-alcohol isotonic beer by interrupted fermentation. *International Journal of Food Science and Technology, 54*(7), 2416–2424.

De Roos, J., & De Vuyst, L. (2019). Microbial acidification, alcoholization, and aroma production during spontaneous lambic beer production. *Journal of the Science of Food and Agriculture, 99*(1), 25–38. https://doi.org/10.1002/jsfa.9291

De Roos, J., Vandamme, P., & De Vuyst, L. (2018). Wort substrate consumption and metabolite production during lambic beer fermentation and maturation explain the successive growth of specific bacterial and yeast species. *Frontiers in Microbiology, 9*(NOV), 1–20. https://doi.org/10.3389/fmicb.2018.02763

Deasy, T., Mahony, J., Neve, H., Heller, K. J., & Van Sinderen, D. (2011). Isolation of a virulent *Lactobacillus brevis* phage and its application in the control of beer spoilage. *Journal of Food Protection, 74*(12), 2157–2161.

Delves-Broughton, J., Blackburn, P., Evans, R., & Hugenholtz, J. (1996). Application of the bacteriocin nisin. *Antonie van Leeuwenhoek, 69*, 193–202.

Dicks, L. M. T., & Holzapfel, W. H. (2015). *Oenococcus*. In *Bergey's manual of systematics of archaea and bacteria* (pp. 1–16). https://doi.org/10.1002/9781118960608.gbm00608

Doyle, C. J., Gleeson, D., Jordan, K., Beresford, T. P., Ross, R. P., Fitzgerald, G. F., et al. (2015). Anaerobic sporeformers and their significance with respect to milk and dairy products. *International Journal of Food Microbiology, 197*, 77–87. https://doi.org/10.1016/j.ijfoodmicro.2014.12.022

Escartin, E. F., Castillo, A., Hinojosa-Puga, A., & Saldaña-Lozano, J. (1999). Prevalence of *Salmonella* in chorizo and its survival under different storage temperatures. *Food Microbiology, 16*(5), 479–486. https://doi.org/10.1006/fmic.1999.0258

EU Regulations. (2016). *Commission implementing regulation (EU) 2016/2023 of 18 November 2016 concerning the authorisation of sodium benzoate, potassium sorbate, formic acid and sodium formate as feed additives for all animal species.*

Farmer, J. J., & Brenner, D. J. (2015). Obesumbacterium. *Bergey's manual of systematics of archaea and bacteria.* https://doi.org/10.1002/9781118960608.gbm01156

Feng, G., Churey, J. J., & Worobo, R. W. (2010). Thermoaciduric *Clostridium pasteurianum* spoilage of shelf-stable apple juice. *Journal of Food Protection, 73*(10), 1886–1890. https://doi.org/10.4315/0362-028X-73.10.1886

Fernandez, J. L., & Simpson, W. J. (1995). Measurement and prediction of the susceptibility of lager beer to spoilage by lactic acid bacteria. *Journal of Applied Microbiology, 78*(4), 419–425.

Feyereisen, M., Mahony, J., Lugli, G. A., Ventura, M., Neve, H., Franz, C. M., et al. (2019). Isolation and characterization of *Lactobacillus brevis* phages. *Viruses, 11*(5), 393.

Figueras, M. J., & Borrego, J. J. (2010). New perspectives in monitoring drinking water microbial quality. *International Journal of Environmental Research and Public Health, 7*(12), 4179–4202. https://doi.org/10.3390/ijerph7124179

Frederiksen, W. (2015). Citrobacter. https://doi.org/10.1002/9781118960608.gbm01143

G-Alegría, E., López, I., Ruiz, J. I., Sáenz, J., Fernández, E., Zarazaga, M., et al. (2004). High tolerance of wild *Lactobacillus plantarum* and Oenococcus oeni strains to lyophilisation and stress environmental conditions of acid pH and ethanol. *FEMS Microbiology Letters, 230*(1), 53–61. https://doi.org/10.1016/S0378-1097(03)00854-1

Gernat, D. C., Brouwer, E., & Ottens, M. (2020). Aldehydes as wort off-flavours in alcohol-free beers—origin and control. *Food and Bioprocess Technology, 13*(2), 195–216. https://doi.org/10.1007/s11947-019-02374-z

Granum, P. E., & Lund, T. (1997). *Bacillus cereus* and its food poisoning toxins. *FEMS Microbiology Letters, 157*(2), 223–228. https://doi.org/10.1016/S0378-1097(97)00438-2

Grenda, T., Jarosz, A., Sapała, M., Grenda, A., Patyra, E., & Kwiatek, K. (2023). *Clostridium perfringens*—opportunistic foodborne pathogen, its diversity and epidemiological significance. *Pathogens, 12*(6), 1–12. https://doi.org/10.3390/pathogens12060768

Grimont, P. A. D., & Grimont, F. (2005a). Enterobacter. *Bergey's manual of systematics of archaea and bacteria* (pp. 1–17). https://doi.org/10.54695/apmc.19.01.1528

Grimont, P. A. D., & Grimont, F. (2005b). Klebsiella. *Bergey's manual of systematics of archaea and bacteria.* https://doi.org/10.1520/STP34835S

Haakensen, M., & Ziola, B. (2008). Identification of novel *horA*-harbouring bacteria capable of spoiling beer. *Canadian Journal of Microbiology, 54*(4), 321–325. https://doi.org/10.1139/W08-007

Haas, G. J., & Barsoumian, R. (1994). Antimicrobial activity of hop resins. *Journal of Food Protection, 57*(1), 59–61. https://doi.org/10.4315/0362-028X-57.1.59

Hawthorne, D. B., Shaw, R. D., Davine, D. F., Kavanagh, T. E., Clarke, B. J., Shaw, R. D., et al. (1991). Butyric acid off-flavors in beer: Origins and control. *Journal of the American Society of Brewing Chemists, 49.* https://doi.org/10.1094/asbcj-49-0004

Hemme, D., & Foucaud-Scheunemann, C. (2004). *Leuconostoc*, characteristics, use in dairy technology and prospects in functional foods. *International Dairy Journal, 14*(6), 467–494. https://doi.org/10.1016/j.idairyj.2003.10.005

Hunter, W. J., Manter, D. K., & Van Der Lelie, D. (2012). Biotransformation of ferulic acid to 4-vinylguaiacol by *Enterobacter soli* and *E. aerogenes*. *Current Microbiology, 65*(6), 752–757. https://doi.org/10.1007/s00284-012-0222-4

Ingram, L. O. N., & Buttke, T. M. (1985). Effects of alcohols on micro-organisms. *Advances in Microbial Physiology, 25*(C), 253–300. https://doi.org/10.1016/S0065-2911(08)60294-5

Ivanov, K., Petelkov, I., Shopska, V., Denkova, R., Gochev, V., & Kostov, G. (2016). Investigation of mashing regimes for low-alcohol beer production. *Journal of the Institute of Brewing, 122*(3), 508–516.

Iwu, C. D., & Okoh, A. I. (2019). Preharvest transmission routes of fresh produce associated bacterial pathogens with outbreak potentials: A review. *International Journal of Environmental Research and Public Health, 16*(22). https://doi.org/10.3390/ijerph16224407

Jääskeläinen, E. L., Häggblom, M. M., Andersson, M. A., & Salkinoja-Salonen, M. S. (2004). Atmospheric oxygen and other conditions affecting the production of cereulide by *Bacillus cereus* in food. *International Journal of Food Microbiology, 96*(1), 75–83. https://doi.org/10.1016/j.ijfoodmicro.2004.03.011

James, T. C., Gallagher, L., Titze, J., Bourke, P., Kavanagh, J., Arendt, E., et al. (2014). In situ production of human β defensin-3 in lager yeasts provides bactericidal activity against beer-spoiling bacteria under fermentation conditions. *Journal of Applied Microbiology, 116*(2), 368–379.

Janda, J. M., & Abbott, S. L. (2021). The changing face of the family *Enterobacteriaceae* (Order: "*Enterobacterales*"): New members, taxonomic issues, geographic expansion, and new diseases and disease syndromes. *Clinical Microbiology Reviews, 28*(February), 1–45.

Jeon, S. H., Kim, N. H., Shim, M. B., Jeon, Y. W., Ahn, J. H., Lee, S. H., et al. (2015). Microbiological diversity and prevalence of spoilage and pathogenic bacteria in commercial fermented alcoholic beverages (beer, fruit wine, refined rice wine, and yakju). *Journal of Food Protection, 78*(4), 812–818. https://doi.org/10.4315/0362-028X.JFP-14-431

Jiang, Z., Yang, B., Liu, X., Zhang, S., Shan, J., Liu, J., et al. (2017). A novel approach for the production of a non-alcohol beer ($\leq$ 0.5% abv) by a combination of limited fermentation and vacuum distillation. *Journal of the Institute of Brewing, 123*(4), 533–536.

Johansson, L., Nikulin, J., Juvonen, R., Krogerus, K., Magalhães, F., Mikkelson, A., et al. (2021). Sourdough cultures as reservoirs of maltose-negative yeasts for low-alcohol beer brewing. *Food Microbiology, 94*, Article 103629.

Jones, P. G., Inouye, M., Wood, R., & Medical, J. (1994). MicroReview the cold-shock response — a hot topic. *Molecular Microbiology, 11*(5), 811–818.

Juneja, V. K., Baker, D. A., Thippareddi, H., Snyder, O. P., & Mohr, T. B. (2013). Growth potential of *Clostridium perfringens* from spores in acidified beef, pork, and poultry products during chilling. *Journal of Food Protection, 76*(1), 65–71. https://doi.org/10.4315/0362-028X.JFP-12-289

Kämpfer, P. (2015). *Rahnella. Bergey's manual of systematics of archaea and bacteria.* https://doi.org/10.1002/9781118960608.gbm01164

Kawacka, I., Olejnik-Schmidt, A., Schmidt, M., & Sip, A. (2020). Effectiveness of phage-based inhibition of *Listeria monocytogenes* in food products and food processing environments. *Microorganisms, 8*(11), 1764.

Keerthirathne, T. P., Ross, K., Fallowfield, H., & Whiley, H. (2019). The combined effect of pH and temperature on the survival of *Salmonella enterica* serovar Typhimurium and implications for the preparation of raw egg mayonnaise. *Pathogens, 8*(4). https://doi.org/10.3390/pathogens8040218

Keevil, C. W., Hough, J. S., & Cole, J. A. (1979). Regulation of respiratory and fermentative modes of growth of *Citrobacter freundii* by oxygen, nitrate and glucose. *Journal of General Microbiology, 113*(1), 83–95. https://doi.org/10.1099/00221287-113-1-83

Kilgour, W., & Smith, P. (1985). The determination of pasteurisation regimes for alcoholic and alcohol-free beer. *Journal of the Institute of Brewing, 91*, 130–130.

Kim, S. A., Kim, N. H., Lee, S. H., Hwang, I. G., & Rhee, M. S. (2014). Survival of foodborne pathogenic bacteria (*Bacillus cereus*, *Escherichia coli* O157:H7, *Salmonella enterica* serovar Typhimurium, *Staphylococcus aureus*, and *Listeria monocytogenes*) and *Bacillus cereus* spores in fermented alcoholic beverages (beer and refined rice wine). *Journal of Food Protection, 77*(3), 419–426. https://doi.org/10.4315/0362-028X.JFP-13-234

Kordialik-Bogacka, E. (2022). Biopreservation of beer: Potential and constraints. *Biotechnology Advances, 58*, Article 107910.

Krogerus, K., Eerikäinen, R., Aisala, H., & Gibson, B. (2022). Repurposing brewery contaminant yeast as production strains for low-alcohol beer fermentation. *Yeast, 39*(1–2), 156–169.

L'Anthoën, N. C., & Ingledew, W. M. (1996). Heat resistance of bacteria in alcohol-free beer. *Journal of the American Society of Brewing Chemists, 54*(1), 32–36. https://doi.org/10.1094/asbcj-54-0032

Labrado, D., Ferrero, S., Caballero, I., Alvarez, C. M., Villafañe, F., & Blanco, C. A. (2020). Identification by NMR of key compounds present in beer distillates and residual phases after dealcoholization by vacuum distillation. *Journal of the Science of Food and Agriculture, 100*(10), 3971–3978.

Lanciotti, R., Sinigaglia, M., Gardini, F., Vannini, L., & Guerzoni, M. E. (2001). Growth/no growth interfaces of *Bacillus cereus*, *Staphylococcus aureus* and *Salmonella enteritidis* in model systems based on water activity, pH, temperature and ethanol concentration. *Food Microbiology, 18*(6), 659–668. https://doi.org/10.1006/fmic.2001.0429

Lawrie, J. (2021). *The brewing of alcohol-free beers and non-intoxicating beers in Scotland.* The Annual Journal of the Scottish Brewing Archive Association.

Liguori, L., De Francesco, G., Russo, P., Albanese, D., Perretti, G., & Matteo, M. D. (2015). Quality improvement of low alcohol craft beer produced by evaporative pertraction. *Chemical Engineering, 43*.

Liu, W. Y., Shi, Y. W., Wang, X. Q., Wang, Y., Wei, C. Q., & Lou, K. (2008). Isolation and identification of a strain producing cold-adapted β-galactosidase, and purification and characterisation of the enzyme. *Czech Journal of Food Sciences, 26*(4), 284–290. https://doi.org/10.17221/31/2008-cjfs

Liu, D., & Zhang, T. (2011). *Listeria*. In *Bergey's manual of systematics of archaea and bacteria* (pp. 279–294). https://doi.org/10.2307/30145199

Logan, N. A., & DeVos, P. (2015a). *Bacillus*. In *Bergey's manual of systematics of archaea and bacteria.* https://doi.org/10.1002/9781118960608.gbm00530

Logan, N. A., & DeVos, P. (2015b). *Brevibacillus. Bergey's manual of systematics of archaea and bacteria* (pp. 1–22). https://doi.org/10.1002/9781118960608.gbm00550

Loredana, L., Giovanni, D. F., Donatella, A., Antonio, M., Giuseppe, P., Marisa, D. M., et al. (2018). Impact of osmotic distillation on the sensory properties and quality of low alcohol beer. *Journal of Food Quality, 2018.*

Lund, B. M., & Gould, G. W. (2000). *The microbiological safety and quality of food.* Aspen Publishers.

Lund, B. M., Graham, A. F., & Franklin, J. G. (1987). The effect of acid pH on the probability of growth of proteolytic strains of *Clostridium botulinum*. *International Journal of Food Microbiology, 4*(3), 215–226. https://doi.org/10.1016/0168-1605(87)90039-0

Müller-Auffermann, K., Grijalva, F., Jacob, F., & Hutzler, M. (2015). Nisin and its usage in breweries: A review and discussion. *Journal of the Institute of Brewing, 121*(3), 309–319.

Magnus, C. A., Ingledew, W. M., & Casey, G. P. (1986). High-gravity brewing: Influence of high-ethanol beer on the viability of contaminating brewing bacteria. *Journal of the American Society of Brewing Chemists, 44*(4), 158–161. https://doi.org/10.1094/asbcj-44-0158

Martens, H., Dawoud, E., & Verachtert, H. (1991). *Wort enterobacteria and other microbial populations involved during the first month of lambic fermentation* (Vol. 97, pp. 435–439).

Massa, S., Altieri, C., Quaranta, V., & De Pace, R. (1997). Survival of *Escherichia coli* O157: H7 in yoghurt during preparation and storage at 4°C. *Letters in Applied Microbiology, 24*(5), 347–350. https://doi.org/10.1046/j.1472-765X.1997.00067.x

McCaig, R., & Morrison, M. (1984). Characterization of *Enterobacter agglomerans* variants and their importance in brewing. *Journal of the American Society of Brewing Chemists, 42*. https://doi.org/10.1094/asbcj-42-0023

Medina, E., Romero, C., Brenes, M., & De Castro, A. (2007). Antimicrobial activity of olive oil, vinegar, and various beverages against foodborne pathogens. *Journal of Food Protection, 70*(5), 1194–1199. https://doi.org/10.4315/0362-028X-70.5.1194

Mellor, D. D., Hanna-Khalil, B., & Carson, R. (2020). A review of the potential health benefits of low alcohol and alcohol-free beer: Effects of ingredients and craft brewing processes on potentially bioactive metabolites. *Beverages, 6*(2), 25.

Menz, G., Aldred, P., & Vriesekoop, F. (2009). Pathogens in beer. In V. Preedy (Ed.), *Beer in health and disease prevention* (pp. 403–413). Academic Press. https://doi.org/10.1016/B978-0-12-373891-2.00039-0

Menz, G., Aldred, P., & Vriesekoop, F. (2011). Growth and survival of foodborne pathogens in beer. *Journal of Food Protection, 74*(10), 1670–1675. https://doi.org/10.4315/0362-028X.JFP-10-546

Menz, G., Vriesekoop, F., Zarei, M., Zhu, B., & Aldred, P. (2010). The growth and survival of food-borne pathogens in sweet and fermenting brewers' wort. *International Journal of Food Microbiology, 140*(1), 19–25. https://doi.org/10.1016/j.ijfoodmicro.2010.02.018

Michel, M., Cocuzza, S., Biendl, M., Peifer, F., Pehl, F., Back, W., et al. (2020). The impact of different hop compounds on the growth of selected beer spoilage bacteria in beer. *Journal of the Institute of Brewing, 8*. https://doi.org/10.1002/jib.624

Mira, N. P., & Teixeira, M. C. (2013). Microbial mechanisms of tolerance to weak acid stress. *Frontiers in Microbiology, 4*, 416. https://doi.org/10.3389/fmicb.2013.0041

Mohanta, Y. K., Biswas, K., Jena, S. K., Hashem, A., Abd_Allah, E. F., & Mohanta, T. K. (2020). Anti-biofilm and antibacterial activities of silver nanoparticles synthesized by the reducing activity of phytoconstituents present in the Indian medicinal plants. *Frontiers in Microbiology, 11*, 1143.

Montanari, L., Marconi, O., Mayer, H., & Fantozzi, P. (2009). Production of alcohol-free beer. In *Beer in health and disease prevention* (pp. 61–75). Academic Press.

Mota, R. V. D., Ramos, C. L., Peregrino, I., Hassimotto, N. M. A., Purgatto, E., & Souza, C. R. De (2018). Identification of the potential inhibitors of malolactic fermentation in wines. *Food Science and Technology (Brazil), 38*, 174–179. https://doi.org/10.1590/1678-457x.16517

Muller, C., Neves, L. E., Gomes, L., Guimarães, M., & Ghesti, G. (2019). Processes for alcohol-free beer production: A review. *Food Science and Technology, 40*, 273–281. https://doi.org/10.1590/fst.32318

Munford, A. R. G., Alvarenga, V. O., Prado-Silva, L. do, Crucello, A., Campagnollo, F. B., Chaves, R. D., et al. (2017). Sporeforming bacteria in beer: Occurrence, diversity, presence of hop resistance genes and fate in alcohol-free and lager beers. *Food Control, 81*, 126–136. https://doi.org/10.1016/j.foodcont.2017.06.003

Nikulin, J., Aisala, H., & Gibson, B. (2022). Production of nonalcoholic beer via cold contact fermentation with Torulaspora delbrueckii. *Journal of the Institute of Brewing, 128*(1), 28–35.

Noots, I., Delcour, J. A., & Michiels, C. W. (1999). From field barley to malt: Detection and specification of microbial activity for quality aspects. *Critical Reviews in Microbiology, 25*(2), 121–153.

Oh, D. H., & Marshall, D. L. (1993). Antimicrobial activity of ethanol, glycerol monolaurate or lactic acid against *Listeria monocytogenes*. *International Journal of Food Microbiology, 20*(4), 239–246. https://doi.org/10.4315/0362-028X-70.5.1194

Pando, J. M., Pfeltz, R. F., Cuaron, J. A., Nagarajan, V., Mishra, M. N., Torres, N. J., et al. (2017). Ethanol-induced stress response of *Staphylococcus aureus*. *Journal: Canadian Journal of Microbiology, 63*(9), 2017. https://doi.org/10.1139/cjm-2017-0221

Peng, J., Xu, Z., Li, L., Zhao, B., & Guo, Y. (2023). Disruption of the sensor kinase *phoQ* gene decreases acid resistance in plant growth-promoting rhizobacterium *Rahnella aquatilis* HX2. *Journal of Applied Microbiology, 134*(2), 1–10. https://doi.org/10.1093/jambio/lxad009

Peyravi, M., Jahanshahi, M., & Banafti, S. (2020). Application of membrane technology in beverage production and safety. In *Safety issues in beverage production* (pp. 271–308). Academic Press.

Pittet, V., Morrow, K., & Ziola, B. (2011). Ethanol tolerance of lactic acid bacteria, including relevance of the exopolysaccharide gene *gtf*. *Journal of the American Society of Brewing Chemists, 69*(1), 57–61. https://doi.org/10.1094/ASBCJ-2011-0124-01

Priest, F. G. (2015). *Paenibacillus*. In *Bergey's manual of systematics of archaea and bacteria* (pp. 1–40). https://doi.org/10.1002/9781118960608.gbm00553

Priest, F. G., Cowbourne, M. A., & Hough, J. S. (1974). Wort enterobacteria - a review. *Journal of the Institute of Brewing, 80*(4), 342–356.

Puligundla, P., Pyun, Y. R., & Mok, C. (2018). Pulsed electric field (PEF) technology for microbial inactivation in low-alcohol red wine. *Food Science and Biotechnology, 27*(6), 1691–1696.

Quain, D. E. (2015). Assuring the microbiological quality of draught beer. In *Brewing microbiology* (pp. 335–354). Woodhead Publishing.

Quain, D. E. (2021). The enhanced susceptibility of alcohol-free and low alcohol beers to microbiological spoilage: Implications for draught dispense. *Journal of the Institute of Brewing, 127*(4), 406–416.

Rachon, G., Rice, C. J., Pawlowsky, K., & Raleigh, C. P. (2018). Challenging the assumptions around the pasteurisation requirements of beer spoilage bacteria. *Journal of the Institute of Brewing, 124*(4), 443–449. https://doi.org/10.1002/jib.520

Rainey, F. A., Hollen, B., & Small, A. M. (2015). *Clostridium*. In *Bergey's manual of systematics of archaea and bacteria* (pp. 1–122). https://doi.org/10.1002/9781118960608.gbm00619

Riu-Aumatell, M., Miró, P., Serra-Cayuela, A., Buxaderas, S., & López-Tamames, E. (2014). Assessment of the aroma profiles of low-alcohol beers using HS-SPME–GC-MS. *Food Research International, 57*, 196–202.

Rocelle, M., Clavero, S., & Beuchat, L. R. (1996). Survival of *Escherichia coli* O157:H7 in broth and processed salami as influenced by pH, water activity, and temperature and suitability of media for its recovery. *Applied and Environmental Microbiology, 62*(8), 2735–2740. https://doi.org/10.1128/aem.62.8.2735-2740.1996

Ross, T., Zhang, D., & McQuestin, O. J. (2008). Temperature governs the inactivation rate of vegetative bacteria under growth-preventing conditions. *International Journal of Food Microbiology, 128*(1), 129–135. https://doi.org/10.1016/j.ijfoodmicro.2008.07.023

Säde, E. (2011). Leuconostoc spoilage of refrigerated, packaged foods. In *Veterinary medicine*.

Sakamoto, K., & Konings, W. N. (2003). Beer spoilage bacteria and hop resistance. *International Journal of Food Microbiology, 89*(2–3), 105–124. https://doi.org/10.1016/S0168-1605(03)00153-3

Salanţă, L. C., Coldea, T. E., Ignat, M. V., Pop, C. R., Tofană, M., Mudura, E., et al. (2020). Non-alcoholic and craft beer production and challenges. *Processes, 8*(11), 1382.

Santamaría, J., & Toranzos, G. A. (2003). Enteric pathogens and soil: A short review. *International Microbiology, 6*(1), 5−9. https://doi.org/10.1007/s10123-003-0096-1

Sarwat, F., Qader, S. A. U., Aman, A., & Ahmed, N. (2008). Production & characterization of a unique dextran from an indigenous *Leuconostoc mesenteroides* CMG713. *International Journal of Biological Sciences, 4*(6), 379−386. https://doi.org/10.7150/ijbs.4.379

Schleifer, K., & Bell, J. A. (2015). *Staphylococcus*. In *Bergey's manual of systematics of archaea and bacteria* (pp. 1−43). https://doi.org/10.1002/9781118960608.gbm00569

Schneider, K. R., Goodrich Schneider, R., Silverberg, R., Kurdmongkoltham, P., & Bertoldi, B. (2017). Preventing foodborne illness: *Bacillus cereus*. *UF IFAS Extension, 2017*(2), 6. https://doi.org/10.32473/edis-fs269-2017

Schulz, L. M., Konrath, A., & Rismondo, J. (2023.02.21). Physiological studies on pH and salt tolerance and the utilization of diverse carbon sources by *Listeria monocytogenes*. *bioRxiv*, Article 529469. http://biorxiv.org/content/early/2023/02/22/2023.02.21.529469.abstract.

Scientific Committee on Veterinary Measures relating to Public Health. (2003). Staphylococcal enterotoxins in milk products, particularly cheeses. In *Scientific committee on veterinary measures relating to public health*. europa.eu/food/fs/sc/scv/out61_en.pdf.

Shimwell, J. L. (1935). The resistance of beer towards *Saccharobacillus pastorianus*. *Journal of the Institute of Brewing, 41*, 245−258.

Simon, M., Vuylsteke, G., & Collin, S. (2022). Flavor defects of fresh and aged NABLABs: New challenges against oxidation. *Journal of the American Society of Brewing Chemists, 0*(0), 1−11. https://doi.org/10.1080/03610470.2022.2142756

Sleha, R., Radochova, V., Mikyska, A., Houska, M., Bolehovska, R., Janovska, S., et al. (2021). Strong antimicrobial effects of xanthohumol and beta-acids from hops against clostridioides difficile infection in vivo. *Antibiotics, 10*(4), 1−11. https://doi.org/10.3390/antibiotics10040392

Smith, G. R. (2012). How RecBCD enzyme and chi promote DNA break repair and recombination: A molecular biologist's view. *Microbiology and Molecular Biology Reviews, 76*(2), 217−228. https://doi.org/10.1128/mmbr.05026-11

Smith, J. L., & Fratamico, P. M. (2015). *Escherichia coli* and other *Enterobacteriaceae*: Food poisoning and health effects. *Encyclopedia of Food and Health*, 539−544. https://doi.org/10.1016/B978-0-12-384947-2.00260-9

Sohrabvandi, S., Mousavi, S. M., Razavi, S. H., Mortazavian, A. M., & Rezaei, K. (2010). Alcohol-free beer: Methods of production, sensorial defects, and healthful effects. *Food Reviews International, 26*(4), 335−352.

Sohrabvandi, S., Razavi, S. H., Mousavi, S. M., & Mortazavian, A. M. (2010). Viability of probiotic bacteria in low alcohol-and nonalcoholic beer during refrigerated storage. *The Philippine Agricultural Scientist, 93*(1), 24−28.

Spitaels, F., Wieme, A. D., Janssens, M., Aerts, M., Van Landschoot, A., De Vuyst, L., et al. (2015). The microbial diversity of an industrially produced lambic beer shares members of a traditionally produced one and reveals a core microbiota for lambic beer fermentation. *Food Microbiology, 49*, 23−32. https://doi.org/10.1016/j.fm.2015.01.008

Stadtman, E. R., Stadtman, T. C., & Barker, H. A. (1949). Tracer experiments on the mechanism of synthesis of valeric and caproic acids by *Clostridium kluyveri*. *Journal of Biological Chemistry, 178*(2), 677−682. https://doi.org/10.1016/s0021-9258(18)56884-8

Storgårds, E. (2000). Process hygiene control in beer production and dispensing. In *VTT publications* (p. 107).

Suzuki, K. (2020). The science of beer emergence of new spoilage microorganisms in the brewing industry and development of microbiological quality control methods to cope with this phenomenon − a review. *Journal of the American Society of Brewing Chemists, 0*(0), 1−15. https://doi.org/10.1080/03610470.2020.1782101

Tantasuttikul, A., & Mahakarnchanakul, W. (2019). Growth parameters and sanitizer resistance of *Raoultella ornithinolytica* and *Raoultella terrigena* isolated from seafood processing plant. *Cogent Food & Agriculture, 5*(1). https://doi.org/10.1080/23311932.2019.1569830

Teuber, M., & Schmalreck, A. F. (1973). Membrane leakage in *Bacillus subtilis* 168 induced by the hop constituents lupulone, humulone, isohumulone, and humulinic acid. *Archives of Microbiology, 94*, 159−171.

Thomas, M., Cole, J. A., & Hough, J. S. (1972). Biochemical physiology of *Obesumbacterium proteus*, a common brewery contaminant. *Journal of the Institute of Brewing, 78*, 332−339.

Threlfall, R. T., & Morris, J. R. (1995). Use of velcorin (DMDC) to control yeast fermentation in juice and wine. In *Proceedings of the viticultural science symposium*.

Tieghem, V. (2015). *Leuconostoc*. In *Bergey's manual of systematics of archaea and bacteria* (pp. 1−23). https://doi.org/10.1002/9781118960608.gbm00607

Tsai, Y. W., & Ingham, S. C. (1997). Survival of *Escherichia coli* O157:H7 and *Salmonella* spp. in acidic condiments. *Journal of Food Protection, 60*(7), 751−755. https://doi.org/10.4315/0362-028X-60.7.751

Uchiyama, H., Hashidoko, Y., Kuriyama, Y., & Tahara, S. (2008). Identification of the 4-hydroxycinnamate decarboxylase (PAD) gene of *Klebsiella oxytoca*. *Bioscience, Biotechnology and Biochemistry, 72*(1), 116−123. https://doi.org/10.1271/bbb.70496

Valero, A., Olague, E., Medina-Pradas, E., Garrido-Fernández, A., Romero-Gil, V., Cantalejo, M. J., et al. (2020). Influence of acid adaptation on the probability of germination of *Clostridium sporogenes* spores against pH, NaCl and time. *Foods, 9*(2), 1−18. https://doi.org/10.3390/foods9020127

Van Doren, J. M., Neil, K. P., Parish, M., Gieraltowski, L., Gould, L. H., & Gombas, K. L. (2013). Foodborne illness outbreaks from microbial contaminants in spices, 1973-2010. *Food Microbiology, 36*(2), 456−464. https://doi.org/10.1016/j.fm.2013.04.014

Van Vuuren, H. J. J., Cosser, K., & Prior, B. A. (1980). Beer flavour was studied . The presence of *E . agglomerans* gave rise to increased levels of in contaminated and control beer column: Column. *Journal of the Institute of Brewing, 86*, 31−33.

Van Vuuren, H. J. J., & Priest, F. G. (2003). Gram-negative brewery bacteria. In *Brewing microbiology* (pp. 219−245).

Vaughan, A., O'Sullivan, T., & Van Sinderen, D. (2005). Enhancing the microbiological stability of malt and beer - a review. *Journal of the Institute of Brewing, 111*(4), 355–371. https://doi.org/10.1002/j.2050-0416.2005.tb00221.x

Virkajärvi, I., Vauhkonen, T., & Storgårds, E. (2001). Control of microbial contamination in continuous primary fermentation by immobilized yeast. *Journal of the American Society of Brewing Chemists, 59*(2), 63–68. https://doi.org/10.1094/asbcj-59-0063

Vriesekoop, F. (2021). Beer and allergens. *Beverages, 7*(4), 79. https://doi.org/10.3390/beverages7040079

Vriesekoop, F., Krahl, M., Hucker, B., & Menz, G. (2012). 125th Anniversary review: Bacteria in brewing: The good, the bad and the ugly. *Journal of the Institute of Brewing, 118*(4), 335–345. https://doi.org/10.1002/jib.49

Wang, Y., Elzenga, T., & van Elsas, J. D. (2021). Effect of culture conditions on the performance of lignocellulose-degrading synthetic microbial consortia. *Applied Microbiology and Biotechnology, 105*(20), 7981–7995. https://doi.org/10.1007/s00253-021-11591-6

Wang, W., Liu, Y., Sun, Z., Du, G., & Li, X. (2017). Hop resistance and beer-spoilage features of foodborne *Bacillus cereus* newly isolated from filtration-sterilized draft beer. *Annals of Microbiology, 67*(1), 17–23. https://doi.org/10.1007/s13213-016-1232-4

Weber, M. H., & Marahiel, M. A. (2003). Bacterial cold shock responses. *Science Progress, 86*(Pt 1–2), 9–75. https://doi.org/10.3184/003685003783238707

Wells-Bennik, M. H. J., Eijlander, R. T., Den Besten, H. M. W., Berendsen, E. M., Warda, A. K., Krawczyk, A. O., et al. (2016). Bacterial spores in food: Survival, emergence, and outgrowth. *Annual Review of Food Science and Technology, 7*, 457–482. https://doi.org/10.1146/annurev-food-041715-033144

Wray, E. (2015). Reducing microbial spoilage of beer using pasteurisation. In *Brewing microbiology* (pp. 253–269). Woodhead Publishing.

Zaika, L. L. (2002). Effect of organic acids and temperature on survival of *Shigella flexneri* in broth at pH 4. *Journal of Food Protection, 65*(9), 1417–1421. https://doi.org/10.4315/0362-028X-65.9.1417

Zaunmüller, T., Eichert, M., Richter, H., & Unden, G. (2006). Variations in the energy metabolism of biotechnologically relevant heterofermentative lactic acid bacteria during growth on sugars and organic acids. *Applied Microbiology and Biotechnology, 72*(3), 421–429. https://doi.org/10.1007/s00253-006-0514-3

Zhao, C., Wen, H., Huang, S., Weng, S., & He, J. (2022). A novel disease (water bubble disease) of the giant freshwater prawn *Macrobrachium rosenbergii* caused by *Citrobacter freundii*: Antibiotic treatment and effects on the antioxidant enzyme activity and immune responses. *Antioxidants, 11*(8). https://doi.org/10.3390/antiox11081491

Chapter 14

Traditional methods of detection and identification of brewery spoilage organisms

Annie E. Hill
The International Centre for Brewing & Distilling, Heriot-Watt University, Edinburgh, United Kingdom

Brewers have been successfully able to reduce incidences of waste batches and product recall due to microbial spoilage through a proactive approach including implementation of good manufacturing practices and HACCP process control systems. However, opportunities for microbiological contaminants to enter the brewing process are available at all stages (Fig. 14.1), and their amazing ability to adapt to seemingly hostile conditions makes them a tenacious threat. The increase in production of no- and low-alcohol beers, and the prevalence of diastatic yeast strains have also raised the requirement to be more vigilant in early detection of spoilage microbes.

The most commonly encountered spoilage microbes are detailed in Fig. 14.2 (Kordialik-Bogacka, 2022; Schneiderbanger, Jacob, & Hutzler, 2020; Suzuki, 2020). Methods employed by breweries to detect and/or identify both yeast and bacteria are constantly changing and vary depending on the scale of operations. Costs for rapid detection methods are still decreasing and most breweries routinely use adenosine triphosphate (ATP) testing as a cleanliness check. However, traditional methods, such as plating, remain a common and cost-effective approach in detection and identification of microbes. In this chapter, the traditional methods available to detect and identify microbes at each stage of the brewing process are detailed.

14.1 Detection of brewery spoilage organisms

14.1.1 Raw materials

14.1.1.1 Cereals

Fungi from the genera *Alternaria*, *Cladosporium*, *Epicoccum*, *Aspergillus*, *Penicillium*, and *Fusarium* are the main hazards in terms of barley and malt infection. Of particular concern is the production of mycotoxins including Aflatoxins (Afs), Ochratoxin A (OTA), Deoxynivalenol (DON), and Zearalenone (ZON). As a result, tolerance of fungal growth on cereals is less than 10 colony forming units (cfu) per gram and zero tolerance of wild yeast. Malt should also be free of mould. Guidelines are available through organisations such as the Home Grown Cereals Authority (HGCA) for the assessment of Fusarium Head Blight on grain, both in the field and in storage. Typically, malt is sampled at each intake and at regular intervals during storage and tested for bacteria, fungi, wild yeast, and mycotoxins. Air samples may also be taken.

The simplest way to evaluate the internal microflora of grain is by direct plating. Grain is immersed in full strength or 50% household bleach for 1 min to kill surface microflora then rinsed in sterile distilled water. Either grind the grain before adding to molten agar (express per gram) or place individual grains on the surface of the agar in a Petri dish (express per grain) and incubate at 25–30°C to allow microbes located in the interior of the grain to grow out. The level of internal infection is an indicator of quality and storability of the grain. If *Aspergillus flavus* and *Aspergillus parasiticus* or Czapek-Dox Iprodione Dichloran Agar (see Table 14.1) are used this technique can also give some information about the safety of the grain by indicating whether or not potentially toxic *A. flavus*, *A. parasiticus*, or *Fusarium* species are present.

Brewing Microbiology. https://doi.org/10.1016/B978-0-323-99606-8.00013-4

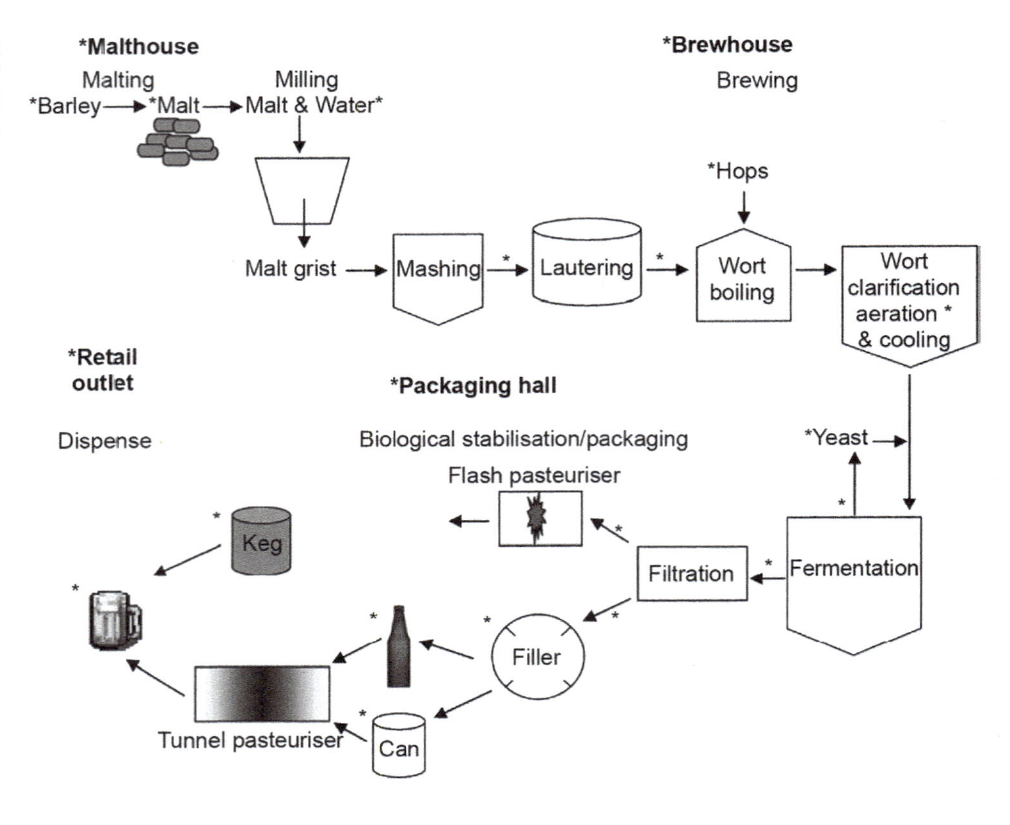

FIGURE 14.1 Schematic of the brewing process. Potential sources of microbiological contamination are indicated by* (Vaughan, O'Sullivan, & van Sinderen, 2005).

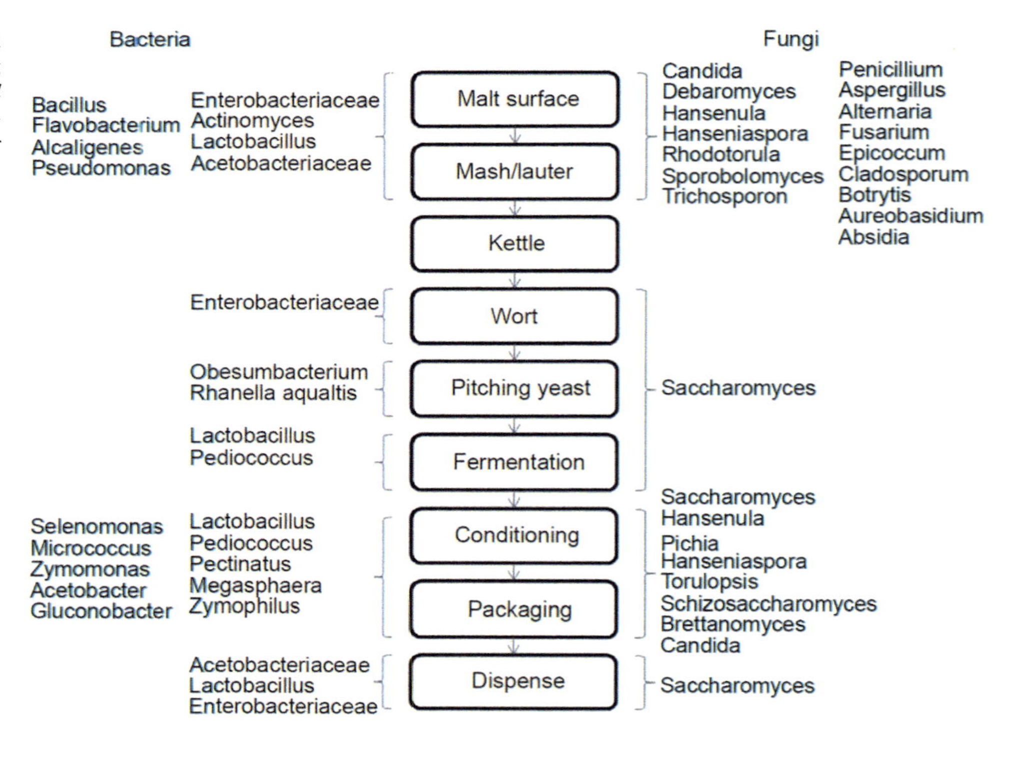

FIGURE 14.2 Common microbial contaminants within the brewing process. *Adapted from Bokulich and Bamforth (2013) using Kordialik-Bogacka (2022), Schneiderbanger et al. (2020), and Suzuki (2020).*

TABLE 14.1 Detection media for grain analysis.

Growth medium		Incubation conditions
Aspergillus flavus and *parasiticus* agar (AFPA)	*Aspergillus flavus* and *Aspergillus parasiticus*	28°C 7 days (morphological analysis), 10 days (toxin production)
Czapek Peptone yeast extract agar (CZPYA)	Actinomycetes	
Czapek-Dox Iprodione Dichloran agar (CZID)	*Fusarium* species	
Dichloran Chloramphenicol Peptone agar (DCPA)	*Fusarium* species	25°C 4 days (initial examination), 6 days (morphological analysis)
Dichloran glycerol agar (DG18)	*Fusarium* species	
Dichloran-Rose Bengal-Chlortetracycline agar (DRBC)	*A. flavus* and *A. parasiticus*	
Pentachloronitrobenzene Peptone agar (PPA)	*Fusarium* species	
Potato Dextrose agar (PDA)	Fungi	

Based on Hocking and Pitt (1980), Mostafa, Barakat, El-Shanawany, and (2005), Thrane (1996).

Grain is susceptible to mycotoxins produced either while the crop is growing by *Fusarium* species or during storage by *Penicillium* species. High quality malting barley should be free from DON resulting from Fusarium head blight, and free from disease. For disease there is no compromise; diseased crops will be rejected and reducing the price will not make them acceptable. Barley from areas with conditions conducive to Fusarium head blight is routinely screened for DON and barley with DON levels over 0.5 ppm will normally be rejected for malting purposes (tolerance is 1 ppm if for human consumption). Unprocessed common wheat and barley are also usually screened for ZON (tolerance is up to 100 ppb). DON and ZON are both included in the HGCA Grain Passport.

Methods for detecting mycotoxins are summarised in Table 14.2. Chromatographic methods are frequently used for detecting, quantifying, and confirming the presence of mycotoxins. These methods include thin-layer chromatography (TLC), high-performance liquid chromatography (HPLC), liquid chromatography combined with mass spectrometry, gas chromatography, and gas chromatography combined with mass spectrometry (GC-MS).

Various detection methods, such as fluorescence, ultraviolet absorption, and others have been combined with chromatographic methods. Methods based on the production of antibodies specific for individual mycotoxins have also been developed and include enzyme-linked immunosorbent assays and immunoaffinity columns. These methods allow for specific and precise detection and quantification of specific mycotoxins. This has led to the development of test kits for mycotoxins, such as VICAM, VERATOX, and REVEALQ, which are rapid and simple to use and can be used in the field and throughout the processing stages. More recently, lateral flow kits have been made available e.g., REVEALQ + MAX·

TABLE 14.2 Methods for detection of mycotoxins.

Mycotoxin(s)	Method(s)
Aflatoxins	TLC, HPLC, ELISA, immunoaffinity column
Deoxynivalenol	GC, HPLC, ELISA, immunoaffinity column
Fumonisins	HPLC, ELISA, immunoaffinity column
Moniliformin	HPLC
Ochratoxin	TLC, HPLC, ELISA, immunoaffinity column
Zearalenone	TLC, HPLC, ELISA, immunoaffinity column

Based on Mirocha and Christensen (1986).

14.1.1.2 Water

As the main ingredient of beer and a utility in the production process, water quality is central to brewing. Algae, protozoa, fungi, yeasts, and bacteria may all be present in water, but fortunately very few waterborne microbes are able to cause serious problems to brewers. Typically, water from boreholes contains fewer micro-microorganisms than surface water, that is, rivers, ponds, and tanks, and public water supplies are of course rigorously tested. Microbiological tests on water predominantly involve detection of an indicator organism. The primary faecal indicator organism is *Escherichia coli* which is abundantly common and has similar survival qualities to *Salmonella*. Water used for human consumption can have no more than one positive sample (>1 coliform/100 mL) in 40 samples tested in a month and the concentration of faecal coliforms must be zero. It is good practice to monitor nonpublic supplies (borehole, spring, etc.) regularly, at a minimum seasonally.

Brewing water should be tested before entering the hot liquor tank, or any mashing vessels. Similarly, water for dilution should be tested prior to use. Clean-in-place (CIP) and rinse water should be checked every cycle. Generally, however, most supplies are checked weekly or upon encountering unstable wort. Sampling points should be uniformly distributed throughout a piped distribution system and the number of sampling points should be proportional to the number of links or branches. The points chosen should generally yield samples that are representative of the system as a whole and of its main components.

The traditional method for detection of waterborne microbes is direct plating. Samples may also be filtered either on- or off-line and filters placed directly on the surface of an agar plate. A range of media for the detection of coliforms is available (Table 14.3) and confirmation of thermotolerant *E. coli* is possible by incubation at 44°C.

Techniques available to rapidly detect bacteria include fluorescence microscopic methods e.g., Epifluorescence microscopy using acridine, detection of specific metabolites, antibody methods, and DNA-based methods. However, many of these methods are expensive, require an enrichment step, sophisticated equipment, and expertise, and/or are not suitable for routine analysis. The determination of ATP with a bioluminescence assay has emerged as the main method for rapid detection of viable bacteria in breweries. Despite ATP being considered a robust monitoring parameter for microbial drinking water quality, a significant increase in ATP should be accompanied by methods for detection of specific bacteria in order to validate whether or not contamination has occurred. Therefore, the best approach for monitoring microbial drinking water quality, in order to enhance water security and safety, is to combine rapid methods with methods targeted for specific bacterial detection (Ferone, Gowen, Fanning, & Scannell, 2020).

14.1.1.3 Yeast

Of all the raw materials, the most likely source of contamination is from yeast because it is added after wort boiling. Yeast handling plants also tend to be very complex and difficult to clean (Briggs, Brookes, Stevens, & Boulton, 2004). Acid washing can be used to reduce or remove bacterial contaminants, but this process does not remove wild yeast. A wild yeast is typically regarded as 'any yeast which is not deliberately used or under full control', which can include both Saccharomyces and non-Saccharomyces strains. Incidences of product recall due to diastatic strains of Saccharomyces (*Saccharomyces cerevisiae* var *diastaticus* or *S. diastaticus*) have risen sharply over recent years, mainly due to higher usage of these yeast in production of sours and Saison's. There is also increased production of specialty products using non-Saccharomyces strains, resulting in a higher risk of cross-contamination. *Brettanomyces bruxellensis*, used in Belgian Lambic Beer production, and *Torulaspora delbrueckii*, used in Wheat Beers, are examples of deliberately used strains, with Brettanomyces, Pichia, Candida, and Hansenula species typically the most frequently detected wild yeast.

The most common bacterial contaminants are the lactic acid bacteria (LAB) Lactobacilli and Pediococci and the microbiological media employed to check pitching yeast reflects their nutritional requirements. *Obesumbacterium proteus* (*Hafnia protea*), Acetic acid bacteria, Zymomonas, and the strict anaerobes Pectinatus and Megasphaera are also frequently problematic, with further species of concern arising with the increase in no- and low-alcohol beer production (see Chapter 13).

Yeast should be checked before pitching (preferably 2–4 days prior to brewing) and monthly checks should also be carried out to test for nonbrewing strains that may populate over time. Tolerance is less than 10 cfu/mL for bacteria and zero tolerance of wild yeast.

Media employed in traditional plate checks typically include inhibitors and/or stimulators (Table 14.4). Such chemicals might include lysine, a nitrogen source that brewing yeast cannot utilise but wild yeast can; copper sulphate, which is also inhibitory to strains of *Saccharomyces*; or plates containing Actidione, which selectively promotes bacterial growth. Such techniques typically require a 2-day 25°C incubation period (Martin & Moll, 1984).

TABLE 14.3 Microbiological media for water analysis.

Medium	Target microorganism(s)	Incubation conditions
Azide	Intestinal enterococci	40–48 h at 36 ± 2°C
Bismuth Sulphite	*Salmonella typhi* and other *salmonellae*	40–48 h at 36 ± 2°C
Cetrimide	*Pseudomonas aeruginosa*	40–48 h at 36 ± 2°C
Chapman	*Staphylococci*	18–72 h at 30–35°C
Chromogenic coliform agar (CCA)	Total coliforms and *Escherichia coli*	21–24 h at 36 ± 2°C
Chromocult coliform agar	Total coliforms and *Escherichia coli*	20–28 h at 36 ± 2°C
ECD	*E. coli*	16–18 h at 44 ± 2°C
Endo	*E. coli* and coliform bacteria	18–24 h at 36 ± 2°C
Glucose Tryptone	Mesophilic and thermophilic bacteria	18–72 h at 30–35°C for mesophilic bacteria; 48–72 h at 55 ± 2°C for thermophilic sporulating microorganisms
LMC broth	Coliforms and *E. coli*	1–2 days at 30–35°C
Lysine	Wild yeast	3–5 days at 30–35°C
MacConkey	Coliform bacteria and other enterobacteriaceae	18–72 h at 30–35°C
Malt extract	Yeasts and moulds	3–5 days at 20–25°C or at 30–35°C depending on the target of the investigation
Meat extract-peptone	Total count	<5 days at 30–35°C
mFC	*E. coli* and faecal coliform bacteria	18–24 h at 36 ± 2°C
MLGA (Membrane lactose Glucuronide agar)	Coliforms and *E. coli*	30°C for 4 h, then 37°C for 14 h
R2A	Heterophilic organisms	>5 days at 30–35°C
Rainbow agar	*E. coli*	18 h 37°C
Sabouraud	Yeast and moulds	<5 days at 20–25°C
Schaufus Pottinger (M Green yeast and mould)	Yeast and moulds	2–5 days at 20–25°C or at 30–35°C depending on the target of the investigation
Soybean-Casein Digest medium (Caso)	Total count	Bacteria: <3 days at 30–35°C Yeasts and moulds: <5 days at 30–35°C
Standard TTC	Total count	<5 days at 30–35°C
Teepol (Lauryl sulphate medium)	*E. coli* and faecal coliform bacteria	18–24 h at 36 ± 2°C
Tergitol TTC	Coliform bacteria and *E. coli*	18–24 h at 36 ± 2°C
Tryptone glucose extract (TGE)	Total count	<5 days at 30–35°C
Wallerstein (WL nutrient)	Microbiological flora of brewing and fermentation processes	2–5 days at 30–35°C aerobic or anaerobic depending on the target of the investigation
Wort	Yeast and moulds	3–5 days at 20–25°C or at 30–35°C depending on the target of the investigation
Yeast extract	Aerobic bacteria	44 ± 4 h at 36 ± 2°C 68 ± 4 h at 22 ± 2°C

When attempting to identify particularly hard-to-culture LAB strains, 'advanced beer-spoiling detection' (ABD) media is the quickest and most effective media, primarily because of the low pH levels which it employs. When seeking to identify LAB, anaerobic incubation at around 28°C is typically employed (Suzuki, 2011). Supplementing deMan Rogosa

TABLE 14.4 Microbiological media for pitching yeast analysis.

Medium	Target microorganism(s)	Incubation conditions
$CuSO_4$	Wild yeast (non-Saccharomyces)	48 h, 25°C
LCSM (Lin's Cupric sulphate medium)	Wild yeast (non-Saccharomyces but some *S. cerevisiae* var. *diastaticus* are able to grow)	4–6 days, 28–30°C; aerobic
Lysine	Enteric, acetic and lactic bacteria, wild yeast	48 h, 25°C
MacConkey + Actidione	Enteric bacteria	48 h, 25°C
MRS (de man, Rogosa, Sharpe)	Enteric, acetic and lactic bacteria, wild yeast	48 h, 25°C
MYGP (yeast extract glucose peptone)	Wild yeast	48–72 h at 37°C; lager strains will not grow
NBB-broth	Lactic acid bacteria	72 h, 25–30°C; aerobic and anaerobic
Nutrient Agar/Broth	General purpose medium for bacteria although many yeasts will grow	48 h, 25°C
Starch	*S. cerevisiae* cannot ferment starch but *S. cerevisiae* var *diastaticus* can	7–10 days or longer as required, 25–30°C; aerobic
WLN (Wallerstein laboratory Differential)	General purpose medium for bacteria	48 h, 25°C
Yeast morphology agar	Assessment of yeast colony morphology	48 h, 25°C

Sharpe (MRS) media with catalase can also potentially speed up growth of LAB (Deng et al., 2014). For very slow growing strains, microcolony methods using carboxyfluorescein diacetate (CFDA; described below), can allow detection within 3 days.

For breweries using flow cytometry to determine yeast count and viability, it is possible to extend use of this method to detect beer spoilers such as *Zygosaccharomyces*, *Dekkera* (*Brettanomyces*), *Pediococcus*, and *Lactobacillus* (Donhauser, Eger, Hubl, Schmidt, & Winnewisser, 1993; Jespersen, Lassen, & Jakobsen, 1993; Takahashi, Kita, Kusaka, Mizuno, & Goto-Yamamoto, 2015; Xu et al., 2022). The principle of flow cytometry is based on fluorescence staining or labelling, and the cells are brought in a fluid stream within a thin capillary where the fluorescence molecules are excited by a laser and the emission is detected. The laser is also used to count the particles and determine the size. All data are collected, and a report is generated with the result of live/dead cells or detection of beer spoilers.

Although PCR is commonly employed for the detection of the *sta*1 gene in diastatic strains of Saccharomyces, the use of culture-dependent techniques is still advised to determine the starch consumption rate, as some strains may contain an inactive form of the gene and grow weakly, or not at all. A more precise test to determine the presence of the promotor sequence for the STA1 gene using multiplex PCR is available (Krogerus, Magalhães, Kuivanen, & Gibson, 2019), and fermentation monitoring can also help in detection of contamination.

14.1.1.4 Hops

The antibacterial properties of hops are one of the main reasons for their use in brewing. Hops are dried down to 8%–10% moisture to prevent spoilage but nonetheless remain susceptible and, as with other raw materials, checks should be made for each batch as a matter of quality control. A number of breweries carry out dry hopping post-brewhouse which increases the risk of introducing contaminants. For hops the most likely microbes are fungi, moulds, and mildew and the tolerance is less than 10 cfu/g, with zero tolerance of wild yeast.

For traditional plating, a weighed sample is rinsed in sterile water and the rinse water is either mixed with molten agar or spread on the surface of an agar plate. Media for the detection of common spoilage organisms is given in Table 14.5.

A range of molecular based techniques are available to detect major fungal crop pathogens, including PCR and DNA microarray methods.

TABLE 14.5 Microbiological media for hops analysis.

Medium	Target microorganism(s)	Incubation conditions
Lysine	"Wild" yeast	3–5 days at 30–35°C
Malt extract agar	Fungi, e.g., *Podosphaera castagnei*, mould, mildew	48 h, 25°C
Sabouraud	Yeast and moulds	5 days at 20–25°C
Schaufus Pottinger	Yeast and moulds	2–5 days at 20–25°C or at 30–35°C
WLD (Wallerstein Differential broth)	Flora of brewing and fermentation processes	2–5 days at 30–35°C; aerobic or anaerobic
Wort agar	Yeast and moulds	3–5 days at 20–25°C or at 30–35°C

14.1.1.5 Sugars and syrups

Noncereal adjuncts and priming sugars are commonly employed in brewing. Irrespective of the point of addition, any materials added to the process should be checked as a matter of quality control. For sugars and syrups the low water potential prevents growth of contaminants, but the main threat is survival of spores from *Bacillus* species. A range of media is available to selectively culture *Bacillus* species (Table 14.6). A sugar solution may be mixed with molten agar or spread directly on the surface of an agar plate.

Rapid methods of analysis tend not to be used for detection of contaminants in sugars and syrups due to the difficulties in extracting genetic material from the complex medium, but PCR primers for *Bacillus* species are readily available.

14.1.2 Brewing process

As we move towards fermentation it is essential that the integrity of the system is maintained and that all raw materials and adjuncts are appropriately stored and transferred under sterile conditions. The first challenge for the microbiologist is to ensure that all vessels and pipework are tested.

14.1.2.1 Brewery surfaces

As the beer progresses through the brewery it comes into contact with vessel surfaces, pipes, and fittings, all of which can harbour infection, particularly on the cold side of the brewery after wort clarification. Any surface that comes into contact with wort, beer, or yeast should be thoroughly cleaned and sterilised, that is, vessels, piping, and implements. Soiled surfaces can support a microbiological growth which can be introduced into the beer. Any recurrent contamination may indicate the presence of a biofilm. Biofilms are particularly difficult to clean as they can bind strongly to the vessel or pipe.

Almost ubiquitously, ATP tests are employed to check plant hygiene. This includes both swab tests of vessels and pipework and liquid tests of CIP final rinse water. Typical sample locations and types are shown in Table 14.7.

TABLE 14.6 Microbiological media for analysis of syrups and sugars.

Medium	Target microorganism(s)	Incubation conditions
BACARA: Chromogenic media	*Bacillus* species	48–72 h at 25–30°C. Combined anaerobic (3 days) followed by aerobic (2 days)
Bacillus Differentiation agar	Differentiation between *Bacillus cereus (colourless)* and *Bacillus subtilis (yellow)*	35–37°C for 18–24 h
Malt extract agar (MXA)	All microorganisms	22–25°C for 3 days
MYP AGAR	*B. cereus*	30°C. 24 h incubation

TABLE 14.7 Brewery ATP testing schedule.

Location	Sample type	Frequency	Target organism(s)
Hot liquor tank	Liquid water sample Final rinse CIP	Weekly Each CIP	All microorganisms
Cereal cooker	Swab Liquid (CIP final rinse)	Each use/cleaning cycle	All microorganisms
Mash tun	Swab Liquid (CIP final rinse)	Each use/cleaning cycle	All microorganisms
Lauter tun	Swab Liquid (CIP final rinse)	Each use/cleaning cycle	All microorganisms
Mash filter	Liquid (CIP final rinse)	Each use/cleaning cycle	All microorganisms
Kettle	Swab Liquid (CIP final rinse)	Each use/cleaning cycle	All microorganisms
Paraflow/chiller	Liquid (CIP final rinse) Liquid (wort)	Each use/cleaning cycle Each batch	All microorganisms

Secondary testing (either traditional plating or rapid methods) tends to only be used if ATP levels are consistently breaching tolerance limits and/or are not reduced by additional cleaning. Malt extract agar or similar may be used for the detection of acetic and lactic bacteria and wild yeast.

14.1.2.2 Air and process gases

Microorganisms are ever-present in the air, often in association with dust particles or airborne moisture droplets. They can also be introduced to the environment by insects and other pests. Every effort must be made to keep the brewing environment as clean as possible and to minimise the ingress of outside contamination. Unless producing a Lambic beer or Coolship ale where open fermenters are required, wherever possible all vessels should be covered to reduce the risk of aerial contamination.

Process gases such as oxygen used during pitching and carbon dioxide applied during packaging may also provide a route for contaminants to enter the system. Any hosing should be inspected for leaks and all gases and air checked. The method involves either exposing an agar plate to the air for a set period of time or filtration of the air through a sterile filter which is then placed on the surface of an agar plate. Typically, a nonspecific medium is used, such as malt extract agar, and incubated for 48 h at 25°C.

14.1.2.3 Wort

As the temperature falls following wort boiling, any bacteria or wild yeast present on surfaces will multiply in the nutrient rich medium. Problems can also arise if wort is saved and allowed to stand (such as with weak wort recycling). Contaminated wort can result in a decreased fermentation rate, off-flavours/odours, and haze and therefore should be checked at each batch. Enterobacteria are the most common contaminants, and the tolerance is less than 10 cfu/mL, with zero tolerance of wild yeast.

Following boiling the likely prevalence of spoilage organisms is low and the wort that is being tested will therefore need to be filtered through a sterile filter and placed on an agar plate. Typical media are given in Table 14.8.

It is from this stage forward that rapid methods of analysis are increasingly likely to be used, simply due to the increased cost of spoilage as we move towards final product. Such methods are described in Chapter 15, but a faster 'traditional' method includes the microcolony method which employs microscopy to detect growing cells that have not yet reached visibly discernible colonies. Several systems are available, including Rapid Micro Biosystem's Growth Direct test, which uses digital imaging technology to automatically enumerate microcolonies. The system captures the native fluorescence (autofluorescence) that is emitted by all living cells. More advanced systems use 96-well microplate formats and automated plate handling systems.

TABLE 14.8 Microbiological media for wort analysis.

Medium	Target microorganism(s)	Incubation conditions
Carr's Bromocresol Green medium	Gram-negative bacteria	27°C for 1 day
Hsu's *Lactobacillus/Pediococcus* medium (HLP)	*Lactobacillus, Pediococcus*	25°C for 2 days
Lee's Multi Differential agar (LMDA)/Schwarz Differential agar (SDA)	Lactic acid bacteria	25°C for 2 days
Lin's Cupric sulphate medium (LCSM)	Wild yeast	25°C for 1–3 days
Lin's wild yeast media (LWYM)	Wild yeast	25°C for 1–3 days
Malt extract agar (MXA)	All microorganisms	22–25°C for 3 days
MRS	Enteric, acetic, and lactic bacteria, wild yeast	25°C for 2 days
MRS + Actidone	Enteric, acetic, and lactic bacteria	25°C for 2 days
Sabouraud Dextrose agar	Yeast	22–25°C for 3 days

14.1.2.4 Fermentation

Fermentation conditions are ideal for microbial growth and contamination can retard or extend fermentation and cause off-flavours and odours. Typically, specific gravity, pH, and flavour are checked while brewing and microbiological analysis is only carried out if issues arise during fermentation. Lactic and acetic bacteria and wild yeast are the main threats with a tolerance of less than 10 cfu/mL and zero count, respectively. Media for their detection is given in Table 14.9.

14.1.3 Product

Once we reach the final stages of the brewing process the sample volume is very low in relation to the batch volume (typically 250 mL from 1000 hL) and levels of beer-spoiling microbes are extremely low. Filtration is therefore needed to improve the likelihood of detecting contaminants, and this can be carried out either in—or off-line.

TABLE 14.9 Microbiological media for fermentation analysis.

Medium	Target microorganism(s)	Incubation conditions
Hsu's *Lactobacillus/Pediococcus* medium (HLP)	*Lactobacillus, Pediococcus*	25°C for 2 days
Lee's Multi Differential agar (LMDA)/Schwarz Differential agar (SDA)	Lactic acid bacteria	25°C for 2 days
Lin's Cupric sulphate medium (LCSM)	Wild yeast	25°C for 1–3 days
Lin's wild yeast media (LWYM)	Wild yeast	25°C for 1–3 days
Malt extract agar (MXA)	All microorganisms	22–25°C for 3 days
MRS	Enteric, acetic, and lactic bacteria, wild yeast	25°C for 2 days
MRS + Actidone	Enteric, acetic, and lactic bacteria	25°C for 2 days

14.1.3.1 Bright beer

Lactic and acetic bacteria present in bright beer cause vinegary, sour astringent off-flavour and odour, excessive gassing, and strong head retention. Every batch should be tested, and the tolerance is less than 10 cfu/mL. Commonly used media for analysis of bright beer are given in Table 14.10.

The complex nature of beer results in difficulties in sensitivity/interference for many rapid methods; however, a number of kits are available and discussed in Chapter 15.

14.1.3.2 Packaging

The packaging process exposes bright beer to a range of new surfaces from the buffer tank through the filling machine and the eventual container. The potential for secondary contamination is high, with this stage representing one of the most common points for entry of spoilage organisms, and as such strict attention should be paid to both hygienic design and cleaning regimes. Process quality control parameters should include turbidity (haze), dissolved oxygen, CO_2 content, original extract or alcohol, and the presence of acetic and lactic acid bacteria. A range of bacteria and yeasts may be present in biofilms, such as Lactic acid bacteria, *Pseudomonas* and *Enterobacteria* species, *Rhodotorula* and *Cryptococcus*, and moulds including *Geotrichum* and *Aureobasidium* (Back, 2005; Maifreni et al., 2015; Riedl, Dünzer, Michel, Jacob, & Hutzler, 2019). However, the most significant spoilage organisms for brewers that enter the product during packaging are *Pectinatus* and *Megasphaera*; improvements in packaging to reduce oxygen in the headspace of bottled beers has led to an increase in incidence of anaerobic beer spoilage bacteria, which are able to survive within filler heads and CO_2 recovery systems (Paradh, Mitchell, & Hill, 2011).

As with other brewing surfaces, ATP testing is the most common method for detection of microbial activity within the packaging area. Media used for traditional plating tests are also identical to those described for bright beer, with Raka Ray and NBB the most common media used within the UK brewing industry (Paradh et al., 2011).

14.1.3.3 Dispense

At the dispense stage product security is usually outside of the control of the brewery. The most common spoilage organisms are acetic and lactic acid bacteria, which cause surface film, haze, and vinegary flavour/odour. Wild yeasts may also proliferate in dispense lines where biofilms are a common hazard. Cask dispensing introduces oxygen into beer, adding further risk, and therefore line cleaning should be carried out at least biweekly. Tolerance is less than 10 cfu bacteria and 0 cfu wild yeast.

As with packaging, ATP testing is commonly employed. Detection media for traditional plating are detailed in Table 14.11 and further information on dispense is given in Chapter 17.

14.2 Identification of brewing spoilage organisms

Once a colony has been isolated from any of the stages above, the next step is to identify it. Identification is necessary to determine whether the contaminant represents a risk in terms of spoilage. Preliminary identification of many of the microbes of significance in brewing has traditionally been made on the basis of the following few simple characteristics of the cells.

TABLE 14.10 Microbiological media for bright beer analysis.

Medium	Target microorganism(s)	Incubation conditions
Beer agar	All microorganisms	22–25°C for 3 days
Malt extract agar (MXA)	All microorganisms	22–25°C for 3 days
MRS	Enteric, acetic, and lactic bacteria, wild yeast	25°C for 2 days
MRS + Actidone	Enteric, acetic, and lactic bacteria	25°C for 2 days
NBB-broth	Lactic acid bacteria, *Pectinatus* and *Megasphaera*	25–30°C for 3 days; aerobic and anaerobic
Raka-Ray	Lactic acid bacteria, *Pectinatus* and *Megasphaera*	30°C for 3 days; anaerobic

TABLE 14.11 Media for the detection of dispense spoilage organisms.

Medium	Target microorganism(s)	Incubation conditions
MRS	Enteric, acetic, and lactic bacteria, wild yeast	25°C for 2 days
Rainbow agar	Enteric, acetic, and lactic bacteria, *E. coli*, *Salmonella*, *Shigella*, and *Aeromonas*	18 h 37°C
WLN (Wallerstein laboratory Differential)	Enteric, acetic, and lactic bacteria, yeast	2–5 days at 30–35°C; aerobic or anaerobic depending on the target of the investigation

- growth requirements,
- ability to grow under aerobic or anaerobic conditions,
- colony characteristics and cell morphology (Fig. 14.3), and
- Gram reaction (for bacterial colonies).

Further identification is made on the basis of biochemical properties such as.

- ability to produce enzymes that can be detected by simple tests,
- ability to metabolise sugars oxidatively or fermentatively, and
- ability to use a range of substrates for growth, for example, glucose, lactose, and sucrose.

These tests can be carried out individually, for example, in broth media containing the specifically required nutrients and/or reagents, but they are more commonly performed using commercial kits or automated systems which have the potential to give a rapid identification based on biochemical profiles.

Bergey's Manual of Determinative Bacteriology is the standard reference for laboratory identification of bacteria. Dichotomous keys, which incorporate information from a variety of identification methods, are also commonly used for the identification of organisms. Dichotomous keys for common brewing fungi and bacteria are given in Fig. 14.4.

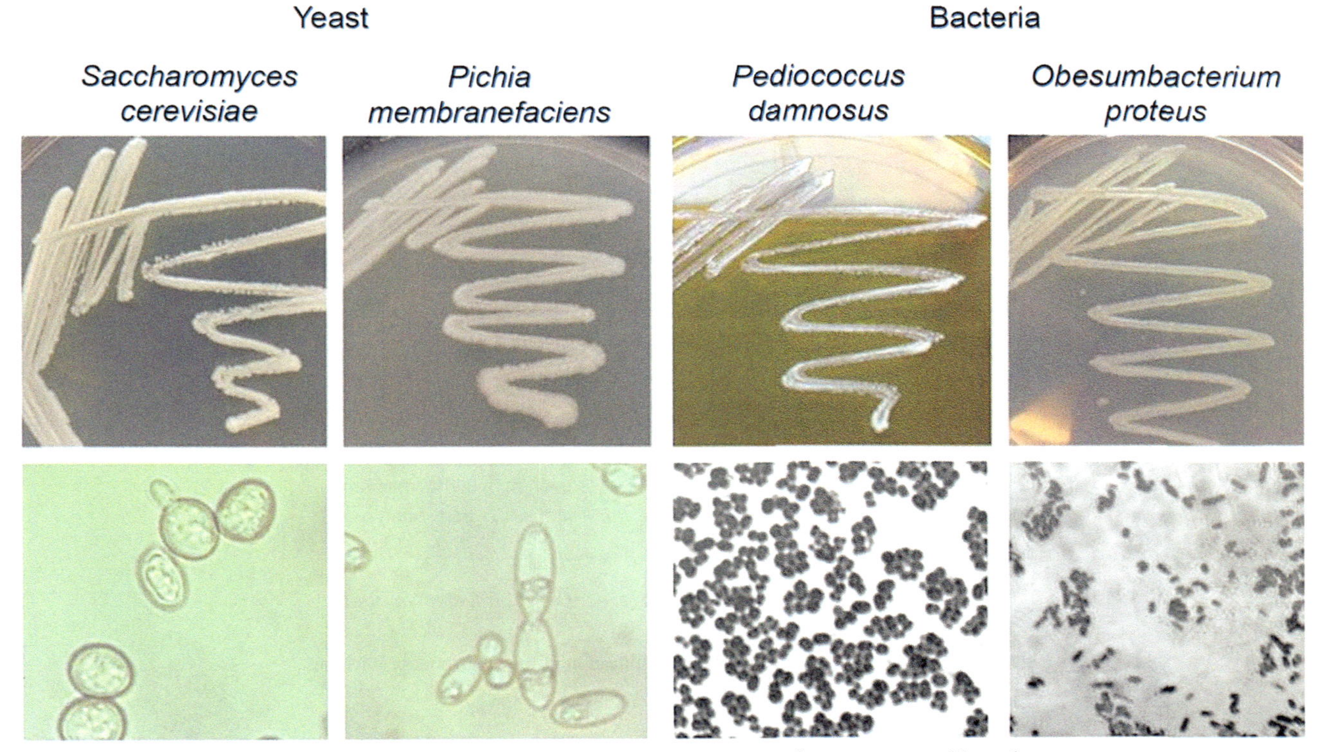

FIGURE 14.3 Colony and cell morphology of common brewery yeast and bacteria.

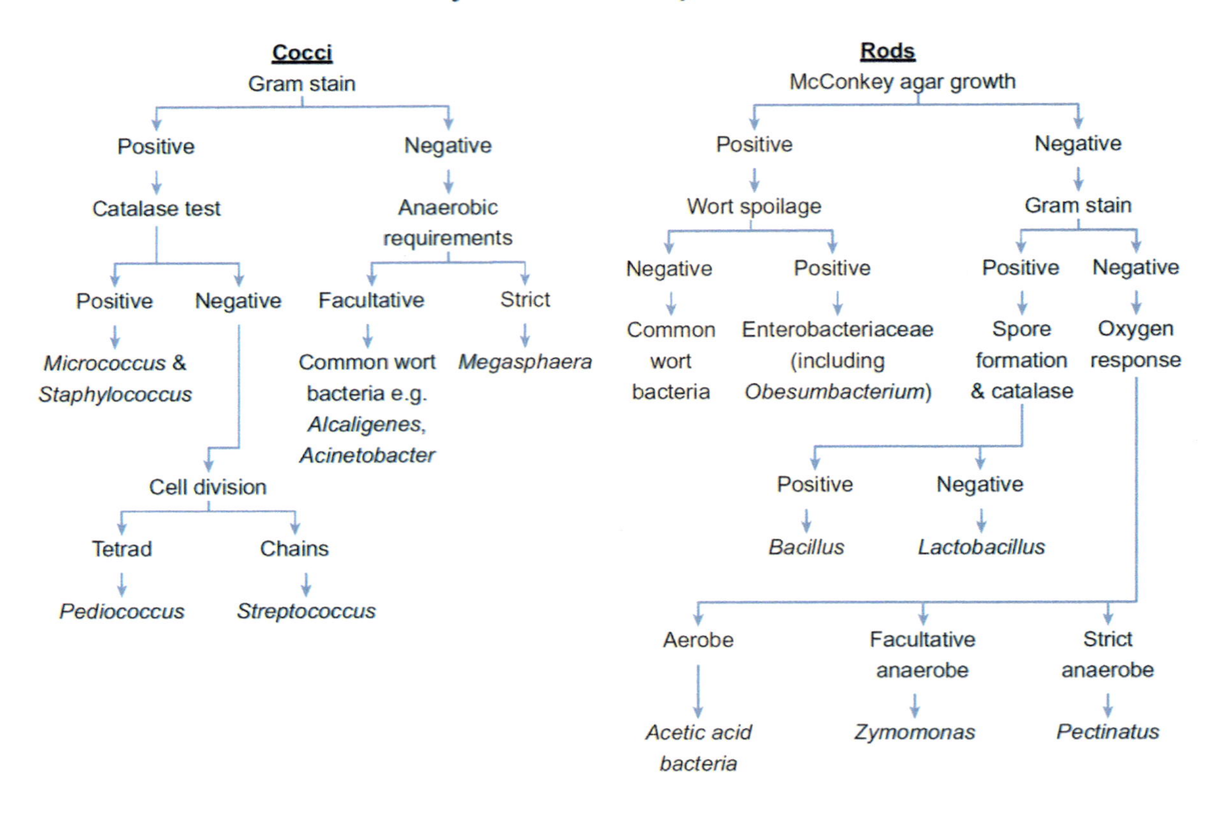

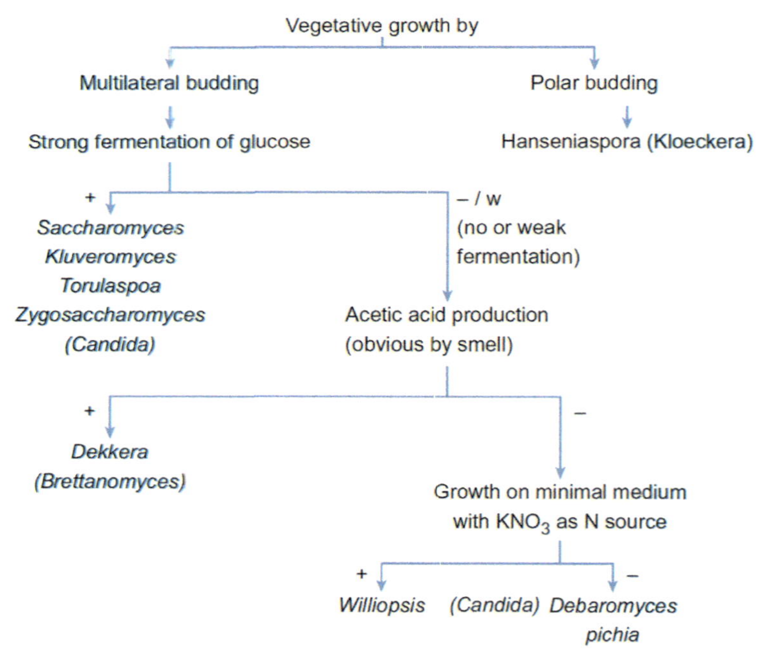

FIGURE 14.4 Dichotomous keys for identification of brewing spoilage microbes.

14.3 Summary

The combination of hygienic plant design, effective CIP, and quality assurance of raw materials represents a sensible strategy for minimising the risk of microbial contamination during the brewing process. However, 'reactive' testing throughout is also essential in quality control. Plate counting and enrichment remain commonly used methods for detection of microbes in breweries during the brewing process and in final product analysis. The wide range of media and methods available may seem overwhelming, but there are common media and tests that may be used for several/all stages, making effective control and maintenance reachable regardless of budget.

References

Back, W. (2005). Winery. In W. Back (Ed.), *Colour atlas and handbook of beverage biology* (pp. 113–135). Nürnberg, Germany: Verlag Hans Carl.

Bokulich, N. A., & Bamforth, C. W. (2013). The microbiology of malting and brewing. *Microbiology and Molecular Biology Reviews, 77*(2), 157–172.

Briggs, D. E., Brookes, P. A., Stevens, R., & Boulton, C. A. (2004). *Brewing: Science and practice.* Elsevier.

Deng, Y., Liu, J., Li, H., Li, L., Tu, J., Fang, H., et al. (2014). An improved plate culture procedure for the rapid detection of beer-spoilage lactic acid bacteria. *Journal of the Institute of Brewing, 120*(2), 127–132.

Donhauser, S., Eger, C., Hubl, T., Schmidt, U., & Winnewisser, W. (1993). *Tests to determine the vitality of yeasts using flow cytometry.* Germany: Brauwelt international.

Ferone, M., Gowen, A., Fanning, S., & Scannell, A. G. (2020). Microbial detection and identification methods: Bench top assays to omics approaches. *Comprehensive Reviews in Food Science and Food Safety, 19*(6), 3106–3129.

Hocking, A. D., & Pitt, J. I. (1980). Dichloran-glycerol medium for enumeration of xerophilic fungi from low-moisture foods. *Applied and Environmental Microbiology, 39*, 488.

Jespersen, L., Lassen, S., & Jakobsen, M. (1993). Flow cytometric detection of wild yeast in lager breweries. *International Journal of Food Microbiology, 17*(4), 321–328.

Kordialik-Bogacka, E. (2022). Biopreservation of beer: Potential and constraints. *Biotechnology Advances, 58*, Article 107910.

Krogerus, K., Magalhães, F., Kuivanen, J., & Gibson, B. (2019). A deletion in the STA1 promoter determines maltotriose and starch utilization in STA1+ *Saccharomyces cerevisiae* strains. *Applied Microbiology and Biotechnology, 103*, 7597–7615.

Maifreni, M., Frigo, F., Bartolomeoli, I., Buiatti, S., Picon, S., & Marino, M. (2015). Bacterial biofilm as a possible source of contamination in the microbrewery environment. *Food Control, 50*, 809–814.

Martin, P. A., & Moll, M. (1984). Ebc analytica microbiologica: Part III. *Journal of the Institute of Brewing, 90*(4), 272–276.

Mirocha, T. J., & Christensen, T. M. (1986). Mycotoxins and the fungi that produce them. *Proceedings of the American Phytopathological Society, 3*, 110–125.

Mostafa, M. E., Barakat, A., & El-Shanawany, A. A. (2005). Relationship between aflatoxin synthesis and *Aspergillus flavus* development. *African Journal of Mycology and Biotechnology, 13*, 35–51.

Paradh, A. D., Mitchell, W. J., & Hill, A. E. (2011). Occurrence of Pectinatus and Megasphaera in the major UK breweries. *Journal of the Institute of Brewing, 117*(4), 498–506.

Riedl, R., Dünzer, N., Michel, M., Jacob, F., & Hutzler, M. (2019). Beer enemy number one: Genetic diversity, physiology and biofilm formation of Lactobacillus brevis. *Journal of the Institute of Brewing, 125*(2), 250–260.

Schneiderbanger, J., Jacob, F., & Hutzler, M. (2020). Mini-Review: The current role of lactic acid bacteria in beer spoilage. *Monatsschrift für Brauwissenschaft, 73*, 2.

Suzuki, K. (2011). 125th Anniversary Review: Microbiological instability of beer caused by spoilage bacteria. *Journal of the Institute of Brewing, 117*(2), 131–155.

Suzuki, K. (2020). Emergence of new spoilage microorganisms in the brewing industry and development of microbiological quality control methods to cope with this phenomenon: A review. *Journal of the American Society of Brewing Chemists, 78*(4), 245–259.

Takahashi, M., Kita, Y., Kusaka, K., Mizuno, A., & Goto-Yamamoto, N. (2015). Evaluation of microbial diversity in the pilot-scale beer brewing process by culture-dependent and culture-independent method. *Journal of Applied Microbiology, 118*(2), 454–469.

Thrane, U. (1996). Comparison of three selective media for detecting *Fusarium* species in foods: A collaborative study. *International Journal of Food Microbiology, 29*, 149–156.

Vaughan, A., O'Sullivan, T., & van Sinderen, D. (2005). Enhancing the microbiological stability of malt and beer – a review. *Journal of the Institute of Brewing, 111*, 355–371.

Xu, Z., Wang, K., Liu, Z., Soteyome, T., Deng, Y., Chen, L., et al. (2022). A novel procedure in combination of genomic sequencing, flow cytometry and routine culturing for confirmation of beer spoilage caused by Pediococcus damnosus in viable but nonculturable state. *LWT, 154*, Article 112623.

Chapter 15

Rapid detection and identification of spoilage microorganisms in beer

Jvo Siegrist[1], Ulf-Martin Kohlstock[2], Kathleen Merx[2], Kathleen Vetter[2] and Annie E. Hill[3]

[1]Merck Group, St Gallen, Switzerland; [2]Scanbec, Bitterfeld-Wolfen, Germany; [3]The International Centre for Brewing & Distilling, Heriot-Watt University, Edinburgh, United Kingdom

15.1 Introduction

Traditional methods in microbiology are primarily culture based, and it often takes several days to detect, identify, and confirm organisms. In the case of the brewing industry, the problem is that the range of spoilage organisms is extensive and growing due to the development of novel products. Often only a very low concentration of contaminants is present, and the medium they are found in is complex. This requires the use of many different classical tests and at minimum, a culture enrichment step. Some may also be viable but non culturable (VBNC). Additionally, in a brewery, large batches are fermented and must be properly stored somewhere in a blocked stock, or they will spoil and eventually end up being recalled. Therefore, the rapid release of a guaranteed high quality new batch of beer saves a lot of money and prevents getting a bad reputation, as well as avoiding an expensive recall. The main drivers for the application of rapid methods are:

1. Product quality and safety.
2. Sustainability: reduction of waste.
3. New products: non-pasteurised beer, low or no-alcohol beer, novel raw materials and adjuncts, and ready-to-drink products.
4. Time and cost savings.

In recent years, diverse rapid methods have been publicised and there are several systems in use today. For some applications and organisms, optimised methods are available which can detect target microbes within only a few hours and most of them can do it within 24 h. Some rapid methods can detect several organisms in one step.

In this Chapter, the methods most commonly used by brewers to rapidly detect spoilage micro-organisms are described first, divided into those based on the physical characteristics of the cell (predominantly phenotypic), and nucleic acid-based methods (predominantly genotypic). Research within medical diagnostics and detection of foodborne pathogens are continuing to advance the development of rapid methods, and emerging techniques with potential application in brewing quality control are also considered.

15.2 Cell-based methods

15.2.1 Hygiene tests (ATP bioluminescence, oxidoreductase)

One of the most used test methods in breweries is analysis of swabs and/or rinse water for the presence of adenosine triphosphate (ATP). ATP is the stored form of energy in microorganisms and is central to many biochemical reactions in cell metabolism. Therefore, it is an excellent indicator for the presence of living microorganisms (Webster, Walker, Ford, & Leach, 1988). It is used in applications where hygiene control is needed, such as food and beverage production and cooking areas. The ATP test is mainly used to check equipment, surface and material sanitation or the effectiveness of treatments from such material or fermenters with biocides and detergents. In the brewing industry the ATP test is a good

Brewing Microbiology. https://doi.org/10.1016/B978-0-323-99606-8.00001-8

indicator to have some evidence that fermenters and used equipment are presumably free of any spoiling organisms and the cleaning process was successful. It does not provide any information on the identity of the micro-organism(s), or indeed if the ATP is derived from other sources, such as human skin cells.

The ATP test is based on the presence of ATP from living cells, which delivers the energy needed for the reaction catalysed by firefly luciferase. In the presence of luciferase, luciferin and ATP react to form luciferyl adenylate and phosphate. Then the luciferyl adenylate and oxygen react, forming oxyluciferin and AMP (adenosine monophosphate). The oxyluciferin is formed in an electronically excited state and therefore a photon is released, which is visible by emitting a yellow-green light. The oxyluciferin returns back to the ground state (luciferin) (Rhodes & McElroy, 1958). This reaction is called bioluminescence and can be seen in fireflies and some other creatures. Light is a positive reaction for living organisms, including bacteria, and is therefore a sign of insufficient cleaning. Approximately 10^2—10^4 colony forming units are needed for detection (Fig. 15.1).

There are a number of ATP systems currently available, including UltraSnap (Hygiena), PocketSwab Plus (Charm Sciences), N-Light™ (NEMIS), and Clean-Trace (3M). The UltraSnap system contains a premoistened swab in a tube and the reaction reagent in a cap on the top. A certain area of surface is wiped over with the swab. Once the cap is closed the reagent moves down to the swab and the tube is shaken a few times. If there is ATP present on the swab it will react with reagent containing luciferyl adenylate and the result of the positive reaction will be bioluminescence, which can be detected in a luminometer. Tolerance or critical limits for relative light units (RLUs) should be determined for the system that is being used and for each sample point within the brewing process. It is also important to ensure consistency in sampling i.e., area swabbed, or volume taken, to enable comparisons to be made between results.

Today's systems are in most cases based on the bioluminescence with ATP and luciferase from the firefly. As an alternative system it is also possible to use a colour test: nicotinamide adenine dinucleotides (NAD/NADH) and nicotinamide adenine dinucleotide phosphates (NADP/NADPH), which are also compounds used for the energy transfer in the metabolism in living cells or compounds found in food debris.

If NAD(P) and/or NAD(P)H is present in the sample, glucose dehydrogenase converts β-D-Glucose into D-glucono-lactone, then diaphorase converts a tetrazolium salt into a coloured formazan salt (HY-RiSE system from Merck). Any colour development on the test strip indicates a positive result (not clean).

It is also possible to use a method with horseradish peroxidase and NAD(P) and/or NAD(P)H. The result of the reaction is the production of H_2O_2 (hydrogen peroxide), which can be detected with the addition of luminol. The luminol reacts with H_2O_2, resulting in bioluminescence (Anand, 2004).

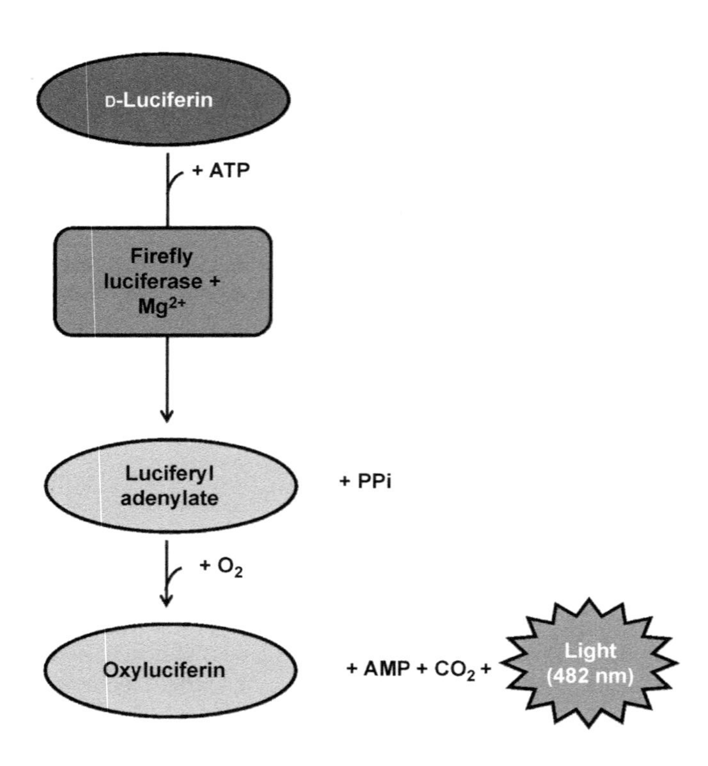

FIGURE 15.1 ATP Luminescence reaction with firefly luciferase.

The oxidoreductase test is based on the detection of NAD/NADH and NADP/NADPH. HY-RiSE (Merck), for example, uses a strip or a card and the result can be directly visible by a colour reaction. The strip is moistened, or the surface must already be wet, or some rinse water is added on the strip. Once the strip has been wiped across the test surface, two further reagents are added. After a 5 min incubation in the dark the reaction zone will appear yellow (negative reaction = clean) or pink/purple to bluish (NAD/NADH and/or NADP/NADPH positive = presence of living cells or food debris). The test has been validated and compared with the ATP test and there was no significant difference (Goll, Kratzheller, & Bülte, 2003).

15.2.2 Fluorescence microscopy and cytometry

Most brewers will be familiar with using a haemocytometer to determine yeast pitching rate. The total viable count is an important parameter in industrial fermentations and traditional methods for total viable count started with the cultivation method of counting the colonies. Later methylene blue and Ponceau S stain came into use for microscopic examinations using a counting chamber (haemocytometer), which give a direct count (Kunkee & Neradt, 1974). Today, systems are available for inline measurement including automated cytometers. Cytometric devices include image cytometers, flow cytometers, and cell sorters. These devices may be used not only for assessment of production yeast strains, but also for the detection and identification of contaminants.

Image cytometry is the oldest form of cytometry and uses an optical microscope. Cells are commonly pre-stained to enhance contrast or to detect specific molecules, and a range of fluorescent stains are available for the specific detection of microbes. In brewing, an image cytometry method has been developed combining fluorescent stains and size exclusion image analysis algorithms to count *Lactiplantibacillus plantarum* and *Saccharomyces cerevisiae* simultaneously in a mixed culture (Williamson et al., 2023).

Direct epifluorescent microscopy (DEM) methods in which both the illuminated and emitted light travel through the same objective lens and the direct epifluorescent filter technique (DEFT) may be used to assess microbial populations in brewery samples.

In DEFT, a homogenised sample is prefiltered through a 12 μm pore filter (Priest & Campbell, 2003) to remove large particles and to avoid blockage in the next filter step (0.2 μm pore black polycarbonate filter) and big numbers of large particles being collected on the filter. The disadvantage is that some organisms do not pass the 5 μm filter and therefore some protocols recommend the use of magnetic particles coated with antibody to fish out the organisms (Boschke, Steingroewer, Ripperger, Klingner, & Bley, 2002). On the 0.2 μm filter the microorganisms are concentrated and remain on the filter. Then the microorganisms are stained with fluorochromes. Acridine orange is used as fluorescence dye in DEFT, which is a nucleic acid intercalator, and stains single-stranded DNA (deoxyribonucleic acids) and RNA (ribonucleic acids) red (650 nm). When acridine orange binds to double-stranded DNA molecules the emitted light is green (526 nm) and so it is possible to differentiate dead and live cells as in dead cells single-stranded nucleic acids are rapidly degraded from active nucleases (Rost, 1995). However, some studies on filtered brewery samples found that differentiation of live and dead cells was not successful, as stained debris was a problem (Barney & Kot, 1992; Kilgour & Day, 1983). A number of other stains are available as alternatives, for example DAPI, primuline, and trypan blue (Table 15.1) (Kregiel & Berlowska, 2009). There are numerous other stains and possible combinations of stains for marking different cell organelles and to differentiate live and dead cells (Report of Subcommittee, 2003).

Following staining, the membrane filter must be examined under a fluorescence microscope and live cells counted to give the total viable count. It is possible to manually perform the counting of the live/dead cells, but for lower numbers of cells and big filters the chance of errors increases. Sorcerer Image Analysis System (Perceptive Instruments) is an example for semiautomatic enumeration by image analysis; the fluorescing cells are viewed in real time by a high sensitivity CCD video camera and viable and nonviable cells readily distinguished by virtue of contrast differences. A completely automated DEFT instrument called COBRA has a high sample throughput, improved reproducibility and lower count limits (Pettipher, Watts, Langford, & Kroll, 1992; Priest & Campbell, 2003).

A faster alternative is to use flow cytometry where the automatisation is easy and the instrument can easily differentiate between debris and yeast (Breeuwer, Drocourt, Rombouts, & Abee, 1994; Bruetschy, Laurent, & Jacquet, 1994). Kits such as BacLight are available to distinguish live and dead cells. BacLight uses SYTO9 and propidium iodide (PI) which differ in their excitation/emission spectra and in their membrane permeability properties; SYTO9 can cross all bacterial cell membranes facilitating a whole cell count when used alone while PI can only enter cells with disrupted membranes (Robertson, McGoverin, Vanholsbeeck, & Swift, 2019). It is even possible to distinguish between stressed and nonstressed yeasts (Edwards, Porter, & West, 1997; Prudêncio, Sansonetty, & Côrte-Real, 1998). Additionally, the detection of beer spoilers such as *Zygosaccharomyces*, *Dekkera* (*Brettanomyces*), *Pediococcus*, and *Lactobacillus* is possible (Bouix,

TABLE 15.1 Examples of fluorescence stains from Sigma–Aldrich (Merck Group) and Life Technologies (Thermo Fisher).

Stain type	Description	Cat. No.	Brand	Optical properties	Application
Damaged cells	Primuline	206865	Sigma-Aldrich	λ_{ex} 340 nm; λ_{em} 425 nm	Primuline is used for visualisation of permeabilized or damaged cells. It binds noncovalently to lipid structures.
	Propidium iodide	P4170	Sigma-Aldrich	λ_{ex} 530 nm; λ_{em} 625 nm	Fluorescent stain for nucleic acids. Cell membrane integrity excludes propidium iodide from staining viable and apoptotic cells. Propidium iodide may be used in flow cytometry to evaluate cell viability when used with other dyes that stain viable cells or cells that are early in the apoptosis process. Propidium iodide is useful for staining dead cells.
	SYTOX blue nucleic acid cell stain	S11348	Molecular probes	λ_{ex} 470 nm; λ_{em} 480 nm	SYTOX blue nucleic acid stain is an excellent blue-fluorescent nuclear and chromosome counterstain that is impermeant to live cells, making it a useful indicator of dead cells within a population.
	SYTOX green nucleic acid stain	S7020	Molecular probes	λ_{ex} 504 nm; λ_{em} 523 nm	green nucleic acid stain is an excellent green-fluorescent nuclear and chromosome counterstain that is impermeant to live cells, making it a useful indicator of dead cells within a population.
	SYTOX orange nucleic acid stain	S11368	Molecular probes	λ_{ex} 532 nm; λ_{em} 547 nm	SYTOX orange dye stains nucleic acids in cells with compromised membranes. This stain is useful as an indicator of cell death.
	SYTOX red dead cell stain	S34859	Molecular probes	λ_{ex} 640 nm; λ_{em} 658 nm	SYTOX red dead cell stain is a simple and quantitative single-step dead cell indicator.
	Trypan blue solution	93595	Sigma-Aldrich	λ_{ex} 488 nm; λ_{em} 675 nm	Trypan blue is a blue acid dye that contains two azo chromophores. It is a large, hydrophilic, tetrasulfonated dye. Trypan blue solution may be used in trypan blue-based cytotoxicity and proliferation assays. It is a vital stain that is not absorbed by healthy viable cells. When cells are damaged or dead, trypan blue can enter the cell, allowing dead cells to be counted. When trypan blue binds to proteins the resulting complex emits red fluorescence. The method is sometimes referred to as the dye exclusion method.

Viable & damaged cells	Acridine orange solution	A9231	Sigma-Aldrich	λ_{ex} 500 nm; λ_{em} 526 nm (bound to DNA); λ_{ex} 460 nm; λ_{em} 650 nm (bound to RNA)	DNA intercalating dye. Suitable for quantitative analysis. Differentially stains double-stranded and single-stranded nucleic acids.
	FUN 1 cell stain	F-7030	Molecular probes	λ_{ex} 480 nm; λ_{em} 560–610 nm (live cells); 510–560 nm (dead cells)	The FUN 1 stain passively diffuses into a variety of cell types and initially stains the cytoplasm with a diffusely distributed green fluorescence. However, in several common species of yeast and fungi, subsequent processing of the dye by live cells results in the formation of distinct vacuolar structures with compact forms that exhibit a striking red fluorescence, accompanied by a reduction in the green cytoplasmic fluorescence. Formation of the intravacuolar structures requires both plasma membrane integrity and metabolic capability. Dead cells fluoresce bright yellow-green, with no discernible red structures.
Viable cells	Bisbenzimide H 33,258	14530	Sigma-Aldrich	λ_{ex} 338 nm; λ_{em} 505 nm (pH 7.0); λ_{ex} 355 nm; λ_{em} 465 nm in TE buffer; DNA	Useful reagent for cytogenetic studies; for the fluorescent staining of DNA in cells; it is membrane-permeable and selectively binds to adenine–thymine regions in the minor groove of B-form DNA with a high binding constant, making it a useful stain for DNA, chromosomes and nuclei. The properties of this dye make it useful in several applications, such as sorting living cells based on DNA content. The methodology of the assay is based on Hoechst 33,258 binding to DNA and is sensitive for a 1000 cells.
	Calcofluor white stain	18909	Sigma-Aldrich	λ_{ex} 355 nm; λ_{em} 433 nm; 0.1 M phosphate pH 7.0	A fluorescent stain for rapid detection of yeasts. Calcofluor white is a nonspecific fluorochrome that binds to cellulose and chitin in cell walls.
	DAPI	32670	Sigma-Aldrich	λ_{ex} 374 nm; λ_{em} 461 nm in 10 mM Tris, 1 mM EDTA, pH 8.0 (DAPI–DNA complex)	DAPI is several times more sensitive than ethidium bromide for staining DNA in agarose gels. It may be used for photofootprinting of DNA to detect annealed probes in blotting applications by specifically visualising the double-stranded complex and to study the changes in DNA and analyse DNA content during apoptosis using flow cytometry. Cell permeable fluorescent minor groove-binding probe for DNA. Binds to the minor groove of double-stranded DNA (preferentially to AT-rich DNA), forming a stable complex which fluoresces approximately 20 times greater than DAPI alone.

Continued

TABLE 15.1 Examples of fluorescence stains from Sigma–Aldrich (Merck Group) and Life Technologies (Thermo Fisher).—cont'd

Stain type	Description	Cat. No.	Brand	Optical properties	Application
	Ethidium bromide solution	46067	Sigma-Aldrich	λ_{ex} 480 nm; λ_{em} 620 nm in H_2O; λ_{ex} 518 nm; λ_{em} 605 nm (bound to DNA); λ_{ex} 530 nm; λ_{em} 600 nm in 50 mM phosphate buffer pH 7.0 (upon binding to DNA)	The fluorescence of EtBr increases 21-fold upon binding to double-stranded RNA and 25-fold on binding double-stranded DNA so that destaining the background is not necessary with a low stain concentration (10 μg/mL). Ethidium bromide has been used in a number of fluorimetric assays for nucleic acids. It has been shown to bind to single-stranded DNA (although not as strongly) and triple-stranded DNA.
	Hoechst 33,258 solution	94403	Sigma-Aldrich	λ_{ex} 355 nm; λ_{em} 465 nm in TE buffer; DNA	These bisbenzimide dyes are blue fluorescence dyes used to stain dsDNA.
	Hoechst 33,342	14533	Sigma-Aldrich	λ_{ex} 355 nm; λ_{em} 465 nm in TE buffer; DNA	
	Hoechst 34,580	63493	Sigma-Aldrich	λ_{ex} 357 nm; λ_{em} 490 ± 10 nm in H_2O (free dye)	
	Nancy-520	01494	Sigma-Aldrich	λ_{ex} 520 nm; λ_{em} 554 nm in TE buffer; DNA	Nancy-520 is a fluorescent stain for dsDNA with higher sensitivity than ethidium bromide and an easy, fast and robust staining procedure. It can be used to determine dsDNA concentrations in solution, with a linear range between 0 and 2 μg/mL of DNA.
	SYBR green I	S-7563	Invitrogen	λ_{ex} 497 nm; λ_{em} 520 nm (DNA–dye complex)	SYBR green I is an asymmetrical cyanine dye used as a nucleic acid stain.

Grabowski, Charpentier, Leveau, & Duteurtre, 1999; Donhauser, Eger, Hubl, Schmidt, & Winnewisser, 1993; Jespersen, Lassen, & Jakobsen, 1993; Xu et al., 2022).

The principle of flow cytometry is also based on fluorescence staining or labelling. The cells are brought in a fluid stream passing a thin capillary where the fluorescence molecules are excited by a laser and the emission is detected. The laser is also used to count the particles and determine the size. All data are collected, and software analyses the data to generate a report with the result of live/dead cells or detection of beer spoilers. A number of other parameters can also be determined (Brown & Wittwer, 2000).

A drawback of DEFT is the inability to detect specific pathogens. The use of special dyes allows only a universal distinction between viable and nonviable cells (Bamforth, 2006), but an extension of the method by the specific detection of microbes is a great advantage. It can help to discover the origin of contaminations during the brewing process (Dodd, Stewart, & Waites, 1990) and enables distinction between different lactic acid bacteria, such as several spoiling species of *Lactobacillus*, and those innocuous or necessary for brewing (Priest & Stewart, 2006). The antibody-DEFT (Ab-DEFT) method combining membrane filtration and pathogen-specific fluorescent antibodies can enumerate spoilage bacteria in food and beverages but with the loss of the evaluation of viability (Tortorello & Gendel, 1993). This method has been applied to the detection and identification of beer spoilage bacteria, including fastidious anaerobic Gram-negative bacteria. Monoclonal antibodies binding to lipopolysaccharide (LPS) of *Megasphaera cerevisiae* have been isolated that also react with the LPS of other brewing-related anaerobic Gram-negative bacteria in the genera *Pectinatus*, *Propionispira*, and *Selenomonas* (Ziola, 2016).

The oligonucleotide-direct epifluorescent filter technique (Oligo-DEFT) represents a further, more sensitive (Uyttendaele & Debevere, 2006) and specific detection method for the direct enumeration of microorganisms. It has been demonstrated for *E. coli* in milk and fruit juices that the oligo-DEFT method can achieve a detection limit of 1 CFU/mL (Tortorello & Reineke, 2000). Fluorescent-labelled oligonucleotides complementary to 16S rRNA were combined with DEFT. Because of the high abundance in cells, the ribosomal RNA represents an optimal target for fluorescence microscopy analysis. After the membrane filtration followed by a short 2 h hybridisation step, it is possible to distinguish between either species or groups of microorganisms (Raybourne & Tortorello, 2003).

15.2.3 MALDI-TOF mass spectrometry

MALDI-TOF MS is a mass spectroscopy method and is an interesting technique for the identification of microbial species. At least 10^3 to 10^6 cells are needed for a determination. The ground principle of MALDI is to look at the mass spectrum, which depends on the protein profile of the cells, and the so-called fingerprints are unique for each microbial species (Fig. 15.2). After the cultivation on an agar plate a colony can be picked and the crude cells or the extracted proteins are spotted on a special slide and covered with a α-cyano-4-hydroxycinnamic acid saturated solution (matrix). Then the drop is dried for a few minutes. Extraction is done by suspending the colony in 80% ethanol followed by a 2 min centrifugation.

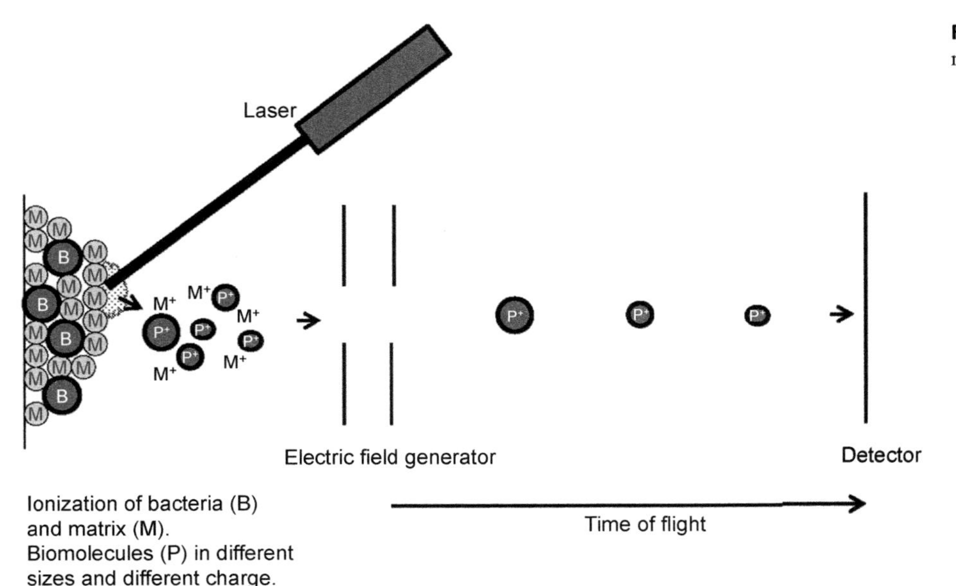

FIGURE 15.2 Principle of MALDI-TOF mass spectroscopy.

The pellet is resuspended in 70% formic acid and the same volume of acetonitrile and again centrifuged for 2 min. One microlitre of the supernatant is then pipetted to the target and covered with the α-cyano-4-hydroxycinnamic acid saturated solution. When the small droplets are dried the target plate is put into the MALDI where the laser (impulse of 1–5 ns) shots the dried droplets which results in positive loaded proteins fly towards the electrode. Based on the weight and the electrical charge they fly faster or slower and with the detector the time of flight is measured and then converted into a mass (Claydon, Davey, Edwards-Jones, & Gordon, 1996; Holland et al., 1996; Krishnamurthy, Ross, & Rajamani, 1996; Lay, 2001). With a simple sample preparation, the profile is measured, and a database checked for reference spectrums and makes a ranking of the hits. The identification is very efficient (about 15 min plus the cultivation) but the machines are still very expensive and the access to the reference database is needed. Newer research studies demonstrated that it is possible to differentiate *L. brevis* strains based on their spoilage potential (Kern, Vogel, & Behr, 2014), and developments in microfluidics have enabled higher throughput for the rapid identification of beer spoilage bacteria (Condina et al., 2019).

15.2.4 Surface enhanced Raman spectroscopy

Raman spectroscopy is based on the interaction of light with the chemical bonds within a material and is used as a non-destructive chemical analysis technique. Surface-enhanced Raman scattering amplifies Raman intensity resulting in improved spectral resolution. This technique has a very wide range of applications and has been successfully used within alcoholic beverage production in counterfeit spirit detection, detection of nutritional components, detection of toxic and harmful substances, and microbiological testing (Li et al., 2022). Raman may be used in fermentation monitoring (Jiang, Xu, Ding, & Chen, 2020) and for the detection and identification of bacteria by comparison of spectra against available databases (Kashif et al., 2021; Shang et al., 2022).

15.2.5 Biosensors

A biosensor is an analytical device that combines a biological component with a chemical or physical transducer that can convert biological, chemical, or biochemical signals into measurable electrical signal. Biological recognition elements include enzymes, whole cells and affinity biomolecules, and the transducer may be electrochemical, optical/optoelectrical, or piezoelectric. The majority of commercially available biosensors are for detection of pathogenic bacteria of significance in medicine and food. Those commonly used in brewing include ATP methods described above but also those used to detect products of microbial spoilage in the brewing process, such as the mycotoxins deoxynivalenol (DON) and ochratoxin A (OTA) (Joshi, Annida, Zuilhof, van Beek, & Nielen, 2016), aflatoxin B1 (Yadav, Yadav, Chhillar, & Rana, 2021), and hydrophobins (Stilman et al., 2022). More recently, an aptamer-based biosensor has been developed to detect *Acetobacter aceti* in beer (Miteu, 2024).

15.3 Nucleic-acid based methods

In 1993, the American Biochemist Kary Mullis was awarded the Nobel Prize for the development of the polymerase chain reaction (PCR; Malmström & Andersson, 2013). PCR is a process used to make a large copy number of a specific DNA fragment from genetic material (DNA) in a relatively short time. The process constituted a major breakthrough because it solved the problem of how to produce multiple copies of any particular piece of DNA using a relatively simple, economical and reliable procedure. Today, all DNA-containing target samples can be analysed by PCR. The applications range from forensic samples, fossil, archaeology, and analysis of metabolic pathways to the identification of plants and animals, trace research, as well as identification and classification of microorganisms (Hutzler, Schuster, & Stettner, 2008). The identification and phylogenetic classification of bacteria is typically carried out by analysis of the rrn operon, especially the 16S rDNA gene and the 16–23S spacer region; it has been the most widely used target for developing PCR tests for beer-spoilage bacteria in various taxonomic ranks (Juvonen, 2009).

PCR has also been instrumental in advances in sequencing, with 'next generation' methods being applied to identification of microbial communities in both brewing and distilling. In this section we examine PCR, sequencing, and other detection and identification methods that are nucleic acid based.

15.3.1 Polymerase chain reaction

A few basic components are needed to perform a PCR. First, two specific oligonucleotides (15–25 nt), called primers, are needed. They are derived from both strands of the target sequence; thus they determine the size and specificity of the resulting PCR product. The other components are a thermo-stable DNA polymerase, deoxyribonucleotides and a defined reaction buffer, which contains magnesium ions as cofactor and estimates the optimal reaction conditions for the polymerisation of DNA (Saiki et al., 1988).

PCR is a cycle reaction and is carried out in a thermocycler. Each cycle consists of three steps. In the first step, denaturation, the initial DNA and primers are heated at 95°C and denatured into single strands. In the second step, called annealing, the primers hybridise to their opposite sequence at both DNA strands. The annealing temperature is usually 3−5°C below the melting temperature of the primer. If the selected annealing temperature chosen is too low the primers may also bind at positions which are not 100% complementary and thus lead to nonspecific products. If the temperature is too high the primers do not bind or bind incompletely and no product, or only a small amount, is formed.

During the third step, the elongation, the new DNA strands are synthesised from the 5′-end to the 3′-end by the polymerase. The elongation temperature depends on the working optimum of the DNA polymerase (68−72°C). The new DNA fragment, which results from the steps 1 to 3 is further multiplied in the following cycles. Ideally, each newly emerging DNA segment is duplicated in 20−50 cycles (Saiki, 1990). Since the original description of PCR as a method to amplify DNA a number of variations of the technologies have been described (Fig. 15.3).

15.3.1.1 Primer design

The specificity of the PCR depends on the quality of the primers. The sequences of primers are derived from the comparison of target sequences from various beer contaminating bacteria. These sequences are available in databases such as the Ribosomal Database Project (http://rdp.cme.msu.edu/) or the Basic Local Alignment Search Tool (BLAST). These sequences have to be compared with each other by creating a sequence alignment. The alignment shows common and different sequence regions between the individual beer spoiling bacteria. Common regions are applied to the design of group specific primers. Species specific primers are derived from different regions. The sequences should comprise 20−25 nt in order to obtain primers with a melting temperature of 50−60°C. The melting temperature of the primers can be predicted according to the formula by Marmur and Doty (1962):

$$T_m = 4 \times GC + 2 \times AT$$

Free primer design software is readily available and potential primer sequences can be tested against databases on their specificity.

15.3.1.2 Endpoint PCR

Based on the PCR product detection, PCR can be categorised into endpoint PCR and time-point, called real-time PCR. In endpoint PCR the PCR product is visualised by agarose or polyacrylamide gel electrophoresis followed by staining with fluorescent dyes, for instance ethidium bromide or SYBR Green I. More recent innovations include Veriflow technology, a vertical flow mediated method for visualisation of PCR products.

Several primer sets have been designed for specific detection of hop resistant lactic acid bacteria, lactic acid-producing bacteria, anaerobic beer spoilage bacteria, and potential spoilage yeast such as *Brettanomyces/Dekkera* species and *Saccharomyces diastaticus* (Bischoff, Bohak, Back, & Leibhard, 2001; DiMichele & Lewis, 1993; Hulin, Harrison, Stratford, & Wheals, 2014; Juvonen & Haikara, 2009; Juvonen, Satokari, Mallison, Haikara, 1999; Satokari, Juvonen, von Wright, & Haikara, 1997; Suzuki, Iijima, Ozaki, & Yamashita, 2005; Suzuki, Sami, Iijima, Ozaki, & Yamashita, 2006; Yamauchi, Yamamoto, Shibano, Amaya, & Saeki, 1998). Veriflow™ technology enables detection of target DNA in less than 3 h with a sensitivity of 10 cells/ml from brewery samples with no enrichment or purification steps required.

15.3.1.3 Real-time or quantitative PCR

The development and application of fluorescent dyes, which are incorporated into the PCR product, and fluorochromes for labelling of oligonucleotides opened the possibility for real-time monitoring of the product formation cycle by cycle. The fluorescence signal increases in proportion to the amount of amplicon. Dyes, such as ethidium bromide and SYBR Green I, that intercalate into the DNA are the simplest way to follow at real time the increase of amplicon. The disadvantage of this method is that distinguishing between different PCR products is not possible.

The other possibility for monitoring of PCR product formation offers the application of fluorescent labelled probes, which emit their signal with the incorporation into the PCR product. Using Förster resonance energy transfer (FRET) probes, the FRET between a donor and an acceptor molecule is exploited. The donor fluorochrome is stimulated by a light

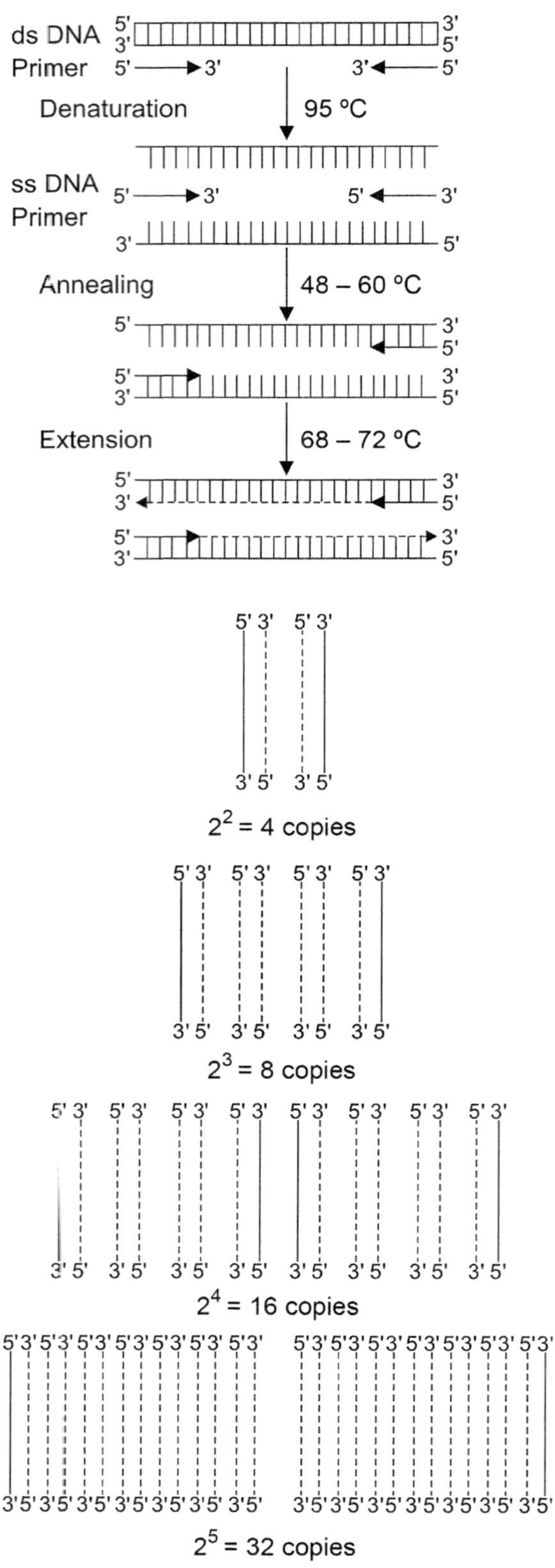

FIGURE 15.3 Basic principle of a polymerase chain reaction.

source and transfers energy to the acceptor fluorochrome that emits a fluorescent signal, which is detected. Therefore, two additional oligonucleotides have to be designed, which contain the adjacent donor and acceptor fluorophore. The FRET signal of the acceptor increases only with the incorporation of the probes into the PCR product. Then the distance between the donor and acceptor only amounts to 1–10 Å. This method provides a high specificity but is expensive.

The FRET principle is applied in different labelled probe systems. TaqMan probes belong to the dual-labelled probes at 5′-end a quencher and at 3′-end a fluorescence molecule. The quencher inhibits the fluorescence signal. During the PCR Taq-polymerase synthesises the DNA strands from 5′-end to the 3′-end and the labelled probe is incorporated into the PCR product. In addition to the polymerisation activity the Taq-polymerase contains a 5′-3′-exonuclease activity to hydrolyse the quencher from the opposite DNA strand. Thus, the fluorophore and the quencher remove from each other, and the rising fluorescence signal can be measured.

Another opportunity of the FRET principle is the application of molecular beacon probes. In such a probe the sequences 5′-end and 3′-end form a stem loop, which is labelled with a reporter fluorophore and quencher. With incorporation of the probe into amplicon, the stem loop opens, the distance between the reporter and the quencher increases and the reporter molecule emits the fluorescence signal that is measured. The possibility to visualise the rise of amplicon, in contrast to the endpoint PCR, was a welcome progress. This expanded the role of PCR from that of a pure research tool to that of a versatile technology, and diagnostic kits such as the foodproof range are available for the detection of beer spoilage bacteria and yeast.

For the quantification of the initial DNA by real-time PCR, a reference gene is usually included in the measurement to perform a relative amount comparison (relative quantification). In the first phase of the product amplification the template amount is limited, because the probability that the template, primer and polymerase meet is suboptimal. When enough amplicon is present, the assay's exponential progress can be monitored as the rate of amplification enters a linear phase (LP). The beginning of the exponential phase, where the fluorescence significantly increases above the background fluorescence, is called the Ct value (cycle threshold) or the Cp (crossing point) value used to describe the cycle. As primers and enzyme become limiting, and product formation has an inhibitory effect to the PCR and is overly competitive to oligoprobe hybridisation accumulate, the reaction slows, entering a transition phase (TP) and eventually reaching a plateau phase (PP) where there is little or no increase in fluorescence. The relative quantification is calculated by comparison with the amplification signal of an internal amplification standard (reference gene) over the Ct value (Fig. 15.4).

An absolute quantification is more demanding and states the exact number of nucleic acid targets in the sample with respect to a particular unit. Absolute quantification may be necessary when there is a lack of sequential specimens to demonstrate a relative change in microbial load or when no suitably standardised reference reagent is available (Mackay, 2004).

The efficiency (E) of a PCR assay is calculated by the gradient (m) of a standard curve. For that purpose cDNA dilutions (e.g., 100%, 10%, 1%, 0.1%) are used as templates for the graphical structure. A linear regression line through the curve has the gradient $-m$ (when plotted with increasing DNA concentration):

$$E = 10^{-1/\mathrm{m}} - 1$$

A gradient of $-3.32\ m$ would thus mean an efficiency of 1 (100%) indicating a doubling of the amplicon per cycle, a gradient of -3.58 and an efficiency of 0.9 (90%). The formula provides meaningful values that are 100% smaller then gradient value -3.32 (Higuchi, Fockler, Dollinger, & Watson, 1993). The efficacy of PCR is determined by its efficiency, fidelity, and specificity, which are in turn influenced by many factors including target length and sequence and primer design.

The first real-time PCR instrument was launched in 1996 and a large selection of systems is currently available. They are normally composed of a fluorescence measuring thermocycler, a computer and software for operation and data analysis. A LightCycler was the first instrument based on rapid cycle PCR. It has the capability to run 30 cycles in 10–15 min (Wittwer et al., 1997). Options for multichannel analyser (MCA) and three to four fluorescence channels are standard features in modern instruments. The configuration with 384-well blocks essentially enables a low-density array setup and nanoplate systems are available that accommodate up to 3072 reactions in a device with the size of a standard microscope slide (Brenan, Roberts, & Hurley, 2009). Downscaling of a PCR thermocycler on a microchip also enables a shortened run time (Juvonen, 2009; Pipper, Zhang, Neuzil, & Hsieh, 2008).

Within brewing, Genesig qPCR Beer Spoilage Detection Kits (rapidmicrobiology.com) have been developed to detect the anaerobic beer spoilage bacteria *Pectinatus* and *Megasphaera*, and lactic acid bacteria belonging to *Pediococcus* and *Lactobacillus* (Haakensen, Dobson, Deneer, & Ziola, 2008; Haakensen et al., 2007; Juvonen, Koivula, & Haikara, 2008).

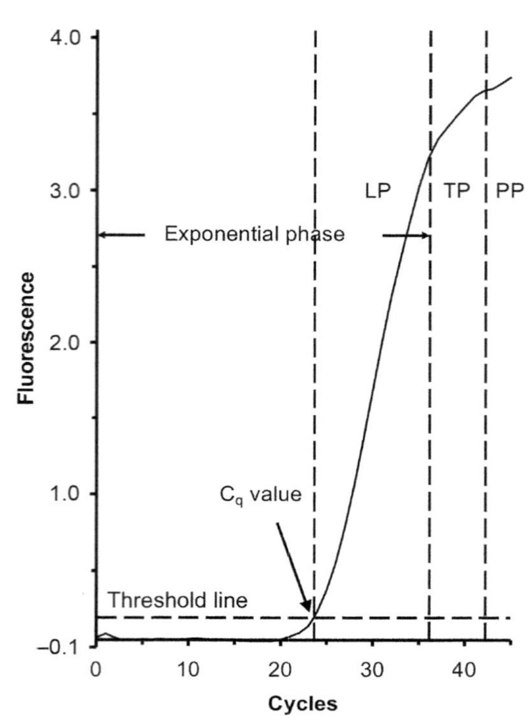

FIGURE 15.4 Kinetic analysis of a PCR reaction. PCR product amount is measured by fluorescence. If enough amplicon is present the rate of amplification and the increase of fluorescence is linear (LP). Under limiting conditions, product formation is in the transition phase (TP) and eventually reaching a plateau phase (PP) where there is little or no increase in fluorescence.

15.3.1.4 Multiplex PCR

A number of modifications of the original PCR basic reaction conditions and technique have been developed to enhance the efficacy. In a multiplex PCR, multiple sets of specific primers are used for the simultaneous amplification of multiple gene targets in a single reaction. It is made more complicated by the development time, since the designed primers have to be adjusted with respect to the melting temperature, and the sequences have to be compared among each other to prevent dimerisation of the primers. These interactions would reduce the sensitivity of the multiplex PCR and must therefore be excluded (Devlin, 2010).

Multiplex PCR kits for the simultaneous detection of beer spoilage yeast and bacteria are available, including Microbiologique's Beer Spoilage Organism Detection Kit (Asano et al., 2008; Haakensen, Schubert, & Ziola, 2008; Iijima, Asano, Suzuki, Ogata, & Kitagawa, 2008). In the Microbiologique method, an enrichment medium is first used before PCR and the amplified products are run on a dipstick (Janagama, Mai, Han, Nadala, Nadala, 2018). The formation of bands on the dipstick indicates a positive result and this method may be used for simultaneous detection of wild yeasts (*Saccharomyces diastaticus*, *Brettanomyces*) and bacteria (hop-resistant Lactic acid bacteria, Acetic acid bacteria, *Megasphaera* and *Pectinatus*). The limit of detection is 1 CFU/mL (Janagama et al., 2018).

A real-time multiplex PCR assay associated with high resolution melting (qPCR-HRM), targeting orthologs sequences, has also been developed to detect beer-spoilage microorganisms from the genus *Acetobacter*, *Bacillus* and *Levilactobacillus* (Machado, 2022). The products of PCR using multiple primer sets all have different melting points, enabling easy differentiation following qPCR (Machado, 2022).

15.3.1.5 Nested PCR

Nested PCR is well suited when only very small amounts of the target DNA are present in comparison with the total sample amount of DNA. Then two PCRs are performed consecutively. The formula provides PCR the target gene is amplified beside unwanted sequence regions as a result of nonspecific binding of the primer. The resulting amplicon is used as a template for a second round of PCR with other primers that bind within the first target region and generate a product with very high specificity. Since the DNA region of choice is amplified a second time, it produces sufficient DNA for further procedures (Busch, 2010). Koivula, Juvonen, Haikara, and Suihko, (2006) and Maugueret and Walker (2002) developed a nested PCR based on primers, which were based on 16S rDNA for the detection of *Obesumbacterium proteus* biotype I and II in beer, wort, and yeast slurry. The detection limit varied in the individual matrices between 10^2 and 10^7cells/sample.

15.3.1.6 Loop-mediated isothermal amplification

Loop-mediated isothermal amplification (LAMP) is a one-step amplification reaction that amplifies a target DNA sequence with high sensitivity and specificity under isothermal conditions (about 65°C). The reaction takes place in three steps: the initial step, a cycling amplification step and an elongation step by a DNA polymerase with strand displacement activity. For the amplification, a set of two outer and two inner primers are required, which are derived from six regions of the target sequence. It provides high amplification efficiency, with DNA being amplified 10^9-10^{10} times in 15−60 min. The increase reaction product can be monitored by turbidity measurement (Mori & Notomi, 2009). The method can also be combined with an RT-PCR. It should be able to amplify a few target copies and be less sensitive to nontarget DNA than PCR. A LAMP-based application for the identification of *L. brevis*, *L. lindneri*, *P. damnosus* and *Pectinatus* from isolated colonies in 1.5 h has been developed (Tsuchiya et al., 2007). The advantage of this technology is that significant investments in equipment are unnecessary.

15.3.1.7 RT-PCR

Besides the use of DNA as a template for PCR, it is possible to convert RNA by a reverse transcriptase into complementary DNA (cDNA). This RT-PCR followed by PCR or real-time PCR is a powerful technique for the qualitative and quantitative detection of messenger RNA. Bergsveinson, Pittet, and Ziola (2012) investigated the expression level of *horA* and *horC* in *L. brevis* and *Pediococcus claussenii* during growth in beer. A deoxyribonuclease (DNase) pretreatment has been shown to be very effective in eliminating DNA contamination when applied prior to RT-PCR analysis, resolving one of the concerns related to this technique in quantification of ribosomal RNA or prerRNA in living cells.

15.3.1.8 Differentiation between viable and nonviable cells

One of the often-stated drawbacks of PCR is the potential for false positives, as the nucleic acid template can be from dead cells. A possibility for differentiating between viable and nonviable cells uses ethidium bromide monoazide (EMA), a DNA binding dye that is used for the differentiation of living and dead cells in flow cytometry and PCR. Dead cells have membrane damage; EMA penetrates and binds covalently to the bacterial DNA. This binding inhibits the amplification of the bound DNA so that the polymerase is sterically hindered (Wang & Levin, 2005). Propidium monoazide (PMA) has also been shown to inhibit the PCR amplification of DNA from dead *Levilactobacillus brevis* cells, but not from live cells (Ma et al., 2017).

Weber, Sahm, Polen, Wendisch, and Antranikian (2008) published an RT-PCR in combination with an oligonucleotide array for the detection and identification of viable beer-spoilage bacteria. In this study a set of primers for the detection of viable bacteria was designed to target the intergenic spacer regions (ISR) between 16 and 23S rRNA. These results suggest that rRNA content is stable and does not necessarily correlate with growth of bacteria. On the other hand, pre-rRNA is a suitable marker of growing bacterial strains. Therefore, RT-PCR targeting the ISR rRNA is a very effective method for detecting growing bacterial cells. Unfortunately, it lacks the evidence that the ISR is also in real beer samples at detectable levels.

15.3.2 In situ hybridisation detection systems

Hybridisation is the process by which two complementary strands of nucleic acid bind together by hydrogen bonds to form a single double-stranded complex. By adjusting the temperature and buffers the most energetically preferred complex is built and this special technique used in laboratories is called annealing. The temperature needed for annealing depends on the number of complementary bases, respectively the number of hydrogen bonds which are formed. In situ hybridisation is used for the detection of specific sequences by using a labelled complementary DNA or RNA strand, called the probe. Since the first in situ hybridisation experiments in 1969 (Gall & Pardue, 1969), many variations of the method have been developed. Results include improved sensitivity and specificity and also different ways of detection and working procedures. The best known in situ hybridisation procedures are fluorescent probes to detect DNA sequences, also called FISH (fluorescence in situ hybridisation) (O'Connor, 2008). Modern methods using RNA as the target nucleotide and new techniques with sandwich hybridisation and detection systems with chromogenic reaction or a biochip (electric measurements) are gaining popularity (Femino, Fay, Fogarty, & Singer, 1998; Pioch et al., 2008; Raj, van den Bogaard, Rifkin, van Oudenaarden, & Tyagi, 2008; Rautio et al., 2003). Radioactive labels have been used, but because of stability and safety issues they are no longer employed (Rudkin & Stollar, 1977). Compared with PCR, the in situ system has no problem with inhibitory effects from the beer matrix as no polymerase enzyme is needed.

The range of probes is wide. The success of this technology relies on finding probes which are highly specific and have an excellent hybridisation rate. Specificity is achieved by targeting conserved or unique rRNA sequences. A probe is normally composed of an oligonucleotide with about 20 nucleotides (Kempf, Trebesius, & Autenrieth, 2000), but some tests are carried out with peptide nucleic acid, which can have some advantages as the molecule is more stable, specific and sensitive (Almeida, Azevedo, Fernandes, Keevil, Vieira, & 2010). For pathogens and beer spoiling organisms, 16S rRNA, 23S rRNA or respectively the corresponding DNA are usually selected as the target (Almeida et al., 2010; Frischer, Floriani, & Nierzwicki-Bauer, 1996; Fuchs, Syutsubo, Ludwig, & Amann, 2001).

There are a number of different probes available, including detection probes, which are labelled with a fluorescence marker or, for example, digoxigenin for a further linkage with an antibody–enzyme complex and then a later colourimetric reaction with a chromogenic substrate (such as nitro blue tetrazolium for alkaline phosphatase) (Helentjaris & McCreery, 1996; Kempf, Trebesius, & Autenrieth, 2000), Capture probes are used to bind the target sequence (RNA or DNA) to a plate or another surface. In most cases the probes are labelled with biotin to react with avidin, which is coated on a plate (Riley, Marshall, & Coleman, 1986).

For the detection of bacteria and other organisms it is interesting using rRNA as a detection target as RNA in normal cases only exists in living cells and also in numerous copies (up to several thousands of ribosomes per bacteria (Kaczanowska & Rydén-Aulin, 2007) and for yeast even more, close to 200,000 (Warner, 1999). PCR methods, for example, also detect DNA of dead cells as double-stranded DNA is quite stable while the single-stranded RNA is decomposed within a few hours from nucleases. Because of the numerous ribosomes in a bacteria or yeast cell there is no need to do a PCR and a direct detection of bacteria or yeasts is possible.

There are three classical FISH detection kits available on the market called VIT®-beer (Vermicon), which detects the beer spoiling organisms. The protocol of the method takes no more than 3 h and can be directly used for isolates or for beer samples after an enrichment step. One kit detects all members of beer-spoilage lactic acid bacteria (red fluorescence) and additional specific *Levilactobacillus brevis* (red and green fluorescence), the most prominent beer spoiling organism. With another kit it is possible to detect *Megasphaera cerevisiae* and *Pectinatus* spp., two obligate beer spoiling organisms. For the detection of obligate and potentially fermentative spoilage yeasts in beer and beer-based drinks another kit is provided.

The principles of these classical FISH kits are quite easy to understand but the procedure takes a while, because several incubation periods are required. The target for the detection probes is the rRNA of the spoiling organisms. First the cells are fixed on a slide and then the cell membrane has to be made permeable with an enzyme mix for Gram-positive organisms. A drop of reagent containing the fluorescent marked probes is added which can penetrate into the cell. During the incubation at 46°C the hybridisation of rRNA and the probes is performed. Then the slide is washed and afterwards the slides are examined under the fluorescence microscope as fluorescent glowing cells (Thelen, Beimfohr, Bohak, & Bac, 2001).

The HybriScan™ system (Sigma–Aldrich/Scanbec) is based on the detection of rRNA through hybridisation events and specific capture and detection probes. The sandwich hybridisation is very sensitive, detecting attomoles of the respective target rRNA molecules. First of all the cell walls are destroyed enzymatically and then the rRNA is centrifuged down, resuspended and used in the assay. The method is highly specific as it uses two probes for the hybridisation: capture probes, which are used to immobilise the bacteria on the microplate (streptavidin coated), and detection probes, which are used for the detection reaction. The capture probe is biotin-labelled while the detection probe is digoxigenin-labelled. After the hybridisation at 50°C the probes and the targets are fixed on the microplate. To the detection probe a horseradish peroxidase is linked by building an anti-DIG-horseradish peroxidase Fab fragment. Then a washing step follows and the bound complex is visualised by horseradish peroxidase substrate TMB (3,3′,5,5′-tetramethylbenzidine). The photometric data are measured at 450 nm and compared with standard solutions, respectively their calibration curve. The measured data and the calibration curve can be used for the calculation of CFU (colony forming units) if no enrichment step was used, but it is also possible just to detect, for example, the beer spoiling organisms (Fig. 15.5).

The HybriScan***D*** Beer kit detects all beer spoiling bacteria of the genera *Lactobacillus*, *Pediococcus*, *Pectinatus* and *Megasphaera*. The sensitivity is 1–10 CFU/L after 24–30 h pre-enrichment in NBB broth, or isolates can directly be used.

The HybriScan***D*** Yeast kit is used for the detection of yeasts in filterable, nonalcoholic drinks. The specificity covers yeasts including the genera *Zygosaccharomyces*, *Saccharomyces*, *Candida*, *Dekkera*, *Torulaspora* and *Pichia*. For direct detection and quantification at least 500 CFU/mL is recommended; after an enrichment step detection of 1–10 CFU/L is possible.

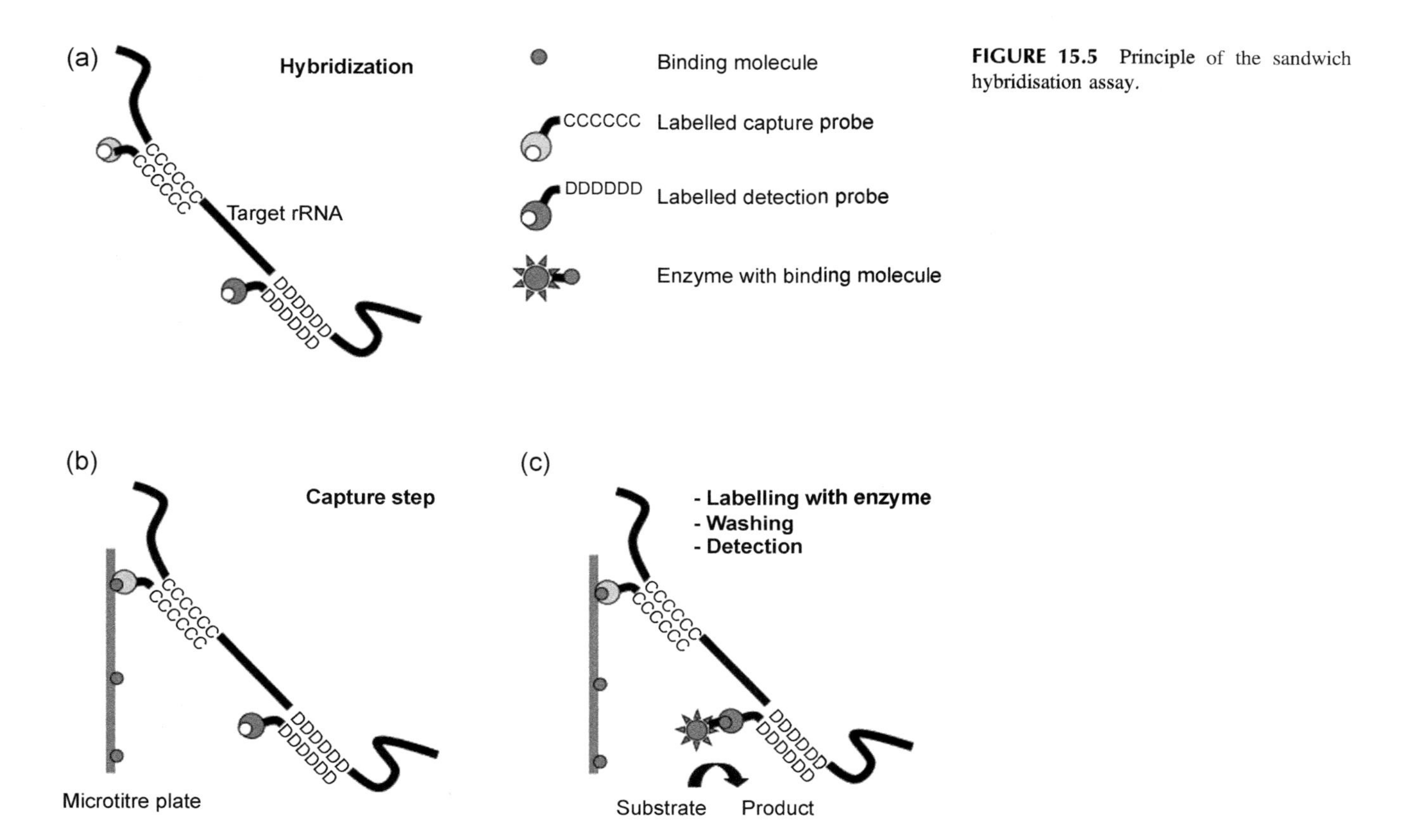

FIGURE 15.5 Principle of the sandwich hybridisation assay.

The test can be performed in 2–2.5 h and as it is in a micro titre plate it is quite economical and can be automated. The work flow is very similar to the ELISA test (Taskila, Tuomola, Kronlöf, & Neubauer, 2010).

15.3.3 DNA Microarrays

A significant extension of nucleic acid-based methods is the development of DNA microarrays (gene chip/DNA chip/biochip). Microarrays are based on the principle that complementary nucleic acid sequences will bind to each other; analysis of hybridisation patterns between capture probes on a DNA chip with labelled fragments in a sample enables gene expression to be characterised. This technique has been widely used to investigate gene expression in brewing yeast and also to monitor structural changes in the yeast genome (Hirasawa, Furusawa, & Shimizu, 2010; Kobayashi et al., 2007).

Weber et al. (2008) developed an oligonucleotide microarray based on the intergenic spacer regions between 16 and 23S rRNA of common beer spoilage bacteria belonging to *Lactobacillus*, *Pediococcus*, *Megasphaera*, and *Pectinatus*. The method was able to detect and discriminate between viable and non-viable species.

15.3.4 Sequencing

Precise identification of microbes can be achieved by DNA sequencing, as the order of nucleotide bases is unique. Sequencing technologies emerged in the 1970s with the first-generation techniques including the Sanger method. By this method, DNA chains are synthesised on a template strand, but chain growth is stopped by incorporation of one of four radio-labelled dideoxy nucleotides (Adenine, Guanine, Cytosine, Thymine). The fragments generated may be visualised through electrophoresis on a high-resolution polyacrylamide gel. Improvements to this method over the years included replacement of radiolabelling with fluorometric based detection, and enhanced separation using capillary-based electrophoresis. Developments in PCR techniques further enhanced speed and scale, with high throughput second (or next) generation methods employing parallelised amplification of DNA and the use of a luminescent method to measure pyrophosphate production as each nucleotide washes through the system.

In terms of identification of beer spoilage microbes, short sequences of highly conserved regions of DNA are sequenced rather than whole genomes. Typically sequencing of the 16S ribosomal RNA (rRNA) gene for bacteria and 18S rRNA gene for yeast are used. As described earlier, rRNA genes contain highly conserved regions that evolve slowly, and these regions may be used for genus identification. There are also regions that evolve faster, and these are used to determine specific species. The next-generation sequencing (NGS) platform Illumina MiSeq (Illumina, Inc., San Diego, CA), has been applied within some breweries as a sequencing tool (Callahan et al., 2019; Sambo et al., 2018; Sobel, Henry, Rotman, & Rando, 2017). This sequencer has a relatively simple workflow and costs for equipment and consumables are continuing to fall.

Even more recently, third generation sequencing technologies have emerged which includes the use of nanopores. Nanopore platforms GridION and MinION produced by Oxford Nanopore Technologies use a process of denaturation of double-stranded DNA by a processive enzyme which ratchets one of the strands to a biological nanopore situated in a synthetic membrane. A voltage is applied across the membrane and the current monitored as strands pass through the pores. Each of the nucleotide bases (Adenine, Guanine, Cytosine, Thymine) impacts the flow of current differently and as a result the sequence of bases in the DNA strand can be determined by the changes in current within each pore.

The MinION platform has been used to generate sequence information in studies of the evolutionary history of brewing yeasts (Kerruish et al., 2024; Salazar et al., 2019; Sobel et al., 2017), in detection of hop tolerance genes, and in discrimination of Saccharomyces strains (Kurniawan et al., 2021, Kurniawan, Shinohara, Sakai, Magarifuchi, & Suzuki, 2022). Further applications in brewing include the rapid identification of yeasts and beer-spoilage bacteria, with protocols already developed (Shinohara, Kurniawan, Sakai, Magarifuchi, & Suzuki, 2021; Suzuki, 2020, Described in Chapter 7 and shown in Fig. 7.9).

15.4 Emerging technologies

Significant advances have been made in 'lab on a chip' technologies in recent years with microfluidic devices available with diverse applications including pharmaceuticals, astrophysics, and manufacturing. Working at micrometric scale enables rapid heat transfer, an increased surface-to-volume ratio, and laminar flow. There are also the significant advantages of using miniscule amounts of samples and reagents, and high resolution and sensitivity.

Microfluidic systems are widely used in procedures such as capillary electrophoresis, isoelectric focussing, immunoassays, flow cytometry, DNA analysis and PCR amplification. One technique that may potentially be applied in brewing microbiology is convective PCR. Convective PCR is a method where the sample/reagent mix is confined in a cylindrical enclosure with the bottom temperature maintained at a higher temperature than the top (Fig. 15.6) and the PCR reagents are shuffled through optimum temperature regimes. Using this method, significantly lower costs are incurred in terms of consumables, and time to result is also shortened.

Third generation PCR builds from real-time/qPCR and is used for absolute quantification. In this method, the PCR reaction system is divided into small volume compartments. Following amplification, the number of positive and negative reactions are counted giving the total copy number. Droplet Digital PCR has been developed for the detection of foodborne pathogens, with significant success in terms of sensitivity of detection from complex foodstuffs without the need for a

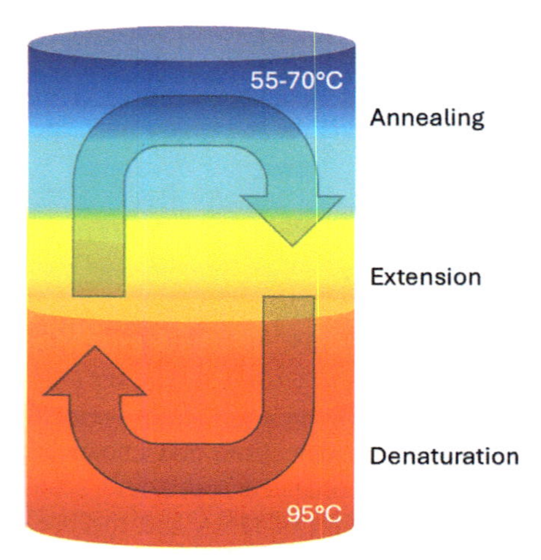

FIGURE 15.6 Convective PCR.

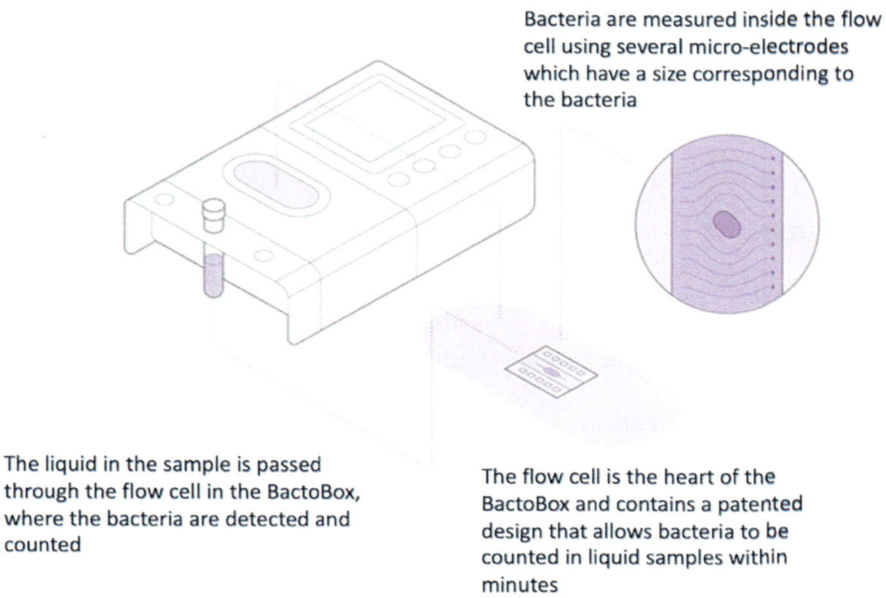

FIGURE 15.7 Direct bacterial cell counting using BactoBox. *With Permission, Gustav Skands, sbtinstruments.com.*

culturing or enrichment step; a detection limit of 2 log CFU/ml of *E. coli* in apple juice has been achieved (Aalto-Araneda, Lundén, Markkula, Hakola, & Korkeala, 2019; Kong et al., 2022), and this technique shows good promise for application in beer quality control.

Another microfluidic device in development is a particle counter from Figura Analytics. In this set up, resistive pulse sensors set in series, are used to characterise particles from 0.1 to 30 μm in diameter. Mixed populations e.g., particles of 10 and 1 μm, can be counted simultaneously in the same device, demonstrating potential detection of bacteria within a mixed yeast-bacteria sample (Pollard, Maugi, & Platt, 2022).

A similar system to detect small particles (0.5—5 μm) and to assess cell viability has been developed to profile probiotic bacteria (Jordal et al., 2023). In the Bactobox system, Impedance Flow Cytometry (IFC; also known as Electrical Impedance Spectroscopy (EIS-FC)) is used. This technique employs a narrow microfluidic channel and an electrode to inspect particle size, membrane integrity, and intracellular content, through changes in impedance (Fig. 15.7). Bactobox may be used as an alternative to tracking growth through optical density measurements or colony forming units, and also to enumerate endospores (Jordal et al., 2023). As yet, it is not suitable for low level contaminants or mixed cultures, but the potential as a rapid, label-free method of detection of bacterial contaminants is clear.

A further advancing field is lipidomics. The importance of lipids in brewing has been widely researched for both process and final product. At the cellular level, investigations of plasma membrane lipid composition of brewing yeast have demonstrated its significance in temperature and ethanol tolerance; changes in composition impact membrane fluidity and functionality (Krogerus, Seppänen-Laakso, Castillo, & Gibson, 2017). Determination of lipid composition may be used as a rapid method of identification for both yeasts and bacteria (Řezanka, Kolouchová, Gharwalová, Palyzová, & Sigler, 2018). Indeed, electrospray ionisation-tandem Mass Spectrometry has been used to identify *Megasphaera* and *Pectinatus* species in contaminated beer with a limit of detection of 830 cells, based on determination of plasmalogen phospholipid molecules (Řezanka, Matoulková, Benada, & Sigler, 2015).

15.5 Conclusions

Brewers have a wide range of tools available for the rapid detection, identification, and quantification of microbes with options available for both common beer spoilage organisms and 'unknowns'. A variety of test kits specifically developed for use in routine beer quality control are commercially available (Table 15.2). Choice of method depends on the requirements of the brewery; some will only use routine hygiene monitoring with specific identification carried out if problems arise/persist. Others track both qualitative and quantitative data.

Most systems described do still require an enrichment step, but advances in sensitivity are continuing to be made along with improvements in terms of useability and cost. ATP determination is well established as the standard for hygiene

TABLE 15.2 Selection of commercially available rapid method kits for the detection of beer spoilage microbes.

Method	Kit/platform	Supplier	Notes
ATP/NAD	SuperSnap, UltraSnap PocketSwab Plus N-Light™ ATP Clean-Trace™ HYRiSE®, HY-LiTE®, MVP ICON®	Hygiena Charm Sciences NEMIS 3M Merck	No identification possible, but very fast and minimal training required.
Fluorescence microscopy/fluorometer/flow cytometry	LIVE/DEAD™ BacLight™ VIT® Test kits	Thermo Fisher Vermicon	Differentiation of live/dead cells. Training required.
Sandwich hybridisation	HybriScan™	Sigma-Aldrich	Differentiation of live/dead cells. Simple equipment requirements. Training required.
PCR	Foodproof Veriflow®, brewPAL/LAP/DEK/BRUX/MAP/STAT, GeneUp BrewPro™ BeerSpoilerAlert™ Beer Spoilage Kit Genesig qPCR Milenia GenLine	Hygiena Invisible Sentinel/ BioMerieux Rheonix Microbiologique Genesig Milenia Biotec	Most kits target traditional beer spoilage organisms. Further kits may be required for no/low alcohol beers and non-traditional products. Training required.
Sequencing	Illumina MiSeq™ MinION, GridION	Illumina Inc Oxford nanopore technologies	Differentiation of closely related species possible. Extensive training required.

control and PCR-based methods are becoming more mainstream. In general, PCR methods take less than 4 h from DNA extraction to result. Sequencing methods are more complex and typically take several hours on top of PCR. Flow cytometry and spectroscopic techniques can be very rapid, but extensive training is required.

Advances in microfluidics represents an exciting development in terms of reducing use of expensive reagents and reducing complexity. Such efforts continue our ambitions to ensure beer quality in the most time and resource efficient way.

References

Aalto-Araneda, M., Lundén, J., Markkula, A., Hakola, S., & Korkeala, H. (2019). Processing plant and machinery sanitation and hygiene practices associate with Listeria monocytogenes occurrence in ready-to-eat fish products. *Food microbiology, 82*, 455–464.

Almeida, C., Azevedo, N. F., Fernandes, R. M., Keevil, C. W., & Vieira, M. J. (2010). Fluorescence in situ hybridization method using a peptide nucleic acid probe for identification of *Salmonella* spp. in a broad spectrum of samples. *Applied and Environmental Microbiology, 76*(13), 4476–4485. https://doi.org/10.1128/AEM.01678-09

Anand, S. K. (2004). Bioluminescence: Realtime indicators of hygiene. In *Committees for organisation of the 17th short course on advances in cleaning and sanitation in food industry* (pp. 137–143).

Asano, S., Suzuki, K., Ozaki, K., Kuriyama, H., Yamashita, H., & Kitagawa, Y. (2008). Application of multiplex PCR to the detection of beer-spoilage bacteria. *Journal of the American Society of Brewing Chemists, 66*, 37–42.

Bamforth, C. W. (2006). Brewing: New technologies. Print Book ISBN: 9781845690038, eBook ISBN: 9781845691738. In E. Storgard, A. Haikara, & R. Juvonen (Eds.), *Brewing control systems: Microbiological analysis*. Cambridge: Woodhead Pub..

Barney, M., & Kot, E. (1992). A comparison of rapid microbiological methods for detecting beer spoilage microorganisms. *MBAA Technical Quarterly, 29*, 91–95.

Bergsveinson, J., Pittet, V., & Ziola, B. (2012). RT-qPCR analysis of putative beer-spoilage gene expression during growth of *Lactobacillus brevis* BSO 464 and *Pediococcus claussenii* ATCC BAA-344T in beer. *Applied Microbiology and Biotechnology, 96*, 461–470.

Bischoff, E., Bohak, I., Back, W., & Leibhard, S. (2001). Schnellnachweis von bierschädlichen Bakterien mit PCR und universellen Primern. *Monatsschr Brauwiss, 54*, 4–8.

Boschke, E., Steingroewer, J., Ripperger, S., Klingner, E., & Bley, T. (2002). Biomonitoring by combination of immunomagnetic separation and direct direct epifluorescence filter technique. *European Cells and Materials, 3*(Suppl. 2), 146–147.

Bouix, M., Grabowski, A., Charpentier, M., Leveau, J., & Duteurtre, B. (1999). Rapid detection of microbial contamination in grape juice by flow cytometrie. *Journal International des Sciences de la Vigne et du Vin, 33*, 25–32.

Breeuwer, P., Drocourt, J. L., Rombouts, F. M., & Abee, T. (1994). Energy-dependent, carrier-medicated extrusion of carboxyfluorescein from *Saccharomyces cerevisiae* allows rapid assessment of cell viability by flow cytometry. *Applied and Environmental Microbiology, 60*, 1467–1472.

Brenan, C. J., Roberts, D., & Hurley, J. (2009). Nanoliter high-throughput PCR for DNA and RNA profiling. *Methods in Molecular Biology, 496*, 161–174.

Brown, M., & Wittwer, C. (2000). Flow cytometry: Principles and clinical applications in hematology. *Clinical Chemistry, 46*(8 Pt 2), 1221–1229.

Bruetschy, A., Laurent, M., & Jacquet, R. (1994). Use of flow cytometry in oenology to analyse yeasts. *Letters in Applied Microbiology, 18*, 343–345. https://doi.org/10.1111/j.1472-765X.1994.tb00885.x

Busch, U. (2010). *Molekularbiologische Methoden in der Lebensmittelanalytik: Grundlegende Methoden und Anwendungen* (Vol 113). Berlin: Springer Verlag.

Callahan, B. J., Wong, J., Heiner, C., Theriot, C. M., Gulati, A. S., McGill, S. K., & Dougherty, M. K. (2019). High-throughput amplicon sequencing of the full-length 16S rRNA gene with single-nucleotide resolution. *Nucleic Acids Research, 18*, e103.

Claydon, M. A., Davey, S. N., Edwards-Jones, V., & Gordon, D. B. (1996). The rapid identification of intact microorganisms using mass spectrometry. *Nature Biotechnology, 14*(11), 1584–1586. https://doi.org/10.1038/nbt1196-1584

Condina, M. R., Dilmetz, B. A., Bazaz, S. R., Meneses, J., Warkiani, M. E., & Hoffmann, P. (2019). Rapid separation and identification of beer spoilage bacteria by inertial microfluidics and MALDI-TOF mass spectrometry. *Lab on a Chip, 19*(11), 1961–1970.

Devlin, T. M. (2010). *269 Textbook of biochemistry with clinical correlations* (Vol. 7). John Wiley and Sons.

DiMichele, L. J., & Lewis, M. J. (1993). Rapid, species-specific detection of lactic acid bacteria from beer using the polymerase chain reaction. *Journal of the American Society of Brewing Chemists, 51*(2), 63–66.

Dodd, C. E. R., Stewart, G. S. B., & Waites, W. M. (1990). Biotechnology-based methods for the detection, enumeration and epidemiology of food poisoning and spoilage organisms. *Biotechnology and Genetic Engineering Reviews, 8*(1). https://doi.org/10.1080/02648725.1990.10647864

Donhauser, S., Eger, C., Hubl, T., Schmidt, U., & Winnewisser, W. (1993). Tests to determine the vitality of yeasts using flow cytometrie. *Brauwelt International, 3*, 221–224.

Edwards, C., Porter, J., & West, M. (1997). Fluorescent probes for measuring physiological fitness of yeast. *Fermentation, 9*, 288–293.

Femino, A., Fay, F., Fogarty, K., & Singer, R. (1998). Visualization of single RNA transcripts in situ. *Science, 280*(5363), 585–590. https://doi.org/10.1126/science.280.5363.585

Frischer, M. E., Floriani, P. J., & Nierzwicki-Bauer, S. A. (1996). Differential sensitivity of 16S rRNA targeted oligonucleotide probes used for fluorescence in situ hybridization is a result of ribosomal higher order structure. *Canadian Journal of Microbiology, 42*(10), 1061–1071. https://doi.org/10.1139/m96-136

Fuchs, B. M., Syutsubo, K., Ludwig, W., & Amann, R. (2001). In situ accessibility of *Escherichia coli* 23S rRNA to fluorescently labeled oligonucleotide probes. *Applied and Environmental Microbiology, 67*(2), 961–968. https://doi.org/10.1128/AEM.67.2.961-968.2001

Gall, J. G., & Pardue, M. L. (1969). Formation and detection of RNA-DNA hybrid molecules in cytological preparations. *Proceedings of the National Academy of Sciences, 63*, 378–383. https://doi.org/10.1073/pnas.63.2.378

Goll, M., Kratzheller, B., & Bülte, M. (2003). Evaluation of the HY-RiSE colour hygiene test strip for checking the cleanness of surfaces. *Fleischwirtschaft, 83*(9), 152–154.

Haakensen, M., Dobson, C. M., Deneer, H., & Ziola, B. (2008). Real-time PCR detection of bacteria belonging to the *Firmicutes* phylum. *International Journal of Food Microbiology, 125*, 236–241.

Haakensen, M., Schubert, A., & Ziola, B. (2008). Multiplex PCR for putative *Lactobacillus* and *Pediococcus* beer-spoilage genes and ability of gene presence to predict growth in beer. *Journal of the American Society of Brewing Chemists, 66*, 63–70.

Haakensen, M. C., Butt, L., Chaban, B., Deneer, H., Ziola, B., & Dowgiert, T. (2007). horA specific real-time PCR for detection of beer-spoilage lactic acid bacteria. *Journal of the American Society of Brewing Chemists, 65*, 157–165.

Helentjaris, T., & McCreery, T. (1996). Utilization of DNA probes with digoxigenin-modified nucleotides in southern hybridizations. *Methods in Molecular Biology, 58*, 41–51. https://doi.org/10.1385/0-89603-402-X:41

Higuchi, R., Fockler, C., Dollinger, G., & Watson, R. (1993). Kinetic PCR analysis: Real-time monitoring of DNA amplification reactions. *Biotechnology, 11*, 1026–1030.

Hirasawa, T., Furusawa, C., & Shimizu, H. (2010). *Saccharomyces cerevisiae* and DNA microarray analyses: What did we learn from it for a better understanding and exploitation of yeast biotechnology? *Applied Microbiology and Biotechnology, 87*, 391–400.

Holland, R. D., Wilkes, J. G., Rafii, F., Sutherland, J. B., Persons, C. C., Voorhees, K. J., et al. (1996). Rapid identification of intact whole bacteria based on spectral patterns using matrix-assisted laser desorption/ionization with time-of-flight mass spectrometry. *PMID: 8759332 Rapid Communications in Mass Spectrometry, 10*(10), 1227–1232. https://doi.org/10.1002/(SICI)1097-0231(19960731)10:10<1227::AID-RCM659>3.0.CO;2-6

Hulin, M., Harrison, E., Stratford, M., & Wheals, A. E. (2014). Rapid identification of the genus Dekkera/Brettanomyces, the Dekkera subgroup and all individual species. *International Journal of Food Microbiology, 187*, 7–14.

Hutzler, M., Schuster, E., & Stettner, G. (2008). Ein Werkzeug in der Brauereimikrobiologie. Real-time PCR in der Praxis. *Brauindustrie, 4*, 52–55.

Iijima, K., Asano, S., Suzuki, K., Ogata, T., & Kitagawa, Y. (2008). Modified multiplex PCR methods for comprehensive detection of *Pectinatus* and beer-spoilage cocci. *Bioscience, Biotechnology, and Biochemistry, 72*, 2764–2766.

Janagama, H. K., Mai, T., Han, S., Nadala, L. M., Nadala, C., & Samadpour, M. (2018). Dipstick assay for rapid detection of beer spoilage organisms. *Journal of AOAC International, 101*(6), 1913–1919.

Jespersen, L., Lassen, S., & Jakobsen, M. (1993). Flow cytometric detection of wild yeast in lager breweries. *International Journal of Food Microbiology, 17*, 321–328. https://doi.org/10.1016/0168-1605(93)90202-R

Jiang, H., Xu, W., Ding, Y., & Chen, Q. (2020). Quantitative analysis of yeast fermentation process using Raman spectroscopy: Comparison of CARS and VCPA for variable selection. *Spectrochimica Acta Part A: Molecular and Biomolecular Spectroscopy, 228*, Article 117781.

Jordal, P. L., Diaz, M. G., Morazzoni, C., Allesina, S., Zogno, D., Cattivelli, D., et al. (2023). Collaborative cytometric inter-laboratory ring test for probiotics quantification. *Frontiers in Microbiology, 14*, Article 1285075.

Joshi, S., Annida, R. M., Zuilhof, H., van Beek, T. A., & Nielen, M. W. (2016). Analysis of mycotoxins in beer using a portable nanostructured imaging surface plasmon resonance biosensor. *Journal of Agricultural and Food Chemistry, 64*(43), 8263–8271.

Juvonen, R. (2009). *DNA-based detection and characterisation of strictly anaerobic beer-spoilage bacteria*. VTT Publications.

Juvonen, R., & Haikara, A. (2009). Amplification facilitators and pre-processing methods for PCR detection of strictly anaerobic beer-spoilage bacteria of the class clostridia in brewery samples. *Journal of the Institute of Brewing, 115*(3), 167–176.

Juvonen, R., Koivula, T., & Haikara, A. (2008). Group-specific PCR-RFLP and real-time PCR methods for detection and tentative discrimination of strictly anaerobic beer-spoilage bacteria of the class Clostridia. *International Journal of Food Microbiology, 125*, 162–169. https://doi.org/10.1016/j.ijfoodmicro.2008.03.042

Juvonen, R., Satokari, R., Mallison, K., & Haikara, A. (1999). Detection of spoilage bacteria in beer by polymerase chain reaction. *Journal of the American Society of Brewing Chemists, 57*, 99–103.

Kaczanowska, M., & Rydén-Aulin, M. (2007). Ribosome biogenesis and the translation process in *Escherichia coli*. *Microbiology and Molecular Biology Reviews, 71*(3), 477–494. https://doi.org/10.1128/MMBR.00013-07

Kashif, M., Majeed, M. I., Nawaz, H., Rashid, N., Abubakar, M., Ahmad, S., et al. (2021). Surface-enhanced Raman spectroscopy for identification of food processing bacteria. *Spectrochimica Acta Part A: Molecular and Biomolecular Spectroscopy, 261*, Article 119989.

Kempf, V. A. J., Trebesius, K., & Autenrieth, I. B. (2000). Fluorescent in situ hybridization allows rapid identification of microorganisms in blood cultures. *PMCID: PMC86216 Journal of Clinical Microbiology, 38*(2), 830–838. https://doi.org/10.1007/0-387-30747-8_26

Kern, C. C., Vogel, R. F., & Behr, J. (2014). Differentiation of *Lactobacillus brevis* strains using matrix-assisted-laser-desorption-ionization-time-of-flight mass spectrometry with respect to their beer spoilage potential. *Food Microbiology, 40*, 18–24. https://doi.org/10.1016/j.fm.2013.11.015

Kerruish, D. W., Cormican, P., Kenny, E. M., Kearns, J., Colgan, E., Boulton, C. A., et al. (2024). The origins of the Guinness stout yeast. *Communications Biology, 7*(1), 68.

Kilgour, W. J., & Day, A. (1983). The application of new techniques for the rapid determination of microbial contamination in brewing. In *The European brewing convention congress* (pp. 177–184). Oxford: IRL Press.

Kobayashi, N., Sato, M., Fukuhara, S., Yokoi, S., Kurihara, T., Watari, J., et al. (2007). Application of shotgun DNA microarray technology to gene expression analysis in lager yeast. *Journal of the American Society of Brewing Chemists, 65*(2), 92–98.

Koivula, T., Juvonen, R., Haikara, A., & Suihko, M.-L. (2006). Characterization of the brewery spoilage bacterium *Obesumbacterium proteus* by automated ribotyping and development of PCR methods for its biotype 1. *Journal of Applied Microbiology, 100*, 398–406.

Kong, J., Fan, C., Liao, X., Chen, A., Yang, S., Zhao, L., et al. (2022). Accurate detection of *Escherichia coli* O157: H7 and *Salmonella enterica* serovar typhimurium based on the combination of next-generation sequencing and droplet digital PCR. *LWT, 168*, Article 113913.

Kregiel, D., & Berlowska, J. (2009). Evaluation of yeast cell vitality using different fluorescent dyes. *Food Chemistry and Biotechnology, 73*.

Krishnamurthy, T., Ross, P. L., & Rajamani, U. (1996). Detection of pathogenic and non-pathogenic bacteria by matrix-assisted laser desorption/ionization time-of-flight mass spectrometry. *Rapid Communications in Mass Spectrometry, 10*(8), 883–888. https://doi.org/10.1002/(SICI)1097-0231(19960610)10:8<883::AID-RCM594>3.0.CO;2-V

Krogerus, K., Seppänen-Laakso, T., Castillo, S., & Gibson, B. (2017). Inheritance of brewing-relevant phenotypes in constructed *Saccharomyces cerevisiae* × Saccharomyces eubayanus hybrids. *Microbial Cell Factories, 16*, 1–22.

Kunkee, R. E., & Neradt, F. (1974). A rapid method for detection of viable yeasts in wines. *Wine Vines, 55*, 36–39. https://doi.org/10.1016/S0167-7012(01)00243-3

Kurniawan, Y. N., Shinohara, Y., Sakai, H., Magarifuchi, T., & Suzuki, K. (2022). Applications of the third-generation DNA sequencing technology to the detection of hop tolerance genes and discrimination of Saccharomyces yeast strains. *Journal of the American Society of Brewing Chemists, 80*(2), 161–168.

Kurniawan, Y. N., Shinohara, Y., Takesue, N., Sakai, H., Magarifuchi, T., & Suzuki, K. (2021). Development of a rapid and accurate nanopore-based sequencing platform for on-field identification of beer-spoilage bacteria in the breweries. *Journal of the American Society of Brewing Chemists, 79*(3), 240–248.

Lay, J. O., Jr. (2001). MALDI-TOF mass spectrometry of bacteria. *Mass Spectrometry Reviews, 20*(4), 172–194. https://doi.org/10.1111/j.1574-695X.2008.00428.x

Li, L., Cao, X., Zhang, T., Wu, Q., Xiang, P., Shen, C., et al. (2022). Recent developments in Surface-Enhanced Raman Spectroscopy and its application in food analysis: Alcoholic beverages as an example. *Foods, 11*(14), 2165.

Ma, Y., Deng, Y., Xu, Z., Liu, J., Dong, J., Yin, H., et al. (2017). Development of a propidium monoazide-polymerase chain reaction assay for detection of viable Lactobacillus brevis in beer. *Brazilian Journal of Microbiology, 48*, 740–746.

Machado, T. I. (2022). *Potential use of multiplex real-time PCR to develop a kit for identification of beer-spoilage microorganisms*.

Mackay, I. M. (2004). Real-time PCR in the microbiology laboratory. *Clinical Microbiology and Infection, 10*(3), 190–212.

Malmström, G., & Andersson, B. (2013). *The Nobel prize in chemistry: The development of modern chemistry*. Available from: Nobelprize.org. Nobel Media AB http://www.nobelprize.org.

Marmur, J., & Doty, P. (1962). Determination of the base composition of deoxyribonucleic acid from its thermal denaturation temperature. *Journal of Molecular Biology, 5*, 109–118.

Maugueret, T. M.-J., & Walker, S. L. (2002). Rapid detection of *Obesumbacterium proteus* from yeast and wort using polymerase chain reaction. *Letters in Applied Microbiology, 35*, 281–284.

Miteu, G. D. (2024). Towards a 2025 global beer market: The use of aptamers in biosensors to detect acetobacter aceti in beer-A biotechnological approach to improve food hygiene and safety in beer developing markets and beyond. *IPS Journal of Nutrition and Food Science, 3*(1), 94–101.

Mori, Y., & Notomi, T. (2009). Loop-mediated isothermal amplification (LAMP): A rapid, accurate, and cost-effective diagnostic method for infectious diseases. *Journal of Infection and Chemotherapy, 15*, 62–69.

O'Connor, C. (2008). Fluorescence in situ hybridization (FISH). *Nature Education, 1*(1), 171.

Pettipher, G. L., Watts, Y. B., Langford, S. A., & Kroll, R. G. (1992). Preliminary evaluation of COBRA, an automated DEFT instrument, for the rapid enumeration of microorganisms in cultures, raw milk, meat and fish. *Letters in Applied Microbiology, 14*, 206–209. https://doi.org/10.1111/j.1472-765X.1992.tb00686.x

Pioch, D., Jürgen, B., Evers, S., Maurer, K. H., Hecker, M., & Schweder, T. (2008). Improved sandwich-hybridization assay for an electrical DNA-chip-based monitoring of bioprocess-relevant marker genes. *Applied Microbiology and Biotechnology, 78*(4), 719–728. https://doi.org/10.1007/s00253-008-1347-z

Pipper, J., Zhang, Y., Neuzil, P., & Hsieh, T. M. (2008). Clockwork PCR including sample preparation. *Angewandte Chemie International Edition in English, 47*(21), 3900–3904.

Pollard, M., Maugi, R., & Platt, M. (2022). Multi-resistive pulse sensor microfluidic device. *Analyst, 147*(7), 1417–1424.

Priest, F. G., & Campbell, I. (2003). *Brewing microbiology* (3rd ed.). New York: Kluwer Academic/Plenum Puplisher. Print ISBN: 0306472880, eBook: ISBN-13: 978-0306472886.

Priest, F. G., & Stewart, G. G. (2006). Handbook of brewing. Microbiology and microbiological control in the brewery, CRC Press. In *Microbiology and microbiological control in the brewery* (2nd ed.). CRC Press. Print ISBN: 978-0-8247-2657-7, eBook ISBN: 978-1-4200-1517-1.

Prudêncio, C., Sansonetty, F., & Côrte-Real, M. (1998). Flow cytometric assessment of cell structural and functional changes induced by acetic acid in the yeasts *Zygosaccharomyces bailii* and *Saccharomyces cerevisiae*. *Cytometry, 31*, 307–313. https://doi.org/10.1002/(SICI)1097-0320(19980401)31:4<307::AID-CYTO11>3.0.CO;2-U

Report of Subcommittee. (2003). Determination of yeast viability by fluorescent staining. *Journal of the American Society of Brewing Chemists, 61*, 231–232. https://doi.org/10.1094/ASBCJ-61-0231

Řezanka, T., Kolouchová, I., Gharwalová, L., Palyzová, A., & Sigler, K. (2018). Lipidomic analysis: From archaea to mammals. *Lipids, 53*(1), 5–25.

Řezanka, T., Matoulková, D., Benada, O., & Sigler, K. (2015). Lipidomics as an important key for the identification of beer-spoilage bacteria. *Letters in Applied Microbiology, 60*(6), 536–543.

Raj, A., van den Bogaard, P., Rifkin, S., van Oudenaarden, A., & Tyagi, S. (2008). Imaging individual mRNA molecules using multiple singly labeled probes. *Nature Methods, 5*(10), 877–879. https://doi.org/10.1038/nmeth.1253

Rautio, J., Barken, K. B., Lahdenperä, J., Breitenstein, A., Molin, S., & Neubauer, P. (2003). Sandwich hybridisation assay for quantitative detection of yeast RNAs in crude cell lysates. *Microbial Cell Factories, 2*, 4. https://doi.org/10.1186/1475-2859-2-4

Raybourne, R., & Tortorello, M. (2003). In T. A. McMeekin (Ed.), *Detecting pathogens in foodMicroscopy techniques: DEFT and flow cytometry*. Cambridge: Woodhead Publishing Limited. Print Book: ISBN: 9781855736702, eBook: ISBN: 9781855737044.

Rhodes, W. C., & McElroy, W. D. (1958). The synthesis and function of luciferyl-adenylate and oxyluciferyl-adenylate. *Journal of Biological Chemistry, 233*, 1528–1537.

Riley, L. K., Marshall, M. E., & Coleman, M. S. (1986). A method for biotinylating oligonucleotide probes for use in molecular hybridizations. *DNA, 5*(4), 333–337. https://doi.org/10.1089/dna.1986.5.333

Robertson, J., McGoverin, C., Vanholsbeeck, F., & Swift, S. (2019). Optimisation of the protocol for the LIVE/DEAD® BacLight™ bacterial viability kit for rapid determination of bacterial load. *Frontiers in Microbiology, 10*, Article 448819.

Rost, F. W. D. (1995). *Fluorescence microscopy* (Vol. 2, p. 202). Cambridge: Cambridge University Press.

Rudkin, G. T., & Stollar, B. D. (1977). High resolution of DNA-RNA hybrids in situ by indirect immunofluorescence. *Nature, 265*, 472–474. https://doi.org/10.1038/265472a0

Saiki, R. K. (1990). Amplification of genomic DNA. In M. A. Innis, D. H. Gelfand, J. J. Sninsky, & T. J. White (Eds.), *PCR protocols, a guide to methods and applications* (pp. 13–20). San Diego: USA Academic Press Inc.

Saiki, R. K., Gelfand, D. H., Stoffel, S., Scharf, S. J., Higuchi, R., Horn, G. T., et al. (1988). Primer-directed enzymatic amplification of DNA with a thermostable DNA polymerase. *Science, 239*, 487–491.

Salazar, A. N., Gorter de Vries, A. R., van den Broek, M., Brouwers, N., de la Torre Cortès, P., Kuijpers, N. G. A., et al. (2019). *Nanopore sequencing and comparative genome analysis confirm lager-brewing yeasts originated from a single hybridization*. bioRxiv, Article 603480.

Sambo, F., Finotello, F., Lavezzo, E., Baruzzo, G., Masi, G., Peta, E., … Di Camillo, B. (2018). Optimising PCR primers targeting the bacterial 16S ribosomal RNA gene. *BMC Bioinformatics, 19*, 1–10.

Satokari, R., Juvonen, R., von Wright, A., & Haikara, A. (1997). Detection of *Pectinatus* beer spoilage bacteria by using the polymerase chain reaction. *Journal of Food Protection, 60*, 1571–1573.

Shang, L., Xu, L., Wang, Y., Liu, K., Liang, P., Zhou, S., et al. (2022). Rapid detection of beer spoilage bacteria based on label-free SERS technology. *Analytical Methods, 14*(48), 5056–5064.

Shinohara, Y., Kurniawan, Y. N., Sakai, H., Magarifuchi, T., & Suzuki, K. (2021). Nanopore based sequencing enables easy and accurate identification of yeasts in breweries. *Journal of the Institute of Brewing, 127*(2), 160–166.

Sobel, J., Henry, L., Rotman, N., & Rando, G. (2017). BeerDeCoded: The open beer metagenome project. *F1000research, 6*, 1676.

Stilman, W., Wackers, G., Sichani, S. B., Khorshid, M., Thesseling, F., Vereman, J., et al. (2022). A table-top sensor for the detection of hydrophobins and yeasts in brewery applications. *Sensors and Actuators B: Chemical, 373*, Article 132690.

Suzuki, K. (2020). Emergence of new spoilage microorganisms in the brewing industry and development of microbiological quality control methods to cope with this phenomenon: A review. *Journal of the American Society of Brewing Chemists, 78*(4), 245–259.

Suzuki, K., Iijima, K., Ozaki, K., & Yamashita, H. (2005). Isolation of hop-sensitive variant from *Lactobacillus lindneri* and identification of genetic marker for beer spoilage ability of lactic acid bacteria. *Applied and Environmental Microbiology, 7*, 5089–5097.

Suzuki, K., Sami, M., Iijima, K., Ozaki, K., & Yamashita, H. (2006). Characterization of *horA* and its flanking regions of *Pediococcus damnosus* ABBC478 and development of more specific and sensitive *horA* PCR methods. *Letters in Applied Microbiology, 42*, 393–399.

Taskila, S., Tuomola, M., Kronlöf, J., & Neubauer, P. (2010). Comparison of enrichment media for routine detection of beer spoiling lactic acid bacteria and development of trouble-shooting medium for *Lactobacillus backi*. *Journal of the Institute of Brewing, 116*(2), 151–156. https://doi.org/10.1002/j.2050-0416.2010.tb00411.x

Thelen, K., Beimfohr, C., Bohak, I., & Back, W. (2001). Spezifischer Schnellnachweis von Bierschädlichen Bakterien mittels fluoreszenzmarkierter Gensonden. *Brauwelt, 141*(38/01), 1596–1603.

Tortorello, M. L., & Gendel, S. M. (1993). Fluorescent antibodies applied to direct epifluorescent filter technique for microscopic enumeration of *Escherichia coli* O157:H7 in milk and juice. *Journal of Food Protection, 56*, 672–677.

Tortorello, M. L., & Reineke, K. F. (2000). Direct enumeration of *Escherichia coli* and enteric bacteria in water, beverages and sprouts by 16S rRNA in situ hybridization. *Food Microbiology, 17*, 305–313. https://doi.org/10.1006/fmic.1999.0317

Tsuchiya, Y., Ogawa, M., Nakakita, Y., Nara, Y., Kaneda, H., Watari, J., et al. (2007). Identification of beer-spoilage microorganisms using the loop-mediated isothermal amplification method. *Journal of the American Society of Brewing Chemists, 65*, 77–80.

Uyttendaele, M., & Debevere, J. (2006). Rapid methods in food diagnostics. In Y. H. Hui (Ed.), *Handbook of food science, technology and engineering (4 volume set)*. CRC Press.

Wang, S., & Levin, R. E. (2005). Discrimination of viable *Vibrio vulnificus* cells from dead cells in real-time PCR. *Journal of Microbiological Methods, 64*(1), 1–8.

Warner, J. R. (1999). The economics of ribosome biosynthesis in yeast. *Trends in Biochemical Sciences, 24*(11), 437–440. https://doi.org/10.1016/S0968-0004(99)01460-7

Weber, D. G., Sahm, K., Polen, T., Wendisch, V. F., & Antranikian, G. (2008). Oligonucleotide microarrays for the detection and identification of viable beer spoilage bacteria. *Journal of Applied Microbiology, 105*, 951–962.

Webster, J. J., Walker, B. G., Ford, S. R., & Leach, F. R. (1988). Determination of sterilization effectiveness by measuring bacterial growth in a biological indicator through firefly luciferase determination of ATP. *Journal of Bioluminescence and Chemiluminescence, 2*(3), 1099–1271. https://doi.org/10.1002/bio.1170020304

Williamson, C., Kennedy, K., Bhattacharya, S., Patel, S., Perry, J., Bolton, J., et al. (2023). A novel image-based method for simultaneous counting of Lactobacillus and Saccharomyces in mixed culture fermentation. *Journal of Industrial Microbiology and Biotechnology, 50*(1), Article kuad007.

Wittwer, C. T., Ririe, K. M., Anrew, R. W., David, D. A., Gundry, R. A., & Balis, U. J. (1997). The LightCycler: A microvolume multisample fluorimeter with rapid temperature control. *Biotechniques, 22*, 176–181.

Xu, Z., Wang, K., Liu, Z., Soteyome, T., Deng, Y., Chen, L., et al. (2022). A novel procedure in combination of genomic sequencing, flow cytometry and routine culturing for confirmation of beer spoilage caused by Pediococcus damnosus in viable but nonculturable state. *LWT, 154*, Article 112623.

Yadav, N., Yadav, S. S., Chhillar, A. K., & Rana, J. S. (2021). An overview of nanomaterial based biosensors for detection of Aflatoxin B1 toxicity in foods. *Food and Chemical Toxicology, 152*, Article 112201.

Yamauchi, H., Yamamoto, H., Shibano, Y., Amaya, N., & Saeki, T. (1998). Rapid methods for detecting *Saccharomyces diastaticus*, a beer spoilage yeast, using the polymerase chain reaction. *Journal of the American Society of Brewing Chemists, 56*(2), 58–63.

Ziola, B. (2016). Monoclonal antibodies binding to lipopolysaccharide from the beer-spoilage bacterium *Megasphaera cerevisiae* exhibit panreactivity with the strictly anaerobic gram-negative brewing-related bacteria. *Journal of the American Society of Brewing Chemists, 74*(4), 267–271.

Chapter 16

Beer packaging: Microbiological—hazards and considerations

Jan Fischer[1], Jan Biering[1] and Ruslan Hofmann[2]
[1]*VLB Berlin e.V., Berlin, Germany;* [2]*PureMalt Products Ltd., Berlin, Germany*

16.1 Introduction

Within the processing of beer, the step of filling and packaging can be considered as the last stage at which a contamination of the product is possible. Any further contamination would be due to manipulation of the packaged product or due to a critical malfunction of the packaging material (e.g., loss of sealing property) and shall not be considered further. Generally, the filling hall can be divided into a 'dry section' that includes the palletising, unpacking and packing and into a 'wet section' that usually starts with cleaning and rinsing of the packaging materials and ends with control of the filled container. The possibilities of stabilising the beer in terms of its microbiological state are subject of Chapters 11 and 12 and will not be specifically addressed within this chapter.

16.2 Microbiological hazards in the filling hall

Microbiological contamination in the packaging hall can also be referred to as secondary contamination. In contrast to primary contamination during beer production, secondary contamination will usually not be noticeable without specific microbiological analysis methods. The risk potential derives from the type of microorganism that is causing the contamination. Referring to the nomenclature by (Back, 1994), the microorganisms may be divided into several categories from which especially the obligate and the potentially beer-spoiling organisms are crucial for the long-term stability and, therefore, quality of the bottled product. If microorganisms get into the filled container and are able to reproduce in the specific product, several negative impacts on the beer's quality may arise, for example,

- Increased turbidity.
- Formation of floating particles or sedimentation.
- Change of flavour (off-flavour).
- Acidification.
- Pressure build-up (increasing CO_2 concentration).
- Deformation of the container.
- Decrease in apparent extract.
- Loss of mouthfeel and body.

Generally, at every stage where the product has contact with any kind of 'new' surface or gets directly exposed to the environment, the risk of contamination is given. In the filling hall, these process stages are limited to the 'wet section'. This chapter will describe the possible risks according to their sources. In particular, the respective filling and capping equipment as well as the process characteristics are reviewed. The packaging materials and their specific risk potentials are addressed. In addition, the supply with water, air and carbon dioxide together with the microbiological hazards in the direct filler periphery (environment) is listed.

Brewing Microbiology. https://doi.org/10.1016/B978-0-323-99606-8.00006-7

16.2.1 Filling machine

Filling machines are first of all characterised by their container that is to be filled. Furthermore, the number of filling organs, shape of filler (circular or line), container size or filling mechanism can be used for characterisation. The purpose of filling can be described as the filling of the product with as low as possible losses and at the desired level within acceptable tolerances. In addition, the preservation of the products quality has to be secured. Preservation of the product quality includes avoiding oxygen pick-up, loss of carbon dioxide and any contamination. As beer is a carbonated drink, the packaging of beer requires a certain pressure level to avoid degassing and thereby loss of carbonation or foaming during the filling process. The necessary pressure level depends on the CO_2 content of the beer and the product temperature during filling. Filling beer at temperatures higher than room temperature is not practice relevant; usually beer is filled at cold temperatures. The temperature difference may lead to water condensation on the container wall. Within this chapter, the reference filling system is the most common circular glass bottle filling. Particular differences and characteristics of filling containers other than glass bottles will be discussed in Section 16.2.3.

16.2.1.1 Filler design

A modern filler design directly reflects the possible measures to reduce contamination risks (hygienic design). Some basic requirements shall be mentioned briefly. For more detailed information on hygienic design, please see Chapter 10.

- All surfaces should be suitable for automated cleaning in place (CIP) and manual cleaning.
- Surfaces with product contact should have an average roughness of ≤ 1.6 μm (better 0.8 μm).
- Surfaces with product contact should be suitable for sterilisation ($T > 80°C$).
- Surfaces should be designed to allow easy run off product residues, cleaning of shards, etc.
- The quality of welding has to be appropriate to the neighbouring surfaces.
- Any kind of niches, gaps or open profiles should be avoided (round, plain surfaces with welded ends preferable).
- Open bores, holes, threads, screws, etc. are to be avoided.
- The floor should be designed in such a way that fluids can easily run off.

Additional measures to minimise the risk of contamination during the filling process will be addressed in Section 16.4.

16.2.1.2 Process steps in the filling machine

Generally, the filling process can be divided into the following steps:

- Transfer of container to the filling valves.
- Cantering and pressing on of the container onto the filling valve.
- Possible pre-evacuation and rinsing as well as pressurising with CO_2.
- Opening of product inlet.
- Filling.
- Closing of product inlet.
- Resting.
- Depressurising and decoupling of container.
- Transfer of filled container to next process step (closing/capping machine).

During the transfer to the filler, the container is usually open. Therefore, the risk is given that microorganisms can enter the container by air or due to aerosol formation. In addition, microorganisms that are attached to construction above the transfer belt, for example, housing, crossing pipes, etc., can fall into the open containers. The longer the open containers are exposed to the environment during transfer, the higher the risk of contamination. Especially longer times of stoppage may have negative effects. During the filling process itself, the container is directly attached to the respective filling valve and can be assumed to be a closed system. Any contamination risk during the filling process arises either from contaminated surfaces within the container and/or product contact or from gas or product flows that are directly involved in the filling process. The filling is followed by the transfer to the capping or closing equipment where again the container can be considered as highly vulnerable to contamination by spraying water or microorganisms in the direct environment. After the container has been closed, the risk of contamination is limited to technical failures only.

16.2.1.3 Media transport

The product is pumped from the pre-filling process step to the filler through a central media distributor and into its buffer vessel. Usually, the product buffer vessel of modern (beer) filling machines has the form of a ring channel. In state-of-the art filling machines, the ring channel is built as a ring tube to ensure the most suitable cleaning conditions and optimal reduction of mass. The media distributor and the ring channel have to be constructed in such a way that any cross-contamination or mixing of media streams can be avoided. In all modern systems, the whole construction can be included in the CIP system. Nevertheless, dead ends and pockets, valves, broken equipment or rough surfaces may increase the potential of contamination due to the growth of microorganisms.

Here not only the transport pipes for the product have to be considered. The supply of CO_2 and vacuum, as well as the pipes for depressurising, may carry substantial microorganism loads or get covered with product residues during the filling process. Any product residue always represents a potential risk for growth of beer contaminants. Therefore, it is absolutely necessary to include all media distribution and transfer pipes in the CIP protocols.

16.2.1.4 Filling process

When beer is filled, usually the glass bottles are evacuated after pressing on onto the filling valve. In some cases, an initial rinsing process with inert gas is applied. After evacuation, the container is pressurised with CO_2 to filling pressure. Depending on the filler design, a second or third evacuation, each followed by CO_2 rinsing up to atmospheric pressure, takes place before the container is pressurised to filling pressure. The purpose is to reduce the oxygen level in the bottle. The filling valves can be designed in several ways. Possible are valves with short, long or without filling tube. The valves can be controlled mechanically, pneumatically, or electro-magnetically. The filling level is adjusted either by height (level) filling or by sensor-based control of mass/volume flow. From the microbiological point of view, filling systems are preferable that enable a 'dry' degassing and a strict separation of gas/fluid flows. Back pressuring into the ring channel should be avoided to decrease the risk of cross-contamination. Valves for degassing or level control that have product contact may supply possible microorganisms with nutrients. Mechanically moving parts may be difficult to clean and may supply spray shadows or niches for microorganism growth and thereby lead to elevated contamination risks.

16.2.1.5 Transfer to closing machine

Before the container is closed, it has to be transferred to the closing machine. During that time, the container is still open to the atmosphere. Usually, a very thin high-pressure water jet is applied to induce a minimal over-foaming of the bottle before it is closed. The high-pressure injection has the purpose to displace the air on top of the product surface in the bottleneck and therefore to reduce the oxygen pick-up during filling.

The water injected into the product has to be treated to avoid contamination. Usually, the water used for the injection gets membrane filtered ($\leq$0.45 μm) and heated to temperatures of 85–90°C. The injection nozzle should be implemented in the CIP programme. Nevertheless, the presence of a nozzle directly above the open container represents a contamination risk, and the nozzle should regularly be checked for absence of microorganisms.

Moreover, the over foaming of the beer usually causes product residues to be spilled at the filling machine and its environment. The spinning carousels of the filling and closing machine lead to a wide spreading of these residues, even into hard to reach niches. Additionally, to the regular CIP operations, a frequent intermediate cleaning, for example, every hour, is mandatory to maintain proper production hygiene. Those intermediate cleanings or flushing steps could also be integrated in phases of unplanned downtime of the filling machines.

16.2.2 Closing machine

For beer filled in glass bottles, crown corks are the dominating form of closure. For reasons of clarity, this chapter focusses on the microbiological risks during the closing with crown corks only. Other possible closures for beer bottles include screw caps, swing tops or corks from various materials. Although technological details may differ, the general microbiological risk potentials are similar for all types of closures.

The closing machine is, on one hand, separating the incoming bottles and transferring them to the closing element. On the other hand, the closing machine has to supply the closures. The closing elements apply the closures and close the bottles mechanically. After the closing procedure is finished, the closed containers are transferred to the following process steps, for example, labelling. It is possible that, after the closing element, a container shower in combination with an air dry rinses product residues to avoid growth of microorganisms on the outside of the bottle or at edges of the crown cork.

Although the growth of microorganisms does not directly represent harm for the product quality, the presence of moulds and other organisms growing on product residues on the outside of the container should be avoided.

The closing elements are mechanically driven systems with several moving parts. Similar to the filling elements, the moving parts and the housing of the moving parts represent microbiological risk potentials. In modern machines, the closing elements are designed to be as 'open' as possible. Thus, the cleaning of the closing elements and the whole machine can be better automated, and inner parts of the closing element may be better implemented in the cleaning procedures. Ideally, the closing element runs grease free. Grease-free elements do not provide possible niches or nutrients to microorganisms and are more cleanable.

Crown corks are transferred magnetically in modern closing machines. Older systems often used pressurised air to transfer the crown corks to the closing element. From a microbiological perspective, the use of pressurised air is not preferable. The use of pressurised air for the crown cork transport leads to possible swirling and aerosol formation. Especially with product residues caused by over foaming, the risk of contamination is elevated.

While using screw caps, for example, on plastic bottles, the risk due to over foaming from the high-pressure injection is even higher. Usually, it is not avoidable that overfoaming beer reaches the screw thread at the bottle's finish. It has to be ensured that these product residues are eliminated entirely before closing by means of a shower. The fact that thermal methods of preservation, such as tunnel pasteurisation, are very limited on plastic bottles, underscores the urgency of removing these product residues.

Regarding swing top closures, it has to be considered that the distance between filling and closing machine is much higher compared to crown corks. Thus, the filled bottles are open longer and exposed to contaminants. Bottles, which are not closed correctly by the closing machine, are often collected on conveyor belts and stand open, before they are closed manually and are processed further.

Similar to the filling machines, the rules of hygienic design should apply for closing machines as well. For more detailed information on hygienic design, please see Chapter 10.

16.2.3 Packaging material

The packaging has, along with others, the purpose of keeping beer quality as good as possible. Of course, this purpose includes microbiological stability. Therefore, the packaging material and its surface with product contact have to be free of pathogenic or product-harming microorganisms.

Depending on the type of packaging, different possibilities exist to clean the packaging material from any harmful microorganisms. This part of the chapter will deal with these actions, as they represent risks if the packaging material is not properly cleaned. Furthermore, depending on the type of packaging, the filling process might differ from the procedures described in Sections 16.2.1 and 16.2.2. Special attention is paid to possible process differences compared to glass bottles.

16.2.3.1 Nonreturnable bottles

Nonreturnable bottles can be considered clean after their manufacturing. Nevertheless, in terms of product safety, it may be useful to rinse the bottles before they are filled. The rinsing can be done either with ionised air or with water. In addition, the bottles get turned upside down to use gravity.

The purpose of the rinsing process is to remove particles that may have fallen into the bottles during transport or handling on site. From a microbiological perspective, only a low-risk potential for beer arises from non-returnable bottles.

Even if rinsing here is mostly done because of non-biological reasons, the water used itself can be a source of contamination if not further treated biologically. Possible methods of water disinfection in this case would be the usage of chloride dioxide (ClO_2) or a treatment with ultraviolet-light. Using ClO_2 has the advantage that due to its reactivity, it may even have a sanitisation effect on possible contaminants in the bottles. In addition, unlike UV light, chlorine dioxide has a depot effect, which largely prevents recontamination of the water.

16.2.3.2 Returnable bottles

Returnable bottles have to be cleaned before they can be refilled. The cleaning has the purpose of removing all particles and fluids from the bottle. Particles and fluids include product residues and grown microorganisms, labels, glue, foils and other material that may be found in the bottles.

After the cleaning process, the bottles shall be clean and bathed all over their surface. All residues or foreign materials shall be removed, and no pathogenic or product harming microorganisms will be found on the bottle surface.

In addition to biological cleanliness, it must be ensured that the bottles leave the bottle washer free of detergent residues (within certain tolerances). To check this, for example, the chemical oxygen demand, the surface tension and the amount of ionic and anionic surfactants in the residual water of the cleaned bottles can be determined.

The cleaning of (glass) bottles is done by applying caustic solution with approximately 1.5%–2.5% NaOH concentration at a temperature of approximately 80°C (range can vary between 70°C to up to ≥ 90°C). The bottle passes several baths to heat up and to achieve the necessary time for soaking about 6–9 min (depending on the design of the machine). Spraying nozzles are installed as well, to achieve an additional mechanical cleaning effect. The spraying nozzles must regularly be checked for organic and inorganic coating that reduces the effectiveness of the rinsing effect and may lead to contamination instead of cleaning in the worst case.

After the cleaning process is finished, the bottle has to be rinsed with fresh water and is cooled down to approximately room temperature. A temperature difference of more than 15 K between glass temperature and product temperature should be strictly avoided; otherwise, the risk of exploding bottles during filling is very high.

The fresh water for the bottle rinsing and cooling is another microbiological risk point. Microbiological control of the fresh water is essential to avoid any re-contamination of the bottle after cleaning. Disinfection with, for example, chlorine dioxide or peracetic acid is used to reduce the risk of microbiological infection in the rinsing zones. Using those chemicals, it has to be ensured not to exceed maximum concentration of residues in the product. The cleaned bottles need to be controlled for microbiological contamination on a regular basis from different positions of the bottle washer.

When the cleaned bottles leave the bottle washing machine, they have to be transported to the filling machine. The time and distance for this transport should be kept as short as possible to avoid any contamination with microorganism from air flow or installation above the transport belt.

Not only for microbiological reasons, but it is also highly recommended to install ceilings above the conveyors of the open bottles, which avoids microbes and other objects, for example, glass splinters, to enter the bottles in a certain amount. In terms of hygiene, it is crucial to clean ceilings on a regular basis. If not, the formation of biofilms is frequently observed there, which is turning a safety measure into a severe hazard.

Once the bottles are in contact with hot caustic for the required soaking time, they can be considered as sterile. However, a re-contamination is possible at several points before finally capped. Inside the bottle washer, especially in the water zones after the main caustic bath, it has to be avoided that microbes get inside the bottles. Thus, using disinfection agents here is strongly recommended. The microbiological state of the fresh water has to be checked frequently, since this last process step of the cleaning procedure.

In terms of machine design, a double-ended bottle cleaning machine is preferable compared to single end machines. Ideally, the bottle infeed is located in another hall than the discharge. Of course, this is not always possible due to means of construction. But it is necessary to set up the machine layout in a way that no cleaned bottle is crossing the conveyer belt for dirty bottles or passing somehow around open doors or gates. When using a single-end machine, it has to taken into account to maximise the distance between dirty and clean bottles. The discharge and all conveyors of the cleaned bottles have to be above the ones from the dirty bottles by any means.

16.2.3.3 Cans

The filling of cans is performed in a manner similar to the filling of glass bottles. Nevertheless, the evacuation cannot take place because the can axial pressure resistance does not allow the application of a vacuum. Therefore, only purging with CO_2 or another inert gas can be used. The fairly high surface-to-volume ratio at the can finish creates difficulties in terms of oxygen pick-up. From a microbiological perspective, the same risks apply as for glass bottles.

Since the diameter of an open can is far bigger than a usual bottle finish, it is more likely, that larger amounts of beer are splashing into the environment of the filling machine. Since these beer residues can act as nutrients for microorganisms, frequent cleaning of the surrounding of the filling machine during production is necessary. In addition, the risk of contamination pick-up by air streams and floating microorganisms is higher because of the increased open surface to the environment.

16.2.3.4 Plastic bottles

When plastic bottles are filled, the bottles themselves are usually produced on site. Polyethylene terephthalate (PET) bottles are blown from preforms. The preforms can either be purchased or produced on site as well.

During the blow moulding process, the PET preforms are treated with pressure up to 40 bar and temperatures up to 240°C. Nevertheless, microorganisms that may be present in the preforms are not totally inactivated, since heat as well as pressure are applied in a dry environment for a comparably short time.

In terms of product safety, it may be useful to rinse the bottles before they are filled. The rinsing can be done either with ionised air or with water. In addition, the bottles are turned upside down to use gravity.

In the special case of aseptic filling, it may be necessary to decontaminate the PET bottles before filling. Usually, hydrogen peroxide or peracetic acid is used for disinfection. Aseptic filling is not necessary for filling beer and beer products.

In contrast to cans and glass bottles, the container walls of PET bottles are permeable. Thus, an increased loss of CO_2 and uptake of oxygen may result. Both factors can be considered as attributes that have a significant impact on growth rates of several microorganisms. Recent research has shown that high permeation rates for plastic bottles may lead to accelerated growth of beer spoilage organisms. Contamination with aerobic bacteria that are not able to grow under low oxygen conditions was seen to increase significantly for samples with elevated permeation rates.

In addition, a tunnel pasteurisation to decrease the contamination risk with a heat treatment after the filling process is limited by the temperature and pressure sensitivity of the plastic material. Generally, a heat treatment of PET containers filled with highly carbonated beverages cannot be recommended and may lead to deformation and loss of sealing properties.

16.2.3.5 Kegs

Kegs in this chapter are to be considered reusable stainless-steel kegs in their various sizes and forms. Various types of fittings are available and used for different markets. Nevertheless, the same basic considerations and hazards apply for all types of stainless-steel kegs.

Before the kegs are emptied and purged with water in the first cleaning station, a pressure control is recommended to select out unpressurised kegs. These kegs may be manipulated by a customer or show leakages and broken gaskets. A cleaning of such pressure-less kegs needs to be done more intensively, and a gasket service is required. The cleaning cycle usually consists of one or two hot caustic and/or acid cleaning steps followed by water steam treatment for sterilisation. Prior to filling, the kegs are pre-pressurised with carbon dioxide to minimise oxygen pick-up during the process.

A challenge in keg cleaning arises from the fact that it must be ensured that the cleaning media reach the entire surface and all internals, for example, fittings. To achieve this, the spray pressure must vary so that the caustic/acid is distributed (high pressure) and flows down (low pressure).

With respect to sufficient cleaning and sanitation of the filling equipment, keg filling is a considerably low-risk procedure for beer filling. The highest microbial risk for kegged beer is usually to be seen at the point of sale (bars, restaurants, etc.) where an unsatisfactory hygienic state of the tapping equipment and low-level trained staff may be the root causes of a secondary contamination of the beer inside the keg. Microbial infection may grow through poorly cleaned hose connections back into the keg and lead to spoilage of the keg content. Assuring the microbiological quality of draft beer is discussed in Chapter 17.

16.2.3.6 Other containers

The group of special containers for beer packaging covers one-way kegs, wooden barrels, bag-in-box systems or growlers. Due to their physiological structure, one-way kegs usually neither withstand a hot caustic cleaning or a steam disinfection step. Therefore, sterilisation on the filling equipment is not possible, which increases the need for a highly hygienic filling process, for example, with a rinsing of the filler head between the single keg fillings with a disinfectant.

In wooden barrels, the natural structure of the wall will enable microorganisms to settle into even the smallest of structural gaps. In addition, modern hygienically designed equipment is most often not suitable for filling of wooden barrels. Therefore, manual handling and filling is necessary in such case individual steam treatment or disinfection with hydrogen peroxide or peracetic acid is possible.

Bag-in-box systems and other possibilities to transport larger volumes of non-carbonated liquids are not very common for beer. Later carbonation at the point of sales or when transferred to the next production step (filling in bottles, cans, etc.) will be necessary.

The absence of carbon dioxide increases the susceptibility of the product and therefore enhances the risk of contamination, especially with microorganisms that, because of their low tolerance of carbon dioxide, would not be considered as beer spoilage organisms.

Growlers as a take away solution for small craft- and pub breweries are more and more common in recent years. These growlers can be made out of glass, PET. Even cans or small stainless-steel containers are available as growlers. The microbiological safe treatment of such growlers is then strongly depending on the used material. Recommendation for the treatment is comparable to the industrial available containers. For example, glass-growlers can be handled as a nonreturnable bottle and rinsed with water or water treated with disinfectant like chlorine dioxide or peracetic acid. However, the main risk for contamination in growlers comes from the filling process as this is done mainly directly from the beer-tap. This underlines again the importance of a hygiene workaround with the tapping equipment as described in Chapter 17 and also limits the microbiological stability of such a package. As the growlers are designed as a 'take away' solution the consumer expectation in regards to shelf-life is also not that long and mainly in-between a few days or weeks.

16.2.3.7 Closures

All types of closures for small beer packages have direct product contact and have to be kept free of possible beer spoilage organisms. In most cases, the closures are transported from the manufacturer as bulk material in large boxes and sealed in plastic bags. Within the dry environment, no microbial growth is possible, except if, due to careless transport or storage conditions, the cardboard and plastic bag material gets physically damaged and moist.

Further risks of the closing procedure are described in Section 16.2.2. In case of approaches towards ultraclean or aseptic filling, the closures may be disinfected before being transferred to the closing machine.

16.2.3.8 Other packaging aids

Other packaging aids, for example, labels, glue, shrink films, carton boxes, trays and crates, are not considered microbial hazards. Nevertheless, a certain standard of hygiene is necessary to deliver an appropriate product appearance to the consumer, and application of a fungicide can be a prudent measure. Beer in packaging that has a dirty or even microbial spoiled look will not be accepted, although the packaged beer may be of best quality.

16.2.4 Water

Within the filling area, several uses of water are necessary. During cleaning and sanitation steps, water is the medium to transport the active ingredients to the surfaces where they are to be acting. All product lines and parts that have direct product content have to be purged with fresh water after cleaning to avoid any carry-over into the product. During product changes, water is used to push out the old product before the new product is run. In addition, a possible high-pressure injection directly brings in hot water to create a controlled overfoaming. All these processes carry a certain risk of microbial product contamination since the used water will have more or less direct product contact.

Due to the direct impact on the microbial state of the product, any process water that is used in the filling area has to be under strict microbial control. Next to a certain water treatment in terms of its mineral and other substance content, the microbial state should be regularly monitored. To reduce the microbial load of process (and product) water, several methods are available that shall not be further discussed here, for example, ultraviolet treatment, microfiltration, chlorine dioxide use, ozone use and many others.

To ensure safe use of the process water after treatment, the water supply pipes have to be part of the regular cleaning regimen. Spray nozzles as well as the high-pressure injection nozzle often show mineral clogging over time and therefore have to be cleaned regularly. Otherwise, the growth of biofilms may be supported, with the effect that microbes are spread over clean surfaces when fresh water is sprayed at the end of the cleaning cycle or may directly enter the container at the high-pressure injection nozzle.

16.2.5 Air

In the filling area, air is used mainly to run pressurised air valves. Next to needs of the valve function itself, the air should be sterile filtered so as not to bring any kind of microorganisms into the filler system. Although the microorganisms will not be able to reproduce in the air itself, the risk is high that, via the pressurised air channels, single cells may get to an environment that enables their growth. For example, in pressure valves of the filling tube, product residues appear during the filling cycles, which may provide the necessary nutrients to start biofilm growth.

16.2.6 Carbon dioxide

Carbon dioxide is used in beer filling to purge the containers. The main purpose is to reduce the air or oxygen content in the package before filling. For can filling, carbon dioxide is additionally used to purge the surface under the lid during closing. The used carbon dioxide gas has direct product contact and may thereby be a direct source of product contamination. The hygienic state of the carbon dioxide gas has to be of highest quality. Especially when carbon dioxide from fermentation is reused, a carry-over of single cells may occur. Usually, the technical needs of the carbon dioxide (purity) demand a certain washing and purification step that may be considered as a stop barrier for the carry-over of microorganisms. Nevertheless, all carbon dioxide pipes should be regularly cleaned to avoid any build-up of biofilms, and a sterile filter should be installed and maintained in the CO_2 Filler inlet line. Especially in the filling and closing machine itself, the formation of aerosols as well as back-pressure may lead to product contamination of the gas supply pipes.

16.2.7 Environment

The environment plays a very important part in terms of microbiological hazards in the packaging hall. Often the source for growth of microorganisms is the direct environment of the filling machine. All kinds of microorganisms may be brought into the direct environment of the filling machine. The transport may, for example, be supported by air flows caused by the fast-moving filling machine, natural air flows in the packaging hall, transport belts, or fork lifter or by operators.

Ideally, the air pressure in the packaging hall should be adjusted to a minimal overpressure. As a consequence, the natural air flow in the packaging hall will always be directed to the outside instead of bringing in possible contaminants with air from outside the packaging hall. Furthermore, the design of the packaging hall should avoid (open) doors or windows close to the filling machine. Ceilings, walls and floors are to be designed in a hygienically appropriate way. Hygienic design is not limited to machines and pipes with product contact but can be applied for the complete packaging hall.

Within the packaging hall, ideally the returned containers (not cleaned) should be strictly separated from cleaned containers and the filling area (wet part of the packaging hall). Any possibility for cross-contaminations should be minimised. Empty as well as filled containers have to be transported to and from the filler. Usually belt transportation systems are used. These transportation systems may bring in microorganisms and therefore have to be cleaned on a regular basis and disinfected constantly. The transport systems should be checked regularly for their hygienic status.

When the packaging process is running, the machines have to be operated and supported with the necessary packaging materials or packaging aids. Operators as well as, for example, fork lifters are moving close to the filling area and may bring in microorganisms from outside the building or another section of the production area.

To avoid or minimise contamination risks, an appropriate hygiene protocol should be followed when acting in the packaging hall.

16.3 Biofilm growth in the packaging hall

In modern microbiology, the biofilm is often referred to as the natural environment microorganisms. Next to the natural environment, technical equipment also may provide conditions that enable biofilm formation (Beckmann, 2010). In fact, many filling devices for beer and other beverages show unwanted growth of biofilms (Timke & et al., 2005). Points of high risk for the development and growth of biofilms can be found especially at the filling and the closing devices. Furthermore, the transport belts, filler housing, filler periphery, bottle washing machine or inspectors can be subject to biofilm growth (Back, 2003) (Fig. 16.1).

FIGURE 16.1 Massive biofilm in filling line.

Biofilms usually consist of water, microorganisms and extracellular polymeric substances (EPS). These EPS react with water to build-up hydrogels. The hydrogels form a slime that coats the microorganisms inside. Within that coating nutrients, metabolites or further substances may be present. The EPS consist of polysaccharides, proteins, lipids and nucleic acids (Beckmann, 2010; Szewzyk, 2003).

The build-up of a biofilm can be divided into several steps. In the first phase, a so-called conditioning film is evolving. The conditioning film is formed by irreversible adsorption of organic macromolecules, for example, proteins and polysaccharides, on suitable surfaces (interfaces). On these conditioning films, microorganisms may settle. At the beginning of the biofilm growth, species that show strong adhesion abilities and the possibility to excrete EPS predominate (Lipski, 2005; Flemming & Wingender, 2001). In the next step, the primary organisms allow the attachment of secondary microorganisms by co-adhesion. As soon as the first microorganisms have completely attached to the respective surface, the phase of microbial growth begins. The phase of microbial growth is dominated by the multiplication of the primary microorganisms. In addition, new microorganisms originating from the closed environment may attach to the growing biofilm.

A general characteristic for the phase of microbial growth during biofilm development is the start or initiation of release of EPS. Prerequisites for the growth as well as the EPS production are the availability of nutrients and water. The evolving biofilm is characterised by a three-dimensional structure.

Therefore, the following prerequisites for biofilm growth can be :

- Surface (interface).
- Sufficient water availability.
- Nutrients.
- Microorganisms.

The filling equipment and its environment provide suitable interfaces as well as microorganisms. When product (substrate) is filled in the filling equipment, the necessary water and nutrients are provided as well. For technical reasons (see Section 16.2.1), a certain spreading of beer residues during the beer filling process can hardly be avoided. Moving parts and residual moisture lead to the formation of aerosols (fine dispersed liquids in the air). The high turning speed of modern filling equipment additionally creates airflows and turbulences that bring the aerosols to practically all surfaces within the filling machine and, therefore, to the suitable interfaces for biofilm development.

Next to the product, residues in returned bottles, grease or lubricants can also serve as substrate for biofilm growth when transferred to suitable surfaces. The mentioned fluids may also serve for the build-up of a conditioning film.

The agglomeration in biofilms provides advantages over single-species colonies. A hydrogel matrix not only provides a certain protection from external stress, but it may also keep water and nutrients that may be degraded by exo-enzymes and metabolised. Several different species of microorganisms may be present in biofilms and create synergies. The respiration of aerobic microorganisms may produce spots of anaerobic conditions within a biofilm structure where anaerobic microorganisms are now enabled to survive and reproduce. Cells of microorganisms are able to communicate via signal molecules (quorum sensing), and genetic information may be exchanged (horizontal gene transfer). The biofilm structure is always in progress and change. Thereby, a permanent adaptation to the environmental conditions is possible.

The presence of biofilms automatically leads to higher risks of product contamination throughout the complete production process. Single cells or whole biofilm fragments may be transferred to other production areas as well. All moving equipment (belt conveyor, fork lifters, etc.), operators, spraying of fluids or airflow may spread the microorganisms and biofilms within the production area (Back, 1994). Especially, the filling area has a high-risk potential for secondary contamination. As mentioned before, the rotation of the equipment, together with high humidity and eventually elevated temperatures, adds up to produce very good conditions for aerosol formation and biofilm development. The risk of cells to enter and contaminate the open containers is high, and contamination reduction measures after filling are limited or unwanted. Next to the risk of contamination, the potential of microbial influenced corrosion (MIC) must be mentioned. Microorganisms that are organised in biofilms may be able to excrete substances that have corrosive properties. Parameters such as pH, redox potential or oxygen concentration may shift and enable the corrosion and damaging of the surface material. Due to the presence of certain microorganisms, for example, *Gallionella, Nitrosomonas* or sulphate-reducing organisms, the electro-chemical potential may shift significantly. Additional acid production may lead to the corrosion of metals, mineral-based substances or coatings. Assumptions relate up to 20% of corrosion damage on metals or other materials to microbial influenced corrosion (MIC) (Beckmann, 2010; Kubik, 2007; Flemming, 1995). However, recent research indicates that the risk of MIC from biofilms in beverage bottling plants is rather low compared to hygienic hazards (Fischer, 2021).

Although there has been a certain gain of knowledge on biofilms and their formation, the diagnosis and targeted counter measures are not very common in practice. The main reasons may be found in the high diversity of biofilm compositions and structures. Different organisms, interfaces and substrate-related aspects lead to differentiated and very specific biofilm structures. Depending on the environmental conditions, the same organisms may build-up a significantly different and unique biofilm structure.

The environmental conditions may vary substantially within a filling machine. Product specifications and changes, cleaning intervals, cleaning procedures and detergents are influencing factors.

The design of the filler defines possible weak spots, such as spray shadows, niches, etc. In addition, weather conditions, seasonal changes, operator hygiene and geographic details influence the growth and structure of microorganisms organised in complex structures such as biofilms.

A research project targeted the brewery-specific identification of microorganisms found in local biofilm structures. As the main inhabitants of the fully grown biofilms that were analysed, various wild yeasts were detected. In single cases, also lactic acid- or acetic acid-producing bacteria were detected (Pahl & Fischer, 2013).

Subsequent research focussed further on the role of non-acid-producing microorganisms during the development of biofilms. As primary organisms, strains of Acinetobacter and Pseudomonas were detected. The development of biofilms at surfaces with indirect product contact may be related to all kinds of ubiquitously existing microorganisms that are slime producing. In particular, Pseudomonas and Enterobacteria are mentioned. In addition, the yeasts *Rhodotorula glutinis* and *Cryptococcus albidus*, as well as the moulds *Geotrichum candidum* and *Aureobasidium pullulans*, are named (Back, 1994).

16.4 Minimisation of risks

16.4.1 General

Minimisation of risk may be achieved by two different approaches. On the one hand, the product stability or susceptibility may be increased or decreased. Processing a less susceptible product does mean reacting on a possibly inadequate level of hygiene to avoid the consequences. Increased levels of toxic or antimicrobial ingredients, for example, ethanol, carbon dioxide and hop bitter acids, may be a possibility or the use of food-grade preservatives such as dimethyldicarbonate or derivatives of sorbate and benzoate.

On the other hand, measures may be taken to avoid contamination in the first place. These measures include hygienic design of the equipment in use and its periphery. In addition, a suitable cleaning and sanitation protocol has to be in place, accompanied by appropriate training of the respective operators and regular revisions. Furthermore, regular control of the hygienic status is a useful tool to track possible weak points within the cleaning and sanitation protocols and to further improve these protocols.

16.4.2 Cleaning and sanitation

Biofilms are usually treated with biocides. However, the use of disinfectants (biocides) has several weaknesses. Even if the disinfection measures successfully inactivated the organisms, the biomass does not leave the system and may serve as nutrient for other microorganisms (Schulte & Flemming, 2006). The development of a new biofilm structure may result. Furthermore, the disinfectant or the kind of disinfection treatment has to pass through the hydrogel matrix without reacting with the matrix. If reactions take place, the result may be lesser disinfection properties of the reaction products or even the supply of additional nutrient to the microorganisms that are organised within the hydrogel matrix (biofilm) (Behmel, 2010).

Therefore, an intense mechanical cleaning success to bring out the organic material is as important as the disinfection or inactivation of microorganisms. Due to the design of the equipment, not all parts and places within the machines can easily be reached and cleaned. Efforts in engineering machines with the best possible hygienic design (Chapter 10) are a possible solution. Nevertheless, biofilms become visible to the human eye only after the structures reach certain size and cell density. Thus, surfaces that seem visibly to be clean are not necessarily free of contaminants and growing biofilms.

There is no general solution available for treating biofilm formation in the filling process; the individual conditions in a particular filling department must always be considered. No cleaning and sanitising strategy is available that would fit each filling machine with its individual risk potentials. The cleaning and disinfection procedures have to be tailored to the (local) demands of the existing equipment. Ideally, this would include a deep knowledge of the local microbial flora.

Within the beer and beverage sector, it is the manufacturer's responsibility to use best practice techniques to ensure a hygienically acceptable product quality. Surfaces, especially when regularly in indirect or direct contact with product, should be easily cleanable and suitable for disinfection. An application of a regular (every few hours) caustic and acid foam cleaning in change is a helpful tool to reduce biofilm formation right in the beginning and remove all sources of nutrient on a regular basis. Also, ensure an appropriate control of hygiene in the production facility by scheduled sampling of surfaces, equipment and product as well as by sufficient safety measures and microbiological control before products are released to the market.

16.4.3 Ultraclean filling

The trend towards beer-mix beverages, low-alcohol and alcohol-free beers is a worldwide driver of product innovations in the brewery sector. The low contents of natural preservatives such as ethanol or hop acids and the increased amounts of (fermentable) sugars have led to a higher microbial susceptibility of these products. If tunnel pasteurisation is avoided or is not possible (with plastic packaging), the filling process should occur under highly controlled hygiene conditions.

The so-called ultraclean filling technology was designed to handle the respective containers under more secure hygiene conditions than standard filler designs would allow. The filler is kept in housing, and sterile filtered air is used to achieve a constant overpressure within the filling compartment. Sanitising of containers and closures is also part of the approach to reduce the microbial risks of filling. In comparison to aseptic filling, the targeted cell count reduction is lower than the five logarithmic steps that are common for aseptic filling. A reduction of target organisms by three logarithmic steps is common for ultraclean fillers.

Aseptic filling is not common for beer or similar fermented beverages with elevated carbonation and lower levels than pH 5.

16.5 Future trends

The ongoing trend towards smaller breweries is also accompanied by new microbiological challenges. Production conditions, for example, narrow facilities, direct contact between areas such as brewhouse, bottling, storage and guest room make it difficult to ensure adequate production hygiene. Sometimes, untrained staff are used, who are not aware of the microbiological hazards of beer production. In addition, due to the variety of beers in these breweries, a much diverse microflora is present, which can lead to increased risks of contamination, for example, lactic acid bacteria in sour beer.

Trends, for the near future, of filling and packaging may be seen in two directions. On one hand, the markets often demand new, innovative products to accompany established brands. In terms of the filling process, the number of single stock-keeping units will increase, and more product or container change will challenge the packaging process and its efficiency.

On the other hand, the productivity of filling lines will undergo constant improvement with better-automated processes (cleaning, product changeover, etc.) and fewer or shorter downtimes. Hygienic design implementation, combined with high-throughput rates, have been the most important improvements over the past years. The biggest challenges for the coming years will be to become even more flexible and to further reduce energy and detergent consumption for a more sustainable process.

References

Back, W. (1994). Sekundärkontaminationen im Abfüllbereich. *BRAUWELT, 16*, 686–695.

Back, W. (2003). Biofilme in der Brauerei und Getränkeindustrie. *BRAUWELT, 24/25*, 766–777.

Beckmann, G. (2010). Biofilme- grenzflächig, grenzwertig, ausgegrenzt. *BRAUWELT, 28-29*, 863–867.

Behmel, U. (2010). Desinfektion Für Trink- Und Brauchwasser in Brauereien (Teil 1). *BRAUWELT, 11*, 314–317.

Fischer, J. (2021). Interaction between microbiology and packaging machinery—Investigations, effects, strategies. In *[Vortrag] BevTech, ISBTs 68th annual meeting—Sanitation and Microbiological Control Technical Committee, 07.05.2021.*

Flemming, H.-C., & Wingender, J. (2001). Biofilme—Die bevorzugte Lebensform der Bakterien. *Biologie in unserer Zeit, 31*, 169–180.

Flemming, H.-C. (1995). Biofouling und Biokorrosion—Die Folgen unerwünschter Biofilme. *Chemie Ingenieur Technik, 67*(11/95), 1425–1430.

Kubik, A. (2007). *Biofilme als mögliche Indikatoren für Umweltbelastungen.* Diss: Friedrich-Schiller-Universität Jena.

Lipski, A. (2005). Ein neuer Fokus bei der Bekämpfung von Biofilmen: Die Pionierorganismen auf gereinigten Flächen. *Brauerei Forum, 7*, 178–179.

Pahl, R., & Fischer, J. (2013). Substrate and environment specific biofilms in beverage filling plants. *Brauerei Forum International, 28*(9), 12–15.

Schulte, S., & Flemming, H.-C. (2006). Ursachen der erhöhten Resistenz von Mikroorganismen in Biofilmen. *Chemie Ingenieur Technik, 78*(11), 1683–1689.

Szewzyk, U. (2003). Biofilem—Die etwas andere Lebensweise. *BIOspektrum, 3*, 253–255.

Timke, et al. (2005). Community Stucture and Diversity of Biofilms from a Beer Bottling Plant as Revealed Using 16S rRNA Gene Clone Libraries. *Applied and Environmental Microbiology*, 6446–6452. Okt.

Further reading

Anger, H.-M. (2008). Aktuelles aus Betriebsrevisionen. Problembereiche und Lösungsansätze in der Brauerei unter besonderer Berücksichtigung der Problematik von Biofilmen - 58. Arbeitstagung des Bundes Österreichischer Braumeister und Brauereitechniker. *Brauwelt, 148*(49), 1485.

Back, W. (1994a). *Farbatlas und Handbuch der Getränkebiologie*. Nümberg: Fachverlag Hans Carl.

Back, W. (1994b). Sekundärkontaminationen im Abfüllbereich. *Brauwelt, 134*(16), 686–695.

Back, W. (2000). Alarm im Flaschenkeller. *Getränkeindustrie*, (4), 212–214.

Back, W. (2003). Biofilme in der Brauerei und Getränkeindustrie. *Brauwelt, 143*(24/25), 766–777.

Beckmann, G. (2010). Biofilme - grenzftächig, grenzwertig, ausgegrenzt. *Brauwelt, 150*(28/29), 863–867.

Behmel, U. (2010). Desinfektion Für Trink- Und Brauchwasser in Brauereien (Teill). *Brauwelt, 150*(11), 314–317.

Brenner, M.w., & Ifftand, H. (1966). Economics of microbiological stabilization of beer. *MBAA TQ, 3*(3), 193–199.

Buchner, N. (1999). *Verpackung von Lebensmitteln*. Heidelberg: Springer-Verlag Berlin.

Ceri, H., Olson, M. E., Morck, D. W., Read, R. R., & Buret, A. G. (July 29, 2003). *Method of growing and analyzing a biofilm. Patent US 6 599 714 B*.

Christiansen, K. (1995). The art and practice of aseptic packaging in a modern brewery. *MBAA TQ, 32*(4), 228–230.

Deutsche, Bundesstiftung Umwelt (2003). *Entwicklung innovativer Strategien zur effizienten und umweltschonenden Bekämpfung von Biofilmen in der Lebensmittelindustrie am Beispiel der Bierabfüllung AZ 13042*.

Dirksen, J. (2005). Monitoring and controlling microbial contamination in the becr filling area. *MBAA TQ, 42*(1), 39–44.

Evers, H., Schmitt, A., & Bechtluft, J. (2009). Tests on rinsing methods for non-refillable PET bottles. *BrewingScience, 62*(5/6), 44–53.

Flemming, H.-C. (1995). Biofouling und Biokorrosion - die Folgen unerwünschter Biofilme. *Chemie Ingenieur Technik, 67*(11), 1425–1430.

Flemming, H.-c., & Wingender, J. (2001). Biofilme - die bevorzugte Lebensform der Bakterien. *Biologie in unserer Zeit, 30*(31), 169–180.

Folz, R. (2010). *Geschmacks- und biologische Stabilität von Bier in KunststojJverpackungen (Diss.)*. Berlin: TU.

Folz, R., Hofmann, R., & Stahl, U. (2011). Impact of permeation of02 and CO2 on the growth behaviour of Saccharomyces diastaticus in Beer. *BrewingScience, 64*(5/6), 52–60.

Gappich, L., & Gipp, M. (June 13, 2002). *Protocol for making a simulated natural biofilm. Patent DE 699 00 632 T2*.

Hays, G. L., Kincaid, C. M., & Warner, R. C. (1964). Microbiological comparison of packaged pre-pasteurized, sub-micron filtered and post-pasteurized beer. *MBAA TQ, 1*(1), 27–35.

Kemmelmeyer, H. (2008). Clean operation. *Brewing and Beverage Industry International, 21*(5), 60–63.

Kubik, A. (2007). *Biofilme als mögliche Indikatoren für Umweltbelastungen (Diss.)*. Jena: Friedrich-Schiller-University.

Lipski, A. (2005). Ein neuer Fokus bei der Bekämpfung von Biofilmen: die Pionierorganismen auf gereinigten Flächen. *Brauerei Forum, 20*(7), 178–179.

Mamvura, T. A., lyuke, S. E., Cluett, J. D., & Paterson, A. E. (2011). Soil films in the, beverage industry - a review. *Journal of Institute of Brewing, 117*(4), 608–616.

Manger, H.-J. (2008). *Füllanlagenfür Getränke*. Berlin: Verlag der VLB.

Mehnert, W. (2014). ~eeping legionella and other germs in check. *Brewing and Beverage Industry InternatIOnal, 27*(2), 30–33.

Odebrecht, E., Schmidt, H.-J., & Franco, B. G. M. (2001). Untersuchungen zur mikrobiologischen Bewertung der Umgebungsluft in den Abfüllräumen von Brauereien. *Brewing_Science - Monatsschriftfür Brauwissenschaf, 54*(7/8), 159–164.

Pahl, R. (2009). Not without residual risk. *Brewing and Beverage Industry International, 22*(1), 33–35.

Papukashvili, N. (2009). *Entwicklung eines in vitro Systems zur Untersuchung der initialen Bakterienadhäsion an selbstanordnenden Monoschichten (Diss.)*. University of Hamburg.

Paradh, A. D., Mitchell, W.1., & Hill, A. E. (2011). Occurrence of Pectinatus and Megasphaera in the major UK breweries. *Journal ofthe Institute of Brewing, 117*(4), 498–506.

Priha, 0., Juvonen, R., Tapani, K., & Storgärds, E. (2011). Acyl homoserine lactone production of brewery process surface bacteria. *Journal ofthe Institute of Brewing, 117*(2), 182–187.

Reed, C. (2001). Sterile filling of beer. In J. Browne, & E. Candy (Eds.), *Excellence in packaging of beverages* (pp. 173–178).

Rumpf, A. K. (2009). *Effekte elektrischer Polarisation leitender Oberflächen auf die akterielle Primäradhäsion und Biojilmentwicklung (Diss.)*. University of Duisburg-Essen.

Schulte, S., & Flemming, H.-c. (2006). Ursachen der erhöhten Resistenz von Mikroorganismen in Biofilmen. *Chemie Ingenieur Technik, 78*(11), 1683–1689.

Szewzyk, U. (2003). Biofilme - die etwas andere Lebensweise. *BIOspektrum, 9*(3), 253–255.

Timke, M., Wang-Lieu, N. Q., Altendorf, K., & Lipski, A. (2005). Community structure and diversity of biofilms from a beer bottling plant as revealed using l6S rRNA gene clone libraries. *Applied and Environmental Microbiology, 7 J*(October), 6446–6452.

University Technologies International fnc. (April 15, 1999). *Vorrichtung zur Inkubation von Biofilmen. Patent DE 297 23 730 U 1*.

Van Den Bogaert, X., Hassan, S. E., & Quesada, E. (1968). Flash pasteurization and aseptic bottling. *MBAA TQ, 5*(4), 218–222.

Vavrova, A., Matoulkova, D., Balazova, T., & Sedo, O. (2014). MALDl-TOF-MS analysis of anaerobic bacteria isolated from biofilm-covered surfaces in brewery bottling halls. *JASBC, 72*(2), 95–101.

Wendler, K. (2001). Hohe Hygiene-Standards bei der Getränkeabfüllung. *Getränkeindustrie, 55*(1), 10–12.

Chapter 17

Draught beer: Hygiene, microbiology and quality

David E. Quain

International Centre for Brewing Science, School of Biosciences, University of Nottingham, Sutton Bonington Campus, Leicestershire, United Kingdom

'We are the Draught Beer Preservation Society'—from the album 'The Village Green Preservation Society' by The Kinks (1968).

17.1 Introduction

Concerns about the quality of draught beer are not new. In 1912, in the Journal of the Institute of Brewing, Mr. G.R. Seton noted that 'it is not possible to find a subject fraught with greater importance to the brewing trade than cellar management'. He further observed that 'the national beverage, as it is served over the counter of many of the public houses in England today, has not the flavour and appearance commensurate with the care bestowed upon its manufacture in the brewery, a fact that often leads the public into the mistaken notion that the beers of to-day are inferior to those of our forefathers'. Obviously not a shrinking violet, Seton hit home with 'it (is) difficult to understand why at the most critical point in its passage from the brewery to the consumer, viz., the public-house, beer is allowed to be treated under conditions which are in direct antithesis to those strictly enforced in the brewery' (Seton, 1912).

So, some 112 years on, with draught beer now predominately in keg rather than cask, Seton's remarks still hold true. Whilst long recognised, draught beer hygiene has received only sporadic attention over the years with publications in the 1950s on cask beer (Hemmons, 1954; Wiles, 1950) and, with the transition to keg, a comparative golden age in the 70 and 80s from the University of Birmingham (Casson 1982, 1985; Harper, 1981; Harper et al. 1980; Hough et al. 1976). Thereafter, the focus has moved to studies on the impact of line composition (Thomas & Whitham, 1996), the use of ATP bioluminescence to validate cleaning (Storgårds & Haikara, 1996) and use of enzymes (Walker et al. 2007) and ozone (Fielding et al. 2007) in line cleaning.

Presciently, a paper by Storgårds in 1996 was titled 'microbiological quality of draught beer—is there reason for concern'? There was and there is! This has driven the work reported here, from studies on the effective cleaning of tap nozzles (Quain, 2016); assessment of draught beer quality (Mallett et al., 2018); quality of lagers and ales in the on-trade (Mallett & Quain, 2019); assessment of biofilm formation by draught beer (Jevons & Quain, 2021); the spoilage of alcohol free beer and the implications for dispense (Quain, 2021); the identification of spoilage microorganisms in different draught beer styles (Jevons & Quain, 2022) and the spoilage of lagers by draught beer microorganisms (Quain & Jevons, 2023). Accordingly, this review is significantly different to that in the first edition (Quain, 2015) of *Brewing Microbiology*.

Brewing Microbiology. https://doi.org/10.1016/B978-0-323-99606-8.00016-X

17.1.1 Global beer market

China has been the biggest producer of beer since 2011. Production volumes in 2020 in China were 341,110 thousand hL (34 mhL) down 8.6% year on year due to the COVID pandemic (British Beer and Pub Association, 2022). Other members of the 'top 10' include the USA (with 61% of the Chinese volume), Brazil (39%), Mexico (37%), Germany (25%), Russia (23%), Japan (14%), Vietnam (12%), Poland (11%) and Spain (10%). In 2020, the UK is the 11th biggest producer with 9% of the volume produced in China.

17.1.2 Draught beer market

Beer is packaged in bottles and cans ('small pack', typically $\leq$0.5 L) or in stainless steel kegs and casks ('large pack', $\geq$20 L). Although both formats are available in the on trade (public houses, bars, hotels, restaurants, etc), the dispense of draught beer from kegs (and casks) predominates. Draught dispense offers 'economies of scale', is more profitable than single serve bottles or cans and stainless steel kegs are environmentally more sustainable. Brands are more visible on the bar with bespoke fonts or tap markers, which drive consumer recognition and purchase decisions.

The draught beer market varies widely by country, although this can reflect the difficulty in obtaining accurate market data on draught beer (no data is available for China, Brazil, Russia or Mexico). With this significant caveat, the top 10 markets by volume of draught beer are reported (Table 17.1) for 2018 and before the global COVID-19 pandemic. Compared to the data from 2013 (Quain, 2015), Ireland, France and the Netherlands are now 11th to 13th in the list and have been replaced by South Korea, Italy and Poland. The USA continues to dominate draught beer volumes with a notable increase in Spain. The major players for market share of draught beer range from 60% or more (Spain, Ireland) through 40% (UK, South Korea and Belgium).

However, in the UK, beer sales are in long term decline (Fig. 17.1). This reflects the demise of draught beer, which in 2000 accounted for 62.3% of the market, 55.2% (2005), 48.2% (2010), 45.1% (2015) and, post COVID-19, 27.7% (2020). Indeed between 2000 and 2020, there is a strong relationship between draught volume (x) and total volume (y) with a linear regression of $y = 0.8084x + 29{,}837$ and $R^2 = 0.916$. To a small extent, the decline in total volume has been offset by an increased contribution from bottled beer.

TABLE 17.1 Top 10 draught beer countries in 2018.

Country	Universe (000 hL)[a]	Market share (%)	Draught (1000 hL)
USA	249,874	17	42,479
Spain	39,889	67	26,726
UK	45,747	46	21,044
Germany	85,042	18	15,308
Japan	49,419	24	11,861
South Korea	23,568	40	9427
Italy	20,318	36	7314
Czech Republic	16,564	36	5963
Poland	39,601	14	5544
Australia	17,220	30	5166

[a]*Universe = production and import volumes—volume exported.*
Data from the BBPA Statistical Handbook (2022), pre-Covid 19.

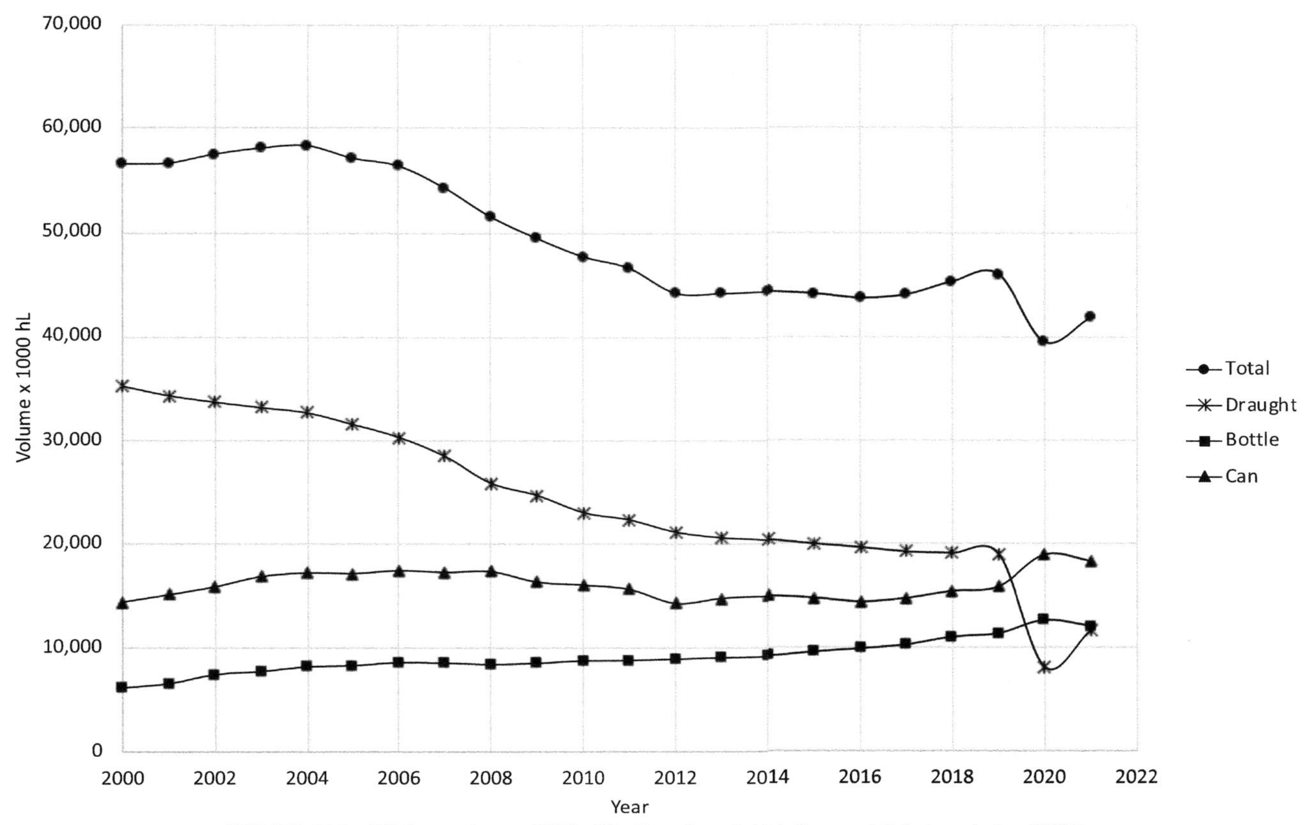

FIGURE 17.1 UK beer volumes (2000–21). *Data from British Beer and Pub Association (2022).*

The UK draught beer market (Fig. 17.2) is dominated by lager with about 60% market share. Not surprisingly, total draught (y) and lager (x) volumes mirror each other with a linear regression of $y = 1.8892x - 3283.4$ ($R^2 = 0.9769$). Since 2000, in a declining market, the share of stout has been stable at 6%–7%, with cask ale increasing marginally to 17%–18%. Keg ale has declined from a market share of 25% (2000), to 20% (2005), 16% (2010), 13% (2015) and 11% (2020). More recently, the COVID-19 pandemic, with the various protracted lockdowns in 2020/21, hit total beer volumes hard in the UK. Not surprisingly with the on-trade shut, draught beer volumes (Fig. 17.1) fell precipitously with an associated uptick in can and bottle volumes, which only marginally compensated for the decline.

Over the last 20 years, the annual UK volumes of draught beer have (pre-COVID) declined from 35 to 19 mhL. Whilst the further decline in 2020 to 12 mhL is directly associated with COVID, there are numerous other factors that have contributed individually and collectively to the demise of draught beer volumes. Table 17.2 updates the 'political', 'economic', 'social' and 'technological' (PEST) analysis first reported by Quain (2007). The further decline in draught volumes can be broadly ascribed to factors impacting on the pub/bar (COVID, Brexit, beer orders, drink driving legislation, move to food) and consumers (ageing demographic, cost of living, drinking less, drinking better, abstinence) or both (smoking ban, pricing). Although undoubtedly a contributor, the role of 'quality' in the downturn of draught beer is harder to quantify. As discussed in Section 17.4.1, many consumers recognise draught beer of compromised quality (aroma, flavour or clarity) and (i) soldier on, (ii) leave the beer/the pub or, preferably (iii) complain. This, together with other issues or complications, undermines the on-trade and drives consumers to purchasing beer from supermarkets.

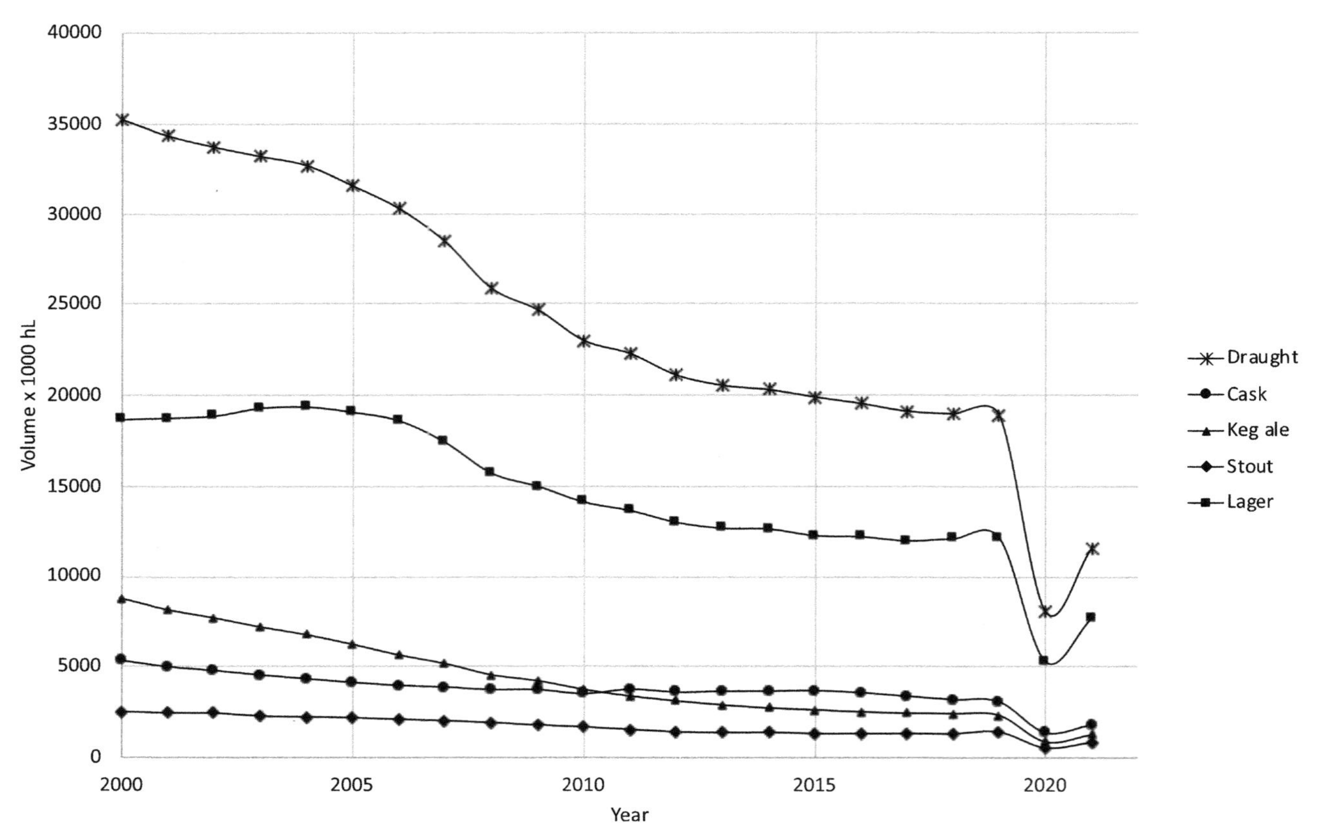

FIGURE 17.2 UK draught beer market (2000–21). *Data from British Beer and Pub Association (2022).*

TABLE 17.2 PEST analysis.

Political	**Economic**
■ 1989 beer orders—cutting the tie between breweries and pubs. ■ Drink driving legislation. ■ Taxation. ■ Licence reform. ■ Smoking ban. ■ Brexit. ■ COVID-19.	■ Off trade predominates. ■ Supermarkets—beer as a loss leader. ■ More food-led pubs than 'wet'. ■ PubCo's consolidation. ■ Cost of living crisis. ■ Jump in running costs and reduced hours/closure of pubs. ■ Price of a pint—'on' versus 'off' trade.
Social	**Technological**
■ Ageing demographic. ■ Wider choice of alcoholic drinks. ■ Drinking less, drinking better. ■ Demise of working men's clubs. ■ COVID—social distancing. ■ 'Sober curious' and growth/availability of alcohol-free beer.	■ Draught RTDs, cocktails, seltzers. ■ Decline of keg ale. ■ Growth of bottled beers. ■ On-trade quality. ■ End to end chilled systems. ■ 3rd party line cleaning.

17.1.3 Beer dispense

The transfer of beer from container to tap is fundamentally simple, exemplified by the now rarely seen direct 'gravity' dispense from cask. Building on this, the simple 'short draw' dispense includes a (disposable) dispense line, a cold cabinet for the keg which is positioned close to (or below) the tap. In the UK, 'short draw' dispense systems are found in low volume (<25 hL/year) licenced cafes and restaurants offering one or two beer brands.

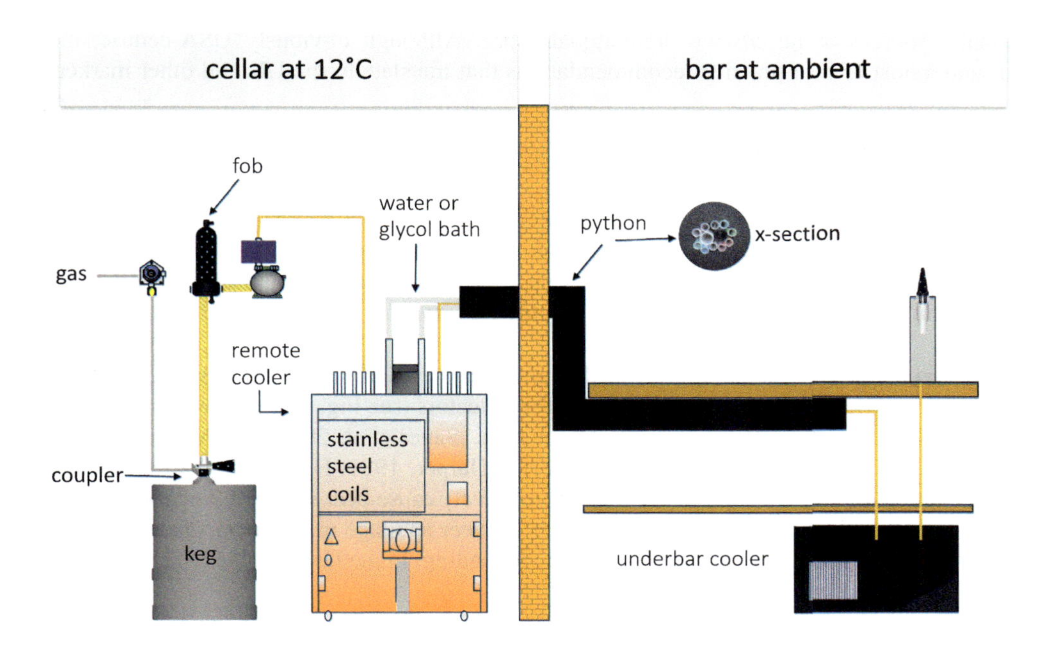

FIGURE 17.3 Standard draught beer dispense configuration in the UK.

Of the 46,800 public houses in the UK (British Beer and Pub Association, 2022), draught beer will typically be dispensed through 'long draw'/'long line' systems (Fig. 17.3). Arguably, cellars at 12°C are too warm for a market dominated by lager dispensed at 4°C (or lower). Accordingly, much of dispense technology is focussed on cooling beer (and keeping it cold) from cellar to tap. This is achieved by passing the beer through stainless steel coils (to increase residence time) in a 'remote cooler' containing ice/water (0°C) or glycol (ca. −2°C). From here, beer is pumped to the tap via dispense tubing ('lines', internal diameter 5.4–7 mm). The beer lines (10–16) are grouped together in an insulated 'python' (10–25 m long or more), which includes two wider bore lines that circulate chilled water or glycol. The proximity of lines to the circulating chilled lines determines beer temperature with lager lines around the core with ale and stout on the periphery. Supplementary trim cooling in the bar is often provided for lager brands with underbar coolers with further trace cooling in the dispense font.

Long draw dispense has innate complexity, with numerous opportunities for microbial contamination and colonisation, including the beer line (Section 17.2.2), keg coupler and fob (Section 17.2.3) and tap nozzle (Section 17.2.4). This is exacerbated by numerous snap-in connectors, flow restrictors and other fittings which have 'dead spots', and are difficult to clean. By comparison, the simplicity of short draw dispense designs out many of the hygienic 'hot spots' particularly where the beer line is short and replaced with each new keg.

Recent innovations in beer dispense by brand owners have focussed on extending cooling and providing cleaning services. These developments reinforce the status quo and do little to remove the complexity or the now inappropriate cellar storage temperature. Whilst repositioning cellars to reduce the distance to the bar is not practicable, building new public houses with cellars (or cold rooms) close to the bar would be a major step in simplifying beer dispense. A more radical change would be to store beer at colder temperatures than 12°C. This temperature harks back to the days when cask beer was the major player in the UK. With lager now being the predominate beer style, storing beer at 4°C (or lower) would simplify the cooling requirements of delivering cold beer to the bar. It would also significantly reduce the risk of microbial spoilage of beer in keg and in the FOB detector.

17.2 Draught beer: hygiene

The quality of draught beer reflects the frequency and, more importantly, the efficacy of line cleaning. Whilst the most significant hygienic intervention, other considerations also play their part including line composition, age and integrity of the line, cleaning of keg couplers and tap nozzles, cellar temperature and throughput of product.

Given the scale of draught beer dispense, resources are surprisingly few and far between. In the UK, the Cellar and Bar Manual has recently been published (Cask Marque, 2023) covering keg and cask dispense. A more exhaustive and detailed guide to the assurance of draught beer quality and dispense hygiene is the 'draught beer quality manual' (Brewers

Association, 2019) with additional resources at http://www.draughtquality.org. Although obviously USA-centric, the manual is a rich source of detail with a host of best practice recommendations that translate to the UK and other markets (see Section 17.6).

17.2.1 Line cleaning

Why—The dispense of beer is not an aseptic process. Accordingly, dispense systems contain microorganisms (Section 17.3.2) that can survive and grow (Section 17.3.1) on the residual nutrients in beer. Spoilage microorganisms reflect the environment and the account 'microbiome'. The direct contribution from beer is negligible as most keg beers are 'commercially sterile' (<1 cfu/L), although primary brewing yeast is contributed by cask and unfiltered/unpasteurised beers.

The diverse mix of yeast and bacteria exist in structured, organised communities or biofilms (Section 17.3.3) that are attached to the surfaces of dispense line, nozzles, connectors, couplers and fob detectors (see Fig. 17.4). However, biofilms need not be attached and are also recognised as free-floating aggregates or flocs (Sauer et al. 2022). With draught beer, biofilms have long been reported, particularly in dispense lines (Casson, 1985; Harper, 1981; Harper et al. 1980; Quain, 2015; Thomas & Whitham, 1996). As these microorganisms grow, they progressively cause beer spoilage (Section 17.3.1), which is characterised by diverse off flavours and aromas. The management of beer spoilage microorganisms in draught dispense systems is achieved by regular line cleaning that reduces the microbial loading. In the UK, line cleaning should be performed every 7 days (see 'when').

FIGURE 17.4 Examples of poor hygiene in draught beer dispense, (a) keg connector (Louis Arnold, Beer Line Wizard Ltd), (b) orifice plate in tap (Kevin Mutch), (c) beer fob detector (Louis Arnold, Beer Line Wizard Ltd), (d) cider fob detector (Justin Lawler, Qualflow Systems Ltd).

What—Almost universally, the cleaning of draught beer lines is performed using proprietary caustic-based solutions of sodium hydroxide or a combination of potassium hydroxide/sodium carbonate. The declared composition of line cleaning solutions can be frustratingly vague but more premium offerings can contain sodium hypochlorite as a biocide and, to improve performance, surfactants, sequestrants and chelating agents. 'Purple' line cleaning solutions contain potassium permanganate which, on pulling though, indicate the presence of biological material (biofilm). As a strong oxidising agent, the colour changes from purple (clean line) to, with a dirty line, blue/green to orange/yellow and clear. If, on pulling through, the solution is not purple, fresh line cleaning solution should be introduced.

Suppliers of line cleaning solutions flex composition to meet different needs. For example, excluding potassium permanganate or, more specifically, removing sodium hypochlorite to avoid concerns about the formation of medicinal TCP taints in dispense tubing (Parker, 2012). Indeed, it is noteworthy that the Brewers Association (Brewers Association, 2019) propose to 'never use solutions that contain any amount of chlorine for regular system maintenance'. The concentration of sodium (or potassium) hydroxide can be elevated for a 'deep clean' or 'bottoming out' of hard to clean or infrequently cleaned dispense lines. However, if used frequently or for a longer period of time, these solutions will damage/age the lines and provide places for microorganisms to attach and colonise (see Section 17.2.2).

How—The steps in the cleaning of cask and keg lines are pictorially described in 'the cellar and bar manual' (Cask Marque, 2023). The process consists of four significant steps (i) chasing beer out of the line by flushing with cold water, (ii) filling with cleaning solution (ca. 2% caustic), (iii) soaking by standing for 30 min and moving (pulling a pint or two) halfway and (iv) and flushing with water before replacing with beer.

Health and safety during line cleaning is of paramount importance. The solutions are corrosive, and their use requires the use of protective personal equipment (goggles and gloves) to assure health and safety. The water flush post cleaning is of especial importance as contamination of beer with line cleaning solution has been reported with awful consequences for a consumer. The use of pH paper to confirm the removal of line cleaning solution provides a simple but effective control point. Equally, the dilution and loss of colour after cleaning with a purple line cleaner provides similar assurance. Further, whilst eliminating any traces of line cleaner, the final rinse must also chase out microorganisms/biofilm detached during cleaning. Whilst the final rinse is the final step of line cleaning, it is important not to abbreviate it. Insight from monitoring beer quality using forcing (Section 17.4.3) in public houses suggests beer quality post line cleaning can decline if the water rinse is insufficient.

When—Of all the variables, meeting the required line cleaning 'frequency' is the most challenging. In the UK, the British Beer and Pub Association (2003) 'recommends the cleaning of beer and cider dispense systems at least every 7 days'. Whilst recognising that weekly line cleaning is a significant commitment, insight from a UK retailer (Quain, 2007) shows that account profitability (as volume growth) is demonstrable (+2%) with weekly line cleaning, breaking even at a frequency of 2 weeks but becoming negative (−2%) with cleaning every two to 4 weeks and increasingly negative thereafter.

A major constraint in the adoption of weekly line cleaning is the volume of beer and associated value that is lost on flushing with water, a concern which is exacerbated by the number of lines and their length. Accordingly, this drives a reduction in the frequency of line cleaning, which results in declining beer quality. Regrettably, the economic pressures identified in the PEST analysis (Table 17.2) increase the mindset that line cleaning wastes beer and 'throws money away'. However, a more enlightened approach on emptying a keg is to dispense the residual beer in the line before flushing with water and cleaning. This approach lends itself to retailers with large accounts with repeated banks/T-bars of beers and ciders. Brands are still then available but provide economies of scale in cleaning groups of taps.

Who—The cleaning of beer dispense lines is performed by publicans, managers, bar staff or external contractors. Operationally, the efficacy of manual line cleaning is strongly influenced by knowledge, skill and understanding. If frequency and other 'corners are cut', the hygiene of the lines and consequently beer quality will be compromised. Indeed, although central to maintaining draught beer quality, line cleaning is a tedious, lengthy process, which is exacerbated by the number of lines/taps. Given this, the growing trend to use third-party cleaning contractors is not surprising and can be recommended, particularly when cleaning with circulation.

Improving the process—The cleaning of draught beer lines is not effective and, in the UK, needs to be repeated every 7 days. Line cleaning reduces the microbial loading in beer (Boulton & Quain, 2001; Storgårds, 1996) and associated biofilm (Fig. 17.5), but regrowth is rapid with cleaning again required. Opportunities for more effective line cleaning would improve the removal of microorganisms, slow recontamination and, potentially, extend the cleaning cycle.

Line cleaning mirrors cleaning in place (CiP) in breweries and involves the balance of four quadrants—time, temperature, chemical action and mechanical action (Fig. 17.6). This 'Sinners Circle' was developed by Herbert Sinner in 1959. If one factor is reduced, other factors need to be increased to achieve the same efficacy of cleaning. In the case of

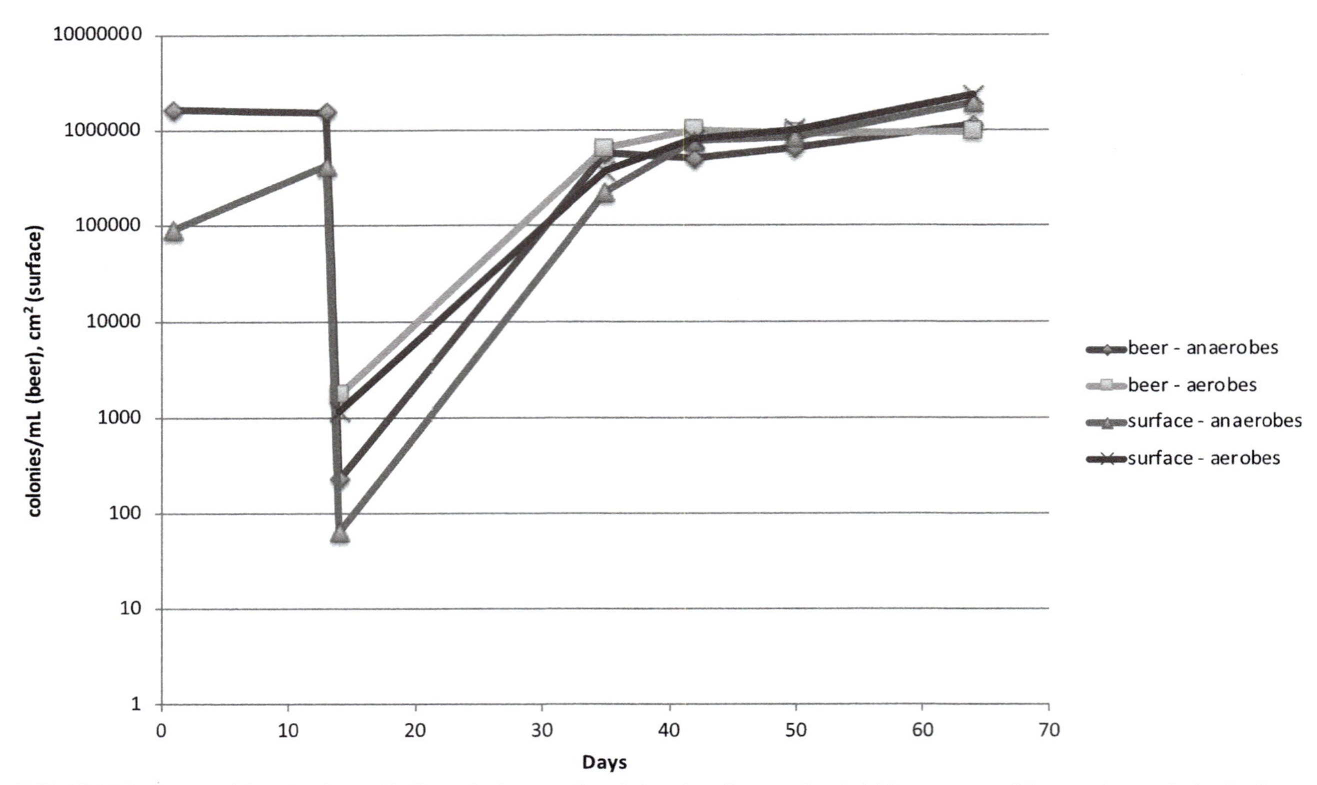

FIGURE 17.5 Impact of line cleaning on biofilm and microorganisms in beer in a dispense rig mimicking a commercial system in complexity, hardware, throughput and length. The microbial loading was monitored using standard methods together with biofilm washing and recovery from short segments of beer line. The data is the average of the results from two lines, sampled after the dispense system 'void volume' was dispensed and — during recontamination — flushed weekly with a total of 25 L of beer (phased to reflect weekly trading pattern).

FIGURE 17.6 Sinners circle (with permission from the Brewers Association)

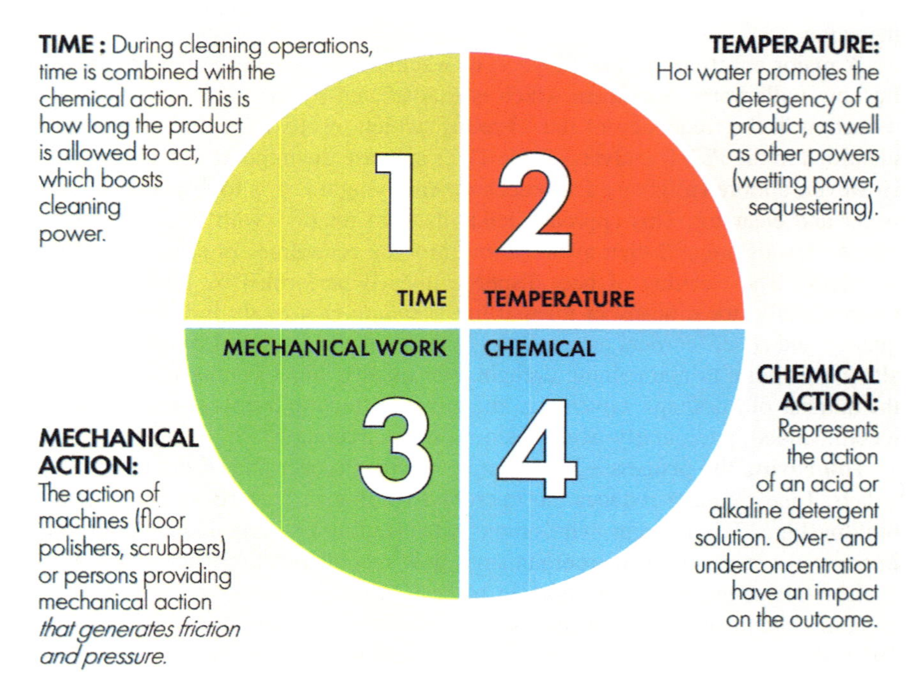

draught beer line cleaning, time is 30 minutes, temperature is 'mains' temperature (5−15°C), with chemical action from caustic (2%) but with little 'mechanical action' other than pulling through water/detergent/water.

Applying the Sinners Circle approach to the assurance of draught beer hygiene strongly suggests that line cleaning in the UK could be optimised to improve performance. The temperature of the line cleaning solution—which reflects the seasons—could increase and mechanical action introduced via recirculation of water and line cleaning solution. Such changes, particularly mechanical action, would improve the balance of Sinners Circle and, it is suggested, would result in a step change in the efficiency of cleaning and in enhanced draught beer quality for longer.

17.2.2 Line composition

The dispense line from keg to tap provides a substantial surface area for microbial colonisation. For example, the surface area of a 6 mm (internal diameter), 25 m dispense line is 471 cm^2 or equivalent of seven and a half A4 sheets of paper (Jevons & Quain, 2021). Accordingly, the dispense line has become an area of commercial innovation aimed at preventing the attachment and growth of microbial biofilms. Casson (1985), Thomas and Whitham (1996) and most recently, Heger and Russell (2021) evaluated the adhesion of yeast and bacteria to plastic lines. All three publications assessed three different line materials (i) polyvinyl chloride/vinyl, (ii) a nylon co-extruded/inner barrier layer and (iii) polyethylene/ polythene (Heger & Russell, 2021) either medium-density polyethylene (MDP) (Thomas & Whitham, 1996) or low-density polyethylene (LDP) (Casson, 1985). Using similar experimental approaches, the results were comparable for the three different line materials in the Casson (1985) report, but—in the more recent publications—the attachment of microorganisms to nylon lined lines was clearly lower than polyethylene lines with PVC lines supporting the greatest microbial attachment. This is nicely demonstrated (Fig. 17.7) in the work reported by Heger and Russell (2021).

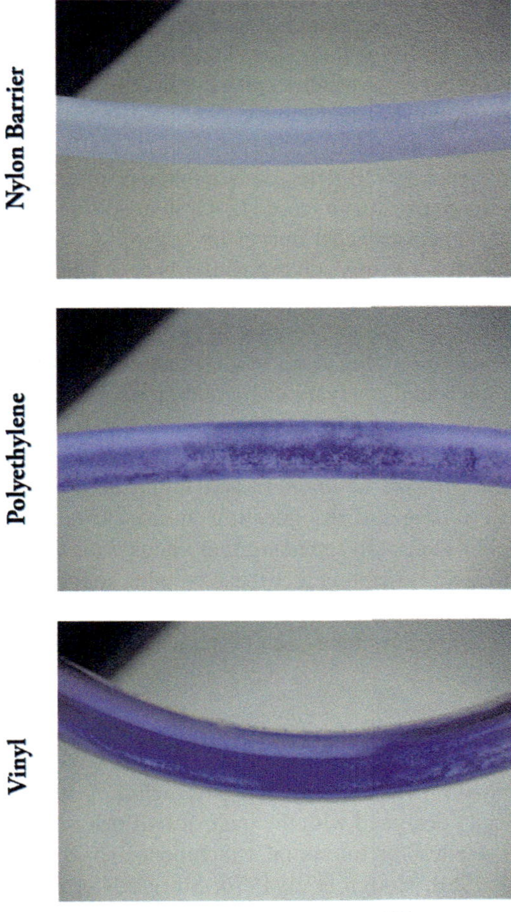

FIGURE 17.7 Crystal violet staining of biofilm in draught beer lines of different composition (Heger & Russell, 2021).

Dispense line tubing used in the UK is mostly MDP or increasingly variants of nylon lined MDP multilayer barrier tubing. Developments include the inclusion of antimicrobials (e.g., silver ions) and lines which are less permeable to gases and consequently reduce egress of carbon dioxide or the ingress of oxygen. These lines are also marketed (mostly without evidence) at reducing the growth of yeast and bacterial biofilms. However, changing the environment in the line may also change the microbial mix. For example, reducing the limited ingress of oxygen may reduce aerobes but select for colonisation by anaerobic microorganisms. Further, dispense lines are invariably in place for 10 years (or more), with erosion and pitting of the line surface through cycles of good and bad line cleaning together with the practice of 'bottoming out' (Section 17.2.1) by using stronger line cleaning solutions. Biofilm attachment would be influenced by surface smoothness, which is greater with nylon than MDP (Thomas & Whitham, 1996) but rougher in older lines from the trade than new (Hough et al. 1976).

17.2.3 Keg contamination (couplers + FOB)

As a potential 'seat' of contamination, the keg coupler gets less of a bad press than the tap (Section 17.2.4). However, the intimate connection between keg and coupler suggests that this interface is worthy of greater attention especially as the cellar environment is often challenging hygienically. As on changing containers, keg couplers are often placed on the floor and other dirty surfaces prior to being reconnected. Consequently, couplers become visibly dirty (Fig. 17.4a). Treatment of the keg head and keg spear with antimicrobial sprays or wipes before connection to the keg is recommended but regrettably is not common practice.

A major contender in the dispense hygiene stakes is the FOB detector (aka cellarbuoy, beer saver, FOB-stop, 'froth on beer', 'fobbing pot' or, less exotically, beer monitor) found in long line dispense systems. The 'foam on beer' detector's role is to eliminate the threat of foam entering the beer dispense line when the container is empty or being changed. The mechanism involves a float control, which in the absence of beer falls and blocks the beer inlet. On connecting to a new keg, the fob detector is bled to drain and then filled with beer enabling dispense to recommence. The bleeding process—though a thin plastic tube—should be performed hygienically into an appropriate receptacle. Regrettably, this is rare and beer is sprayed onto kegs and the cellar floor, providing sites for microbial colonisation. Over time this will encourage the growth of spoilage microorganisms which are distributed around the environment by movement of air (bar staff, cooling fans, open doors, container delivery and removal).

FOB detectors are a potent reservoir of microbial contamination (Fig. 17.4c and d). Each dispense line has a FOB which is positioned a metre or so from the keg. FOB detectors are made of plastic and provide an early visual indication of microbial contamination which will be more obvious in cool (12°C) than cold (3°C) cellars. This reflects the inefficiency of cleaning FOB detectors (especially the upper surfaces) during line cleaning.

Whilst 'commercially sterile' leaving the brewery, on broaching beer in keg becomes progressively contaminated with microorganisms, which can compromise quality with kegs that are on sale for weeks rather than days. The evidence for this is two-fold. Firstly, direct aseptic sampling of kegs and forcing from an account with hygiene issues showed the beers (2× lager, keg ale and stout) to be contaminated ranging from low ('excellent') through marked ('acceptable') to heavy (2× 'poor') (Quain, unpublished). Direct microbial analysis using multiplex PCR GeneDisc technology (Jevons & Quain, 2022) showed the beers ex keg to contain yeasts (*Brettanomyces* species, *Saccharomyces* including a diastatic *Saccharomyces* and *Candida/Pichia*) and bacteria (*Levilactobacillus brevis*, *Fructilactobacillus lindneri*, *Secundilactobacillus paracollinoides; formerly Lactobacillus brevis*, *L. lindneri* and *L. paracollinoides*, *respectively*). Other work—much repeated—demonstrated that post line cleaning and reconnecting 'in use' kegs, the dispensed beer was microbiologically compromised (Mallett and Quain, unpublished). On replacing these kegs with new unbroached kegs, the beer ex dispense was 'excellent' and clean microbiologically. Taken together these insights suggest that on dispense, a small volume of beer from the line enters the keg. The section of dispense line between the FOB detector and keg coupler is vulnerable to microbial contamination which is exacerbated by being at (12°C) and not chilled like the rest of the line in the python (Section 17.2.5).

17.2.4 Taps and nozzles

The tap/faucet is susceptible to contamination via a variety of routes including staff handling, air, dried beer and, ironically, cleaning. Further, taps are not hygienically designed with the inclusion of orifice plates (Fig. 17.4b), restrictors or diffusers, sparklers and flow straighteners adding sites for microbial attachment and colonisation. Accordingly, the tap is a rich source of contamination (Harper et al. 1980; Hough et al. 1976; Storgårds & Haikara, 1996).

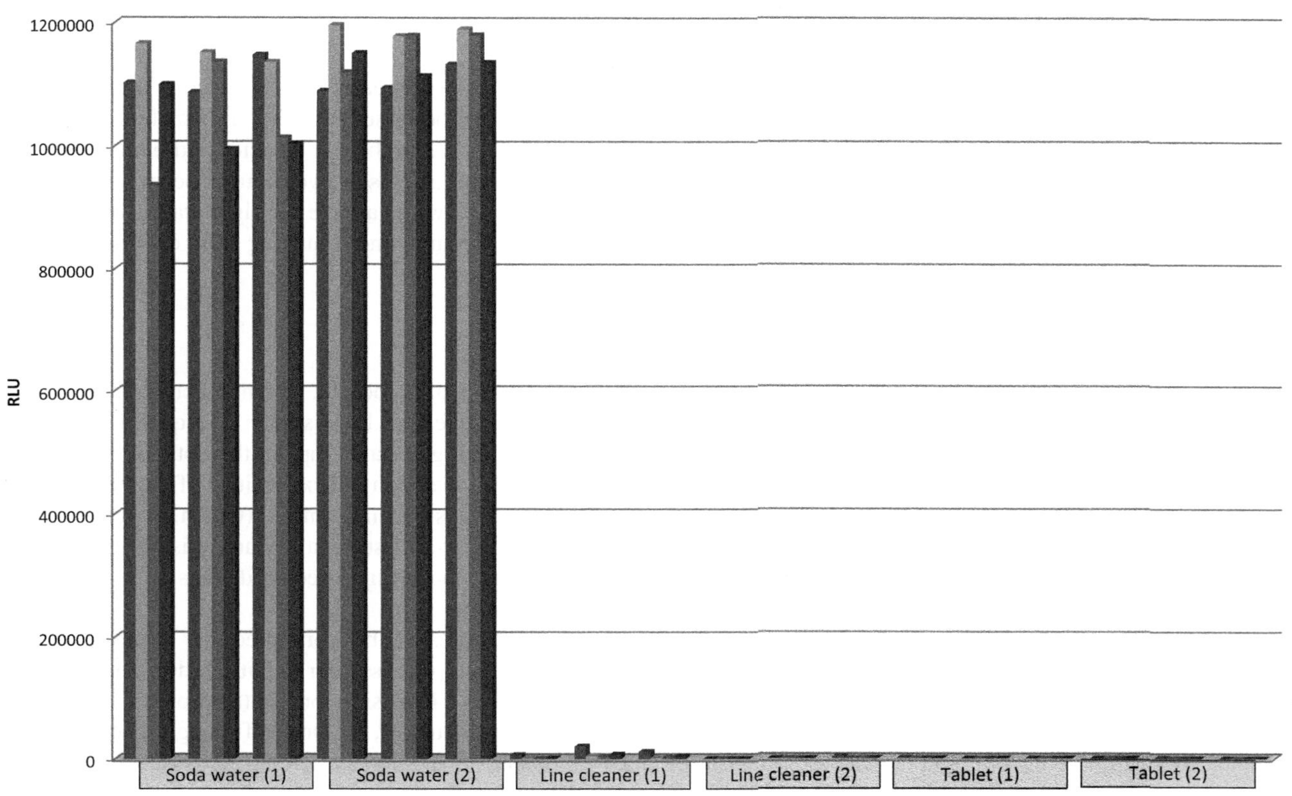

FIGURE 17.8 Impact on nozzle hygiene of soaking nozzles in soda water, line-cleaning solution and a solution of hypochlorous acid (sanitizing tablet) (Quain, 2016).

In the UK, tap spouts or nozzles fall into two categories—removable (either plastic or stainless steel) or one piece as part of the tap, which are cleaned, in situ by line cleaning. Removable spouts, which are either 'straight through' or containing diffusers and straighteners, together with sparklers are 'cleaned' using a bizarre daily ritual! At the end of a day's trading, all removable nozzles (and any internal plasticware) are soaked overnight—together with sugary cider and soft drink nozzles—in carbonated (soda) water. Next morning, they are (hopefully) rinsed with tap water and returned to a (different) tap. Given that the nozzles and associated plastic 'furniture' are soaked in beery/sugary water overnight at room temperatures, it is no surprise that this process markedly exacerbates microbial contamination. Indeed, after soaking nozzles, the liquid in the jug or pint glass can be visibly fermenting.

Soaking nozzles in soda water is bad practice, and more effective approach to their hygienic management has been investigated (Quain, 2016). Nozzles were incubated in spoilt beer to encourage the growth of microorganisms. Groups of contaminated nozzles were soaked overnight in soda water and solutions of diluted line cleaner, and hypochlorous acid (sanitising agent). The nozzles were recovered, washed with sterile water, swabbed internally four times (north, south, east and west) and the microbial contamination determined using ATP bioluminescence. Of the treatments, sanitising tablets achieved 'commercial sterility' and a 4-log reduction in bioluminescence compared with carbonated water (Fig. 17.8). Line cleaning solution was less effective and, from the perspective of health and safety, cannot be recommended as this should only be used in the cellar and not in the bar. However, sanitising tablets can be used safely and are highly effective in reducing the threat of the microbial contamination of the dispense line from dirty nozzles.

17.2.5 Temperature

Brand owners recommend different serving temperatures for draught beer from 2 to 6°C (lager), 4–8°C (stout), 6–12°C (ales) and 11–14°C (cask ale). Dispense temperature is flexed by the position in the python of dispense lines relative to circulating ice water/glycol coupled—for extra cold beers—with supplementary trim cooling under the bar (Section 17.1.3). Cask beers (and some keg ales) at 12°C are delivered via a dedicated mini-remote or an ale python cooler (Cask

Marque, 2023). Dispense temperature of lagers will invariably flex with throughput and will increase during busy trading with the erosion of the ice bank 'reserve' in remote coolers.

The mixed microorganisms found in draught beer are mesophiles that grow between 5 and 40°C at an increasing rate with higher temperature. Consideration of the temperature gradient across the process suggests the microbial risk is greatest in the cellar at 12°C. As noted in 17.2.3, microbial contamination can be visible in the fob detector and demonstrated in kegs, particularly with slow throughput. Comparison of the growth of microorganisms in two draught beers incubated for 8 days at 2 and 12°C. (Quain & Jevons, 2023) showed little or no growth at 2°C with variable but significant growth (20- to 50-fold greater) at 12°C. This is justification for reviewing keg storage temperature and switching from cellars at 12°C to cold rooms at 2–4°C (Section 17.6.3).

17.2.6 Throughput

The size of the bar is determined by the dimensions of the account and anticipated footfall. The number of dispense taps will reflect the demands of busy trading periods where banks of taps with repeating brands may be required. Invariably there will be a 'hotspot' where, in quieter trading, customers and bar staff will gravitate. Taps in the hotspot will be used preferentially and throughput through them will be greater. Conversely taps located in the extremities of the bar will only be used as trading picks up towards the end of the week. Quality from these taps is compromised (Quain, unpublished) by the lack of the movement, with the beer stagnating in the line. Although by no means straightforward, this can be managed by the lines being emptied over the weekend, cleaned and (preferably) gas blocked until required. This approach is also necessary in function rooms where beer dispense (and cleaning) is infrequent.

Having more taps on the bar than is required or 'overfonting' requires to be controlled by removal and—if required—the use of 'dummy' fonts. Too many taps would be expected to compromise beer quality. Indeed, an extensive study of draught beer quality in 57 accounts (Mallett & Quain, 2019; Section 17.4.3) using forcing (Section 17.4.2) showed that quality declined as the number of dispense taps increased across the bar. Further, the price of draught beer would be anticipated to impact on throughput. Indeed, analysis of price compared to quality by forcing showed that the most expensive price band had the lowest quality (Mallett & Quain, 2019).

17.3 Draught beer: microbiology

17.3.1 Spoilage

As has long been recognised (Mossel & Ingram, 1955), microbial spoilage reflects the growth of a small proportion of microorganisms exposed to a food or beverage. Further, the fingerprint of microbial spoilage is further defined by the composition of the food (termed 'intrinsic factors'). However, the threat of microbiological spoilage during production of beer is minimised or, preferably, eliminated by hygienic practices (closed vessels, wort boiling, CiP, cold processing, filtration, pasteurisation/sterile fill). Accordingly, packaged beer in cans, bottle or kegs is 'commercially sterile' with a microbial loading which is low/barely detectable (e.g., <1 colony/L) and consequently in-pack microbiological spoilage is a rare event.

However, as dispense is not an aseptic process, draught beer is exposed to microorganisms growing in multicellular biofilms (Section 17.3.3) as free-floating flocs or attached to internal surfaces. Accordingly, 'commercially sterile' beer from the keg is pulled through the dispense system becoming contaminated with yeast and bacteria. The number of microorganisms in draught beer will depend on the effective application of hygienic processes (line cleaning, sanitising keg couplers, tap nozzles), beer throughput and temperature (storage, dispense).

The German DIN standard (6650 Part 6) provides a useful guideline (Table 17.4a) for the number of microorganisms in different draught beverages including beer, wine and water (Anonymous, 2006). Using conventional microbiological plate methods, four bands are described with loadings of (i) < 1000, (ii) 1–10,000, (iii) 10–50,000 and (iv) > 50,000 cfu/mL. The standard suggests that cleaning is recommended if the count is 10,000 or higher, and loadings of 50,000 cfu/mL are considered unacceptable. Although laudable, this is not a practical approach for managing beer quality as microbiological testing requires three to 7 days incubation before quantifying the number of microorganisms. However, these bands are relevant and reflect the range of microbial loading found in draught beer (Quain, 2016, Table 17.4a).

Microbiological spoilage is unpredictable reflecting the mixed consortia of microorganisms, their interaction and the complexity of nutrients in beer. Candidate microorganisms, spoilage notes, threshold and concentrations in beer have been extensively reviewed (Spedding & Aiken, 2024). In terms of damage to product quality, archetypal indicators of microbiological growth and spoilage include turbidity/haze, acidification (lactic and acetic acids), phenolic aromas (medicinal, 'barnyard'), diacetyl/butterscotch and super attenuation. Other less common aromas include 'eggy' sulphur aromas (hydrogen sulphide), fruity characters (esters and higher alcohols) and 'sweaty socks' (short chain fatty acids). Unless characteristic of a beer type or style, such notes will never normally be found by consumers in beer in can or bottle but, unfortunately, can experience them in draught beer.

In the literature, beer is often described as being 'hostile' and 'inhospitable' to microorganisms and an unfavourable medium for microbial growth. This is suggested to be due to a lack of nutrients, the presence of ethanol (4−8% ABV), carbonation (ca. 0.5% w/v) together with antimicrobial hop compounds (17−55 mg iso-alpha acids/L) and an acidic pH (3.8−4.7) (Sakamoto & Konings, 2003). Other yeast metabolites, sulphur dioxide (5−30 mg/L) and organic acids (<500 mg/L), also contribute to the antimicrobial mix. Given this, it is surprising that such a robust beverage can be spoilt by some '30 microbial species belonging to several genera' (Suzuki, 2020). These include aerotolerant Gram-positive bacteria (*Lactobacillus*, *Pediococcus*), aerobic/microaerophilic Gram-negative bacteria (*Acetobacter*, *Gluconobacter*), facultatively aerobic yeasts (*Saccharomyces*), aerobic yeasts (*Brettanomyces*, *Candida*, *Pichia*, *Rhodotorula*) and strictly anaerobic bacteria (*Megasphaera*, *Pectinatus*). Fortuitously, there is no evidence that pathogens grow in conventional (containing ethanol) beers, although some (*Escherichia coli*, *Salmonella typhimurium*) may be able to survive in beer (Menz et al. 2011).

The 'cost of failure' by microbial spoilage of beer has not been determined but on a global basis is suggested to be 'high' (Suiker & Wösten, 2022). Although the threat of microbial spoilage is managed by 'building in quality' through the quality assurance of hygienic practices, quality control is still practiced either in house or contracted out through—for example—the Research Centre Weihenstephan for Brewing and Food Quality in Germany. Here, this long term programme of analysis of beer samples across the process and in final package reflects breweries across Europe and beyond. A report from the analysis of more than 13,000 samples processed by PCR between 2010 and 2016 noted over a 1000 bacterial 'incidents', dominated by identification of *Levilactobacillus brevis* and other *Lactobacillus* species (Schneiderbanger et al., 2018).

As noted above, the philosophy in breweries is to minimise or better, eliminate the threat of microbial spoilage. However, draught beer is inevitably exposed to microorganisms through dispense and, if not managed through hygienic practices, spoilage is increasingly real. Given this, it is of interest whether beers vary in spoilability and, if so, does this reflect differences in composition. This would be useful in identifying beers that are either more susceptible or resistant to microbial spoilage. Such insights could then be used to design beers that are less susceptible to spoilage.

Assessment of spoilage of different beers has received sporadic attention with reports using lactic acid bacteria (Dolezil & Kirsop, 1980; Fernandez & Simpson, 1995; Geissler et al. 2016). These 'challenge tests' involve the inoculation of low concentrations of pure cultures of *Lactobacillus* and *Pediococcus* species into different beers with incubation at 25°C for 60 or more days. Spoilage is determined visually or by measurement of absorbance and, where possible, related to compositional analysis of the beers.

Dolezil and Kirsop (1980) assessed the growth of 16 lactic acid bacteria in 31 beers (lager, ale and stout, 2.1−8.1% ABV). The extent of spoilage varied with three beers spoilt by all the inoculated bacteria whereas five beers exhibited no spoilage. Notably, susceptibility to spoilage showed no correlation with pH, specific gravity, amino nitrogen, fermentable sugars, colour or sulphur dioxide. However, no spoilage was observed with unpasteurised, bottle conditioned beers suggesting that resistance was associated with a heat labile yeast metabolite.

Fernandez and Simpson (1995) reported relationships between spoilage and beer composition in 17 lagers (3.5−6.1% ABV) challenged with 14 strains of lactic acid bacteria. Spoilage varied widely with two beers being spoilt by two bacterial strains, three by more than 10 with the remainder in the middle ground. Contrary to the conclusions of Dolezil and Kirsop (1980), relationships were found between susceptibility to spoilage and beer composition. A negative correlation was found between with the 'resistance to spoilage value' and pH, free amino and total soluble nitrogen, maltotriose and a positive correlation with colour and the undissociated forms of sulphur dioxide and hop bitter acids. Fernandez and Simpson (1995) suggest that the adaptation to growth in beer of their inocula may explain the different conclusion as the inocula used by Dolezil and Kirsop (1980) were not pre-grown in beer.

The involvement of beer pH in spoilage was also flagged in a more recent study (Geissler et al. 2016) which explored the utilisation of sugars, organic acids and amino acids in three beer styles (pilsner, lager and wheat, 5.1−5.5% ABV) by six species (26 strains) of lactic acid bacteria. Although spoilage varied with beer and the inoculated microorganism, spoilage was uniformly greater where the pH of the beers was adjusted to pH 5 from 4.3/4.4.

An alternative approach to evaluating the susceptibility of spoilage of lagers was reported by Quain and Jevons (2023). Rather than using pure cultures of lactic acid bacteria in challenge testing, mixed microbiota from different draught beers was inoculated under controlled conditions into 10 different lagers. After static incubation at 30°C for 4 days, the increase in absorbance at A_{660} was determined. The inoculum in this study were from four draught beer styles, cask ale and keg

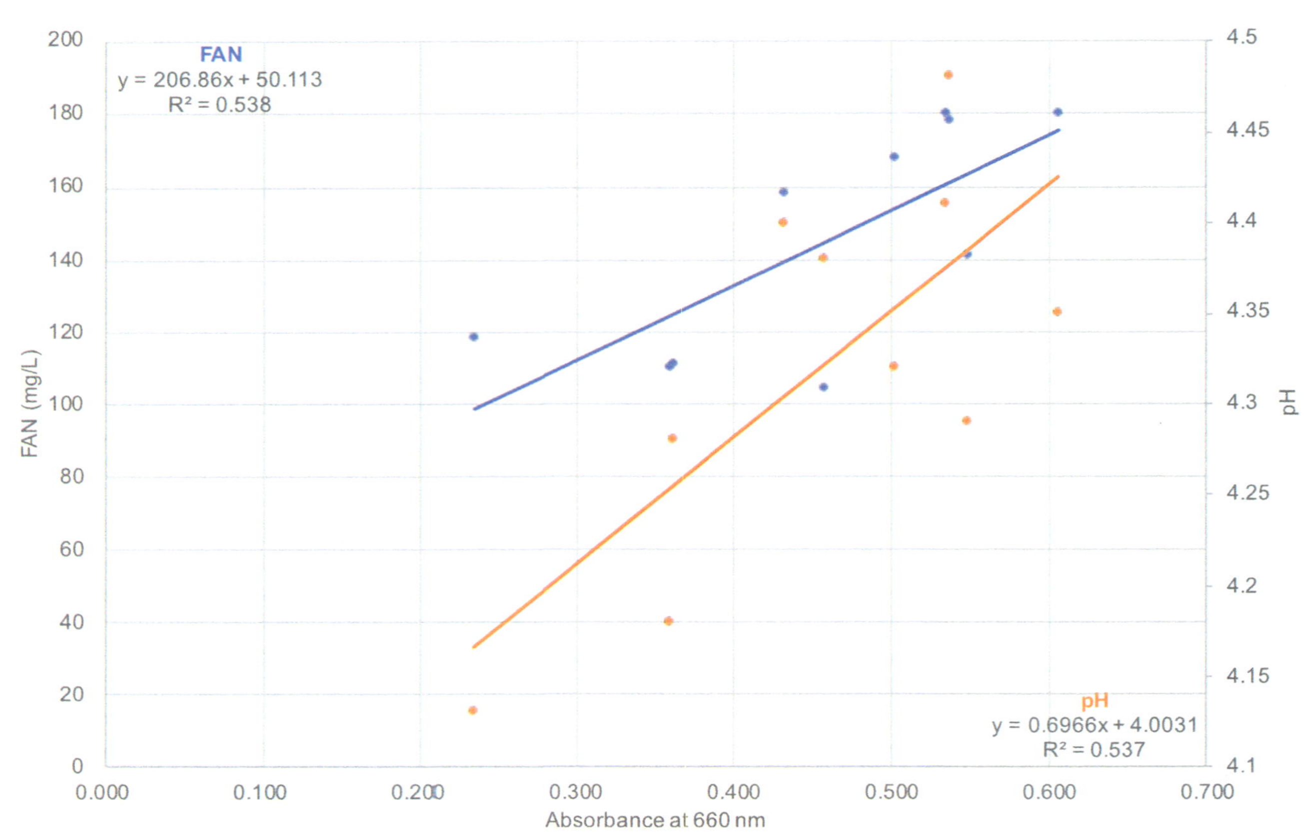

FIGURE 17.9 Spoilage of 10 lagers vs lager pH and FAN (Quain & Jevons, 2023).

lager, ale and stout—sampled twice in five different accounts in four different locations. Analysis of the culture-dependent microbiota of these samples revealed a core group of microorganisms, with some microorganisms associated with specific beer styles (Section 17.3.2; Table 17.3; Jevons & Quain, 2022).

With 40 challenge tests per individual brand, the 10 commercial lagers (4–5.1% ABV) varied three-fold in susceptibility to spoilage. This extends the observation of Cosbie (1943) who noted that 'beers vary considerably in their resistance to infection and the organisms which apparently flourish with ease in one beer appear to be suppressed in another'. The ranking of spoilage of the different lagers was broadly the same irrespective of the source of microorganisms (Quain & Jevons, 2023), suggesting that composition drives spoilage rather than the mix of microorganisms. In agreement with Fernandez and Simpson (1995), spoilage negatively correlated with pH and free amino nitrogen (FAN) (Fig. 17.9), suggesting that beers with a lower pH and FAN level provide some protection to spoilage.

Although 'a lack of nutrients' is one of the factors that are considered to limit the microbial spoilage of beer. Work by Hucker et al. (2017) reported that the addition of thiamine (vitamin B_1) or riboflavin (vitamin B_2) enhanced the growth in minimal media of spoilage microorganisms. Although, replication of this approach in challenge testing with the least and the most spoilable of the 10 lagers was without great effect, the addition of yeast extract resulted in almost a three-fold uplift in spoilage by microorganisms from draught beer. Accordingly, the least spoilable lager of the 10 brands switched on supplementation with yeast extract from 'excellent' to 'unacceptable' on challenge testing (Section 17.4.2). This suggests that the rate and extent of spoilage is impacted by the availability of trace nutrients that support the growth of diverse microorganisms in draught beer. These nutrients may not have been assimilated during fermentation, derived from yeast autolysis in fermenter or—together with bacteria – autolysis within dispense biofilms. Autolysis or 'self-digestion' of yeast results in the release of polysaccharides, organic acids, amino acids, nucleic acid degradation products and lipids (Hernawan & Fleet, 1995). It has been reported that extreme autolysis of yeast may increase the susceptibility of beer to spoilage, by lactic acid bacteria (Kulka, 1953).

Generally, the presence of ethanol in beer is considered to provide protection against microbial spoilage. Certainly, the converse—an absence of alcohol—is suggested (Hill, 2009; Kordialik-Bogacka, 2022; Suiker & Wösten, 2022; Suzuki, 2020; Vaughan et al. 2005) to make alcohol-free beers more vulnerable to microbiological spoilage. This idea has been investigated (Quain, 2021) by challenge testing of six alcohol free beers (AFBs, ≤0.05% ABV) and two low alcohol beers (LABs, ≤1.2% ABV) with mixed microbiota from draught beers. These AFB/LABs were markedly more spoilable (2.5×) than the two most and least spoilable lagers (4.4% ABV) reported above (Quain & Jevons, 2023). The enhanced spoilage of AFB/LABs correlated ($R^2 = 0.9554$) with the level of 'fermentables' (glucose + fructose + maltose) which were 4–9× greater than in the two premium lagers. It is noteworthy that challenge testing of the AFB/LABs supplemented with ethanol (2–8% ABV) resulted only in the modest inhibition of spoilage (collectively 24% at 8% ABV), suggesting that the compositional matrix of these products provides protection and/or supports microbial growth.

17.3.2 Biofilms

Line cleaning (Section 17.2.1) is front and centre of assuring draught beer quality. This process reduces but does not eliminate, biofilm on surfaces in the dispense system. Biofilms are communities of heterogeneous microorganisms (yeast and bacteria) that are organised in three dimensional structures. Biofilms are how microorganisms live in the 'real world' and importantly are involved in chronic medical infections and industrial biofouling. Indeed, since 1995, there has been exponential growth in research to better understand, manage—or better still—eliminate biofilm.

Contrary to earlier studies, biofilms are diverse in microbiota and physical size, structure and shape. The formation of biofilm is currently considered to involve three steps: (i) aggregation and attachment, (ii) growth and accumulation and (iii) disaggregation and detachment (Sauer et al. 2022). Importantly, this model extends the view that biofilms are attached to surfaces, but the microbial aggregates can also be unattached, free-floating flocs which aids dispersal and colonisation of new sites. Unsurprisingly, microbial metabolism within biofilms is unpredictable reflecting complex environmental gradients such as oxygen tension, macro and micro-nutrients, pH and extracellular enzymes. These multicellular communities are increasingly recognised as being highly sophisticated with cell-to-cell communication within genera ('quorum sensing'), which triggers a collective response to environmental stimulus across the population. Further, biofilms protect themselves against the wider world by laying down an outer glycoprotein/slime layer ('extracellular polymeric substance'). Pertinently, it is recognised that the loading of microorganisms in a biofilm greatly outnumbers the planktonic loading. Consequentially, measurement of loading in the aqueous phase (e.g., draught beer) is the tip of the microbial iceberg.

In brewing, biofilms have been described as 'the real enemy of process and product hygiene' (Quain, 1999). A review of biofilms in brewing (Storgårds & Priha, 2009) details reports of biofilms on conveyors, the filler and pasteuriser in the packaging of bottled beer. It is noteworthy that the primary bacterial colonisers (e.g., *Acinetobacter* spp.) are unable to spoil beer but are required for the involvement of beer spoilage and secondary colonising microorganisms. Similar observations have been reported with yeasts where attachment by *Candida* spp. provides the foundation for beer spoilage yeasts (*S. diastaticus*, *Brettanomyces*) (Suiker & Wösten, 2022).

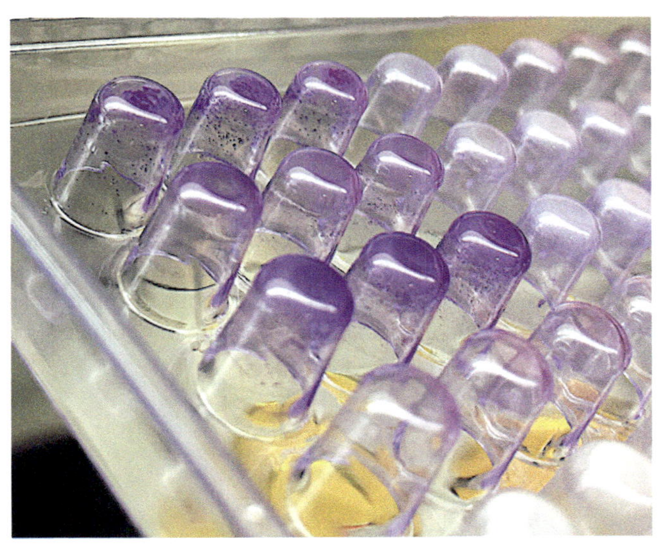

FIGURE 17.10 Crystal violet-stained biofilms in microplate wells. In triplicate from (top to bottom) – lager, stout, keg ale and cask ale (Jevons & Quain, 2021).

Dispense biofilms have long been recognised by dispense service technicians and can be readily seen by applying light behind an exposed beer line or FOB detector. Although predating the term 'biofilms', publications from the University of Birmingham reported the attachment of microorganisms to draught beer dispense tubing (Casson, 1985; Harper, 1981; Harper et al. 1980). The attachment of yeast and bacteria to the surface of dispense lines has been visualised using scanning electron microscopy (Casson, 1985; Harper, 1981; Storgårds & Priha, 2009; Thomas & Whitham, 1996; Walker et al. 2007) and confocal microscopy (Heger & Russell, 2021).

As noted above (Fig. 17.5), line cleaning is not fully effective, reducing but not eliminating attached or planktonic (floating) microorganisms in beer. On resuming dispense, both populations regrow to levels comparable with those before cleaning (Quain, 2015). This reflects in part the lack of mechanical action in line cleaning. However, there will be other factors—known and unknown—that contribute to the intractability of line cleaning. In addition to the surface of the line being damaged (Section 17.2.2), Casson (1985) suggested that the exopolysaccharide matrix that protects biofilm hardens and is not removed by line cleaning resulting in sites for microbial colonisation. To add further complexity, there is not a 'universal' biofilm in dispense systems but one with a diverse history, age and microbial composition (Walker et al. 2007).

Attachment of draught beer microorganisms to PVC has been assessed in 96-well microplates (Jevons & Quain, 2021). The method was found to be reproducible using crystal violet to visualise and quantify biofilm formed by microorganisms from four draught beer styles, keg lager, ale, stout and cask ale (Fig. 17.10). From this, several insights align with dispense parameters (dispense temperature, oxygen ingress) for the beers. For example, the rate of biofilm formation by microbiota from keg beers decreased with increasing temperature, whereas with microorganisms from cask ale, it increased. Oxygen enhanced biofilm formation with microorganisms from cask ale but not keg. Simulation of line cleaning in microplates with a proprietary alkaline solution failed to kill all microorganisms and the microorganisms regrew in all four beer styles. Further, the line cleaning process was increasingly ineffective with older biofilms.

17.3.3 Microbiota

Despite evolving methods of microbial identification, market decline and switch from cask to keg with the associated reduction in the availability of oxygen, the microbial genera (Section 17.3.1) reported in the last 50 years or so in draught beer (Quain, 2015) remain as Gram-positive bacteria (*Lactobacillus* and *Pediococcus*), Gram-negative acetic acid bacteria (*Acetobacter*, *Gluconobacter*) and wild yeasts (*Saccharomyces*, *Brettanomyces* and, less so, *Pichia*, *Candida*).

Recent studies (Jevons & Quain, 2021; Quain & Jevons, 2023) described in Section 17.3.1 reported the diverse culture-based microbiota found in four styles of draught beer sampled twice in five different accounts in four different locations. Using culture-dependent molecular techniques with identification of DNA sequences in forced samples (Table 17.3), 28 different microorganisms were identified at species level in the four beer styles, with *Brettanomyces bruxellensis*, *B. anomalus* and *Acetobacter fabarum* predominating in all four beer styles. Some microorganisms—*Pichia manshurica*, *Secundilactobacillus paracollinoides*—were exclusive to lager with *Rhodotorula mucilaginosa* only found in stout and keg ale. However, the advent of holistic, culture-independent methods (removing selection) will invariably identify microorganisms that are new to draught beer. For example, using such an approach, diastatic strains of *S. cerevisiae* (Andrews & Gilliland, 1952) were found in draught lager, keg and cask ale (Jevons & Quain, 2021).

It is noteworthy that the microbiota in forced samples of draught beer sampled on two occasions four or so weeks apart were found on 49 occasions. This suggests that the microbiota of draught beer is consistent, presumably reflecting beer composition and the microflora in the biofilm attached to the dispense line. Indeed, using Principal Coordinates Analysis to compare the two rounds of challenge testing of the 10 lagers with microbiota from the four beer styles revealed four independent clusters (Fig. 17.11) that were separated by style.

It is not perhaps surprising that of the 10 most abundant microorganisms found in draught beer (Table 17.3), nine have been reported in the Belgian sour beer Lambic (Jevons & Quain, 2021). There are parallels between the spoilage of draught beer and Lambic, produced by spontaneous fermentation over 3 years with microorganisms from the air and surfaces in wooden barrels in which Lambic is matured. Perhaps semantics, but both are products of microbial spoilage albeit 'good' spoilage (Lambic) or 'bad' (draught beer).

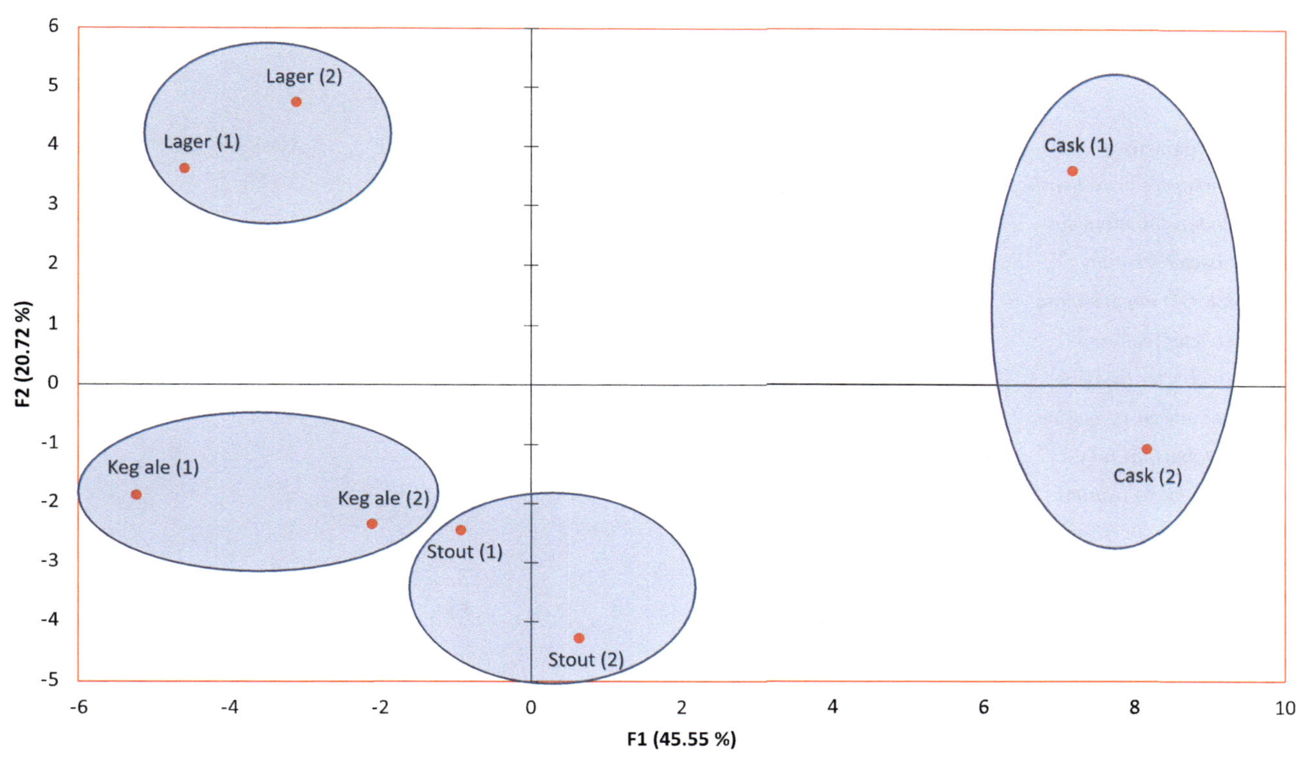

FIGURE 17.11 PCA plot of spoilage of 10 lagers by paired samples from 20 accounts (Quain & Jevons, 2023)

17.4 Draught beer: quality

17.4.1 Consumers

Draught beer is more expensive than beer in the off trade. Poor or 'so-so' quality coupled with high purchase price, are two of many drivers for consumers to switch from the on-trade to drinking beer in small pack at home (Table 17.2). Regrettably quality issues are all too common with draught beer, which in the consumer's eyes, reflects initially on the brand and then the account. The consumer experiences several cues—taste, aroma, visual, emotional, environmental—and they are adept at picking up things that are not (in their experience) right! This can include 'temperature' particularly as brand owners and retailers talk-up 'cold' dispense. Although, in the case of flavour and aroma, they are unlikely to articulate 'diacetyl', 'phenolic' and 'lactic' but may spot 'vinegar' or, more likely, that the beer is 'off'. Appearance though is a little more straightforward especially if the beer is fobbing or flat. Beer in a dirty or wrong branded glass is a frustrating and irritating fault, which undermines the 'quality' offer! In terms of clarity, issues are invariably more obvious in lager and ales where haze readily confirms a problem. That said, clarity is not always a good measure of compromised quality *ex* dispense with wheat beers, dark ales, stouts, porters and the growing category of 'murky' unfined or unfiltered beers.

17.4.2 Microbiological testing

The isolation and quantification of microbial colonies on agar plates has long been the standard method for assessing the microbiological status of wort, beer and other liquids in the brewing process. Indeed, plate tests have used to quantify microorganisms in draught beer (Table 17.4a). The approach takes time (days), is tuned to be selective for specific types of microorganisms (therefore missing some) and provides a 'solid' environment for microorganisms from a liquid. Further, quantifying the number of colonies can be 'hit and miss', depending on sample dilution with some viable microorganisms being dormant or non-culturable and fail to grow on agar plates.

Sampling draught beer in the on-trade and assessment on agar plates requires consumables and technical expertise. The results are retrospective and as a quality control tool not actionable as they reflect beer quality three (or more days) ago. The DIN standard (Section 17.3.1) provides guidelines for acceptability (or not) of microbial loading of beverages (including beer) which align with levels reported in draught beer (Table 17.4a).

TABLE 17.3 Microorganisms found in different draught beer styles.

	Beer style				
Microorganism	Lager (SL3)	Stout (ST1)	Ale (KA1)	Cask ale (SC1)	Total
Brettanomyces bruxellensis	19 (5)*	18 (1)	10 (2)	30 (5)	77
Brettanomyces anomalus	13 (3)	13 (3)	32 (4)	15 (2)	73
Acetobacter fabarum	16 (2)	15	8 (2)	16 (2)	55
Rhodotorula mucilaginosa	–	13	24 (2)	–	37
Acetobacter malorum	2	6	2 (1)	19 (4)	29
Gluconobacter oxydans	11 (3)	7	1	5	24
Saccharomyces cerevisiae	2	3	8 (1)	4 (1)	17
Levilactobacillus brevis	13 (2)	2	–	1	16
Saccharomyces uvarum	3 (1)	1	5	–	9
Pichia membranifaciens	2	1	2	1	6
Leuconostoc fallax	3 (1)	2	–	–	5
Acetobacter lovaniensis	–	1	3	1	5
Secundilactobacillus paracollinoides	4	–	–	–	4
Acetobacter pasteurianus	–	–	–	4	4
Pichia manshurica	4 (1)	–	–	–	4
Acetobacter cerevisiae	–	1	1	1	3
Acetobacter persici	–	–	3	–	3
Gluconobacter frateurii	2 (1)	1	–	–	3
Acetobacter sicerae	2	–	–	–	2
Acetobacter tropicalis	2	–	–	–	2
Candida boidinii	–	1	–	–	1
Saccharomyces bayanus	–	1	–	–	1
Pichia fermentans	–	1	–	–	1
Acetobacter estunensis	–	–	–	1	1
Acetobacter indonesiensis	1	–	–	–	1
Gluconobacter albidus	1	–	–	–	1
Pediococcus damnosus	–	1	–	–	1
Weissella cibaria	–	1	–	–	1
Total	100	89	99	98	386

From Jevons and Quain (2022).
Indicates the number of occasions where the microorganism was found in both samples from the account.

17.4.3 Forcing

The use of microbiological counts to measure draught beer quality is difficult and has limitations (Table 17.4a). An alternative approach uses 'forcing', first used in breweries in Burton-on-Trent in the 1870s to assess beers produced in winter and spring for sale in summer. Here, samples of beer from racking tank (or better) casks were stored 'under such conditions of temperature as would hasten the development of any of the adverse bacterial changes to which the beer was liable when stored under the ordinary conditions which rule in practice' (Brown, 1916). After forcing for up to 3 weeks at

TABLE 17.4A Microbiological testing of draught beer—agar plates.

Insight	Comment
Long established analytical methodology for quantifying microbiota across the brewing process.	Requires skill, training, laboratory consumables and hardware (incubator, anaerobic jars etc).
Selective agars (often containing inhibitors—cycloheximide, copper, etc) are used to quantify different groups of microorganisms ('aerobes', 'anaerobes', 'wild yeast' etc).	Selective agars select some microorganisms (and 'miss' others). Difficult to differentiate beer spoilage from environmental microbiota.
Quantifies the number of microorganisms in the sample.	Incorrect dilutions can result in inaccuracy (confluent colonies or 'too numerous too count').
German quality standard—Deutsches Institut für Normung (DIN 6650)—'dispense systems for draught beverages'- published in seven parts. Specifically, part 6 of the standard covers 'requirements for cleaning and disinfection'. Generic for draught beverages (beer, wine, carbonates etc.) using plate tests (which are not defined), four categories of microbial loading (colonies/mL) are described, (a) < 1000 as a 'positive result'; (b) 1–10,000 as 'acceptable'; (c) 10–50,000 'cleaning is necessary'; and (d) > 50,000 as 'unacceptable.	Results from commercial accounts suggests good quality beer contain around 1000 colonies per mL of beer. Loadings can be lower (<100/mL) or substantially higher (>10,000/mL) (Boulton & Quain, 2001; Quain, 2015; Storgårds, 1996, Storgårds & Haikara, 1996). Subsequently, a survey of 12 licenced premises (accounts) showed aerobes to range (per mL) from <1000 to >50,000 with anaerobes from <1000 to <50,000 (Quain, 2016). Similar results were reported for 52 samples from 24 accounts (Mallett et al. 2018).
Takes time for microbial colonies to grow at 28–30°C.	Aerobes = 3 days, anaerobes = 7 days.
Microorganisms are removed from their environment to grow on solid, selective agars. Conditions do not reflect those of draught beer dispense.	Dilution of the sample for countable counts removes any cell-to-cell trading of nutrients and communication (quorum sensing). The growth of nutritionally fastidious microorganisms and viable but non-culturable organisms will be compromised.
Microorganisms can be identified by recovery of the colony.	Traditional methods use microscopy, assimilation of sugars, growth on different sources of nitrogen or, increasingly, DNA based methods.
Sample may not be representative.	'First runnings' from dispense taps will have a higher microbial load. Ideally, sample when the line volume has been pulled through (account open for some hours and busy). Clumps and flocs of microorganisms can compromise obtaining a homogeneous sample to process.

TABLE 17.4B Microbiological testing of draught beer—forcing.

Insight	Comment
Developed in the 1870s by Horace Brown to predict the microbiological stability of cask beers brewed in the Autumn and Spring for sale in the summer. Samples were stored at 24–27°C for 10 days to 3 weeks and examined for changes in clarity and flavour, specific gravity and acidity. Any deposit was quantified and examined microscopically.	Beers were stored 'under such conditions of temperature as would hasten the development of any of the adverse bacterial changes to which the beer was liable when stored under the ordinary conditions which rule in practice' (Brown, 1916). A decision can be taken as to 'which beer will stand the longest storage, is best for bottling and which, on the contrary, is best to send out without delay' (Bailey, 1907). The forcing approach continues to find application with the assessment of the 'spoilability' of (alcoholic and non-alcoholic) beers.
Draught beer samples (250 mL) in sterile containers are kept cold and processed within 8 hours. After thorough mixing, 2 × 25 mL are transferred to plastic universal bottles, the cap located on top (but not tightened, to allow gas transfer) and incubated statically at 30°C for 4 days. Cycloheximide (4 mg/L) is added to cask, unfiltered and unfined beers to suppress the growth of primary *Saccharomyces* yeasts. Forced samples are thoroughly mixed by inversion. The absorbance of the samples was measured (in duplicate) at 660 nm with the dispensed beer and incubated. The increase in absorbance is used to classify the samples into four bands: A (0–0.3), B (>0.3–0.6), C (>0.6–0.9) and D (>0.9). The change in absorbance reflected the microbiological 'quality' at dispense, such that the A category ('excellent') with relatively little change in absorbance was superior to B ('acceptable'), which in turn was better than C ('poor'), with D being of 'unacceptable' quality.	Method development and details reported in Mallett et al. (2018). Quality index (%) $= \frac{\Sigma\ quality\ score}{number\ of\ samples \times 4} \times 100$ The 'Quality index' calculated from the sum of the individual scores for each quality band (where A = 4, B = 3, C = 2, D = 1) divided by (number of samples × 4) × 100. The method has been used in two PhD projects (Jevons, 2022; Mallett, 2019) and publications (Mallett et al., 2018; Mallett & Quain, 2019; Jevons & Quain, 2021; Quain, 2021; Jevons & Quain, 2022; Quain & Jevons, 2023, Werner et al. 2023).
Although retrospective, the method selects for microflora that can grow in/spoil beer.	The environment during forcing with a mixed microflora gradually changes and becomes a 'better medium for growth, allowing development of some organisms at the end of the forcing period which were initially incapable of growth' (Kulka, 1960).
Takes time.	Incubation for 96 h.
Sample may not be representative.	As above for 'counts on agar plates'.
Population heterogeneity can compromise obtaining a homogeneous sample to measure turbidity.	On mixing/resuspension, microorganisms can be adhesive (attaching to the universal surface), sedimentary/flocculent (dropping out) or trapped in the foam.
Forcing does not reflect the conditions of draught beer dispense.	Static, oxygen ingress from headspace, temperature.

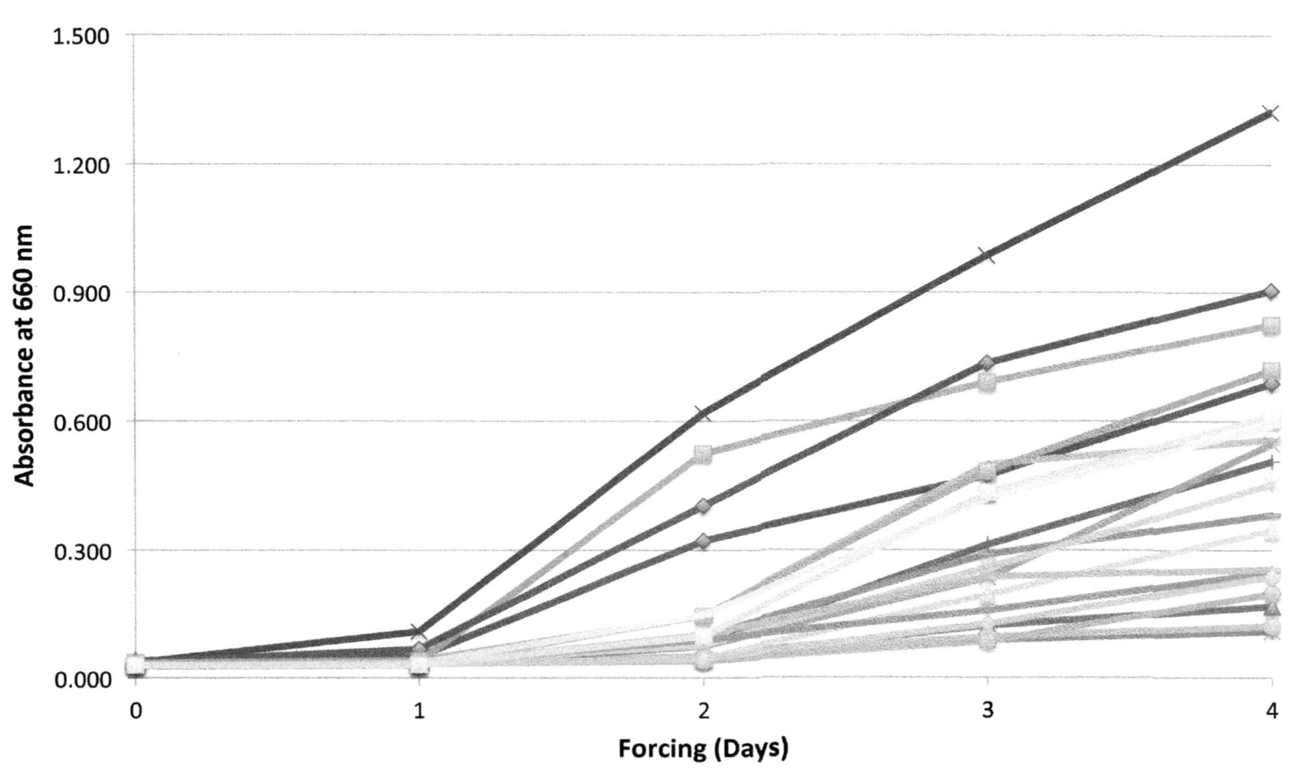

FIGURE 17.12 Changes in absorbance of draught beer samples on forcing at 30°C (Mallett et al, 2018).

24–27°C, a decision can be taken as to 'which beer will stand the longest storage, is best for bottling and which, on the contrary, is best to send out without delay' (Bailey, 1907).

The method was reimagined for assessing the quality of draught beer (Mallett et al., 2018) by incubating beer statically (25 mL) at 30°C for 96 h (see Table 17.4b for details). Microbial growth—and, retrospectively, beer quality—is measured by determining the increase in turbidity (A_{660}). Spoilage of good quality draught beer with a low microbial loading is minimal so the increase in A_{660} is small. The more contaminated the beer the greater the increase in turbidity (Fig. 17.12). Depending on the increase in turbidity post forcing, four quality categories are recognised ranging from 'excellent', 'acceptable', 'poor' to 'unacceptable'.

The upsides and downsides of using forcing for the measurement of draught beer quality are reviewed in Table 17.4b. Importantly, compared to agar plates, the indigenous microorganisms are not subject to selection on a solid menstruum but allowed to grow in situ in beer. Accordingly, forcing reflects the microbiological quality retrospectively of beer at dispense, enabling tracking over time and a cumulative 'quality index' (Table 17.4b) for individual brands or collectively for an account (Jevons & Quain, 2022; Mallett et al., 2018; Mallett & Quain, 2019). This is a useful metric, where if all samples are measured as excellent (quality band A), the quality index is 100%, whereas if all samples are in quality band B (acceptable), the index is 75%.

17.4.4 Trade quality

Other than microbiological testing, the quantitative assessment of draught beer quality has received little attention. The forcing test lends itself to the measurement of (retrospective) quality and was used to assess the quality of draught ales and lager draught beers purchased on two occasions from 57 on-trade licenced premises in 10 locations in the UK Midlands Mallett & Quain, 2019). Nothing was known of line cleaning frequency or hygienic practices in the accounts. Of the 149 samples of standard lager (abv ≤4.2%), 44% were in the 'excellent' quality band compared with 16% of 88 samples of keg ale (abv ≤4.2%). This was reinforced by measurement of the quality index with lager SL3 at 84% and 68% for keg ale KA1. Of the 237 samples, 20.3% were 'poor' and 2.7% were 'unacceptable'. As keg beer should be of 'excellent' quality, 65% of the keg beers sampled in this study had suffered some microbiological damage through dispense.

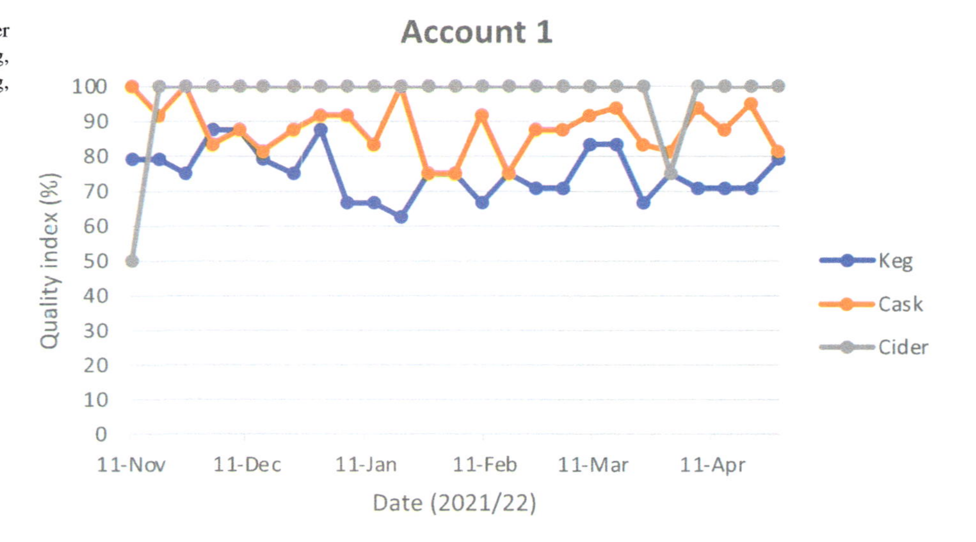

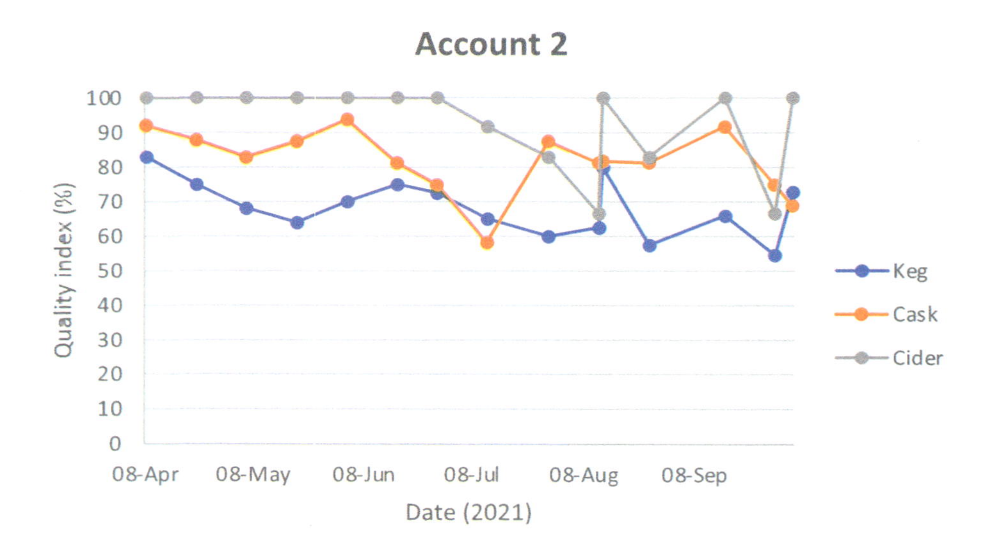

FIGURE 17.13 Quality of beer and cider in two public houses (account 1 – 6× keg, 4× cask, 1× cider; account 2 –10× keg, 4× cask, 3× cider) (Quain, unpublished).

For a host of reasons, draught beer quality will vary over time. For example, the quality by forcing of keg and cask beers along with cider was monitored in two accounts over several months. Quality was broadly consistent in Account 1 but less so in Account 2 (Fig. 17.13). This was borne out by the quality index with Account 1 (n = 25) performing better than Account 2 (n = 15) in all categories; keg 75 versus 68%, cask 88 versus 81% and cider 97 versus 93%. With a quality index of 75% being of 'acceptable' quality, it is disappointing that in both accounts keg beer was at or below this benchmark. It is perhaps surprising that the quality of cask beer is better in both accounts. In Account 1, this reflects line cleaning—rather than rinsing with water—on changing casks. Conversely in Account 2, the collective quality of keg beer was compromised by three taps at the far end of the bar, away from the hotspot (14.2.6). These beers were regularly found to be either 'poor' or 'unacceptable'.

17.4.5 Impact of best hygienic practice

Effective and regular line cleaning is at the heart of hygienic best practice for draught beer. Indeed, the profitability of accounts (as volume growth) (Section 17.2.1) is directly linked to the frequency of line cleaning with a 2% uplift with weekly line cleaning, breaking even at 2 weeks and declining thereafter. Optimisation of the line cleaning process (Section 17.6) would be expected to drive growth in quality and (potentially) throughput.

Profit is measurable, whereas beer quality is subjective and harder to define. The forcing approach (Section 17.4.2) considers quality from the perspective of microbiological loading in beer at dispense. It is of note that the quality index in two studies of trade quality with the same brands were similar (Jevons & Quain, 2022; Mallett & Quain, 2019) with respectively a QI for lager of 75% (109 and 10 samples) and with keg ale of 68.3% (63 samples) and 67.5% (10 samples).

The quality index has been used to quantify the effect of best practice on the quality of draught products in two successful 'gastro pubs' (A—3× lager, stout, keg ale, cider and B—4× lager, 3× cask, stout, cider) (Mallett and Quain, unpublished). All the products were assessed by forcing weekly for 3 weeks. Best practice was limited to line cleaning together with nozzle and keg coupler hygiene. Lines were cleaned weekly with a premium line cleaner, keg coupler and spear sprayed with sanitiser on changing containers (Section 17.2.3) and nozzles soaked daily with a sanitising tablet (Section 17.2.4) and beer quality monitored weekly for a further 3 weeks. Prior to introduction of best practice, the average ($n = 3$) quality index pre- and post-cleaning increased from 74.1% to 81.5% (A) and 75.8%–84.7% (B). The switch to more hygienic practices resulted in an improvement in microbiological quality of beer both pre- and post-cleaning in the two (accounts A [88.4/93.2%] and B [84.2/94.2%]). Similarly, the hygiene of dispense nozzles was found (using ATP bioluminescence) to switch from heavily contaminated to 'commercially sterile' with similar results to those previously reported (Quain, 2016).

Better hygienic practices would be expected to have a positive impact on beer quality. Although not quantified, feedback from the owner of the two accounts indicated an uplift in sales from the intervention. Despite the benefit, the transition in behaviours from a trial to a routine is difficult. Whilst the frequency of line cleaning is more embedded, the take up of additional hygienic actions is more challenging requiring training, coaching and reinforcement. In particular, a routine of sanitising keg couplers can be lost in the pressures of busy trading. Change management is always difficult which—in the on-trade—is exacerbated by the high turnover of bar staff.

17.5 Other draught beverages

17.5.1 Alcohol-free beers

Alcohol-free and low-alcohol beers are in growth in Western Europe and elsewhere (Bellut & Arendt, 2019). Currently, alcohol-free beers produced in the UK contain $\leq$0.05% ABV with low alcohol $\leq$1.2% ABV. To add confusion, alcohol-free beers produced in Europe and on sale in the UK contain $\leq$0.5% ABV. Alcohol-free beers are produced using biological, physical or hybrid methods (Bellut & Arendt, 2019), and their composition is markedly different to alcoholic beers. As has long been recognised (Adams et al.1989; L'Anthoen & Ingledew, 1996), spoilage bacteria exhibit greater heat resistance in alcohol free beer, and, accordingly, these products require more pasteurisation than those containing ethanol. As noted in Section 17.3.1, AFBs and LABs are more vulnerable to microbial spoilage reflecting the liquid matrix and a lack of ethanol and high levels of fermentable sugars (Quain, 2021).

In the UK, the introduction of draught alcohol-free brands on the bar is gaining momentum. Whilst some are delivered using innovative stand-alone systems (Quain, 2021), others tap into existing long draw dispense. This is of concern as these products are more susceptible to microbial spoilage and are more vulnerable to contamination by hygienically compromised conventional dispense systems. This will result in poor product quality, undermining the alcohol-free draught beer 'promise' for consumers.

However, a bigger concern for draught alcohol-free beer is one of food safety as pathogenic microorganisms have been reported to grow in low- and alcohol-free beers. L'Anthoen and Ingledew (1996) reported (without data) that pathogens grew in alcohol free (0.5% ABV) but not in beers at 5% ABV. Importantly, Menz et al. (2011) demonstrated the growth of Gram-negative pathogens (*Escherichia coli* O157:H7, *Salmonella typhimurium*) in alcohol free beer (0.5% ABV) but not of Gram-positive microorganisms (*Listeria monocytogenes* and *Staphylococcus aureus*). More recently, Çobo et al. (2023) reported similar conclusions in challenge tests with *E.coli*, *S. enterica* and *L. monocytogenes* in alcohol-free beer.

Whilst pH ($\leq$4) limits the growth of these organisms (Menz et al. 2011), most AFB/LABs have a pH > 4 (Quain, 2021) and are potentially vulnerable to pathogens through the handling by bar staff of nozzles and keg couplers. Given these insights, it was strongly recommended (Quain, 2021) that 'low or alcohol-free beer styles are not offered to consumers in the on-trade by conventional long line dispense systems' and that the risk of pathogens is minimised by the use of 'hygienically designed stand-alone dispense systems'.

17.5.2 Cider

Total UK cider volumes pre-COVID were 6.8 mhL in the UK in 2019 (British Beer and Pub Association, 2022). Increasingly, public houses offer two draught ciders, one apple and one fruit. With a lower pH (<4, Qin et al. 2018) and added sulphur dioxide (<200 mg/L), cider has a reputation in the trade for being microbiologically robust. Consequently, in some accounts, line cleaning of cider brands can be less regular than with beers.

Although more robust than beer, cider is susceptible to microbial spoilage in public houses (FOB detector, Fig. 17.4c). Trade data reported in Fig. 17.13 show the apple cider in Account 1 to be broadly of excellent quality with occasional examples of microbial spoilage. However, three draught ciders (2× apple, fruit) in Account 2 exhibited more variation in quality, particularly in the second half of monitoring. Similarly, laboratory studies show apple cider microbiota forming biofilm in a model system (Jevons & Quain, 2021) and, in an experiment with extended challenge testing up to 14 days, microorganisms from draught lager and keg ale spoiling fruit cider and, interestingly but slowly, seltzer (Fig. 17.14).

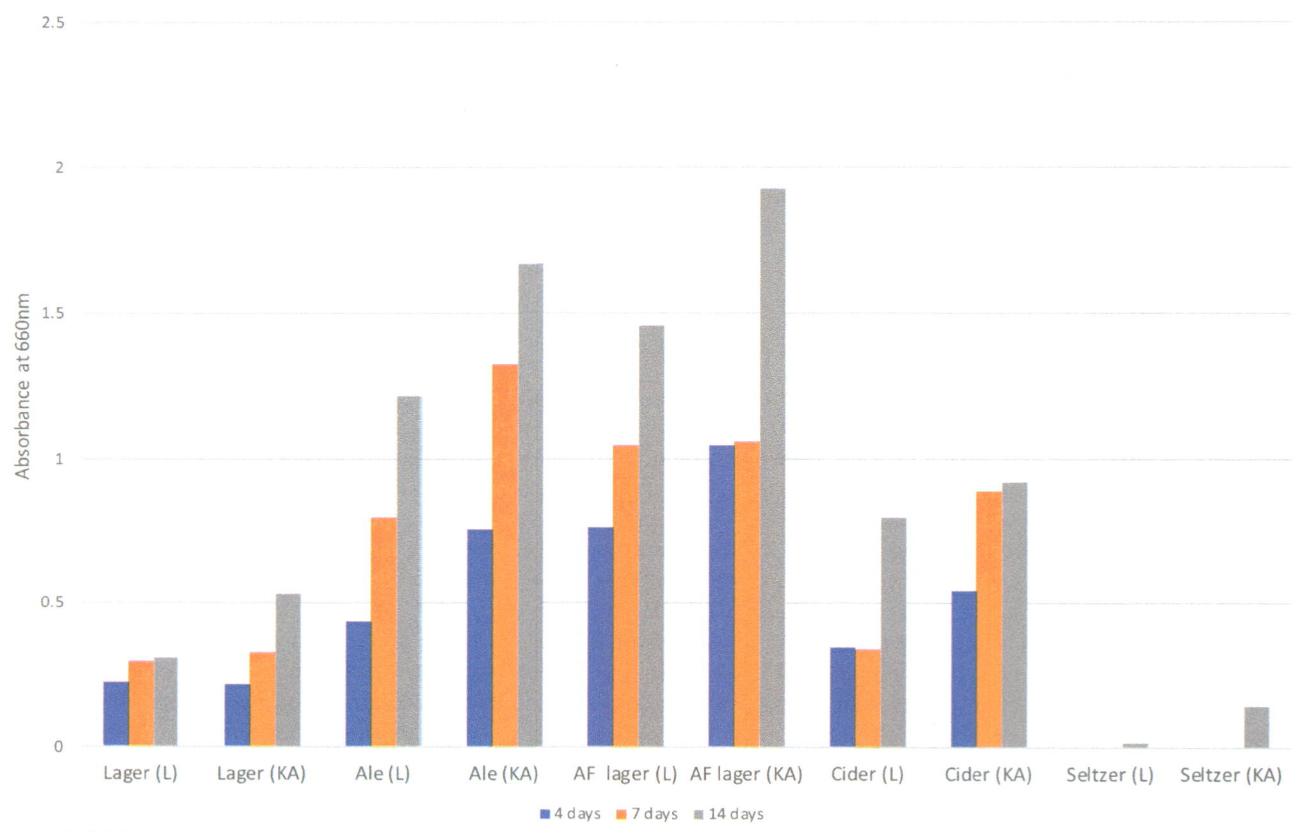

FIGURE 17.14 Extended challenge test with microbiota (KA, keg ale and L, lager) of lager, ale, cider and seltzer (Quain, unpublished).

17.6 Improving quality

17.6.1 Low hanging fruit

- Clean draught dispense lines every 7 days.
- Use a quality 'purple' line cleaning solution that visually indicates the presence of dirt.
- Soak tap nozzles daily in sanitising solution.
- Clean keg couplers every week.
- Sanitise keg coupler and spear on changing containers.
- Throughputs—sell keg beer in 5 days, cask beer in 3 days. Avoid 'overfonting', where necessary trade down the size of containers.

- Avoid damage to dispense lines by eliminating practices such as soaking overnight/using overly strong line cleaning solutions.
- On installing a new python, use lines which minimise biofilm formation (nylon lined, reduced ingress of oxygen and egress of carbon dioxide).
- Challenge suppliers who by—using technology—encourage the relaxation of line cleaning frequency. Request information on long-term performance of beer quality in several accounts rather than focus on the suggested financial savings on reducing the frequency of line cleaning.

17.6.2 Step change

- Get stake holders (owner, manager, bar staff) on board to understand the 'why' and 'how' of assuring draught beer quality through a continuing programme of education and training.
- Minimise the risks to product quality and food safety by using 'stand-alone' dispense solutions and not long line dispense for alcohol free beer.
- Explicitly flag to consumers, the link between hygiene/cleaning and beer quality. Communicate the date the lines were last cleaned on.
- Sign up for 'Beer Marque', quality accreditation for draught products (https://cask-marque.co.uk/product/beer-marque-new-member/)
- Introduce 'mechanical action' in line cleaning. Circulate (ideally in both directions) using an 'electric recirculating pump', which is the 'preferred method for nearly all systems' in the USA (Brewers Association, 2019).
- Chill the dispense line from keg to FOB and add trace cooling to the back of the tap.
- Use warm line cleaning solutions which are actively circulated (see Section 17.2.1, 'improving the process').
- Quarterly—'all FOB-stop devices should be completely disassembled and hand-cleaned' (Brewers Association, 2019).
- Quarterly—use a phosphoric acid clean to remove any oxalate and scale build up in dispense lines (Brewers Association, 2019).
- Quarterly—disassemble and clean taps.

17.6.3 Cultural change

- With new public houses change the traditional model to (i) locate the cellar adjacent to the bar to reduce line lengths and (ii) instal a cold room at 2–4°C. This configuration would allow the frequency of line cleaning to be comfortably extended from every 7 days to every 2 weeks.

Acknowledgements

Big thanks to my former colleagues at the University of Nottingham; Ph.D. students—Dr. James Mallett (funded by the Brewers' Research and Education Fund, Diageo GB, Molson Coors Beverage Company) and Dr. Alex Jevons (BBSRC Doctoral Training Programme), together with Mel Stuart, Dr. Steph Brindley and Honours students Sam Garvey, Vicki Humphries, Sean Sargent and Zoe Chapman. As ever, thanks to Dr. Chris Boulton for friendship, insight and debate. Finally, to our labradoodle Indy for daily walks and reflection. She was with us for far too short a time (January 2011–June 2022).

References

Adams, M. R., O'Brien, P. J., & Taylor, G. T. (1989). Effect of ethanol content of beer on the heat resistance of spoilage *Lactobacillus*. *Journal of Applied Bacteriology, 66*, 491–495.

Andrews, J., & Gilliland, R. B. (1952). Super-attenuation of beer: A study of three organisms capable of causing abnormal attenuations. *Journal of the Institute of Brewing, 58*, 189–196.

Anonymous. (2006). *DIN 6650-6 Dispense systems for draught beverages—Part 6: Requirements for cleaning and disinfection.*

Bailey, D. R. (1907). *The Brewer's Analyst.* Available from: https://archive.org/details/brewersanalystsy00bailrich.

Bellut, K., & Arendt, E. K. (2019). Chance and challenge: Non-*Saccharomyces* yeasts in non-alcoholic and low alcohol beer brewing—A review. *Journal of the American Society of Brewing Chemists, 77*, 77–91.

Boulton, C., & Quain, D. (2001). *Brewing yeast and fermentation*. Oxford, UK: Blackwell.

Brewers Association. (2019). *Draught beer quality manual* (4th ed., pp. 1–119).

British Beer and Pub Association. (2003). *Hygienic management of cellars and cleaning of beer dispense lines.* https://beerandpub.com/policy-campaigns/product-assurance/.

British Beer and Pub Association. (2022). In N. Fish (Ed.), *Statistical handbook*. London: Brewing Publications Ltd.

Brown, H. T. (1916). Reminiscences of fifty years' experience of the application of scientific method to brewing practice. *Journal of the Institute of Brewing, 22*, 267–354.

Çobo, M., Charles-Vegdahl, A., Kirkpatrick, K., & Worobo, R. (2003). Survival of foodborne pathogens in low and nonalcoholic craft beer. *Journal of Food Protection, 86*, 100183. Article.

Cask Marque. (2023). *The Cellar and Bar manual. Driving profit through quality*. Essex: Cask Marque Trust.

Casson, D. (1982). Beer dispense lines (pp. 447–453). *The Brewer*. November.

Casson, D. (1985). Microbiological problems of beer dispense (pp. 417–421). *The Brewer*. November.

Cosbie, A. J. C. (1943). Beer disease organisms. *Journal of the Institute of Brewing, 49*, 288–295.

Dolezil, L., & Kirsop, B. H. (1980). Variations amongst beers and lactic acid bacteria relating to beer spoilage. *Journal of the Institute of Brewing, 86*, 122–124.

Fernandez, J. L., & Simpson, W. J. (1995). Measurement and prediction of the susceptibility of lager to spoilage by lactic acid bacteria. *Journal of Applied Bacteriology, 78*, 419–425.

Fielding, L. M., Hall, A., & Peters, A. C. (2007). An evaluation of ozonated water as an alternative to chemical cleaning and sanitisation of beer lines. *Journal of Food Service, 18*, 59–68.

Geissler, A. J., Behr, J., von Kamp, K., & Vogel, R. F. (2016). Metabolic strategies of beer spoilage lactic acid bacteria in beer. *International Journal of Food Microbiology, 216*, 60–68.

Harper, D. R. (1981). A microbiologist looks at beer dispense. August (pp. 23–28). *Brewers' Guardian*, 31.

Harper, D. R., Hough, J. S., & Young, T. W. (1980). Microbiology of beer dispensing systems (pp. 24–28). *Brewers' Guardian*. January.

Heger, P., & Russell, A. (2021). Nylon oxygen barrier tubing reduces biofouling in beer draught lines. *Fine Focus, 7*, 25–35.

Hemmons, L. M. (1954). Wild yeasts in draught beer 1. An exploratory survey. *Journal of the Institute of Brewing, 60*, 288–291.

Hernawan, T., & Fleet, G. (1995). Chemical and cytological changes during the autolysis of yeasts. *Journal of Industrial Microbiology, 14*, 440–450.

Hill, A. E. (2009). Microbiological spoilage of beer. In C. W. Bamforth (Ed.), *Beer—A quality perspective* (pp. 163–183). London, UK: Academic Press.

Hough, J. S., Young, T. W., Braund, A. M., Longstaff, D., Weeks, R. J., & White, M. A. (1976). Keg and cellar tank beer in public houses—A microbiological survey (pp. 179–183). *The Brewer*. June.

Hucker, B., Christophersen, M., & Vriesekoop, F. (2017). The influence of thiamine and riboflavin on various spoilage microorganisms commonly found in beer. *Journal of the Institute of Brewing, 123*, 24–30.

Jevons, A. J. (2022). *The impact of process, environment and hygiene on the microbiome and metabolome of draught beer*. PhD thesis. University of Nottingham.

Jevons, A. J., & Quain, D. E. (2021). Draught beer hygiene: Use of microplates to assess biofilm formation, growth and removal. *Journal of the Institute of Brewing, 127*, 176–188.

Jevons, A. J., & Quain, D. E. (2022). The culture-dependent spoilage microflora of four styles of draught beer. *Journal of Applied Microbiology, 133*, 3728–3740.

Kordialik-Bogacka, E. (2022). Biopreservation of beer: Potential and constraints. *Biotechnology Advances, 58*, Article 107910.

Kulka, D. (1953). Yeast autolysis and the biological stability of beers. *Journal of the Institute of Brewing, 59*, 285–293.

Kulka, D. (1960). Bacterial spoilage under anaerobic conditions. *Journal of the Institute of Brewing, 66*, 28–35.

L'Anthoen, N. C., & Ingledew, W. M. (1996). Heat resistance of bacteria in alcohol-free beer. *Journal of the American Society of Brewing Chemists, 54*, 32–36.

Mallett, J. R. (2019). *Assessing the factors affecting draught beer hygiene and validation of a novel dispense system*. PhD thesis. University of Nottingham.

Mallett, J. R., & Quain, D. E. (2019). Draught beer hygiene: A survey of on-trade quality. *Journal of the Institute of Brewing, 125*, 261–267.

Mallett, J. R., Stuart, M. S., & Quain, D. E. (2018). Draught beer hygiene: A forcing test to assess quality. *Journal of the Institute of Brewing, 124*, 31–37.

Menz, G., Aldred, P., & Vriesekoop, F. (2011). Growth and survival of foodborne pathogens in beer. *Journal of Food Protection, 74*, 1670–1675.

Mossel, D. A. A., & Ingram, M. (1955). The physiology of the microbial spoilage of foods. *Journal of Applied Microbiology, 18*, 232–268.

Parker, D. K. (2012). Beer: Production, sensory characteristics and sensory analysis. In J. Piggott (Ed.), *Alcoholic beverages—Sensory evaluation and consumer research* (pp. 133–158). Cambridge, UK: Woodhead Publishing.

Qin, Z., Petersen, M. A., & Bredie, W. L. P. (2018). Flavor profiling of apple ciders from the UK and Scandinavian region. *Food Research International, 105*, 713–723.

Quain, D. (1999). The 'new' microbiology. In *Proceedings of the European Brewery Convention Congress*, Cannes. (pp. 239–248). Nürnberg, Germany: Fachverlag Hans Carl.

Quain, D. (2007). Draught beer quality—Challenges and opportunities. In *Proceedings of the European Brewery Convention Congress*, Venice. (pp. 791–801). Nürnberg, Germany: Fachverlag Hans Carl.

Quain, D. E. (2015). Assuring the microbiological quality of draught beer. In A. Hill (Ed.), *Brewing microbiology—Managing microbes, ensuring quality and valorising waste* (pp. 333–352). Cambridge, UK: Woodhead Publishing.

Quain, D. E. (2016). Draught beer hygiene: Cleaning of dispense tap nozzles. *Journal of the Institute of Brewing, 122*, 388–396.

Quain, D. E. (2021). The enhanced susceptibility of alcohol-free and low alcohol beers to microbiological spoilage: Implications for draught dispense. *Journal of the Institute of Brewing, 127*, 406–416.

Quain, D. E., & Jevons, A. J. (2023). The spoilage of lager by draught beer microbiota. *Journal of the Institute of Brewing, 129*, 307–320.

Sakamoto, K., & Konings, W. N. (2003). Beer spoilage bacteria and hop resistance. *International Journal of Food Microbiology, 89*, 105–124.

Sauer, K., Stoodley, P., Goeres, D. M., Hall-Stoodley, L., Burmølle, M., Stewart, P. S., et al. (2022). The biofilm life cycle: Expanding the conceptual model of biofilm formation. *Nature Reviews Microbiology, 20*, 608–620.

Schneiderbanger, J., Grammer, M., Jacob, F., & Hutzler, M. (2018). Statistical evaluation of beer spoilage bacteria by real-time PCR analyses from 2010 to 2016. *Journal of the Institute of Brewing, 124*, 173–181.

Seton, G. R. (1912). Cellar management. *Journal of the Institute of Brewing, 18*, 389–406.

Spedding, G., & Aiken, T. (2024). Sensory analysis as a tool for beer quality assessment with emphasis on its use for microbial control in the brewery. In A. Hill (Ed.), *Brewing microbiology—Managing microbes, ensuring quality and valorising waste* (2nd). Cambridge, UK: Woodhead Publishing.

Storgårds, E. (1996). Microbiological quality of draught beer—Is there a reason for concern? (pp. 92–103). *European Brewery Convention Monograph XXV, Fachverlag Hans Carl, Nürnberg*.

Storgårds, E., & Haikara, A. (1996). ATP bioluminescence in the hygiene control of draught beer. *Ferment, 9*, 352–360.

Storgårds, E., & Priha, O. (2009). Biofilms and brewing. In P. M. Fratamico, B. A. Annous, & N. W. Gunther (Eds.), *Biofilms in the food and beverage industries* (pp. 432–454). Cambridge, UK: Woodhead.

Suiker, I. M., & Wösten, H. A. B. (2022). Spoilage yeasts in beer and beer products. *Current Opinion in Food Science, 44*, Article 100815.

Suzuki, K. (2020). Emergence of new spoilage microorganisms in the brewing industry and development of microbiological quality control methods to cope with this phenomenon—A review. *Journal of the American Society of Brewing Chemists, 78*, 245–259.

Thomas, K., & Whitham, H. (1996). Improvements in beer line technology (pp. 124–137). *European Brewery Convention Monograph XXV, Fachverlag Hans Carl, Nürnberg*.

Vaughan, A., O'Sullivan, T., & van Sinderen, D. (2005). Enhancing the microbiological stability of malt and beer—A review. *Journal of the Institute of Brewing, 11*, 355–371.

Walker, S. L., Fourgialakis, M., Cerezo, B., & Livens, S. (2007). Removal of microbial biofilms from dispense equipment: The effect of enzymatic pre-digestion and detergent treatment. *Journal of the Institute of Brewing, 113*, 61–66.

Werner, R., Hoi, K. A., & Becker, T. (2023). Microorganisms: Life in the hoses of a beverage dispensing unit. *Brauwelt International, 41*, 28–31.

Wiles, A. E. (1950). Studies of some yeasts causing spoilage of draught beer. *Journal of the Institute of Brewing, 56*, 183–193.

Part IV

Impact of microbiology on sensory quality

Chapter 18

Biotransformation of wort components for appearance, flavour and health

Mei Z.A. Chan[1] and Shao Quan Liu[1,2]
[1]National University of Singapore, Singapore; [2]National University of Singapore (Suzhou) Research Institute, Suzhou, Jiangsu, China

18.1 Introduction

Beer is a product of microbial transformation of cereal-based substrates, predominantly yeast and barley malt. Microorganisms are inevitably associated with the whole beer production process, from barley grains, germination, malting, wort preparation, fermentation, post-fermentation processing, to the finished product (beer). As a result, microorganisms are expected to have an impact on beer quality in a positive or negative manner at different stages of the brewing process. This chapter focuses on the impact of yeast and bacteria associated with beer fermentation on beer appearance, flavour and health properties. In this context, beer appearance refers to haze, turbidity, sedimentation and foam; flavour refers to taste and aroma (and compounds that cause these sensations) and health properties refer to bioactive/toxicological compounds released. Off-flavour (off-taste, off-odour) elicited by spoilage yeast and bacteria is also covered where appropriate.

This chapter is divided into seven sections including an introduction (Section 17.1), impact of yeast on beer appearance (Section 17.2), impact of yeast on beer flavour (Section 17.3), impact of bacteria on beer appearance and flavour (Section 17.4), impact of yeast on the health properties of beer (Section 17.5), impact of bacteria on the health properties of beer (Section 17.6) and future trends (Section 17.7). In sections on beer appearance and flavour, the emphasis is on the impact of brewing yeasts (*Saccharomyces cerevisiae* and *Saccharomyces pastorianus*) on barley malt-based beers (ale and lager respectively) because more information is available on this topic. Nevertheless, references are also made to other yeasts involved in brewing speciality beers derived from both barley malt and other cereal malts. The effect of lactic acid bacteria (LAB) on beer appearance and flavour in spoiled and acidic beers are also discussed. In sections relating to fermentative roles of yeasts and bacteria on the health properties of beer, the emphasis is on the production of bioactive metabolites, in particular by probiotic microorganisms.

18.2 Impact of yeast on beer appearance

Growth of yeasts in finished beers obviously brings about turbidity changes and cell sedimentation that are undesirable for most beer types. In particular, turbidity arises from the release of proteins and β-glucans from autolysed yeasts, which also contributes to haze formation (Mastanjević et al., 2018). Besides turbidity and haze formation, brewing yeasts also have a significant impact on foam formation, stability and head retention as reviewed by Blasco, Vinas, and Villa (2011), and a summary of this review is provided below.

Yeasts can either stabilise or destabilise beer foam during the fermentation process, depending on the yeast factors involved. For example, carbon dioxide facilitates foam generation, while mannoproteins derived from yeast cell walls minimise haze formation and stabilise foam by adhering to the gas−liquid interface of foam bubbles (i.e., the bio-emulsification effect of yeast mannoproteins). On the other hand, ethanol destabilises foam, and vigorous yeast fermentation may lead to overfoaming and loss of foam-active substances (e.g., proteins), which prevents beer head formation and retention. Proteinase-A, an enzyme that is secreted by yeasts under nutrient-stress conditions, degrades malt proteins involved in foaming and stabilising beer foam. Additionally, autolysed yeasts can release β-glucanases that hydrolyse

Brewing Microbiology. https://doi.org/10.1016/B978-0-323-99606-8.00003-1

β-glucans, resulting in viscosity reduction and liquid drainage from the foam. Lipids and longer-chain fatty acids released from autolysed yeasts also damage beer foam by promoting the coalescence of foam bubbles through a film-bridging mechanism (Bravi, Perretti, Buzzini, Della Sera, & Fantozzi, 2009). Moreover, a foam-negative protein, thioredoxin, has also been identified by proteomics, presumably as a result of yeast proteolysis upon autolysis (Iimure & Sato, 2013).

18.3 Impact of yeast on beer flavour

Beer is a product of transformation of the wort and hops by yeasts under brewing conditions. Although the hops added principally contribute to the bitter taste and aroma to a certain extent, it is the yeast used in brewing that makes a significant impact on beer flavour, especially aroma, through yeast autolysis, catabolism of sugars, assimilable nitrogen, organic acids and other substances, then generation of acids, alcohols, aldehydes, esters, ketones, volatile phenolic compounds, terpenoids and volatile sulphur compounds (VSCs).

18.3.1 Impact of yeast on beer taste

Brewing yeasts contribute positively to beer mouthfeel by producing carbon dioxide and foam bubbles. The impact of yeasts on beer taste varies with the yeast species involved in beer fermentation. For example, *Brettanomyces* spp. and *Saccharomyces kudriavzevii* yeasts involved in the spontaneous fermentation of acidic beers such as Lambics can degrade exopolysaccharides (EPS; produced by LAB) and malto-oligosaccharides that are unfermentable by conventional *S. cerevisiae* brewing yeasts. The degradation of beer polysaccharides by non-conventional brewing yeasts eliminates ropiness, while generating thinner beers with higher ethanol content (Bongaerts, De Roos, De Vuyst, & Björkroth, 2021). *Schizosaccharomyces pombe* produced better foam consistency and foam persistence due to its greater capacity for CO_2 production and release of mannoproteins and polysaccharides upon autolysis (Callejo et al., 2019). If a lactic acid-producing yeast such as *Lachancea thermotolerans* is involved in beer production, the resultant beer would have a higher acidity and a sour taste (Callejo et al., 2019), a trait suitable for acidic beers.

Glycerol, another product of yeast fermentation in the glycolytic pathway, is highly desired in beers as it improves mouthfeel and body while imparting a slight sweetness and suppressing roughness derived from ethanol production (Zhao, Procopio, & Becker, 2015). In beers, glycerol concentrations typically range from 1 to 3 g/L, but yields can be enhanced by adopting high-gravity wort fermentation to shift metabolic pathways towards the re-oxidation of NADH to favour glycerol production. Yields can also be enhanced by other factors such as yeast strain, pH, temperature and mild heat shocks (Zhao et al., 2015).

Flavour-active metabolites such as peptides (e.g., glutathione), nucleotides and amino acids may be released during yeast ageing and autolysis. For example, sweet tasting amino acids (Pro and Ala) and bitter tasting amino acids (Phe, Ile, Trp) were prominent in unfiltered beers (e.g., craft beers) rich in yeast sediments. Glutamic acid (umami) was also abundant in the unfiltered beers, which interestingly, exhibited umami synergy with nucleotide rich foods (e.g., oysters, tuna) (Vinther Schmidt, Olsen, & Mouritsen, 2021).

18.3.2 Impact of yeast on beer aroma

Beer contains numerous volatile compounds, depending on ingredients such as malt type and hop variety, yeast strain, brewing and maturation conditions, as well as storage conditions. However, not all volatile compounds contribute to beer aroma, and only a small number of volatiles in beer are aroma-active (Olaniran, Hiralal, Mokoena, & Pillay, 2017). Furthermore, some of the beer volatiles (both aroma-active and aroma-inactive) are not of yeast origin. In addition, among the aroma-active volatiles, some impart positive aromas to beer such as fruity, floral and fragrant flavour notes, whereas others may cause off-odours such as green, butter-like, sulphury and phenolic. In this section, the impact of yeasts on beer aroma compound formation is discussed with regard to their potential positive and negative influences on beer aroma, covering the following classes of odourants: alcohols, aldehydes, acids, esters, ketones, volatile phenolic compounds, terpenoids and VSCs.

18.3.2.1 Alcohols

Ethanol is known to be the dominant alcohol formed via the glycolytic pathway by brewing yeasts, which not only gives off an alcoholic odour but also acts as a carrier of other odour-active higher alcohols (also known as fusel alcohols or fusel oils). These higher alcohols include aliphatic higher alcohols (e.g., n-propanol, isobutanol, active amyl alcohol, isoamyl alcohol) and aromatic alcohols (e.g., 2-phenylethyl alcohol). Depending on the concentration and type of alcohols, aliphatic alcohols predominantly contribute to alcoholic, solvent-like aromas, while aromatic alcohols contribute to sweet and rose-like aromas in beers (Ferreira & Guido, 2018; Olaniran et al., 2017). The aroma importance of higher alcohols extends to other facets of beer flavour by serving as ester precursors (elaborated in Section 17.3.2.4). The biogenesis of higher alcohols in beer in relation to yeast metabolism has been reviewed by several authors (Cordente, Schmidt, Beltran, Torija, & Curtin, 2019; Ferreira & Guido, 2018; Pires, Teixeira, Brányik, & Vicente, 2014), and a summary is given below.

Higher alcohols are biosynthesised by yeasts from sugars and selected amino acids (typically branched-chain and aromatic amino acids) via the anabolic pathway and Ehrlich pathway, respectively. In the anabolic pathway, α-keto acids are generated from carbohydrates via de novo biosynthesis of amino acids, whereas in the Ehrlich pathway, α-keto acids are formed from amino acid breakdown by way of transamination. The α-keto acids are then decarboxylated with the formation of aldehydes, which are subsequently reduced to higher alcohols. In the Ehrlich pathway, the type of higher alcohols produced is determined by the type of amino acids present, commonly threonine (n-propanol), valine (isobutanol), leucine (isoamyl alcohol), isoleucine (active amyl alcohol) and phenylalanine (2-phenylethyl alcohol). Tyrosol and tryptophol, which impart bitterness and 'taste sharpening' effects in beer, can also be produced from tyrosine and tryptophan through the Ehrlich pathway, respectively.

18.3.2.2 Aldehydes

There are a number of aldehydes present in beer such as acetaldehyde, hexanal, (*E*)-2-nonenal, furfural, 2-methylpropanal (isobutanal), 2-methylbutanal (amyl aldehyde), 3-methylbutanal (isoamyl aldehyde), 3-methylthiopropanal (methional), 2-phenylacetaldehyde and benzaldehyde (Filipowska et al., 2021). These aldehydes affect beer flavour by imparting organoleptic notes ranging from green apple—like (acetaldehyde), malty (branched-chain aldehydes), cooked-potato (methional), flowery (2-phenylacetaldehyde) and almond—or cherry-like (benzaldehyde), which are concentration-dependent. Naturally, not all aldehydes are of yeast origin, as malt is the main brewing raw material which delivers bound aldehydes to the final brew, with the bound aldehydes subsequently released during beer ageing. A brief description of the chemical and biological origins of aldehydes in beer are described below, with reference to recent reviews and other reports (Filipowska et al., 2021; Lehnhardt, Gastl, & Becker, 2019; Stewart, 2017).

Acetaldehyde is a precursor to ethanol in the glycolytic pathway by brewing yeasts, formed via the decarboxylation of pyruvate. Small amounts of acetaldehyde are excreted under normal physiological conditions, with yields dependent on the yeast strain and fermentation conditions (e.g., oxygen, pitching rate, wort gravity, temperature) (Stewart, 2017). Small amounts of branched-chain aldehydes and 2-phenylacetaldehyde are also excreted by brewing yeasts during fermentation from the catabolism of respective amino acids by way of the Ehrlich pathway mentioned in Section 17.3.2.1 (Cordente et al., 2019). Although the contribution of aldehydes to the final beer in this fashion may be quantitatively rather limited as most aldehydes are directly reduced to their corresponding alcohols, aldehydes have very low odour detection thresholds and can still impact beer aromas individually, additively or synergistically (Filipowska et al., 2021).

Some aldehydes formed either biologically during wort fermentation and/or chemically during beer ageing can cause aged or stale beer aroma (beer staling), for instance, cardboard aroma ((*E*)-2-nonenal), green/grassy (hexanal), cooked potato-like (methional) and honey-like (2-phenylacetaldehyde). As the staling aldehydes are a result of oxidation or lipid oxidation, aged beer aroma can be decreased considerably by taking advantage of the reducing activity of the yeasts through bottle re-fermentation (also known as bottle conditioning) such that aldehydes are reduced to their corresponding alcohols (e.g., reduction of (*E*)-2-nonenal to nonenol) (Silva Ferreira, Bodart, & Collin, 2019).

18.3.2.3 Acids

The acids in beer consist of inorganic acids (mainly phosphoric acid) and organic acids (non-volatile and volatile acids); together they contribute to the total acidity of beer. The non-volatile acids include, but not restricted to, malic, citric, pyruvic, α-ketoglutaric, succinic and lactic acids. The volatile acids mainly comprise acetic (C2), butyric (C4), caproic (C6), caprylic (C8), capric (C10) and lauric (C12) (Horák, Čulík, Jurková, Čejka, & Kellner, 2008; Rodrigues et al., 2010). Other minor but relatively potent volatile acids found in beer are 3-methylbutanoic (cheesy, sweaty) and 2-phenylacetic acids (honey-like, sweet) (Hazelwood, Daran, Maris, Pronk, & Dickinson, 2008; Olšovská et al., 2019). Whereas the non-volatile acids contribute to the sour taste of beers, the volatile acids of C2 to C12 exert unpleasant aromas such as rancid, cheesy, soapy and fatty flavours if present in high concentrations (Olšovská et al., 2019). Some of the acids originate from the wort (e.g., mashing process), whereas others are derived from yeast autolysis and metabolism.

Brewing yeasts are known to produce both non-volatile and volatile acids during fermentation and beer ageing. The non-volatile acids produced by the yeasts are pyruvic, α-ketoglutaric, succinic and lactic acids, although usually in small quantities, whereas most of the volatile acids are formed by the yeasts. Yeast autolysis also produces the longer-chain fatty acid due to membrane lipid breakdown, especially the unsaturated fatty acids, which, upon oxidation, would affect beer flavour adversely. Quantitatively the most significant odour-active volatile acids derived from yeast metabolism are acetic, caprylic, capric and lauric acids (Amata & Germain, 1990; Clapperton, 1978). Production of non-volatile and volatile acids by yeasts is predominantly associated with glycolysis (e.g., C2), the TCA cycle (e.g., the non-volatile organic acids), amino acid metabolism (e.g., branch-chained and aromatic acids) and fatty acid metabolism (e.g., C4–C18 fatty acids).

18.3.2.4 Esters

Esters of short-chain and branched-chain fatty acids, which are the most aroma-active, are arguably the most important volatile compounds in beer. They have a positive impact on the overall beer flavour, especially fruity aromas, but excessive levels of esters can lead to overly fruity, fermented off-flavour. Esters found in beer can be categorised into two main groups: acetate (acetyl) esters (e.g., ethyl acetate, isoamyl acetate and 2-phenylethyl acetate) and ethyl esters of medium-chain fatty acids (e.g., ethyl hexanoate, ethyl octanoate, ethyl decanoate). These esters possess sensory descriptions ranging from fruity and solvent-like (ethyl acetate, depending on concentration), banana- and pear-like (isoamyl acetate), rose- and honey-like (2-phenylethyl acetate) or apple-like and sweet (ethyl hexanoate and ethyl octanoate). Of these esters, ethyl acetate is produced in the highest concentration (threshold 30 mg/L) by brewing yeasts, although the other esters also synergistically contribute to fruity aromas due to lower threshold concentrations (<4 mg/L) (Olaniran et al., 2017; Stewart, 2017).

Esters can be synthesised chemically or biologically in beer, with brewing yeasts recognised as the principal ester producers in beer fermentation. Esters and their formation mechanisms in brewing *S. cerevisiae* yeast strains have attracted much research attention and have been reviewed several times (Olaniran et al., 2017; Pires et al., 2014; Stewart, 2017). Therefore, it is beyond the scope of this section to elaborate the details of ester biosynthesis, and interested readers are referred to these review articles for further information. However, a summary of ester biosynthesis based on these reviews are provided below.

Ester formation is associated with yeast growth in the early phase of fermentation. Acetate (acetyl) esters are produced via the reaction between an alcohol and acetyl Co-A, which is catalysed by the enzyme alcohol acetyl transferases (ATF1 and ATF2). Ethanol, branched-chain alcohols and 2-phenylethanol are the common moieties of acetate (acetyl) esters. Ethyl esters of medium-chain fatty acids are formed through the reaction between ethanol and respective fatty acyl Co-A, which is catalysed by the enzyme alcohol acyl transferases. *Saccharomyces cerevisiae* strains also produce esterases that hydrolyse esters, and thus the final concentration of esters in beers is the net balance between ester synthesis and hydrolysis. In brewing yeasts, esters are synthesised intracellularly, which then passively diffuse out into the medium due to their lipophilic nature. However, the diffusion of ethyl esters of medium-chain fatty acids decreases with increasing chain length, resulting in lower yields in beers. Ester production in beer is regulated by a number of factors such as yeast strain, temperature, hydrostatic pressure, wort composition, sugar type and concentration, type and amount of yeast-assimilable nitrogen, aeration and unsaturated fatty acids.

18.3.2.5 Ketones

Beer does not seem to contain many ketones. Among the few ketones found in beer, such as β-damascenone, β-ionone and oct-1-en-3-one, the vicinal diketones diacetyl and 2,3-pentanedione are the most important ketones with respect to their impact on beer flavour (Féchir, Reglitz, Mall, Voigt, & Steinhaus, 2021; Gonzalez Viejo, Fuentes, Torrico, Godbole, & Dunshea, 2019; Stewart, 2017). Diacetyl is a very potent, volatile odour-active compound with a butter- or butterscotch-like odour, causing a flavour defect in most beer styles, especially lager-style beers. Reported diacetyl thresholds in lagers and ales are 0.1–0.2 mg/L and 0.1–0.4 mg/L, respectively. On the other hand, 2,3-pentanedione has a toffee-like aroma (threshold ~1 mg/L) but it is not as potent as diacetyl and can also bring about off-flavour in beer. The vicinal diketones are generally undesirable but may constitute a special feature in some beer styles such as Czech Pilsners and some English ales (Stewart, 2017).

Brewer's yeasts are the indirect producers of the vicinal diketones in beer in the sense that they produce the precursors to the diketones during metabolism. The formation pathways of diacetyl and 2,3-pentanedione, as well as their control, have been reviewed by Krogerus and Gibson (2013a). Briefly, their precursors α-acetolactate and α-acetohydroxybutyrate (both are the respective intermediates in the biosynthesis of valine and isoleucine) are secreted out of the cells during yeast growth and metabolism, respectively, due to certain rate-limiting steps in the biosynthetic pathways. The α-acetohydroxy acids, α-acetolactate and α-acetohydroxybutyrate, in the fermenting wort are then non-enzymatically and oxidatively decarboxylated into diacetyl and 2,3-pentanedione, respectively.

Yeast cells also have the ability to reduce the vicinal diketones to acetoin, 2,3-butanediol and 2,3-pentanediol, which have much higher odour detection thresholds and which are the biological basis of the industrial practice of 'diacetyl rest' to decrease or even remove the vicinal diketones. Vicinal diketones can additionally be controlled by regulating process and fermentation conditions, modifying wort composition, applying yeast immobilisation techniques or by improving/modifying yeast strains. For example, restricting oxygen ingress, reducing pitching rates, introducing α-acetolactate decarboxylase genes into brewing yeasts and addition of selected amino acids such as branched-chain amino acids (e.g., valine) are effective measures of reducing vicinal diketone production in beer (Krogerus & Gibson, 2013b; Stewart, 2017).

18.3.2.6 Volatile phenolic compounds

A number of odour-active volatile phenolic compounds are present in different beers, including guaiacol, 4-vinylsyringol, 4-vinylguaiacol, 4-vinylphenol, 4-ethylguaiacol, 4-ethylphenol and vanillin. Most volatile phenols impart various odour notes such as clove, spicy, smoky, medicinal or phenolic (collectively known as 'phenolic odour' or 'phenolic off-flavour'). Phenolic flavour is undesirable in most beer types but is regarded as an attribute or even as essential in certain speciality beers such as lambic and wheat beers (Lentz, 2018).

In beers brewed with phenolic off flavour (POF+) *Saccharomyces* yeasts, 4-vinylguaiacol and 4-vinylphenol are the most abundant, due the enzymatic decarboxylation of ferulic and *p*-coumaric acids respectively. The latter hydroxycinnamic acids are primarily derived from malted grains and hops, which are enzymatically converted by phenyl acrylic acid decarboxylase (PAD1) and ferulic acid decarboxylase (FDC1) to their corresponding vinylphenols. However, POF + yeasts are unable to further reduce vinylphenols to their corresponding ethylphenols, a trait more commonly observed in *Brettanomyces* spp. in speciality beers (e.g., acidic beers) (Lentz, 2018).

18.3.2.7 Terpenoids

The terpenes and terpenoids found in hops and beer include myrcene, limonene, farnesene, humulene, β-caryophyllene, β-citronellol, nerol, α-terpineol, linalool, geraniol, citronellyl acetate, geranyl 2-methyl propanoate, geranyl acetate, etc (Dietz, Cook, Huismann, Wilson, & Ford, 2020; Holt, Miks, de Carvalho, Foulquié-Moreno, & Thevelein, 2019). These terpenes and terpenoids contribute to bitterness, as well as fruity, citrusy, woody and floral aromas while imparting trigeminal-type and mouthfeel sensations (Dietz et al., 2020). Some of the terpenes are carried over from the hops, whereas others such as terpenoid esters and terpene alcohols are produced or released from other monoterpene alcohols and glycoside precursors during yeast fermentation (Fig. 18.1). The biotransformation of beer terpenes and terpenoids by yeasts has been reviewed by Holt et al. (2019) and is briefly summarised below.

Monoterpene alcohols undergo a series of complex biotransformation by yeasts. As illustrated in Fig. 18.1, geraniol (most abundant in wort) is converted into β-citronellol (almost absent in wort, by old yellow enzyme 2, Oye2), while linalool is produced from both geraniol and nerol; then α-terpineol is produced from nerol and linalool. Most of the monoterpene alcohols can be transformed into esters, especially acetate (acetyl) esters. In addition, *Saccharomyces* yeasts are able to release small amounts of monoterpenoids (e.g., geraniol and linalool), either de novo or via terpene glycosides to release their aglycones.

It is expected that these biotransformations would have an impact on beer aroma, since oxygenated terpenes possess subtle odour differences, for example, rose-like/floral for geraniol, fresh, coriander or lavender for linalool, citrusy (lemon or lime) for β-citronellol, floral, fresh, or green for nerol and lilac for α-terpineol. Moreover, the coexistence of geraniol,

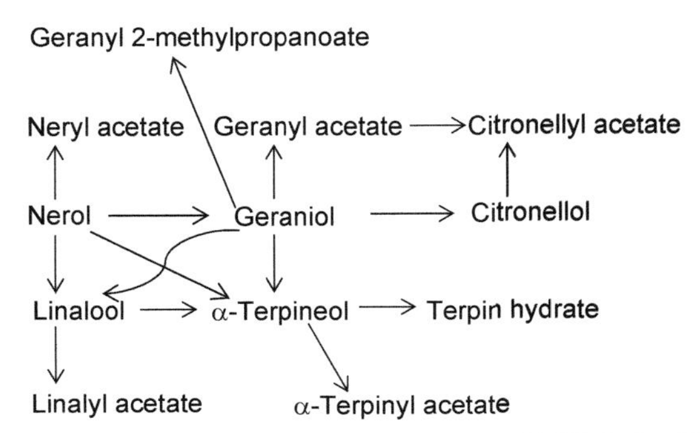

FIGURE 18.1 Proposed pathways of terpenoid biotransformation by yeasts. *Modified after King and Dickinson (2000, 2003).*

β-citronellol and excess linalool have an additive or synergistic effect on the total flavour impression by maximising the citrusy character (Holt et al., 2019). One strategy to realise the coexistence of the three monoterpene alcohols is to control the degree of yeast biotransformation by delaying hop addition so as to retain some geraniol in the finished beer (Takoi et al., 2014). Besides the flavour impact of terpene alcohols, terpenoid acetate esters add more fruitiness (Dietz et al., 2020).

18.3.2.8 Volatile sulphur compounds

Numerous odour-active VSCs have been detected in beers. Examples of beer VSCs encompass SO_2, methanethiol, ethanethiol, H_2S, dimethyl sulfide, dimethyl disuphide, methional, methionol, 3-(methylthio)propyl acetate and 2-mercapto-3-methyl-1-butanol (Ferreira & Guido, 2018; Stewart, 2017). Small amounts of VSCs are acceptable, although excessive amounts cause off-odours such as rotten egg-like, cabbage-like, onion-like and garlic-like. However, some VSCs have a positive impact on beer flavour by accentuating blackcurrant aromas (e.g., 3-mercaptohexanol and 3-mercaptohexyl acetate) (Morimoto et al., 2010).

Brewing yeasts contribute to the genesis of some VSCs in two modes: direct production of VSCs and further chemical and/or a combination of chemical and yeast-mediated conversions of these VSCs into other potent odourous VSCs. For example, H_2S and SO_2 are intermediates of methionine and cysteine de novo synthesis in yeasts, especially under nitrogen limitation (Stewart, 2017). The contents of SO_2 and H_2S in beer usually peak within 2–4 days of fermentation, with H_2S subsequently decreasing to trace levels towards the final stage of fermentation, possibly due to re-assimilation or binding by yeast cells (Oka, Hayashi, Matsumoto, & Yanase, 2008; Stewart, 2017). Nevertheless, contents of SO_2 and H_2S are also influenced by factors such as nitrogen limitation, cysteine and methionine availability, oxygen, type and geometry of fermentation vessels. These factors can be manipulated to produce small amounts of SO_2, which reacts with carbonyl compounds and improve flavour stability by removing cardboard flavour during ageing (Stewart, 2017).

Methional is known to be a key contributor to the worty flavour of alcohol-free beers (Piornos et al., 2020). This sulphur-containing aldehyde and methionol are also precursors to dimethyl trisulphide in aged beers (giving off onion- and garlic-like off-odours) (Gijs, Perpète, Timmermans, & Collin, 2000). Metabolism of methionine via the Ehrlich pathway in yeasts leads to the production of a variety of VSCs, including methional, methionol, methionic acid, methionyl acetate and ethyl 3-methylthio-1-propanoate (Fig. 18.2). Methional is mostly reduced to methionol, which is the VSC of methionine metabolism by yeasts. There is evidence that trace amounts of the highly potent methional (raw potato-like odour) with a very low detection threshold at a few ppb (0.2–40 ppb) can be secreted out of the yeast cells (Liu & Crow, 2010; Quek, Seow, Ong, & Liu, 2011; Seow, Ong, & Liu, 2010; Tan, Lee, Seow, Ong, & Liu, 2012).

Cysteine and homocysteine are also important precursors of VSCs, as shown in Fig. 18.2. A number of thiols can be generated from cysteine, cysteine conjugate and homocysteine transformation during yeast fermentation, including 2-mercaptoethanol, 3-mercaptoethanol, 3-mercaptohexanol, 4-mercapto-4-methyl-2-pentanone and acetate esters such as 3-mercaptohexyl acetate (Gros, Peeters, & Collin, 2012; Kishimoto, Morimoto, Kobayashi, Yako, & Wanikawa, 2008; Nizet et al., 2013).

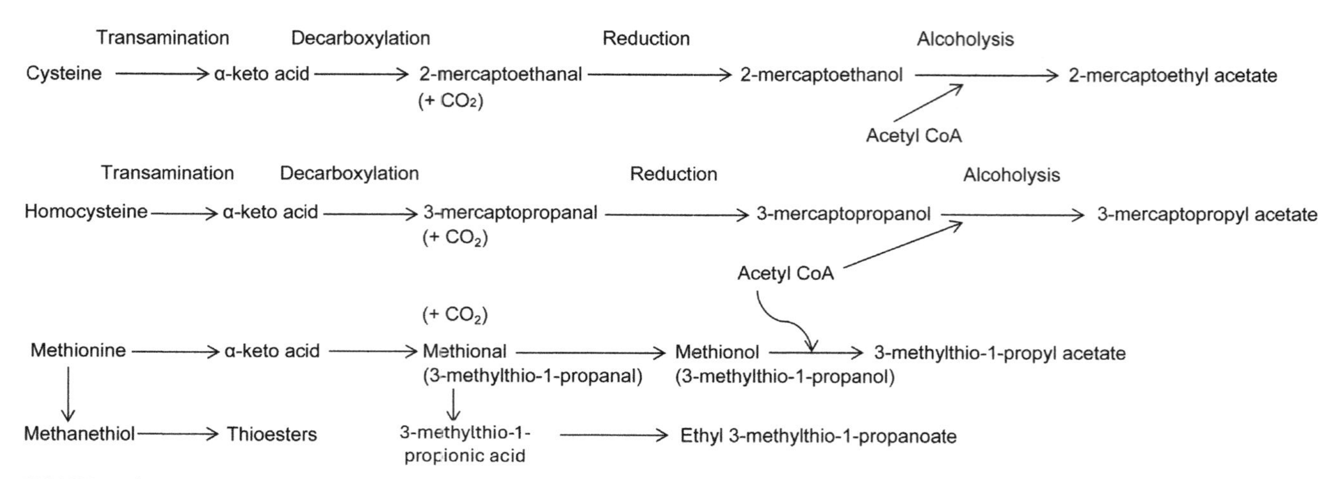

FIGURE 18.2 Proposed pathways of bioproduction of volatile sulphur compounds from cysteine, homocysteine and methionine by yeasts. *Modified after Tan et al. (2012) and Vermeulen, Lejeune, Tran, and Collin (2006).*

18.4 Impact of bacteria on beer appearance and flavour

Relative to the impact of yeast on beer appearance and flavour as discussed above, information is scarce on the impact of bacteria on beer appearance and flavour. In conventional lagers and ales, bacteria are undesirable, mainly causing turbidity, acidification, off-flavour formation (e.g., diacetyl) and ropiness. Spoilage bacteria and their negative influence on beer appearance and flavour have been recently reviewed (Rodhouse & Carbonero, 2019; Suzuki, 2020). Briefly, spoilage bacteria include *Pectinatus* and *Megasphaera*, which cause turbidity and produce rotten egg odour compounds such as H_2S and short-chain fatty acids (e.g., butyric). Acetic acid bacteria such as *Gluconobacter* and *Acetobacter* elicit vinegar odours by producing acetic acid, besides producing ropy textures in beers. Enterobacteria such as *Rahnella aquatilis* and *Obesumbacterium proteus* generate acetaldehyde and methyl acetate to elicit fruity, milky or sulphury aromas. Firmicutes such as *Pediococcus*, lactobacilli (e.g., *L. brevis*, *L. lindneri*) and *Leuconostoc* can elicit ropiness (due to β-glucan production) and produce acidic off-flavour and buttery aromas arising from lactic and diacetyl formation. New beer spoilage microorganisms that cause turbidity have also emerged as a result of rising trends in non-traditional brewing practices and ingredients, including *L. curtus*, *L. cerevisiae*, *L. acetotolerans* and *Prevotella cerevisiae*.

In contrast, LAB are beneficial in the production of acidulated malt, acidified wort and acidic beers (Dysvik et al., 2021; Prado, Gastl, & Becker, 2021). Lactic acid is formed from the Embden-Meyerhof-Parnas (EMP) or phosphogluconate pathway by homofermentative and heterofermentative LAB, respectively. Acetic acid may also be produced by LAB via the phosphogluconate pathway but are predominantly produced by acetic acid bacteria in acidic beers via the oxidation of ethanol to acetic acid (Bongaerts et al., 2021). Besides giving acidic beers their distinctive acidic tastes, the benefits of biological acidification include lower risk of protein haze formation and microbial contamination (and associated microbial turbidity, sediments and off-flavour), finer foam bubbles and stable, longer-lasting foam, fresher mouth-feel, smoother bitterness and fuller and smoother flavour profile, as summed up in the review by Vriesekoop, Krahl, Hucker, and Menz (2012). It should be stressed that strains selected for biological acidification must not produce diacetyl.

18.5 Impact of yeast on the health properties of beer

Besides its impact on flavour, non-alcoholic metabolites produced by brewing yeasts exhibit bioactivities (e.g., antioxidant, anti-inflammatory), which may be responsible for some beneficial health outcomes associated with moderate beer consumption (e.g., cardiovascular disease) (Osorio-Paz, Brunauer, & Alavez, 2020). For example, SO_2 (Section 17.3.2.8) and glutathione (Section 17.3.1) secreted by yeasts are effective antioxidants in beer (Vilela, 2019; Yang & Gao, 2021). Tryptophol and tyrosol (Section 17.3.2.1) also elicit bioactivities ranging from anti-oxidant, anti-carcinogenic, cardioprotective and anti-microbial properties (Cordente et al., 2019; Vilela, 2019). In particular, tyrosol is further converted to hydroxytyrosol in the human host after beer consumption, where it may exert antioxidant effects and confers protection against low density lipoprotein oxidation (Soldevila-Domenech et al., 2019).

In addition to tryptophol, tryptophan is also catabolised to other bioactive metabolites by brewing yeasts. For example, tryptophan is a precursor to serotonin and melatonin (tryptophan → 5-hydroxytryptophan → serotonin → *N*-acetylserotonin → melatonin). Serotonin and melatonin are neurohormones that regulate the circadian rhythm, in addition to exerting anti-oxidant, anti-carcinogenic, immunomodulatory and neuroprotective effects (Cordente et al., 2019; Vilela, 2019). The intermediate, 5-hydroxytryptophan, may also exert neuroprotective effects as it inhibited amyloid β-peptide aggregation in vitro (Hornedo-Ortega et al., 2018). Moreover, tryptophan is a precursor of indole-3-acetic acid via the Ehrlich pathway, where it exerts immunomodulating, anti-thrombotic and microglial activation inhibitory effects in an aryl hydrocarbon receptor (AhR)-dependent manner (Fernández-Cruz et al., 2020; Liu, Hou, Wang, Zheng, & Hao, 2020). In fact, indole-3-acetate and intermediates of the melatonin biosynthesis pathway (5-hydroxytryptophan, serotonin, *N*-acetylserotonin and melatonin) have been detected in commercial beers, although low concentrations detected (pg-ng/mL range) may not be physiologically relevant (Fernández-Cruz et al., 2020).

Volatile phenolic compounds produced by brewing yeasts (Section 17.3.2.6) may also contribute to beer antioxidant and physiological activities (e.g., cancer, gut microbiota) (Ambra, Pastore, & Lucchetti, 2021). For example, enzymatic decarboxylation of hydroxycinnamic acids to vinylphenols may improve antioxidant activities since the latter are more efficient antioxidants in o/w emulsions (Terpinc et al., 2011). Therefore, enriching the wort with hydroxycinnamic acid precursors prior to fermentation may enhance levels of vinylphenols produced by POF + yeasts. This may be achieved via optimising mashing parameters, barley and wheat malt ratios or brewing with adjuncts or ingredients (e.g., herbs, spices and fruits) rich in phenolic compounds (Kalb, Seewald, Hofmann, & Granvogl, 2021; Yang & Gao, 2021). Nevertheless, it must be cautioned that boosting the level of phenolic acids such as ferulic acid in the wort may increase the content of volatile phenols in the finished product, and a flavour defect may ensue. Moreover, excessive cinnamic acid levels in the wort may undergo decarboxylation by yeasts to generate styrene, a class 2b carcinogen (Kalb et al., 2021).

Amid inclinations towards craft beers and probiotic functional foods, beers brewed with the probiotic yeast, *Saccharomyces cerevisiae* var. *boulardii*, are becoming popular and have been reviewed by Chan, Toh, & Liu (2021). Briefly, *S. boulardii* is the only yeast that qualifies as a probiotic, having been shown to alleviate several gastrointestinal disorders such as antibiotic associated and travellers' diarrhoea, irritable bowel syndrome and *Helicobacter pylori* infection (McFarland, Evans, & Goldstein, 2018). Despite being an undomesticated brewing strain, *S. boulardii* demonstrated satisfactory growth (>six to seven Log colony forming units/mL) and fermentation characteristics (e.g., sugar utilisation, ethanol tolerance) in beers. Lower ethanol yields, greater antioxidant capacities and higher acidification rates seem to be characteristic traits in beers fermented by this probiotic yeast (Chan et al., 2021). In particular, greater acidification rates due to acetic acid production by *S. boulardii* was commonly reported (Pereira de Paula et al., 2021; Silva et al., 2021), which, although might increase sourness in beers, may also prevent contamination or food-borne pathogenesis by inhibiting *E. coli* (Chan & Liu, 2022). Besides acetic acid, *S. boulardii* may also produce bioactive metabolites such as polyamines (trophic effects), decanoic acid (antimicrobial), anti-microbial and toxin degrading enzymes, glutathione, γ-aminobutyric acid (GABA), B vitamins, etc (Chan & Liu, 2022), although their production in beers has not been elucidated.

18.6 Impact of bacteria on the health properties of beer

Besides *S. boulardii*, probiotic LAB have also been used to brew beers, primarily for sour beer production. However, hop iso-α-acids adversely impair the viability of Gram-positive probiotic bacteria, necessitating modifications to brewing processes (e.g., co-culturing with yeasts, reduction/removal of hops) to retain probiotic viability (Chan et al., 2021). Retaining probiotic viabilities in foods is paramount, as it is generally recommended that a dosage of 9 Log colony forming units per serving of food product is consumed daily to confer health benefits (Hill et al., 2014). By modifying conventional brewing processes, several studies have successfully brewed beers with co-cultures of probiotic LAB and brewing yeasts. For example, unhopped beers fermented with co-cultures of *L. paracasei* DTA 81 (potential probiotic) and *S. cerevisiae* S-04 promoted anti-depressant behaviours in mice. However, the mechanisms and substances responsible for the psychobiotic properties were not elucidated (Silva et al., 2021).

In unhopped wheat beer fermented by co-cultures of *L. paracasei* Lpc-37 and *S. boulardii*, amino acid catabolites such as (*S*)-(−)-2-hydroxyisocaproic acid, phenylacetic acid (PLA), *p*-hydroxyphenyllactic acid (OH-PLA) and indolelactic acid (ILA) were produced by the probiotic bacteria (Loh, Ng, Toh, Lu, & Liu, 2021). These catabolites are formed from leucine, phenylalanine, tyrosine and tryptophan via the Ehrlich pathway, respectively. Indoleacrylic acid (IA) was also produced, postulated to be via indolelactate dehydratase from ILA. The abovementioned catabolites are antimicrobial compounds, and some are also immunomodulators via the hydroxycarboxylic acid receptor 3 (HCA_3; PLA, ILA) and the AhR (ILA, IA) (Laursen et al., 2021; Liu, Chang, Chang, & Mou, 2020, Liu, Hou et al., 2020). Intriguingly, co-culturing *L. paracasei* Lpc-37 with the probiotic yeast *S. boulardii* enhanced levels of (*S*)-(−)-2-hydroxyisocaproic acid, PLA, OH-PLA and ILA (Loh et al., 2021), which may enhance the health-promoting effects of beers. Indeed, *S. boulardii* similarly enhanced the production of indolic compounds (indole-3-acetic acid and indole) by gut lactobacilli of mice, with the increased production translating to improved gastrointestinal motility and reduced anxiety (Constante et al., 2021).

Nonprobiotic lactic and acetic acid bacteria in mixed and spontaneously fermented sour beers also produce valuable bioactive compounds such as short chain fatty acids (lactate, acetate, succinate) (Tyakht et al., 2021). Lactic and acetic acids in particular exhibit antimicrobial activities, which prevent fungal contamination in acidulated malts (Section 17.4). Lactic and acetic acids also downregulate pro-inflammatory responses in gut, while acting as substrates for butyrate producing members of the gut microbiome (Pascari, Ramos, Marín, & Sanchís, 2018). Moreover, LAB are capable of biotransforming hydroxycinnamic acids into volatile phenols. Unlike yeasts, however, LAB are not capable of producing styrene, and specific strains of lactobacilli may even reduce hydroxycinnamic acids to hydroxyphenylpropionic acids (e.g., phloretic, dihydrocaffeic and dihydroferulic acids). Dihydrocaffeic acid, for example, exhibited stronger in vitro bioactivities than its precursor, caffeic acid, in areas relating to anti-oxidant capacities, platelet and cardiovascular modulating effects, preventing retinal degeneration and anti-inflammatory effects in endothelial cells (Chan & Liu, 2022). However, the production of hydroxyphenylpropionic acids by LAB in beer has not been elucidated.

EPS may also be produced in beers intentionally fermented by bacteria. For example, *Weissela cibara* MG1 was used to produce EPS and oligosaccharides for the development of a prebiotic functional wort beverage (Zannini et al., 2013). However, it should be noted that EPS may cause turbidity and ropiness reminiscent of beer spoilage bacteria (Bongaerts et al., 2021). Other bioactive metabolites that may be derived from LAB-induced wort fermentations include vitamins, minerals, peptides, bacteriocins and enzymes (e.g., bile salt hydrolase) (Barros et al., 2020; Peluzio, Martinez, & Milagro, 2021), although their production in beers has not been demonstrated.

Nonetheless, biogenic amines may also be produced by certain bacterial species, such as *Bifidobacterium*, *Pediococcus* and lactobacilli (Chan et al., 2021; Poveda, Ruiz, Seseña, & Palop, 2017). Biogenic amines such as histamine and tyramine are formed from the decarboxylation of histidine and tyrosine respectively, and excessive amounts may cause migraines, vomiting and hypertension. Biogenic amines are especially relevant in beers since monoamine oxidase, the enzyme responsible for biogenic amine detoxification in humans, is suppressed by ethanol (Chan et al., 2021; Poveda et al., 2017).

18.7 Future trends

Beer has been the subject of intensive biotechnological research to improve appearance and flavour, as well as to add functional and technological benefits. With the advent and maturation of molecular biology over the past few decades, genetic engineering has been exploited extensively to modify brewing yeast strains for the production of beers with specific functional attributes. Commercial examples include genetically modified *S. cerevisiae* strains designed to produce lactic acid and aromatic monoterpenes in beers (Alperstein, Gardner, Sundstrom, Sumby, & Jiranek, 2020). Although not explored, genetically modified strains of *S. boulardii* could also be used to enhance the health-promoting effects of beers. For example, the probiotic yeast has been engineered to secrete bioactive metabolites such as an antibody that neutralised *C. difficile* endotoxins (Chen et al., 2020), β-carotene (Durmusoglu et al., 2021), antilisterial peptide leucocin C (Li, Wan, Takala, & Saris, 2021), neoagaro oligosaccharides (Jin et al., 2021) and atrial natriuretic peptide (Liu, Chang, et al., 2020).

Nonetheless, genetic modification still faces consumer resistance and regulatory challenges (Alperstein et al., 2020). This has prompted the application of non-genetically modified approaches to improve brewing efficiency and quality, including bioprospecting, hybridisation and adaptive laboratory evolution (Gibson et al., 2020; Iattici, Catallo, & Solieri, 2020. For example, bioprospecting for non-conventional brewing microorganisms has led to the usage of *Starmerella bacillaris* (isolated from grape/wine environments) for low-alcohol beer production (Alperstein et al., 2020) and potentially probiotic yeasts (e.g., *Kazachstania unispora* isolated from sourdough) for probiotic beers (Canonico, Zannini, Ciani, & Comitini, 2021). Hybridisation has also led to the development of *S. cerevisiae* × *S. eubayanus* hybrids for low-temperature lager brewing (Gibson et al., 2020) and an *S. jurei* × *S. cerevisiae* hybrid that eliminates hyper attenuation and produce beers with a more complex flavour profile (Giannakou et al., 2021). Adaptive evolution, which applies defined cultivation conditions and selective pressures (e.g., mutagens, base alkylating agents) to select for favourable traits within microbial populations, has produced yeasts with superior environmental resistance (e.g., ethanol), fermentation performance (e.g., maltotriose utilisation, flocculation) and favourable metabolite production (e.g., eliminating excessive acetaldehyde, H_2S, SO_2 productions) (Gibson et al., 2020).

In this 'omics' age, more research is being directed at the genomics, metabolomics, lipidomics and proteomics of brewing yeasts with a view to better understand their impact on beer appearance, flavour and health properties. Untargeted metabolomics, for example, has been employed to characterise bioactive components (e.g., aromatic amino acid catabolites) produced by probiotic yeast and bacteria in beers (Section 17.6) (Loh et al., 2021). Untargeted metabolomics and comparative transcriptomics have also been used to classify and identify differential metabolites among *Saccharomyces* brewing yeasts (e.g., *S. cerevisiae* and *S. pastorianus*) (Behr, Kliche, Geißler, & Vogel, 2020; Seo et al., 2020). Proteomics, on the other hand, has been applied to identify foaming proteins in beers (including yeast origins) (Gonzalez Viejo et al., 2020). Thus, it is not difficult to envision that the 'omics' approach will continue to be taken in this field.

18.8 Further information

Journal of the Institute of Brewing has published a series of reviews (125th Anniversary Reviews) on various aspects of brewing and beer, including bacterial and yeast impact on beer appearance and flavour. The 125th Anniversary Reviews Virtual Issue is recommended for further reading. The Leuven Institute for Beer Research (http://libr.be/) is a source of information on brewing yeast and beer fermentation in relation to impact on appearance and flavour, in addition to the International Centre for Brewing and Distilling (http://www.icbd.hw.ac.uk/).

Information covered in this chapter can also be found in several journals devoted to beer and brewing listed below:

The Journal of the American Society of Brewing Chemists (http://www.asbcnet.org/journal/default.htm)
Journal of the Institute of Brewing (http://onlinelibrary.wiley.com/journal/10.1002/(ISSN)2050-0416)
Master Brewers Association of the Americas Technical Quarterly. (http://www.mbaa.com/publications/tq/Pages/default.aspx)
Brewing Science — Monatsschrift für Brauwissenschaft (http://www.brewingscience.de/)
Cerevisia (http://www.journals.elsevier.com/cerevisia)

References

Alperstein, L., Gardner, J. M., Sundstrom, J. F., Sumby, K. M., & Jiranek, V. (2020). Yeast bioprospecting versus synthetic biology—which is better for innovative beverage fermentation? *Applied Microbiology and Biotechnology, 104*(5), 1939–1953. https://doi.org/10.1007/s00253-020-10364-x

Amata, B. I. A., & Germain, P. (1990). The effect of pitching yeast aeration on the production of acetic acid during fermentations with brewers' yeast: An enzymatic approach. *Journal of the Institute of Brewing, 96*(3), 131–134. https://doi.org/10.1002/j.2050-0416.1990.tb01023.x

Ambra, R., Pastore, G., & Lucchetti, S. (2021). The role of bioactive phenolic compounds on the impact of beer on health. *Molecules, 26*(2), 486. https://doi.org/10.3390/molecules26020486

Barros, C. P., Guimarães, J. T., Esmerino, E. A., Duarte, M. C. K. H., Silva, M. C., Silva, R., et al. (2020). Paraprobiotics and postbiotics: Concepts and potential applications in dairy products. *Current Opinion in Food Science, 32*, 1–8. https://doi.org/10.1016/j.cofs.2019.12.003

Behr, J., Kliche, M., Geißler, A., & Vogel, R. F. (2020). Exploring the potential of comparative de novo transcriptomics to classify *Saccharomyces* brewing yeasts. *PLoS One, 15*(9), Article e0238924. https://doi.org/10.1371/journal.pone.0238924

Blasco, L., Vinas, M., & Villa, T. G. (2011). Proteins influencing foam formation in wine and beer: The role of yeast. *International Microbiology, 14*, 61–71. https://doi.org/10.2436/20.1501.01.136

Bongaerts, D., De Roos, J., De Vuyst, L., & Björkroth, J. (2021). Technological and environmental features determine the uniqueness of the lambic beer microbiota and production process. *Applied and Environmental Microbiology, 87*(18), Article e00612-21. https://doi.org/10.1128/AEM.00612-21

Bravi, E., Perretti, G., Buzzini, P., Della Sera, R., & Fantozzi, P. (2009). Technological steps and yeast biomass as factors affecting the lipid content of beer during the brewing process. *Journal of Agricultural and Food Chemistry, 57*(14), 6279–6284. https://doi.org/10.1021/jf9007423

Callejo, M. J., García Navas, J. J., Alba, R., Escott, C., Loira, I., González, M. C., et al. (2019). Wort fermentation and beer conditioning with selected non-*Saccharomyces* yeasts in craft beers. *European Food Research and Technology, 245*(6), 1229–1238. https://doi.org/10.1007/s00217-019-03244-w

Canonico, L., Zannini, E., Ciani, M., & Comitini, F. (2021). Assessment of non-conventional yeasts with potential probiotic for protein-fortified craft beer production. *LWT, 145*, Article 111361. https://doi.org/10.1016/j.lwt.2021.111361

Chan, M. Z. A., & Liu, S.-Q. (2022). Fortifying foods with synbiotic and postbiotic preparations of the probiotic yeast, *Saccharomyces boulardii*. *Current Opinion in Food Science, 43*, 216–224. https://doi.org/10.1016/j.cofs.2021.12.009

Chan, M. Z. A., Toh, M., & Liu, S.-Q. (2021). In A. Gomes da Cruz, C. S. Ranadheera, F. Nazzaro, & A. Mortazavian (Eds.), *Chapter 10 - beer with probiotics and prebiotics* (pp. 179–199). Academic Press. https://doi.org/10.1016/B978-0-12-819662-5.00004-5

Chen, K., Zhu, Y., Zhang, Y., Hamza, T., Yu, H., Saint Fleur, A., et al. (2020). A probiotic yeast-based immunotherapy against *Clostridioides difficile* infection. *Science Translational Medicine, 12*(567), Article eaax4905. https://doi.org/10.1126/scitranslmed.aax4905

Clapperton, J. F. (1978). Fatty acids contributing to caprylic flavour in beer. The use of profile and threshold data in flavour research. *Journal of the Institute of Brewing, 84*(2), 107–112. https://doi.org/10.1002/j.2050-0416.1978.tb03849.x

Constante, M., De Palma, G., Lu, J., Jury, J., Rondeau, L., Caminero, A., et al. (2021). *Saccharomyces boulardii* CNCM I-745 modulates the microbiota–gut–brain axis in a humanized mouse model of irritable bowel syndrome. *Neuro-Gastroenterology and Motility, 33*(3), Article e13985. https://doi.org/10.1111/nmo.13985

Cordente, A. G., Schmidt, S., Beltran, G., Torija, M. J., & Curtin, C. D. (2019). Harnessing yeast metabolism of aromatic amino acids for fermented beverage bioflavouring and bioproduction. *Applied Microbiology and Biotechnology, 103*(11), 4325–4336. https://doi.org/10.1007/s00253-019-09840-w

Dietz, C., Cook, D., Huismann, M., Wilson, C., & Ford, R. (2020). The multisensory perception of hop essential oil: A review. *Journal of the Institute of Brewing, 126*(4), 320–342. https://doi.org/10.1002/jib.622

Durmusoglu, D., Al'Abri, I. S., Collins, S. P., Cheng, J., Eroglu, A., Beisel, C. L., et al. (2021). In situ biomanufacturing of small molecules in the mammalian gut by probiotic *Saccharomyces boulardii*. *ACS Synthetic Biology, 10*(5), 1039–1052. https://doi.org/10.1021/acssynbio.0c00562

Dysvik, A., La Rosa Sabina, L., De Rouck, G., Rukke, E.-O., Westereng, B., Wicklund, T., et al. (2021). Microbial dynamics in traditional and modern sour beer production. *Applied and Environmental Microbiology, 86*(14), Article e00566-20. https://doi.org/10.1128/AEM.00566-20

Féchir, M., Reglitz, K., Mall, V., Voigt, J., & Steinhaus, M. (2021). Molecular insights into the contribution of specialty barley malts to the aroma of bottom-fermented lager beers. *Journal of Agricultural and Food Chemistry, 69*(29), 8190–8199. https://doi.org/10.1021/acs.jafc.1c01846

Fernández-Cruz, E., Carrasco-Galán, F., Cerezo-López, A. B., Valero, E., Morcillo-Parra, M.Á., Beltran, G., et al. (2020). Occurrence of melatonin and indolic compounds derived from L-tryptophan yeast metabolism in fermented wort and commercial beers. *Food Chemistry, 331*, Article 127192. https://doi.org/10.1016/j.foodchem.2020.127192

Ferreira, I. M., & Guido, L. F. (2018). Impact of wort amino acids on beer flavour: A review. *Fermentation, 4*(2), 23. https://doi.org/10.3390/fermentation4020023

Filipowska, W., Jaskula-Goiris, B., Ditrych, M., Bustillo Trueba, P., De Rouck, G., Aerts, G., et al. (2021). On the contribution of malt quality and the malting process to the formation of beer staling aldehydes: A review. *Journal of the Institute of Brewing, 127*(2), 107–126. https://doi.org/10.1002/jib.644

Giannakou, K., Visinoni, F., Zhang, P., Nathoo, N., Jones, P., Cotterrell, M., et al. (2021). Biotechnological exploitation of *Saccharomyces jurei* and its hybrids in craft beer fermentation uncovers new aroma combinations. *Food Microbiology, 100*, Article 103838. https://doi.org/10.1016/j.fm.2021.103838

Gibson, B., Dahabieh, M., Krogerus, K., Jouhten, P., Magalhães, F., Pereira, R., et al. (2020). Adaptive laboratory evolution of ale and lager yeasts for improved brewing efficiency and beer quality. *Annual Review of Food Science and Technology, 11*(1), 23–44. https://doi.org/10.1146/annurev-food-032519-051715

Gijs, L., Perpète, P., Timmermans, A., & Collin, S. (2000). 3-Methylthiopropionaldehyde as precursor of dimethyl trisulfide in aged beers. *Journal of Agricultural and Food Chemistry, 48*(12), 6196–6199. https://doi.org/10.1021/jf0007380

Gonzalez Viejo, C., Caboche, C. H., Kerr, E. D., Pegg, C. L., Schulz, B. L., Howell, K., et al. (2020). Development of a rapid method to assess beer foamability based on relative protein content using robobeer and machine learning modeling. *Beverages, 6*(2), 28. https://doi.org/10.3390/beverages6020028

Gonzalez Viejo, C., Fuentes, S., Torrico, D. D., Godbole, A., & Dunshea, F. R. (2019). Chemical characterization of aromas in beer and their effect on consumers liking. *Food Chemistry, 293*, 479–485. https://doi.org/10.1016/j.foodchem.2019.04.114

Gros, J., Peeters, F., & Collin, S. (2012). Occurrence of odorant polyfunctional thiols in beers hopped with different cultivars. first evidence of an s-cysteine conjugate in hop (*Humulus Lupulus* l.). *Journal of Agricultural and Food Chemistry, 60*(32), 7805–7816. https://doi.org/10.1021/jf301478m

Hazelwood, L. A., Daran, J., Van Maris, A. J., Pronk, J. T., & Dickinson, J. R. (2008). The Ehrlich pathway for fusel alcohol production: A century of research on *Saccharomyces cerevisiae* metabolism. *Applied and Environmental Microbiology, 74*(8), 2259–2266. https://doi.org/10.1128/AEM.02625-07

Hill, C., Guarner, F., Reid, G., Gibson, G. R., Merenstein, D. J., Pot, B., et al. (2014). The international scientific association for probiotics and prebiotics consensus statement on the scope and appropriate use of the term probiotic. *Nature Reviews Gastroenterology and Hepatology, 11*(8), 506–514. https://doi.org/10.1038/nrgastro.2014.66

Holt, S., Miks, M. H., de Carvalho, B. T., Foulquié-Moreno, M. R., & Thevelein, J. M. (2019). The molecular biology of fruity and floral aromas in beer and other alcoholic beverages. *FEMS Microbiology Reviews, 43*(3), 193–222. https://doi.org/10.1093/femsre/fuy041

Horák, T., Čulík, J., Jurková, M., Čejka, P., & Kellner, V. (2008). Determination of free medium-chain fatty acids in beer by stir bar sorptive extraction. *Journal of Chromatography A, 1196–1197*, 96–99. https://doi.org/10.1016/j.chroma.2008.05.014

Hornedo-Ortega, R., Da Costa, G., Cerezo, A. B., Troncoso, A. M., Richard, T., & Garcia-Parrilla, M. C. (2018). In vitro effects of serotonin, melatonin, and other related indole compounds on amyloid-β kinetics and neuroprotection. *Molecular Nutrition and Food Research, 62*(3), Article 1700383. https://doi.org/10.1002/mnfr.201700383

Iattici, F., Catallo, M., & Solieri, L. (2020). Designing new yeasts for craft brewing: When natural biodiversity meets biotechnology. *Beverages, 6*(1), 3. https://doi.org/10.3390/beverages6010003

Iimure, T., & Sato, K. (2013). Beer proteomics analysis for beer quality control and malting barley breeding. *Food Research International, 54*(1), 1013–1020. https://doi.org/10.1016/j.foodres.2012.11.028

Jin, Y., Yu, S., Liu, J.-J., Yun, E. J., Lee, J. W., Jin, Y.-S., et al. (2021). Production of neoagarooligosaccharides by probiotic yeast *Saccharomyces cerevisiae* var. *boulardii* engineered as a microbial cell factory. *Microbial Cell Factories, 20*(1), 160. https://doi.org/10.1186/s12934-021-01644-w

Kalb, V., Seewald, T., Hofmann, T., & Granvogl, M. (2021). Investigations into the ability to reduce cinnamic acid as undesired precursor of toxicologically relevant styrene in wort by different barley to wheat ratios (grain bill) during mashing. *Journal of Agricultural and Food Chemistry, 69*(32), 9443–9450. https://doi.org/10.1021/acs.jafc.1c03018

King, A., & Dickinson, J. R. (2000). Biotransformation of monoterpene alcohols by *Saccharomyces cerevisiae, Torulaspora delbrueckii* and *Kluyveromyces lactis*. *Yeast, 16*, 499–506. https://doi.org/10.1002/(SICI)1097-0061(200004)

King, A. J., & Dickinson, J. R. (2003). Biotransformation of hop aroma terpenoids by ale and lager yeasts. *FEMS Yeast Research, 3*(1), 53–62. https://doi.org/10.1111/j.1567-1364.2003.tb00138.x

Kishimoto, T., Morimoto, M., Kobayashi, M., Yako, N., & Wanikawa, A. (2008). Behaviors of 3-mercaptohexan-1-ol and 3-mercaptohexyl acetate during brewing processes. *Journal of the American Society of Brewing Chemists, 66*(3), 192–196. https://doi.org/10.1094/ASBCJ-2008-0702-01

Krogerus, K., & Gibson, B. R. (2013a). Diacetyl and its control during brewery fermentation. *Journal of the Institute of Brewing, 119*(3), 86–97. https://doi.org/10.1002/jib.84

Krogerus, K., & Gibson, B. R. (2013b). Influence of valine and other amino acids on total diacetyl and 2,3-pentanedione levels during fermentation of brewer's wort. *Applied Microbiology and Biotechnology, 97*(15), 6919–6930. https://doi.org/10.1007/s00253-013-4955-1

Laursen, M. F., Sakanaka, M., von Burg, N., Mörbe, U., Andersen, D., Moll, J. M., et al. (2021). *Bifidobacterium* species associated with breastfeeding produce aromatic lactic acids in the infant gut. *Nature Microbiology, 6*(11), 1367–1382. https://doi.org/10.1038/s41564-021-00970-4

Lehnhardt, F., Gastl, M., & Becker, T. (2019). Forced into aging: Analytical prediction of the flavor-stability of lager beer. A review. *Critical Reviews in Food Science and Nutrition, 59*(16), 2642–2653. https://doi.org/10.1080/10408398.2018.1462761

Lentz, M. (2018). The impact of simple phenolic compounds on beer aroma and flavor. *Fermentation, 4*(1), 20. https://doi.org/10.3390/fermentation4010020

Li, R., Wan, X., Takala, T. M., & Saris, P. E. J. (2021). Heterologous expression of the leuconostoc bacteriocin leucocin c in probiotic yeast *Saccharomyces boulardii*. *Probiotics and Antimicrobial Proteins, 13*(1), 229–237. https://doi.org/10.1007/s12602-020-09676-1

Liu, C.-H., Chang, J.-H., Chang, Y.-C., & Mou, K. Y. (2020a). Treatment of murine colitis by *Saccharomyces boulardii* secreting atrial natriuretic peptide. *Journal of Molecular Medicine, 98*(12), 1675–1687. https://doi.org/10.1007/s00109-020-01987-8

Liu, S.-Q., & Crow, V. L. (2010). Production of dairy-based, natural sulphur flavor concentrate by yeast fermentation. *Food Biotechnology, 24*(1), 62–77. https://doi.org/10.1080/08905430903562724

Liu, Y., Hou, Y., Wang, G., Zheng, X., & Hao, H. (2020b). Gut microbial metabolites of aromatic amino acids as signals in host–microbe interplay. *Trends in Endocrinology and Metabolism, 31*(11), 818–834. https://doi.org/10.1016/j.tem.2020.02.012

Loh, L. X., Ng, D. H. J., Toh, M., Lu, Y., & Liu, S. Q. (2021). Targeted and nontargeted metabolomics of amino acids and bioactive metabolites in probiotic-fermented unhopped beers using liquid chromatography high-resolution mass spectrometry. *Journal of Agricultural and Food Chemistry, 69*(46), 14024–14036. https://doi.org/10.1021/acs.jafc.1c03992

Mastanjević, K., Krstanović, V., Lukinac, J., Jukić, M., Vulin, Z., & Mastanjević, K. (2018). Beer—the importance of colloidal stability (non-biological haze). *Fermentation, 4*(4), 91. https://doi.org/10.3390/fermentation4040091

McFarland, L. V., Evans, C. T., & Goldstein, E. J. C. (2018). Strain-specificity and disease-specificity of probiotic efficacy: A systematic review and meta-analysis. *Frontiers of Medicine, 7*(5), 124. https://doi.org/10.3389/fmed.2018.00124

Morimoto, M., Kishimoto, T., Kobayashi, M., Yako, N., Iida, A., Wanikawa, A., et al. (2010). Effects of bordeaux mixture (copper sulfate) treatment on blackcurrant/muscat-like odors in hops and beer. *Journal of the American Society of Brewing Chemists, 68*(1), 30−33. https://doi.org/10.1094/ASBCJ-2009-1118-01

Nizet, S., Gros, J., Peeters, F., Chaumont, S., Robiette, R., & Collin, S. (2013). First evidence of the production of odorant polyfunctional thiols by bottle refermentation. *Journal of the American Society of Brewing Chemists, 71*(1), 15−22. https://doi.org/10.1094/ASBCJ-2013-0117-01

Oka, K., Hayashi, T., Matsumoto, N., & Yanase, H. (2008). Decrease in hydrogen sulfide content during the final stage of beer fermentation due to involvement of yeast and not carbon dioxide gas purging. *Journal of Bioscience and Bioengineering, 106*(3), 253−257. https://doi.org/10.1263/jbb.106.253

Olšovská, J., Vrzal, T., Štěrba, K., Slabý, M., Kubizniaková, P., & Čejka, P. (2019). The chemical profiling of fatty acids during the brewing process. *Journal of the Science of Food and Agriculture, 99*(4), 1772−1779. https://doi.org/10.1002/jsfa.9369

Olaniran, A. O., Hiralal, L., Mokoena, M. P., & Pillay, B. (2017). Flavour-active volatile compounds in beer: Production, regulation and control. *Journal of the Institute of Brewing, 123*(1), 13−23. https://doi.org/10.1002/jib.389

Osorio-Paz, I., Brunauer, R., & Alavez, S. (2020). Beer and its non-alcoholic compounds in health and disease. *Critical Reviews in Food Science and Nutrition, 60*(20), 3492−3505. https://doi.org/10.1080/10408398.2019.1696278

Pascari, X., Ramos, A. J., Marín, S., & Sanchís, V. (2018). Mycotoxins and beer. Impact of beer production process on mycotoxin contamination. A review. *Food Research International, 103*, 121−129. https://doi.org/10.1016/j.foodres.2017.07.038

Peluzio, M. do C. G., Martinez, J. A., & Milagro, F. I. (2021). Postbiotics: Metabolites and mechanisms involved in microbiota-host interactions. *Trends in Food Science and Technology, 108*, 11−26. https://doi.org/10.1016/j.tifs.2020.12.004

Pereira de Paula, B., de Souza Lago, H., Firmino, L., Fernandes Lemos Júnior, W. J., Ferreira Dutra Corrêa, M., Fioravante Guerra, A., et al. (2021). Technological features of *Saccharomyces cerevisiae* var. *boulardii* for potential probiotic wheat beer development. *LWT, 135*, Article 110233. https://doi.org/10.1016/j.lwt.2020.110233

Piornos, J. A., Balagiannis, D. P., Methven, L., Koussissi, E., Brouwer, E., & Parker, J. K. (2020). Elucidating the odor-active aroma compounds in alcohol-free beer and their contribution to the worty flavor. *Journal of Agricultural and Food Chemistry, 68*(37), 10088−10096. https://doi.org/10.1021/acs.jafc.0c03902

Pires, E. J., Teixeira, J. A., Brányik, T., & Vicente, A. A. (2014). Yeast: The soul of beer's aroma—a review of flavour-active esters and higher alcohols produced by the brewing yeast. *Applied Microbiology and Biotechnology, 98*(5), 1937−1949. https://doi.org/10.1007/s00253-013-5470-0

Poveda, J. M., Ruiz, P., Seseña, S., & Palop, M. L. (2017). Occurrence of biogenic amine-forming lactic acid bacteria during a craft brewing process. *LWT - Food Science and Technology, 85*, 129−136. https://doi.org/10.1016/j.lwt.2017.07.003

Prado, R., Gastl, M., & Becker, T. (2021). Aroma and color development during the production of specialty malts: A review. *Comprehensive Reviews in Food Science and Food Safety, 20*(5), 4816−4840. https://doi.org/10.1111/1541-4337.12806

Quek, J. M. B., Seow, Y.-X., Ong, P. K. C., & Liu, S.-Q. (2011). Formation of volatile sulfur-containing compounds by *Saccharomyces cerevisiae* in soymilk supplemented with L-methionine. *Food Biotechnology, 25*(4), 292−304. https://doi.org/10.1080/08905436.2011.617254

Rodhouse, L., & Carbonero, F. (2019). Overview of craft brewing specificities and potentially associated microbiota. *Critical Reviews in Food Science and Nutrition, 59*(3), 462−473. https://doi.org/10.1080/10408398.2017.1378616

Rodrigues, J. E. A., Erny, G. L., Barros, A. S., Esteves, V. I., Brandão, T., Ferreira, A. A., et al. (2010). Quantification of organic acids in beer by nuclear magnetic resonance (NMR)-based methods. *Analytica Chimica Acta, 674*(2), 166−175. https://doi.org/10.1016/j.aca.2010.06.029

Seo, S.-H., Kim, E.-J., Park, S.-E., Park, D.-H., Park, K. M., Na, C.-S., et al. (2020). GC/MS-based metabolomics study to investigate differential metabolites between ale and lager beers. *Food Bioscience, 36*, Article 100671. https://doi.org/10.1016/j.fbio.2020.100671

Seow, Y.-X., Ong, P. K. C., & Liu, S.-Q. (2010). Production of flavour-active methionol from methionine metabolism by yeasts in coconut cream. *International Journal of Food Microbiology, 143*(3), 235−240. https://doi.org/10.1016/j.ijfoodmicro.2010.08.003

Silva, L. C., de Souza Lago, H., Rocha, M. O. T., de Oliveira, V. S., Laureano-Melo, R., Stutz, E. T. G., et al. (2021). Craft beers fermented by potential probiotic yeast or lacticaseibacilli strains promote antidepressant-like behavior in swiss webster mice. *Probiotics and Antimicrobial Proteins, 13*(3), 698−708. https://doi.org/10.1007/s12602-020-09736-6

Silva Ferreira, C., Bodart, E., & Collin, S. (2019). Why craft brewers should be advised to use bottle refermentation to improve late-hopped beer stability. *Beverages, 5*(2), 39. https://doi.org/10.3390/beverages5020039

Soldevila-Domenech, N., Boronat, A., Mateus, J., Diaz-Pellicer, P., Matilla, I., Pérez-Otero, M., et al. (2019). Generation of the antioxidant hydroxytyrosol from tyrosol present in beer and red wine in a randomized clinical trial. *Nutrients, 11*(9), 2241. https://doi.org/10.3390/nu11092241

Stewart, G. G. (2017). The production of secondary metabolites with flavour potential during brewing and distilling wort fermentations. *Fermentation, 3*(4), 63. https://doi.org/10.3390/fermentation3040063

Suzuki, K. (2020). Emergence of new spoilage microorganisms in the brewing industry and development of microbiological quality control methods to cope with this phenomenon: A review. *Journal of the American Society of Brewing Chemists, 78*(4), 245−259. https://doi.org/10.1080/03610470.2020.1782101

Takoi, K., Itoga, Y., Takayanagi, J., Kosugi, T., Shioi, T., Nakamura, T., et al. (2014). Screening of geraniol-rich flavor hop and interesting behavior of β-citronellol during fermentation under various hop-addition timings. *Journal of the American Society of Brewing Chemists, 72*(1), 22−29. https://doi.org/10.1094/ASBCJ-2014-0116-01

Tan, A. W. J., Lee, P.-R., Seow, Y.-X., Ong, P. K. C., & Liu, S.-Q. (2012). Volatile sulphur compounds and pathways of L-methionine catabolism in *Williopsis* yeasts. *Applied Microbiology and Biotechnology, 95*(4), 1011–1020. https://doi.org/10.1007/s00253-012-3963-x

Terpinc, P., Polak, T., Šegatin, N., Hanzlowsky, A., Ulrih, N. P., & Abramovič, H. (2011). Antioxidant properties of 4-vinyl derivatives of hydroxycinnamic acids. *Food Chemistry, 128*(1), 62–69. https://doi.org/10.1016/j.foodchem.2011.02.077

Tyakht, A., Kopeliovich, A., Klimenko, N., Efimova, D., Dovidchenko, N., Odintsova, V., et al. (2021). Characteristics of bacterial and yeast microbiomes in spontaneous and mixed-fermentation beer and cider. *Food Microbiology, 94*, Article 103658. https://doi.org/10.1016/j.fm.2020.103658

Vermeulen, C., Lejeune, I., Tran, T. T. H., & Collin, S. (2006). Occurrence of polyfunctional thiols in fresh lager beers. *Journal of Agricultural and Food Chemistry, 54*, 5061–5068. https://doi.org/10.1021/jf060669a

Vilela, A. (2019). The importance of yeasts on fermentation quality and human health-promoting compounds. *Fermentation, 5*(2), 46. https://doi.org/10.3390/fermentation5020046

Vinther Schmidt, C., Olsen, K., & Mouritsen, O. G. (2021). Umami potential of fermented beverages: Sake, wine, champagne, and beer. *Food Chemistry, 360*, Article 128971. https://doi.org/10.1016/j.foodchem.2020.128971

Vriesekoop, F., Krahl, M., Hucker, B., & Menz, G. (2012). Bacteria in brewing: The good, the bad and the ugly. *Journal of the Institute of Brewing, 118*(4), 335–345. https://doi.org/10.1002/jib.49

Yang, D., & Gao, X. (2021). Research progress on the antioxidant biological activity of beer and strategy for applications. *Trends in Food Science and Technology, 110*, 754–764. https://doi.org/10.1016/j.tifs.2021.02.048

Zannini, E., Mauch, A., Galle, S., Gänzle, M., Coffey, A., Arendt, E. K., et al. (2013). Barley malt wort fermentation by exopolysaccharide-forming *Weissella cibaria* MG1 for the production of a novel beverage. *Journal of Applied Microbiology, 115*(6), 1379–1387. https://doi.org/10.1111/jam.12329

Zhao, X., Procopio, S., & Becker, T. (2015). Flavor impacts of glycerol in the processing of yeast fermented beverages: A review. *Journal of Food Science and Technology, 52*(12), 7588–7598. https://doi.org/10.1007/s13197-015-1977-y

Chapter 19

Sensory analysis as a tool for microbial quality control in the brewery

Gary Spedding[1] **and Tony Aiken**[2]
[1]*Brewing and Distilling Analytical Services, LLC, Lexington, KY, United States;* [2]*Data Collection Solutions, Lexington, KY, United States*

[Understanding Key Beer Flavours Associated With Bacteria, Mould, Saccharomyces *and Wild Yeast, Hybrid and Engineered Yeasts, and Bacterial and Yeast Metabolism in Brewing. Emphasis on the use of new microbes selected for brewing, and use of sensory analytics in flavour profile evaluation]*

19.1 Introduction

Sensory analysis of beer forms a complex topic, the mechanics of which have been well documented in entire volumes. A single chapter here is not realistically adequate to thoroughly explore the subject. However, an attempt is made to outline some key sensory terms and approaches to investigating, from an organoleptic perspective, key flavours produced by microorganisms. Ones that are either regarded as unwanted contaminants, causing beer spoilage or desired organisms used to produce distinctive flavour profiles in speciality beers. In this regard, we believe this chapter to be unique. The methods outlined will indeed apply to testing other flavour attributes in beer, though the focus is on directly or indirectly microbially generated flavour notes that might typically form only a part of full sensory training in the brewery; that said references will be provided for those wishing to set up a full sensory evaluation programme in the modern brewery environment. Following some definitions on taints and off-flavours, a brief overview is presented regarding microbial spoilage organisms and the distinct stages at which they either provoke the damage to flavour quality and therefore, desirability for consumption, or provide desirable aroma/flavour properties to unique beer styles such as traditional sour beers. These days the fermentation of beer worts may be performed by a growing number of wild yeasts, *Saccharomyces* and non-*Saccharomyces* species/strains and bacteria found via bioprospecting or as obtained via natural selection or via genetic alteration. Such strains are covered in some detail here – with a growing knowledge accumulating regarding their metabolomes – the machinery of production of their volatile flavour metabolites. The desired flavour notes found in beer as produced by these alternate microbial fermentative organisms are outlined here, and then those flavour notes are described from a point of understanding how, and when they are produced and can be controlled and defined. Then Part 2 follows on with a brief review of setting up sensory evaluation programmes. Sensory analysis using human subjects, rather than via electronic noses, etc., require a full understanding of how to describe the flavour notes and their origins – with a growing lexicon now available for training purposes and allowing for more meaningful reporting of actionable information back to brewers.

Brewing Microbiology. https://doi.org/10.1016/B978-0-323-99606-8.00005-5

19.2 Part 1: Microbes, flavours, novel flavour profiles, off-flavours and taints in brewing

19.2.1 Microbial spoilage overview

Other chapters in this volume, and other reference works, cover the general safety aspects of beer based upon microbial presence — both in terms of bacteria and wild yeast species that might be unknowingly and undesirably introduced into the product at some stage of the process. Some preliminary factors are noted here with bearing on more coverage that appears in later sections. Issues of note include recent incidents of brewing product recalls due to the rising appearance of the yeast species *Saccharomyces* var. *diastaticus* as a case example (see later in this section), the increasing consumer demand for non- or low-alcoholic beers (NAB/LAB); avoiding worty, bland beer flavour profiles (Section 5.2) and an increasing understanding of the concerns over the production and control of the buttery flavour note, diacetyl (Section 5.3).

Fermentation and spoilage yeasts and bacteria, reviews of beer safety detailing challenges, trends and issues within the craft and large-scale sector brewery environment have been detailed by others (Ciont et al., 2022; Hutzler, Riedl, Koob, & Jacob, 2012; Schneiderbanger, Grammer, Jacob, & Hutzler, 2018; Schneiderbanger et al., 2020). Regarding bacterial issues, lower alcohol content beers and reduced hop bittering can be a cause of concern for beer spoilage. Some Gram-positive bacteria, *Lactobacillus* and *Pediococcus* species, can grow in beer with some LAB useful for sour beer production and showing resistance to hop iso-alpha acids. Other Gram-positive bacteria are susceptible to the antibacterial action of the hop acids and other hop oil components. Sensory evaluation — evidenced often by foul odours — can be a leading indicator of microbial infection with recent cases noted in market recalls (Ciont et al., 2022). Sensory training to enable the recognition of off-flavours and taints (defined below) is thus useful in this regard. Another case in point being that of *Staphylococcus xylosus* shown to be a cause of turbid and off-flavoured craft beer (Ciont et al., 2022). Unwanted beer sourness is linked to acetic acid and lactic acid bacteria with *Pediococcus damnosus* and *Levilactobacillus brevis* noted as most widely occurring in beer; diacetyl off-flavour is also associated with *P. damnosus* contamination (detailed further in Section 5.3). Other spoilage bacteria detailed from earlier research are covered in Table 19.1.

Most spoilage yeasts and bacteria are non-pathogenic, and the properties of beer ensure that no pathogens can survive the process and in the packaged product (Hill, 2009; Suzuki, 2011; Vaughan, O'Sullivan, & Van Sinderen, 2005; Vriesekoop, Krahl, Hucker, & Menz, 2012). However, spoilage can occur early in the production of beer and any taint flavours (defined below) will often remain leading to a 'spoilt' product. That said mycotoxin-forming moulds can pose a health risk from tainted raw materials, though again issues would be more obvious via gushing of beer contents or mouldy flavour notes (taints) rather than through any illnesses caused by such organisms (Hill, 2009; Vaughan et al., 2005). While beer contaminated with spoilage organisms will normally not be harmful the results of the spoilage may be noticeable by the consumer and will be rejected by them. Microbial spoilage of beer is, therefore, defined as growth of the spoilage organisms to a sufficient level as to promote an alteration in that beer perceptible to a consumer and liable to cause dissatisfaction, complaint or rejection of that beer (Stratford, 2006). Visual cues: distorted cans (due to over-carbonation), beer fobbing or gushing, hazes or visible yeast colonies, blooms (surface films, biofilms) or pellicles, etc., can provide clues to spoilt product (Boulton & Quain, 2001; Hill, 2009; Romano, Capece, & Jesperson, 2006; Stratford, 2006) as can beer that pours in a viscous-oily way. With respect to mycotoxins, three recent papers address the current understanding of the issues concerning product quality and safety (Pascari, Ramos, Marín, & Sanchís, 2018; Pascari, Marin, Ramos, & Sanchis, 2022; Peters et al., 2017). From a sensory evaluation viewpoint — the mycotoxins issue needs to be addressed with the raw materials and mouldy aromas — a taints issue (see Section 2.4 for more on typical mould taint issues).

Detectable spoilage requires a substantial number of yeasts (and/or bacteria) in the order of $1 \times 10^5 - 1 \times 10^6$ cells (Stratford, 2006), and it is, therefore, only noted or detected through the continued growth, within a contaminated product, of the yeast or bacterial population. Such strains can be isolated and examined but, with the low threshold of detection for many of the diverse metabolic or autolysis-derived components produced, spoilage may often be determined by sensory means well before the physical detection of deterioration of a product or the actual identification of specific microbial species present. Two excellent summaries, now a decade on, are still of note by way of nice introductions to the good, the bad and the ugly of bacteria in brewing (Suzuki, 2011; Vriesekoop et al., 2012). Good follow up reports by Esmaeili and group and Paradh and Hill deal with common spoilage microorganisms of beer and a review of Gram-negative bacteria in brewing, respectively (Esmaeili et al., 2015; Paradh & Hill, 2016). Spoilage lactic acid bacteria and detection methods seeing recent coverage by Xu and colleagues (Xu et al., 2020) and others (Rodríguez-Saavedra et al., 2020; Tsekouras et al., 2021). The identification of spoilage organisms in draught beer by using culture-dependent methods is also of current note (Jevons & Quain, 2022). New media formulations are needed to isolate previously 'unnoticed' organisms such as *S. diastaticus* (see below) and *Zygosaccharomyces* species strains as noted before. The works of Ciont et al. (2022) also

TABLE 19.1 Bacteria associated with beer spoilage with specific reference to typical flavour notes produced and general sensory flavour changes.

Group, species, or genera—wort and beer spoiling bacteria (some useful for flavour production in speciality beers)	Spoilage/flavour notes produced
Acetobacter species: *A. aceti, A. hansenii, A liquefaciens, A. pasteurianus.*	Produce acetic acid (vinegar) and ropiness. Aerobes so limited to certain process points – can be used to impart notable levels of acid in certain speciality beers.
Enterobacteriaceae ('termo' – original designation to a loose group of wort spoiling bacteria) (see Back, 2005) Includes: *Enterobacter* *Citrobacter (Citrobacter freundii)* *Hafnia** *Rahnella (Rahnella aquatilis)* *Klebsiella* *Serratia* **Hafnia protea* (formerly *Obesumbacterium proteus*)	Wort: Acetate, celery-like, parsnip, phenols and cooked cabbage and DMS (dimethyl sulphide) notes. Acetaldehyde, diacetyl, iso-amyl alcohol (fusel alcohols) and volatile organo-sulphur compounds are also reported[a] and phenolic notes. *Hafnia protea* and *Rahnella aquatilis* can produce. Excessive amounts of diacetyl and DMS (dimethylsulfide). *Hafnia* contamination leads to a parsnip-like odour. *Hafnia protea* spoils beer and wort by producing acetoin, DMS, isobutanol, lactic acid, propanol and 2, 3 butanediol[b] and phenols[c]. *Citrobacter freundii* and *Rahnella aquatilis* produce various off-notes and aromas; acetaldehyde, acetoin diacetyl, DMS, lactic acid and 2,3-butanediol. *Parsnip-like refers to early descriptors for wort spoilage flavour. Strong sulphury and vegetal notes indicate spoilage along with turbidity. Sensory evaluation would deal with specific flavour terms such as DMS and with research and training other specific sulphur notes and standards for the other flavours noted above. In general terms are vague for wort spoilage flavours.*
Gluconobacter (*Acetomonas*) *G. oxydans*	Produce acetic acid in beer sometimes giving a cidery note.
Lactobacillus species	Turbidity and souring of beer via lactic and acetic acid formation some strains produce diacetyl and ropiness.
Megasphaera (*M. cerevisiae*)	Leads to turbidity and production of a variety of fatty acids, butyric, caproic and valeric and isovaleric acids along with acetic acid and H_2S; manure-like aromas also described.
Micrococcus spps.	Hazes and fruity esters.
Pectinatus (*P. cerevisiiphilus*)	Acetic acid, acetoin, propionic acid, succinic acid and turbidity. Sour and rotten egg aromas due to the H_2S and variety of acids produced. Other sulphur notes – methyl mercaptans (sewer-like notes also possible) and manure-like aromas also described[d]. Pectinatus mainly spoils unpasteurised beers.
Pediococcus species (*P. damnosus* – 'Beer sarcina' – early term)	Sours beer: lactic acid and some strains produce diacetyl (esp. *P. damnosus*). Sediments and reduced foam stability may also result. Sarcina sickness referred to major *Pediococcus* infections.
Selenomonas	Acetic, lactic and propionic acids.
Zymomonas species (*Z. mobilis*)	Causes fruity and sulphidic (H_2S) characters and acetaldehyde (rotten apples) during fermentation. Also, turbidity. Higher alcohols and acetic esters, DMS (dimethylsulphide) and dimethyl disulphide also reported[e].
Zymophilus	Acetic, lactic and propionic acids.

*The data in the table represent summaries of information culled from many of the references cited in the text and so for clarity specific points are not referenced in the table itself. Certain details are also included in the text. Other details may also be found elsewhere in this volume (see Ashtavinayak & Elizabeth, 2016; Rodríguez-Saavedra et al., 2020; Storgårds, 2000; Storgårds, Haikara, & Juvonen, 2006; Tsekouras, Tryfinopoulou, & Panagou, 2021).
[a]*Harrison, Webb, and Martin (1974).*
[b]*Middlekauff (1995).*
[c]*Back (2005).*
[d]*Paradh et al. (2011) and Rodríguez-Saavedra et al. (2021).*
[e]*Back (2005) and Dennis and Young (1982).*

provide information on key spoilage organisms being found today as uncovered by modern molecular approaches to detection and elucidation of strains. With works by Suzuki, Takahashi et al., and other current researchers providing updates with coverage of the emergence of new spoilage microorganisms in brewing, their generation of off-odours/flavours and how to deal with the issues or the discovery of many new microorganisms in post-boiled worts that can impact fermentation (Suzuki, 2020; Takahashi, Kita, Kusaka, Mizuno, & Goto-Yamamoto, 2015).

As for both yeast and beer spoilage, the paper by Ciont et al. (2022) also provides a wealth of information which supplements that in Tables 19.1 and 19.2. Other organisms of note for non-alcoholic and low-alcoholic beer (NA, NAB/LA, or LAB) and spoilage being *Pichia*, *Rhodotorula*, *Alternaria*, *Hansenia*, *Wickerhamomyces* and *Cladosporium species* and strains (Ciont et al., 2022). Though, as seen later, many of these so-called 'spoilage organisms' are now finding increasing use for novel brewing operations.

Recently, a silent brewery resident awoke — an organism likely to have been present lurking in the wings of most breweries since brewing began, but which came to bear on a few recalls is the wild yeast *Saccharomyces cerevisiae* var. *diastaticus* (Meier-Dörnberg, Jacob, Michel, & Hutzler, 2017, Meier-Dörnberg, Kory, Jacob, Michel, & Hutzler, 2018; Suiker, Arkesteijn, Zeegers, &Wösten, 2021; Suiker & Wösten, 2022). This organism evaded detection based upon lack of growth on earlier established standardised selective growth media. *S. cerevisiae* var. *diastaticus* is an obligatory wild yeast spoilage organism with high spoilage impact based on its glucoamylolytic (enzymatic) ability to ferment residual sugars, dextrins and starches in beer and other malt beverages (especially in naturally conditioned beers) (Meier-Dörnberg et al., 2017). Said to accumulate in breweries as mixed biofilms (alongside *Candida* and/or *Pichia* species) (Suiker & Wösten, 2022). Both vegetative cells and spores can spoil all tested beer products. In addition to promoting changes in taste (with dry and winey body characteristics and noticeable phenolic off-notes, including clove like/spicy qualities (that might fit German wheat beers — noted below — but not for other beer styles). Several other detrimental issues include over-carbonation (leading to package swelling and to bottle or can rupture), the appearance of sediments and turbidity and the gushing of beer from containers. A sizeable number of cases of positive results for this organism were noted in Europe from 2008 to 2017, and around 2015, an increase was recorded in the rate of contaminations (Meier-Dörnberg et al., 2017), which saw an impact with major brand product recalls in the American craft brewing industry in 2014 and 2016. These issues and potential resolutions are detailed elsewhere (Meier-Dörnberg et al., 2018).

TABLE 19.2 Yeast associated with beer spoilage with specific reference to typical flavour notes produced and general sensory flavour changes.

Yeast strain—wort and beer spoiling yeast (some useful for flavour production in speciality beers)	Spoilage/flavour notes produced
Brettanomyces (*B. bruxellensis*) Also classified as *Dekkera* — see text for definitions/nomenclature details.	Imparts typical acetic acid ester aroma (high ethyl acetate fruity-solvent notes). Produce copious quantities of acids, lactic and consequently ethyl lactate, and acetic acid. 4-Ethylphenol (4-EP) and 4-ethylguaiacol (4-EG) also associated with Brett beers (see text).
Candida Alternate names exist for *Candida* species. Some species noted in the text.	Cloudiness and off-flavours; esters (ethyl acetate), acids (acetic) and phenols (4-vinylguaiacol — 4-VG).
Hansenula (now merged with genus *Pichia*, Kurtzman, 1984, 1996)	Cloudiness and off-flavours (high ethyl acetate ester production; solvent-like odour) (see *Pichia*).
Kloeckera	Produces acid; acetic and lactic, esters (ethyl acetate: Fruity odour) and cloudiness in beer.
Pichia (*P. anomala*) *P. anomala* now renamed *Wickerhamomyces anomalus* Kurtzman (2011).	Cloudiness and off-flavours. Volatile phenols (4-vinyl guaiacol), ethyl acetate, amyl acetate (higher alcohols). Aerobic: spoilage potential limited to beers stored in the presence of air. However, under suitable conditions, they grow rapidly and often give rise to films on the surface of the beer as well as resulting in the production of hazes and off-flavours.
Saccharomyces (wild strains)	Phenolic off-flavours and contamination can lead to over carbonation of beer via over attenuation.

The data in the table represent summaries of information obtained from many of the references cited in the text: for clarity specific points are not referenced in the table itself. Certain details are also included in the text. See Figs. 19.1 and 19.2.

Of current and further interest is a statement that '*Saccharomyces cerevisiae*, var. *diastaticus* can lead to changes in flavour, but does not give an overall unpleasant taste to beer' (cited in Meier-Dörnberg et al., 2018). This point led to the suggested possible use of specific strains of this organism, with super-attenuating properties, to produce carbohydrate- and calorie reduced de-alcoholysed beverages. Thus, *S. cerevisiae* var. *diastaticus* is now also now being used to produce the phenolic accented German wheat beer style (Meier-Dörnberg et al., 2018) and is suspected to have unknowingly been used as production strains in some breweries. Thus, in revealing some positive traits, and if understood and controlled, this organism might then be considered, depending upon the situation, as both a useful and a spoilage organism (Krogerus & Gibson, 2020).

19.2.2 Off-flavours and off-odours

In general, consumers are not sufficiently aware as to what constitutes an off-flavour or taint (defined below) in beer unless, it is present in a very pronounced way. However, they are becoming more educated as to several significant issues such as diacetyl (Section 5.3). Yeast and bacteria produce a multitude of organoleptically powerful (low threshold detection) metabolic by-products – many quite volatile, which means they present both as off-odours and as off-tastes = off flavour (flavour being largely odour based – olfactory, but a combination of odour, taste – gustation and tactile/trigeminal mouthfeel and 'irritating' – spicy, cooling/camphoraceous and warming/balsamic sensations) (Barwich & Smith, 2022). This chapter deals with those organisms, given the right conditions that can grow and either spoil the beer organoleptically in the main or provide those desirable but atypical flavours and, these days, novel or enhanced flavour profiles and flavours in speciality beers. It is to be noted that trained panelists can detect flavour issues at levels below which consumers may notice them.

19.2.3 Taints and off-flavours

To follow the approaches to understanding the microbial metabolites that have a sensory impact, and to generating sensory training programs (outlined in Part 2), a definition of taints and of off-flavours is in order (Hughes, 2009; Kilcast, 1996; Saxby, 1996). A taint or an off flavour is caused by the presence of a chemical that imparts a flavour that is unacceptable/unusual (or is atypical) in a food or beverage product. A taint is often defined further as the presence of a substance totally alien to all foods (and may include components imparting atypical flavours or odours from external sources such as air, water, packaging materials, processing lines, etc.). An off-flavour is defined as arising from a chemical reaction of a naturally occurring component in the food or beverage (or through internal deteriorative changes), giving rise to an atypical compound with an undesirable or unexpected taste. For purposes of microbiologically derived flavours – a metabolic by-product (or autolytic components) leading to atypical or unwanted flavours or odours in beer would be regarded as an off-flavour rather than a taint.

Microbes can, however, metabolise certain compounds derived from disinfectants and sanitisers, for example, to generate undesirable flavours such as chloroanisoles (mouldy or musty accents) from the methylation of chlorophenols (in water supplies) or the production, by cyanobacteria, of low threshold detectable compounds such as geosmin (trans-1, 10-dimethyl-trans-9-decalol, conveying an earthy, musty or beetroot-like aroma detected in air at five parts per trillion!) and (−)2-methylisoborneol (2-MIB) (earthy or musty odour - odour detection threshold of 2–20 parts per trillion – ppt) (Jüttner & Watson, 2007; Liato & Aïder, 2017). With respect to water supplies it is seasonal algal blooms that often invoke the flavour taint of geosmin (Westerhoff, Rodriguez-Hernandez, Baker, & Sommerfeld, 2005). Such aromatics would be undesirable in any beverage and would be considered as odour complaint level taints rather than off-flavours. Other sources of taints may be from mould derived flavours carried into beer production from contaminated raw materials and are difficult to remove (see notes on mycotoxin producing fungi below). We will be referring to off-flavours and not taints through most of this article regarding spoilage or unwanted flavours. Part 2 of the chapter deals with both desirable and undesirable flavour notes as assessed through sensory evaluation.

19.2.4 Air, water, minerals and raw materials

The potential for beer spoilage occurs prior to the production of wort (see below). Water supplies and all raw materials as well as any exposure to air or surfaces that come in contact with wort or beer may carry wild-yeast, moulds and/or bacteria that may cause downstream processing problems. These sources can also be addressed in sensory programs if their key 'microbial contamination indicators' are based on sensory perception. Air will also carry a multitude of microorganisms that are taken advantage of for spontaneously fermented beer production. Interestingly, while water quality needs to be considered by brewers in attempts to avoid unwanted microorganisms leading to subsequent beer spoilage, or to flavour taints as discussed above, certain components of water – the minerals that brewers often play with to affect different flavour qualities for the different beer styles can, in fact, be inhibitory to key microorganisms involving 'Brett' beer brews. This issue was reported upon most recently (Witrick & Pitts, 2021). The bicarbonate content and the fermentation activity of *Brettanomyces bruxellensis* within beer has been evaluated in this regard, with a statistical analysis showing a significant difference between fermentation activity and, most importantly, with flavour compound production within different batches brewed at different bicarbonate levels (Witrick & Pitts, 2021). Revealing again that many facets and mysteries remain to be solved to provide a more complete understanding of brewing and to take it to the next level. Major factors dealing with and affecting *Brettanomyces*-brewed beers are covered in the introductory speciality beers Sections 4.5 and 4.6 and in more detail in Sections 5.4, 5.5 and 5.6).

19.3 The microbiology of 'atypical flavour' production in brewing—An overview

19.3.1 Moulds—Additional factors for consideration

Several notable fungi and moulds may infect barley and stored malt with some fungi of the *Fusarium* genera associated with gushing (the violent spontaneous ejection of beer from containers) (Hill, 2009; Mastanjević, Krstanović, Mastanjević, & Šarkanj, 2018). This is clearly noted visually, and any beer showing gushing should be examined for its cause including microbiological contamination. Although moulds are not directly spoilers of wort or beer, their presence in barley may negatively impact on the quality of the malt, wort and beer (Vaughan et al., 2005). Observations have been made that mould growth on malt can be responsible for strong off flavours in beer produced from it. These off flavours ranged from 'burnt molasses' to 'unclean,' 'winey' and 'harsh' (Kneen, 1963, p. 51, cited in Vaughan et al., 2005; the latter terms too vague today and such tainted beers needing more technical evaluation to clearly define sensory terms. Beer brewed with malt contaminated with *Aspergillus fumigatus* had pronounced roughness (again a vague term Kneen, 1963, p. 51; cited in Vaughan et al., 2005) and a stale flavour (defined stale flavour notes are better understood today and covered on the beer flavour wheels and sensory maps discussed in Part 2). Suffice to say, moulds can carry through to beer in the form of a range of off-tastes and odours and raw materials, such as barley malt and other cereal grains, should be taste evaluated to ensure they are free of mouldy, musty or earthy taints. Further studies on the sensory impact of microorganisms found on and within grain raw materials, not specifically discussed here, have been made recently with one example involving an organism that is discussed later in this chapter: *Wickerhamomyces anomalus* (aka. *Pichia anomala*) (Kurtzman, 2011; Laitila et al., 2011). Mycotoxins from moulds can survive the brewing process and end up in the final product, and while causing other issues, rather than major flavour problems, some mycotoxins have been shown to cause genetic damage in cells and cancer in animals. Two recent articles shed some important light on this with respect to beer spoilage and provide good overviews of the literature on this topic (Pascari et al., 2022; Peters et al., 2017). A further reference provides descriptive notes on the fungal volatile organic compounds, that when present can affect raw materials and resultant products (Pennerman, Al-Maliki, Lee, & Bennett, 2016). Such notes to be added to the sensory evaluation flavour lexicon.

19.3.2 Beer and wort—Overview, with some sensory evaluation considerations

Beer and wort are both perishable liquids prone to microbial attack. While beer provides a less hospitable environment for microbial growth than the initially non-sterile beer wort (Hill, 2009; Suzuki, 2011; Vriesekoop et al., 2012) the solution still retains residual sugars, nitrogenous compounds, minerals, and vitamins that can provide nutrients for bacterial and wild-yeast contaminants to thrive (see Takahashi et al., 2015). The pH of beer, normally around 4 to 5, is also favourable for the survival and growth of certain species of bacteria. Overall, however, a limited number of the multitude of known microorganisms is responsible for beer spoilage – a few species of bacteria and wild yeasts as detailed below. Moreover, things are changing with respect to many of the so-called 'spoilage organisms' with several yeast species and their strains or newer hybrid strains—as already mentioned – now being beneficial to brewing operations, as will be seen throughout the rest of the chapter. (Notes considering spoilage—earlier references still relevant: Ault, 1965; Back, 2005; Boulton & Quain, 2001; Hill, 2009; Linske & Weygandt, 2013, pp. 33–37; Manzano et al. 2011; Middlekauf, 1995; Campbell, 1987; Priest & Campbell, 1987; Rainbow, 1981; Spedding & Lyons, 2001; Storgårds, 2000; Storgårds et al., 2006; Suzuki, 2011; Vaughan et al., 2005; Vriesekoop et al., 2012). Moulds are not regarded as beer spoilage organisms as they require oxygen to grow but can, as noted above, cause flavour issues through their contamination of raw materials (Pennerman et al., 2016). See Tables 19.1–19.3 for summary descriptions of many key organisms involved in spoilage, together with the details of a number of the flavour volatile metabolites causing detrimental sensory qualities to beer. The properties of the

TABLE 19.3 Characteristic flavour notes found in beer and associated with yeast and bacterial metabolism.

Flavour note/indicator compound (generic and specific names)	General flavour Descriptors[a]	Typical concentration (ppm) found in beer[a]	Typical threshold mg/L (ppm.)[b] (Thresholds vary with beer style)	General notes/and example associated microorganisms capable of producing the flavour note (see also Tables 19.1 and 19.2 and Figs. 19.1 and 19.2) Note: 1 ppm = 1 mg/L
Acidic (Generic –see specific acids)	Sour cream (*sourness and apples); - see also acetic and lactic.	30–280	Varies with acid; see individual acids	*Lactobacillus* spp. (see details in Table 19.1) (**Acetomonas*) microorganisms may produce various acids – give a sour/tart note to beer.
Acetaldehyde (Ethanal)	Apples, emulsion paint, grassy (green/Bruised apple), avocado, green leaves, melon, and pumpkin.	2–15	5–15	*Acetomonas/Gluconobacter/Zymomonas*
Acetoin (3-Hydroxy-2-butanone)	Buttery, sweet-buttery, creamy aroma, dairy, milk, fatty. Fruity nuances.	2.9–19.3	Detected at 150 mg/L in aq. ethanol 8–20 in beer	Lactic acid bacteria, *Pediococcus* and *Megasphaera*
Acetic acid (Ethanoic acid) (see acidic)	Sour, vinegar, acidic, acetic.	30–200	130–200	*Acetobacter*
Autolysis of yeast	Yeasty, sulphurous, broth/bouillon or meaty-like.	–	–	Notes associated with yeast autolysis also include caproic, caprylic and other medium chain fatty acids.
Butyric acid (Butanoic acid)	Rancid, sharp cheese, baby vomit, pungent/putrid, sour spent grains.	0.6–3.3 (0.5–1.5 ppm more typical in beer?)	2–3	Produced by wort spoiling bacteria and will not volatilise away – carries through to finished beer. Occasionally formed during bacterial spoilage of packaged beer (*Megasphaera* and *Pectinatus*)
Caproic acid (hexanoic) acid	Sour, fatty, sweat, cheese.	1–5.8	2.3?	*Megasphaera, Clostridium* spp.
Caprylic acid (octanoic) acid	Goaty, waxy, fatty, rancid, cheesy, tallow.	2–14.7	4–6	Part of the caprylic flavour (several medium chain fatty acids associated (low levels) with pale lager beers (released during autolysis of yeast)) (Tressl, Bahri, & Kossa, 1980) May be produced by *Brettanomyces*. (wet leather, goat-like, wet dog notes associated with caprylic, capric, and caproic acids).
Cheesy (General note)	Old cheese, sweaty, rancid fat, old hops, stale (see isovaleric acid).	–	–	General descriptor associated with fatty acids – see e.g., isovaleric acid.
Diacetyl (2,3-butanedione)	Butter, butterscotch, movie popcorn (toffee – but usually suggestive based on butteriness), may also give an oily mouthfeel sensation.	0.08–0.6	0.08 (Varies with the beer)	*Lactobacillus/Pediococcus* Referred to in early days of brewing as 'Sarcina sickness'. yeast strains too (culture yeast and brewing process related (Krogerus & Gibson, 2013) – yeast mutations and wild yeast).

Continued

TABLE 19.3 Characteristic flavour notes found in beer and associated with yeast and bacterial metabolism.—cont'd

Flavour note/indicator compound (generic and specific names)	General flavour Descriptors[a]	Typical concentration (ppm) found in beer[a]	Typical threshold mg/L (ppm.)[b] (Thresholds vary with beer style)	General notes/and example associated microorganisms capable of producing the flavour note (see also Tables 19.1 and 19.2 and Figs. 19.1 and 19.2) Note: 1 ppm = 1 mg/L
Dimethyl sulphide (DMS) (Methylsulfanylmethane)	Sweet corn/creamed corn, asparagus, parsnip, tomato juice/ketchup, tinned beans, oysters, sea-spray.	0.05–0.15	0.03–0.08	*Hafnia protea* (*Obesumbacterium proteus*) – other enteric bacteria and *Zymomonas* (see Table 19.1) Flavour nuances change with concentration.
4-Ethyl guaiacol (4-Ethyl-2-methoxyphenol)	Phenolic, clove, smoky, ash-like, bacon, smoked bacon/cheese.	Low	Less was known about this in beer – compared to wine until recently – see text.	Ethyl phenol and ethyl guaiacol said to be the characteristic odours of *Brettanomyces* spp.
4-Ethyl phenol	Band-aid, contaminated with *Brettanomyces*, plastic medicinal, horsey.	0.006–0.02?	Threshold and concentrations in beer not well known, though current citations are covering more detail.	Ethyl phenol and ethyl guaiacol are said to be the characteristic odours of *Brettanomyces* spp. (esp. in wine, little research on beers until recently) (see Licker, Acree, & Henick-Kling, 1998; Romano, Perello, Lonvaud-Funel, Sicard, & de Revel, 2009; Suarez et al., 2007, and text for updates.
Ethyl acetate [Most common ester in all beers]	Acetone, (nail varnish remover), estery, paint thinner.	10–50	30–50	A typical component of all beer; can be elevated due to microbial contamination (incl. *Brettanomyces*)
Ethyl butyrate (Ethyl butanoate)	Tropical fruit, mangoes, canned pineapple.	0.05–0.25	0.4	Common ester in beer but elevated levels can indicate bacterially contaminated worts. As for all esters, it derives from condensation of alcohol and an acid (here butyric acid). Alcohols are solvent and acids sharp and sometimes cheesy/goaty but esters formed from them are often fruity or floral.
Ethyl lactate (ethyl(S)-2-hydroxypropanoate)	Fruity, strawberry.	0.1–0.8	250	High lactic acid could lead to high ethyl lactate levels.
Eugenol (4-Allyl-2-methoxyphenol)	See spicy.	Zero - low	0.013? (No reference)	This compound is often used as a standard for spicy-clove phenolic notes but other phenolic compounds such as 4-vinyl guaiacol (4-VG) are more often found in beers at threshold levels.
Geosmin	Earthy, beetroot.	Not normally present	Detected at parts per trillion levels. See text	*Cyanobacteria*. A taint rather than an off flavour. Text also describes methylisoborneol – (MIB) another earthy/musty note.
Indole (2,3-Benzopyrrole)	Farmyard, like pigs on a farm, faecal, coliform, jasmine.	<0.005	0.015	*Coliform* bacteria during early fermentation. Some beers will exhibit a slight note of this based on author judging experience.

TABLE 19.3 Characteristic flavour notes found in beer and associated with yeast and bacterial metabolism.—cont'd

Flavour note/indicator compound (generic and specific names)	General flavour Descriptors[a]	Typical concentration (ppm) found in beer[a]	Typical threshold mg/L (ppm.)[b] (Thresholds vary with beer style)	General notes/and example associated microorganisms capable of producing the flavour note (see also Tables 19.1 and 19.2 and Figs. 19.1 and 19.2) Note: 1 ppm = 1 mg/L
Iso-Amyl acetate (3-Methylbutyl-acetate) *(banana oil, pear essence)*	Fruity, banana and pear drops. [US: Circus peanuts.]	0.5–1.5 ppm (higher in wheat beers)	1.4–2	A typical component of certain beers at detectable levels (wheat beers); it can be elevated in other beers as an off-flavour due to wild yeast carrying the phenolic off-flavour gene (POF) (Tressl et al., 1980 and extended details in text.)
Iso-amyl alcohol (3-Methyl-1-butanol)	Fusel oil (higher alcohol), whiskey-like; represents the main higher alcohol known as 'fusel oil'.		Thresholds vary depending on specific fusel alcohol 50 –800 ppm*	Produced by *Brettanomyces* or other organisms in association with Brett beer production. *(Another higher alcohol: n-propanol with 600–800 ppm. threshold may be found in beers at 7–45 ppm. depending upon beer style and is produced by certain contaminating microorganisms)
Isovaleric acid (3-methyl butanoic acid) (May be confused with butyric flavour)	Cheesy, old hop-like, sweaty, sweat socks. Rancid, putrid, stale cheese.	0.1–3.4	0.1–1.5	Usually from old/aged hops. May be produced by *Brettanomyces and Megasphaera (see* Tables 19.1 and 19.2).
Lactic acid (2-hydroxypropanoic acid; see acidic)	Sour, sour milk, yogurt. (No odour). Dulls the sensation of beer.	–	170–180	*Lactobacillus/Pediococcus* contamination – also deliberately encouraged in acidification of malt and wort and in sour or 'wild-beer' (see text).
Meaty or broth-like (aka yeasty)	Yeast extract, meat extract, broth, old yeast.	–	–	From autolysis of yeast
Medicinal (An older general term.)	TCP, antiseptic, phenolic.	–	–	*Enterobacter/Klebsiella associated with chlorophenol taints and defined phenolic compounds.*
Methyl mercaptan (methanethiol)	Rotten cabbage, garlic, sulphurous, eggy.	0.001 Typically, extremely low.	–	*Pectinatus* Mercaptan (ethanethiol) with similar sulphury notes may also be involved with some contaminating organisms.
Musty [General term]	Musty, mouldy, earthy.	–	–	More a taint than off-flavour from mould contaminated grains, or water supplies (see also geosmin)
Phenolic [General term]	Herbal, cloves, medicinal.	–	–	Wild yeast. See specific phenolic notes, e.g., eugenol/4-vinyl guaiacol (4-VG) and text for new research details.
Propionic acid (more correctly propanoic acid)	Acidic, rancid, dairy, nutty flavour, pungent, cheesy vinegar.	0.5–5	–	*Pectinatus/Clostridium* spp.

Continued

TABLE 19.3 Characteristic flavour notes found in beer and associated with yeast and bacterial metabolism.—cont'd

Flavour note/indicator compound (generic and specific names)	General flavour Descriptors[a]	Typical concentration (ppm) found in beer[a]	Typical threshold mg/L (ppm.)[b] (Thresholds vary with beer style)	General notes/and example associated microorganisms capable of producing the flavour note (see also Tables 19.1 and 19.2 and Figs. 19.1 and 19.2) Note: 1 ppm = 1 mg/L
Spicy (General term)	Clove, eugenol, nutmeg, allspice.	–	–	A more general term but could include eugenol and see 4-vinylguaiacol (4-VG).
Styrene (ethenylbenzene)	Polystyrene, plastic, burning plastic, styrene, modeller's aeroplane glue.	<0.005 not detectable in normal beer	0.02	Off-flavour produced by contaminant wild yeast during fermentation; or a taint from raw materials/packaging. Styrene has a mechanism of production like those of traditional wheat beer phenolics. Related to the POF phenolic off-flavour gene, it may be found in bottle re-fermented beers if POF + strains are present. (Schwarz, Stübner, & Methner, 2012)
Succinic acid (butanedioic acid)	Odour: None, sour acidic flavour.	–	–	*Pectinatus/Selenomonas/Zymophilus*
Sulphidic (hydrogen sulphide) (sulphur) H_2S	Rotten eggs, sewer drains, mercaptans, onions/garlic.	0.004 (<4 ppb)	0.004	Wild yeast, *Zymomonas*
Valeric acid (pentanoic acid)	Fatty, earthy, putrid acidic, sweaty, cheesy odour, sharp, acidic, milky cheese, sl. fruity.	–	–	*Megasphaera/Brettanomyces/Clostridium* spp.
Vinyl guaiacol (4-VG)	Spicy, clove, herbal, phenolic.	0.05–0.55	0.3	A typical wheat beer note (Coghe, Benoot, Delvaux, Vanderhaegen, & Delvaux, 2004). Wild yeasts or speciality yeasts.

*This table forms a key part of instruction for sensory evaluation programs in general, and specifically for flavours associated with atypical fermentation activities of bacteria and or wild yeast. Today atypica or enhanced flavours are desired from fermentation activities utilising alternative yeast species in single/pure culture or co-fermentation systems. The data in the table represent summaries of information obtained from many of the references cited in the text, product summaries from suppliers of sensory standards (specifically FlavorActiv and Cara Technologies – AROXA Sensory Standards) and collected notes of the authors over many years. So, for clarity, not all specific points are referenced in the table itself.

[a]*Sensory panels will decide upon the desired term for their team but should arrive at a consensus as to how best to describe the flavour notes (suppliers of standards and sensory training such as Cara Technologies and FlavorActiv can advise here). The American Society of Brewing Chemists (ASBC) flavour wheel and new style flavour maps provide tools as described in part 2. The ASBC provides technical material as to standards to use as do several references listed in Part 2 of this article.*

[b]*See the text for details on approximate concentrations found in beer and threshold values and definitions (consensus values are provided based in part on subjective author opinion and experience working with standards – and are beer style dependent). For an extensive set of threshold data see Angelino (1991) and for the most extensive lists see Engan (1981) and de la Torre-González et al. (2017). A concise discussion as to how sensory threshold values are obtained was published recently (Spedding, 2023). A few references not mentioned in the text are included in this table to keep some facts localised. More research and publications are adding new results and updated information to such tables of sensory data, and as such should be used in conjunction with this table to maintain a current lexicon of flavour terms and the most up-to-date threshold data and values. Moreover, synergistic, and antagonistic interactions are at play in how the human senses and brain interpret flavour information with such implications described recently by Barwich (2020), Barwich and Smith (2022). With the complex nature of flavour and both machine and human sensory evaluations covered by Chambers and Koppel (2013). This table represents only a small fraction of key beer volatiles – a little more information is to be found in Figs. 19.1 and 19.2. Two very recent additions to the literature discuss esters, higher alcohols and other odour-active compounds produced by various conventional and non-conventional yeasts (Roberts et al., 2023; Satora & Pater, 2023). Many varied and flavourful esters are particularly relevant to this topic for enhanced and novel flavourful beverage production.*

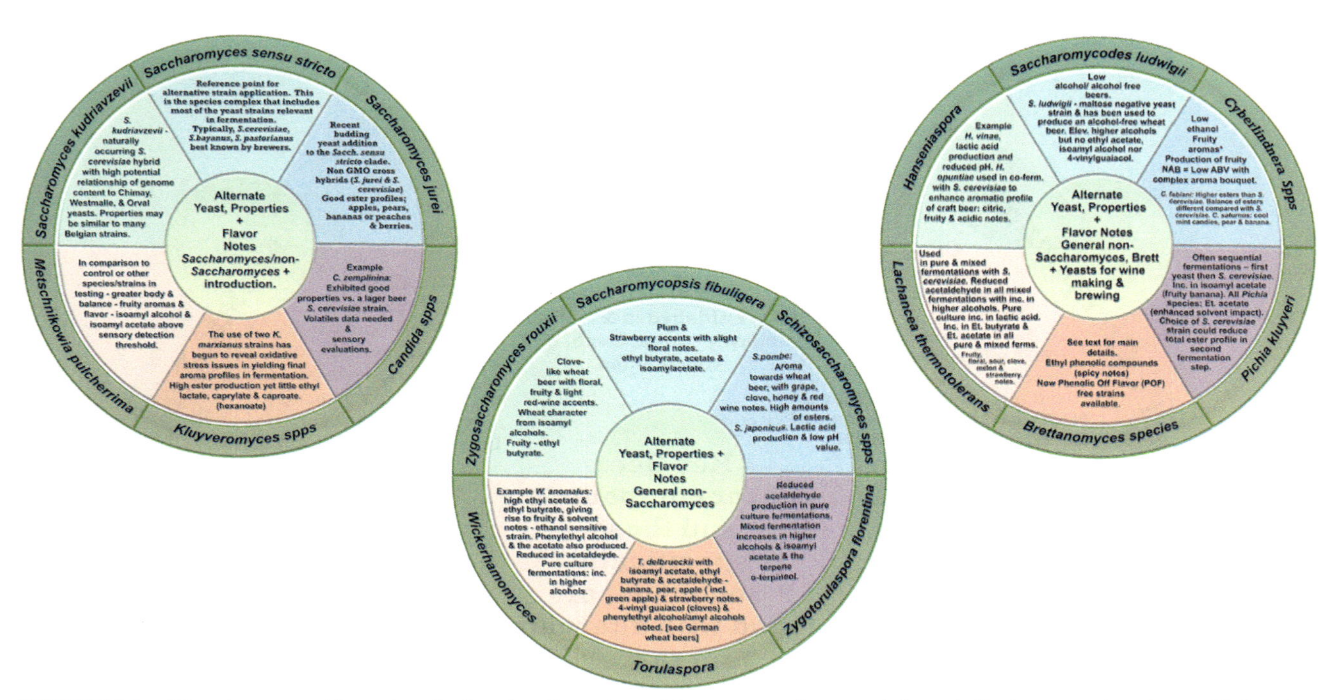

FIGURE 19.1 Alternate yeast used in modern day brewing: Properties and flavour notes. A representative set of notes on yeast species used in brewing. Some once considered spoilage yeast now being utilised to enhance flavour profiles in beer ranging from sour beer to traditional styles and with low- and non-alcohol beers included. Notes taken from many of the references cited in the text. See flavour notes in Table 19.2. Further details on each organism and others that are not illustrated here are covered in the body text.

indicator compounds addressed in Table 19.3 include threshold of detection values, sensory descriptors and potential changes in sensory qualities imparted to beer, via their presence, which then lead us on to evaluating the key flavour notes in a sensory programme as discussed in Part 2. Figs. 19.1 and 19.2 provide summary details complementing that in the tables. The typical levels of odour/flavour compounds in beer and threshold values presented in Table 19.3 are obtained from various sources (Anderson et al., 2000; Angelino, 1991; Engan, 1981; Kunze, 2010; O'Rourke, 2000, pp. 29–31; Spedding, 2013a; Taylor & Organ, 2009). Additional references published since the first edition of this chapter appear in the sub-heading sections with details most relevant to each specific topic. All such references to be regarded as guides only; threshold values for volatiles vary by type and style of beer and production methods with a lot of synergistic and antagonistic factors in play (see Bossaert, Kocijan, Winne, Schlich, et al., 2022; Bossaert, Kocijan, Winne, Van Opstaele, et al., 2022; Bossaert, Winne, et al., 2022). A trained sensory person or panel often needs retraining when moving to a new brewery or onto new product formulations and to learn how to avoid many ingrained or 'house biases'. A threshold value is the concentration at which an aroma or taste can be detected as here in beer or beer wort, with a recognition threshold being a concentration at which a compound can be positively identified. As threshold values in beer and other food items are dependent upon several other variables the simple definitions above will have to suffice here. Threshold values may best be determined through the sensory approaches outlined in Part 2 and the references cited in that section. Recent articles cover some updated information also on thresholds and sensory properties of many compounds discussed in this chapter and with relevance to beer sensory evaluation in general (de la Torre-González, Narváez-Zapata, & Larralde-Corona, 2017; Romero-Rodríguez, Durán-Guerrero, Castro, Díaz, & Lasanta, 2022).

19.3.3 Beer wort spoiling bacteria

Wort is particularly susceptible to contamination by bacteria and wild yeasts as it provides an ideal nutrient medium for many organisms; enteric bacteria, acetic and lactic acid bacteria, and some wild yeast strains (Back, 2005; Hill, 2009). Bacterial species found are often Gram-negative asporogenous rods and were originally (now archaically) termed 'termo bacteria' – they are unable to develop in beer, but the off-flavours produced often carry through to finished beer (Paradh & Hill, 2016). Poor attention to wort production leads to spoilage with sewer like, parsnip and celery notes (related to dimethyl sulfide – DMS and other sulphur metabolites). In general, termo bacteria may be regarded as mixed populations containing

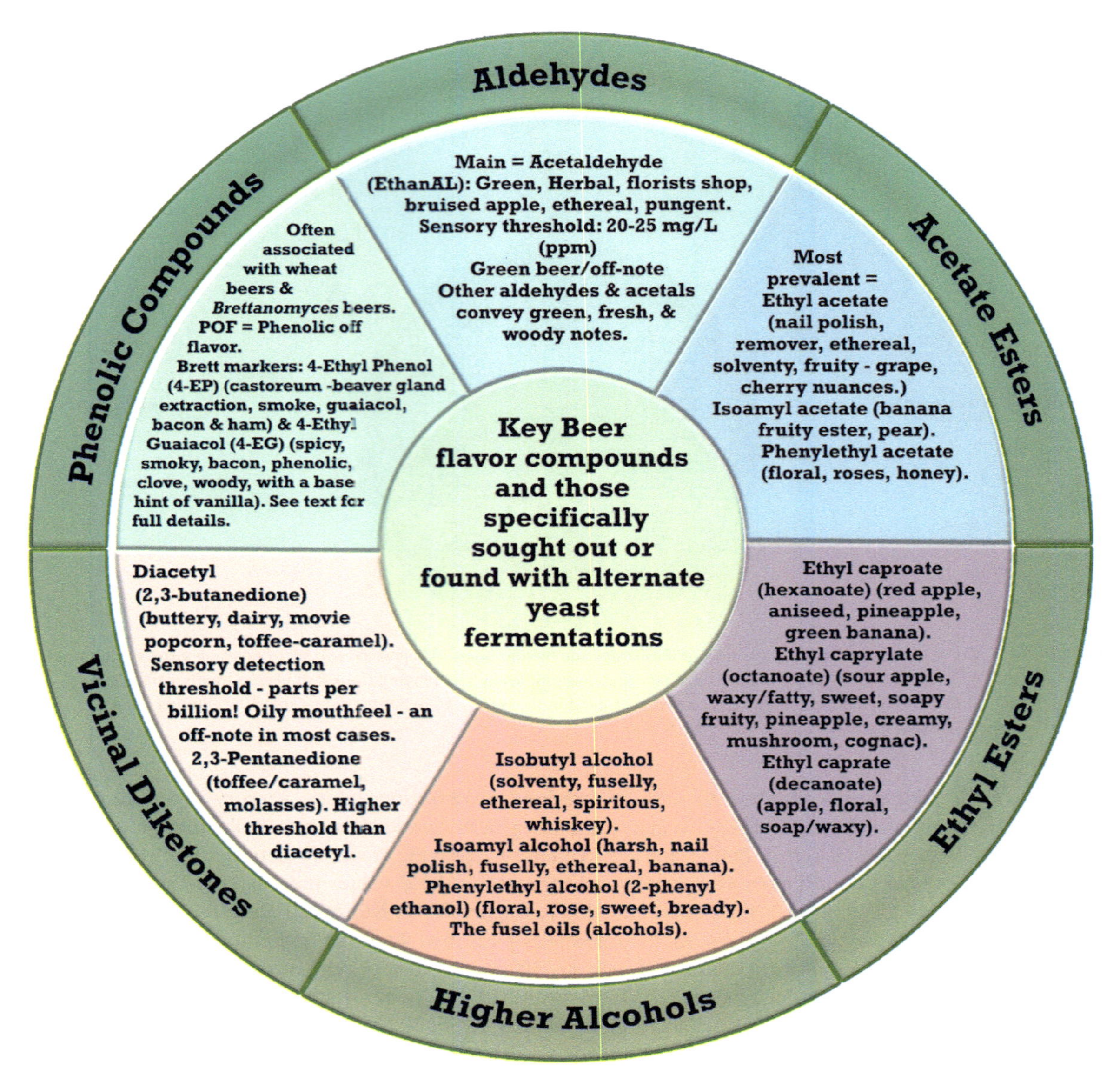

FIGURE 19.2 Key classes of flavour notes found in beers. With examples of those sought out through the use of alternate yeast in fermentation practice. Beer flavours with descriptors. Many are found in traditionally fermented beer using *S. cerevisiae* and that are sought after in the production of novel beer styles, or for enhancement of flavour in other beers via the use of alternative choices of yeast species and strains. This includes the production of non-alcoholic and low alcohol beers to mask worty flavours for example. See also Fig. 19.1 in context here.

representatives of the Enterobacteriaceae and Pseudomonadaceae (Back, 2005). Predominantly found are species of *Enterobacter*, *Citrobacter*, *Hafnia*, *Klebsiella*, *Serratia*, *Pseudomonas* and *Xanthomonas*. Some of these strains and off-flavour production are described further in Tables 19.1 and 19.3 and elsewhere (Ault, 1965; Back, 2005; Boulton & Quain, 2001; Hill, 2009; Linske & Weygandt, 2013, pp. 33–37; Manzano et al., 2011; Middlekauf, 1995; Priest & Campbell, 1987; Priest, Cowbourne, & Hough, 1974; Rainbow, 1981; Spedding & Lyons, 2001; Storgårds, 2000; Storgårds et al., 2006; Suzuki, 2011; Vaughan et al., 2005; Vriesekoop et al., 2012). *Enterobacteria* may grow in warm wort causing unwanted flavours such as hydrogen sulphide, acetaldehyde and vinegar – other acidic flavours, diacetyl, phenolic compounds and fruity off-flavours from ester formation. These flavours are produced in quantities greater than typically produced by culture yeast during normal fermentations. Culture or pitching yeast may also be a source of both bacteria and wild yeast, which can impact beer flavour and stability. As stated above, wort spoilage flavours often persist through the brewing process.

A second group of wort bacteria includes *Bacillus* and *Clostridium* species – sporogenic species with high heat resistant spores that may enter beer processes via raw materials (malt, sugar syrups and hops). These are associated with the production of butyric acid and sulphur compounds (Back, 2005). A third group includes the genera *Lactobacillus*, *Enterococcus*, *Lactococcus* and *Pediococcus*, many producing lactic acid and diacetyl (Back, 2005; Schneiderbanger et al., 2020). Such issues as noted here are not as much of a concern these days with a few cases cropping up now and again, within the smaller-operation craft-side of brewing. Though current research is uncovering more organisms within the brewery environment and in brewing vats.

19.3.4 Beer spoiling bacteria

Beer spoiling bacteria are characterised as microorganisms capable of multiplying in beer, resulting in product deterioration. Gram-positive bacteria are in general inhibited by hop bittering components, but the growth of Gram-negative bacteria is unaffected (see elsewhere in this volume for details and Suzuki (2011) for an extensive review of hop resistance in microbes associated with beer spoilage. Gram-negative bacteria are undesirable – they include acetic acid bacteria, *Zymomonas* and certain members of the Enterobacteriaceae (*Rahnella*, *Hafnia*) and Acidaminococcaceae (*Pectinatus*, *Megasphaera*, *Selenomonas and Zymophilus*.) Acetic acid and lactic bacteria can grow in stored beer (Ashtavinayak & Elizabeth, 2016). Minimising oxygen can assist in keeping acetic acid bacteria at bay and from spoiling beer. In packaged beer, anaerobic spoilers include *Lactobacilli*, *Pectinatus* spp., *P. frisingensis*, *P. cerevisiiphilus* and the less frequently found organism, *Megasphaera cerevisiae* (Helander et al., 2004; Juvonen, 2015; Rodriguez-Saavedra et al., 2020). These organisms can produce foul odour metabolites such as methyl mercaptan, dimethyl sulfide (DMS) and hydrogen sulfide (H_2S) along with turbidity. A microbially contaminated beer may also convey lactic and acetic notes, diacetyl (buttery, butterscotch), liquid manure odour, rotten egg, cooked vegetable, phenolic aromas, fusel alcohols (propanol and isobutanol) and ropiness. (See the extended set of references under the wort section above for detailed accounts.) For beer at the point of sale, acetic and lactic bacteria, and sometimes *coliforms*, may plague dispense systems and bar drains and food and beverage contact points (Spedding, 2013a). As for wort spoiling bacteria, a summary of the major beer spoilage organisms is presented in Table 19.1 along with their associated spoilage flavour notes. Some of these issues again now only show up occasionally and more so in craft brewing operations, dispense systems in pub-breweries and especially with unpasteurised beers (Paradh & Hill, 2016; Spedding, 2013a).

19.3.5 Beer spoiling yeast—An early broad introduction

A spoilage yeast species is one with the ability to cause spoilage and as such, yeasts simply isolated from foods/beverages are not necessarily spoilage yeasts (Stratford, 2006). Yeasts not deliberately used in the brewery but find their way into beer production are designated wild yeasts and form a diverse group including both non-*Saccharomyces* and *Saccharomyces* species. Many species are described in the remainder of the text. Naturally, culture yeasts will be present in most beers and these, through their fermentative metabolic activities, are the source of most of the chemical species found in beer. These flavour notes are described in detail in several works (Anderson et al., 2000; Angelino, 1991; Engan, 1981; Hammond, 1993; Romero-Rodríguez et al., 2022), many of these flavours understood by trained panelists. Major types of wild or spoilage yeasts found in the brewery include: *Brettanomyces/Dekkera*, *Candida*, *Debaryomyces*, *Pichia*, *Hanseniaspora*, *Kluyveromyces*, *Torulaspora*, *Williopsis* and non-brewing strains of *Saccharomyces*. Many of these are also described in terms of spoilage below and elsewhere in this volume and have been discussed in earlier publications and in general by others (extensively by Back, 2005, and see Boulton & Quain, 2001; Hill, 2009; Priest & Campbell, 1987; Russell & Stewart, 1998; Spedding & Lyons, 2001). Many of these yeasts compete with culture yeast and can produce copious amounts of metabolites rising to levels above their threshold concentrations, and this leads to their sensory perception as spoilage off-flavours. In speciality beer production, they may be harnessed to generate high concentrations of some of the same flavour notes, which are then considered desirable for the special style intended by the brewer. The details as to which of several yeast genera and species cause specific flavour notes are presented in Table 19.2 and in Figs. 19.1 and 19.2 and in a listing below (Section 5.1.5). In this chapter, many of these organisms, as listed above in terms of being spoilage agents, are now justifiably described in a much more positive light in terms of their use today as favourable, beneficial fermentative agents involved in generating respected flavourful beer profiles.

19.4 Speciality beer production and processes

In addition to looking at the sensory properties of beer from a microbiological contamination issue, microbes have historically been used selectively, or from spontaneous inoculations, to create some artisanal and interesting highly flavoured beer styles; such beers provide unique or unusual complex flavour profiles with their production requiring much mindfulness to ensure safety, wholesomeness and consistency of product flavour and quality. Some of these beers require blending which also requires trained individuals to be able to flavour-match each production run. Such being the case with wheat beers, so-called 'sour-beers', Lambic beers and *Brettanomyces* inoculated beer wort ('Brett beers'), for example, and where brewers use *Lactobacilli* species in the mash to lower the pH through lactic acid production.

19.4.1 Biological acidification

From a sensory aspect, the deliberate use of microorganisms to attain acidification of malt and beer wort warrants a few sentences. Acidification of brewing raw materials, mash and wort can result in beer of a 'superior' flavour quality (Kunze, 2010; Spedding, 2012a). The process deals with careful use of *Lactobacillus* strains and lactic acid production (Back, 2005; Kunze, 2010; Spedding, 2012a; Vaughan et al., 2005). Sensory panels might have a role to play in seeing how the effects of biological acidification play on the overall beer flavour profiles resulting from its use, see also Sections 5.6 and 5.8.

19.4.2 Speciality beers—Wheat beers

Wheat beers utilise raw or malted wheat, rather than malted barley as their primary raw material base. Fermentation may be spontaneous (natural flora of the brewery – Belgian Lambic beers – see below) or with a top fermentation yeast strain. South German wheat beers (Weissbier) utilise a top fermenting yeast strain, which produces a characteristic phenolic flavour (4-vinyl-guaiacol – clove-like – see Table 19.3 (Anderson et al., 2000; Coghe et al., 2004; Jung, Karabín, Jelínek, & Dostálek, 2023; Lentz, 2018; McMurrough et al., 1996; Russell & Stewart, 1998; Spedding, 2012c; Vanbeneden, Gils, Delvaux, & Delvaux, 2008). Certain wild yeasts can produce phenolic off-flavours (POF) in non-wheat beer styles and as such would be regarded as contaminants. Though, as seen later, genetically altered yeast strains or alternative yeast species are also finding use in wheat beer production under controlled conditions. Phenolic off flavour is considered in detail in Section 4.5.

19.4.3 Speciality beers—Berliner Weisse

Berliner Weisse is a regional variation of the wheat beer (white beer/Weissbier) style from Northern Germany and a beer, which is very pale in colour. Lactic fermentation is carried out following the addition of a starter culture resulting in a very low pH, 3.2 to 3.4. *Levilactobacillus brevis*, and *Brettanomyces bruxellensis* (aka. *lambicus*) are used in fermentation resulting in pure lactic acid flavour (Anderson et al., 2000; Back, 2005; Dysvik et al., 2019; Vriesekoop et al., 2012). Quite recently studies on the use of lactic acid bacteria, varied species and strains and optimal ratio of organisms to use for the preliminary acidification of the wort with ultimate taste characteristic dependencies in mind for such beers have taken place (Chervina, Geide, & Selezneva, 2019). Sensory panels would need to ensure the quality of the pure lactic acid flavour and lack of any contaminant-generated off-flavours in such highly acidic/tart beers.

19.4.4 Speciality beers—Sour beers—An introduction

Craft brewers have experimented with sour beer production, creating some unique styles many of which are still being categorised. Brewers of such beers need to be careful of contaminating their other traditional yeast-fermented beers, and both the production of the sour beers and traditional brands need careful monitoring for inappropriate sensory changes and determination as to trueness of desired flavour qualities and therefore, trueness-to-style and to brand. Such types of fermentations are more difficult to control, and consistency may vary with each batch. Sensory panels can determine the range over which the flavours can be allowed to vary in such cases and provide the brewer with instructions on safety concerns and on blending or ageing regimens as appropriate. The acid beers of Roeselare (Rodenbach) would belong here as an example of the use of *Lactobacillus* and *Pediococcus* strains for lactic acid fermentation yielding 500–600 ppm lactic acid! And the recently rediscovered Gose beer style with *Lactobacillus* spp., *as* the souring agent in the kettle souring process. *Brettanomyces* species also play a key role in sour beer production (Verachtert & Iserentant, 1995). See Section 5.8 for more on sour beers.

19.4.5 Speciality beers—Brett beers—And introducing phenolic off-flavour

Brettanomyces, an organism with a solid history with regard to beer production, presents us with an interesting topic. The *Brettanomyces/Dekkera* species can cause characteristic changes in beverages; sediments, turbidity, off-aromas (mousy, rodent cage litter, clove, spice, medicinal, wet wool, cedar, horse, farmyard and sewage notes) with notable acetic acid and a heavy content of ethyl acetate. Whilst most early dated research into spoilage caused by *Brettanomyces* (*Dekkera*) strains had been done with wine as a substrate, many beers have also had quite a past with respect to this organism, including beers with 'English character' (Licker et al., 1998; Varela & Borneman, 2022) — porters and stouts (organisms likely acquired through contact with wooden vats used during fermentation or storage of such beers) (Eliodorio et al., 2019). Belgian beers are also noted for the production by *Brettanomyces* species, of strong fruity, estery-like aromas and metabolites that produce the flavour notes known as 'horse sweat' (Eliodorio et al., 2019; Licker et al., 1998). A multitude of flavours are thus associated with 'Brett' character and a lot of very recent research and more sensory work, discussed below has been reported on which helps to pin down a definition as to all the nuances of Brett beers. The key flavours are said to be from 4-ethylphenol (smoky, medicinal) and 4-ethylguaiacol (woody, smoky, spicy, vanilla) though octanoic and decanoic acids, acetic, isobutyric, isovaleric and several other compounds have all been ascribed to 'Brett character' (Anderson et al., 2000; Heresztyn, 1986; Suarez et al., 2007; Van Nedervelde & Debourg, 1995). In addition, the compound 4-vinyl guaiacol (clove or spice-like) is now noted as being both a positive and a negative flavour component, depending upon beer style context and expectation. The POF volatiles detailed and reviewed lately (Serra Colomer, Funch, & Forster, 2019) and with a new, carefully constructed and sensorially evaluated, *Brettanomyces* flavour wheel (again accented towards wines, rather than beers, yet fully applicable to other beverages) proposed to assist in the newer developments with respect to this class of wild yeasts (Joseph, Albino, & Bisson, 2017). Most interestingly, there are strains that are 'sensorially neutral', causing no undesirable aromas and that are being used in the production of lower alcohol containing beers. These issues are addressed at greater length below (Section 5.2).

As phenolic off-flavour is an important sensory topic for brewing with several phenolic compounds responsible in creating the complex impression of cloves/spices, bacon, smoky, woody and vanilla notes, with yeast strains involved in the metabolic chemistry, a few defining notes are provided here ahead of several further mentions of POF flavour below (Lentz, 2018). 4-Vinyl guaiacol (4-VG) is a member of a group of aromatics, monomeric phenols that include vanillin. Found naturally in most beers (especially ales) but usually well below threshold level. However, 4-VG is present in Weissbiers (wheat beers), rauchbiers (smoked beers) and some speciality Belgian beers in notable concentrations, imparting those desirable spicy, pungent and clove-like notes (Jung et al., 2023). At higher concentrations, and under certain conditions, however, it may be present as a medicinal, off-flavour note. With a low-flavour threshold of detection at about 200 parts per billion (ppb), 4-VG can exert an enormous influence on the flavour of beer. The main production of the compound is yeast strain specific (requiring a specific gene originally called POF; phenolic off-flavour, and now renamed PAD1+) but also depends upon other process factors involved in malting and in the brewhouse. 4-VG can be produced by wild yeasts and some bacteria (which harbour the POF or PAD gene with variable degrees of expression). Unless a specific wheat beer yeast strain or a Belgian strain has been used, the analysis of 4-VG in beer can often be regarded as an indicator of contamination caused by wild yeast and bacteria in the brewery. Today, however, 4-VG might be a positive marker for some beers created by alternative yeast species and altered strains. Finally, because 4-VG breakdown is related to vanilla flavour production, it is now being suggested that the clove-like aroma in fresh speciality beers such as wheat beers and top-fermented Belgian-style ales may shift to sweeter, more vanilla-like flavour impressions as these beers age (Lentz, 2018; Spedding, 2012c; Vanbeneden et al., 2008). Full accounts of the history of research, flavour volatiles production and health effects of 4-vinyl guaiacol and 4-vinylphenol (the pair produced by the older named 'killer strain' of *S. cerevisiae*), and the pair of compounds, 4-ethylguaiacol and 4-ethyl phenol primarily associated with Brettanomyces species have been regaled by Ambra, Pastore and Lucchetti, who also make note that *B. bruxellensis* was the first microorganism to be patented for beer production (Ambra, Pastore, & Lucchetti, 2021).

19.4.6 Speciality beers—Lambic and gueuze

Lambic and gueuze are Belgian beer styles of profound acidity (pH 3.3) and complex flavour (Spitaels et al., 2014, 2015). Wort is cooled overnight in shallow open trays and picks up a variety of microorganisms from the air (they are not usually inoculated with added yeast). Whilst the breweries are within a tight geographical area for this 'style', there is a possibility for some local differences based on the local microflora/brewery structure essentially making such breweries the first in which a more global 'house character' might have been investigated (Winter 2014). Lambics are not blended, while the gueuze style blends old and new Lambics, which are re-fermented in the bottle. In emulating this type of spontaneously fermented, beer American craft brewers are producing a style known now as Coolship beers or American coolship ale (ACA) as detailed below (Bokulich, Bamforth, & Mills, 2012; Carriglio, Budner, & Thompson-Witrick, 2022). This topic and the relationship to *Brettanomyces* yeasts and to maturation are detailed in comprehensive fashion under three separate but related headings below, Sections 5.4, 5.5 and 5.6.

19.4.7 Speciality beers—Wood and barrel-aged beers

Another new class of beers emerging as well as an understanding of early traditional wooden and cask-aged beers will require sensory studies based on the effect of the microflora present in such barrels (Bossaert et al., 2021). This topic involves understanding a unique set of flavour notes derived also from the wood itself. New flavour wheels describing wood- and barrel-aged flavours are available, and a lot of current research into the production of such beers is underway. Sensory programs as described in Part 2 can be initiated to study these unique beer styles. A full section on the latest developments on liquid in wood ageing appears below – Section 5.7.

19.4.8 Pasteurisation, dry hopping and introducing bottle conditioning

Of concern to craft brewers is the issue of pasteurisation, a concomitant loss of top-note flavour characters (some fruity notes), increases in beer colour and a decrease in bitterness along with an increase in the formation of perceivable stale flavour compounds (Štulíková, Bulíř, et al., 2020). Some beers are dry hopped, and some bottle conditioned, and as a result, many beers today in craft as in the past with cask-conditioned ales are 'live products' with some desirable and sometimes undesirable organisms in tow that can lead to sensory degradation of the beer's quality if not properly packaged. Heat-induced flavour changes can occur with pasteurisation, and issues also arise via the use of priming yeast during bottle conditioning – continued metabolic activities or with autolysis of yeast over long maturation or ageing periods, and these features would also bear upon sensory evaluation to monitor such changes (Milani & Silva, 2022; Xu, Wang, & Li, 2014). The storage stability of non-filtered beer, which had been pasteurised, with respect to thiol content, sulphite levels and volatile profiles has been covered by Hoff and others (Hoff, Lund, Petersen, Frank, & Andersen, 2013, and with related work studies cited therein). Two other processing issues are then to be to be noted as points for consideration in any decisions concerning pasteurisation. These being the production of the volatile polyfunctional thiols during bottle refermentation (Nizet, Peeters, Gros, & Collin, 2013) as noted by Hoff and cited above and changes in Belgian beer flavour profiles during bottle conditioning (Dekoninck, Mertens, Delvaux, & Delvaux, 2013). These topics being considered further in Section 5.6. Brewers of today have a much wider range of yeast strains to choose from, enabling them to create exciting new beers, as discussed throughout this chapter. As such, judicious choices will need to be made by them regarding novel beer production and the details of both the desirable and potentially undesirable flavour notes produced by that wider spectrum choice of yeast. Decisions to thermally protect such wares via pasteurisation will also need a careful assessment with sensory evaluation providing a key to mark that decision making. A caveat here is that many pasteurisation studies as reported on to date are pilot scale, in-lab tests. Scale up to production in many cases is needed to see how well these new and novel strains could perform in industrial capacity production both with and without pasteurisation.

19.5 New developments in the uses and applications of microorganisms for flavourful beer production: Strains and hybrids as used in pure culture or within in-sequence fermentations

19.5.1 Introduction—A wild world of fermentation

An increased knowledge base of flavours that are present in many beers, and other fermented beverages, has led to the adoption of many newer organisms – mainly yeasts, with both *Saccharomyces* and non-*Saccharomyces* species and hybrid organisms included for beer production. The goal stemming from this new research, and alternate strain implementation, being the enhancement or production of novel or unique flavour profiles for an audience ever hungry for something new in the way of flavour complexity (Ciani, Canonico, & Comitini, 2020). Such naturally generated, selected or genetically engineered strains garnering both criticism and favour, and with much very recent literature published within the field and on the topics. With the caveat that genetically modified organisms are still not often accepted for modern food/beverage production. That must and will change eventually. However, in that vein, a more natural approach has been touted as a 'Yeast Flavour Diversity Screening' strategy (Carrau, Gaggero, & Aguilar, 2015). This system attempting to integrate sensory profiling work, natural genome evolution and selection of suitable strains, along with flavour insights from the metabolic networks of such microbes and learning how such biochemical networks are regulated (see Metabolome below) (Carrau et al., 2015). Regulation of course dependent in part on actual fermentation conditions, nutrient supplementation and raw materials presented to the organisms for fermentation, plus presence or absence of preservatives and a host of other factor variables. Such developments, in search of new flavour spectrum tapestries, also include a reversion to 'spontaneously inoculated fermentations' or at least to wild-species hunting in local or even new environments, to mixed culture fermentations and to dealing with natural and artificial diversity (Libkind et al., 2020; Molinet & Cubillos, 2020; Petruzzi, Rosaria Corbo, Sinigaglia, & Bevilacqua, 2016; Steensels & Verstrepen, 2014; Steensels et al., 2014). Though brewers might have been unaware of the true microbiome involved in their so called 'pure culture' inoculated worts based on presence or availability of ambient environment and 'in-house' (autochthonous) microbe residents. In that same manner, recent discoveries of beers found in old shipwrecks, etc. are revealing secrets of microorganisms that were engaged in producing beers in much earlier days, along with long-lost-to memory flavour profiles for some classic beer styles (Londesborough et al., 2015; Thomas, Ironside, Clark, & Bingle, 2021, and now see Pieczonka et al., 2022 for more on historical findings with significant implications for the resurrection of old beer styles — that might not be so 'old and lost' after all).

Now turning to engineering and strain selection. While focussed more on wine, the article by Carrau et al. (2015) provides a very succinct and yet powerful overview of the topic and the way forwards. The team posits both the arguments for and against, and the advantages and disadvantages of the various approaches to selection or engineering of new culture strains. Designer approaches for creating superior brewers' yeasts are also covered elsewhere and that includes hybridisation of species, strains or gene swapping (see below for more on this) (Gibson et al., 2017; Iattici, Catallo, & Solieri, 2020; Krogerus, Magalhães, Vidgren, & Gibson, 2017; Petruzzi et al., 2016; Sipiczki, 2018; Steensels & Verstrepen, 2014; Steensels et al., 2014; Van Wyk, Kroukamp, & Pretorius, 2018). Noting that the organisms created by genetic modifications, and non-GM techniques, including hybridisation, mutagenesis and evolutionary engineering, however promising as fermentative agents, have typically only been assessed out in laboratory-scale trials. With that caveat in mind and this basic introduction on the topic now providing a lead-in to the details contained in the following sections.

19.5.1.1 Saccharomyces or non-Saccharomyces?—Of yeasts, alternate yeast species, hybrids, mutants and genomic variation

Saccharomyces sensu stricto covers the most important industrial yeast strains including *S. bayanus*, *S. cerevisiae and S. pastorianus* (synonym *S. carlsbergensis* — better known as the bottom fermenting lager yeast), and with this important topic well-reviewed and referenced by Barrio et al., and others (Barrio, González, Arias, Belloch, & Querol, 2006; Mertens et al., 2015). ('*Saccharomyces sensu stricto* is a species complex that includes most of the yeast strains relevant in the fermentation industry as well as in basic science. The taxonomy of these yeasts has always been controversial, particularly at species level' [Rainieri, Zambonelli, & Kaneko, 2003]). It can thus be difficult keeping up with the naming and renaming of such organisms, especially more so today with increasingly rapid genetic and metabolomic findings (Kidd, Abdolrasouli, & Hagen, 2023). Many of the genetic changes noted in those well-known strains indicated above being highly detailed in the article by Barrio et al., 2006.

It should be made clear that the earlier focus on *S. cerevisiae* (for ales) and *S. pastorianus* (for lagers) limited genotypic and phenotypic variation and thus also limited aroma and flavour compound production. Leading on from this understanding, as detailed herein, several strains of yeasts, including those from within the *Brettanomyces*, *Hanseniaspora*, *Lachancea* and *Pichia* genera have now, therefore, found application as alternative yeasts for beer production (more species noted below) (Iorizzo, Coppola, Letizia, Testa, & Sorrentino, 2021; Osburn et al., 2017). Afterall, the production of most of the aroma-active compounds is dependent on the yeast strain chosen — the metabolic profile of that organism, and the fermentation conditions employed. The addition of many other organisms — *Saccharomyces* and non-*Saccharomyces* yeasts to the spectrum or band of fermentation players is further detailed in a listing below — Section 5.1.5 and in Fig. 19.1.

Thus, as now seen, non-*Saccharomyces* yeasts represent attractive alternatives for production of beers, enabling the enhancement of flavour. This includes roles for producing non- or low-alcohol versions of fermented malt and beer-style beverages and to issues of diacetyl reduction in beers, as well as to the maturation and bottle conditioning of sour beers. The production of beers with low ethanol content is made possible due to the weak fermentative capacity of a substantial percentage of the non-*Saccharomyces* species (Bruner & Fox, 2020; Postigio et al., 2022; Salanță et al., 2020). Gluten-free beer and nutritionally or natural flavour-enhanced beverages are also possible through the application of such factors as discussed above. Taking all properties in view, there are crosses or hybrids created between pairs of the *Saccharomyces* strains, between pairs of non-*Saccharomyces* strains and between pairs involving both *Saccharomyces* and non-*Saccharomyces* species and their strains (see more under Section 5.1.2; Species hybridisation — seeking desired traits for flavour production). Fig. 19.1 provides a summary listing of 18 yeast species or strains and brief sensory properties conveyed via their use. Table 19.3 and Fig. 19.2 provide sensory descriptors for the main flavour compounds of interest to brewers and for the creation of new ever more flavour-rich (and off-flavour free) beers.

Natural selection and adaptation (domestication) of brewing yeasts clearly took place over the hundreds of years of more industrialised and modern beer production throughout the world (Barrio et al., 2006; Gallone et al., 2016). Modern molecular methods and clearer understanding of genetics have allowed scientists, through mechanistic knowledge, to alter metabolic activities of yeasts and bacteria and allowed for the selection of other yeast species and bacteria now proving useful in modern brewery fermentations. Such knowledge required a new field term – metabolomics. Metabolomics is defined as a comprehensive study of small molecules (metabolites), within cells, biofluids, tissues or organisms and now to beer (Cavallini, Savorani, Bro, & Cocchi, 2021; Ellis, Kerr, Schenk, & Schulz, 2022; Pieczonka, Lucio, Rychlik, & Schmitt-Kopplin, 2020; Pieczonka, Rychlik, & Schmitt-Kopplin, 2021; Roullier-Gall et al., 2020; Spevacek, Benson, Bamforth, & Slupsky, 2016). Collectively, these small molecules and their interactions within a biological system are known as the metabolome (Færgestad et al., 2009 – with descriptions of the genome, transcriptome, proteome and metabolome – covering the machinery needed by cells to live and, incidentally produce most of the useful volatiles described herein. The metabolome is defined with further detail below). Such work seeks to unravel the complexities of whole picture metabolism and, increasingly, those activities associated with flavour production, identifying and quantifying a selected number of metabolites – providing Carrau's yeast flavour diversity screening maps (Carrau et al., 2015) noted above. Then, the engineering or selection of suitable strains to enhance or produce desired and novel flavours and flavour profiles in beers begins. Noting that it is not possible to characterise yeast strains solely based on their genome but also with a need to comprehend and understand the basis of their specific metabolism. With even more complexity involved when mixed-culture species are intertwined in fermentation and flavour production. Calling upon the needs of a trained sensory panel again.

19.5.1.2 Species hybridisation—Seeking desired traits for flavour production

A few definitions:

Genus is a taxonomic classification that includes closely related species. A *species* is defined as a group of organisms made up of similar individuals capable of interbreeding or exchanging genes. *Hybridisation* is the most widely used breeding approach to develop phenotypes which have desirable traits. *Interspecific hybridisation* is the crossing of two distinct species from the same genus. This allows the exploitation of useful genes from wild, unimproved species for the benefit of the cultivated species. *Intraspecific hybridisation* is the mating of two individuals from the same species that are genetically distinct. In other words, intraspecific hybridisation is the sexual reproduction that occurs within the same species. Therefore, it can occur between different sub-species within a species. Selective breeding is another name for intraspecific hybridisation. Both types result in a hybrid that is genetically different from the parents, and mating is done artificially in both methods.

Hybridisation then is a molecular mechanism occurring naturally, or that can be done in the laboratory, whereby the exchange of one or more genes occurs or is driven from one organism to another adding to beneficial diversity (Bendixsen, Frazão, & Stelkens, 2022; Gabaldón, 2020; Naseeb et al., 2021; Stelkens & Bendixsen, 2022). Thus, alleles (genes/genetic material) from one genetic matrix or background may be integrated into another if favoured by selection. Moreover, as stated by Sipiczki: 'Interspecies hybridisation – effectively genetically re-addressing yeast has provided a powerful approach for transferring genes between *Saccharomyces* species without the resultant hybrids being genetically modified organisms'. (Sipiczki, 2018). Hybrid strains often outperform both parents and have proven useful in providing new, novel or enhanced flavour profiles when used in beer production (Mertens et al., 2015; Winans, 2022).

The innovations in brewing fermentation practices, using complex hybridisations of wild and domesticated yeast, are discussed by Langdon et al. (2019). As a case in point, POF genes have been inactivated or eliminated from lager brewing yeasts by multiple types of mutations, whereas these genes have been retained in yeasts that ferment products where POF is prized. As a reminder here, 4-vinyl guaiacol (4-VG) is perceived as a clove-like, phenolic or smoky flavour and considered an undesirable off-flavour in most beers (see discussion above on 4-VG; Section 4.5). Lager beers are known for their crisp flavour profiles that lack appreciable 4-VG, while wild strains of *S. eubayanus* and another 12 species, noted by Langdon et al., produce 4-VG (Langdon et al., 2019; see also Diderich, Weening, van den Broek, Pronk, & Daran, 2018 for details on reduced 4-VG formation).

Again, understanding that the limited genetic variation within available lager yeasts only leads to limited aromatic diversity, an ambitious plan led to the generation of novel interspecific yeast hybrids between selected strains of *S. cerevisiae* and *S. eubayanus* (natural mating so non-genetically modified). The programme led to an increase in lager yeast biodiversity. While many hybrids created and discussed by Mertens et al. (2015) had some off-flavour issues, one hybrid, H29, showed good fermentation characteristics, yielding a high ethanol concentration and delivered a desirably complex and fruity sensory profile to the lager beers so produced. The chemical composition including isoamyl acetate (sweet, fruity, banana, with a green ripeness and pear), ethyl acetate (ethereal/solventy, fruity, nuances of cherry and grape), isoamyl alcohol (fusel, fermented, fruity banana, cognac and ethereal) and phenylethyl acetate (floral roses, honey-like), see Fig. 19.2 for a quick glimpse summation on chemical classes and flavour descriptors.

The publication by Sipiczki (2018) presents two highly detailed tables of data covering interspecies hybridisation with the pairs of different *Saccharomyces* strains and resultant phenotypes largely expressing positive attributes with respect to beer production. Several phenotypic expressions leading to enhanced higher alcohol and ester formation, positive aroma profiles, reduction of certain off-notes (hydrogen sulfide, phenolic characteristics), decreased acid production and better fermentation performance (Sipiczki, 2018). Overall, the generation of novel hybrids has led to improvements in the fitness of yeast and to the diversity of aroma profiles in beers and, to more complex and unique sensory experiences and consumer enjoyment. *S. jurei* related hybrids with tropical and floral notes illustrating well the flavour complexity of such new hybrids (see Section 19.5.1.5). With increases in modern molecular genetic approaches and development of new techniques, the generation of many more hybrid yeast strains has been a feature of much discussion within the past decade beyond that of the *S. jure*i x *S. cerevisiae* story (Krogerus, Magalhães, Vidgren, & Gibson, 2015, Krogerus et al., 2017). The craft beverage producer today have a vast number of choices in the flavours they wish to elicit and display in their brews, and the appropriate strains to choose from, to effectively produce those desired beer profiles (see '*Saccharomyces* yeast hybrids on the rise' by Bendixsen et al., 2022).

19.5.1.3 The taming of the yeasts and bioprospecting

The text discussion above has considered *Saccharomyces* yeast strain hybridisation. Now onto a more notable mention of 'taming of wild yeast' — the rounding up and use of nonconventional yeasts in fermentation (Steensels & Verstrepen, 2014). Early brewing and some other alcoholic beverages relied upon, and brewing styles produced today still rely upon, spontaneous fermentation of naturally present microorganisms with such processes being inconsistent, inefficient and often producing off-flavours. This of course is what led to the use of defined starter cultures — usually the domesticated strains of *S. cerevisiae*, *bayanus* or *pastorianus*. However, as seen for other styles, such as Belgian Lambic beers (noted in detail below; Section 5.4) the selection of non-conventional yeasts such as *Brettanomyces*, *Hanseniaspora* and *Pichia* species, and several others, lend sensorial complexity and diversity to beer styles and flavour profiles (Gibson et al., 2017; Steensels & Verstrepen, 2014; Steensels et al., 2014, 2015).

So, what are some of the more unusual or unexpected flavourful results and desirable fermentation features that result from the adoption and use of alternative yeast strains and hybrid yeast in brewing? Let us go to the hop — first off. Increasing knowledge of plant terpenes such as derived from hops, hemp, cannabis and other plant materials (along with their biotransformation), together with brewers gaining a better understanding of volatile thiol compounds and their imparting desirable flavour notes has led to further diversification of beer flavour (Bruner & Fox, 2020; Dufour, Zimmer, Thibon, & Marullo, 2013; King & Dickinson, 2003; Molitor et al., 2022; Ramírez & Viveros, 2021; Svedlund, Evering, Gibson, & Krogerus, 2022; Swiegers, Saerens, & Pretorius, 2016). Now yeast interspecies hybrids have been constructed with the enhanced ability to release or manipulate hop-derived flavours through various enzymatic activities (King & Dickinson, 2003; see Dufour et al., 2013 for related details). Such hybrid construction also eliminates the undesirable POF trait and allows for 'boosted hop aroma with less hops' (Denby et al., 2018; Krogerus, Rettberg, & Gibson, 2022). Full sensory properties and chemical changes occurring in beer brewed with yeast modified to release polyfunctional thiols from malt and hops have been detailed by researchers at Oregon State University (Molitor et al., 2022). The sensory evaluation work dealing with descriptive terminology, which included fruity, resinous, melon, tropical, sweaty, stone fruit, guava, passionfruit, mango pineapple, herbal, vegetal, grainy, citrus floral, nutty, caramel, grassy and earthy. Noting here decidedly tropical flavour overtones. Interestingly many of these terms also apply to the wider spectrum of plant terpenes.

As for several other non-*Saccharomyces* yeast, once considered a microbial spoilage organism, *Torulaspora delbrueckii* is playing a role in innovative practices not only in brewing but in the baking industry with a comprehensive review covered by De La Cruz Pech-Canul, Ortega, and Solís-Oviedo (2019). The authors' discussion focussing on the higher stress tolerances of such organisms, that make them useful at least in co-fermentations, which enhance product qualities, and then noting some commercial sources for such yeasts and the flavour production potential, including an enhanced clove-like aroma from 4-vinyl guaiacol production (see also Table 19.3) (De La Cruz Pech-Canul et al., 2019; see also Burini, Eizaguirre, Loviso, & Libkind, 2021; Canonico, Ciani, Galli, Comitini, & Ciani, 2020 — the latter in relation to mixed fermentation activities at a Microbrewery plant and this as an application at an industrial level unlike many other laboratory-only-level trials).

Non-conventional yeast species for producing these diverse, novel flavourful beverages may be utilised first in musts and worts followed by inoculation of *S. cerevisiae* to complete the fermentations or a consort may be used. Such fermentation programmes or protocols now also go by the term bioflavouring. By using multiple yeast species, occasionally together with a few bacteria in the mix, this might be a case of science catching up with what nature intended, practices or works towards with respect to selective advantage and survival. One organism, or a group of organisms working in succession or together within a commensal ecological environment or matrix adding to variability in beer production processes. Leading brewers also to a selective advantage in creating novel products for an ever more competitive marketplace. Of course, this is not totally unlike the situation for Belgian beers for example, whereby a diverse ecological spectrum of organisms engages in the overall process leading to finished beer, including those organisms that enter at the time of bottling or casking for in-package or cask-conditioning of certain beers (see Section 5.6). Some of the details concerning the relevant organisms leading to desirable or alternative or enhanced flavours are presented in Fig. 19.1 (Callejo et al., 2019; Holt, Mukherjee, Lievens, Verstrepen, & Thevelein, 2018). Many of the cited studies revealing back-up to their assertions and results from sensory evaluations — thus confirming flavour changes, preferences and elimination of unwanted flavour notes. Indeed, further understanding of sensory perception — especially for odours (responsible for 'taste' — flavour) also plays a role in driving the consumer to seek out new sensory experiences (see van Wyk et al., 2018 on this point). With an increasing demand for new flavour experiences and the high price of obtaining and purifying flavours from natural plant sources, the use of microorganisms to produce them is helping reduce the costs of preparing flavours and flavour additives, which can be used in foods and beverages. Moreover, these then do not have to be claimed as 'added flavours'. For brewing purposes, specific flavour notes can be enhanced, 'more naturally' — examples include raspberry, cinnamon, roses and vanilla, plus strawberry or tropical fruit accents (van Wyk et al., 2018).

19.5.1.4 Beer fermented with yeasts more commonly associated with other fermented beverages

From wine to mezcal, to sourdough bread, to kombucha, exotic rotting fruits on the vine or a mead-makers honey-bee hive and everywhere in between — if an organism has played a role in fermentation of food or other beverages, there is a chance it is finding use in the brewing of beer these days. Details of some of these cross-beverage-fermentation possibilities — a modern non-human take on the synergistic 'friends with benefits' approach follow. Several recent key works come to mind here (Bellut et al., 2018; Borren & Tian, 2020; De Simone et al., 2021; Postigo, García, Cabellos, & Arroyo, 2021; Postigo, Sánchez-Arroyo, Cabellos, & Arroyo, 2022; Postigo, Sanz, García, & Arroyo, 2022; Roullier-Gall et al., 2020). In addition to the consideration of sensory attributes to final products arising from raw materials ingredients — which also relates to features linked to specific geographical regions — terroir/provenance, De Simone et al. (2021) present an interesting discussion of the microorganisms involved in beer fermentation. The topic includes the use of traditional culture yeast and hybrids — covered elsewhere in this chapter, yet also discusses sequential or contemporary fermentation by different microbes such as for Lambic beers or *Saccharomyces* yeast and LAB, and from more spontaneous fermentations initiated by the introduction of local air borne organisms. *S. cerevisiae* and non-*Saccharomyces* strains (including *Hanseniaspora*, *Lachancea*, *Pichia* and *Torulaspora strains*, *Lactobacillus* and *Pediococcus* bacteria) isolated from grape musts, wine bakeries, sourdough and wine and apple stillage, and the fermented tea beverage Kombucha just to mention a few noted examples (Bellut, Krogerus, & Arendt, 2020; Bellut et al., 2018; De Simone et al., 2021; Postigo, et al., 2021; Postigo, Sánchez-Arroyo, et al., 2022; Postigo, Sanz, et al., 2022). The flavour impact on beer quality is also addressed in those publications and illustrated in Table 19.3 and Fig. 19.2 (see Cioch, Cichoń, Satora, Skoneczny, 2022; Cioch, Królak, Tworzydło, Satora, Skoneczny, 2022; Postigo et al., 2021; Postigo, Sánchez-Arroyo, et al., 2022; Postigo, Sanz, et al., 2022). In the case of kombucha the isolation from the complex; microbiome, known as an SCOBY (for symbiotic collection — or 'cohort' — of bacteria and yeast) of several non-*Saccharomyces* yeast (*Hanseniaspora*, *Torulaspora* and *Zygosaccharomyces* strains) were investigated for non-alcoholic beer production. A sensory panel not able to discriminate between non-*Saccharomyces* alcohol free beer (AFB) and one produced with a commercial AFB strain (Bellut et al., 2018, see Section 19.5.1.5 for more on *Zygosaccharomyces*). In another major study with non-*Saccharomyces* species (including *Torulaspora*, *Metschnikowia*, *Wickerhamomyces and Hanseniaspora* strains) isolated from the wine industry were seen to be useful for low ethanol beer production and, in co-fermentation with a *Saccharomyces* yeast an improvement in the organoleptic characteristics of beer was noted (Postigo, Sánchez-Arroyo, et al., 2022). Once again, it is seen that the development of complex microbial starters, comprising different microbial strains, in either co- or sequential fermentation offers opportunities for emerging regional styles, unique flavour profiles and market segmentation (De Simone et al., 2021). For more on NA and LA beers, see Section 5.2, see also Figs. 19.1 and 19.2.

19.5.1.5 Of species, strains and flavour tales. A brief listing of some key alternate brewery fermentative yeasts

A brief introduction now to a few of the many *Saccharomyces* and non-*Saccharomyces* strains currently being used or considered for beer production — either in single or co-culture fermentations. In addition to the details and strains noted in the list, and numerous times through the chapter, *Lachancea fermentati*, *Schizosaccharomyces japonicus and*

Wickerhamomyces anomalus (Pichia anomala) species showed promising fermentation aptitude and sensory features to produce sour beer. Yeasts belonging to *Cyberlindnera fabianii*, *Pichia kudriavzevii* and *Pichia kluyveri* have been evaluated with a view to tailor the aroma production and the ethanol content in beer. With *C. saturnus* producing apple/cooked apple and red-berry-like notes in non-alcoholic beer partly through (E)-β-damascenone production, another compound of growingly notable importance in beer and distilled spirits beverages (sweet, fruity, rose, plum, grape, raspberry, sugar and with spicy tobacco nuances in the taste) (Methner, Dancker, et al., 2022). See additional sections and Fig. 19.1 for more insights into the value of using these yeasts and noting interest there also in the application of *Hanseniaspora vineae/L. thermotolerans/Torulaspora delbrueckii*. Quick overviews dealing with the expanding number of wild yeast strains useable in craft beer production with desirable brewing characteristics has been provided by Baiano (2020). With notes on yeasts in tradition and innovation noted by others (Iorizzo et al., 2021; Maicas, 2020). A question now being asked more frequently is how *S. cerevisiae* interacts with other microbes in natural habitats (Liti, 2015; see also Sicard & Legras, 2011). Indeed, with new developments opening more avenues of discovery, and multimicrobe fermentation models being explored, this can only mean exciting discoveries ahead for food and beverage flavour.

A selection of those yeasts finding ever more application in modern brewing operations

Candida zemplinia. Despite the suspected infectious and health issues with respect to *Candida* yeasts, some strains have been considered for craft brewery use (Estela-Escalante, Rosales-Mendoza, Moscosa-Santillán, & González-Ramírez, 2016). This organism often found on the surfaces of grapes and overripe fruits for example has not been assessed at any significant level for its flavour potential for brewery work or at least is not well recorded yet. Noted species also include *C. stellata*, and *C. pulcherrima*. Ethanol, glycerol and a balanced production of volatile compounds of sensory importance are positives for these strains, however, with limited experiments in beer — more currently understood in the wine sphere.

Cyberlindnera yeasts. Finding application in producing fruity non-alcoholic beers (NAB) which compete well with commercial NABs. Noting *C. subsufficiens* here (Bellut, Michel, Zarnkow, et al., 2019). See text above for more on red-berry flavours produced with this yeast (Methner, Hutzler, et al., 2022).

Dekkera/Brettanomyces. Extensively covered in the text — Lambic/Sour beers. Also, bioflavouring and calorie reduction potential (see Capece, Romaniello, Siesto, & Romano, 2018 and other notable references cited in this chapter).

Hanseniaspora strains. *H. guilliermondii*, *H. opuntiae* — aroma profiles — fruity and toffee notes, respectively. Such yeasts (often found on the skins of red grapes) seeing co-fermentation application with S. *cerevisiae* to enhance the aromatic profile of craft beer (Bourbon-Melo et al., 2021; Escalante, 2018; Piraine et al., 2022; Rocha, 2019). Both strains; mixed culture fermentations — reduced ethyl ester amounts, yet *H. guilliermondii* use led to increased concentrations of (other) acetate esters — especially when sequentially inoculated, with large increases in phenethyl acetate with the characteristic rose and honey notes present in the final beverages. Though both organisms in single-cultures led to excessive sweetness and unbalanced aroma rendering them unsuitable for non- or low-alcohol beers (Bourbon-Melo et al., 2021). One might wonder if the noted influence on terpene components, in interesting wine studies of late, could bode well for the potential use of *Hanseniaspora* spp. with respect to the flavour of hoppy beers (del Fresno et al., 2022).

KVIEK — A Consortium. The consortia of yeast known as Kveik, primarily used for farmhouse ale production in Nordic countries have recently been employed, along with non-conventional yeasts for targeted aroma modulation in beers. Once again showing the power of using non-*Saccharomyces* yeast in brewing with resultant desired flavour characteristics — for beer and distilled spirits (Dippel et al., 2022). The Norwegian kveik yeasts represent a genetically distinct group of domesticated beer yeasts with unique properties relevant to the wider brewing sector. Kveik yeasts obtained from several diverse sources — have been shown to have lost 4-VG production capacities and to exhibit unique flavour metabolite production profiles (Preiss, Tyrawa, Krogerus, Garshol, & van der Merwe, 2018; Preiss, Tyrawa, & van der Merwe, 2017). Kveik strains produce fruity ethyl esters at above detection thresholds; ethyl caproate (hexanoate — pineapple, fruity, waxy, banana green and estery), caprylate (octanoate — waxy, fruity, pineapple, creamy, fatty, mushroom and cognac notes), caprate (decanoate — waxy, sweet fruit, apple, brandy-like), and the more floral phenethyl acetate (honey, floral, rose, green, fruity and yeasty). Additional distinct populations of kveik domesticated or 'landrace yeasts' may exist thus increasing the potential flavour spectrum diversity for beers produced with them (Preiss et al., 2018).

Lachancea thermotolerans. This organism, as a low lactic-acid producer, is finding use as a new starter for beer production (Zdaniewicz, Satora, Pater, & Bogacz, 2020).The organism shows a high tolerance to hop-derived compounds and, as noted under the NA Section (5.2) — like for *Lachancea fermentati* produces a low level of ethanol (Bellut et al.,

2020). This organism has been the subject of a major work dealing with biomodulation of the physicochemical, aromatic and sensory aspects of craft beers made using it as a main or co-fermentative organism (Peces-Pérez, Vaquero, Callejo, & Morata, 2022). The organism produces high concentrations of lactic acid leading to the formation of ethyl lactate (ester: sweet, fruity, creamy, butterscotch, pineapple-like and caramellic notes). Moreover, and interestingly, bottle conditioning was additionally studied with other organisms also included in the mix. Banana fruity notes were a noted feature, as is the case with many alternative yeasts used in novel brewing scenarios. It is thus now posited that, due to high lactic acid production, this organism could play useful roles in the acidification of beer without the need for lactic acid use or lactic acid bacterial production and thus avoiding some other unwanted flavours (Peces-Pérez et al., 2022, see Section 5.8). The interplay of the various metabolomes at work in mixed culture interactions and brewing 'ecosystems' worthy of further investigation and sensory evaluation.

Saccharomyces jurei. As a new budding yeast species, *Saccharomyces jurei* was initially discovered in oak bark (*Quercus robur*) in the French Alps (Giannakou et al., 2021; Naseeb et al., 2018). See the section concluding text in the following, this listing for a more complete account of and discussion of the uses of this organism in current brewing practice.

Saccharomyces paradoxus. An organism stated to be commonly isolated from environmental samples in both Europe and America, though rarely found in association with fermentation, is showing recent potential for industrial brewing application. Beer produced in pilot scale; full-bodied, clean flavour profile though with clove-like 4-VG present (Nikulin et al., 2020). Noted here is the review of wild yeast for the future by Molinet and Cubillos, which makes mention of *S. paradoxus* and notes the allele origin for boxwood and black currant aromas in wines. Moreover, their publication emphasises that, via bioprospecting, useful strains of yeast have been isolated from wasps, flies, oak trees and bark, soil and flowers (Molinet & Cubillos, 2020). Related to the latter point on bioprospecting is an interesting account of the sourcing of wild Ecuadorian *S. cerevisiae* strains together with extended flavour details from the fermentation of many of the isolates (Simbaña, Portero-Barahona, & Carvajal Barriga, 2022). And the details concerning the isolation of wild yeasts from Olympic National Park, including the strain *Moniliella megachiliensis*, with coverage of its physiological characterisation and potential for fermented beverage production also recently presented (Piraine et al., 2022, see also Ciani et al., 2020; Lengeler, Stovicek, Fennessy, Katz, & Förster, 2020).

Saprochaete suaveolens. A strain isolated from dragon fruit (formely known as *Geotrichum fragrans*) with characterisation of mutant strains showing a higher capacity for producing flavour compounds — notably rare in other yeast strains — isobutyl, isoamyl and ethyl tiglate (fruity nutty, caramel, herbal) (Tan, Caro, Shum-Cheong-Sing, et al., 2021; Tan, Caro, Sing, et al., 2021). This work represents a first case example of volatile organic compound enhancement in a microbial strain via the UV-mutagenesis approach. An evaluation, by the same team, of mixed fermentations using *S. cerevisiae* and *Saprochaete suaveolens* yielded a natural fruity beer from an industrial wort. The use of *S. suaveolens* alone led to fruity and low-ethanol content beers thus showing potential for high- or low-ethanol fruity flavoured beers. The metabolism, production of the flavour compounds, including ethyl tiglate, and flavour properties and attributes described elsewhere (Grondin et al., 2015).

Schizosaccharomyces pombe. Research with this organism in mixed-culture fermentations, with a more complete attenuation of high gravity wort reveals an interesting prospect for brewers. Such yeast and fermentations can result in the over yield of certain desirable flavour notes, thus invoking the need to perform batch-mixing or dilution of such often higher gravity brewed ferments to produce desirably attenuated, flavoured and ethanol content beers. This allows for another format of pairing different strains or creating/blending multiply fermented microorganism ferments for novel beer flavour production. Moreover, some authors discuss a concept of recent, though growing importance — the search for, and the application of organisms sourced from different branches of the fermented/distilled beverage industry (Pownall, Reid, Hill, & Jenkins, 2022, see body text Section 5.1.4 for more on this conceptualisation).

Torulaspora delbrueckii. *T. delbrueckii* — (formely *Saccharomyces delbrueckii* or *S. rosei* — of long-term oenological interest). Summation flavour; ethyl lactate (coffee, strawberry), 2-phenylethyl acetate (roses), 3-ethoxypropanol (fruity/blackcurrant) and isoamyl acetate (banana/pear) (De La Cruz Pech-Canul et al., 2019). Several discussions pertain to this organism (see elsewhere in this chapter for even more detail) including its ability to enhance bioflavour and reduce ethanol content along with performing transformations on hop monoterpenoids (Canonico, Agarbati, Comitini, & Ciani, 2016; Canonico, Comitini, & Ciani, 2017; Canonico et al., 2020; De La Cruz Pech-Canul et al., 2019; Eliodório et al., 2019; Michel et al., 2016). *T. delbrueckii* was evaluated for beer production, in both pure and in mixed cultures with a *Saccharomyces cerevisiae* starter strain — also serving as the control. Interactions were noted between *S. cerevisiae* and *T. delbrueckii*, with enhanced ethyl hexanoate and ethyl octanoate production observed with levels depending upon relevant inoculation ratios of the two organisms. *T. delbrueckii* use also resulted in reduced 2-phenyl ethanol and isoamyl acetate levels — compounds noted for floral and fruity aroma impressions. Beer produced with pure cultures of *T. delbrueckii* had a low alcohol content <3.0% ABV, while also showing a characteristic analytical and overall useful aromatic profile. Qualifying flavour descriptors included fruity/estery, fruity/citric, alcoholic/solvent, hop, DMS, sulfidic, cereal, malt, caramel, toasted, body/mouthfeel, astringent, acid, bitter, sweet and oxidised/aged (Canonico et al., 2016). The behaviour of this organism also noted by others (Postigo, Sánchez-Arroyo, et al., 2022; Postigo, Sanz, et al, 2022) with an illuminating note that 'the perception of aroma compounds depends on multiple factors, known as the 'matrix effect', as well as synergisms or antagonisms. For this reason, it is necessary to study the sensory profile of the beers brewed (with such organisms in single or co-culture) to evaluate the positive or negative effect of the high concentrations of some compounds, such as ethyl acetate, produced in the different experiments' (Postigo, Sánchez-Arroyo, et al., 2022; Postigo, Sanz, et al, 2022). Brewers should therefore not simply rely on anticipated, or yeast-supplier advertised flavour notes, but on full sensory evaluation of trials and successive production runs to be sure things are working in harmony.

Wickerhamomyces anomalus. (Formerly *Pichia anomala, Hansenula anomala, Candidia pelliculosa* – Kurtzman, 2011). Accepted by the wine industry since noting its exhibition of interesting metabolic characteristics and flavour production, though often seen as an uncontrollable risk (Padilla, Gil, & Manzanares, 2018), it is now frequently reported in the beer brewing literature and receives mention throughout this chapter. Seen as a bioflavouring, low alcohol and sour beer co-inoculation microbe with noted pear, apple and peach flavour contributions (Burini et al., 2021; Capece et al., 2018; Gibson et al., 2017; Postigo, Sanz, et al, 2022; Ravasio et al., 2018). Spoilage issues and sources in a brewery recounted by Sohlberg, Sarlin, and Juvonen (2022).

Zygosaccharomyces with *Z. rouxii* and sister species *Z. bailii* noted for food and beverage spoilage even with food preservatives present (Solieri, 2021). Now transitioning from being regarded as mere culprits of spoilage to finding useful roles as 'ZygoFactories' for production of useful flavours and flavourings. See the concluding part of this section for more details regarding this species and its strains and the NA section (19.5.2) for roles in production of such brews.

Zygotorulaspora florentina. A new find is the maltotriose-positive organism *Z. florentina* (formely *Zygosaccharomyces florentinus*) isolated from oak (Nikulin et al., 2020). With good fermentation performance and high volatile production in a composition and with component concentrations that masks the 4-VG produced by this 'Finnish' strain. *Torulaspora delbrueckii* (see above) is its closest brewing related species. Beers produced with *Z. florentina* have been characterised by an increase in the isoamyl acetate (fruity banana, pear) and α-terpineol content (citrus, lemon/lime, woody, soapy) and is tolerant of some stress conditions and certain preservatives used in other beverage production (Canonico, Galli, Ciani, Comitini, & Ciani, 2019).

In concluding this section, two of the listed organisms from the aforementioned are deemed worthy of additional commentary. Deservingly so based upon their broad-range application to brewing and flavour production. In *Saccharomyces jurei* – we see a representative example of a recently discovered yeast species finding notable use in modern craft brewing. *S. jurei*, as a new budding yeast species, is a recent addition to the *Saccharomyces* sensu stricto yeast clade, initially discovered in oak bark (*Quercus robur*) in the French Alps (Giannakou et al., 2021; Naseeb et al., 2018). Simply defined, *Saccharomyces* sensu stricto is a species complex that includes most of the yeast strains relevant in the fermentation industry as well as in basic science (Rainieri et al., 2003; and see details in Sipiczki, *2018*). Two strains of *S. jurei* have also been isolated from the European ash (*Fraxinus excelsior*) in Germany (Hutzler et al., 2021). Strains from both sources showed brewing potential and this species and its non-GMO derived hybrids (*S. jurei* × *S. cerevisiae*) have seen successful application in the production of a British brewery made beer with two different hybrid fermented beers exhibiting a good estery profile – notes of apple, pear and banana. A third hybrid leading to a beer with a tropical fruity expression; notes of peaches and berries but with some astringency and light sour character. A more expanded list of flavour compounds and their descriptors was also supplied by the research group (Giannakou et al., 2021). An *S. jurei* strain isolated in Germany also showed potential to produce beer with enhanced apple and aniseed like aromatics from ethyl hexanoate, and with higher amounts of medium chain fatty acid esters in comparison with *S. pastorianus* brewed beers (fruity, berry-like ascribed to higher levels of ethyl esters), but also with some less desirable phenolic spice/clove accents (Hutzler et al., 2021). Beers produced from the *S. jurei* × *S. cerevisiae* hybrids perform better from an overall desirable flavour profile viewpoint as judged by trained sensory panels (Giannakou et al., 2021; Hutzler et al., 2021).

In addition to *S. jurei,* the other microorganism worthy of mention, and one that is little known about, and yet one that has plagued the beverage industry from beer to lower alcohol-content-packaged spirit cocktails – in the form of product spoilage or package damage is the species *Zygosaccharomyces.* An organism can be considered as friend or foe. With *Z. rouxii* and sister species *Z. bailii* noted for food and beverage spoilage even with food preservatives present (Solieri, 2021). However, while classed as a spoilage yeast, as for other non-conventional yeast (NCY, i.e., other non-classical *S. cerevisiae* yeast), the *Zygosaccharomyces* yeast species noted here, based on tolerance to certain stress conditions, are transitioning from being regarded as mere culprits of spoilage to finding useful roles as 'ZygoFactories'. Such systems are capable of carrying out bioconversion of highly sugary and salty worts, poorly fermentable by other NCYs, leading to the production of secondary metabolites for use as flavour additives (Solieri, 2021). For *Z. bailii* and *Z. rouxii*, their ability to utilise fructose preferentially in glucose/fructose media (a condition known as fructophily) may extend their usefulness. Low alcohol beer production and aroma diversification in wine have been noted features of the use of the ZygoFactories. *Z. rouxii* strains for ethanol free beer production, due to inability to ferment maltose, can also consume ethanol under aerobic conditions while simultaneously producing active flavour compounds impacting favourably beer aroma profiles. *Z. bailii* has been used to produce a Pilsner style beer with reduced ethanol content in co-culture fermentation with *S. cerevisiae* (Solieri, 2021). Further research and understanding needed for scalability and cost-effective production is required.

19.5.2 Non- and low-alcoholic beers

An increasingly important segment of current brewing sales includes low- and non-alcoholic beers (see Chapter 13). The removal of alcohol and early packaging in cans or bottles sometimes led to a poor reception on flavour for such beers. Beers, characterised as bland, tasted worty and potato-like without the alcohol or overly malty and sometimes with metallic taints (if in cans) and with an overall lack of enticing aroma/flavour – quite unlike their alcohol-laden counterparts (Strejc, Kyselová, Karabin, Silva, & Brányik, 2013). Lager yeast mutants (*S. pastorianus*) with elevated amounts of flavouring compounds were then noted for production of alcohol-free beer (Strejc et al., 2013). Key flavour notes contributed by elevated levels of isoamyl alcohol and isoamyl acetate (fruity banana – more on banana to come!) and palate fullness sensations.

Non-*Saccharomyces* yeasts are now seeing much more application in the production of LAB (0.5%–1.2% ABV) and NAB (<0.5% ABV) (Bellut et al., 2018, Bellut, Michel, Hutzler, et al., 2019, Bellut, Michel, Zarnkow, et al., 2019; Vaštík et al., 2022). Suitable strains of *Zygosaccharomyces rouxii* and especially *Saccharomycodes ludwigii* are seeing some favour (Callejo, Gonzalez, & Morata, 2017; Callejo et al., 2019; De Francesco, Turchetti, Sileoni, Marconi, & Perretti, 2015). Good flavour profiles are being noted and often with reduced diacetyl content. The pilot-scale production of a fruity NAB has quite recently been ascribed to *Cyberlindnera* yeast strains (*C. subsufficiens*) with good masking of the wort-like aromatics. Further studies are, however, suggested as needed on the masking effect (Bellut, Michel, Zarnkow, et al., 2019).

The application of non-Saccharomyces yeasts isolated from the complex zoogloeal mat (SCOBY) from kombucha has also been considered for producing alcohol free beers. Many of the yeasts involved there representative of the common ones discussed in this chapter. Strains of *Lachancea fermentati* isolated from kombucha for example may have potential application in LAB brewing (Bellut & Arendt, 2019; Bellut et al., 2018, 2020; Bellut, Michel, Hutzler, et al., 2019). Sensory evaluation programs were implemented to compare the flavour of such beers – with the application of terms such as fruity, wort-like and floral, sweet and acidic/sour taste descriptors on the ballots and included evaluation of the body or mouthfeel of the beers. The alcohol-free beers, made with kombucha-derived microorganisms then being characterised with attributes such as worty/wort-like and with bready and honey terms, but also with some cereal-like character and with diacetyl noted by several panelists in one study instance (Bellut et al., 2018). Principal component analysis (PCA) has been applied to transform and combine copious amounts of sensory data collected from non-alcohol containing beers into new 'mapped' components, based on variation and correlations within the data set (Bellut et al., 2018). Such mapping has shown, based on the sensory evaluations and analyses of secondary metabolites, that alcohol-free beers are not well-discriminated (set apart from one another), but are usually all honed-in on the aforementioned sweet taste and wort-like aroma characteristics. Some other noted descriptors within the literature for LA and NA beers include black tea, caramel, slightly grassy, fruity and white wine. Beers at less than 1.3% ABV were produced using *Lachancea fermentati* strains and exhibited a balance of sweetness and lactic acidity though were plagued with an elevated level of diacetyl off-flavour (Bellut et al., 2020). As mentioned above some cases of diacetyl reduction have been noted, though as here this is not always the case. Worty character, overly sweet impressions and potential diacetyl issues, could be the significant issues to contend with here with sensory detection likely influenced by the actual alcohol content. The still problematic issue of worty-like as an accented off-note in alcohol free beers is given due commentary in several of the cited works. Maltose-negative yeasts known to be useful in non-alcoholic and low-alcoholic beer production have also been reported on in the literature (Yabaci Karaoglan et al., 2022). However, once again, the results of sensory evaluations indicate that none of the investigated maltose-negative strains were able to mask the worty off-flavours. The aldehyde 3-methylthiopropionaldehyde is now noted as being the key compound responsible for the worty off-flavour (aka. methional, extensive notes; musty, potato {skin and French fries}, tomato, vegetal & vegetable oil, earthy, creamy, yeasty, bready, limburger cheese, mouldy, onion, savoury, meaty, brothy and seafood – and *fishy/fish like* – according to Spedding, one of this chapter's authors. The roles of ethanol and yeast metabolism in the reduction of worty character are important considerations here (Bellut et al., 2018; Ramsey et al., 2020). The absence of ethanol, the lack of aldehyde reduction due to shortened fermentation times and the higher levels of residual sugars – mono and disaccharides such as maltose intensify undesirable worty flavours, and these issues should be at the forefront of the brewers' mind when creating alcohol free-beers and the choices of organisms to use for their chosen fermentation conditions.

Overall, the performance of such wild non-conventional yeasts operating under different wort conditions and the selection of suitable starter strains resulting in low-alcohol beers with pleasing organoleptic flavour profiles – has recently been summarily described (Capece, De Fusco, Pietrafesa, Siesto, & Romano, 2021). For sensory evaluation purposes, extensive coverage of flavours produced from various strains with threshold data has been provided by Vaštik and colleagues and Romero-Rodríguez et al. (Vaštik et al., 2022; Romero-Rodríguez et al., 2022). Most of the cited works provide important directives for producing NABLABs via the use of non-conventional yeasts, with respect to how flavour production is encouraged while also steering alcohol levels (van Rijswijck, Wolkers–Rooijackers, Abee, & Smid, 2017).

Ending this section with hints of mints and bananas. In a paper by Methner, Dancker, et al. (2022), the authors showed the potential for *Cyberlindnera* species strains to yield cool mint, pear and banana notes, whilst yeasts of the species *Kluyveromyces marxianus* and *Saccharomycopsis fibuligera* produce a wide range of fruity flavours. Of sensory significance here was the finding that key secondary metabolites (the flavour volatiles) conveying the notes mentioned above were produced below odour detection thresholds as discoverable in regular strength alcoholic beers but that their thresholds are significantly lower in non-alcoholic beers (Methner, Dancker, et al., 2022). Oh, and about the banana notes. A key flavour note associated with many alternate yeast fermentations is isoamyl acetate – oh then of bananas and circus-peanuts (Quilter, Hurley, Lynch, & Murphy, 2003; Vaštík et al., 2022). See Fig. 19.2 (The Flavours Graphic).

19.5.3 The problem with diacetyl—Avoiding buttery beers and the situation with dry hopping and yeast-based bio-transformations

As a massive problem for brewers since the beginning days of brewing, the topic of diacetyl (2,3 butanedione – a vicinal diketone) was the subject of a 125th Anniversary Review in 2013 (Krogerus & Gibson, 2013) and has been a focus of intense research (Bruner, Marcus, & Fox, 2022; Gibson, Vidgren, Peddinti, & Krogerus, 2018; Lodolo, Kock, Axcell, & Brooks, 2008). The 125th year review provides the historical context and a picture of the issue with respect to diacetyl production, plus limitations as to an understanding of how the yeast reduces the concentration of this off-flavour volatile during the post-fermentation process known as the diacetyl rest. Diacetyl conveys a trigeminal slick/oily mouthfeel and a cloying buttery, buttered popcorn or butterscotch aroma and flavour if present in too high of an amount in finished beer. Furthermore, its precursor can exist in beer and form more diacetyl with time in package, as can the oxidation of the diacetyl end products of acetoin and 2,3-butanediol. Interestingly, and often overlooked, is the production of diacetyl via the complex system known as the Maillard reaction (Cha, Debnath, & Lee, 2019). Yeast strain development through genetic engineering or adaptive evolution, with the aim of diacetyl amelioration, is fully discussed in the paper by Krogerus and Gibson (2013). Control factors, the use of enzymes (including enzyme treatments of fermented beer, immobilised yeast – on columns for passage of continuously produced beer – or for high gravity brewing scenarios) and genetic modifications to yeast to reduce final diacetyl levels are introduced in the review by Krogerus and discussed by others (Bruner et al., 2022; Gibson et al., 2018; Lodolo et al., 2008). The issue of acceptance of genetically modified organisms is also a topic of debate here. Contaminating bacteria, *Lactobacillus* and *Pediococcus* can (more directly) produce diacetyl (see Table 19.1). The ratio of diacetyl to another vicinal diketone, pentanedione, can be used to indicate if elevated diacetyl concentrations are due to microbial contaminants or to the fermentation by-products (Lodolo et al., 2008). Adaptive laboratory evolution, via mutation, selection and use, of a lager yeast for diacetyl control was explored further by Gibson and group, leading to a 60% reduction in diacetyl levels compared to the control strain (Gibson et al., 2018). Most recently, an issue, known for a long time as the 'freshening power of hop' (since ca. 1893) – but with the new term now known as hop creep, whereby dry-hopping of beer may lead to an increase in alcohol content, concomitant higher carbonation levels and diacetyl production has been addressed (Cottrell, 2022). Such work involved examination of several *Saccharomyces* species and strains (Bruner & Fox, 2020; Werrie, Deckers, & Fauconnier, 2022). Certain experimental yeast strains were in accord with production of diacetyl at levels below detection threshold values as explained by Bruner et al. (2022). The production of diacetyl from hop creep is associated with yeast health, and the yeast health during fermentation is related to free amino nitrogen content (FAN; amino acids and short peptides as nutrients) (Bruner et al., 2022; Spedding, 2013b). Thus, an increased understanding of FAN content, diacetyl levels and yeast selection in relation to hop creep was needed and addressed (Bruner et al., 2022). Several *Saccharomyces* yeasts were chosen and included *S. cerevisiae*, *S. cerevisiae* var. *diastaticus* (modified to prevent the formation of diacetyl) *and S. bayanus*, and as such used to produce several classic beer styles. The study inferred that there is a link between amino acid assimilation by the yeast strains and diacetyl production or final levels present in the fermented beers. Further analysis of amino acid content during fermentation in dry-hopped beers is now warranted (Bruner et al., 2022).

It was noted above that in the presence of dry-hops, certain selected or modified yeast strains can create unique flavours through biotransformation of the terpenes in the hop oils. Those bioconversions lead to the production of more pleasing aromatic terpenic compounds. Moreover, this leads to a reduction in amounts, or elimination of the need for dry hopping while still producing 'hoppy flavour notes' (Denby et al., 2018). In addition, biotransformation utilising yeast-encoded enzymes can convert flavourless precursor molecules, found in barley and hops, into volatile thiols that impart a variety of desirable flavours and aromas in beer (Molitor et al., 2022). Such transformation components are being investigated as a potential means to eliminate the currently popular practice of dry hopping, as already noted above in the case of terpenes, while, at the same time, reducing diacetyl or masking the perception of the buttery characteristics as associated with hop creep (Molitor et al., 2022). Two volatile thiols produced by such biotransformation conveying notes of guava and passionfruit, with sensory perception inferences to pineapple and tropical fruity qualities (Molitor et al., 2022). The work of Molitor's group shows the potential of genetic modification in dramatically enhancing yeast biotransformation to produce beers without off-flavours and with full fermentation ability. Sensory evaluations, including detection and terminal threshold value determinations, were of utmost importance for evaluating the results of such engineered yeast and their biotransformation capabilities with respect to the volatile thiol compounds (Molitor et al., 2022).

19.5.4 Lambic beers

Note the brief introduction to this topic above; Section 4.6. Scientific efforts have increased our understanding of Lambic beer microbiology and chemistry. A neat review has most recently been prepared by De Roos and De Vuyst (2022). Wherein they posit the details for Flanders red ale, Flanders brown ale — the red/red-brown acidic ales and especially Lambic beer and provide a deep dive into the main literature on the topic. A summary of the process, showing the action of the unique blend of microorganisms illustrates: Phase one — Enterobacterial/AAB phase with enterobacteria, acetic acid bacteria (AAB) and oxidative yeast involvement; Phase 2 — main fermentation phase; *S. cerevisiae* and the cryophilic *S. kudriavzevii* as the active performers; Phase 3 — acidification phase; lactic acid bacteria: *Pediococcus damnosus*, acetic acid bacteria: *Acetobacter pasteurianus* and the mesophilic bacterium, *Acetobacter lambici* and, Phase 4 — the maturation phase; cycloheximide resistant yeasts: *Brettanomyces bruxellensis*, *B. anomalus* and *B. custersianus* (De Roos, Verce, Weckx, & De Vuyst, 2020; De Roos & De Vuyst, 2022; see also Straka & Hleba, 2022 who purport 100 species of yeasts and more than 50 species of bacteria to be involved in spontaneously fermented beers! Such facts regarding the microbiota/microbiome and ecological complexity were initially reported in two important papers (Spitaels et al., 2014; Vriesekoop et al., 2012).

The *Brettanomyces* species continue to play a role in cask and bottle (see also below). Most Lambics are blends and one blend component may include a beer produced by a mixed population of *Brettanomyces*, lactic and acetic acid bacteria (Anderson et al., 2000). *Brettanomyces* may produce a host of components which give beers made with its involvement a very complex sensory profile; alluded to below — see also Sections 4.5 and 5.5 (Suarez et al., 2007). An improved knowledge of the flavours of Lambic beers is noted by Witrick et al., whereby it is stated, through the extended flavour results, that the range of typical values for many of the aroma compounds and the acids which present in Belgian Lambic beers could help clarify the situation with respect to variation in flavour characteristics noted in commercial products and may assist the craft brewer to attain better consistency production of their own beers, especially if combined with a better understanding of the quality and nature of the barrels used for any ageing (Witrick, Duncan, Hurley, & O'Keefe, 2017; see also De Roos, Vandamme, & De Vuyst, 2018). Several well-recognised brands of Lambic beer were assessed for the content of several key flavour-associated acids, esters, phenols as well as for lactic and acetic acids (Witrick et al., 2017). With respect to aged bottled gueuze beers, the flavour profile has been followed over time with suggestions that the preferred ageing period is less than 10 years. A spectrum of volatile flavour compounds was followed, with some produced and others lost over time (Spitaels et al., 2015). The age time for gueuze beers further detailed under the section on Coolships (koelschip), blending and bottle conditioning (Section 5.6).

The justification for the need, in commercial beer production today, for normally unwanted/undesirable microorganisms (beer spoiling bacteria and phenolic compound-generating yeasts) and as to which organisms produce the key flavour components nicely described by De Roos and De Vuyst, together with factors that elevate such spontaneously fermented acidic beers to the higher level of acceptance as a potentially health-promoting benefit beverage (De Roos & De Vuyst, 2022). Key molecular and biochemical details of *Brettanomyces* in brewing covered by Menoncin and Bonatto, with notes on potential biotransformation of herb and hop (leading to terpenes and thiols from their precursors) and with those authors also presenting brewers with a new aroma/flavour wheel (Menoncin & Bonatto, 2019). The distribution and roles of acetic acid bacteria in wooden casks used during fermentation and ageing of Lambic beer detailed also by De Roos et al. (2018) and the variations in practice, technologies applied and the environmental conditions affecting the complexity of taste and the aroma profiles of Lambic-derived beer products such as gueuze detailed by Bongaerts, Roos, and De Vuyst (2021, see also Tyrawa, Preiss, Armstrong, and Merwe, 2019 regarding strain variability and temperature-dependent variations in processing to effect product aroma and flavour diversity). Descriptions of the key phenolic compounds that persist in 'Brett' and other beers comprehensively dealt with by Lentz (2018). This latter account covers the clove-like, curry, spice, smoky, bacon, phenolic, medicinal leathery, horse stable/barnyard, creosote and vanilla notes, conveyed by the vinyl- and ethylguaiacols, vinyl- and ethylphenols, guaiacol, vanillin, and 4-vinyl syringol and how such compounds can degrade to give sweeter (more vanillin enhanced qualities) and stale old beer character (Lentz, 2018; Spedding, 2012c).

An extensive volatiles data list, stemming from recent gas chromatography-mass spectrometry analyses of ageing Lambic beers is also now available (Joseph, Albino, Ebeler, & Bisson, 2015; Joseph et al., 2017; Witrick, Pitts, & O'Keefe, 2020) together with additional new flavour wheels, which add to the arsenal of tools for craft brewers and their sensory teams. Thus, enabling them to better assess quality aspects of their production processes and for qualifying such complex flavoured beers for blending purposes, and for 'Go-No Go' decisions on release of final packaged or kegged beers (see Part 2) (Flavour wheels: Joseph et al., 2017; Menoncin & Bonatto, 2019; Serra Colomer et al., 2019 – Styles wheel).

19.5.5 Brettanomyces—Organisms with multiple complex roles to play

Following the opening of the session on Lambic beers above, important developments relevant to a better understanding of the main 'Brett' character organisms are noted here. *Dekkera/Brettanomyces* are synonyms, with some authors choosing *Dekkera* and some *Brettanomyces* (de Barros Pita et al., 2019). While many details for the *Dekkera/Brettanomyces* wild yeasts apply to the Lambic beers section, these organisms are playing increasingly important roles in other areas of brewing with great flavour impact and implications.

The yeast as isolated from Belgian Lambic beer was assigned as *Brettanomyces bruxellensis*, with the genus Dekkera later introduced to describe strains able to produce ascospores. *Brettanomyces* then is the asexual budding form known as an anamorph and Dekkera, the sexual reproducing form/stage in the life cycle of a fungus known as a teleomorph – the same organism in different forms. The genus of *Brettanomyces* has five species, two of which are better known to brewers as *Brettanomyces bruxellensis* and *Brettanomyces anomalus*. A third strain sometimes noted in the literature with potential applications in brewing being *Brettanomyces custersianus*. The other two species: *B. nanus* and *B. naardenensis*. Based on changing nomenclature, it should thus be noted that strains known as *Brettanomyces lambicus* and *Brettanomyces claussenii* are *Brettanomyces bruxellensis* and *Brettanomyces anomalus*, respectively (de Barros Pita et al., 2019; Yakobson, 2012). The discussion below presents data using the nomenclature of the cited authors, and the above definitions should hopefully assist in understanding which organism is referred to when reading different works.

The production of flavours and the beer spoilage potential varies depending upon specific *Dekkera/Brettanomyces* species (Crauwels et al., 2017; Shimotsu et al., 2015). With respect to the genetic selection and manipulation of yeast strains for specific or directed enhanced aroma volatiles production, there is a need to provide the lexicon for the sensory evaluation of such specific-strain-fermented beers. A case in point for *Brettanomyces* yeasts is the creation and usage of the new *Brettanomyces* aroma/flavour/styles wheels as noted earlier (Joseph et al., 2017; Menoncin & Bonatto, 2019; Serra Colomer et al., 2019).

Aroma characteristics pertaining to these organisms were determined via the growth of several *Brettanomyces* yeast strains in a synthetic growth medium, with and without supplementation of specific precursor compounds. These studies, while focused on wine, showed the generation of an array of aroma compounds, by *Brettanomyces*, in addition to the well-known aromatic phenol derivatives referred to as the phenolic off flavour (Di Canito, Foschino, Mazzieri, & Vigentini, 2021; Suárez, Suárez-Lepe, Morata, & Calderón, 2007). The properties of a POF negative *B. anomalus* strain with the loss of ability to convert ferulic acid to 4-ethylguaiacol and 4-ethylphenol and thus loss of the main spoilage character being described recently by Colomer et al. (2020). A recent interesting finding is the potential for *Brettanomyces* yeast to release sweet and floral monoterpene alcohols from hops with the promotion of floral and citrus notes in beer. Nicely adding to flavour complexity and for bioflavouring of beers and foods as discussed for other yeasts in this chapter (Crauwels et al., 2017). More specifically, the noted terpenes being β-citronellol (floral, rose, sweet, citrus, green, fatty/waxy, leathery, terpenic) and/or γ-terpineol (pine, woody/resinous, terpenic, lilac, citrus {lemon, lime}, floral, camphoraceous) (Colomer et al., 2020; Serra Colomer et al., 2019). Recalling here that Lambic brewers incorporate aged or old hops (surannés) primarily over fresh hops; further research may be justified here. In that light, and as an aside, industrial brewing yeast has also been metabolically engineered to assist in the generation of primary flavour determinants in hopped beer, hoppier than via traditionally dosed hop according to sensory panel evaluations (Denby et al., 2018). In this instance, we now add to the flavour spectrum the monoterpenes linalool (citrus — orange/lemon, floral, rose, sweet, woody) and geraniol (floral, rosy, fruity, waxy, perfumed with a fruity, peach-accented nuance); two more key flavour determinants of final beer taste and aroma. The monoterpene synthase-engineered *S. cerevisiae* yeast strains biosynthesise the monoterpenes and give rise to the hoppy flavour of the beer, thus reducing or eliminating the late addition of flavour hops to the wort boil (Denby et al., 2018). One must wonder if this is the reason that old/aged hops were used in the early days of brewing such Belgian beers — fresher hop flavours perhaps not in balance with the microbially generated phenols and other flavour notes.

The volatiles produced by this yeast genus thus include both potent, undesirable off-notes and some desirably pleasing components, or ones said to add to the unique complexity of certain beer styles, new beer flavour profiles and to more suitably flavoured (less worty and bland) low alcohol containing beers. Beers with savoury, woody, spicy, floral, earthy, chemical (ethyl-estery) and the animalic notes (Colomer et al., 2020; Di Canito et al., 2021; Joseph et al., 2017; Steensels et al., 2015). The topic most cogently summed up by Colomer, Funch and Forster, 'The wide range of existing *Brettanomyces* allows the straight-forward establishment of novel-to-beer-flavours'. (Serra Colomer et al., 2019).

19.5.6 Coolships, blending and bottle conditioning/refermentation

The history, metabolic aspects and traditional and more modern approaches for bottle conditioning with mixed yeast/bacterial cultures and the impact on properties of the final beer, including the flavour aspects are detailed by Štulíková, Bulíř, et al. (2020), Štulíková, Novák, et al. (2020). Many of the organisms and the flavours produced as discussed in this chapter are further covered here (Štulíková, Bulíř, et al., 2020; Štulíková, Novák, et al., 2020, see also Callejo et al., 2019). Flavour attributes vary depending upon the use of the primary yeast used for the main fermentation and for bottle re-fermentation or if distinct species or strains are used for the two fermentations. Test fermentations have been performed using *Schizosaccharomyces pombe*, *Torulaspora delbrueckii*, *Saccharomycodes ludwigii* and *Lachancea thermotolerans* and *with S. cerevisiae* as the control. Additionally analyses of the metabolic activities and sensory contributions to beer taste included *Brettanomyces bruxellensis*. Such experiments have been conducted with several of the noted strains (not the *Brettanomyces*) or with *S.cerevisiae* as pure culture fermentations and to seek out their influences and effects upon and during the bottle conditioning. Or the beer worts were first fermented with *S. cerevisiae* with the four non-*Saccharomyces* strains used respectively for bottle conditioning (Callejo, Gonzalez, & Morata, 2017; Callejo et al., 2019). Differences in flavour profiles were seen when the stains were used for both primary fermentation and bottle conditioning (re-fermentation) compared to the first fermentation conducted by *S. cerevisiae* and then the other strains used for the bottle conditioning (higher alcohols and esters of some note here). While some acetaldehyde and diacetyl issues were noted in some cases, overall favourable reactions were observed with these yeast species. Higher levels of diacetyl are possible due to low capacity to reduce the compound during bottle or cask conditioning with no potential for a traditional diacetyl rest (Dekoninck et al., 2013). Active dry yeast and the performance of such yeast in primary and secondary fermentations under stresses imposed by the high lactic and acetic acid content present in sour beers have also received recent attention (Shayevitz, Abbott, Van Zandycke, & Fischborn, 2022).

From these studies, the sensory attributes evaluated included colour, malt, caramel, yeast, hop, banana/fruity, bitterness, sweetness, acidity body/mouthfeel, astringency and aftertaste (Callejo et al., 2019, see also Callejo et al., 2017 for some more related coverage here). More recently, a physicochemical analysis of aroma compounds with the sensory profiling of craft beers made using *L. thermotolerans*, *S. cerevisiae*, *Hanseniaspora vineae* and *Schizosaccharomyces pombe* has been reported, along with the chemical details involved in the evolution of bottle conditioning; 20 key aroma volatile components detailed in the report (Peces-Pérez et al., 2022).

Many sour beers spend a long time resting in wood (see below Section 4.7) or undergo sometimes quite lengthy re-fermentation periods in the bottle. As already noted in this chapter, mixed microbial cultures (microbiota) are used to create sour beers during the primary fermentation period. However, it might be considered that the complexity of microbiology carries over to effectively form a new microbiome when such beers are in the bottle during secondary or conditioning fermentation. At this point things can go well or badly — resulting in beer for sale or that tainted or spoilt and ready for dumping.

Even so-called pure culture fermentations (known pure single organism or known multiply species inoculated) are more complex than thought or often desired with genetic sequencing efforts directed at establishing the identities of all organisms in a sour beer community and how they interact — synergistically or antagonistically (Dysvik, La Rosa, 2020; Dysvik, Rosa, et al., 2020; Piraine, Leite, & Bochman, 2021; Piraine, Nickens, et al., 2021; Roullier-Gall et al., 2020). Such efforts aiding in the understanding of American coolship ales, wild ales, sour ales and Belgian Lambic beers (Witrick et al., 2017; Carriglio et al., 2022).

The emerging understanding of the release of volatile and odorant polyfunctional thiols created during fermentation as noted above shows that this also occurs during bottle refermentation. The topic addressed in two earlier dated papers (Nizet et al., 2013, 2014) as were the changes noted as occurring to Belgian beer flavour profiles during bottle conditioning (Dekoninck et al., 2013). So, in relation to dry hopping of beers, and owing to a better understanding of the flavour impacts of terpenic compounds, a cautionary note regarding beer stability and flavour issues has been made. It is now being suggested to use bottle re-fermentation to improve such late-hopped beer stability (Silva Ferreira, Bodart, & Collin, 2019). (See the speculative statement on the potential origin for the use of old hops in Belgian brewing in Section 5.5).

Another cogent discussion of bottle conditioning this time expanding on the bioflavouring issue was provided by Štulíková et al., wherein prevailing microbiota, *Brettanomyces*, *Pediococcus* and acetic acid bacteria are noted as playing important roles in Belgian Flanders region red and brown acidic ales production leading to the flavour active components lactic acid, ethyl acetate, isoamyl acetate, ethyl hexanoate and ethyl octanoate (Štulíková, Novák, et al., 2020). The principles of flavour enrichment by such bottle conditioning, as well as the technology and practices involved to produce shelf-stable Belgian acidic beers are fully considered by these authors (Štulíková, Novák, et al., 2020). Then, the issue of sitting on lees might also play a role in the taste profile of the beers, via the release of the amino acids during yeast autolysis. It is known that long-term contact with yeast during ageing of various beverages leads to an increase in the level of the amino acid glutamate, one component that is known to elicit the fifth taste sensation — umami. This has been noted for some sakes, beers, wines, and champagnes — other beverages that may remain in contact with yeasts for lengthy periods (Diepeveen, Moerdijk-Poortvliet, & van der Leij, 2022; Vinther Schmidt, Olsen, & Mouritsen, 2021). This would most likely also impact flavour via the continued chemical activities associated with the Maillard reaction. The potential for synergistic processes to also occur and influence flavour sensations. Publications as noted above provide valuable references for those considering bottle conditioning of their own complex, microbially fermented beer creations.

How long should this bottle conditioning go on? Research efforts have also followed the progression of organisms (microbial markers) and development, appearance and disappearance of various flavour volatiles for gueuze beers (blended Lambic) aged between a few months up to 17 years. Interestingly all yeast could continue to be grown and isolated over time, yet bacteria were not cultivatable after 5 years. Lactic acid (odourless but sour acidic, yogurt-like) and ethyl lactate (tart, creamy, pineapple, caramellic) concentrations increased during ageing, with ethanol remaining constant. Isoamyl acetate (sweet, fruity, estery, banana- and pear-like) and ethyl decanoate (waxy, fruity, apple, grape, oily, brandy) decreased during ageing. It was thus concluded that ethyl lactate and ethyl decanoate can be used as positive and negative ageing markers, respectively, (ethyl decanoate paradoxically as the negative is based on the loss of desired flavour; Spitaels et al., 2015). Other markers were also followed, leading to an overall suggestion that gueuze beers should be aged for less than 10 years for best flavour and character. This is based on the low perceived fruitiness of the oldest gueuze beers evaluated. In addition, as also noted above, there are issues of the ageing of the yeast during extended periods of such activity (or inactivity during the 'ecological' progression of the microbiome) (Wauters, Britton, & Verstrepen, 2021). Craft brewers are not likely to be waiting more than a few years to release product and would be advised to learn the key sensory attributes and follow these flavour markers over their time of ageing such or similar beers and release them at their estimated prime (Carriglio et al., 2022; Witrick et al., 2017). Frequent sensory panel profiling recommended.

19.5.7 Barrel ageing: Sorting out some issues with the use of wood

Considerable advances have been made in looking at maturation of liquids in wood. With a growing number of brewers experimenting with wood ageing of many different beer styles, there is a need to better understand the microbiology and flavour complexity of such beers as well as extending the knowledge base for traditionally matured-in-wood products such as Belgian Lambic beers or for open fermentation coolship ales. Several recent papers have bearings on this topic. A keen understanding of the impact of wood species on the microbial community composition not only deals with potential desirable flavour production during liquid—wood contact but also contaminating microbe generated off-odours/flavours. This is important and with a need for careful sensory evaluation, as wood-ageing of beers remains, for craft brewers, a trial-and-error process in the main (Bossaert, Kocijan, Winne, Schlich, et al., 2022; Bossaert, Kocijan, Winne, Van Opstaele, et al., 2022). Different woods harbour different microbial species, with *Pediococcus damnosus*, *Brettanomyces bruxellensis* and *Acetobacter* plus *Paenibacillus* species (see Grady, MacDonald, Liu, Richman, & Yuan, 2016; Munford et al., 2017) noted for oak and acacia-aged beers and in various percentages. Moreover, these bacteria and yeast species modulated the sensory perceptions of such matured beers — eugenol, oak lactones (coconut or celery — isomer dependent) and vanillin that allow for future potentially noteworthy flavour profile complexity. The relationships between wood chemical components and microbial metabolites and the selection for and stimulation of the growth of such organisms and relationships involved in overall flavour production needs to be better understood. Talks with Coopers involved in barrel making for the distilled spirits industry might be contacted for their expertise on this issue. The same research group of Bossaert, Kocijan, Winne, Schlich, et al. (2022) are also investigating sour beers produced by barrel ageing of otherwise traditionally fermented beers. Proving that beer ethanol content, and iso-alpha-acid levels affect the microbial community (microbiota), the ecosystem (microbiome) and the metabolic chemistry (flavour profiles – metabolome) through the beer-in-wood maturation period. Nothing that a variety of wood-derived flavour profiles are possible (Bossaert, Kocijan, Winne, Schlich, et al., 2022; Bossaert, Winne, et al., 2022). Details on the temporal dynamics in microbial community composition, understanding of wooden surface borne microbial inoculation sources and beer chemistry in sour beer production, via barrel ageing of beers, can be found in several recently published materials. Along with additional commentary on acetic acid and other bacterial resident communities in wooden casks used for maturation of lambic beers, the quality of craft-brewed barrel-aged beers and sanitation of wooden barrels for ageing beer (Bossaert et al., 2021; Costa, Sierra-Garcia, & Cunha, 2022; De Roos, Van der Veken, & De Vuyst, 2019; De Roos et al., 2018; Kocijan et al., 2021; Shayevitz, Harrison, & Curtin, 2021). Such issues are also of concern to distillers and brewers who use wooden or open vessels for fermentation and for brewer's barrel ageing more traditional lager or ale styles. Those brewers recreating some older styles may need to note that in earlier times Porter production in the UK carried flavour impressions imparted to them through 'contaminating' or 'unknown at the time' *Brettanomyces* strains (Kocijan et al., 2021). Brewers then need to ensure that barrels are microbiologically clean for certain other ageing operations and for avoiding cross contamination if producing different beer styles in the same facility. Today brewers are using used spirits barrels for imparting both wood and prior content flavours to their beers. In turn, distillers are then often secondarily maturing their spirits in prior use beer barrels! The study of De Roos et al. (2019) might be considered as a house character study resulting in the finding of organisms that might well have been part of the individual brewery's identity – be that a positive or negative connection. Thus, understanding the population of organisms in barrels used for ageing, as well as in the local environment, is now an important part of quality control and for ensuring product flavour consistency moving forwards. Especially when wild/sour ales are being made alongside traditional more 'single-culture-inoculated' brews (if such a thing truly exists). Microbiological testing and sensory evaluation sessions going hand in hand as part of solid brewery quality assurance/quality control and good manufacturing practice programme. Sensory teams and brewers learning how to pick sweet spot barrels as done by distillery master blenders and knowing which ones to blend for a reliably consistent or desirably aged flavourful beer.

As a tool to aid in such sensory programme quality control evaluations, a new sensory assessment flavour wheel on barrel maturation of beer is now available for brewers (Silvello, Bortoletto, & Alcarde, 2020). A cogent review of microbial spoilage and conditions necessary for such spoilage and dealing with health consequences, along with notes on detrimental flavour changes was recently published that will also provide sage advice for all brewers and especially those choosing more complex fermentation routes for their brewing of beers that will stand out in an ever more competitive marketplace (Shankar et al., 2021).

19.5.8 Ending on a sour note: As beer goes sour by intention. A bio-prospecting addendum note and a resounding summation

The desire for thirst quenching tartness has led to a rapid growth of sour beers of design beyond those earlier 'styles' noted in this Chapter, with extended details on them provided recently (Bossaert, Crauwels, De Rouck, & Lievens, 2019; Osburn et al., 2017). Many varied fruit-flavoured 'thirst-quenching' versions are now on store shelves, alongside seltzers. As a result of such interest, scientists and enthusiastic brewers have again gone 'bio-prospecting' for wild yeasts. Several isolates of five yeast species that produce lactic acid and ethanol were discovered with good potential for sour beer production. *Hanseniaspora vineae*, *Lachancea fermentati*, *Lachancea thermotolerans*, *Schizosaccharomyces japonicus* (close relation to *S. pombe*) and *Wickerhamomyces anomalus* were demonstrated to exhibit good fermentative abilities and desirable sensory characteristics, thus providing alternative microbes to LAB for sour beer production. Such yeast species use eliminates the need for kettle souring and the use of mixed culture fermentations (Osburn et al., 2017). The production process is dubbed primary souring and this, along with a history of sour beers and other souring methods are covered elsewhere (Bossaert et al., 2019; Osburn et al., 2017). See also the Section 4.4 above introducing the process of microbial acidification of sour beers. The use of traditional lactic acid and acetic acid bacterial and *Pediococcus* strains plus *Brettanomyces* in such brewing practices is now also well covered in a neat new review (Bossaert et al., 2021).

Ending this short section on a 'sour note' then reminds us that an understanding of the brewing of beer can be both simple and complex, with many microbiological and chemical details still be revealed. More microorganisms are being found to be playing a role in the 'background' so to speak, occasionally, adapting to the nutritious milieu – feeding on the remaining sugars and buoyed up by 'nutritious' metabolites and pH conditions produced by the main fermentative organism, which is *S. cerevisiae* in most cases. Such either occasionally spoiling the broth or leading to the mystery of 'house character' – (terroir/provenance?). However, brewers are learning more about complex, previously 'unseen' microbiota and learning how to better control them to prevent spoilage or to harness them in order to create ever more novel and flavourful beers and beer styles (Berg et al., 2020; Ellis et al., 2022; Luan et al., 2023; Marchesi & Ravel, 2015; Rodríguez-Saavedra, González de Llano, & Moreno-Arribas, 2020). In the end, the finish is either sweet, bitter or sour then! Oh, add a pinch of salt if the Gose style is included. And that is only a small part of the flavour.

19.6 Conclusion—Part 1

Part 1 of this paper has presented: An overview of microbial spoilage of beers as it relates to the organoleptic senses, some key sensory definitions (taints, off-flavours, thresholds) and flavour note descriptors for many typical and atypical flavours produced by both wild yeast and bacteria; A detailed look at the new drive for alternative, novel and enhanced flavour profiles in beers, including the issue of flavour in de-alcoholysed or low-alcohol beers and how this is achieved, and then, coverage of the vastly increased knowledge of the flavours of multi-organism-fermented beers such as Belgian Lambic/gueuze beers, increasingly popular sour beers and including issues of barrel ageing and bottle conditioning of such beers. Short descriptions and discussions presented on many of the varied yeast species (and some acetic and lactic acid bacteria), and their strains now used to produce flavourful beers – some of which used to be regarded as beer spoilage organisms. Now, via bioprospecting, genetic enhancements and under careful brewing management, many are proving to be decidedly good at producing an ever-increasing spectrum of beer styles and flavour profiles. Other chapters in this volume (and references cited herein) describe in depth the microbiology of many of the organisms discussed in lesser and general detail here. The emphasis of this chapter dealing with sensory attributes of their use and the reasons such organisms are being employed in modern-day and especially craft breweries. The discussion above, along with the flavour notes/descriptors (Table 19.3, Fig. 19.2) provides the lead now into an outline discussion of setting up Sensory Evaluation Programs in Part 2. Until very recently specific works had not addressed in detail sensory programs, for brewers, from a perspective of understanding all the flavours associated with atypical (special beer production) or contaminant microbial fermentation activities. Wine and distilled spirits had been covered in depth with more research with wine flavour and microbiology published. Thankfully, a great deal of published research from within the past decade has vastly improved our overall understanding of the situation with respect to the brewing of superior quality beers. The material from part 1, used in conjunction with the discussions and cited references in part 2, will hopefully address this earlier imbalance in coverage of a key area of discovery. Sensory programs can form a powerful approach to our understanding of beverages including beer. Perhaps more so needed today with an ever-increasing number of organisms to choose from for fermentation and from the viewpoint of an audience ever hungrier for new and exciting flavour experiences.

19.7 Part 2: Sensory evaluation

19.7.1 Getting started with sensory

A sensory panel can be a cost-effective tool for quality control, although it is often seen as a luxury, affordable only to large, well-established breweries. While it is true a sensory programme can represent a significant investment, even a modest programme can deliver significant returns to safeguard the brand and retain customers. At a minimum, a Go/No-Go (aka Ship/Don't Ship) sensory panel should approve each beer prior to packaging and shipping. A Go/No-Go ballot can be as simple as the words 'Ship' and 'Don't Ship' written on a piece of paper. Panelists, familiar with the brand profile, will circle their response and if negative, will then write down on paper or enter into the computer the reasons for the findings. Fig. 19.3.

A Go/No-Go ballot is the last chance for a brewery to prevent a flawed batch of beer from reaching the market and perhaps negatively affecting the perception of the brand and brewery. This sort of issue can easily result in costly returns/buy-backs. Because of its importance, Go/No-Go panels should ideally be staffed by the best trained most, proficient sensory panelists available but when just starting a programme, a grasp of the brand profile, enthusiasm and an appreciation of the work's importance may have to do.

This description of a Go/No-Go panel is an illustration of the basic philosophy behind sensory panels and their mission. Find problems before they become big problems. A taint found in raw ingredients is an easier and cheaper way to fix a problem before brewing a spoilt batch of beer. A problem detected in the brewhouse is easier and cheaper to fix than finding it once the product has shipped. For issues discovered when the product is in market, a goal is to learn about the mistake and fix the

Brand: ________________
Date: ________________
Panelist: ________________

Should this beer batch of beer be packaged and shipped:

YES NO

Why: ________________________
____

FIGURE 19.3 Go – No-Go ballot.

TABLE 19.4 Sensory evaluation in the brewhouse can (should) encompass the following.

Phase/Purpose	Focus	Details in this chapter
All raw materials and the brewing water	Tainted components	See air, water and raw materials and taints and off flavours
Wort and beer at all stages of process	Spoilage and off-flavours	See: Table 19.3, Beer, and wort - overview
Final Ship/Don't Ship	Batches that do not fit the brand profile	Getting started
Buy-back programme that periodically purchases beer from retailers	Assess product longevity by looking at oxidation, non-biological haze, and biological issues	Shelf-life evaluation
Field and customer support	Evaluate all product returned for any reason	Proactive sensory

Areas of sensory focus.

causes, so the next batch is better. This process is true for most all Quality Assurance programs regardless of the product. Find and fix problems as early as possible and do not repeat the same mistakes. This philosophy lends itself to every stage of production and can/should be extended beyond the brewery to product shelf-life evaluation (Table 19.4).

19.7.2 Creating an effective sensory team

A reliable sensory programme requires commitment from management. A sensory team must have sufficient resources (both money and time) along with an understanding of the complex nature of establishing, maintaining an ongoing programme and evaluating product. Paramount is management's understanding that the sensory team is a tool. As with any tool, it is of no use if its work product is not utilised. Management has a responsibility to take the findings of the sensory team into account when making applicable business decisions. Demands from the revenue generating side of a product's life cycle are not cause for the QA team's finding to be ignored. At the same time, the quality team is not typically in charge of making business decisions.

The concept/idea of having a sensory team is often met with enthusiasm, in practice; however, convincing managers and accountants that people stopping work in the middle of the day to taste beer is a good idea can be problematic but doing so is paramount to the success of a panel. How to accomplish this is beyond the scope of this chapter.

When building a programme, the first question is *who will be on the team*? What are the basic qualifications of panelists? Of course, those who do not imbibe or anyone that has issues with alcohol will be precluded. Availability is also a factor. A forklift driver might not be a suitable candidate for a panel that meets during his/her workday. The rigid time requirements of the panel might not be compatible with the dynamic schedule of the head brewer. Though not specific to breweries or beer an in-depth treatment of the subject of organising a sensory panel has been presented by Kilcast (2010), Stone, Bliebaum, and Thomas (2012), Meilgaard, Carr, and Carr (2007) and most recently by Habschied, Krstanović, and Mastanjevićal (2022).

It is desirable to have 8 to 12 panelists. Productive sensory sessions can be run with as few as five or six participants, but it is a great advantage to have more available personnel in case some are out sick, on vacation, leave the business or are otherwise unavailable.

While having 8 to 12 panelists is desirable, a startup brewery might only have half that number of employees in total. Having few employees is no reason to forego a program. The rule of sensory is to work with the resources at hand.

A shortage of employees raises the question of who else you might recruit? The short answer is *everyone*. Are there tap room, gift shop, or maintenance staff available that might be amenable to devoting the time and effort needed? Are there people looking to get into professional brewing that are willing to intern? Consider local distributors and retailers to augment the team and build brand loyalty. Are there BJCP certified (US) or other recognised judge program, home brewers that might work for little to no remuneration? Are there any folks that have or would like to take part in any of the Cicerone Certification Programs? Do not overlook customers that have exhibited above average beer and tasting knowledge or have shown special interest in the brewery operation (Craine, Bramwell, Ross, Fisk, & Murphy, 2021). It is not at all unusual to find a very proficient taster working in the gift shop, mowing the lawn or working as an intern. By the same token, individuals who are principals in the business and brewers are not always the most gifted. These decisions are unique to each operation and have no set rules; just don't limit the search to the brewery.

19.7.3 New panelists

Having selected panelists, training might seem in order, but if the situation permits, consider giving candidates an assessment period. Have them attend some sessions, evaluate beers like a trained panelist. Are they attending regularly and willing to learn? After a suitable trial period, ask the candidate if they are still interested in participating. People often volunteer envisioning a sensory panelist's job as sitting around drinking beer like they would in a bar or tasting room. It does not suit some to devote the time, concentration and learning required. Training can be time-consuming and expensive. An assessment period might be in the brewery's best business interest.

Once willing participants are on board, the next step is assessing their basic level of sensory acuity to determine aptitude. An effective means of this testing is the use of aroma bottles where a person sniffs and then describes an aroma instead of tasting the flavour in a beer. A brief explanation of this test's administration can be found in Lawless (2013a). The goal is to screen out those with impairments or who are blind to basic beer elements. For untrained candidates, correctly describing 60% of common beer aromas is considered acceptable.

19.7.4 Building a vocabulary

Once it is time for training, the initial goals should be to have panelists consistently identify beer attributes and to build a common vocabulary. Specific flavour notes and qualities must be understood and described in technical terms. The importance of sensory lexicons for research, evaluation and product development is emphasised by Suwonsichon (2019).

Often new panelists have never evaluated a beer beyond saying, 'I like it', 'Too hoppy', 'Its skunky'. A training exercise designed to push panelists to think in more detail is to limit the descriptive terms panelists are permitted. An example might be to have trainees describe a piece of toast using no form of the words 'toast' or 'crunch'. Vague terms such as 'typical of style', 'smooth', 'skunky', etc. should almost always be avoided. The goal is for a panelist to use specific, defined terms wherever possible.

Tactile sensations — mouthfeel such as warming, cooling, oily, stinging (carbonation) may be acceptable, though should be related to the correct attributes causing such stimuli when possible (examples camphoraceous and balsamic, peppery, minty, spicy). A buttery flavour indicative of diacetyl (see above) may also be sensed as an oily slickness or increased viscosity on the palette. Increased viscosity is not a certain pointer to diacetyl, rather a flag that signals it is a possibility. Did other panelists find indications of diacetyl? Is the panelist reporting the increased viscosity prone to detecting diacetyl with this descriptor? What does the history of the brand that is being sampled show about changes in viscosity and with respect to issues of diacetyl contamination in the past?

These sensations show the complexity of defining a food or beer in technical and quantitative terms. The panel will need to use technical terms here such as dimethyl sulfide (DMS) or geosmin (see Table 19.3, Fig. 19.2) for, respectively, the less technical though still descriptive terms like cooked corn/vegetal or musty. It is of little use for a problem to be identified if it is not clearly communicated to those in the brewery and lab, responsible for identifying and correcting the underlying production and microbial issues.

A common place to start building the required vocabulary is with the *Beer Flavour Wheel*. First published in 1979 by Dr. Morten Meilgaard in cooperation with the European Brewery Convention, the Master Brewers Association of the

Americas and the American Society of Brewing Chemists. The wheel as a tool provides a concise list of beer attributes and a broad categorisation as to odour and taste (Aiken & Spedding, 2023; Spedding, 2012b, 2022 and see Fig. 19.4). A further breakdown of the wheel's defined attributes can be found in a Practical Guide for Beer Quality: *Flavour*, by Bamforth (2014). This work relates an attribute's relevance to odour, taste, mouthfeel, warming, and aftertaste, as well as giving accepted descriptors and reference standards for the wheel's elements.

In recent years, some sensory experts have broken down (split) the beer flavour wheel into a set of more manageable sized tools. New flavour wheels at various levels of professional activity have included ones for wood-aged beers. For example, in 2017, a Brettanomyces flavour wheel was published (https://www.asevcatalyst.org/content/1/1/12) Descriptions of other flavour wheels appears in Part 1.

An alternative to the *Beer Flavour Wheel* is the *Beer Flavour Map*, Created by Lindsay Barr and Dr. Nicole Garneau (Fig. 19.5). First published by DraughtLab in 2016, the map builds upon Dr. Meilgaard's wheel. The map updates and organises attributes in hierarchies that reflect contemporary understandings in sensory science. Like wheels, *Flavour Maps* have been created for base malts, hops and speciality malts.

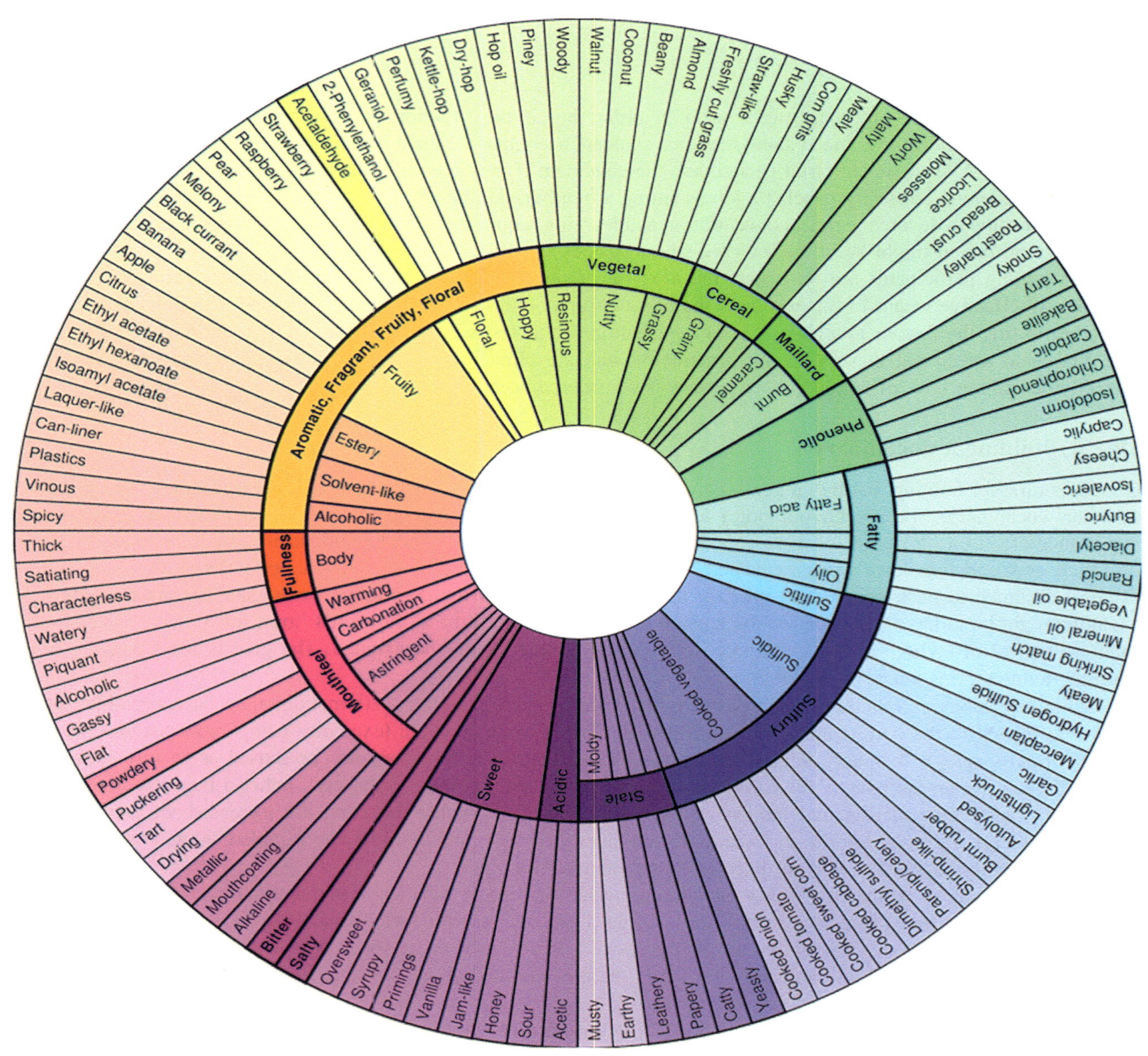

FIGURE 19.4 Beer flavour wheel (American Society of Brewing Chemists).

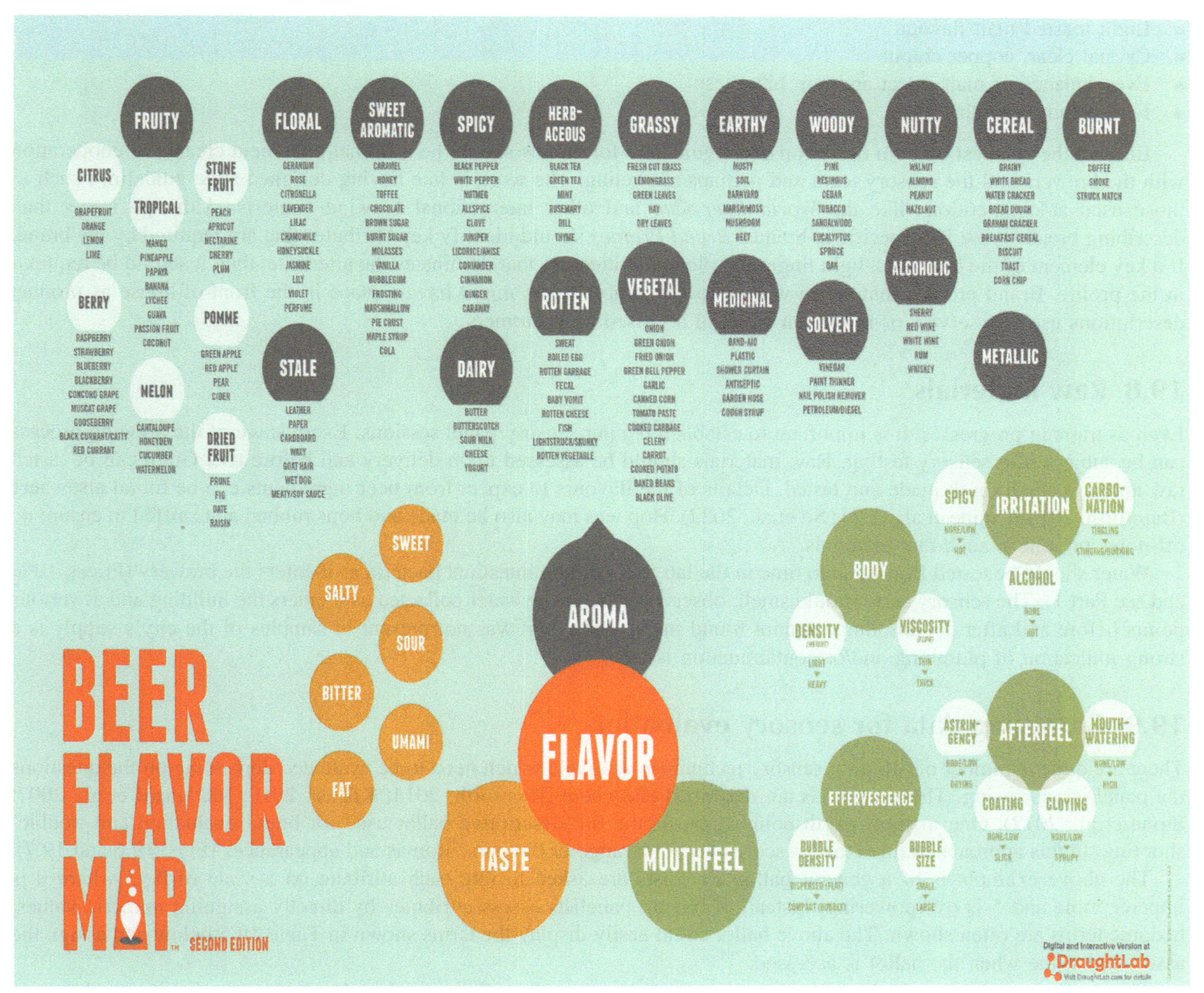

FIGURE 19.5 Beer flavour map. *Courtesy of DraughtLab.*

When using any of these tools, there may be elements which never come into play in an operation and other attributes that are missing from the work. A sensory programme administrator is free to build their own tools, be they wheels, maps or simple lists of descriptors.

The vernacular used in an established brewery that only produces light lagers might differ from that of a start-up brewpub specialising in American-style barley wines. Additional information available to assist in building the required sensory vocabulary have been detailed in the following noted papers (de la Torre-González et al., 2017; Habschied et al., 2022; Martins, Brandão, Almeida, & Rocha, 2020; Ravasio et al., 2018; Su et al., 2022). See also Table 19.3 and Fig. 19.2. In addition, major new findings ascribing aroma to molecular features of aroma volatiles as detectable by both human and electronic noses and sensory systems have been reviewed elsewhere (Spedding, 2024).

19.7.5 Brand profiles

An in-house sensory programme should also have a *brand profile* ('aroma-print'/'taste-print') for every beer in a brewery's regular portfolio. These profiles describe the key attributes of each brand. That is, documentation that describes which qualities must always be present in the brand. Examples of *brand profile* terms are.

- Light toasted malt flavour
- Crystal clear, copper colour
- Even balance of malt sweet and hop bitterness
- Full, creamy mouth feel

Beyond the smallest start-up operations, this guideline for brands should be a formal document created in cooperation with the brewers and the sensory team and perhaps marketing. The work is like having designed style guidelines such as those used at competitions like the *World Beer Cup* and other international brewing contests. Although rather than describing broad styles, it is specific to brands. *Brand Profiles* should identify key attributes that are required by the brand. If a key element of the brand is a light lingering, pleasant bitterness that dominates the aftertaste, then it should be required in the profile. Brand profiles that are re-written to use simple terms might have a place in the front-of-house as product descriptions used by servers or in written material accessed by customers.

19.8 Raw materials

Even as training progresses, it is important to establish regular sensory panel sessions. Every stage of the brewing process can be subjected to sensory testing. Raw materials should be assessed upon delivery and before use. Grain can be tasted raw and/or a small mash made and tasted. Details of the flavours to expect from beer ingredients can be found elsewhere (Bamforth, 2014; Craine et al., 2021; Su et al., 2022). Hop teas may also be made and hops rubbed and sniffed to ensure no off-notes present in such raw materials.

Water should be tested from time to time in the lab for common intestinal bacteria as it enters the brewery (Priest, 1990 and see Part 1). The sensory crew should smell, observe and taste the water collected as it enters the building and at various points before and after conditioning. A taint found in the water that was not present in samples of the city's supply is a strong indication of plumbing and/or contamination issues.

19.9 Gathering data for sensory evaluation

There are quite a number of different sensory techniques available which need to be evaluated depending on the questions the panel is addressing. These methods are described elsewhere (Bamforth, 2014; Kilcast, 2010; Meilgaard et al., 2007; Stone et al., 2012). One method worth pointing out is the full descriptive ballot used for brand evaluation and trouble-shooting. In this approach, trained tasters score beer for a range of flavours, aromas and appearances (Figs. 19.6 and 19.7).

The above example is of a general ballot. Panelists are asked to rate each attribute on a scale of 0–5 where 0 is imperceptible and 5 is overpowering. Instead of having panelists assess attributes by directly assigning numeric values, hedonic terms are often shown. The above ballot could easily display the terms shown in Table 19.5 below and assign the associated value when the ballot is accessed.

The comment section at the bottom allows for an explanation of anything perceived or lacking in the sample that does not fit the format of the ballot. Many of the off-flavour attributes on the ballot can be traced to microbial issues and taints introduced during the brewing, packaging or handling processes. Of course, the scale mentioned is arbitrary and could just as easily range from −5 to +5 or 0 to 10 or 1 to 100 or whatever the organisation finds most useful. More specific ballots might be created for particular brands, training or specialised panels.

While sensory typically focusses on aroma and taste, all other senses can come into play. Haze and gushing are visual perceptions but can indicate microbial issues. Tactile impressions of the beer on the tongue or the sound of a can or bottle being opened may point to a packaging concern. It will be noticed, however, that the ballot shown in Fig. 19.2 does not include visual attributes. It was designed to be used in a setting where a beer's appearance is concealed from the panelists; visual cues can bias panelists. A discussion of the effects of appearance on flavour perception can be found in Bamforth (2014). If, on the other hand, the sensory panel is a brewery's in-house team, then panelists will be familiar with the brands being assessed and deviations in appearance should likely be considered by the sensory team.

19.9.1 Paper or electronic?

The same attributes presented in Fig. 19.6 ballot could be included on a paper ballot. Paper has been used for generations and represents the lowest start-up cost. The issue to consider, however, is the time and expense of handling, analysing, summarising and reusing the data recorded. Transcribing data into an analysis tool like Microsoft Excel are time-consuming and error prone. Strong consideration should be given to a computerised sensory application. The cost of

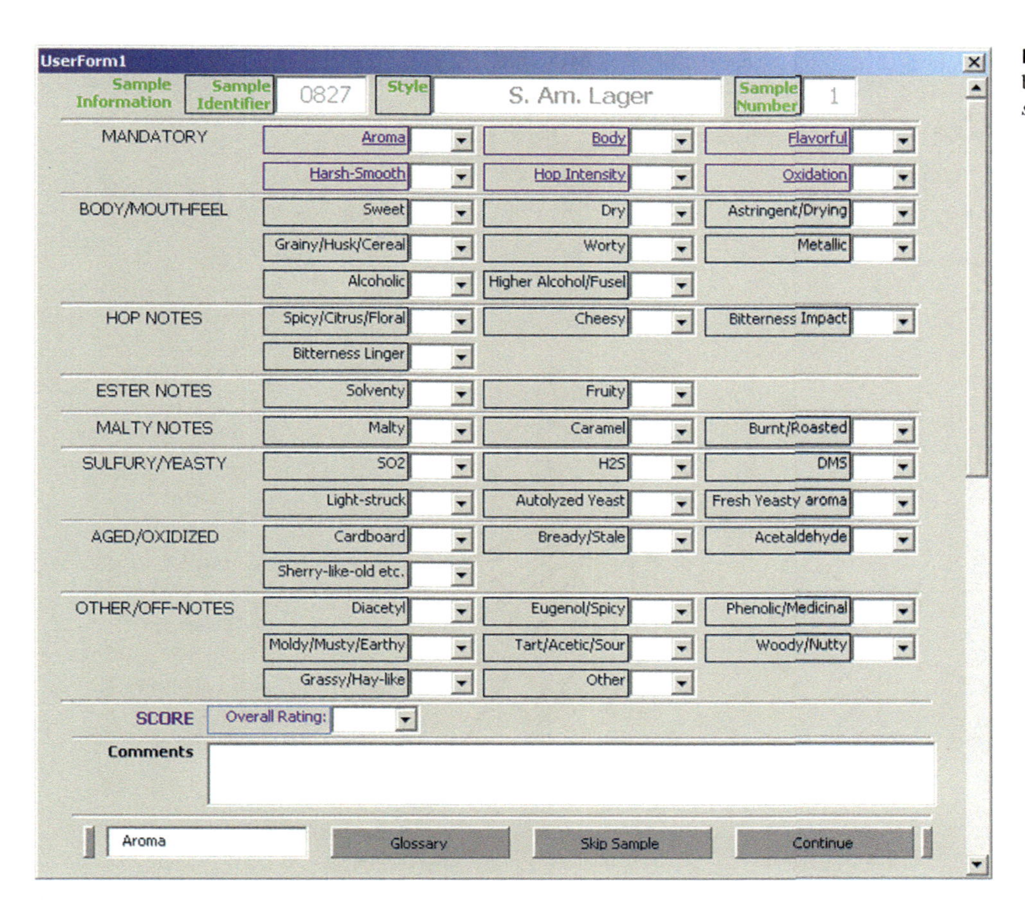

FIGURE 19.6 General diagnostic ballot. *Courtesy of data collection solutions.*

such software will be recovered in the person-hours saved, and the increased usability will likely result in more frequent sensory sessions and greater productivity. There are free 'survey' tools available online. These tools are generic, require some learning and protection of data may not but up to your organisation's security standards. Other tools, specifically designed for sensory testing, are also available online for purchase.

19.9.2 Presenting sensory panel data

Sensory data may contain many important details that will remain just a collection of results unless they can be presented in a concise, understandable manner. The illustration below is a spider chart (aka Radar Profile) showing the results from a descriptive ballot evaluating a brewery's IPA that meets the requirements of the brand profile and is considered True to Brand. [Full statistical approaches can be implemented for data analysis and discussions may be found in the main references cited above and extensively in Lawless, 2013b; O'Mahoney, 1986.]

19.9.3 Shelf-life evaluation

Buy-back programs have two forms. One is when retailers or distributors return product that has not sold and now exceeds its *Best-By* date and likely no longer fits its *brand profile*. Another type is a proactive programme that periodically purchases beer from retailers and has the sensory team evaluate how well the product is holding up. A beer returned to the brewery might be handled as a sample from the proactive buyback programme. That is, have it tasted as a suspect sample, without detailing the customer complaint. Once the sensory team has done its work, the differences can be easily contrasted with a sample that is considered True to Brand.

A quick look at the contrasting spider chart, Fig. 19.8, shows an increase in diacetyl, oxidation and sweetness levels and a decrease in bitterness and overall flavour in the aged sample. Depending on the brand and the sample's age, the changes noted in oxidation, sweetness, caramel notes and bitterness may be expected transformations because of the ageing process. The increase in diacetyl may be a cause for concern that warrants further investigation.

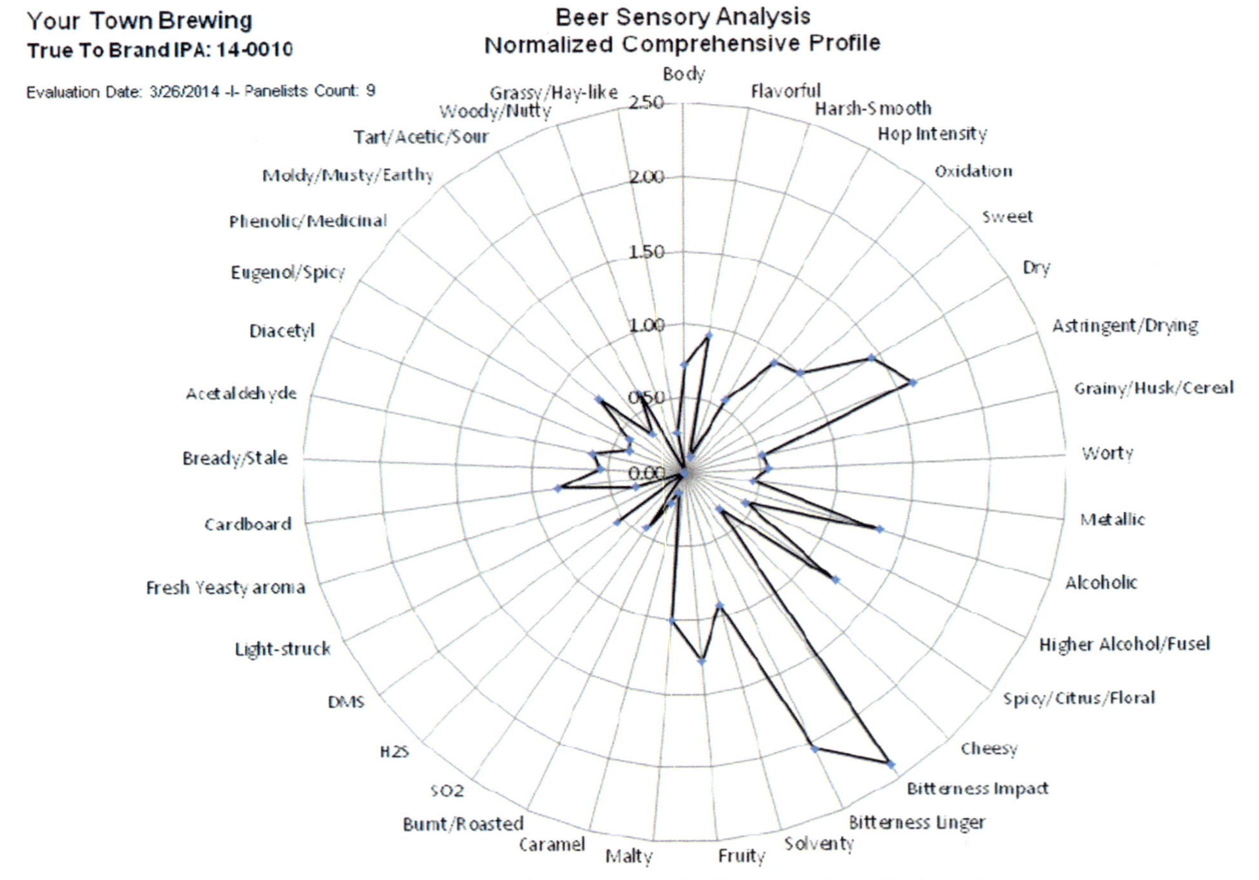

FIGURE 19.7 Graph of beer fitting the brand profile. *Courtesy of data collection solutions.*

TABLE 19.5 Sensory ballot using hedonic terms.

Term displayed	Value assigned
Imperceptible	0
Just perceptible	1
Easily perceived	2
Strong	3
Very strong	4
Overpowering	5

A succinct introduction to shelf-life Evaluation that fuses sensory and chemometric techniques can be found in a most recent publication [de Lima, Aceña, Mestres, Boqué, 2022: An Overview of the Application of Multivariate Analysis to the Evaluation of Beer Sensory Quality and Shelf-Life Stability].

19.9.4 Proactive sensory

A buyback programme is one way a sensory programme may extend beyond the brewery, but there is no reason others that oversee the brand cannot be involved in monitoring the product's shelf life. If amenable, distributors and sales reps can be trained to evaluate beer. Business is better for everyone when the beer is fresh and remains within its brand profile. If beer

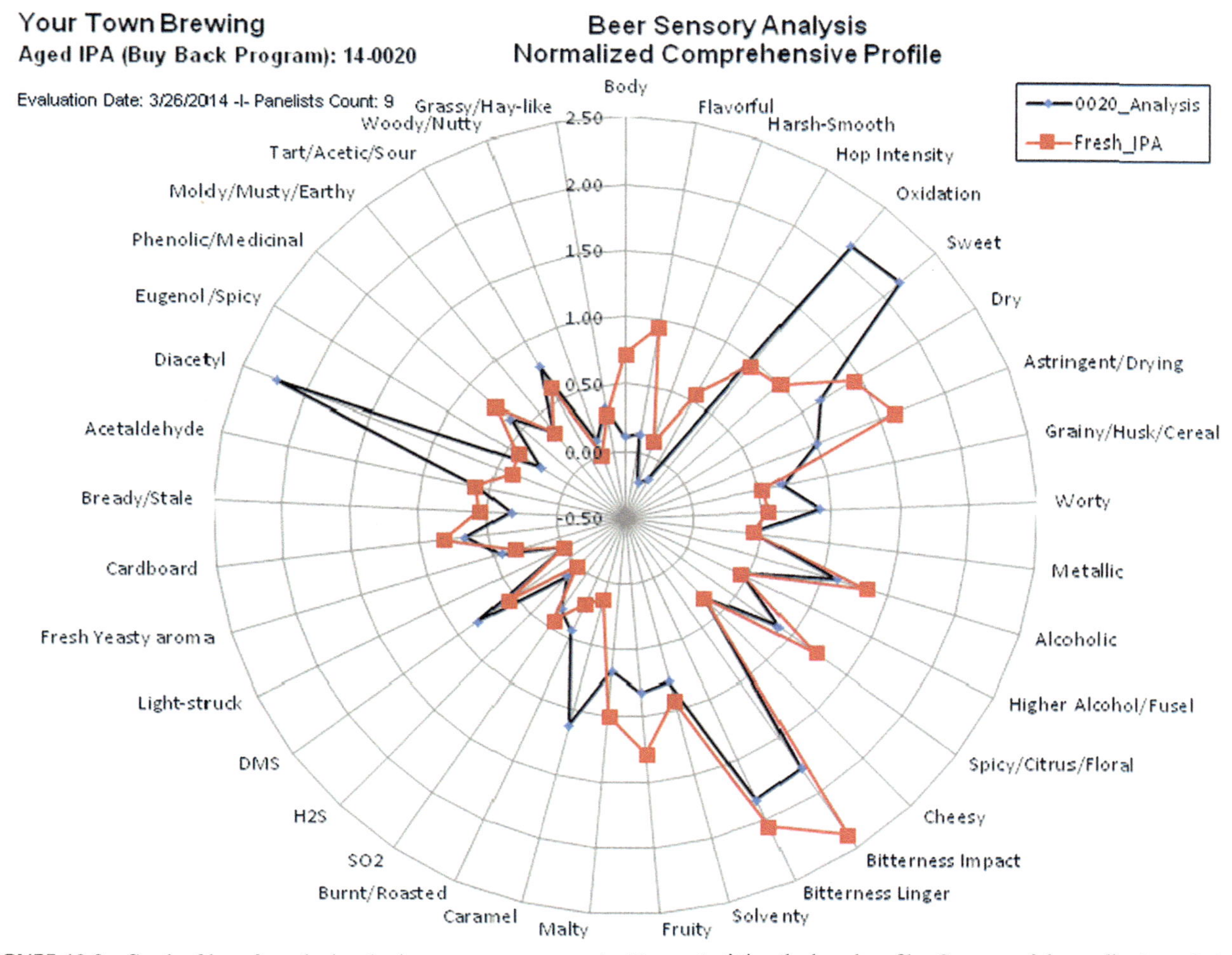

FIGURE 19.8 Graph of beer from the buy back programme contrasted with sample fitting the brand profile. *Courtesy of data collection solutions.*

in trade is not handled correctly, microbiological issues may ensue – growth of microorganisms in the package or at the location of dispense (Spedding, 2013a). Such issues discussed in full in Part 1.

19.9.5 Sensory training

Of course, sensory panelists must be trained on the flavour and aroma characteristics of the key metabolites of microorganisms to make informed decisions about the desirability for release of tainted or contaminated beer or to recall product from the market. Or, these days, to assess the flavour profiles of new style beers created via the fermentations of wort, or in secondary conditioning protocols using many of the alternate yeast species and strains outlined in Part 1. An effective means of introducing new flavour notes is the use of aroma/odour bottles. Training bottles are normally created by dosing a small amount of a standard reference compound into a brown glass bottle (approx. 30 mL) containing a cotton ball. Optionally the cotton can be soaked in a light beer for background aromas (see: the Flavor Sub-Committee Sensory Analysis Manual, 1995 for more on basic training with aroma bottles and Aiken & Spedding, 2023).

Training can then be extended to actual tasting of off-flavours in the brewery's own beers. For example, it may be unlikely that the beer will ever have dimethyl sulfide (DMS) levels above threshold, but the panelist should be capable of identifying the defect if it ever occurs. This is where sensory training kits can be especially useful tools. These kits are for creating aroma bottles and spiking samples. Kits are available from several sources including the following.

- FlavorActiv (www.flavoractiv.com)
- AROXA (www.aroxa.com)
- The Siebel Institute of Technology (www.siebelinstitute.com)

Typically, panel members are presented spiked beers (with particular flavour characteristics added to a sample) alongside a standard beer (a control with no perceived defect or other added training flavour) to understand the flavours versus a true to brand sample. Any material added to beer intended for human consumption must be food-grade quality and safe in order to avoid any health hazard. The threshold values mentioned in Table 19.3 were based on light style lagers and as such levels may have to be increased to accommodate their perception in more flavourful craft beers (Spedding, 2023). Indeed, threshold tests will become a part of routine sensory training practice, especially with more flavourful beers, with details on such testing covered in many of the references cited in this section.

Sensory training must be an ongoing process. Validation of panelist performance is important to determine proficiency. Keeping in mind, when testing panelists, results should be used to better the panel and not to punish underperformers. A common way of assessing panelist accuracy is by introducing spiked samples alongside the panel's regular work. Validation should be designed to track not only team members that correctly identify off-flavours but to illuminate patterns in the incorrect responses. Failure to describe correctly the added character is indicative of a training requirement or a blind spot. For example, if multiple panelists are misidentifying acetaldehyde (green/bruised apple – See Table 19.3) as ethyl hexanoate (red apple/aniseed) there is a need for additional training. Panelist assessments may also point out that certain individuals are extremely sensitive to some attributes and below par with some others. This does not mean that the panelist is not of use to the team but is important to note when faults are found by only some in the group. A good overview of motivating and evaluating panel members is provided by Meilgaard et al. (2007).

Difference testing can be a useful tool for assessing panelist's skills. Methods such as the duo-trio, triangle and tetrad tests are communally known as forced choice discrimination tests because panelists are presented with multiple samples and must decide, which examples are the same and which are different. These tests can also be used in assessing the impact of changes in the supply chain and production techniques. Succinct descriptions of these and other discrimination tests can be found in Bamforth (2014), Simpson (2006) and Aiken and Spedding (2023). More in-depth discussions of these techniques are covered by Ennis and Rousseau (2020), Kilcast (2010), Meilgaard et al. (2007), Stone et al. (2012) and most recently by Habschied et al. (2022).

19.10 Conclusion—Part 2

Part 2 of this paper has presented an overview of the sensory analysis of beer and beer wort which also extends to brewing raw materials. Part 1 alluded to some areas where sensory programs could provide answers to some interesting questions. In particular, the growing role of alternate yeasts and bacteria in the production of new styles, enhanced or specific flavours and unique flavour profiles will require more careful sensory panel scrutiny. Furthermore, many more sophisticated programs are being established for the assessment of sensory data and relationships to the levels of specific flavour volatiles present. Tools like the beer flavour wheel and flavour maps can be instrumental in providing clues and a memory jogger for many important beer odour and flavour notes, yet increasingly more data is being added to the flavour lexicon. Using the knowledge gained from the brief account of sensory programs here will help lead readers into setting up their own programs for both fundamental research and for quality control of their beers as they face an ever-growing competitive field and a consumer base evermore seeking out something new and a little different and knowing that little bit of difference when they smell and taste it!

References

Aiken, T., & Spedding, G. (2023). Brewery sensory panels and practice: A brief overview. *Master Brewers Association Americas – Technical Quarterly (MBAA-TQ), 60*(2), 32–36. https://doi.org/10.1094/TQ-60-2-0609-01

Ambra, R., Pastore, G., & Lucchetti, S. (2021). The role of bioactive phenolic compounds on the impact of beer on health. *Molecules, 26*(2). https://doi.org/10.3390/molecules26020486

Anderson, R., Sanchez, A. B., Devreux, A., Due, J., Hammond, J. R. M., Martin, P. A., et al. (2000). *Fermentation and maturation: Manual of good practice. European brewery convention.* Fachverlag Hans Carl.

Angelino, S. A. G. F. (1991). 16 - beer. In H. Maarse (Ed.), *Volatile compounds in foods and beverages* (pp. 581–616). Marcel Dekker, Inc.

Ault, R. G. (1965). Spoilage bacteria in brewing—a review. *Journal of the Institute of Brewing, 71*(5), 376–391. https://doi.org/10.1002/j.2050-0416.1965.tb06362.x

Back, W. (2005). *Color Atlas and handbook of beverage biology.* Germany: Fachverlag Hans Carl. Nurnberg.

Baiano, A. (2020). Craft beer: An overview. *Comprehensive Reviews in Food Science and Food Safety, 20.* https://doi.org/10.1111/1541-4337.12693

Bamforth, C. W. F. (2014). *ASBC handbook series, practical guides for beer quality.* American Society of Brewing Chemists.

Barrio, E., González, S. S., Arias, A., Belloch, C., & Querol, A. (2006). Molecular mechanisms involved in the adaptive evolution of industrial yeasts. In A. Querol, & G. Fleet (Eds.), *Yeasts in food and beverages* (pp. 153–174). Springer Berlin Heidelberg.

Barwich, A.-S. (2020). *Smellosophy: What the nose tells the mind.* https://doi.org/10.4159/9780674245426

Barwich, A.-S., & Smith, B. (2022). From molecules to perception: Philosophical investigations of smell. *Philosophy Compass, 17.* https://doi.org/10.1111/phc3.12883

Bellut, K., & Arendt, E. K. (2019). Chance and challenge: Non-Saccharomyces yeasts in ionalcoholic and low alcohol beer brewing – a review. *Journal of the American Society of Brewing Chemists, 77*(2), 77–91. https://doi.org/10.1080/03610470.2019.1569452

Bellut, K., Krogerus, K., & Arendt, E. K. (2020). Lachancea fermentati strains isolated from kombucha: Fundamental insights, and practical application in low alcohol beer brewing. *Frontiers in Microbiology, 11*, 764. https://doi.org/10.3389/fmicb.2020.00764

Bellut, K., Michel, M., Hutzler, M., Zarnkow, M., Jacob, F., De Schutter, D. P., et al. (2019). Investigation into the potential of Lachancea fermentati strain KBI 12.1 for low alcohol beer brewing. *Journal of the American Society of Brewing Chemists, 77*(3), 157–169. https://doi.org/10.1080/03610470.2019.1629227

Bellut, K., Michel, M., Zarnkow, M., Hutzler, M., Jacob, F., Atzler, J. J., et al. (2019). Screening and application of Cyberlindnera yeasts to produce a fruity, non-alcoholic beer. *Fermentation, 5*(4), 103. https://doi.org/10.3390/fermentation5040103

Bellut, K., Michel, M., Zarnkow, M., Hutzler, M., Jacob, F., De Schutter, D., et al. (2018). Application of non-Saccharomyces yeasts isolated from kombucha in the production of alcohol-free beer. *Fermentation, 4*, 66. https://doi.org/10.3390/fermentation4030066

Bendixsen, D. P., Frazão, J. G., & Stelkens, R. (2022). Saccharomyces yeast hybrids on the rise. *Yeast, 39*(1–2), 40–54. https://doi.org/10.1002/yea.3684

Berg, G., Rybakova, D., Fischer, D., Cernava, T., Vergès, M.-C. C., Charles, T., et al. (2020). Microbiome definition re-visited: Old concepts and new challenges. *Microbiome, 8*(1), 103. https://doi.org/10.1186/s40168-020-00875-0

Bokulich, N. A., Bamforth, C. W., & Mills, D. A. (2012). Brewhouse-resident microbiota are responsible for multi-stage fermentation of American coolship ale. *PLoS One, 7*(4), Article e35507. https://doi.org/10.1371/journal.pone.0035507

Bongaerts, D., Roos, J., & De Vuyst, L. (2021). Technological and environmental features determine the pniqueness of the lambic beer microbiota and production process. *Applied and Environmental Microbiology, 87*, Article AEM0061221. https://doi.org/10.1128/AEM.00612-21

Borren, E., & Tian, B. (2020). The important contribution of non-Saccharomyces yeasts to the aroma complexity of wine: A review. *Foods, 10*(1). https://doi.org/10.3390/foods10010013

Bossaert, S., Crauwels, S., De Rouck, G., & Lievens, B. (2019). The power of sour - a review: Old traditions, new opportunities. *BrewingScience, 72*, 78–88. https://doi.org/10.23763/BrSc19-10bossaert

Bossaert, S., Kocijan, T., Winne, V., Schlich, J., Herrera-Malaver, B., Verstrepen, K. J., et al. (2022b). Beer ethanol and iso-α-acid level affect microbial community establishment and beer chemistry throughout wood maturation of beer. *International Journal of Food Microbiology, 374*, Article 109724. https://doi.org/10.1016/j.ijfoodmicro.2022.109724

Bossaert, S., Kocijan, T., Winne, V., Van Opstaele, F., Schlich, J., Herrera-Malaver, B., et al. (2022a). Development of a tractable model system to mimic wood-ageing of beer on a lab scale. *bioRxiv.* https://doi.org/10.1101/2022.03.11.483928

Bossaert, S., Winne, V., Opstaele, F., Buyse, J., Verreth, C., Herrera, B., et al. (2021). Description of the temporal dynamics in microbial community composition and beer chemistry in sour beer production via barrel ageing of finished beers. *International Journal of Food Microbiology.* https://doi.org/10.1016/j.ijfoodmicro.2020.109030

Bossaert, S., Winne, V., Van Opstaele, F., Buyse, J., Verreth, C., Herrera-Malaver, B., et al. (2022). Impact of wood species on microbial community composition, beer chemistry and sensory characteristics during barrel-ageing of beer. *International Journal of Food Science and Technology, 57*(2), 1122–1136. https://doi.org/10.1111/ijfs.15479

Boulton, C., & Quain, D. (2001). 8 – microbiology. In *Brewing yeast & fermentation* (pp. 510–585). Blackwell.

Bourbon-Melo, N., Palma, M., Rocha, M. P., Ferreira, A., Bronze, M. R., Elias, H., et al. (2021). Use of Hanseniaspora guilliermondii and Hanseniaspora opuntiae to enhance the aromatic profile of beer in mixed-culture fermentation with Saccharomyces cerevisiae. *Food Microbiology, 95*, Article 103678. https://doi.org/10.1016/j.fm.2020.103678

Bruner, J., & Fox, G. (2020). Novel non-cerevisiae Saccharomyces yeast species used in beer and alcoholic beverage fermentations. *Fermentation, 6*, 116. https://doi.org/10.3390/fermentation6040116

Bruner, J., Marcus, A., & Fox, G. (2022). Changes in diacetyl and amino acid concentration during the fermentation of dry-hopped beer: A look at bwelve Saccharomyces species and strains. *Journal of the American Society of Brewing Chemists*, 1–13. https://doi.org/10.1080/03610470.2022.2078946

Burini, J. A., Eizaguirre, J. I., Loviso, C., & Libkind, D. (2021). [Non-conventional yeasts as tools for innovation and differentiation in brewing]. *Revista Argentina de Microbiología, 53*(4), 359–377. https://doi.org/10.1016/j.ram.2021.01.003

Callejo, M. J., García Navas, J. J., Alba, R., Escott, C., Loira, I., González, M. C., et al. (2019). Wort fermentation and beer conditioning with selected non-Saccharomyces yeasts in craft beers. *European Food Research and Technology, 245*(6), 1229–1238. https://doi.org/10.1007/s00217-019-03244-w

Callejo, M. J., Gonzalez, C., & Morata, A. (2017). *Use of non-Saccharomyces yeasts in bottle fermentation of aged beers.* https://doi.org/10.5772/intechopen.68793

Campbell, I. (1987). 7 - wild yeasts in brewing and distilling. In F. G. Priest, & I. Campbell (Eds.), *Brewing microbiology* (pp. 187–205). Springer US.

Canonico, L., Agarbati, A., Comitini, F., & Ciani, M. (2016). Torulaspora delbrueckii in the brewing process: A new approach to enhance bioflavour and to reduce ethanol content. *Food Microbiology, 56*, 45–51. https://doi.org/10.1016/j.fm.2015.12.005

Canonico, L., Ciani, E., Galli, E., Comitini, F., & Ciani, M. (2020). Evolution of aromatic profile of Torulaspora delbrueckii mixed fermentation at Microbrewery plant. *Fermentation, 6*, 7. https://doi.org/10.3390/fermentation6010007

Canonico, L., Comitini, F., & Ciani, M. (2017). Torulaspora delbrueckii contribution in mixed brewing fermentations with different Saccharomyces cerevisiae strains. *International Journal of Food Microbiology, 259*, 7–13. https://doi.org/10.1016/j.ijfoodmicro.2017.07.017

Canonico, L., Galli, E., Ciani, E., Comitini, F., & Ciani, M. (2019). Exploitation of three non-conventional yeast species in the brewing process. *Microorganisms, 7*(1). https://doi.org/10.3390/microorganisms7010011

Capece, A., De Fusco, D., Pietrafesa, R., Siesto, G., & Romano, P. (2021). Performance of wild non-conventional yeasts in fermentation of wort based on different malt rxtracts to uelect novel starters for low-alcohol beers. *Applied Sciences, 11*, 801. https://doi.org/10.3390/app11020801

Capece, A., Romaniello, R., Siesto, G., & Romano, P. (2018). Conventional and non-conventional yeasts in beer production. *Fermentation, 4*, 38. https://doi.org/10.3390/fermentation4020038

Carrau, F., Gaggero, C., & Aguilar, P. S. (2015). Yeast diversity and native vigor for flavor phenotypes. *Trends in Biotechnology, 33*(3), 148–154. https://doi.org/10.1016/j.tibtech.2014.12.009

Carriglio, J., Budner, D., & Thompson-Witrick, K. A. (2022). Comparison review of the production, microbiology, and sensory profile of lambic and American coolship ales. *Fermentation, 8*, 646. https://doi.org/10.3390/fermentation8110646

Cavallini, N., Savorani, F., Bro, R., & Cocchi, M. (2021). A metabolomic approach to beer characterization. *Molecules, 26*, 1472. https://doi.org/10.3390/molecules26051472

Cha, J., Debnath, T., & Lee, K.-G. (2019). Analysis of α-dicarbonyl compounds and volatiles formed in Maillard reaction model systems. *Scientific Reports, 9*(1), 5325. https://doi.org/10.1038/s41598-019-41824-8

Chambers, E.t., & Koppel, K. (2013). Associations of volatile compounds with sensory aroma and flavor: The complex nature of flavor. *Molecules, 18*(5), 4887–4905. https://doi.org/10.3390/molecules18054887

Chervina, N. M., Geide, I. V., & Selezneva, I. S. (2019). The study of possible use of lactic bacteria in brewing. *AIP Conference Proceedings, 2174*(1), Article 020099. https://doi.org/10.1063/1.5134250

Ciani, M., Canonico, L., & Comitini, F. (2020). 19 - beer between tradition and innovation. In A. Sankaranarayanan, N. Amaresan, & D. Dhanasekaran (Eds.), *Fermented food products* (1st ed., pp. 313–328). CRC Press. https://doi.org/10.1201/9780429274787-19

Cioch, M., Cichoń, N., Satora, P., & Skoneczny, S. (2022). Physicochemical characteristics of beer with grape must addition produced using non-Saccharomyces yeasts. *European Food Research and Technology*. https://doi.org/10.1007/s00217-022-04182-w

Cioch, M., Królak, K., Tworzydło, Z., Satora, P., & Skoneczny, S. (2022). Characteristics of beer brewed with unconventional yeasts and addition of grape must, pulp and marc. *European Food Research and Technology*. https://doi.org/10.1007/s00217-022-04166-w

Ciont, C., Epuran, A., Kerezsi, A. D., Coldea, T. E., Mudura, E., Pasqualone, A., et al. (2022). Beer safety: New challenges and future trends within craft and large-scale production. *Foods, 11*, 2693. https://doi.org/10.3390/foods11172693

Coghe, S., Benoot, K., Delvaux, F., Vanderhaegen, B., & Delvaux, F. R. (2004). Ferulic acid release and 4-Vinylguaiacol formation during brewing and fermentation: Indications for Feruloyl msterase activity in Saccharomyces cerevisiae. *Journal of Agricultural and Food Chemistry, 52*(3), 602–608. https://doi.org/10.1021/jf0346556

Colomer, M. S., Chailyan, A., Fennessy, R. T., Olsson, K. F., Johnsen, L., Solodovnikova, N., et al. (2020). Assessing population diversity of Brettanomyces yeast species and identification of strains for brewing applications. *Frontiers in Microbiology, 11*. https://doi.org/10.3389/fmicb.2020.00637

Costa, J., Sierra-Garcia, I. N., & Cunha, A. A. (2022). Culture-independent comparison of microbial communities of two maturating craft beers styles. *Microbiology and Biotechnology Letters, 50*(3), 404–413. https://doi.org/10.48022/mbl.2204.04002

Cottrell, M. T. (2022). A search for diastatic enzymes endogenous to Humulus lupulus and produced by microbes associated with pellet hops driving "hop creep" of dry hopped beer. *Journal of the American Society of Brewing Chemists*, 1–13. https://doi.org/10.1080/03610470.2022.2084327

Craine, E. B., Bramwell, S., Ross, C. F., Fisk, S., & Murphy, K. M. (2021). Strategic malting barley improvement for craft brewers through consumer sensory evaluation of malt and beer. *Journal of Food Science, 86*(8), 3628–3644. https://doi.org/10.1111/1750-3841.15786

Crauwels, S., Van Opstaele, F., Jaskula-Goiris, B., Steensels, J., Verreth, C., Bosmans, L., et al. (2017). Fermentation assays reveal differences in sugar and (off-) flavor metabolism across different Brettanomyces bruxellensis strains. *FEMS Yeast Research, 17*(1). https://doi.org/10.1093/femsyr/fow105

de Barros Pita, W., Teles, G. H., Peña-Moreno, I. C., da Silva, J. M., Ribeiro, K. C., & de Morais Junior, M. A. (2019). The biotechnological potential of the yeast Dekkera bruxellensis. *World Journal of Microbiology and Biotechnology, 35*(7), 103. https://doi.org/10.1007/s11274-019-2678-x

De Francesco, G., Turchetti, B., Sileoni, V., Marconi, O., & Perretti, G. (2015). Screening of new strains of Saccharomycodes ludwigii and Zygosaccharomyces rouxii to produce low-alcohol beer. *Journal of the Institute of Brewing, 121*(1), 113–121. https://doi.org/10.1002/jib.185

De La Cruz Pech-Canul, Á., Ortega, D., & Solís-Oviedo, A. G. R. L. (2019). Torulaspora delbrueckii: Towards innovating in the legendary baking and brewing industries. In R. L. Solís-Oviedo, & Á. de la Cruz Pech-Canul (Eds.), *Frontiers and new trends in the science of fermented food and beverages*. IntechOpen. https://doi.org/10.5772/intechopen.83522

de la Torre-González, F. J., Narváez-Zapata, J. A., & Larralde-Corona, C. P. (2017). 5 - microbial diversity and flavor quality of fermented beverages. In A. M. Holban, & A. M. Grumezescu (Eds.), *Microbial production of food ingredients and additives* (pp. 125–154). Academic Press. https://doi.org/10.1016/B978-0-12-811520-6.00005-2

de Lima, A. C., Aceña, L., Mestres, M., & Boqué, R. (2022). An overview of the application of multivariate analysis to the evaluation of beer sensory quality and shelf-life stability. *Foods, 11*, 2037. https://doi.org/10.3390/foods11142037

De Roos, J., & De Vuyst, L. (2022). Lambic beer, a unique blend of tradition and good microorganisms. In F. J. de Bruijn, H. Smidt, L. S. Cocolin, M. Sauer, D. Dowling, & L. Thomashow (Eds.), *Good microbes in medicine, food production, biotechnology, bioremediation, and agriculture* (pp. 225–235). https://doi.org/10.1002/9781119762621.ch18

De Roos, J., Van der Veken, D., & De Vuyst, L. (2019). The interior surfaces of wooden barrels are an additional microbial inoculation source for lambic beer production. *Applied and Environmental Microbiology, 85*(1). https://doi.org/10.1128/aem.02226-18

De Roos, J., Vandamme, P., & De Vuyst, L. (2018). Wort substrate consumption and metabolite production during lambic beer fermentation and maturation explain the successive growth of specific bacterial and yeast species. *Frontiers in Microbiology, 9*. https://doi.org/10.3389/fmicb.2018.02763

De Roos, J., Verce, M., Weckx, S., & De Vuyst, L. (2020). Temporal shotgun metagenomics revealed the potential metabolic capabilities of specific microorganisms during lambic beer production. *Frontiers in Microbiology, 11*. https://doi.org/10.3389/fmicb.2020.0169

De Simone, N., Russo, P., Tufariello, M., Fragasso, M., Solimando, M., Capozzi, V., et al. (2021). Autochthonous biological resources for the production of regional craft beers: Exploring possible contributions of cereals, hops, microbes, and other ingredients. *Foods, 10*, 1831. https://doi.org/10.3390/foods10081831

Dekoninck, T. M. L., Mertens, T., Delvaux, F., & Delvaux, F. R. (2013). Influence of beer characteristics on yeast refermentation performance during bottle conditioning of Belgian beers. *Journal of the American Society of Brewing Chemists, 71*(1), 23–34. https://doi.org/10.1094/ASBCJ-2013-0118-01

del Fresno, J. M., Escott, C., Carrau, F., Herbert-Pucheta, J. E., Vaquero, C., González, C., et al. (2022). Improving aroma complexity with Hanseniaspora spp.: Terpenes, acetate esters, and Safranal. *Fermentation, 8*, 654. https://doi.org/10.3390/fermentation8110654

Denby, C., Li, R., Vu, V., Costello, Z., Lin, W., Chan, L.-J., et al. (2018). Industrial brewing yeast engineered for the production of primary flavor determinants in hopped beer. *Nature Communications, 9*. https://doi.org/10.1038/s41467-018-03293-x

Dennis, R. T., & Young, T. W. (1982). A simple, rapid method for the detection of Subspecies of *Zymomonas mobilis*. *Journal of the Institute of Brewing, 88*(1), 25–29. https://doi.org/10.1002/j.2050-0416.1982.tb04065.x

Di Canito, A., Foschino, R., Mazzieri, M., & Vigentini, I. (2021). Molecular tools to Exploit the biotechnological potential of Brettanomyces bruxellensis: A review. *Applied Sciences, 11*(16), 7302.

Diderich, J. A., Weening, S. M., van den Broek, M., Pronk, J. T., & Daran, J.-M. G. (2018). Selection of Pof-Saccharomyces eubayanus variants for the construction of S. cerevisiae × S. Eubayanus hybrids with reduced 4-vinyl guaiacol formation. *Frontiers in Microbiology, 9*. https://doi.org/10.3389/fmicb.2018.01640

Diepeveen, J., Moerdijk-Poortvliet, T. C. W., & van der Leij, F. R. (2022). Molecular insights into human taste perception and umami tastants: A review. *Journal of Food Science, 87*(4), 1449–1465. https://doi.org/10.1111/1750-3841.16101

Dippel, K., Matti, K., Muno-Bender, J., Michling, F., Brezina, S., Semmler, H., et al. (2022). Co-fermentations of Kveik with non-conventional yeasts for targeted aroma modulation. *Microorganisms, 10*(10). https://doi.org/10.3390/microorganisms10101922

Dufour, M., Zimmer, A., Thibon, C., & Marullo, P. (2013). Enhancement of volatile thiol release of Saccharomyces cerevisiae strains using molecular breeding. *Applied Microbiology and Biotechnology, 97*(13), 5893–5905. https://doi.org/10.1007/s00253-013-4739-7

Dysvik, A., La Rosa, S. L., Liland, K. H., Myhrer, K. S., Østlie, H. M., De Rouck, G., et al. (2020). Co-Fermentation involving Saccharomyces cerevisiae and Lactobacillus species tolerant to brewing-related stress factors for controlled and rapid production of sour beer. *Frontiers in Microbiology, 11*. https://doi.org/10.3389/fmicb.2020.00279

Dysvik, A., Liland, K. H., Myhrer, K. S., Westereng, B., Rukke, E.-O., de Rouck, G., et al. (2019). Pre-fermentation with lactic acid bacteria in sour beer production. *Journal of the Institute of Brewing, 125*(3), 342–356. https://doi.org/10.1002/jib.569

Dysvik, A., Rosa, S. L. L., Rouck, G. D., Rukke, E.-O., Westereng, B., & Wicklund, T. (2020). Microbial dynamics in traditional and modern sour beer production. *Applied and Environmental Microbiology, 86*(14). https://doi.org/10.1128/AEM.00566-20

Eliodório, K. P., Cunha, G., Müller, C., Lucaroni, A. C., Giudici, R., Walker, G. M., et al. (2019). Advances in yeast alcoholic fermentations for the production of bioethanol, beer and wine. *Advances in Applied Microbiology, 109*, 61–119. https://doi.org/10.1016/bs.aambs.2019.10.002

Ellis, D. J., Kerr, E. D., Schenk, G., & Schulz, B. L. (2022). Metabolomics of non-Saccharomyces yeasts in fermented beverages. *Beverages, 8*, 41. https://doi.org/10.3390/beverages8030041

Engan, S. (1981). 3 - beer composition: Volatile substances. In J. R. Pollock (Ed.), *Brewing science* (Vol. 2, pp. 93–165). London: Academic Press.

Ennis, D. M., & Rousseau, B. (2020). *Tools and applications of sensory and consumer science*. Virginia: The Institute for Perception, 21-21, 26-31, 34.

Escalante, W. D. E. (2018). Perspectives and uses of non-Saccharomyces yeasts in fermented beverages. In R. L. Solís-Oviedo, & Á. de la Cruz Pech-Canul (Eds.), *Frontiers and new trends in the science of fermented food and beverages*. IntechOpen. https://doi.org/10.5772/intechopen.81868

Esmaeili, S., Mogharrabi, M., Safi, F., Sohrabvandi, S., Mortazavian, A. M., & Bagheripoor-Fallah, N. (2015). The common spoilage microorganisms of beer: Occurrence, defects, and determination-a review. *Carpathian Journal of Food Science and Technology, 7*, 68–73.

Estela-Escalante, W. D., Rosales-Mendoza, S., Moscosa-Santillán, M., & González-Ramírez, J. E. (2016). Evaluation of the fermentative potential of Candida zemplinina yeasts for craft beer fermentation. *Journal of the Institute of Brewing, 122*(3), 530–535. https://doi.org/10.1002/jib.354

Færgestad, E. M., Langsrud, Ø., Høy, M., Hollung, K., Sæbø, S., Liland, K. H., et al. (2009). 4.08 - analysis of megavariate data in functional Genomics. In S. D. Brown, R. Tauler, & B. Walczak (Eds.), *Comprehensive chemometrics* (pp. 221–278). Elsevier.

Flavor Sub-Committee. (1995). *Sensory analysis manual*. The Institute of Brewing.

Gabaldón, T. (2020). Hybridization and the origin of new yeast lineages. *FEMS Yeast Research, 20*(5). https://doi.org/10.1093/femsyr/foaa040

Gallone, B., Steensels, J., Prahl, T., Soriaga, L., Saels, V., Herrera-Malaver, B., et al. (2016). Domestication and divergence of Saccharomyces cerevisiae beer yeasts. *Cell, 166*(6). https://doi.org/10.1016/j.cell.2016.08.020

Giannakou, K., Visinoni, F., Zhang, P., Nathoo, N., Jones, P., Cotterell, M., et al. (2021). Biotechnological exploitation of Saccharomyces jurei and its hybrids in craft beer fermentation uncovers new aroma combinations. *Food Microbiology, 100*, Article 103838. https://doi.org/10.1016/j.fm.2021.103838

Gibson, B., Geertman, J. A., Hittinger, C. T., Krogerus, K., Libkind, D., Louis, E. J., et al. (2017). New yeasts-new brews: Modern approaches to brewing yeast design and development. *FEMS Yeast Research, 17*(4). https://doi.org/10.1093/femsyr/fox038

Gibson, B., Vidgren, V., Peddinti, G., & Krogerus, K. (2018). Diacetyl control during brewery fermentation via adaptive laboratory engineering of the lager yeast Saccharomyces pastorianus. *Journal of Industrial Microbiology and Biotechnology, 45*(12), 1103–1112. https://doi.org/10.1007/s10295-018-2087-4

Grady, E. N., MacDonald, J., Liu, L., Richman, A., & Yuan, Z.-C. (2016). Current knowledge and perspectives of Paenibacillus: A review. *Microbial Cell Factories, 15*(1), 203. https://doi.org/10.1186/s12934-016-0603-7

Grondin, E., Shum Cheong Sing, A., Caro, Y., Billerbeck, G. M., François, J. M., & Petit, T. (2015). Physiological and biochemical characteristics of the ethyl tiglate production pathway in the yeast Saprochaete suaveolens. *Yeast, 32*, 57–66. https://doi.org/10.1002/yea.3057

Habschied, K., Krstanović, V., & Mastanjević, K. (2022). Beer quality evaluation—a sensory aspect. *Beverages, 8*, 15. https://doi.org/10.3390/beverages8010015

Hammond, J. R. M. (1993). 2 - brewer's yeast. In A. Rose, & J. Harrison (Eds.), *The yeasts* (2nd ed., Vol. 5, pp. 7–67). Academic Press.

Harrison, J., Webb, T. J. B., & Martin, P. A. (1974). The rapid detection of brewery spoilage micro-organisms. *Journal of the Institute of Brewing, 80*(4), 390–398. https://doi.org/10.1002/j.2050-0416.1974.tb03637.x

Helander, I. M., Haikara, A., Sadovskaya, I., Vinogradov, E., & Salkinoja-Salonen, M. S. (2004). Lipopolysaccharides of anaerobic beer spoilage bacteria of the genus Pectinatus– lipopolysaccharides of a Gram-positive genus. *FEMS Microbiology Reviews, 28*(5), 543–552. https://doi.org/10.1016/j.femsre.2004.05.001

Heresztyn, T. (1986). Metabolism of volatile phenolic compounds from hydroxycinnamic acids byBrettanomyces yeast. *Archives of Microbiology, 146*(1), 96–98. https://doi.org/10.1007/BF00690165

Hill, A. E. (2009). 5 - microbiological stability of beer. In C. W. Bamforth (Ed.), *Beer* (pp. 163–183). Academic Press. https://doi.org/10.1016/B978-0-12-669201-3.00005-1

Hoff, S., Lund, M. N., Petersen, M. A., Frank, W., & Andersen, M. L. (2013). Storage stability of pasteurized non-filtered beer. *Journal of the Institute of Brewing, 119*(3), 172–181. https://doi.org/10.1002/jib.85

Holt, S., Mukherjee, V., Lievens, B., Verstrepen, K. J., & Thevelein, J. M. (2018). Bioflavoring by non-conventional yeasts in sequential beer fermentations. *Food Microbiology, 72*, 55–66. https://doi.org/10.1016/j.fm.2017.11.008

Hughes, P. (2009). 2 - beer flavor. In C. W. Bamforth (Ed.), *Beer* (pp. 61–83). Academic Press. https://doi.org/10.1016/B978-0-12-669201-3.00002-6

Hutzler, M., Michel, M., Kunz, O., Kuusisto, T., Magalhães, F., Krogerus, K., et al. (2021). Unique brewing-relevant properties of a strain of Saccharomyces jurei isolated from ash (Fraxinus excelsior). *Frontiers in Microbiology, 12*. https://doi.org/10.3389/fmicb.2021.645271

Hutzler, M., Riedl, R., Koob, J., & Jacob, F. (2012). Fermentation and spoilage yeasts and their relevance for the beverage industry - a review. *BrewingScience, 65*, 33–52.

Iattici, F., Catallo, M., & Solieri, L. (2020). Designing new yeasts for craft brewing: When natural biodiversity meets biotechnology. *Beverages, 6*, 3. https://doi.org/10.3390/beverages6010003

Iorizzo, M., Coppola, F., Letizia, F., Testa, B., & Sorrentino, E. (2021). Role of yeasts in the brewing process: Tradition and innovation. *Processes, 9*, 839. https://doi.org/10.3390/pr9050839Bottle

Jevons, A. L., & Quain, D. E. (2022). Identification of spoilage microflora in draught beer using culture-dependent methods. *Journal of Applied Microbiology, 133*(6), 3728–3740. https://doi.org/10.1111/jam.15810

Joseph, C. M. L., Albino, E., & Bisson, L. F. (2017). Creation and use of a Brettanomyces aroma wheel. *Catalyst: Discovery Into Practice, 1*(1), 12. https://doi.org/10.5344/catalyst.2016.16003

Joseph, C. M. L., Albino, E. A., Ebeler, S. E., & Bisson, L. F. (2015). Brettanomyces bruxellensis aroma active compounds determined by SPME GC-MS olfactory analysis. *American Journal of Enology and Viticulture, 66*. https://doi.org/10.5344/ajev.2015.14073

Jung, R., Karabín, M., Jelínek, L., & Dostálek, P. (2023). Balance of volatile phenols originating from wood- and peat-smoked malt during the brewing process. *European Food Research and Technology, 249*(1), 33–45. https://doi.org/10.1007/s00217-022-04130-8

Jüttner, F., & Watson, S. B. (2007). Biochemical and ecological control of geosmin and 2-methylisoborneol in source waters. *Applied and Environmental Microbiology, 73*(14), 4395–4406. https://doi.org/10.1128/AEM.02250-06

Juvonen, R. (2015). 9 - Strictly anaerobic beer-spoilage bacteria. In A. E. Hill (Ed.), *Brewing microbiology* (pp. 195–218). Woodhead Publishing. https://doi.org/10.1016/B978-1-78242-331-7.00009-5

Kidd, S. E., Abdolrasouli, A., & Hagen, F. (2023). Fungal nomenclature: Managing change is the name of the game. *Open Forum Infectious Diseases, 10*(1). https://doi.org/10.1093/ofid/ofac559

Kilcast, D. (1996). Sensory evaluation of taints and off-flavours. In M. J. Saxby (Ed.), *Food taints and off-flavours* (pp. 1–40). Springer US. https://doi.org/10.1007/978-1-4615-2151-8_1

Kilcast, D. (2010). *Sensory analysis for food and beverage quality control.* Woodhead Publishing/CRC Press.

King, A. J., & Dickinson, J. R. (2003). Biotransformation of hop aroma terpenoids by ale and lager yeasts. *FEMS Yeast Research, 3*(1), 53–62. https://doi.org/10.1111/j.1567-1364.2003.tb00138.x

Kneen, E. (1963). *Proceedings of the Irish Maltsters technical conference*. Dublin: Irish Maltsters Association.

Kocijan, T., Bossaert, S., Van Boeckel, G., De Rouck, G., Lievens, B., & Crauwels, S. (2021). Sanitation of wooden barrels for ageing beer - a review. *BrewingScience, 74*, 51–62. https://doi.org/10.23763/BrSc21-04kocijan

Krogerus, K., & Gibson, B. R. (2013). 125th Anniversary review: Diacetyl and its control during brewery fermentation. *Journal of the Institute of Brewing, 119*(3), 86–97. https://doi.org/10.1002/jib.84

Krogerus, K., & Gibson, B. (2020). A re-evaluation of diastatic Saccharomyces cerevisiae strains and their role in brewing. *Applied Microbiology and Biotechnology, 104*. https://doi.org/10.1007/s00253-020-10531-0

Krogerus, K., Magalhães, F., Vidgren, V., & Gibson, B. (2015). New lager yeast strains generated by interspecific hybridization. *Journal of Industrial Microbiology & Biotechnology, 42*. https://doi.org/10.1007/s10295-015-1597-6

Krogerus, K., Magalhães, F., Vidgren, V., & Gibson, B. (2017). Novel brewing yeast hybrids: Creation and application. *Applied Microbiology and Biotechnology, 101*(1), 65–78. https://doi.org/10.1007/s00253-016-8007-5

Krogerus, K., Rettberg, N., & Gibson, B. (2022). Increased volatile thiol release during beer fermentation using constructed interspecies yeast hybrids. *European Food Research and Technology*. https://doi.org/10.1007/s00217-022-04132-6

Kunze, W. (2010). *Technology brewing and malting* (4th Intl). Berlin: VLB.

Kurtzman, C. P. (1984). Synonomy of the yeast genera Hansenula and Pichia demonstrated through comparisons of deoxyribonucleic acid relatedness. *Antonie van Leeuwenhoek, 50*(3), 209–217. https://doi.org/10.1007/bf02342132

Kurtzman, C. P. (1996). *Transfer of* Hansenula Ofunaensis *to the genus* Pichia *Mycotaxon, LIX* (pp. 85–88).

Kurtzman, C. P. (2011). Phylogeny of the ascomycetous yeasts and the renaming of Pichia anomala to Wickerhamomyces anomalus. *Antonie van Leeuwenhoek, 99*(1), 13–23. https://doi.org/10.1007/s10482-010-9505-6

Laitila, A., Sarlin, T., Raulio, M., Wilhelmson, A., Kotaviita, E., Huttunen, T., et al. (2011). Yeasts in malting, with special emphasis on Wickerhamomyces anomalus (synonym Pichia anomala). *Antonie van Leeuwenhoek, 99*(1), 75–84. https://doi.org/10.1007/s10482-010-9511-8

Langdon, Q., Peris, D., Baker, E., Opulente, D., Huu-Vang, N., Bond, U., et al. (2019). Fermentation innovation through complex hybridization of wild and domesticated yeasts. *Nature Ecology & Evolution, 3*. https://doi.org/10.1038/s41559-019-0998-8

Lawless, H. T. (2013a). *Laboratory exercises for sensory evaluation (food science text series)*. NY: Springer.

Lawless, H. T. (2013b). *Quantitative sensory analysis: Psychophysics: Models and intelligent design*. Wiley Blackwell.

Lengeler, K., Stovicek, V., Fennessy, R., Katz, M., & Förster, J. (2020). Never change a brewing yeast? Why not, there are plenty to choose from. *Frontiers in Genetics, 11*. https://doi.org/10.3389/fgene.2020.582789

Lentz, M. (2018). The impact of simple phenolic compounds on beer aroma and flavor. *Fermentation, 4*. https://doi.org/10.3390/fermentation4010020

Liato, V., & Aïder, M. (2017). Geosmin as a source of the earthy-musty smell in fruits, vegetables and water: Origins, impact on foods and water, and review of the removing techniques. *Chemosphere, 181*. https://doi.org/10.1016/j.chemosphere.2017.04.039

Libkind, D., Peris, D., Cubillos, F., Steenwyk, J., Opulente, D., Langdon, Q., et al. (2020). Into the wild: New yeast genomes from natural environments and new tools for their analysis. *FEMS Yeast Research, 20*. https://doi.org/10.1093/femsyr/foaa008

Licker, J., Acree, T., & Henick-Kling, T. (1998). What is "Brett" (Brettanomyces) flavor?: A preliminary investigation. *ACS Symposium Series, 714*, 96–115. https://doi.org/10.1021/bk-1998-0714.ch008

Linske, M., & Weygandt, A. (Nov-Dec). *Brewers digest quality control series: troubleshooting tips: Basic microbiological identification techniques and interpretation*. Brewers Digest.

Liti, G. (2015). The fascinating and secret wild life of the budding yeast S. cerevisiae. *Elife, 4*. https://doi.org/10.7554/eLife.05835

Lodolo, E. J., Kock, J. L., Axcell, B. C., & Brooks, M. (2008). The yeast Saccharomyces cerevisiae- the main character in beer brewing. *FEMS Yeast Research, 8*(7), 1018–1036. https://doi.org/10.1111/j.1567-1364.2008.00433.x

Londesborough, J., Dresel, M., Gibson, B., Juvonen, R., Holopainen, U., Mikkelson, A., et al. (2015). Analysis of beers from an 1840s' shipwreck. *Journal of Agricultural and Food Chemistry, 63*(9), 2525–2536. https://doi.org/10.1021/jf5052943

Luan, C., Cao, W., Luo, N., Tu, J., Hao, J., Bao, Y., et al. (2023). Genomic insights into the adaptability of the spoilage bacterium Lactobacillus acetotolerans CN247 to the beer microenvironment. *Journal of the American Society of Brewing Chemists, 81*(1), 171–180. https://doi.org/10.1080/03610470.2021.1997280

Maicas, S. (2020). The role of yeasts in fermentation processes. *Microorganisms, 8*(8). https://doi.org/10.3390/microorganisms8081142

Manzano, M., Iacumin, L., Vendrames, M., Cecchini, F., Comi, G., & Buiatti, S. (2011). Craft beer microflora identification before and after a cleaning process. *Journal of the Institute of Brewing, 117*(3), 343–351. https://doi.org/10.1002/j.2050-0416.2011.tb00478.x

Marchesi, J. R., & Ravel, J. (2015). The vocabulary of microbiome research: A proposal. *Microbiome, 3*, 31. https://doi.org/10.1186/s40168-015-0094-5

Martins, C., Brandão, T., Almeida, A., & Rocha, S. M. (2020). Enlarging knowledge on lager beer volatile metabolites using multidimensional gas chromatography. *Foods, 9*, 1276. https://doi.org/10.3390/foods9091276

Mastanjević, K., Krstanović, V., Mastanjević, K., & Šarkanj, B. (2018). Malting and brewing industries encounter Fusarium spp. related problems. *Fermentation, 4*, 3. https://doi.org/10.3390/fermentation4010003

McMurrough, I., Madigan, D., Donnelly, D., Hurley, J., Doyle, A.-M., Hennigan, G., et al. (1996). Control of ferulic acid and 4-vinyl guaiacol in brewing. *Journal of the Institute of Brewing, 102*(5), 327–332. https://doi.org/10.1002/j.2050-0416.1996.tb00918.x

Meier-Dörnberg, T., Jacob, F., Michel, M., & Hutzler, M. (2017). Incidence of Saccharomyces cerevisiae var. diastaticus in the beverage industry: Cases of contamination, 2008–2017. *Technical Quarterly, 54*. https://doi.org/10.1094/TQ-54-4-1130-01

Meier-Dörnberg, T., Kory, O. I., Jacob, F., Michel, M., & Hutzler, M. (2018). Saccharomyces cerevisiae variety diastaticus friend or foe?—spoilage potential and brewing ability of different Saccharomyces cerevisiae variety diastaticus yeast isolates by genetic, phenotypic and physiological characterization. *FEMS Yeast Research, 18*(4). https://doi.org/10.1093/femsyr/foy023. . (Accessed 31 December 2022)

Meilgaard, M. C., Carr, B. T., & Carr, B. T. (2007). *Sensory evaluation techniques* (4th ed.). CRC Press. https://doi.org/10.1201/b16452

Menoncin, M., & Bonatto, D. (2019). Molecular and biochemical aspects of Brettanomyces in brewing. *Journal of the Institute of Brewing, 125*(4), 402–411. https://doi.org/10.1002/jib.580

Mertens, S., Steensels, J., Saels, V., De Rouck, G., Aerts, G., & Verstrepen, K. J. (2015). A large set of newly created interspecific Saccharomyces hybrids increases aromatic diversity in lager beers. *Applied and Environmental Microbiology, 81*(23), 8202–8214. https://doi.org/10.1128/aem.02464-15

Methner, Y., Dancker, P., Maier, R., Latorre, M., Hutzler, M., Zarnkow, M., et al. (2022). Influence of varying fermentation parameters of the yeast strain Cyberlindnera saturnus on the concentrations of selected flavor components in non-alcoholic beer focusing on (E)-β-Damascenone. *Foods, 11*(7). https://doi.org/10.3390/foods11071038

Methner, Y., Hutzler, M., Zarnkow, M., Prowald, A., Endres, F., & Jacob, F. (2022). Investigation of non-Saccharomyces yeast strains for their suitability for the production of non-alcoholic beers with novel flavor profiles. *Journal of the American Society of Brewing Chemists, 80*(4), 341–355. https://doi.org/10.1080/03610470.2021.2012747

Michel, M., Kopecká, J., Meier-Dörnberg, T., Zarnkow, M., Jacob, F., & Hutzler, M. (2016). Screening for new brewing yeasts in the non-Saccharomyces sector with Torulaspora delbrueckii as model. *Yeast, 33*(4), 129–144. https://doi.org/10.1002/yea.3146

Middlekauff, J. E. (1995). 17 - sanitation and pest control, Part B: Microbiological aspects. In W. A. Hardwick (Ed.), *Handbook of brewing* (pp. 480–499). Marcel Dekker, Inc.

Milani, E. A., & Silva, F. V. M. (2022). Pasteurization of beer by non-thermal technologies. *Frontiers in Food Science and Technology, 1*. https://doi.org/10.3389/frfst.2021.798676

Molinet, J., & Cubillos, F. (2020). Wild yeast for the future: Exploring the use of wild strains for wine and beer fermentation. *Frontiers in Genetics, 11*. https://doi.org/10.3389/fgene.2020.589350

Molitor, R. W., Roop, J. I., Denby, C. M., Depew, C. J., Liu, D. S., Stadulis, S. E., et al. (2022). The sensorial and chemical changes in beer brewed with yeast genetically modified to release polyfunctional thiols from malt and hops. *Fermentation, 8*, 370. https://doi.org/10.3390/fermentation8080370

Munford, A. R. G., Alvarenga, V. O., Prado-Silva, L.d., Crucello, A., Campagnollo, F. B., Chaves, R. D., et al. (2017). Sporeforming bacteria in beer: Occurrence, diversity, presence of hop resistance genes and fate in alcohol-free and lager beers. *Food Control, 81*, 126–136. https://doi.org/10.1016/j.foodcont.2017.06.003

Naseeb, S., Alsammar, H., Burgis, T., Donaldson, I., Knyazev, N., Knight, C., et al. (2018). Whole Genome sequencing, de Novo assembly and phenotypic profiling for the new budding Yeast Species Saccharomyces jurei. *G3 (Bethesda), 8*(9), 2967–2977. https://doi.org/10.1534/g3.118.200476

Naseeb, S., Visinoni, F., Hu, Y., Hinks Roberts, A. J., Maslowska, A., Walsh, T., et al. (2021). Restoring fertility in yeast hybrids: Breeding and quantitative genetics of beneficial traits. *Proceedings of the National Academy of Sciences, 118*(38), Article e2101242118. https://doi.org/10.1073/pnas.2101242118

Nikulin, J., Vidgren, V., Krogerus, K., Magalhães, F., Valkeemäki, S., Kangas-Heiska, T., et al. (2020). Brewing potential of the wild yeast species Saccharomyces paradoxus. *European Food Research and Technology, 246*. https://doi.org/10.1007/s00217-020-03572-2

Nizet, S., Gros, J., Peeters, F., Chaumont, S., Robiette, R., & Collin, S. (2013). First evidence of the production of odorant polyfunctional thiols by bottle refermentation. *Journal of the American Society of Brewing Chemists, 71*(1), 15–22. https://doi.org/10.1094/ASBCJ-2013-0117-01

Nizet, S., Peeters, F., Gros, J., & Collin, S. (2014). Chapter 43 - odorant polyfunctional thiols issued from bottle beer refermentation. In V. Ferreira, & R. Lopez (Eds.), *Flavour science* (pp. 227–230). Academic Press.

O'Mahoney, M. (1986). *Sensory evaluation of food: Statistical methods and procedures*. Marcel Dekker Inc.

O'Rourke, T. (2000). *Flavour quality*. Brewer's Guardian Dec.

Osburn, K., Amaral, J., Metcalf, S., Nickens, D., Rogers, C., Sausen, C., et al. (2017). Primary souring: A novel bacteria-free method for sour beer production. *Food Microbiology, 70*. https://doi.org/10.1016/j.fm.2017.09.007

Padilla, B., Gil, J. V., & Manzanares, P. (2018). Challenges of the non-conventional yeast Wickerhamomyces anomalus in Winemaking. *Fermentation, 4*, 68. https://doi.org/10.3390/fermentation4030068

Paradh, A., & Hill, A. E. (2016). Review: Gram negative bacteria in brewing. *Advances in Microbiology, 06*, 195–209. https://doi.org/10.4236/aim.2016.63020

Paradh, A. D., Mitchell, W. J., & Hill, A. E. (2011). Occurrence of Pectinatus and Megasphaera in the major UK breweries. *Journal of the Institute of Brewing, 117*(4), 498–506. https://doi.org/10.1002/j.2050-0416.2011.tb00497.x

Pascari, X., Marin, S., Ramos, A. J., & Sanchis, V. (2022). Relevant Fusarium mycotoxins in malt and beer. *Foods, 11*(2). https://doi.org/10.3390/foods11020246

Pascari, X., Ramos, A. J., Marín, S., & Sanchís, V. (2018). Mycotoxins and beer. Impact of beer production process on mycotoxin contamination. A review. *Food Research International, 103*, 121–129. https://doi.org/10.1016/j.foodres.2017.07.038

Peces-Pérez, R., Vaquero, C., Callejo, M. J., & Morata, A. (2022). Biomodulation of physicochemical parameters, aromas, and sensory profile of craft beers by using non-Saccharomyces yeasts. *ACS Omega, 7*(21), 17822–17840. https://doi.org/10.1021/acsomega.2c01035

Pennerman, K. K., Al-Maliki, H. S., Lee, S., & Bennett, J. W. (2016). Chapter 7 - fungal volatile organic compounds (VOCs) and the genus Aspergillus. In V. K. Gupta (Ed.), *New and future developments in microbial biotechnology and bioengineering* (pp. 95–115). Elsevier.

Peters, J., van Dam, R., van Doorn, R., Katerere, D., Berthiller, F., Haasnoot, W., et al. (2017). Mycotoxin profiling of 1000 beer samples with a special focus on craft beer. *PLoS One, 12*(10). https://doi.org/10.1371/journal.pone.0185887

Petruzzi, L., Rosaria Corbo, M., Sinigaglia, M., & Bevilacqua, A. (2016). Brewer's yeast in controlled and uncontrolled fermentations, with a focus on novel, nonconventional, and superior strains. *Food Reviews International, 32*(4), 341–363. https://doi.org/10.1080/87559129.2015.1075211

Pieczonka, S. A., Lucio, M., Rychlik, M., & Schmitt-Kopplin, P. (2020). Decomposing the molecular complexity of brewing. *Npj Science of Food, 4*(1), 11. https://doi.org/10.1038/s41538-020-00070-3

Pieczonka, S. A., Rychlik, M., & Schmitt-Kopplin, P. (2021). 08 - metabolomics in brewing research. In A. Cifuentes (Ed.), *Comprehensive foodomics* (pp. 116–128). Elsevier. https://doi.org/10.1016/B978-0-08-100596-5.22790-X

Pieczonka, S. A., Zarnkow, M., Diederich, P., Hutzler, M., Weber, N., Jacob, F., et al. (2022). Archeochemistry reveals the first steps into modern industrial brewing. *Scientific Reports, 12*(1), 9251. https://doi.org/10.1038/s41598-022-12943-6

Piraine, R. E. A., Leite, F. P. L., & Bochman, M. L. (2021). Mixed-culture metagenomics of the microbes making sour beer. *Fermentation, 7*, 174. https://doi.org/10.3390/fermentation7030174

Piraine, R., Nickens, D., Sun, D., Leite, F., & Bochman, M. (2021). *Isolation of wild yeasts from Olympic National Park and Moniliella megachiliensis ONP131 physiological characterization for beer fermentation.* https://doi.org/10.1101/2021.07.21.453216

Piraine, R. E. A., Retzlaf, G. M., Gonçalves, V. S., Cunha, R. C., Conrad, N. L., Bochman, M. L., et al. (2022). Brewing and probiotic potential activity of wild yeasts Hanseniaspora uvarum PIT001, Pichia kluyveri LAR001 and Candida intermedia ORQ001. *European Food Research and Technology.* https://doi.org/10.1007/s00217-022-04139-z

Postigo, V., García, M., Cabellos, J. M., & Arroyo, T. (2021). Wine Saccharomyces yeasts for beer fermentation. *Fermentation, 7*, 290. https://doi.org/10.3390/fermentation7040290

Postigo, V., Sánchez-Arroyo, A., Cabellos, J. M., & Arroyo, T. (2022). New approaches for the fermentation of beer: non-Saccharomyces yeasts from wine. *Fermentation, 8*, 280. https://doi.org/10.3390/fermentation8060280

Postigo, V., Sanz, P., García, M., & Arroyo, T. (2022). Impact of non-Saccharomyces wine yeast strains on improving healthy characteristics and the sensory profile of beer in sequential fermentation. *Foods, 11*(14). https://doi.org/10.3390/foods11142029

Pownall, B., Reid, S. J., Hill, A. E., & Jenkins, D. (2022). Schizosaccharomyces pombe in the brewing process: Mixed-culture fermentation for more complete attenuation of high-gravity wort. *Fermentation, 8*, 643. https://doi.org/10.3390/fermentation8110643

Preiss, R., Tyrawa, C., Krogerus, K., Garshol, L. M., & van der Merwe, G. (2018). Traditional Norwegian Kveik are a genetically distinct group of domesticated Saccharomyces cerevisiae brewing yeasts. *Frontiers in Microbiology, 9*, 2137. https://doi.org/10.3389/fmicb.2018.02137

Preiss, R., Tyrawa, C., & van der Merwe, G. (2017). Traditional Norwegian Kveik yeasts: Underexplored domesticated Saccharomyces cerevisiae yeasts. *bioRxiv.* , Article 194969. https://doi.org/10.1101/194969

Priest, F. G. (1990). 1 - contamination. In *An Introduction to brewing Science and technology (Series II)* (Vol. 3, pp. 1−15). The Institute of Brewing. Quality.

Priest, F. G., & Campbell, I. (1987). *Brewing microbiology*. Elsevier.

Priest, F. G., Cowbourne, M. A., & Hough, J. S. (1974). Wort enterobacteria—a review. *Journal of the Institute of Brewing, 80*(4), 342−356. https://doi.org/10.1002/j.2050-0416.1974.tb03629.x

Quilter, M. G., Hurley, J. C., Lynch, F. J., & Murphy, M. G. (2003). The production of isoamyl acetate from Amyl alcohol by Saccharomyces cerevisiae. *Journal of the Institute of Brewing, 109*(1), 34−40. https://doi.org/10.1002/j.2050-0416.2003.tb00591.x

Rainbow, C. (1981). 9 - beer spoilage organisms. In J. R. Pollock (Ed.), *Brewing science* (Vol. 2, pp. 491−550). London: Academic Press.

Rainieri, S., Zambonelli, C., & Kaneko, Y. (2003). Saccharomyces sensu stricto: Systematics, genetic diversity and evolution. *Journal of Bioscience and Bioengineering, 96*, 1−9. https://doi.org/10.1016/S1389-1723(03)90089-2

Ramírez, A., & Viveros, J. M. (2021). Brewing with cannabis sativa vs. Humulus lupulus: A review. *Journal of the Institute of Brewing, 127*(3), 201−209. https://doi.org/10.1002/jib.654

Ramsey, I., Dinu, V., Linforth, R. S. T., Yakubov, G. E., Harding, S. E., Yang, Q., et al. (2020). Understanding the lost functionality of ethanol in non-alcoholic beer using sensory evaluation, aroma release and molecular hydrodynamics. *Scientific Reports, 10.*

Ravasio, D., Carlin, S., Boekhout, T., Groenewald, M., Vrhovsek, U., Walther, A., et al. (2018). Adding flavor to beverages with non-conventional yeasts. *Fermentation, 4*, 15. https://doi.org/10.3390/fermentation4010015

Roberts, R., Khomenko, I., Eyres, G. T., Bremer, P., Silcock, P., Betta, E., et al. (2023). Online monitoring of higher alcohols and esters throughout beer fermentation by commercial Saccharomyces cerevisiae and Saccharomyces pastorianus yeast. *Journal of Mass Spectrometry, 58*(10), Article e4959. https://doi.org/10.1002/jms.4959

Rocha, M. P. (2019). Use of Hanseniaspora opuntiae in co-fermentation with Saccharomyces cerevisiae to enhance the aromatic profile of craft beer. In *IBB − Institute for Bioengineering and Biosciences* (pp. 1−11). Lisboa, Portugal: Instituto Superior Técnico, Universidade de Lisboa.

Rodríguez-Saavedra, M., González de Llano, D., Beltran, G., Torija, M. J., & Moreno-Arribas, M. V. (2021). Pectinatus spp. - unpleasant and recurrent brewing spoilage bacteria. *International Journal of Food Microbiology, 336*, Article 108900. https://doi.org/10.1016/j.ijfoodmicro.2020.108900

Rodríguez-Saavedra, M., González de Llano, D., & Moreno-Arribas, M. V. (2020). Beer spoilage lactic acid bacteria from craft brewery microbiota: Microbiological quality and food safety. *Food Research International, 138*(Pt A), Article 109762. https://doi.org/10.1016/j.foodres.2020.109762

Romano, P., Capece, A., & Jespersen, L. (2006). Taxonomic and ecological diversity of food and beverage yeasts. In A. Querol, & G. Fleet (Eds.), *Yeasts in food and beverages* (pp. 13−53). Springer Berlin Heidelberg. https://doi.org/10.1007/978-3-540-28398-0_2

Romano, A., Perello, M. C., Lonvaud-Funel, A., Sicard, G., & de Revel, G. (2009). Sensory and analytical re-evaluation of "Brett character." *Food Chemistry, 114*(1), 15−19. https://doi.org/10.1016/j.foodchem.2008.09.006

Romero-Rodríguez, R., Durán-Guerrero, E., Castro, R., Díaz, A. B., & Lasanta, C. (2022). Evaluation of the influence of the microorganisms involved in the production of beers on their sensory characteristics. *Food and Bioproducts Processing, 135*, 33−47. https://doi.org/10.1016/j.fbp.2022.06.004

Roullier-Gall, C., David, V., Hemmler, D., Schmitt-Kopplin, P., & Alexandre, H. (2020). Exploring yeast interactions through metabolic profiling. *Scientific Reports, 10*(1), 6073. https://doi.org/10.1038/s41598-020-63182-6

Russell, I., & Stewart, G. G. (1998). *An introduction to brewing science and technology. Series III, brewer's yeast.* The Institute of Brewing.

Satora, P., & Pater, A. (2023). The influence of different non-conventional yeasts on the odour-active compounds of produced beers. *Appl. Sci., 13*, 2872. https://doi.org/10.3390/app13052872

Štulíková, K., Bulíř, T., Nešpor, J., Jelínek, L., Karabín, M., & Dostálek, P. (2020). Application of high-pressure processing to assure the storage stability of unfiltered lager beer. *Molecules, 25*(10). https://doi.org/10.3390/molecules25102414

Štulíková, K., Novák, J., Vlček, J., Šavel, J., Košin, P., & Dostálek, P. (2020). Bottle conditioning: Technology and mechanisms applied in refermented beers. *Beverages, 6*(3), 56. https://doi.org/10.3390/beverages6030056

Salanță, L., Coldea, T., Ignat, M., Tofana, M., Pop, C., Mudura, E., et al. (2020). Non-alcoholic and craft beer production and challenges. *Processes, 8*, 1382. https://doi.org/10.3390/pr8111382

Saxby, M. J. (1996). A survey of chemicals causing taints and off-flavours in food. In M. J. Saxby (Ed.), *Food taints and off-flavours* (pp. 41–71). Springer US. https://doi.org/10.1007/978-1-4615-2151-8_2

Schneiderbanger, J., Grammer, M., Jacob, F., & Hutzler, M. (2018). Statistical evaluation of beer spoilage bacteria by real-time PCR analyses from 2010 to 2016. *Journal of the Institute of Brewing, 124*(2), 173–181. https://doi.org/10.1002/jib.486

Schneiderbanger, J., Jacob, F., Hutzler, M., Schneiderbanger, J., Jacob, F., & Hutzler, M. (2020). Mini-review: The current role of lactic acid bacteria in beer spoilage. *BrewingScience, 73*, 1–5. https://doi.org/10.23763/BrSc19-28schneiderbanger

Schwarz, K. J., Stübner, R., & Methner, F.-J. (2012). Formation of styrene dependent on fermentation management during wheat beer production. *Food Chemistry, 134*(4), 2121–2125. https://doi.org/10.1016/j.foodchem.2012.04.012

Serra Colomer, M., Funch, B., & Forster, J. (2019). The raise of Brettanomyces yeast species for beer production. *Current Opinion in Biotechnology, 56*, 30–35. https://doi.org/10.1016/j.copbio.2018.07.009

Shankar, V., Mahboob, S., Al-Ghanim, K., Ahmad, Z., Almulhim, N., & Govindarajan, M. (2021). A review on microbial degradation of drinks and infectious diseases: A perspective of human well-being and capabilities. *Journal of King Saud University Science, 33*, Article 101293. https://doi.org/10.1016/j.jksus.2020.101293

Shayevitz, A., Abbott, E., Van Zandycke, S., & Fischborn, T. (2022). The impact of lactic and acetic acid on primary beer fermentation performance and secondary Re-fermentation during bottle-conditioning with active dry yeast. *Journal of the American Society of Brewing Chemists, 80*(3), 258–269. https://doi.org/10.1080/03610470.2021.1952508

Shayevitz, A., Harrison, K., & Curtin, C. D. (2021). Barrel-induced variation in the microbiome and Mycobiome of aged sour ale and Imperial porter beer. *Journal of the American Society of Brewing Chemists, 79*(1), 33–40. https://doi.org/10.1080/03610470.2020.1795607

Shimotsu, S., Asano, S., Iijima, K., Suzuki, K., Yamagishi, H., & Aizawa, M. (2015). Investigation of beer-spoilage ability of Dekkera/Brettanomyces yeasts and development of multiplex PCR method for beer-spoilage yeasts. *Journal of the Institute of Brewing, 121*(2), 177–180. https://doi.org/10.1002/jib.209

Sicard, D., & Legras, J.-L. (2011). Bread, beer, and wine: Yeast domestication in the Saccharomyces sensu stricto complex. *Comptes Rendus Biologies, 334*(3), 229–236. https://doi.org/10.1016/j.crvi.2010.12.016

Silva Ferreira, C., Bodart, E., & Collin, S. (2019). Why craft brewers should Be advised to use bottle refermentation to improve late-hopped beer stability. *Beverages, 5*, 39. https://doi.org/10.3390/beverages5020039

Silvello, G. C., Bortoletto, A. M., & Alcarde, A. R. (2020). The barrel aged beer wheel: A tool for sensory assessment. *Journal of the Institute of Brewing, 126*(4), 382–393. https://doi.org/10.1002/jib.626

Simbaña, J., Portero-Barahona, P., & Carvajal Barriga, E. J. (2022). Wild Ecuadorian Saccharomyces cerevisiae strains and their potential in the malt-based beverages industry. *Journal of the American Society of Brewing Chemists, 80*(3), 286–297. https://doi.org/10.1080/03610470.2021.1945366

Simpson, W. J. (2006). 20 - brewing control systems: Sensory evaluation. In C. W. Bamforth (Ed.), *Brewing* (pp. 427–460). Woodhead Publishing. https://doi.org/10.1533/9781845691738.427

Sipiczki, M. (2018). Interspecies hybridisation and genome Chimerisation in Saccharomyces: combining of gene pools of species and its biotechnological perspectives. *Frontiers in Microbiology, 9*, 3071. https://doi.org/10.3389/fmicb.2018.03071

Sohlberg, E., Sarlin, T., & Juvonen, R. (2022). Fungal diversity on brewery filling hall surfaces and quality control samples. *Yeast, 39*(1–2), 141–155. https://doi.org/10.1002/yea.3687

Solieri, L. (2021). The revenge of Zygosaccharomyces yeasts in food biotechnology and applied microbiology. *World Journal of Microbiology and Biotechnology, 37*(6), 96. https://doi.org/10.1007/s11274-021-03066-

Spedding, G. (2012a). Acidification. In G. Oliver (Ed.), *The Oxford Companion to beer* (pp. 6–7). Oxford University Press.

Spedding, G. (2012b). In G. Oliver (Ed.), *Flavor wheel in the Oxford Companion to beer* (pp. 362–363). Oxford University Press.

Spedding, G. (2012c). 4-vinyl guaiacol. In G. Oliver (Ed.), *The Oxford Companion to beer* (p. 372). Oxford University Press.

Spedding, G. (2013a). *Best practices guide to quality craft beer: Delivering optimal flavor to the consumer*. Brewers Association Publications.

Spedding, G. (2013b). The World's most popular Assay? A review of the Ninhydrin-based free amino nitrogen reaction (FAN Assay) emphasizing the development of newer methods and conditions for testing alcoholic beverages. *Journal of the American Society of Brewing Chemists, 71*(2), 83–89. https://doi.org/10.1094/ASBCJ-2013-0411-0

Spedding, G. (2022). A brief history and use of sensory flavour wheels. *Artisan Spirit, 39*, 86–92.

Spedding, G. (2023). Sensory thresholds. Do you know yours? *Artisan Spirit, 43*, 83–87.

Spedding, G. (2024). Something to sniff about. *Artisan Spirit, 45*, 92–98.

Spedding, G., & Lyons, T. P. (July/August 2001). Microbiological media for bacteria and wild yeast detection in the brewery. In *Brewers digest* (pp. 66–70).

Spevacek, A. R., Benson, K. H., Bamforth, C. W., & Slupsky, C. M. (2016). Beer metabolomics: Molecular details of the brewing process and the differential effects of late and dry hopping on yeast purine metabolism. *Journal of the Institute of Brewing, 122*(1), 21–28. https://doi.org/10.1002/jib.291

Spitaels, F., Van Kerrebroeck, S., Wieme, A. D., Snauwaert, I., Aerts, M., Van Landschoot, A., et al. (2015). Microbiota and metabolites of aged bottled gueuze beers converge to the same composition. *Food Microbiology, 47*, 1–11. https://doi.org/10.1016/j.fm.2014.10.004

Spitaels, F., Wieme, A. D., Janssens, M., Aerts, M., Daniel, H. M., Van Landschoot, A., et al. (2014). The microbial diversity of traditional spontaneously fermented lambic beer. *PLoS One, 9*(4), Article e95384. https://doi.org/10.1371/journal.pone.0095384

Steensels, J., Daenen, L., Malcorps, P., Derdelinckx, G., Verachtert, H., & Verstrepen, K. J. (2015). Brettanomyces yeasts — from spoilage organisms to valuable contributors to industrial fermentations. *International Journal of Food Microbiology, 206*, 24–38. https://doi.org/10.1016/j.ijfoodmicro.2015.04.005

Steensels, J., Snoek, T., Meersman, E., Nicolino, M. P., Voordeckers, K., & Verstrepen, K. J. (2014). Improving industrial yeast strains: Exploiting natural and artificial diversity. *FEMS Microbiology Reviews, 38*(5), 947–995. https://doi.org/10.1111/1574-6976.12073

Steensels, J., & Verstrepen, K. J. (2014). Taming wild yeast: Potential of conventional and nonconventional yeasts in industrial fermentations. *Annual Review of Microbiology, 68*, 61–80. https://doi.org/10.1146/annurev-micro-091213-113025

Stelkens, R., & Bendixsen, D. P. (2022). The evolutionary and ecological potential of yeast hybrids. *Current Opinion in Genetics & Development, 76*, Article 101958. https://doi.org/10.1016/j.gde.2022.101958

Stone, H., Bliebaum, R. N., & Thomas, H. A. (2012). *Sensory evaluation practices* (4th ed.). Elsevier/Academic Press.

Storgårds, E. (2000). *Process hygiene control in beer production and dispensing*. VTT Publications. https://www.vttresearch.com/sites/default/files/pdf/publications/2000/P410.pdf.

Storgårds, E., Haikara, A., & Juvonen, R. (2006). 19 - brewing control systems: Microbiological analysis. In C. W. Bamforth (Ed.), *Brewing* (pp. 391–426). Woodhead Publishing. https://doi.org/10.1533/9781845691738.391

Straka, D., & Hleba, L. (2022). Microbiological phases of spontaneously fermented beer. *J microb biotech food sci, 12*, Article e9624.

Stratford, M. (2006). Food and beverage spoilage yeasts. In A. Querol, & G. Fleet (Eds.), *Yeasts in food and beverages* (pp. 335–379). Springer Berlin Heidelberg. https://doi.org/10.1007/978-3-540-28398-0_11

Strejc, J., Kyselová, L., Karabin, M., Silva, J., & Brányik, T. (2013). Production of alcohol-free beer with elevated amounts of flavouring compounds using lager yeast mutants. *Journal of the Institute of Brewing, 119*. https://doi.org/10.1002/jib.72

Suárez, R., Suárez-Lepe, J. A., Morata, A., & Calderón, F. (2007). The production of ethylphenols in wine by yeasts of the genera Brettanomyces and Dekkera: A review. *Food Chemistry, 102*(1), 10–21. https://doi.org/10.1016/j.foodchem.2006.03.030

Su, X., Yu, M., Wu, S., Ma, M., Su, H., Guo, F., et al. (2022). Sensory lexicon and aroma volatiles analysis of brewing malt. *NPJ Science of Food, 6*(1), 20. https://doi.org/10.1038/s41538-022-00135-5

Suiker, I. M., Arkesteijn, G. J. A., Zeegers, P. J., & Wösten, H. A. B. (2021). Presence of Saccharomyces cerevisiae subsp. diastaticus in industry and nature and spoilage capacity of its vegetative cells and ascospores. *International Journal of Food Microbiology, 347*, Article 109173. https://doi.org/10.1016/j.ijfoodmicro.2021.109173

Suiker, I. M., & Wösten, H. A. B. (2022). Spoilage yeasts in beer and beer products. *Current Opinion in Food Science, 44*, Article 100815. https://doi.org/10.1016/j.cofs.2022.100815

Suwonsichon, S. (2019). The importance of sensory lexicons for research and development of food products. *Foods, 8*(1). https://doi.org/10.3390/foods8010027

Suzuki, K. (2011). 125th Anniversary review: Microbiological instability of beer caused by spoilage bacteria. *Journal of the Institute of Brewing, 117*(2), 131–155. https://doi.org/10.1002/j.2050-0416.2011.tb00454.x

Suzuki, K. (2020). Emergence of new spoilage microorganisms in the brewing industry and development of microbiological quality control methods to cope with this phenomenon — a review. *Journal of the American Society of Brewing Chemists, 78*, 1–15. https://doi.org/10.1080/03610470.2020.1782101

Svedlund, N., Evering, S., Gibson, B., & Krogerus, K. (2022). Fruits of their labour: Biotransformation reactions of yeasts during brewery fermentation. *Applied Microbiology and Biotechnology, 106*(13–16), 4929–4944. https://doi.org/10.1007/s00253-022-12068-w

Swiegers, J. H., Saerens, S. M. G., & Pretorius, I. S. (2016). Novel yeast strains as tools for adjusting the flavor of fermented beverages to market specifications. In D. Havkin-Frenkel, & N. Dudai (Eds.), *Biotechnology in flavor production* (pp. 62–132). https://doi.org/10.1002/9781118354056.ch3

Takahashi, M., Kita, Y., Kusaka, K., Mizuno, A., & Goto-Yamamoto, N. (2015). Evaluation of microbial diversity in the pilot-scale beer brewing process by culture-dependent and culture-independent method. *Journal of Applied Microbiology, 118*(2), 454–469. https://doi.org/10.1111/jam.12712

Tan, M., Caro, Y., Shum-Cheong-Sing, A., Robert, L., François, J.-M., & Petit, T. (2021). Evaluation of mixed-fermentation of Saccharomyces cerevisiae with Saprochaete suaveolens to produce natural fruity beer from industrial wort. *Food Chemistry, 346*, Article 128804. https://doi.org/10.1016/j.foodchem.2020.128804

Tan, M., Caro, Y., Sing, A., Reiss, H., Francois, J., & Petit, T. (2021a). Selection by UV mutagenesis and physiological characterization of mutant strains of the yeast Saprochaete suaveolens (former Geotrichum fragrans) with higher capacity to produce flavor compounds. *Journal of Fungi, 7*, 1031. https://doi.org/10.3390/jof7121031

Taylor, B., & Organ, G. (2009). Sensory evaluation. In H. M. Eßlinger (Ed.), *Handbook of brewing* (pp. 675–701). Wiley-VCH. https://doi.org/10.1002/9783527623488.ch29

Thomas, K., Ironside, K., Clark, L., & Bingle, L. (2021). Preliminary microbiological and chemical analysis of two historical stock ales from Victorian and Edwardian brewing. *Journal of the Institute of Brewing, 127*(2), 167–175. https://doi.org/10.1002/jib.641

Tressl, R., Bahri, D., & Kossa, M. (1980). Formation of off-flavor components in beer. In G. Charalambous (Ed.), *The analysis and control of less desirable flavors in foods and beverages* (pp. 293–318). Academic Press.

Tsekouras, G., Tryfinopoulou, P., & Panagou, E. Z. (2021). Detection and identification of lactic acid bacteria in semi-finished beer products using molecular techniques. *Biology and Life Sciences Forum, 6*, 122. https://doi.org/10.3390/Foods2021-11046

Tyrawa, C., Preiss, R., Armstrong, M., & Merwe, G. (2019). The temperature dependent functionality of Brettanomyces bruxellensis strains in wort fermentations. *Journal of the Institute of Brewing, 125*. https://doi.org/10.1002/jib.565

Van Nedervelde, L., & Debourg, A. (1995). Properties of Belgian acid beers and their microflora II. Biochemical properties of Brettanomyces yeasts. *Cerevisia, 1*, 43–48.

van Rijswijck, I. M. H., Wolkers–Rooijackers, J. C. M., Abee, T., & Smid, E. J. (2017). Performance of non-conventional yeasts in co-culture with brewers' yeast for steering ethanol and aroma production. *Microbial Biotechnology, 10*(6), 1591–1602. https://doi.org/10.1111/1751-7915.12717

Van Wyk, N., Kroukamp, H., & Pretorius, I. S. (2018). The smell of synthetic biology: Engineering strategies for aroma compound production in yeast. *Fermentation, 4*(3), 54. https://doi.org/10.3390/fermentation4030054

Vanbeneden, N., Gils, F., Delvaux, F., & Delvaux, F. (2008). Formation of 4-vinyl and 4-ethyl derivatives from hydroxycinnamic acids: Occurrence of volatile phenolic flavour compounds in beer and distribution of Pad1-activity among brewing yeasts. *Food Chemistry, 107*, 221–230. https://doi.org/10.1016/j.foodchem.2007.08.008

Varela, C., & Borneman, A. (2022). Molecular approaches improving our understanding of Brettanomyces physiology. *FEMS Yeast Research, 22*. https://doi.org/10.1093/femsyr/foac028

Vaštík, P., Rosenbergová, Z., Furdíková, K., Klempová, T., Šišmiš, M., & Šmogrovičová, D. (2022). Potential of non-Saccharomyces yeast to produce non-alcoholic beer. *FEMS Yeast Research, 22*(1). https://doi.org/10.1093/femsyr/foac039

Vaughan, A., O'Sullivan, T., & Van Sinderen, D. (2005). Enhancing the microbiological stability of malt and beer — a review. *Journal of the Institute of Brewing, 111*(4), 355–371. https://doi.org/10.1002/j.2050-0416.2005.tb00221.x

Verachtert, H., & Iserentant, D. (1995). Properties of Belgian acid beers and their microflora. I. The production of gueuze and related refreshing acid beers. *Cerevisia, 1*, 37–41.

Vinther Schmidt, C., Olsen, K., & Mouritsen, O. G. (2021). Umami potential of fermented beverages: Sake, wine, champagne, and beer. *Food Chemistry, 360*, Article 128971. https://doi.org/10.1016/j.foodchem.2020.128971

Vriesekoop, F., Krahl, M., Hucker, B., & Menz, G. (2012). 125th Anniversary review: Bacteria in brewing: The good, the bad and the ugly. *Journal of the Institute of Brewing, 118*(4), 335–345. https://doi.org/10.1002/jib.49

Wauters, R., Britton, S. J., & Verstrepen, K. J. (2021). Old yeasts, young beer—the industrial relevance of yeast chronological life span. *Yeast, 38*(6), 339–351. https://doi.org/10.1002/yea.3650

Werrie, P.-Y., Deckers, S., & Fauconnier, M.-L. (2022). Brief insight into the underestimated role of hop Amylases on beer aroma profiles. *Journal of the American Society of Brewing Chemists, 80*(1), 56–74. https://doi.org/10.1080/03610470.2021.1937453

Westerhoff, P., Rodriguez-Hernandez, M., Baker, L., & Sommerfeld, M. (2005). Seasonal occurrence and degradation of 2-methylisoborneol in water supply reservoirs. *Water Research, 39*(20), 4899–4912. https://doi.org/10.1016/j.watres.2005.06.038

Winans, M. J. (2022). Yeast hybrids in brewing. *Fermentation, 8*, 87. https://doi.org/10.3390/fermentation8020087

Winter, I. (2014). *Investigations of possible location dependence of unique microflora in spontaneously fermented lambic beer.*

Witrick, K., Duncan, S., Hurley, K., & O'Keefe, S. (2017). Acid and volatiles of commercially-available lambic beers. *Beverages, 3*(4), 51. https://doi.org/10.3390/beverages3040051

Witrick, K., & Pitts, E. (2021). Bicarbonate inhibition and its impact on Brettanomyces bruxellensis ability to produce flavor compounds. *Journal of the American Society of Brewing Chemists, 80*, 1–9. https://doi.org/10.1080/03610470.2021.1940654

Witrick, K., Pitts, E. R., & O'Keefe, S. F. (2020). Analysis of lambic beer volatiles during aging using gas chromatography–mass spectrometry (GCMS) and gas chromatography–olfactometry (GCO). *Beverages, 6*, 31. https://doi.org/10.3390/beverages6020031

Xu, Z., Luo, Y., Mao, Y., Peng, R., Chen, J., Soteyome, T., et al. (2020). Spoilage lactic acid bacteria in the brewing industry. *Journal of Microbiology and Biotechnology, 30*(7), 955–961. https://doi.org/10.4014/jmb.1908.08069

Xu, W., Wang, J., & Li, Q. (2014). Comparative proteome and transcriptome analysis of lager brewer's yeast in the autolysis process. *FEMS Yeast Research, 14*(8), 1273–1285. https://doi.org/10.1111/1567-1364.12223

Yabaci Karaoglan, S., Jung, R., Gauthier, M., Kinčl, T., & Dostálek, P. (2022). Maltose-negative yeast in non-alcoholic and low-alcoholic beer production. *Fermentation, 8*, 273. https://doi.org/10.3390/fermentation8060273

Yakobson, C. M. (2012). Brettanomyces. In G. Oliver (Ed.), *The Oxford Companion to beer* (pp. 157–159). Oxford University Press.

Zdaniewicz, M., Satora, P., Pater, A., & Bogacz, S. (2020). Low lactic acid-producing strain of Lachancea thermotolerans as a new starter for beer production. *Biomolecules, 10*, 256. https://doi.org/10.3390/biom10020256

Part V

The recycling and valorisation of brewing residues

Chapter 20

Anaerobic treatment of brewery wastes

Joseph C. Akunna
Abertay University, Dundee, United Kingdom

20.1 Introduction

The anaerobic digestion process is a biochemical process that occurs in the presence of readily biodegradable organic carbon and the absence of oxygen. This is similar to the process that occurs naturally in stomachs of ruminants, marshes or sanitary landfill. The process results in the production of biogas, a complex mixture of carbon dioxide, methane and other gases and a slurry or liquid by-product called digestate. The biogas is an important source of energy, which can be converted to electrical or mechanical energy for municipal or industrial use. The digestate is rich in nutrients that can enrich the soil. Anaerobic treatment technology is widely used throughout the world as a cost-effective treatment solution for biodegradable organic wastes and wastewater, from both municipal and industrial sources. The industrial wastes commonly treated using anaerobic digestion technologies include wastewater from food, meat and beverage production and processing, alcohol distilling, dairy and cheese processing, fish processing, fruit and vegetable processing, pulp and paper production, sugar processing, chemical manufacturing and brewing wastes. This chapter discusses key principles important to the anaerobic digestion process and how these apply specifically to brewing waste.

The accepted biochemical pathway for the process is shown in Fig. 20.1. It involves four main stages, namely, hydrolysis, acidogenesis, acetogenesis and methanogenesis, with the last three stages catalysed by acidogenic, acetogenic and methanogenic microbes, respectively, as shown in the figure. Acetogens and methanogens are strict anaerobes, while acidogens are mainly facultative microbes. The final products of each of these stages serve as substrates for other stages as shown in the figure, with the final gaseous product comprised mainly of methane gas and carbon dioxide and trace gases such as hydrogen sulphide and hydrogen.

Some of the common key microorganisms involved in the different degradation stages of the anaerobic digestion process can be seen on Table 20.1.

In comparison with aerobic treatment processes, where biodegradation occurs in the presence of oxygen, the advantages and disadvantages of anaerobic digestion are summarised in Table 20.2. This table shows that anaerobic treatment can play an important role in cost-effective waste management and environmental protection.

20.2 Key factors affecting the anaerobic digestion process

20.2.1 Organic content

Anaerobic treatment is most suitable for solid residues, slurries and wastewaters with chemical oxygen demand (COD) concentrations in the intermediate to high strength range, i.e., from 2000 mg COD/l (Akunna, 2018). Organic removal efficiencies tend to increase with increasing organic strength of the wastewater. However, in general, about 80%–90% COD removal is achievable in an efficiently operated anaerobic digestion system. To achieve higher COD removal, anaerobically pretreated effluent must be further treated with aerobic biological processes. If the wastewater is dilute (i.e., with COD <2000 mg/L), treatment using aerobic processes will be more cost effective.

The types of compounds present in waste or wastewater are one of the primary indicators of the potential bioavailability of the organic matter for the anaerobic microbial population. Fig. 20.2 shows the relative biodegradation rates and reaction times of various types of organic compounds. Biodegradability may be limited by the chemical structure of common

Brewing Microbiology. https://doi.org/10.1016/B978-0-323-99606-8.00008-0

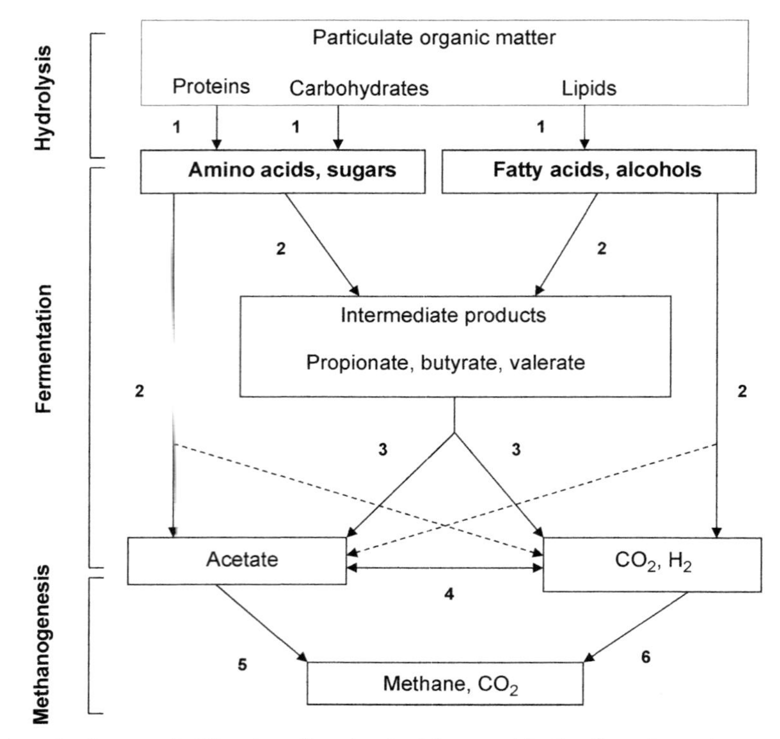

FIGURE 20.1 Simplified schematic diagram of different reactions involved in anaerobic digestion of complex organic matter. Fig. 20.1 Key: (1) Hydrolysis of complex polymers by extracellular enzymes to simpler soluble products. (2) Fermentative or acedogenic bacteria convert simpler compounds to short chain fatty acids, alcohols, ammonia, hydrogen, sulphides and carbon dioxide. (3) Break-down of short chain fatty acids to acetate, hydrogen and carbon dioxide, which act as substrates for methanogenic bacteria. (4) Reaction carried out by acetogenic bacteria. (5) About 70% of methane is produced by acetoclastic methanogens using acetate as substrate. (6) Methane production by hydrogenophilic methanogens using carbon dioxide and hydrogen. *Adapted from Kasper and Wuhrmann (1978), Gujer and Zehnder (1983).*

TABLE 20.1 Anaerobic digestion stages and typical associated microbial species (Akunna, 2018; Stronach, Rudd, & Lester, 1986).

Stage	Associated microbial species
Hydrolysis	*Bacillus, Clostridium, Acetovibrio, Peptococcus, Streptococcus, Staphylococcus, Micrococcus, Eubacterium, Butyrivibrio*, etc.
Acidogenesis	*Veillonella, Bacillus, Desulfobacter, Selenomonas, Clostridium, Desulforomonas, Staphylococcus, Butyrivibrio*, etc.
Acetogenesis	*Clostridium, Methanobacillus omelianskii, Syntrophomonas buswelli, Syntrophomonas wolinii, Syntrophomonas wolfei*, etc.
Methanogenesis	Acetoclastic methanogens: *Methanosarcina, Methanosaeta*, etc. Hydrogenophilic methanogens: *Methanospirillum, Methanobacterium, Methanobrevibacter, Methanoplanus*, etc.

compounds, such as lignin, cellulose and hemicellulose, which are not readily amenable to enzymatic hydrolysis. These compounds may require other types of treatment (termed pre-treatment) before treatment by anaerobic processes. Pre-treatment requirements for anaerobic digestion are discussed later in this chapter.

If the wastewater contains biodegradable particulate organic matter, usually expressed in total solids (TS) or volatile solids (VS), these must be hydrolysed in the first stage of anaerobic digestion process as shown in Fig. 20.1. TS is a measure of all solids in the wastewater while VS measures only the organic fraction (i.e., both biodegradable and non-biodegradable) of the TS. Hydrolysis of particulate biodegradable organic matter is a relatively slow biological reaction. Therefore, if the solids content of the waste is high, effective anaerobic treatment will require relatively long periods of contact (i.e., retention time) between the substrate and the anaerobic microbial consortium. Conversely, if the organic constituents of the waste are primarily soluble in nature, shorter retention times will be required.

TABLE 20.2 Advantages and disadvantages of anaerobic waste treatment processes compared to aerobic treatment (Akunna, 2018; Hall, 1992; Malina, 1992).

Advantages	Disadvantages
• Low sludge production • Low nutrient (nitrogen and phosphorus) requirement • Low capital cost and operating costs • Methane production, a source of energy • Production of liquid and solids residues that may be used as a soil conditioner • Inactivation of pathogens present in the waste • Microbial biomass in anaerobic treatment reactors can survive long periods of little or no feeding.	• Long start-up and retention times • Requires high temperatures for effective operation • Requires monitoring for smooth operation • Shock and variable load can upset microbial balance • Usually used as a pre-treatment stage. Aerobic 'polishing' may be required before discharge to the aquatic environment.

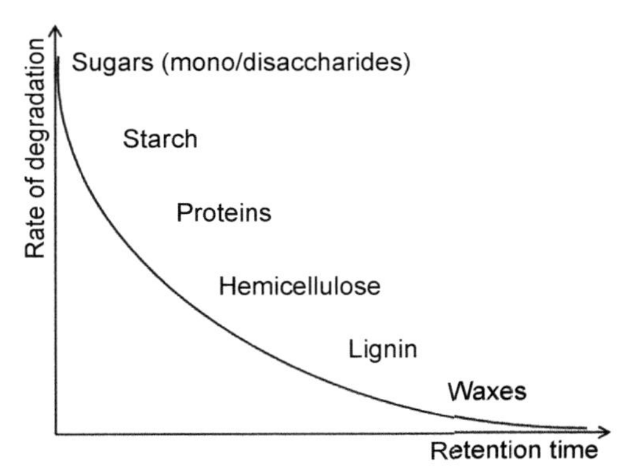

FIGURE 20.2 Effect of substrate type and rate of biodegradation. *Adapted from Elder and Schulz (2006).*

20.2.2 Nutrients

The ideal feedstock composition ratio for carbon (C), nitrogen (N), phosphorus (P) and sulphur (S) for hydrolysis and acidogenesis phases (the C:N:P:S ratio) is considered to be 500:15:5:3, and for methanogenesis, the ultimate ratio is theoretically assumed to be 600:15:5:3 (Weiland, 2001). These elements are called macro-nutrients. The sulphur and phosphorous requirements are very low compared to the carbon and nitrogen requirements and carbon is naturally abundant in organic waste streams. Therefore the limiting nutrient for anaerobic digestion process is considered to be mainly nitrogen. The carbon/nitrogen (C/N) ratio is used to measure nitrogen suitability of the waste to be treated by the anaerobic digestion process, with appropriate values ranging from 20 to 30 (Deublein & Steinhauser, 2008; Polprasert, 2007). Higher C/N ratios can lead to decreased bacterial growth due to nitrogen deficiency, while low ratios may result in ammonia toxicity on the microbial population. For example, high protein content waste with low C/N ratio can produce high ammonia nitrogen during hydrolysis, which can lead to microbial toxicity at higher pH values. Where there is a nitrogen deficiency, nutrient supplement may be needed and this is commonly achieved using urea, sewage sludge or animal manure. Where a phosphorus deficiency exists, phosphorous can be added as phosphate salt or phosphoric acid where necessary.

Another vital parameter in effective digestion is the availability of trace elements (or micro-nutrients), notably iron, cobalt, nickel and zinc, in the feedstock. These elements, when available in relatively small amounts, can stimulate methanogenic activities. The exact amount needed can vary for different wastewater, and prior trials are needed before they are added. More information on the importance of micro-nutrients on the anaerobic digestion process can be found in the literature (Banks, Zhang, Jiang, & Haven, 2012; Demirel & Scherer, 2011; Facchin et al., 2013).

20.2.3 pH and alkalinity

The stability of the anaerobic digestion process is highly dependent on the pH. Whilst the acidogenic bacteria are more tolerant to pH values below 6.0, the optimum pH values for methanogenic bacteria lie between 7 and 8 (Akunna, 2018;

Angelidaki & Sanders, 2004; Raposo, De la Rubia, Fernandez-Cegri, & Boja, 2012). Therefore, the pH range of 6.5–7.8 is suitable for the main microbial groups involved in the process. Acidic pH can occur in anaerobic digestion systems where the methanogenesis rate is slower than the acidogenic rate, thereby bringing about accumulation of the volatile fatty acids. This situation commonly occurs where there is a sudden or excessive increase in the wastewater addition to the anaerobic system. On the other hand, alkaline pH can result in the treatment of wastes containing high amount of nitrogenous compounds, such as proteins. These compounds hydrolyse to produce ammonia, which can bring about alkaline pH. When pH value rises higher than 8.5, it begins to exert a toxic effect on the methanogenic bacteria (Hartmann & Ahring, 2006).

Wastewater alkalinity is also an important parameter in process control. It is a measure of the potential resistance of the digestion process to pH fluctuations. High alkalinity thus ensures process stability. Alkalinity concentrations in the range of 2000–5000 mg/L as $CaCO_3$ are typically required to maintain the pH at or near neutral (Akunna, 2018; Metcalf & Eddy, 2014; Polprasert, 2007).

20.2.4 Temperature

Like other biological processes, anaerobic biodegradation is also affected by temperature as shown in Fig. 20.3. The process can be operated at psychrophilic (<20°C), mesophilic (25–40°C) or thermophilic (45–60°C), temperature ranges, with the optimum temperatures for the mesophilic and thermophilic process at about 37 and 55°C, respectively (Abbasi, Tauseef, & Abbasi, 2012; Akunna, 2018; Raposo et al., 2012). Psychrophilic temperatures are rarely employed due to the resulting relatively low reaction rate. The choice of either mesophilic or thermophilic digestion is dependent on net economic gain that each can provide. In practice, most commercial anaerobic digestion plants operate at mesophilic range.

20.2.5 Solid and hydraulic retention times

The solid retention time (SRT) refers to the average dwelling time of microorganisms within the reactor. This time depends on the growth rate of the microbes and also on the rate at which the microbial biomass is removed from the treatment system. The appropriate SRT varies from one microbial group to another and is dependent on the degradation rate, which itself is dependent on the nature of organic compound in the waste as discussed in Section 1.2.1 above. Methanogenic bacteria have significantly slower growth rates than other microbial groups in the anaerobic digestion process, as shown in Table 20.3. Consequently, the appropriate retention time of anaerobic digestion systems is controlled by the need to reduce the rate of removal of the methanogenic microorganisms from the anaerobic digestion system.

Operational temperature also plays a vital role in the microbial regeneration time and hence the retention time. The higher the operating temperature, the lower the retention time (Akunna, 2018; Metcalf & Eddy, 2014). Hence, thermophilic systems tend to operate at shorter retentions times than mesophilic systems.

The hydraulic retention time (HRT) is defined as the theoretical amount of time that the liquid is resident within the reactor. For completely mixed systems without biomass recycling, the HRT is same as the SRT. For systems designed to

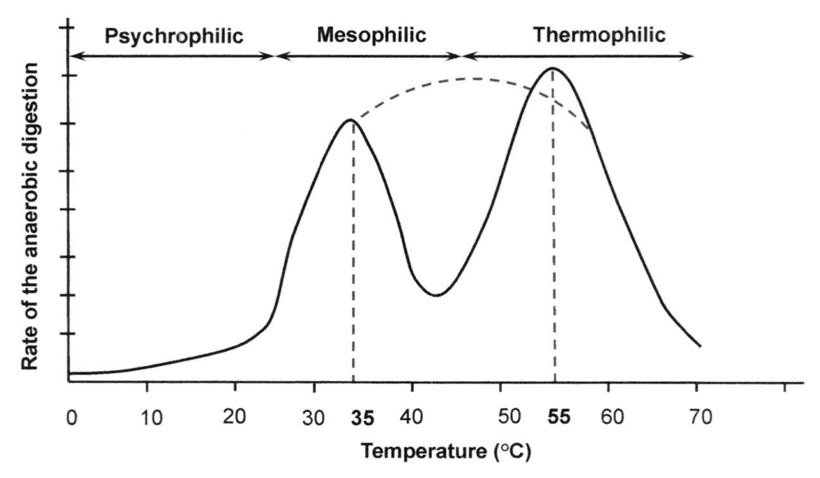

FIGURE 20.3 Effect of temperature on the rte of anaerobic digestion. *Adapted from Mata-Alvarez (2003).*

TABLE 20.3 Average time of regeneration of some microbial groups (Deublein & Steinhauser, 2008).

Microorganisms	Time of regeneration
Acidogenic bacteria	Less than 36 h
Acetogenic bacteria	80–90 h
Methanogenic bacteria	15–16 days

encourage greater biomass retention, the SRT is generally longer that the HRT. Separating both parameters in this manner improves process stability and efficiency.

20.2.6 Organic loading rate

The organic loading rate (OLR) describes the relationships between the rate of treatment of the organic matter and the size of the treatment system or reactor. It is expressed as weight of organic matter in terms of COD or VS (or TS) per volume of reactor per day. The higher the OLR that a system can treat efficiently, the greater the cost-effectiveness of the application. Anaerobic treatment technologies or systems that can handle relatively high OLR are usually referred to as high-rate systems.

20.2.7 Toxic compounds

The anaerobic digestion process can be inhibited by substances in the waste that are toxic to anaerobic microorganisms. The common inhibitors include ammonia, sulphide, long chain fatty acids, salts, heavy metals, phenolic compounds and xenobiotics (Chen, Chang, Guoa, Hong, & Wu, 2016; Chen, Cheng, & Creamer, 2008; Mata-Alvarez, 2003; Speece, 2008).

Sulphide is produced during the hydrolysis of sulphates contained in the waste, and its microbial inhibitory effect is likely to occur where the COD/SO_4^{2-} ratio of the waste is less than 7.7. Where this is the case, the inhibition can be minimised by the following measures (Akunna, 2018; Pohland, 1992):

- Dilution of the influent
- Addition of iron salts to precipitate sulphide from solution
- Stripping the reactor liquid or scrubbing and recirculation of the reactor biogas
- Biological sulphide oxidation and sulphide recovery

In general, the level of toxicity of a substance to microorganisms will depend on its nature, concentration and the degree to which the process has become acclimated to it. As with most microorganisms, anaerobic bacteria can develop a tolerance to a wide variety of inhibitors following an adequate acclimation period.

20.2.8 Treatment configuration: Single- and multi-stage systems

In single-stage systems, all the processes outlined in Fig. 20.1 take place in a single reactor. The main advantage of a single reactor system is its relatively lower capital and maintenance costs. A major drawback is that the system cannot take sufficiently into account the differences in substrate and environmental requirements, and kinetic properties of the major microbial groups involved in the process, notably, the acidogenic and methanogen micro-organisms. Providing and ensuring optimal conditions for both acidogenic and methanogenic microorganisms in a single reactor usually entail long retention times (SRT and HRT) and low treatment rates (i.e., low OLR).

Multi-stage systems involve separation of mainly the hydrolysis and acidogenesis stages from the methanogenesis stage, in different reactors connected in series or in a single compartmentalised reactor. The latter is also known as an anaerobic baffled reactor (ABR) system (Akunna, 2018; Akunna & Clark, 2000; Baloch & Akunna, 2003; Barber & Stuckey, 1999; Shanmugam & Akunna, 2008). Stage or phase separation enables each of the different processes to be maintained at their optimal conditions, to promote greater process stability and biogas yield. The main drawback associated with staged operation is the increased technical complexity and relatively higher capital and maintenance costs, which may not always lead to higher process rates and subsequent biogas yields (Akunna, 2018; Weiland, 1992).

20.3 Factors affecting the application of anaerobic digestion in waste treatment

20.3.1 Pre-treatment of wastes

As explained in Section 1.2.1, the hydrolysis stage is usually the limiting stage for the anaerobic digestion of organic solids. Increasing the rate of hydrolysis can lead to an increase in the rate of biogas production. A range of hydrolysis-enhancing pre-treatment methods for organic solids have been developed. These include physical (e.g., thermal, ultrasound, mechanical, etc.), chemical (alkaline, acid, ozone, etc.), biological (enzymes, aerobic, etc.) and their various combinations. Determining the most appropriate pre-treatment method(s) will depend on the characteristics of the waste. More information on pre-treatment can be found in the literature (e.g., Akunna, 2018; Al-Alm Rashed, Akunna, El-Halwany, & Abou Atiaa, 2010; Delgenes, Penaaud, & Moletta, 2003; Hendriks & Zeeman, 2009; Mallick, Akunna, & Walker, 2008; Taherzadeh & Karimi, 2008).

20.3.2 Co-digestion

Co-digestion or co-treatment refers to the digestion of a mixture of two or more types of waste. The practice can provide the following benefits (Akunna, 2018; Akunna, Abdullahi, & Stewart, 2007; Hartmann, Angelidaki, & Ahring, 2003):

- Waste dilution to reduce the inhibitory levels of certain constituents of the waste streams.
- Increase in readily biodegradable organic matter and vital nutritional balance needed for increased biogas production and effective digestion.
- Mitigation against seasonal and diurnal variations in quantity and quality of waste.

In essence, co-digestion can lead to greater process stability and improved economic gains.

20.3.3 Technology selection

Various types of anaerobic technologies (or reactor systems) have been developed and some are only suitable for certain types of wastes. Where the waste contains significant amounts of organic solids, a suitable technology will involve a longer SRT to allow sufficient time for effective hydrolysis of the solids. Long retention times result in large reactors, operating at relatively low OLR. Alternatively, the solids can be separated in a pre-treatment stage, using physical or physic-chemical methods, and the liquid and semi-solid fractions treated separately. This option can lead to a smaller overall reactor size. Solid—liquid separation pre-treatment is a common practice in the anaerobic digestion of waste from the food and beverage processing wastewaters.

Technologies that can (a) provide phase separation and/or (b) ensure simultaneous high SRT and low HRT are usually referred to as high-rate systems. Other technological variations include operating at different temperatures (i.e., mesophilic, thermophilic or a combination of both in staged systems), use of inorganic or poorly biodegradable organic media support to enhance biomass retention (biofilm systems) and use of various methods to provide reactor mixing (use of biogas or treated effluent, fluidisation, etc.). More information on the range of available technologies can be found in the literature (e.g., Akunna, 2018; Akunna & Clark, 2000; Baloch & Akunna, 2003; Barber & Stuckey, 1999; Hall, 1992; Letting & Hulshoff, 1992; Metcalf & Eddy, 2014; Stamatelatou, Antonopoulou, & Michailides, 2014; van Lier, Mahmoud, & Zeeman, 2008; Werkneh, Beyene, & Osunkunle, 2019).

20.3.4 Biogas production and use

Irrespective of the technology adopted, the performance of an effectively operated anaerobic digestion system treating is similar to values shown in Table 20.4.

Table 20.5 shows a typical composition of biogas from a good functioning anaerobic treatment process.

Biogas may be used directly in an appropriate gas boiler for heating or burned in an engine to produce combined heat and power (CHP). In CHP units, about 70% of the energy contained in biogas is converted to heat and the rest to electricity. The heat can be used to maintain the digester to mesophilic or thermophilic operating temperature and also in supplementing industrial and residential heating requirements. The biogas can also be further cleaned to upgrade the methane content to up to 95%, for injection into the district gas supply network for domestic and industrial use. Upgrading of biogas to pure methane (termed biomethane) creates an option for the biogas to be used as transport fuel for vehicles. More information on options for biogas use can be obtained elsewhere (Akunna, 2018, 2010; DGS & Ecofys, 2005; IEA Bioenergy, 2010, 2014; Polprasert, 2007).

TABLE 20.4 Typical anaerobic treatment performance levels (Pohland, 1992).

Treatment parameter	Typical value
BOD removal (%)	80%–90%
COD removal (mg/L)	1.5 × BOD removed
Biogas production	0.5 m^3/kg COD removed
Methane production	0.35 m^3/kg COD removed
Sludge production	0.05–0.10 kg VS/kg COD removed

TABLE 20.5 Typical biogas composition of a normal functioning anaerobic digestion process (Akunna, 2018; DGS and Ecofys, 2005; IEA Bioenergy, 2006).

Component	Typical range (% volume)
Methane (CH_4)	50–75
Carbon dioxide (CO_2)	25–50
Nitrogen (N_2)	0–10
Hydrogen (H_2)	0.01–5
Oxygen (O_2)	0.1–2
Water vapour	0–10
Ammonia (NH_3)	Less than 1%
Hydrogen sulphide (H_2S)	0.01–3

20.3.5 Digestate handling and disposal

The anaerobic digestion process produces a semi-solid by-product, referred to as digestate, whose properties will depend on the C/N ratio and the solid content of the raw waste. In the absence of significant amounts of toxic compounds such as heavy metals, the digestate is a good source of fertiliser (Akunna, 2018; IEA Bioenergy, 2010). Low waste C/N ratios will produce relative high ammonia content digestate and vice versa. Where the conversion of the solids is relatively low, the moisture content of the digestate will be lower than where the solids are made up of readily biodegradable organic matter. Consideration must therefore be made as to the potential end uses of digestate and whether additional treatment is required such as dewatering for volume reduction or aerobic digestion (or composting) (Abdullahi, Akunna, White, Hallett, & Wheatley, 2008; Akunna, 2018).

20.4 Anaerobic treatment of brewery wastes

This section addresses some of the specific factors pertinent to the use of anaerobic processes as part of a treatment regime for the management of brewery waste. A range of treatment methods applicable to brewery wastewaters can be found elsewhere (Werkneh et al., 2019).

20.4.1 Waste production and collection

The brewery process is a significant water consumer and wastewater producer. It has been estimated that the volume of wastewater discharged from a brewery is within the range of 2.5–10 times the volume of beer being produced (Chen et al., 2016; Fakoya & van der Poll, 2013; Götz, Geißen, Ahrens, & Reimann, 2014; Janhom, Wattanachira, & Pavasant, 2009; Reed, 2006; Simate et al., 2011).

Brewing involves unit processes, some of which are carried out daily in batch or discontinuous operations, with some breweries operating only a 5-day week and weekends reserved for plant maintenance. Consequently, there are wide variations in the rates of wastewater production over a given period. If the wastewater is fed directly to an anaerobic treatment system, the variation in quality and quantity can upset the balance between the key microbial communities, notably the acidogenic and methanogenic microorganisms, thereby causing instability in the process. This problem can be prevented by collecting the wastewater in a balancing tank from where a consistent wastewater quality can be fed to the anaerobic digestion reactor.

20.4.2 Constituents of wastes

The basic beer production process involves mashing, boiling and fermentation followed by cooling, clarification, pasteurisation and packaging (Baloch, Akunna, & Collier, 2007; Mussatto, Dragone, & Roberto, 2006). Each of these stages produces wastewater with varying concentrations of organic compounds.

The particulate organic constituents of brewery wastes consist mainly of spent grain, spent hops and yeast. Spent grains or brewers' spent grains (BSG) is the most abundant brewing by-product, corresponding to about 85% of total by-products generated (Mussatto et al., 2006). BSG is typically more than 75% moisture, and the dry matter composed of about 20% protein and about 70% fibre (Mussatto et al., 2006). It has a C/N ratio of less than 25 and hence amenable to the anaerobic digestion, either as a sole substrate or in co-digestion with other readily biodegradable organic compounds (Kuzmanova & Akunna, 2013; Sturm, Butcher, Wang, Huang, & Roskilly, 2012; Thomas & Rahman, 2006). Other sources of waste include waste liquor from the mashing process and 'out of specification' beer. The various waste streams in the brewing process can each therefore be characterised by one or more of the following properties:

- Highly variable (continuous or intermittent) flow and composition
- High or low organic strength, expressed in term of biological oxygen demand (BOD) or COD
- High or low organic solids content
- High or low nitrogen content
- High concentration of sulphates
- High pH or very low pH values

The characteristics of brewery waste will also be affected by the collection system and by housekeeping practice. Where the waste streams are separated according to their concentrations of organic contents, the wastewater from the fermentation and maturation vessels are likely to contain high levels of COD and suspended solids. Baloch et al. (2007) reported concentrations of wastewater from the tank bottoms, which are the remains of the fermentation and maturation vessels after the beer has been held for a period of time in order to improve its flavour and allow any remaining yeast to settle, to be in the range of 115,000–125,000 mg/L for COD, 1400–16,000 mg/L for TS (composed of over 90% VS), and pH of about 4.2.

Where more mixing of various wastewater streams are carried out, the compositions of brewery waste can contain concentration ranges as reported in Tables 20.6 and 20.7. The yeast-enriched wastewater is usually the residues containing excess yeast used in the process.

As shown by the values reported in Tables 20.6 and 20.7, brewery waste is extremely variable in composition. In general, the waste can be classified (a) intermediate to high strength, (b) made up mainly of organic compounds and (c) does not contain compounds known to be toxic to the anaerobic treatment microorganisms. However, adequate

TABLE 20.6 Types and composition of brewery wastewaters (Akunna, 2010).

	Concentration(mg/L)[a]	
Parameter	Brewery effluent[b]	Yeast-enriched wastewater
COD	3000–6000	50,000–11 0000
Solids	50–1000	2000–3000
TKN	24–200	500–10000
SO_4^{2-}	35	160
pH	5–11	8.3

[a]*All parameters are in mg/L except pH.*
[b]*Refer to literature for more information (Werkneh et al., 2019; Simate et al. 2011).*

TABLE 20.7 Characteristics of wastewater from a local brewery (Akunna, 2003).

Parameter	Range	Typical
COD (mg/L)	1800–50,000	10,000
BOD (mg/L)	2700–38,000	16,000
Solids (mg/L)	50–6000	500
TKN (mg/L)	20–600	50
$P_{(total)}$ (mg/L)	4–103	10
SO_4^{2-} (mg/L)	20–50	35
pH	5–11	9

attention should always be paid to ensure that the COD/SO_4^{2-} ratio is not within the range that can bring about sulphide toxicity. Mixing excess yeast residues with other waste streams can help dilute the sulphate concentration to appropriate levels (Akunna, 2010, 2018).

20.4.3 Treatment and pre-treatment requirements

Depending on the composition of the waste and its discharge or re-use requirements, the following treatment processes can be applied, as standalone or as part of a treatment regime:

- Solids separation (e.g., filter cake washing):
  - Sedimentation (with or without use of chemical to enhance the process)
  - Filtration or floatation
- Neutralisation (e.g., for caustic wash)
- Biological treatment for medium/high strength streams (e.g., wort washings, tank bottoms, surplus yeast, kegging), consisting of aerobic or anaerobic processes, as standalone or combined

Pre-treatment to enhance the rate of the hydrolysis of the brewery solids is necessary in order to accelerate the digestion process and increase biogas yield. In their study (Kuzmanova & Akunna, 2013), a VS reduction of about 24% was obtained after 40 days of batch digestion of the BSG. A common hydrolytic pre-treatment operation for BSG is mechanical disintegration to reduce the size of the particles, which increases the available surface area available for biological process. Other methods include alkaline and acid, ultrasound, thermal and enzymatic treatments. Although some of these processes have been trialled in the laboratory with variable levels of success, full-scale application is hampered by their current high operational costs.

Nutrient correction and pH balancing may be part of pre-treatment steps before anaerobic treatment. Nutrient correction may involve dosing with a solution of mainly nitrogen and phosphorus, micronutrients or compounds to provide alkalinity. Alkaline compounds (e.g., lime, caustic soda) could also be used for pH correction, dosed in the balancing tank and/or directly inside the reactor whenever the need arises.

20.4.4 Co-digestion

For some breweries, co-digestion with other easily available organic substrates may be the only way in which the operation will be economically viable. Co-digestion of brewery waste with animal slurry can bring about C/N ratio correction, which would have been more difficult to achieve with either of the substrates alone. Other appropriate substrates include source separated food wastes, agricultural and other food and beverage processing wastes, domestic wastewater treatment sludges, crop residues and grasses.

Effective co-digestion relies on proper determination of suitable blends that can provide the desired objectives, taking into consideration the biochemical pathways and kinetics of the individual components of the mixture. It is therefore important to be able to predict the outcome of digestion of a chosen waste mixture and to manage the process carefully. Where one or more of the substrates have no history of successful co-digestion with brewery wastes, it is important to conduct laboratory- and pilot-scale trials to establish suitable design and operational data before embarking on a commercial-scale operation. When a full-scale plant is operational, it is always advisable to maintain laboratory scale models for quicker and efficient assessment of the various combinations of new substrates. Anaerobic digestion models combined with experimental results can be used to build and validate co-digestion models to support effective feedstock management (Akunna et al., 2007; Hartmann et al., 2003; Hierholtzer & Akunna, 2012, 2014).

20.4.5 Biogas production

Biogas yield reported in the literature for brewery wastes are very variable and depend on the wastewater characteristics. Values such as 0.25–0.3 L CH_4/g VS, 60–100 m^3/wet tonne of 20% TS and 311 mL CH_4/gTS_{added}) for BSG have been reported (Agler, Aydinkaya, Cummings, Beer, & Angenent, 2010; Kuzmanova & Akunna, 2013).

20.4.6 Digestate management

Digestates from the digestion of brewery wastewater have generally been found suitable for land application, preferably following post-treatment such as dewatering for volume reduction and aerobic digestion (composting) for further breakdown (or stabilisation). The latter is usually carried out alone or in combination with plant biomass (green wastes). Where co-digestion is practised, depending on the sources of the other constituting wastes, post-treatment for land application may include disinfection for pathogen reduction and reduction of potential toxic compounds (such as organic micro-pollutants and heavy metals). Other common disinfection methods include thermal treatment and lime addition. Further information on digestate handling and disposal can be found in elsewhere (e.g., Akunna, 2018).

20.5 Conclusion and perspectives

Anaerobic digestion is now widely regarded as a sustainable management approach for those high organic strength wastes that cannot be re-used for other purposes due to economic and public health reasons. The current worldwide quest for renewable sources of energy and the restrictions now in place in many countries on the disposal of organic wastes to landfill have also contributed to the uptake of anaerobic digestion for the management of municipal, agricultural and industrial organic residues. The brewery industry is one of the sectors that are embracing anaerobic digestion, both as a method of complying with environmental regulations and as a vital source of energy to offset its high dependence on fossil-fuel-based energy sources.

Being environmentally friendly is now considered by many organisations as a tool to improve the corporate image, particularly for those sectors such as the food and beverage producing industry that are heavy consumers of natural resources and energy and heavy producers of wastes.

In the brewery sector, one of the potential challenges to the uptake of anaerobic digestion for waste treatment is the economy of scale particularly for breweries that do not produce sufficient organic waste to make the process viable. For these breweries, co-digestion can be the solution. There is therefore a need for more research on co-digestion of brewery organic waste with other amenable organic materials.

More research is also needed in finding cost effective ways of enhancing the hydrolysis of the solids residues (BSG).

Finally, most of the current anaerobic digestion plants operate on mesophilic temperatures. More research is needed on process optimisation at lower temperatures and also at higher temperatures, to explore if any of these or their combinations can result in greater net energy gains than existing practices.

References

Abbasi, T., Tauseef, S. M., & Abbasi, S. A. (2012). Anaerobic digestion for global warming control and energy generation: An overview. *Renewable and Sustainable Energy Reviews, 16*(5), 3228–3242.

Abdullahi, Y. A., Akunna, J. C., White, N. A., Hallett, P. D., & Wheatley, R. (2008). Investigating the effects of anaerobic and aerobic post-treatment on the quality and stability of organic fraction of municipal solid waste as soil amendment. *Bioresource Technology, 99*(18), 8631–8636.

Agler, M. T., Aydinkaya, Z., Cummings, T. A., Beers, A. R., & Angenent, L. T. (2010). Anaerobic digestion of brewery primary sludge to enhance bioenergy generation: A comparison between low- and high-rate solids treatment and different temperatures. *Bioresource Technology, 101*, 5842–5851.

Akunna, J. C. (2003). *Feasibility study of anaerobic treatment of brewery wastewater*. Dundee, UK: Abertay University.

Akunna, J. C. (2010). Anaerobic treatment of distillery and brewery wastewaters. In G. Walker, & P. Hughes (Eds.), *Distilled spirits - proceedings of the worldwide distilled spirits conference. September 2008* (pp. 111–114). Edinburgh, Nottingham: Nottingham University Press.

Akunna, J. C. (2018). *Anaerobic waste-wastewater treatment and biogas plants: A practical guide*. Boca Baton: CRC Press.

Akunna, J. C., Abdullahi, Y. A., & Stewart, N. A. (2007). Anaerobic digestion of municipal solid wastes containing variable proportions of waste types. *Water Science and Technology, 56*(8), 143–151.

Akunna, J. C., & Clark, M. (2000). Performance of a granular-bed anaerobic baffled reactor (GRABBR) treating whisky distillery wastewater. *Bioresource Technology, 73*(3), 257–261.

Al-Alm Rashed, I. G., Akunna, J., El-Halwany, M. M., & Abou Atiaa, A. F. F. (2010). Improvement in the efficiency of hydrolysis of anaerobic digestion in sewage sludge by the use of enzymes. *Desalination and Water Treatment, 21*, 280–285.

Angelidaki, I., & Sanders, W. (2004). Assessment of the anaerobic biodegradability of macropollutants. *Reviews in Environmental Science and Biotechnology, 3*(2), 117–129.

Baloch, M. I., & Akunna, J. C. (2003). Granular bed baffled reactor (GRABBR): A solution to a two-phase anaerobic digestion system. American Society of Civil Engineers (ASCE). *Journal of Environmental Engineering, 29*(11), 1015–1021.

Baloch, M. I., Akunna, J. C., & Collier, P. J. (2007). The performance of a phase separated granular bed bioreactor treating brewery wastewater. *Bioresource Technology, 98*, 1849–1855.

Banks, C. J., Zhang, Y., Jiang, Y., & Haven, S. (2012). Trace element requirements for stable food waste digestion at elevated ammonia concentrations. *Bioresource Technology, 104*, 127–135.

Barber, W. P., & Stuckey, D. C. (1999). The use of anaerobic baffled reactor for wastewater treatment: A review. *Water Research, 33*(11), 1559–1578.

Chen, H., Chang, S., Guoa, Q., Hong, Y., & Wu, P. (2016). Brewery wastewater treatment using an anaerobic membrane bioreactor. *Biochemical Engineering Journal, 105*, 321–331.

Chen, Y., Cheng, J., & Creamer, K. S. (2008). Inhibition of anaerobic digestion process: A review. *Bioresource Technology, 99*, 4044–4064.

Delgenes, J. P., Penaaud, V., & Moletta, R. (2003). Pre-treatment for the enhancement of anaerobic digestion of solid wastes. In J. Mata-Alvarez (Ed.), *Biomethanization of organic fraction of municipal solid waste* (pp. 202–228). Cornwall: IWA Publishing.

Demirel, B., & Scherer, P. (2011). Trace element requirements of agricultural biogas digesters during biological conversion of renewable biomass to methane: Review. *Biomass and Bioenergy, 35*, 992–998.

Deublein, D., & Steinhauser, A. (2008). *Biogas from waste and renewable resources*. Weinheim Willey-VCH Verlag GmbH & Co. KGaA.

Eder, B., & Schulz, H. (2006). *Biogas-praxis, Grundlagen, Planung. Anlagenbau, Beispiele Wirtschaftlichkeit* (3rd ed.). Freiburg: Staufen bei.

Facchin, V., Cavinato, C., Fatone, F., Pavan, P., Cecchi, F., & Bolzonella, D. (2013). Effect of trace element supplementation on the mesophilic anaerobic digestion of foodwaste in batch trials: The influence of inoculum origin. *Biochemical Engineering Journal, 70*, 71–77.

Fakoya, M. D., & van der Poll, H. M. (2013). Integrating ERP and MFCA systems for improved waste-reduction decisions in a brewery in South Africa. *Journal of Cleaner Production, 40*, 136–140.

German Solar Energy Society (DGS) and Ecofys. (2005). *Planning and installing bioenergy systems: A guide for installers, architects and engineers* (pp. 53–100). London: Earthscan.

Götz, G., Geißen, S., Ahrens, A., & Reimann, S. (2014). Adjustment of the wastewater matrix for optimization of membrane systems applied for water reuse in breweries. *Journal of Membrane Science, 465*, 68–77.

Gujer, W., & Zehnder, A. J. B. (1983). Conversion processes in anaerobic digestion. *Water Science and Technology, 15*, 127–167.

Hall, E. R. (1992). Anaerobic treatment of wastewater in suspended growth and fixed film processes. In J. F. Malina, & F. G. Pohland (Eds.), *Design of anaerobic processes for the treatment of industrial and municipal wastes* (Vol. 7, pp. 41–118). Lancaster, PA: Technomic Publishing Company Inc.

Hartmann, H., & Ahring, B. K. (2006). Strategies for the anaerobic digestion of the organic fraction of municipal solid waste: An overview. *Water Science and Technology, 53*(8), 7–22.

Hartmann, H., Angelidaki, I., & Ahring, B. K. (2003). Co-digestion of the organic fraction of municipal waste with other waste types. Fundamentals of the anaerobic digestion process. In J. Mata-Alvarez (Ed.), *Biomethanization of organic fraction of municipal solid waste* (pp. 181–197). Cornwall: IWA Publishing.

Hendriks, A. T. W. M., & Zeeman, G. (2009). Pretreatments to enhance the digestibility of lignocellulose biomass. *Bioresource Technology, 100*, 10–18.

Hierholtzer, A., & Akunna, J. C. (2012). Modelling sodium inhibition on the anaerobic digestion process. *Water Science and Technology, 66*(7).

Hierholtzer, A., & Akunna, J. C. (2014). Modelling start-up performance of anaerobic digestion of saline-rich macro-algae. *Water Science and Technology, 69*(10), 2059–2064.

International Energy Agency (IEA) Bioenergy. (2006). In M. Persson, O. Jönsson, & A. Wellinger (Eds.), *IEA Bioenergy Task 37—Energy from BiogasBiogas upgrading to vehicle fuel standards and grid injection*. IEA Bioenergy.

International Energy Agency (IEA) Bioenergy. (2010). In C. T. Lukehurst, P. Frost, & T. Al Seadi (Eds.), *IEA Bioenergy Task 37—Energy from BiogasUtilisation of digestate from biogas plants as biofertiliser*. IEA Bioenergy.

International Energy Agency (IEA) Bioenergy. (2014). In T. Persson, J. Murphy, A.-K. Jannasch, E. Ahern, J. Liebetrau, M. Trommler, et al. (Eds.), *IEA Bioenergy Task 37— Energy from BiogasA perspective on the potential role of biogas in smart energy grids*. IEA Bioenergy.

Janhom, T., Wattanachira, S., & Pavasant, P. (2009). Characterisation of brewery wastewater with spectrofluorometry analysis. *Journal of Environmental Management, 90*, 1184–1190.

Kasper, H. F., & Wuhrann, K. (1978). Kinetic parameters and relative turn-overs of some important catabolic reactions in digesting sludge. *Applied and Environmental Microbiology, 36*, 1–7.

Kuzmanova, E., & Akunna, J. C. (2013). *Evaluation of the biochemical methane potential (BMP) from brewers' spent grain*. Project Report (Unpublished).

Letting, G., & Hulshoff, L. W. (1992). UASB process design for various types of wastewaters. In J. F. Malina, & F. G. Pohland (Eds.), *Design of anaerobic processes for the treatment of industrial and municipal wastes* (Vol. 7, pp. 119–146). Lancaster, PA: Technomic Publishing Company Inc.

Malina, J. F. (1992). Anaerobic digestion. In J. F. Malina, & F. G. Pohland (Eds.), *Design of anaerobic processes for the treatment of industrial and municipal wastes* (Vol. 7, pp. 167–212). Lancaster, PA: Technomic Publishing Company Inc.

Mallick, P., Akunna, J. C., & Walker, G. M. (2008). Benefits of enzymatic pre-treatment of intact yeast cells for anaerobic digestion of distillery pot ale. In G. Walker, & P. Hughes (Eds.), *Distilled spirits - proceedings of the worldwide distilled spirits conference. September 2008* (pp. 197–201). Edinburgh, Nottingham: Nottingham University Press.

Mata-Alvarez, J. (2003). Fundamentals of the anaerobic digestion process. In J. Mata-Alvarez (Ed.), *Biomethanization of organic fraction of municipal solid waste* (pp. 1–20). IWA Publishing.

Metcalf and Eddy. (2014). *Wastewater engineering: Treatment, disposal and reuse* (5th ed.). New York: McGraw-Hill.

Mussatto, S. I., Dragone, G., & Roberto, I. C. (2006). Brewers' spent grain: Generation, characteristics and potential applications. *Journal of Cereal Science, 43*, 1–14.

Pohland, F. G. (1992). Anaerobic treatment: Fundamental concepts, applications, and new horizons. In J. F. Malina, & F. G. Pohland (Eds.), *Design of anaerobic processes for the treatment of industrial and municipal wastes* (Vol. 7, pp. 1–33). Lancaster, PA: Technomic Publishing Company Inc.

Polprasert, C. (2007). *Organic waste recycling: Technology and management*. London: IWA Publishing.

Raposo, F., De la Rubia, M. A., Fernández-Cegrí, V., & Borja, R. (2012). Anaerobic digestion of solid organic substrates in batch mode: An overview relating to methane yields and experimental procedures. *Renewable and Sustainable Energy Reviews, 16*(1), 861–877.

Reed, R. (2006). Waste handling in brewing industry. In C. W. Bamforth (Ed.), *Brewing: New technologies* (pp. 335–357). Cambridge: Woodhead Publishing.

Shanmugam, A., & Akunna, J. C. (2008). Comparison of the performance of GRABBR and UASB for the treatment of low strength wastewaters. *Water Science & Technology, 58*(1), 225–232.

Simate, G. S., Cluett, J., Iyuke, S. E., Musapatika, E. T., Ndlovu, S., Walubita, L. F., et al. (2011). The treatment of brewery wastewater for reuse: State of the art. *Desalination, 273*(2–3), 235–247.

Speece, R. E. (2008). *Anaerobic biotechnology and odor/corrosion control* (pp. 246–249). Nashville: Archae Press.

Stamatelatou, K., Antonopoulou, G., & Michailides, P. (2014). Biomethane and bihydrogen production via anaerobic digestion/fermentation. In K. Waldron (Ed.), *Advances in biorefineries* (pp. 478–524). Cambridge: Elsevier.

Stronach, S. M., Rudd, T., & Lester, J. N. (1986). *Anaerobic digestion processes in industrial wastewater treatment*. Berlin: Springer-Verlag.

Sturm, B., Butcher, M., Wang, Y., Huang, Y., & Roskilly, T. (2012). The feasibility of the sustainable energy supply from bio wastes for a small-scale brewery-A case study. *Applied Thermal Engineering, 39*, 45–52.

Taherzadeh, M. J., & Karimi, K. (2008). Pretreatment of lignocellulosic wastes to improve ethanol and biogas production: A review. *International Journal of Molecular Sciences, 9*, 1621–1651.

Thomas, K. R., & Rahman, P. K. S. M. (2006). Brewery wastes. Strategies for sustainability. A review. *Aspects of Applied Biology, 80*, 147–153.

van Lier, J. B., Mahmoud, N., & Zeeman, G. (2008). Anaerobic wastewater treatment. In M. Henze, M. C. M. Van Loosdrecht, G. A. Ekama, & D. Brdjanovic (Eds.), *Biological wastewater treatment: Principles, modelling and design* (pp. 415–456). London: IWA Publishing.

Weiland, P. (1992). One and two-step anaerobic digestion of solid agro-industrial residues. In F. Cecchi, J. Mata-Alvarez, & F. G. Pohland (Eds.), *Proceedings of the international symposium on anaerobic digestion of solid waste* (pp. 193–199), 14-17th April, 1992, Venice, Italy.

Weiland, P. (2001). Grundlagen der Methangärung – Biologie und Substrate (Fundamentals of methane fermentation – biology and substrates). *VDI-Berichte, 1620*, 19–32.

Werkneh, A. A., Beyene, H. D., & Osunkunle, A. A. (2019). Recent advances in brewery wastewater treatment; approaches for water reuse and energy recovery: A review. *Environmental Sustainability, 2*, 199–209.

Chapter 21

Water treatment and reuse in breweries

Geoff S. Simate
School of Chemical and Metallurgical Engineering, University of the Witwatersrand, Johannesburg, South Africa

21.1 Introduction

Brewing is a multibillion-dollar industry that creates jobs, generates taxes, supports agriculture and attracts tourism globally. However, one of the major challenges faced by the brewing industry is water consumption (Simate, 2012). The use of water is increasingly pressing as a result of climate change, with rising incidences of water scarcity. Grain to glass, it is estimated that up to 60 L of water are used for every litre of beer produced, with 3−5 L per litre of beer used within the brewhouse in various activities including as a raw material, heating and cooling, cleaning and sanitation (Olajire, 2020). However, most of this water is allowed to flow out to the drains after its intended use (Olajire, 2020). Furthermore, some of the water is lost with spent grains and some is lost through evaporation during wort boiling (Olajire, 2020). Besides large water consumption, the brewing industry is also a large producer of wastewater (Baloch, Akunna, & Collier, 2007). Though there are variations in the composition of brewery wastewater (Brito et al., 2007; Rao et al., 2007), typically, it has high organic components (Brewers of Europe, 2002; Goldammer, 2008). If the brewery wastewater is discharged into waterways, high levels of organic compounds can deplete dissolved oxygen needed for survival of aquatic species. Some organic compounds also cause serious physiological and neurological damage to the human body when ingested. Thus, it is imperative that brewery wastewater is subjected to some degree of pre-treatment before discharge. Generally, the wastewater generated is pre-treated within the brewery before being discharged into the waterway or municipal sewer system (Goldammer, 2008; Huige, 2006).

As a result of public perception and the possibility of the quality of beer deteriorating, beer brewing is characterised by the use of high-quality fresh water (Janhom, Wattanachira, & Pavasant, 2009). However, the use of fresh water is unsustainable because of an increase in water demand from various other sectors of society and a significant dwindling of fresh water sources. Therefore, it is important that appropriate processes are developed that should not only remove macro-, micro- and nano-pollutants from brewery wastewater but also purify it to a suitable level so that it may be used in primary and/or secondary applications. This chapter focusses on the current and future processes for treating brewery wastewater including prospective applications for reuse. The chapter is divided into five themes: (1) production and composition of brewery wastewater, (2) pre-treatment of brewery wastewater, (3) advanced treatment of brewery wastewater, (4) challenges and future prospects and (5) conclusions.

21.2 Production and composition of brewery wastewater

Beer is one of the oldest alcoholic beverages humans have ever produced (Arnold, 1911; Hornsey, 2003; McGovern et al., 2004; Wyatt, 1900). In fact, the antecedents of our modern-day beer existed many years ago in several places, including Asia, Africa and Europe (Poelmans & Swinnen, 2011). Throughout the years, the brewing industry has developed systematically to include several new process developments and genetic inventions (Linko, Haikara, Ritala, & Penttilä, 1998). Batch-type operations are predominantly employed to process raw materials for the final beer product (van der Merwe & Friend, 2002). The five steps shown in Fig. 21.1 dominate the brewing process (Harrison, 2009), though production methods will differ from brewery to brewery as well as according to the type of beer, brewery equipment and national legislation (Brewers of Europe, 2002). Mashing and fermentation are the two vital processes (Phiarais & Arendt, 2008).

Brewing Microbiology. https://doi.org/10.1016/B978-0-323-99606-8.00021-3

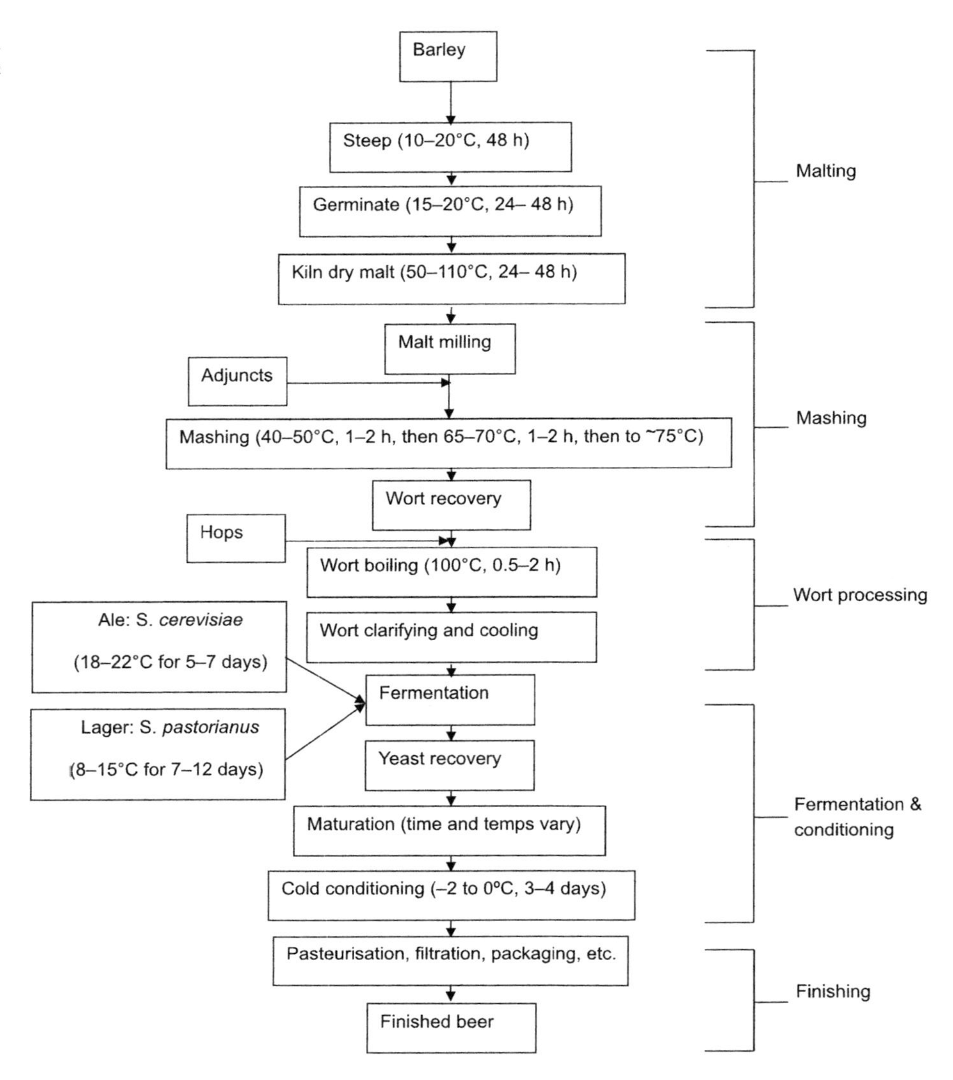

FIGURE 21.1 The brewing process. *Reprinted with permission from Harrison (2009).*

Mashing involves the breaking down of starch to sugar and fermentation is the conversion of the sugars to alcohol and carbon dioxide.

Due to a large number of steps in the brewing process and because of its batch-wise nature, an enormous amount of water is utilised in beer making itself, washing, cleaning and destruction of bacteria and other microorganisms from various units after completion of each and every batch (van der Merwe & Friend, 2002). Consequently, large volumes of brewery wastewater are produced.

The quantity and characteristics of brewery wastewater can differ significantly from time to time and location to location since it is dependent on several different processes that occur within the brewery (Driessen & Vereijken, 2003). Table 21.1 is a summary of some of the physicochemical characteristics of brewery wastewater (Driessen & Vereijken, 2003; Rao et al., 2007). As shown in Table 21.1, the composition of brewery wastewater is highly variable. However, the major component of brewery effluent is organic material (Brewers of Europe, 2002; Goldammer, 2008), as evidenced from high chemical oxygen demand (COD) and biological oxygen demand (BOD). Both of these parameters (i.e., COD and BOD) are important diagnostic parameters for determining the quality of water in natural waterways and waste streams (Mantech, 2011). BOD is a measure of the amount of oxygen required by microorganisms to degrade organic matter, whilst COD is a measure of the total quantity of oxygen needed to oxidise organic as well as inorganic matter present in the wastewater into carbon dioxide and water (Metcalf & Eddy, 1991; ReVelle & ReVelle, 1988). Nevertheless, brewery wastewater is non-toxic, does not carry a considerable amount of heavy metals and is easily biodegradable (Brewers of Europe, 2002; Olajire, 2020). This implies that if properly treated, brewery wastewater may be reused as primary water and/or secondary water without harming public health and the environment. Primary water is the water that is utilised in producing beer itself, whereas secondary water is

TABLE 21.1 Physicochemical characteristics of brewery wastewater.

Parameter	Value
pH	3–12
Temperature (°C)	18–40
COD (mg/L)	2000–6000
BOD (mg/L)	1200–3600
COD:BOD ratio	1.667
VFA (mg/L)	1000–2500
Nitrogen (mg/L)	25–80
Phosphates as PO_4 (mg/L)	10–50
TKN (mg/L)	25–80
TS (mg/L)	5100–8750
TSS (mg/L)	2901–3000
TDS (mg/L)	2020–5940

Reprinted with permission from Rao et al. (2007).

water that does not have any physical contact with beer (Simate et al., 2011), for example, water used for cooling utilities, water utilised in the packaging process and water used for general purpose cleaning.

21.3 Pre-treatment of brewery wastewater

Brewery wastewater is recognised as a significant environmental problem because of the considerable amount of impurities, particularly organic load, created by the brewing process. The disposal of wastewater with high organic load into water bodies can have severe consequences for the biota. This is because during the decomposition of organic pollutants, the dissolved oxygen in the receiving waterways may be used at a faster rate than it can be replenished, exhausting oxygen and thus depriving biota of oxygen needed for survival (Rashed, 2011). Furthermore, wastewater with high organic pollutants contains a large amount of suspended solids that minimise the light accessible to photosynthetic organisms and, on settling out, significantly change the characteristics of the river bed, making it an inappropriate habitat for many invertebrates. More importantly, the disposal of untreated (or partially treated) brewery wastewater directly into waterways or municipal sewers costs more in municipal fees because there are environmental restrictions limiting the amount of contaminants in solution sent through to the municipal reticulation system. This section discusses the physicochemical and biological processes that are commonly used to remove as much particulate and colloidal contaminants from brewery wastewater as possible before it enters the waterways or municipal sewer systems. Table 21.2 lists the unit operations that are included within each category (Simate et al., 2011).

21.3.1 Physical pre-treatment methods

Physical methods encompass all processes in which contaminants are removed by means of or through the application of physical forces. These are the first treatment methods that separate coarse solid matter, instead of dissolved pollutants (Simate et al., 2011). Large solids and grit are removed first so that they do not restrain treatment processes or cause excessive mechanical wear and increased maintenance on subsequent wastewater treatment equipment (EPA, 2003). In most cases, preliminary treatment consists of flow equalisation, screening, grit removal and gravity sedimentation (EPA, 2003). In general, physical pre-treatment requires the least energy but is also the least effective in removing contaminants.

21.3.2 Chemical pre-treatment methods

Chemical methods are wastewater treatment processes in which contaminants are removed by means of or through chemical reactions (Metcalf & Eddy, 1991). Thus, in this pre-treatment method, different chemicals are mixed with the

TABLE 21.2 Wastewater treatment unit operations and processes.

Physical unit operations	• Screening • Comminution • Flow equalisation • Sedimentation • Flotation • Granular-medium filtration
Chemical unit operations	• Chemical precipitation • Adsorption • Disinfection • Chlorination • Other chemical applications
Biological unit operations	• Activated sludge processes • Aerated lagoons • Trickling filters • Rotating biological contactors • Pond stabilisation • Anaerobic digestion • Biological nutrient removal

Reprinted with permission from Simate et al. (2011).

brewery wastewater to adjust the water chemistry (Huang, Schwab, & Jacangelo, 2009). Coagulation and flocculation and/or pH adjustment are some of the most commonly used chemical treatment methods for removing toxic materials and colloidal impurities at breweries (Olajire, 2020; Simate et al., 2011).

Chemical pre-treatment methods have an advantage of being easily applied as soon as it is required (Mohan, 2008). However, one of the inherent disadvantages of chemical treatment methods, as compared to physical methods, is that they are additive processes (Metcalf & Eddy, 1991). As a result, there is usually a positive increase in the dissolved constituents in the wastewater. This additive aspect is in contrast to physical and biological treatment methods that may be described as being subtractive because material is removed from the wastewater. Another drawback of chemical methods is that they are all intensive in operating costs (Metcalf & Eddy, 1991). The costs of some of the chemicals are tied to the cost of energy and thus can be expected to increase similarly.

21.3.3 Biological pre-treatment methods

The objective of a biological pre-treatment method is to eliminate or reduce the concentration of organic and inorganic compounds (Metcalf & Eddy, 1991). It is hinged on the activity of a variety of microorganisms, breaking down the biodegradable organic and inorganic pollutants in the wastewaters (Simate et al., 2011). The principal applications of these processes, also identified in Table 21.3 are (Metcalf & Eddy, 1991): (1) the removal of the carbonaceous organic matter, (2) nitrification, (3) denitrification and (4) stabilisation.

Biological methods of treating wastewater can be either anaerobic (without oxygen) or aerobic (with air/oxygen supply) (Goldammer, 2008). Another process, which is not very different from anaerobic, is anoxic process (used for the removal of nitrogen from wastewater). The individual processes are subdivided further, depending on whether treatment is performed in suspended-growth systems, attached-growth systems or combinations thereof (Metcalf & Eddy, 1991). Table 21.4 compares aerobic and anaerobic biological treatment systems such as activated sludge (Driessen & Vereijken, 2003; Simate et al., 2011). Compared with physicochemical or chemical methods, biological treatment methods possess three main advantages (Dai, Yang, Dong, Ke, & Wang, 2010): (1) the treatment technology is fully developed, (2) high COD and BOD removal efficiency (80%–90%) and (3) low cost of investment. Whilst biological treatment processes are highly effective in reducing conventional pollutants, they also require a high energy input (Feng, Wang, Logan, & Lee, 2008).

21.4 Advanced treatment of brewery wastewater

Water of drinking quality is one of the most important resources in breweries (Blomenhofer, Groß, Procelewska, Delgado, & Becher, 2013). This water is required for brewing, rinsing or cooling purposes (Braeken, Van der Bruggen,

TABLE 21.3 Major biological treatment processes.

Type	Common name	Use
Aerobic processes		
Suspended growth	Activated sludge process Suspended growth nitrification Aerated lagoons Aerobic digestion High-rate aerobic algal ponds	Carbonaceous BOD removal; nitrification Nitrification Carbonaceous BOD removal; nitrification Stabilisation; carbonaceous BOD removal Carbonaceous BOD removal
Attached growth	Trickling filters Roughing filters Rotating biological contactors Packed bed reactors	Carbonaceous BOD removal; nitrification Carbonaceous BOD removal Carbonaceous BOD removal; nitrification Nitrification
Combined processes	Trickling filter, activated sludge Activated sludge, trickling filter	Carbonaceous BOD removal; nitrification Carbonaceous BOD removal; nitrification
Anoxic processes		
Suspended growth	Suspended growth denitrification	Denitrification
Attached growth	Fixed film denitrification	Denitrification
Anaerobic processes		
Suspended growth	Anaerobic digestion Anaerobic contact process	Stabilisation; carbonaceous BOD removal Carbonaceous BOD removal
Attached growth	Anaerobic filter Anaerobic lagoon (ponds)	Carbonaceous BOD removal; stabilisation Denitrification
Aerobic/anoxic/anaerobic processes		
Suspended growth	Single stage	Carbonaceous BOD removal; nitrification; denitrification
Attached growth	Nitrification–denitrification	Nitrification–denitrification
Combined processes	Facultative lagoons (ponds) Maturation or tertiary ponds Anaerobic facultative-lagoons Anaerobic facultative-aerobic lagoons	Carbonaceous BOD removal Carbonaceous BOD removal; nitrification Carbonaceous BOD removal Carbonaceous BOD removal

Metcalf and Eddy (1991).

TABLE 21.4 Anaerobic treatment as compared to aerobic treatment.

	Aerobic systems	Anaerobic systems
Energy consumption	High	Low
Energy production	No	Yes
Biosolids production	High	Low
COD removal (%)	90–98	70–85
Nutrients (N/P) removal	High	Low
Space requirement	High	Low
Discontinuous operation	Difficult	Easy

Driessen and Vereijken (2003).

Vandecasteele, 2004; Fakoya & van der Poll, 2013). Brewing water is utilised during the brewing process itself; rinsing water is needed for the cleaning of bottles, vessels and installations, and cooling water is applied at different stages of the brewing process (Braeken et al., 2004). However, with a growing human population combined with climate challenges, water resources are under stress both quantitatively and qualitatively (Manios, Gaki, Banou, Ntigakis, & Andreadakis, 2006), making operations in the brewery industry very difficult. Therefore, the perpetual necessity for high-quality, but ever insufficient water in the brewery industry has continued to drive the need to find other sources of water (Simate, 2012). One option that needs serious consideration is wastewater reclamation and reuse (Simate, 2012). In fact, the future reuse of water appears to be inescapable, as the concern of water shortage has become a grave global and environmental problem (Janhom et al., 2009).

It must be noted, however, that due to expected high standards, reuse of treated water in breweries is considered unacceptable and would thus require that drinking water standards are complied with (Braeken et al., 2004). In many countries, standards for wastewater reuse have been influenced by the World Health Organisation (WHO) health guidelines (WHO, 1989) and the United States Environmental Protection Agency (US−EPA/USAID) guidelines for water reuse (EPA, 1992). The WHO health guidelines focus mainly on the presence of pathogens, while the EPA guidelines also include physiochemical parameters such as organic load (BOD or COD), total suspended solids (TSS) and residual chlorine concentration (Manios et al., 2006). Table 21.5 shows some of the vital quality requirements for rinsing, cooling and drinking water. Amongst the parameters in Table 21.5, the most important parameter for recycling water is the COD; this is also the most important parameter for measuring (Braeken et al., 2004; Ince, Ince, Sallis, & Anderson, 2000).

Because of the tighter water quality regulations coupled with unsatisfactory results from conventional or pre-treatment processes, the use of intensive treatment processes is necessary if brewery wastewater is to be reused. Therefore, after the brewery wastewater has been subjected to physical, chemical and biological treatments, the wastewater can then go through advanced treatment. This section will discuss some of the current and future advanced treatment processes needed to improve the overall water quality.

21.4.1 Membrane filtration technologies

In the last few decades, various membrane filtration technologies have been used in water and wastewater treatment because of proven solid−liquid separation efficiency and more importantly because of drastic cost reduction of manufacturing membrane materials (Xie, Zhou, Chong, & Holbein, 2008). In addition, this technology has a lot of other advantages including stable and quality effluent, a small area (Hua et al., 2007), low-energy requirements, a small volume of retentate to be handled and selective removal of pollutants (Wu, Li, Wang, Xue, & Li, 2012). Moreover, no chemical addition is required. According to Mallevialle, Odendall, and Wiesner (1996), membrane filtration is a process that uses a semipermeable membrane to separate the feed stream into two portions: a permeate that contains the species passing through the membrane and a retentate consisting of materials left behind. In other words, a membrane is a semipermeable barrier existing between two homogeneous phases and has the ability to transport one component more readily than another because of differences in physical and/or chemical properties between the membrane and the permeating components (Mulder, 1997).

TABLE 21.5 Quality standards for rinsing and cooling water and aimed value for drinking water.

	Quality standard rinsing water	Quality standard cooling water	Quality standard drinking water
COD (mg O_2/L)	0−2	0−2	0−2
Na^+ (mg/L)	0−200	/	20
Cl− (mg/L)	50−250	/	25
pH	6.5−9.5	6.5−9.5	6.5−9.5
Conductivity (μS/cm)	/	/	400

/, Not specified.
Reprinted with permission from Braeken et al. (2004).

There are four groups of membrane filtration that depend on the effective pore size of the membrane (Gregory, 2006). In the order of decreasing pore size, the four groups are as follows: microfiltration (MF), ultrafiltration (UF), nanofiltration (NF) and hyperfiltration (HF) or reverse osmosis (RO). Apart from the size range of permeating species, membrane filtration can be classified further in terms of the mechanisms of rejection of permeating species, the driving forces employed, the chemical structure and composition of membranes and the geometry of construction (Zhou & Smith, 2002). Table 21.6 summarises the essential features of the common membrane filtration processes.

Practically, membranes should have a high permeate flux, high contaminant rejection, great durability, good chemical resistance and low cost (Zhou & Smith, 2002). The other property that is also important in the selection and/or classification of a membrane process is pore size or molecular weight cutoff (MWCO) (Zhou & Smith, 2002). The MWCO expresses the retention characteristics of the membrane in terms of molecules of known sizes (Brock, 1983) and defines the maximum molecular weight of a solute to be rejected (Zhou & Smith, 2002).

The performance of membrane processes also relies on the use of correct module configurations (Zhou & Smith, 2002). Typical commercial membrane geometries are flat sheet and tubular. There are five module types: plate-and-flame and spiral-wound modules, based on flat membranes, and tubular, capillary and hollow-fibre modules, based on tubular membrane geometries (Basile, 2013). A qualitative comparison amongst some of the different model configurations is presented in Table 21.7.

Typically, NF and RO are of spiral-wound configuration so as to promote turbulence, thereby reducing concentration polarisation fouling and particle cake deposition (Zhou & Smith, 2002). However, this type of membrane configuration is vulnerable to biofouling. The weakness of seals and glue lines also prevents the use of vigorous backwashing and may lead to loss of module integrity. In contrast, MF and UF usually use hollow-fibre geometry to facilitate backwash and yield a high surface area-to-volume ratio. A major drawback is the high energy consumption necessary to maintain high crossflow velocity (CFV).

As stated at the beginning of this section, various membrane technologies have been used successfully for water and wastewater treatment applications. Fakhru'l-Razi (1994) used UF membranes of 10,000 nominal molecular weight limit in conjunction with an anaerobic reactor to treat wastewater from a brewery. The percentages of COD removal achieved were above 96%. The results indicated that the UF membranes are capable of efficient biomass—effluent separation, thus preventing any biomass loss from the reactor, and have potential for treating industrial wastewaters. In an attempt to treat brewery wastewater for recycling, Braeken et al. (2004) used NF. Four different water streams (wastewater after biological treatment, bottle rinsing water, rinsing water of the brewing room and rinsing water of the bright beer reservoir) were filtered with four different NF membranes. The results for the biologically treated wastewater were the most promising with removal of COD, Na^+ and $Cl-$ averaging 100%, 55% and 70%, respectively. The other three wastewater streams were not suitable for recycling using NF. These results clearly show the significance of pre-treatment processes.

RO membranes have been used to remove both organics and inorganics in various wastewaters for wastewater reclamation. Compared to other processes, RO offers several advantages (Williams, 2003): (1) high removal rates for many contaminants and pollutants, and can remove both inorganic and organic pollutants simultaneously, (2) simple to design and operate with low maintenance costs and (3) often consume less energy. As a result of these advantages and many others, RO has been employed for treating wastewaters in chemical, textile, petrochemical, electrochemical, pulp and

TABLE 21.6 Typical characteristics of common membrane filtration processes.

Process	Operating pressure (bar)	Pore size (nm)	Molecular weight cut-off range	Size cut-off range (nm)	Main mechanisms	Permeate flux
Microfiltration (MF)	<4	100–3000	>500,000	50–3000	Sieving	High
Ultrafiltration (UF)	2–10	10–200	1000–01,000,000	15–200	Sieving	High
Nanofiltration (NF)	5–40	1–10	100–20,000	1–100	Diffusion + exclusion	Medium
Reverse osmosis (RO)	15–150	<2	<200	<1	Diffusion + exclusion	Low

Gregory (2006), Zhou and Smith (2002).

TABLE 21.7 Comparison of different membrane configurations.

Criteria	Spiral wound	Hollow fibre	Tubular	Plate and frame	Rotating disc
Packing density (m^2/m^3)	++	+++	−	+	−
Wall shear rate	++	+	+++	+	+++
Permeate flux ($L/(m^2\ h)$)	++	++	+++	+	+++
Holdup volume	+	+	−	+	−
Cost per area	+++	+++	−	−	−
Replacement cost	++	++	−	+++	−
Energy consumption	+	++	−	+	++
Fouling tendency	+	++	+++	++	+++
Ease of cleaning	−	+	++	+	+
Pretreatment requirement	−	+	+++	+	+++

The configurations are ranked from clear disadvantage (−) to clear advantage (+++).
Zhou and Smith (2002).

paper, mining and food industries as well as municipal wastewater (Ghabris, Abdel-Jawad, & Aly, 1989; Williams, 2003). A review of RO applications has shown that COD of the effluent may decrease by 90% or may be completely removed (Madaeni & Mansourpanah, 2006; Williams, 2003). Madaeni and Mansourpanah (2006) evaluated various polymeric RO and NF membranes for COD (900−1200 mg/L) removal from biologically treated wastewater from an alcohol manufacturing plant. A complete COD removal (100%) and high flux (33 kg/m^2 h) were obtained from the hydrophilic polyethylene terephthalate PVD RO membrane. These results illustrate that RO is the best method for separating organics from water.

RO systems can also replace or be integrated with other treatment processes such as oxidation, adsorption, stripping or biological treatment to produce a high-quality product water that can be reused or safely discharged (Simate et al., 2011; Williams, 2003). For example, a combination of UF and RO resulted in very high removals of COD (98%−99%), colour and conductivity from the pulp and paper industry effluents (Koyuncu, Yalcin, & Ozturk, 1999; Yalcin, Koyuncu, Oztürk, & Topacik, 1999). Shao, Wei, Yo, and Levy (2009) applied UF and RO for mine wastewater reuse. A study was carried out in two plants treating copper and coal mine wastewater that were characterised by high levels of total dissolved solids (TDS), COD, hardness and TSS as well as high concentrations of sulphates, silica, iron and other metals. The results showed that by integrating UF with RO, suspended solids, bacteria and colloids could be removed effectively. UF membranes could provide feed water for RO with low silt density index (SDI) and turbidity, even in difficult applications where raw water quality fluctuates. The study demonstrated that careful design of a multistage treatment process, and especially the combination of UF and RO membranes, can allow efficient and cost-effective reuse of wastewater that otherwise would be discharged to the environment. Actually, there are many advantages of using UF membrane technology as a pre-treatment for RO (Shao et al., 2009; Yeung, Chu, Rosenberg, & Tong, 2008): (1) stable quality of UF permeate independent of raw water quality, (2) low SDI and turbidity of the UF permeate and (3) reliable removal of bacteria and viruses by UF, thus reducing biofouling of the RO membranes.

Many studies of membrane separation have also been reported for oily wastewater treatment from various industries such as oil fields, petrochemical, metallurgical, pharmaceutical and others. Oil concentrations in wastewater generated in such industries may go up to 1000 mg/L or above (Chakrabarty, Ghoshal, & Purkait, 2008); however, the acceptable discharge limit is only 10−15 mg/L (Maphutha, Moothi, Meyyappan, & Iyuke, 2013). Using ceramic MF membrane, Hua et al. (2007) studied the effects of transmembrane pressure (TMP), CFV, oil concentration in feed, pH and salt concentration on the permeate flux and total organic carbon (TOC) removal efficiency during the separation of oily wastewater. The high permeate flux was achieved under high TMP, high CFV and low oil concentration. The TOC removal efficiencies were higher than 92.4% for all experimental conditions. The results also indicated that the permeate flux decreased either under high salt concentration or under low pH value in the feed solution. Maphutha et al. (2013) used a carbon nanotube integrated polymer composite membrane with a polyvinyl alcohol barrier layer to treat oil-containing wastewater. The permeate through the membrane contained oil concentrations below the acceptable 10 mg/L limit with an excellent throughput and oil rejection of over 95%.

Since brewery wastewater contains high levels of organic impurities, the results discussed in this section show that membrane technologies may be considered as preferred treatment methods for the brewing industry because of their environmentally friendly results, simplicity regarding design, user-friendly aspects in terms of operations and the small amount of space they require. Furthermore, no regenerating chemicals are required, which means no additional salts have to be added for wastewater neutralisation. Membrane technologies have been applied successfully to the treatment and reuse of brewery wastewater, and a range of commercially available products developed by companies such as Half Moon Bay Brewing Company, USA (Tunnel Vision IPA), Village Brewery, Canada (Village Blonde Ale), New Carnegie Brewery, Sweden (PU:REST), Stone Brewing, USA (Full Circle Pale Ale) and Berliner Wasserbetriebe, Germany (Reuse Brew).

21.4.2 Membrane bioreactor technologies

Membrane bioreactor (MBR) technology combines biological-activated sludge processes and membrane filtration technologies, as shown in Fig. 21.2. This technology has become more popular, abundant and accepted for the treatment of many types of wastewaters where the conventional-activated sludge (CAS) process cannot cope with either composition of wastewater or fluctuations of wastewater flow rate (Radjenovic, Petrovic, & Barceló, 2007). Depending on how the membrane is integrated with the bioreactor, the process may be carried out either by pressure-driven filtration in side-stream MBRs or with vacuum-driven membranes immersed directly into the bioreactor in submerged MBRs (Radjenović et al., 2007; Simate et al., 2011). Fig. 21.3 shows the two MBR process configurations (Simate et al., 2011), and Table 21.8 gives a comparison of the two process configurations (Côté & Thompson, 2000). As can be seen in Table 21.8, the side-stream MBRs are more energy intensive compared to submerged MBRs due to higher operational TMPs and the elevated volumetric flow required to achieve the desired CFV (Jeison, 2007). However, submerged MBRs use more membrane area and operate at lower flux levels (Seneviratne, 2007).

Several studies have investigated the efficiencies of MBR and CAS processes operating under comparable conditions, and results have shown significantly improved performance for an MBR in terms of COD, NH_3-N and suspended solids (SS) removal (Bailey, Hansford, & Dold, 1994; Muller, Stouthamer, van Verseveld, & Eikelboom, 1995; Ng & Hermanowicz, 2005; Yamamoto, Hiasa, Mahmood, & Matsuo, 1989). In fact, MBR has been studied not only for wastewater but also for drinking water treatment (Fan & Zhou, 2007; Li & Chu, 2003) and is applied to municipal wastewater treatment at full scale (Lyko et al., 2007). It has also been successfully applied to brewery wastewater treatment at Lagunitas Brewing Company, California, USA, removing over 90% of pollutants.

Li and Chu (2003) used MBR to treat raw water supply that was contaminated by domestic sewage discharge. The results showed that nearly 60% of influent TOC was removed by MBR, accompanied by more than 75% reduction in trihalomethanes formation potential (THMFP). The MBR was also highly effective in removing turbidity, microorganisms and UV_{254} absorbance. The MBR technology was also applied to the brewery wastewater for the purpose of reuse (Dai et al., 2010). Dai et al. (2010) investigated various operating parameters during the process of brewery wastewater treatment in an MBR. The COD reduction in MBR influent (500–1000 mg O_2/L) of up to an average of 96% was achieved. Ammonium and phosphorus impurities were also reduced by 92% and 98%, respectively. Treatment of brewery wastewater in MBR was also conducted by various other researchers (Fakhru'l-Razi, 1994; Kimura, 1991; Nagano, Arikawa, & Kobayashi, 1992). In most of these studies, significant amounts of COD removals (~90%) were reported. Improved COD removal in MBR applications is attributed to the prevention of biomass washout problems commonly encountered in activated sludge processes as well as to complete particulate retention by the membrane (Côté, Buisson, Pound, & Arakaki, 1997, Côté, Buisson, & Praderie, 1998). In another study, brewery bioeffluent was obtained using an internal aerobic MBR (internal MEMBIOR) which was superior to a conventional wastewater treatment plant (Cornelissen,

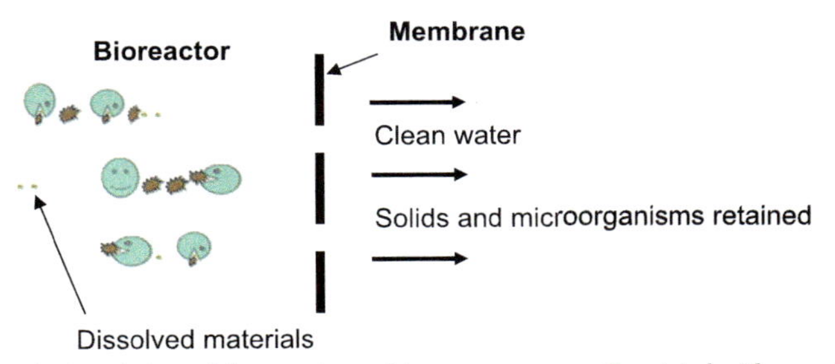

FIGURE 21.2 Simplified schematic description of the membrane bioreactor process. *Reprinted with permission from Simate et al. (2011).*

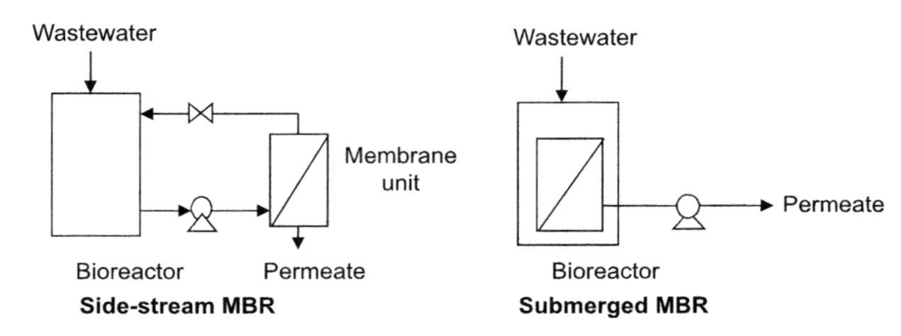

FIGURE 21.3 Membrane bioreactor configurations. *Reprinted with permission from Simate et al. (2011).*

Janse, & Koning, 2002). In this study, the COD of brewery wastewater varied from 1500 to 3500 mg/L, but after the internal MEMBIOR treatment, the COD was reduced to around 30 mg/L regardless of the COD fluctuations of the influent. The suspended solids were also completely retained by the flat-plate membrane. This made the effluent perfectly suited for reuse via RO as process water, omitting the need for expensive pre-treatment methods. This demonstrates that proper process design can provide a visible and feasible solution to the treatment of brewery wastewater.

With these promising results, it can be seen that the MBR process is an attractive option for the treatment and reuse of brewery wastewaters.

21.4.3 Electrochemical technologies

Electrochemical methods of treating wastewaters have gained increasing interest due to their outstanding technical characteristics for eliminating a wide variety of pollutants such as refractory organic matter, nitrogen species and microorganisms (Anglada, Urtiaga, & Ortiz, 2009). Furthermore, electrochemical methods of treatment are favoured because they are neither subject to failure due to variation in wastewater strength nor due to the presence of toxic substances and require less hydraulic retention time (Simate et al., 2011). This method of treating wastewater came into existence when it was first used to treat sewage generated onboard by ships (Bockris, 1977), but extensive investigation of this technology commenced in the 1970s, when Nilsson and others investigated the anodic oxidation of phenolic compounds (Nilsson, Ronlan, & Parker, 1973). Fig. 21.4 shows a conceptual diagram of an electrochemical reactor for wastewater electro-oxidation (Anglada et al., 2009).

Electrochemical oxidation of pollutants can take place directly or indirectly. In direct oxidation (or anodic oxidation), the pollutants are destroyed (or oxidised) at the anode surface. The anodic oxidation can take place through two different pathways (Anglada et al., 2009; Drogui, Blais, & Mercier, 2007): electrochemical conversion where organic compounds are only partially oxidised; therefore, a subsequent treatment may be required or electrochemical combustion where organic compounds are transformed into water, carbon dioxide and other inorganic or biodegradable components. In indirect oxidation, the mediator species (e.g., HClO, $H_2S_2O_8$ and others) are electrochemically generated to carry out the oxidation (Anglada et al., 2009). As far as the indirect oxidation is concerned, the most used electrochemical oxidant is

TABLE 21.8 Comparison of filtration conditions for side-stream and submerged membrane bioreactors.

	Side-stream tubular membrane	Submerged membrane
Manufacturer	Zenon	Zenon
Model	PermaFlow Z-8	ZeeWeed ZW-500
Surface area (m^2)	2	46
Permeate flux ($L/(m^2$ h))	50–100	20–50
Pressure (bar)	4	0.2–0.5
Air flow rate (m^3/h)	–	40
Energy for filtration (kWh/m)	4–12	0.3–0.6

Côté and Thompson (2000).

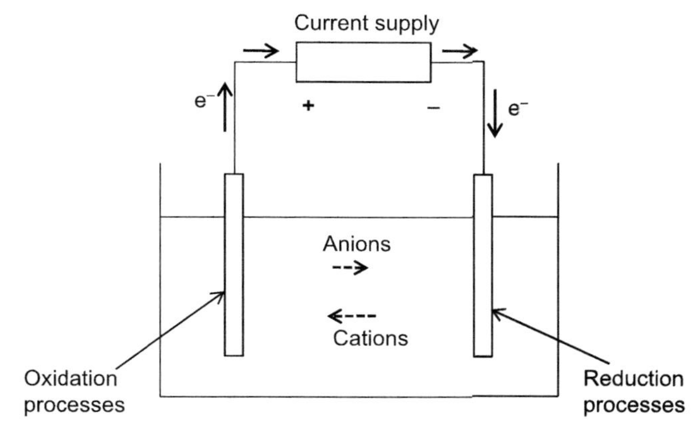

FIGURE 21.4 Conceptual diagram of an electrochemical reactor. *Reprinted with permission from Anglada et al. (2009).*

probably chlorine (HClO, in neutral or alkaline media), as a result of the ubiquitous character of Cl− species in wastewaters and due to their effective action (Martínez-Huitle & Ferro, 2006).

The efficiency and flexibility of electrochemical methods have been studied in a wide spectrum of effluents (e.g., in chemical industry, textile industry, tannery industry, food industry, agro-industry, landfill leachate and urban wastewater) (Anglada et al., 2009). Overall, the aim of these studies was mainly to eliminate non-biodegradable and/or toxic organic pollutants and ammonia nitrogen contained in the effluent. In the food industry, Vijayaraghavan, Ramanujam, and Balasubramanian (1999) studied electrochemical oxidation of high-strength organic waste of distillery spentwash in the presence of sodium chloride. The COD removal from the spentwash was found to be 99% for an initial COD concentration of 15,000 mg/L within 4 h. Because the graphite anode and stainless steel cathode were kept in an undivided electrolytic reactor, chlorine produced from sodium chloride during electrolysis underwent a disproportionation reaction, forming hypochlorous acid. The hypochlorous acid formed thus oxidised the organic matter present in the wastewater. Later, Vijayaraghavan, Ahmad, and Lesa (2006) developed a novel brewery wastewater treatment method also based on in situ hypochlorous acid generation. The generated hypochlorous acid served as an oxidising agent that destroyed organic compounds present in the brewery wastewater. An influent COD value of 2470 mg/L was reduced to only 64 mg/L (i.e., over 97% COD reduction). In the same period, Piya-areetham, Shenchunthichai, and Hunsom (2006) investigated the removal of colour and COD from distillery wastewater by using electro-oxidation processes. The commercial Ti/RuO_2 grid was used as the cathode, and two voluminous surface area materials including graphite particles and the commercial titanium sponge were used as the anode. Effects of several parameters including the initial pH of wastewater (1−5), dilution factor, current intensity (1−10 A), type of additive (H_2O_2 or NaCl) and additive concentration were investigated. The results showed that the optimum condition for treating effluent from distillery wastewater with 10 times dilution was found at the current intensity of 9 A at an initial pH of 1 with a titanium sponge anode in the presence of 1.0 M NaCl. At this condition, approximately 92.24% and 89.62% of colour and COD were removed, respectively.

Another electrochemical method that has the potential to be an effective alternative to the various traditional techniques employed for the distillery and/or brewery effluent treatment is electrocoagulation. Electrocoagulation is based on the in situ formation of the coagulant as the sacrificial anode dissolves due to the applied current, while the simultaneous evolution of gases at the electrodes allows for organic pollutant removal by flotation (Khandegar & Saroha, 2012).

Electrocoagulation cells consist of pairs of parallel metal plate electrodes separated by a few millimetres with a low voltage applied at high current densities, as shown in Fig. 21.5 (Global Advantech, 2011; Xu & Zhu, 2004). The current flowing between the electrodes destabilises electrical charges, which maintain suspensions of particulates, for example, clays and emulsions/microemulsions of hydrocarbons and insoluble organic compounds. The particulates coagulate together into flocs. The hydrocarbons and insoluble organic compounds coalesce into larger droplets and rise in the cells. Electrochemical reactions at the electrodes produce very fine H_2 and O_2 gas bubbles and highly chemically reactive hydroxyl (OH^-) and superoxide (HO_2^-) radicals. The gas bubbles promote the flotation of coagulated solids and coalesced hydrocarbons, etc. The hydroxyl and superoxide radicals cause precipitation of hydroxides of heavy metals and the breakdown of many soluble organic molecules.

A few studies have been reported in the literature on the use of electrocoagulation for the treatment of distillery and/or brewery wastewater. Manisankar, Rani, and Viswanathan (2004) used electrocoagulation to remove COD, BOD and colour from distillery effluent using graphite electrodes and studied the effect of pH, current density and the halides

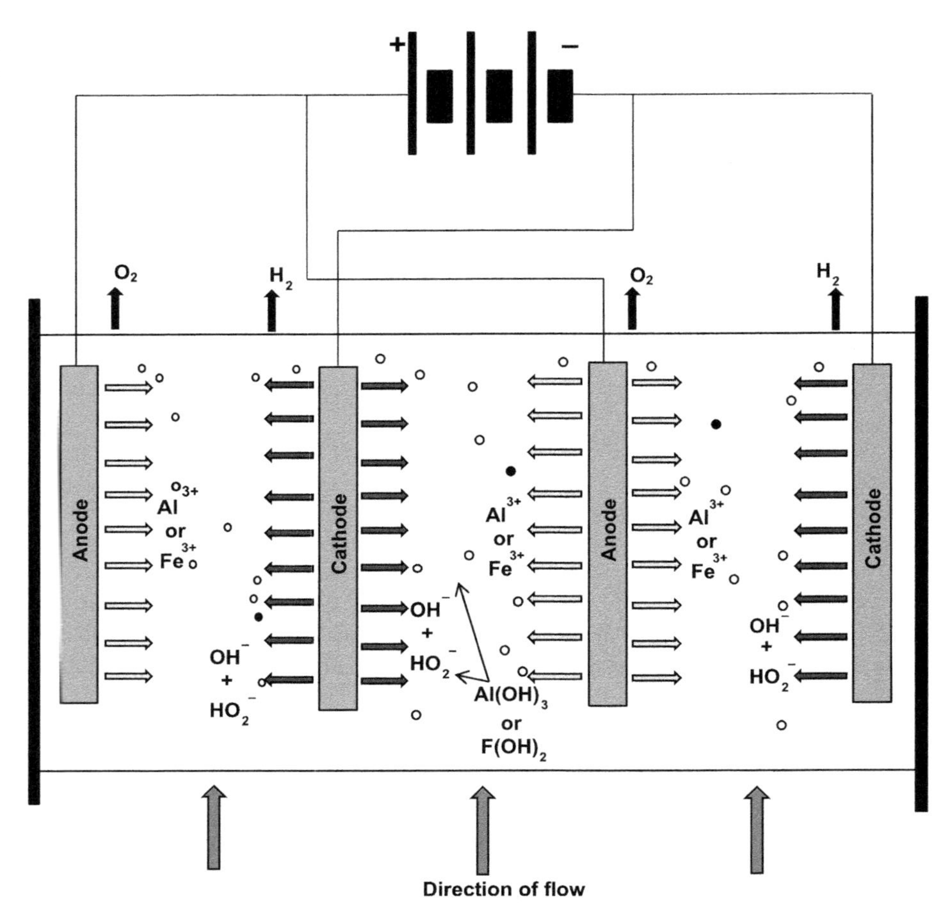

FIGURE 21.5 Conceptual diagram of an electrocoagulation reactor. *Adapted with permission from Xu and Zhu (2004).*

(sodium fluoride, sodium chloride and sodium bromide) as supporting electrolytes on the treatment of distillery effluent. An influent COD value of 12,000 mg/L was reduced by 85.2% in the presence of sodium chloride electrolyte. Colour and BOD were also reduced by 98% and 93.5%, respectively. Thakur, Srivastava, and Mall (2009) investigated the removal of COD (9310 mg/L) and colour from a two-stage aerobic treatment effluent using stainless-steel electrodes and studied the effect of pH, current density, inter-electrode distance and electrolysis time. At the optimum conditions (pH = 6.75; current density = 146.75 A/m^2; interelectrode distance = 1 cm and electrolysis time = 130 min), 61.6% and 98.4% COD and colour were removed, respectively. The results also showed that for pH < 6, the primary mechanism for COD and colour removal was charge neutralisation by monomeric cationic iron species, while sweep coagulation with amorphous iron hydroxide was the dominant mechanism for higher pH.

Khandegar and Saroha (2012) employed different combinations of electrodes in the treatment of alcohol distillery spentwash having very high COD (120,000 mg/L). The tests were performed to study the effect of current density, pH of the spentwash, agitation speed, electrolysis time and the distance between the electrodes on the COD removal efficiency. It was observed that aluminium electrodes were more suitable for treatment of distillery spentwash as compared to iron electrodes. The maximum COD removal efficiency of 81.3% was obtained with aluminium anode and cathode electrodes at the current density of 0.187 A/cm and pH 3 for an electrolysis time of 2 h. In this study, the COD reduction of the distillery spentwash happened due to two mechanisms (Khandegar & Saroha, 2012). Firstly, the coagulants [$Al(OH)_3$ and $Fe(OH)_2$] were generated in situ in the electrocoagulation process (see Fig. 21.5), which helped in coagulation of the organic content. The coagulants were generated through the electrochemical reactions occurring at the aluminium and iron electrodes as follows (Khandegar & Saroha, 2012).

For aluminium electrodes:

$$\text{Anode: Al} \rightarrow Al^{3+} + 3e^- \quad (21.1)$$

$$\text{Cathode: } 3H_2O + 3e^- \rightarrow 1\tfrac{1}{2}H_2 + 3OH^- \quad (21.2)$$

$$\text{Overall: Al} \rightarrow \text{Al}^{3+} \rightarrow \text{Al(OH)}_n^{(3-n)} \rightarrow \text{Al(OH)}_2{}^{4+} \rightarrow \text{Al(OH)}_4{}^{5+} \rightarrow \text{Al}_{13}\ \text{(complex)} \rightarrow \text{Al(OH)}_3 \quad (21.3)$$

For iron electrodes:

$$\text{Anode: Fe} \rightarrow \text{Fe}^{2+} + 2\text{e}^- \quad (21.4)$$

$$\text{Fe}^{2+} + 2\text{OH}^- \rightarrow \text{Fe(OH)}_2 \quad (21.5)$$

$$\text{Cathode: } 2\text{H}_2\text{O} + 2\text{e}^- \rightarrow \text{H}_2 + 2\text{OH}^- \quad (21.6)$$

$$\text{Overall: Fe} + 2\text{H}_2\text{O} \rightarrow \text{Fe(OH)}_2 + \text{H}_2 \quad (21.7)$$

Secondly, the presence of chlorides in the distillery spentwash and the application of electric current led to the generation of chlorine and hypochlorite ions, which reacted with the organic molecules and oxidised them. The hypochlorous acid and hypochlorite ions can decompose organic matter due to their high oxidative potentials (Krishna et al., 2010; Vijayaraghavan et al., 1999, 2006). The reactions at anode and cathode were as follows:

$$\text{Anode: } 2\text{Cl}^- \rightarrow \text{Cl}_2{}^+\ 2\text{e}^- \quad (21.8)$$

$$\text{Cathode: } 2\text{H}_2\text{O} + 2\text{e}^- \rightarrow \text{H}_2 + 2\text{OH}^- \ (21.9)$$

$$\text{Bulk solution: Cl}_2 + \text{H}_2\text{O} \rightarrow \text{HOCl} + \text{HCl} \quad (21.10)$$

$$\text{HOCl} \rightarrow \text{OCl}^- + \text{Hl}^+ \quad (21.11)$$

This section of the chapter has shown that electrochemical methods are efficient and versatile processes that are able to handle a wide variety of wastewaters. The coupling of electron-driven reactions (direct oxidation) with in situ generation of oxidants (indirect oxidation) makes this technique a valuable treatment alternative (Anglada et al., 2009). It must be noted that if any chlorinated organics are formed during electrolytic treatment of wastewater, they can be removed by passing the treated effluent through activated carbon before the discharge (Vijayaraghavan et al., 1999). Moreover, any excess concentration of chlorine can be reduced by the addition of bisulphite (Vijayaraghavan et al., 2006).

21.4.4 Microbial fuel cell technologies

Microbial fuel cells (MFCs) have gained a lot of attention as a means for converting organic waste, including low-strength wastewaters and lignocellulosic biomass, into electricity (Pant, van Bogaert, Diels, & Vanbroekhoven, 2010). Actually, MFCs are considered to be the major type of bioelectrochemical systems that convert biomass spontaneously into electricity through the metabolic activity of the micro-organisms (Pant et al., 2010). In other words, MFCs allow a direct conversion of chemical energy from the biodegradable organic matter into electricity via microbial catalysis (Liu, Liu, Zhang, & Su, 2009). Though the idea of using micro-organisms as catalysts in an MFC has been explored since the 1970s (Roller et al., 1984; Suzuki, 1976), the MFCs used to treat domestic wastewater were only introduced relatively recently by Habermann and Pommer (1991).

An MFC typically consists of a porous anode chamber, a porous cathode chamber and a membrane (or an electrolyte) sandwiched between the two. In a two-chamber setup, the anode and cathode compartments are separated by a proton exchange membrane (PEM) that allows proton transfer from anode to cathode but prevents oxygen diffusion to the anode chamber (Pant et al., 2010). In the single-chamber MFC, the cathode is exposed directly to the air (Pant et al., 2010). The basic operation of an MFC is as follows: micro-organisms oxidise organic matters in the anode chamber under anaerobic conditions and produce electrons and protons (Köroğlu, Özkaya, & Çetinkaya, 2014; Liu et al., 2010; Pant et al., 2010). Electrons transfer via the external circuit to the cathode chamber (thus generating electric current) where electrons, protons and electron acceptors (mainly oxygen) combine to form water (Köroğlu et al., 2014; Liu et al., 2009; Logan, 2008; Pant et al., 2010). Essentially there are three configurations amongst MFCs with a PEM (Fig. 21.6): (a) bioreactor separated from the MFC: the micro-organisms generate hydrogen that is then used as fuel in a fuel cell, (b) bioreactor integrated into the MFC: the micro-organisms generate hydrogen that is converted into electricity in a single cell and (c) MFC with direct electron transfer: microbiological electricity generation and direct transfer to the anode (Alzate-Gaviria, 2011; Rabaey, Lissens, & Verstraete, 2005).

MFCs can be monitored via electrochemical parameters, such as power density, generated electrical current and voltage (Alzate-Gaviria, 2011). Equally, a very important parameter is the organic load of the substrate to be used (Rabaey,

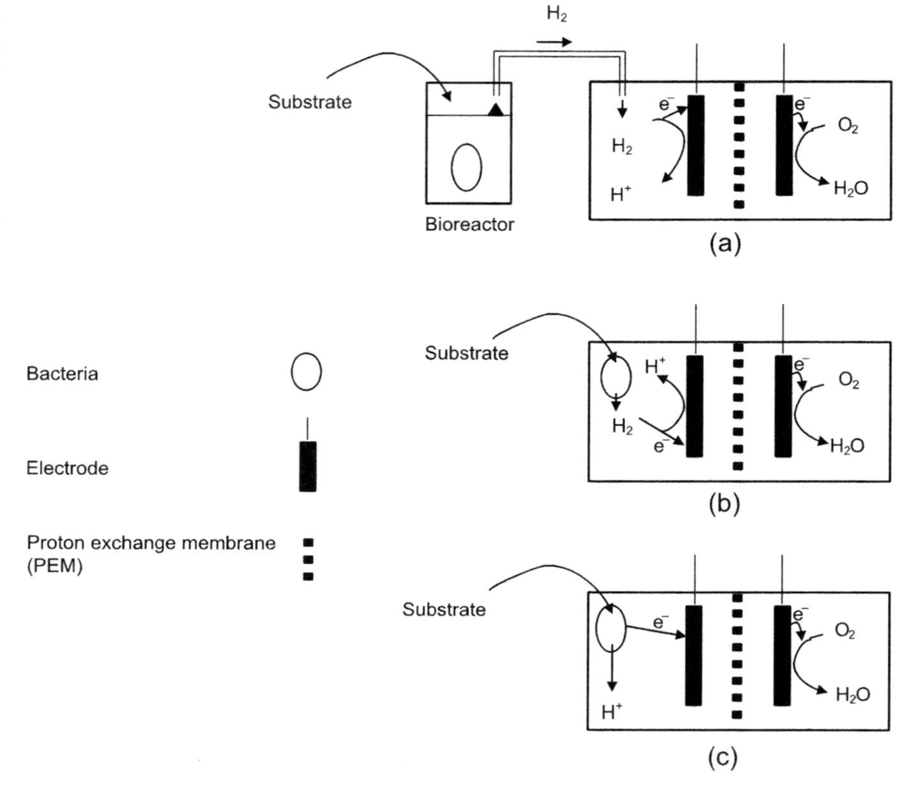

FIGURE 21.6 Three typical microbial fuel cell configurations: (a) Bioreactor separate from the MFC: The micro-organisms generate hydrogen that is then used as fuel in a fuel cell, (b) bioreactor integrated into the MFC: The micro-organisms generate hydrogen that is converted into electricity in a single cell and (c) MFC with direct electron transfer: Microbiological electricity generation and direct transfer to the anode. *Reprinted with permission from Rabaey et al. (2005).*

Lissens, Siliciano, & Verstraete, 2003). Actually, in MFCs, a substrate is regarded as one of the most important biological factors affecting electricity generation (Liu et al., 2009; Pant et al., 2010). As already stated, brewery wastewater is characterised by the presence of organic material; therefore, it is a suitable substrate for MFCs. In fact, wastewater from breweries has been a favourite substrate for MFCs among researchers, primarily because of its low strength (Pant et al., 2010) compared to other organic substrates. Besides, it is suitable for electricity generation in MFCs due to the food-derived nature of the organic matter and the lack of high concentrations of inhibitory substances (Feng et al., 2008; Pant et al., 2010). In other words, it is an ideal substrate for MFCs due to its nature of high carbohydrate content and low ammonium nitrogen concentration (Pant et al., 2010).

Feng et al. (2008) investigated the treatment of beer brewery wastewater using air cathode MFC. In this study, the efficiency of wastewater treatment was examined in terms of maximum power densities, Coulombic efficiencies (CEs) and COD removal as a function of temperature. It was found that with an influent COD of beer brewery wastewater of 2250 ± 418 mg/L, the COD removal efficiency was 85% and 87% at 20 and 30°C, respectively. Decreasing the temperature from 30 to 20°C reduced the maximum power density from 205 to 170 mW/m^2, while CEs decreased only slightly with temperature. The performance of electricity production from beer brewery wastewater in a single-chamber membrane-free MFC was investigated by Wang, Feng, and Lee (2008). Experimental results showed that the MFCs could generate electricity with the maximum power density of 483 mW/m^2 at 30°C and 435 mW/m^2 at 20°C, from wastewater with influent COD of 2239 mg/L. Wen, Wu, Zhao, Sun, and Kong (2010) used an MFC model based on a polarisation curve to investigate the performance of brewery wastewater treatment in conjunction with electricity generation. With influent COD of 1250 ± 100 mg/L, this sequential anode–cathode MFC achieved COD removal efficiency of more than 90%. This study also showed that the most important factors which influenced the performance of the MFC with brewery wastewater were reaction kinetic loss and mass transport loss. However, these can be avoided by increasing the concentration of brewery wastewater and by increasing the reaction temperature and using a rough electrode to provide more reaction sites (Pant et al., 2010). Mathuriya and Sharma (2010) also used an MFC to simultaneously treat brewery wastewater and produce electricity. This study reported 93.8% COD removal efficiency and up to 10.89 mA electric current generation. The study also showed that the addition of readily utilisable substrates like glucose and sucrose to the wastewater can enhance the electricity production and COD removal.

Since high COD removal efficiencies were achieved in these studies, it can be ascertained that MFCs can provide a new approach for brewery wastewater treatment while offering a valuable alternative to energy generation. Thus, the use of brewery wastewater as a substrate for MFCs has great development potential not only in terms of wastewater treatment, but also in terms of energy self-sufficiency as well as reducing competition with food production as is the apprehension with conventional biofuels.

21.4.5 Electric discharge plasma technologies

One of the vital developments of advanced oxidation processes is concerned with using electrical discharges to generate very powerful and non-selective oxidising agents. The class of advanced oxidation processes involving electric discharge that will be discussed in this section is referred to as gliding arc discharge plasma. Plasma is a highly ionised gas that occurs at high temperatures (Moreau, Orange, & Feuilloley, 2008; Simate et al., 2011). The gas consists of positive and negative ions and electrons as well as neutral species (Kaunas, 2012). Plasmas are energetically the strongest and are characterised by their acidic, oxidising and complexing properties (Abba, Gongwala, Laminsi, & Brisset, 2014). The plasma gas can be directly cooled and projected onto the target or quenched by a reaction with water. In both cases, highly reactive oxidative species are formed locally and can react with the macromolecules of contaminants (Moreau et al., 2008). Like gas, plasma does not have a definite shape or a definite volume unless enclosed in a container; unlike gas, in the influence of a magnetic field, it may form structures such as filaments, beams and double layers (Simate et al., 2011).

There are two categories of plasma: those in thermal equilibrium and those not in thermal equilibrium (Kaunas, 2012), defined according to the conditions in which they are created (Moreau et al., 2008). Thermal equilibrium implies that the temperatures of active species (electrons, ions and neutrals) are the same. In the case of non-thermal equilibrium plasmas, the temperatures of active species are not the same. To be more precise, electrons are characterised by much higher temperatures compared to heavy ions (Kaunas, 2012). Thermal plasmas are obtained at high pressure ($\geq$105 Pa) and need substantial power (up to 50 MW) to be observed. This type of plasma is found, for example, in plasma torches and in electric arcs. Non-thermal plasmas are obtained at lower pressures and use less power. Such plasma can be generated by electric discharges in lower pressure gases.

A third category of plasmas is an intermediate between thermal and non-thermal discharges (Moreau et al., 2008). Usually, these plasmas are included in the category of the non-thermal plasmas because they are formed near atmospheric pressure and ambient temperature. These low-temperature and medium-pressure plasmas are of particular interest technically and industrially because they do not require extreme conditions. Typical examples of these plasmas are the corona discharge and the gliding arc discharge. As already mentioned, of particular interest in this chapter is the gliding arc discharge plasma.

A gliding arc is an electrical discharge formed between two or more thin diverging electrodes with a high-velocity (>1 m/s) gas flowing between the electrodes (Burlica, Kirkpatrick, & Locke, 2006). An arc forms at the narrowest gap between the electrodes (Djepang, Laminsi, Njoyim-Tamungang, Ngnintedem, & Brisset, 2014). A gas flow directed along the axis of the electrodes gently pushes the arc feet along the conductors, so that the arc length increases until breaking in a plasma plume. Its temperature decreases, as does its energy, when the arc is short-circuited by a new one (Brisset et al., 2008; Djepang et al., 2014). In other words, the gliding arc plasma generator consists of two divergent electrodes, where the arc starts at the shortest distance between the electrodes and then moves with the gas flow. The length of the arc column increases together with the voltage (Kaunas, 2012). The arc discharge disappears when arc maintenance voltage exceeds input voltage. This process of generating an arc, movement (gliding), and disappearance is repeated continuously (Burlica & Locke, 2008). The gliding arc generates regions of both thermal and non-thermal plasma at the conditions of atmospheric pressure and ambient temperature (Fridman, Nester, Kennedy, Saveliev, & Mutaf-Yardimci, 1995).

Gliding arc discharges have been investigated as a potential treatment technology for gas-phase pollution treatment (Krawczyk & Motek, 2001) and for liquid-phase pollution treatment (Moussa & Brisset, 2003). Ghezzar, Abdelmalek, Belhadj, Benderdouche, and Addou (2007) used a non-thermal gliding arc at atmospheric pressure to remove anthraquinonic acid green 25 (AG 25) from an aqueous solution. The removal of the dye was carried out in the absence and presence of TiO_2 as a photocatalyst. The gaseous species formed in the discharge (particularly OH* radicals) induced strong oxidising effects in the target solution. At the optimum concentration, the dye (80 μM) was totally decolourised within 15 min of plasma treatment time, and 93% of the initial COD was removed after 180 min of plasma treatment time. In the absence of catalyst, colour removal was 46% after 15 min, while COD abatement reached 84% after 180 min. The results have shown that the combined plasma–TiO_2 method is a rapid and cost-effective means that might prove well adapted to the removal of other organic pollutants such as those in brewery wastewater.

The plasma chemical treatment of wastewater has also been applied to brewery wastewater. Doubla et al. (2007) reported the use of humid air plasma created by an electric gliding arc discharge in humid air to lower organic pollutants in brewery wastewater. The gliding arc discharge in humid air generates NO and OH radicals, which have strong oxidising characteristics. The OH radical is a very powerful oxidising agent [$E^0(OH/H_2O) = 2.85$ V/SHE] and is responsible for oxidation reactions with organic targets, both due to its own properties and to its derivative and/or parent molecule H_2O_2, as shown in Eq. (20.12) (Doubla et al., 2007):

$$H_2O_2 \leftrightarrow 2OH \tag{21.12}$$

The nitrate ions also participated in the oxidising characteristics of the humid air plasma. It must be noted that, initially, NO in humid air led to the formation of nitrite in neutral mediums but was further oxidised to stable nitrate ion species. As can be seen, the high standard oxidation–reduction potentials of the HNO_2/NO (1.00 V) and NO_3-/HNO_2 (1.04 V) systems reflect the oxidising power of the nitrate ions (Doubla et al., 2007). In the study by Doubla et al. (2007), the BOD removal efficiency of the gliding arc discharge process with brewery wastewaters of BOD values of 385 and 1018 mg/L were 74% and 98%, respectively. The alkaline wastewaters were also rapidly neutralised due to the pH-lowering effect of the plasma treatment emanating from the production of nitrate ions (Benstaali, Moussa, Addou, & Brisset, 1998). This process can be coupled with other methods, such as biological processes, to further lower the organic pollutant concentration more easily and rapidly to an acceptable level for reuse (Doubla et al., 2007).

This section has shown that non-thermal plasma technology (or the gliding arc discharge specifically) is one of the most attractive of the advanced oxidation techniques for treating wastewaters because of low equipment and energy costs and greater efficiency (Benstaali et al., 1998). In summary, the rapid interest in the application of gliding arc discharges results from the unusual chemical properties and enhanced reactivity of the activated species (atoms, radicals and excited molecules) produced in the plasma. These activated species formed are responsible for acid and oxidising effects in the target solution (Abba et al., 2014).

21.5 Challenges and future prospects

This section explores the existing and emerging challenges in relation to water treatment and reuse in breweries. The section will also discuss future prospects. Water reuse has been dubbed the 'greatest challenge of this century' as water supplies continue to dwindle and water demands increase because of the increase in population (Fatta et al., 2005). In the brewery industry, this statement is exacerbated by public perception and possible product quality deterioration problems (Janhom et al., 2009). In fact, most studies investigating public acceptance of recycled water come to the same conclusion – that people are very open to using recycled water for uses with low personal contact, such as watering trees and shrubs in their garden, but are reluctant to adopt recycled water for uses with high personal contact, such as drinking or bathing (Dolnicar, Hurlimann, & Grün, 2011). Moreover, concerns for human health and the environment are the most important constraints in the reuse of wastewater (Fatta et al., 2005). Nevertheless, the main problem preventing the safe reuse of treated wastewater in breweries is the nonexistence of the reuse criteria related to hygiene, public health and product quality control.

Notwithstanding the obstacles, several promising results have shown that new wastewater treatment processes such as gliding arc discharge plasma, electrochemical methods and MBRs have great capacity to be used for the treatment of brewery wastewater for reuse. In other words, there have been several technological advances and innovations that can achieve significant improvements in the treatment of brewery wastewater to guarantee its reuse. Furthermore, integrating these processes together as two or more stage processes would be more suitable thus giving the brewery wastewater treatment processes good economics and a high degree of energy efficiency (Simate et al., 2011).

Previous research has shown that integration of several processes may be able to partially or completely eliminate a wide range of contaminants (Dobias, 1993; Harrelkas, Azizi, Yaacoubi, Benhammou, & Pons, 2009). For example, integrated anaerobic and aerobic processes in brewery wastewater treatment (Driessen & Vereijken, 2003) have resulted in up to 98% COD and nutrient removal (Biothane, 2014). Besides wastewater treatment, the other benefit from the anaerobic–aerobic system is the production of biogas. When biogas is burned in brewery boilers or in a combined heat and power unit, the whole treatment can create a positive energy balance (Biothane, 2014). Thus, the combination of wastewater treatment together with power production may help in reducing the cost of wastewater treatment.

In view of environmental problems accompanied by the use of non-renewable fossil fuels and an urgent need for renewable energy, it is suggested that MFCs are used as the first pre-treatment stage of every integrated process, particularly with membrane filtration techniques. Besides generating electricity, MFCs would substantially reduce the organic load, thus minimising membrane fouling. It must, however, be noted that despite the potential to treat brewery wastewater as well as produce electricity, MFCs have a lot of limitations. A major drawback associated with MFCs is the start-up time that may vary from just days to months depending on the inoculum, electrode materials, reactor design and operating conditions (Pant et al., 2010). Furthermore, scaleup is still a big challenge; the high cost of cation exchange membranes, the potential for biofouling and associated high internal resistance restrain the power generation and limit the practical applications of MFCs (Hu, 2008; Pant et al., 2010). Therefore, before the potential of MFCs is fully realised additional research and development is needed.

Electrochemical methods can be well suited to be coupled in the latter stages of the integrated process (Simate et al., 2011). Sanitising agents (often called disinfectants), which are present in brewery wastewater, contain chlorine compounds. These compounds produce chlorine during electrolysis and, thereafter, chlorine generates hypochlorous acid, which may oxidise organic compounds. Furthermore, chlorine produced may also deactivate pathogenic microorganisms. Therefore, electrochemical methods coupled in the latter stages can serve as an organic oxidation and disinfecting stage. Nevertheless, to achieve an efficient and cost competitive electrochemical treatment process, the wastewater should have relatively high conductivity (Anglada et al., 2009). The main obstacles that require attention before the full-scale implementation of electrochemical oxidation are the high operating cost and lack of efficient and stable electrode materials (Anglada et al., 2009). Therefore, a major area for future research is the improvement of the electrocatalytic activity and electrochemical stability of the electrode materials, which will result in lower operational and capital costs (Anglada et al., 2009).

Though Doubla et al. (2007) recommend integrating gliding arc discharge plasma techniques with other treatment processes in order to lower the organic pollutant concentrations more easily and rapidly to an acceptable level for reuse, the processes can be very expensive (Simate et al., 2011). This is because of the high ionisation energy requirements and the cost of energy sources such as lasers (Simate et al., 2011). However, the gliding arc discharge can easily be powered by a DC or AC power supply source. The DC gliding arc plasma generator is characterised by stability of discharge and a simple design. The main advantages of the AC gliding arc discharge plasma generator are simplicity of the power supply system and its low cost (Lie, Bin, Chi, & Chengkang, 2006).

Membrane filtration processes, particularly RO, have been demonstrated to be very effective in removing organic and inorganic materials. The COD removal efficiencies up to 99%, TSS removal efficiencies up to 100% and complete removal of pathogens have been reported. Nevertheless, in order to improve the operations of membrane processes, the following measures are required (Zhou & Smith, 2002): (1) Better understanding of membrane fouling mechanisms; (2) effective fouling control strategies; (3) better membrane materials and module designs; and (4) membrane integrity management. The high cost of membranes is still a significant issue impeding a faster commercialisation (Skouteris, Hermosilla, López, Negro, & Blanc, 2012). So, even though the membrane costs have been dramatically reduced over time, they are still a critical issue.

To date, much progress has been achieved in research and applications of both anaerobic and aerobic MBRs. Just like conventional membranes, fouling is one of the main disadvantages of MBRs because it hinders the operation of the systems in a constant, reliable way (Skouteris et al., 2012). The deposition of solids on anaerobic MBR membrane surfaces is lower than on aerobic MBR membrane surfaces. However, since anaerobic MBRs are usually operated at lower membrane permeate fluxes, they are characterised by lower sludge filterabilities, which favour membrane fouling (Skouteris et al., 2012). Therefore, it is imperative that further research is carried out to mitigate this problem.

The energy produced from biogas in anaerobic MBRs could be used to cover the energy required for membrane filtration and the excess energy could be used elsewhere. However, more research is required to investigate in detail to what extent the biogas produced in anaerobic MBR can lead to sustainable energy operations. For example, little information is available regarding the energy that is consumed by anaerobic MBRs as a whole or by each of their components (Skouteris et al., 2012).

21.6 Conclusions

Though the conventional or pre-treatment processes for getting rid of many pollutants from brewery wastewater have long been established, their effectiveness is limited. This chapter discussed various processes that can be used individually or coupled with others to treat brewery wastewater for reuse. The chapter has shown that most of the processes studied could be successfully implemented for high-level treatment of brewery wastewater for reuse. In summary, these processes may have the much-needed solution for the future because, if properly utilised, they can give the most efficacious and cost-effective approach to treating brewery wastewater for reuse. A number of hybrid treatment methods have also been proposed that are formed by integrating these processes with other traditional treatment processes or amongst themselves.

However, there is a need to carry out extensive research so as to understand both synergistic and antagonistic effects of the suggested hybrid processes.

References

Abba, P., Gongwala, J., Laminsi, S., & Brisset, J. L. (2014). The effect of the humid air plasma on the conductivity of distilled water: Contribution of ions. *International Journal of Research in Chemistry and Environment, 4*(1), 25–30.

Alzate-Gaviria, L. (2011). Microbial fuel cells for wastewater treatment. In F. S. G. Einschlag (Ed.), *Waste water: Treatment and utilisation* (pp. 152–170). In Tech. Available fromAccessed February 2014 http://cdn.intechopen.com/pdfs-wm/14554.pdf.

Anglada, A., Urtiaga, A., & Ortiz, I. (2009). Contributions of electrochemical oxidation to waste-water treatment: Fundamentals and review of applications. *Journal of Chemical Technology and Biotechnology, 84*, 1747–1755.

Arnold, J. P. (1911). *Origin and history of beer and brewing*. Chicago: Alumni Association of the Wahl-Henius Institute of Fermentology.

Bailey, A. D., Hansford, G. S., & Dold, P. L. (1994). The use of crossflow microfiltration to enhance the performance of an activated sludge reactor. *Water Research, 28*(2), 297–301.

Baloch, M. I., Akunna, J. C., & Collier, P. J. (2007). The performance of a phase separated granular bed bioreactor treating brewery wastewater. *Bioresource Technology, 98*, 1849–1855.

Basile, A. (2013). *Handbook of membrane reactors: Fundamental materials science, design and optimisation*. Cambridge: Woodhead Publishing Limited.

Benstaali, B., Moussa, D., Addou, A., & Brisset, J.-L. (1998). Plasma treatment of aqueous solutes: Some chemical properties of a gliding arc in humid air. *The European Physical Journal - Applied Physics, 4*, 171–179.

Biothane. (2014). *Biogas production in breweries*. Available fromAccessed March 2014 . Available fromAccessed March 2014 http://www.biothane.com/en/news_and_media/articles-publications/biogas-breweries.htm.

Blomenhofer, V., Groß, F., Procelewska, J., Delgado, A., & Becher, T. (2013). Water quality management in the food and beverage industry by hybrid automation using the example of breweries. *Water Science and Technology: Water Supply, 13*(2), 427–434.

Bockris, J. O. H. (1977). *Environmental chemistry*. New York: Plenum Press.

Braeken, L., Van der Bruggen, B., & Vandecasteele, C. (2004). Regeneration of brewery waste water using nanofiltration. *Water Reserach, 38*(13), 3075–3082.

Brewers of Europe. (2002). *Guidance note for establishing BAT in the brewing industry*. Brussels: Brewers of Europe Accessed January 2014 http://www.brewersofeurope.org/docs/publications/guidance.pdf.

Brisset, J. L., Moussa, D., Doubla, A., Hnatiuc, E., Hnatiuc, B., Youbi, G. K., et al. (2008). Chemical reactivity of discharge and temporal post-discharge in plasma treatment of aqueous media: Examples of gliding discharge treated solutions. *Industrial and Engineering Chemistry Research, 47*, 5761–5781.

Brito, A. G., Peixoto, J., Oliveira, J. M., Oliveira, J. A., Costa, C., Nogueira, R., et al. (2007). Brewery and winery wastewater treatment: Some focal points of design and operation. In V. Oreopoulous, & W. Russ (Eds.), *Utilisation of byproducts and treatment of waste in the food industry* (Vol. 3). New York: Springer.

Brock, T. D. (1983). *Membrane filtration: A user's guide and reference manual*. New York: Springer-Verlag.

Burlica, R., Kirkpatrick, M. J., & Locke, B. R. (2006). Formation of reactive species in gliding arc discharges with liquid water. *Journal of Electrostatics, 64*, 35–43.

Burlica, R., & Locke, B. R. (2008). Confined plasma gliding arc discharges. *Waste Management, 2*(4–5), 484–498.

Côté, P., Buisson, H., Pound, C., & Arakaki, G. (1997). Immersed membrane activated sludge for the reuse of municipal wastewater. *Desalination, 113*(2–3), 189–196.

Côté, P., Buisson, H., & Praderie, M. (1998). Immersed membranes activated sludge process applied to the treatment of municipal wastewater. *Water Science and Technology, 38*(4–5), 437–442.

Côté, P., & Thompson, D. (2000). Wastewater treatment using membranes: The north American experience. *Water Science and Technology, 41*(10–11), 209–215.

Chakrabarty, B., Ghoshal, A. K., & Purkait, M. K. (2008). Ultrafiltration of stable oil-in water emulsion by polysulfone membrane. *Journal of Membrane Science, 325*, 427–437.

Cornelissen, E. R., Janse, W., & Koning, J. (2002). Wastewater treatment with the internal MEMBIOR. *Desalination, 146*, 463–466.

Dai, H., Yang, X., Dong, T., Ke, Y., & Wang, T. (2010). Engineering application of MBR process to the treatment of beer brewing wastewater. *Modern Applied Science, 4*(9), 103–109.

Djepang, S. A., Laminsi, S., Njoyim-Tamungang, E., Ngnintedem, C., & Brisset, J. L. (2014). Plasma-chemical and photo-catalytic degradation of bromophenol blue. *Chemical and Materials Engineering, 2*(1), 14–23.

Dobias, B. (1993). *Coagulation and flocculation*. New York: Marcel Dekker.

Dolnicar, S., Hurlimann, A., & Grün, B. (2011). What affects public acceptance of recycled and desalinated water? *Water Research, 45*, 933–943.

Doubla, A., Laminsi, S., Nzali, S., Njoyim, E., Kamsu-Kom, J., & Brisset, J. L. (2007). Organic pollutants abatement and biodecontamination of brewery effluents by a non-thermal quenched plasma at atmospheric pressure. *Chemosphere, 69*, 332–337.

Driessen, W., & Vereijken, T. (2003). Recent developments in biological treatment of brewery effluent. In *The institute and guild of brewing convention, livingstone, Zambia, march 2–7*. Available fromAccessed October 2013 http://www.environmental-expert.com/Files%5C587%5Carticles%5C3041%5Cpaques24.pdf.

Drogui, P., Blais, J. F., & Mercier, G. (2007). Review of electrochemical technologies for environmental applications. *Recent Patents on Engineering, 1*, 257–272.

EPA. (1992). *Guidelines for agricultural reuse of wastewater*. http://water.epa.gov/aboutow/owm/upload/Water-Reuse-Guidelines-625r04108.pdf. Available fromAccessed January 2014 . Available fromAccessed January 2014.

EPA. (2003). *Wastewater technology fact sheet*. Washington: Municipal Technology Branch U.S. EPA. http://water.epa.gov/aboutow/owm/upload/2004_07_07_septics_final_sgrit_removal.pdf. Available fromAccessed March 2014.

Fakhru'l-Razi, A. (1994). Ultrafiltration membrane separation for anaerobic wastewater treatment. *Water Science and Technology, 30*(12), 321–327.

Fakoya, M. D., & van der Poll, H. M. (2013). Integrating ERP and MFCA systems for improved waste-reduction decisions in a brewery in South Africa. *Journal of Cleaner Production, 40*, 136–140.

Fan, F. S., & Zhou, H. D. (2007). Interrelated effects of aeration and mixed liquor fractions on membrane fouling for submerged membrane bioreactor processes in wastewater treatment. *Environmental Science and Technology, 41*(7), 2523–2528.

Fatta, D., Alaton, A., Gockay, C., Rusan, M. M., Assobhei, O., Mountadar, M., et al. (2005). Wastewater reuse: Problems and challenges in Cyprus, Turkey, Jordan and Morocco. *European Water, 11*(12), 63–69.

Feng, Y., Wang, X., Logan, B. E., & Lee, H. (2008). Brewery wastewater treatment using air-cathode microbial fuel cells. *Applied Microbiology and Biotechnology, 78*, 873–880.

Fridman, A., Nester, S., Kennedy, L., Saveliev, A., & Mutaf-Yardimci, O. (1995). Gliding arc gas discharge. *Progress in Energy and Combustion Science, 25*, 211–231.

Ghabris, A., Abdel-Jawad, M., & Aly, J. (1989). Municipal wastewater renovation by reverse osmosis: State of the art. *Desalination, 75*, 213–240.

Ghezzar, M. R., Abdelmalek, F., Belhadj, M., Benderdouche, N., & Addou, A. (2007). Gliding arc plasma assisted photocatalytic degradation of anthraquinonic acid green 25 in solution with TiO_2. *Applied Catalysis B: Environmental, 72*, 304–313.

Global Advantech. (2011). *Electrocoagulation and advanced electrochemical oxidation*. Available fromAccessed December 2013 . Available fromAccessed December 2013 http://www.globaladvantech.com/Decontamination/TDS801%20Electrocoagulation%20and%20Advanced%20Electrochemical%20Oxidation%20EN%2009.pdf.

Goldammer, T. (2008). *The brewers' handbook*. Clifton: Apex Publishers.

Gregory, J. (2006). *Particles in water: Properties and processes*. London: IWA Publishing/CRC Press.

Habermann, W., & Pommer, E. H. (1991). Biological fuel cells with sulphide storage capacity. *Applied Microbiology and Biotechnology, 35*, 128–133.

Harrelkas, F., Azizi, A., Yaacoubi, A., Benhammou, A., & Pons, M. N. (2009). Treatment of textile dye effluents using coagulation-flocculation coupled with membrane processes or adsorption on powdered activated carbon. *Desalination, 235*, 330–339.

Harrison, M. A. (2009). Beer/brewing. In M. Schaechter (Ed.), *Encyclopedia of microbiology* (3rd ed., pp. 23–33). Oxford: Elsevier.

Hornsey, I. S. (2003). *A history of beer and brewing*. Cambridge: The Royal Society of Chemistry.

Hu, Z. (2008). Electricity generation by a baffle-chamber membraneless microbial fuel cell. *Journal of Power Sources, 179*, 27–33.

Hua, F. L., Tsang, Y. F., Wang, Y. J., Chan, S. Y., Chua, H., & Sin, S. N. (2007). Performance study of ceramic microfiltration membrane for oily wastewater treatment. *Chemical Engineering Journal, 128*, 169–175.

Huang, H., Schwab, K., & Jacangelo, J. G. (2009). The pretreatment for low pressure membranes in water treatment: A review. *Environmental Science and Technology, 43*(9), 3011–3019.

Huige, N. J. (2006). Brewery by-products and effluents. In F. G. Priest, & G. G. Stewart (Eds.), *Handbook of brewing*. Boca Raton: CRC Press.

Ince, B. K., Ince, O., Sallis, P. J., & Anderson, G. K. (2000). Inert COD production in a membrane anaerobic reactor treating brewery wastewater. *Water Research, 34*(16), 3943–3948.

Janhom, T., Wattanachira, S., & Pavasant, P. (2009). Characterisation of brewery wastewater with spectrofluorometry analysis. *Journal of Environmental Management, 90*, 1184–1190.

Jeison, D. (2007). *Anaerobic membrane bioreactors for wastewater treatment: Feasibility and potential applications*. PhD Thesis. Wageningen, The Netherlands: Wageningen University.

Köroğlu, E. O., Özkaya, B., & Çetinkaya, A. Y. (2014). Microbial fuel cells for energy recovery from waste. *International Journal of Energy Science, 4*(1), 28–30.

Kaunas. (2012). *Report on the different plasma modules for pollution removal MO 03*. Available fromAccessed March 2014 . Available fromAccessed March 2014 http://www.plastep.eu/fileadmin/dateien/Outputs/Report_on_the_different_Plasma_Modules_for_Pollution_Removal.pdf.

Khandegar, V., & Saroha, A. K. (2012). Electrochemical treatment of distillery spent wash using aluminum and iron electrodes. *Chinese Journal of Chemical Engineering, 20*(3), 439–443.

Kimura, S. (1991). Japan's aqua renaissance '90 project. *Water Science and Technology, 23*(7–9), 1573–1582.

Koyuncu, I., Yalcin, F., & Ozturk, I. (1999). Color removal of high strength paper and fermentation industry effluents with membrane technology. *Water Science and Technology, 40*(11–12), 241–248.

Krawczyk, K., & Motek, M. (2001). Combined plasma-catalytic processing of nitrous oxide. *Applied Catalysis B: Environmental, 30*, 233–245.

Krishna, B. M., Murthy, U. N., Manoj, K. B., & Lokesh, K. S. (2010). Electrochemical pretreatment of distillery wastewater using aluminum electrode. *Journal of Applied Electrochemistry, 40*, 663–667.

Li, X. Y., & Chu, H. P. (2003). Membrane bioreactor for drinking water treatment of polluted surface water supplies. *Water Research, 37*(19), 4781–4791.

Lie, L., Bin, W., Chi, Y., & Chengkang, W. (2006). Characteristics of gliding arc discharge plasma. *Plasma Science and Technology, 8*(6), 653–655.

Linko, M., Haikara, A., Ritala, A., & Penttilä, M. (1998). Recent advances in the malting and brewing industry. *Jounral of Biotechnology, 65*, 85–98.

Liu, Z., Liu, J., Zhang, S., & Su, Z. (2009). Study of operational performance and electrical response on mediator-less microbial fuel cells fed with carbon- and protein-rich substrates. *Biochemical Engineering Journal, 45*, 185–191.

Liu, M., Yuan, Y., Zhang, L. X., Zhuang, L., Zhou, S. G., & Ni, J. R. (2010). Bioelectricity generation by a gram-positive *Corynebacterium* sp. strain MFC03 under alkaline condition in microbial fuel cells. *Bioresource Technology, 101*, 1807–1811.

Logan, B. E. (2008). *Microbial fuel cells*. New York: John Wiley & Sons.

Lyko, S., Al-Halbouni, D., Wintgens, T., Janot, A., Hollender, J., Dott, W., et al. (2007). Polymeric compounds in activated sludge supernatant-characterisation and retention mechanisms at full scale-scale municipal membrane bioreactor. *Water Research, 41*(17), 3894–3902.

Madaeni, S. S., & Mansourpanah, Y. (2006). Screening membranes for COD removal from dilute wastewater. *Desalination, 197*, 23–32.

Mallevialle, J., Odendall, P. E., & Wiesner, M. R. (1996). *Water treatment membrane processes*. New York: McGraw-Hill.

Manios, T., Gaki, E., Banou, S., Ntigakis, D., & Andreadakis, A. (2006). Qualitative monitoring of a treated wastewater reuse extensive distribution system: COD, TSS, EC and pH. *WaterSA, 32*(1), 99–104.

Manisankar, P., Rani, C., & Viswanathan, S. (2004). Effect of halides in the electrochemical treatment of distillery effluent. *Chemosphere, 57*, 961–966.

Mantech. (2011). *PeCOD aplicaion note # 1*. Available fromAccessed October 2013. Ontario: Mantech Inc. Accessed October 2013 http://www.titralo.hu/WEBSET_DOWNLOADS/613/PeCOD-Wastewater%20Industry.pdf.

Maphutha, S., Moothi, K., Meyyappan, M., & Iyuke, S. E. (2013). A carbon nanotube-infused polysulfone membrane with polyvinyl alcohol layer for treating oil-containing waste water. *Scientific Report, 3*, 1509.

Martínez-Huitle, C. A., & Ferro, S. (2006). Electrochemical oxidation of organic pollutants for the wastewater treatment: Direct and indirect processes. *Chemical Society Reviews, 35*, 1324–1340.

Mathuriya, A. S., & Sharma, V. N. (2010). Treatment of brewery wastewater and production of electricity through microbial fuel cell technology. *International Journal of Biochemistry and Biotechnology, 6*(1), 71–80.

McGovern, P. E., Zhang, J., Tang, J., Zhang, Z., Hall, G. R., Moreau, R. A., et al. (2004). Fermented beverages of pre-and proto-historic China. *Proceedings of the National Academy of Sciences, 101*(51), 17593–17598.

Metcalf, L., & Eddy, H. P. (1991). *Wastewater engineering – treatment, disposal, and reuse*. Singapore: McGraw-Hill.

Mohan, S. V. (2008). Fermentative hydrogen production with simultaneous wastewater treatment: Influence of pretreatment and system operating conditions. *Journal of Scientific and Industrial Research, 67*, 950–961.

Moreau, M., Orange, N., & Feuilloley, M. G. J. (2008). Non-thermal plasma technologies: New tools for bio-decontamination. *Biotechnology Advances, 26*, 610–617.

Moussa, D., & Brisset, J. L. (2003). Disposal of spent tributylphosphate by gliding arc plasma. *Journal of Hazardous Materials, 102*(2–3), 189–200.

Mulder, M. (1997). *Basic principles of membrane technology*. Dordrecht: Kluwer Academic Publishers.

Muller, E. B., Stouthamer, A. H., van Verseveld, H. W., & Eikelboom, D. H. (1995). Aerobic domestic waste water treatment in a pilot plant with complete sludge retention by cross-flow filtration. *Water Research, 29*(4), 1179–1189.

Nagano, A., Arikawa, E., & Kobayashi, H. (1992). The treatment of liquor wastewater containing high-strength suspended solids by membrane bioreactor system. *Water Science and Technology, 26*(3–4), 887–895.

Ng, H. Y., & Hermanowicz, S. W. (2005). Membrane bioreactor operation at short solids retention times: Performance and biomass characteristics. *Water Research, 39*(6), 981–992.

Nilsson, A., Ronlan, A., & Parker, V. D. (1973). Anodic hydroxylation of phenols: A simple general synthesis of 4-alkyl-4-hydroxycyclo-hexa-2,5-dienones from 4-alkylphenols. *Journal of the Chemical Society, Perkin Transactions, 1*, 2337–2345.

Olajire, A. A. (2020). The brewing industry and environmental challenges. *Journal of Cleaner Production, 256*, Article 102817.

Pant, D., van Bogaert, G., Diels, L., & Vanbroekhoven, K. (2010). A review of the substrates used in microbial fuel cells (MFCs) for sustainable energy production. *Bioresource Technology, 101*, 1533–1543.

Phiarais, B. P. N., & Arendt, E. K. (2008). Malting and brewing with gluten-free cereals. In *Gluten-free cereal products and beverages* (pp. 347–372).

Piya-areetham, P., Shenchunthichai, K., & Hunsom, M. (2006). Application of electrooxidation process for treating concentrated wastewater from distillery industry with a voluminous electrode. *Water Research, 40*, 2857–2864.

Poelmans, E., & Swinnen, J. F. M. (2011). A brief economic history of beer. In J. F. M. Swinnen (Ed.), *The economics of beer*. Oxford: Oxford University Press.

Rabaey, K., Lissens, G., Siliciano, S., & Verstraete, W. (2003). A microbial fuel cell capable of converting glucose to electricity at high rate and efficiency. *Biotechnology Letters, 25*, 1531–1535.

Rabaey, K., Lissens, G., & Verstraete, W. (2005). Microbial fuel cells: Performances and perspectives. In P. Lens, P. Westerman, A. Haberbauer, & A. Moreno (Eds.), *Biofuels for fuel cells: Renewable energy from biomass fermentation*. London: IWA Publishing.

Radjenovic, J., Petrovic, M., & Barceló, D. (2007). Analysis of pharmaceuticals in wastewater and removal using a membrane bioreactor. *Analytical and Bioanalytical Chemistry, 387*, 1365–1377.

Rao, A. G., Reddy, T. S. K., Prakash, S. S., Vanajakshi, J., Joseph, J., & Sarma, P. N. (2007). pH regulation of alkaline wastewater with carbon dioxide: a case study of treatment of brewery wastewater in UASB reactor coupled with absorber. *Bioresource Technology, 98*, 2131–2136.

Rashed, M. N. (2011). Adsorption technique for the removal of organic pollutants from water and wastewater. Available fromAccessed March 2014. In M. N. Rashed (Ed.), *Organic pollutants – monitoring risk and treatment* (pp. 167–194). In-Tech. Available fromAccessed March 2014 http://library.umac.mo/ebooks/b28046055.pdf.

ReVelle, P., & ReVelle, C. (1988). *The environment: Issues and choices for society*. Boston: Jones and Bartlett Publishers.

Roller, S., Bennetto, H., Delaney, G., Mason, J., Stirling, J., & Thurston, C. (1984). Electrontransfer coupling in microbial fuel cells. Comparison of redox-mediator reduction rates and respiratory rates of bacteria. *Journal of Chemical Technology and Biotechnology, 34*, 3–12.

Seneviratne, M. (2007). *A practical approach to water conservation for commercial and industrial facilities*. Oxford: Elsevier.

Shao, E., Wei, J., Yo, A., & Levy, R. (2009). Application of ultrafiltration and reverse osmosis for mine waste water reuse. Available fromAccessed January 2014. In *Water in mining conference, perth, 15 – 17 september* http://www.nirosoft.com/files/CollahausiChileMiningEffluents(1).pdf. Available fromAccessed January.

Simate, G. S. (2012). *The treatment of brewery wastewater using carbon nanotubes synthesized from carbon dioxide carbon source*. PhD thesis. University of the Witwatersrand.

Simate, G. S., Cluett, J., Iyuke, S. E., Musapatika, E. T., Ndlovu, S., Walubita, L. F., et al. (2011). The treatment of brewery wastewater for reuse: State of the art. *Desalination, 273*(2–3), 235–247.

Skouteris, G., Hermosilla, D., López, P., Negro, C., & Blanc, Á. (2012). Anaerobic membrane bioreactors for wastewater treatment: A review. *Chemical Engineering Journal, 198–199*, 138–148.

Suzuki, S. (1976). Fuel cells with hydrogen forming bacteria. *Hospital-Hygiene: Gesundheitswesen und Desinfektion, 68*, 159.

Thakur, C., Srivastava, V. C., & Mall, I. D. (2009). Electrochemical treatment of a distillery wastewater: Parametric and residue disposal study. *Chemical Engineering Journal, 148*, 496–505.

van der Merwe, A., & Friend, J. F. C. (2002). Water management at a malted barley brewery. *WaterSA, 28*(3), 313–318.

Vijayaraghavan, K., Ahmad, D., & Lesa, R. (2006). Electrolytic treatment of beer brewery wastewater. *Industrial and Engineering Chemistry Research, 45*, 6854–6859.

Vijayaraghavan, K., Ramanujam, T. K., & Balasubramanian, N. (1999). In situ hypochlorous acid generation for the treatment of distillery spentwash. *Industrial and Engineering Chemistry Research, 38*, 2264–2267.

Wang, X., Feng, Y. J., & Lee, H. (2008). Electricity production from beer brewery wastewater using single chamber microbial cell. *Water Science and Technology, 57*(7), 1117–1121.

Wen, Q., Wu, Y., Zhao, L., Sun, Q., & Kong, F. (2010). Electricity generation and brewery wastewater treatment from sequential anode-cathode microbial fuel cell. *Journal of Zhejiang University - Science B, 11*(2), 87–93.

WHO. (1989). Health guidelines for the use of wastewater in agriculture and aquaculture. In *Technical report series 778*. Geneva: World Health Organisation.

Williams, M. E. (2003). *A review of wastewater treatment by reverse osmosis*. Available fromAccessed March 2014. EET Corporation and Williams Engineering Services Company, Inc. Accessed March 2014 http://www.eetcorp.com/heepm/RO_AppsE.pdf.

Wu, B., Li, X., Wang, H., Xue, X., & Li, J. (2012). Full-scale application of an integrated UF/RO system for treatment and reuse of electroplating wastewater. *Journal of Water Sustainability, 2*(3), 185–191.

Wyatt, F. (1900). The influence of science in modern beer brewing. *Journal of the Franklin Institute, 150*(3), 190–214.

Xie, X., Zhou, H., Chong, C., & Holbein, B. (2008). Coagulation assisted membrane filtration to treat high strength wastewater from municipal solid waste anaerobic digesters. *Journal of Environmental Engineering and Science, 7*, 21–28.

Xu, X., & Zhu, X. (2004). Treatment of refectory oily wastewater by electro-coagulation process. *Chemosphere, 56*, 889–894.

Yalcin, F., Koyuncu, I., Oztürk, I., & Topacik, D. (1999). Pilot scale UF and RO studies on water reuse in corrugated board industry. *Water Science and Technology, 40*(4–5), 303–310.

Yamamoto, K., Hiasa, M., Mahmood, T., & Matsuo, T. (1989). Direct solid-liquid separation using hollow fiber membrane in an activated sludge aeration tank. *Water Science and Technology, 21*(4–5), 43–54.

Yeung, A., Chu, R., Rosenberg, S., & Tong, T. (2008). Integrated membrane operations in difficult waters, commercial case studies of UF + RO. In *Proceedings of the water environment Federation (WEFTEC)* (pp. 2575–2590).

Zhou, H., & Smith, D. W. (2002). Advanced technologies in water and wastewater treatment. *Journal of Environmental Engineering and Science, 1*, 247–264.

Index

'*Note:* Page numbers followed by "f" indicate figures and "t" indicate tables.'

C

D

X

Y

Z

9780323996068